VLSI: SYSTEMS ON A CHIP

IFIP - The International Federation for Information Processing

IFIP was founded in 1960 under the auspices of UNESCO, following the First World Computer Congress held in Paris the previous year. An umbrella organization for societies working in information processing, IFIP's aim is two-fold: to support information processing within its member countries and to encourage technology transfer to developing nations. As its mission statement clearly states,

IFIP's mission is to be the leading, truly international, apolitical organization which encourages and assists in the development, exploitation and application of information technology for the benefit of all people.

IFIP is a non-profitmaking organization, run almost solely by 2500 volunteers. It operates through a number of technical committees, which organize events and publications. IFIP's events range from an international congress to local seminars, but the most important are:

- The IFIP World Computer Congress, held every second year;
- open conferences;
- working conferences.

The flagship event is the IFIP World Computer Congress, at which both invited and contributed papers are presented. Contributed papers are rigorously refereed and the rejection rate is high.

As with the Congress, participation in the open conferences is open to all and papers may be invited or submitted. Again, submitted papers are stringently refereed.

The working conferences are structured differently. They are usually run by a working group and attendance is small and by invitation only. Their purpose is to create an atmosphere conducive to innovation and development. Refereeing is less rigorous and papers are subjected to extensive group discussion.

Publications arising from IFIP events vary. The papers presented at the IFIP World Computer Congress and at open conferences are published as conference proceedings, while the results of the working conferences are often published as collections of selected and edited papers.

Any national society whose primary activity is in information may apply to become a full member of IFIP, although full membership is restricted to one society per country. Full members are entitled to vote at the annual General Assembly, National societies preferring a less committed involvement may apply for associate or corresponding membership. Associate members enjoy the same benefits as full members, but without voting rights. Corresponding members are not represented in IFIP bodies. Affiliated membership is open to non-national societies, and individual and honorary membership schemes are also offered.

VLSI: SYSTEMS ON A CHIP

IFIP TC10 WG10.5 Tenth International Conference on Very Large Scale Integration (VLSI '99) December 1-4, 1999, Lisboa, Portugal

Edited by

Luis Miguel Silveira
INESC
Portugal

Srinivas Devadas
Massachusetts Institute of Technology
USA

Ricardo Reis
Universidade Federal do Rio Grande do Sul
Brazil

KLUWER ACADEMIC PUBLISHERS
BOSTON / DORDRECHT / LONDON

Distributors for North, Central and South America:
Kluwer Academic Publishers
101 Philip Drive
Assinippi Park
Norwell, Massachusetts 02061 USA
Telephone (781) 871-6600
Fax (781) 871-6528
E-Mail <kluwer@wkap.com>

Distributors for all other countries:
Kluwer Academic Publishers Group
Distribution Centre
Post Office Box 322
3300 AH Dordrecht, THE NETHERLANDS
Telephone 31 78 6392 392
Fax 31 78 6546 474
E-Mail <services@wkap.nl>

Electronic Services <http://www.wkap.nl>

Library of Congress Cataloging-in-Publication Data

IFIP TC10/WG10.5 International Conference on Very Large Scale Integration (10th : 1999 : Lisbon, Portugal)
VLSI : systems on a chip : IFIP TC10 WG10.5 Tenth International Conference on Very Large Scale Integration (VLSI'99), December 1-4, 1999, Lisboa, Portugal / edited by Luis Miguel Silveira, Srinivas Devadas, Ricardo Reis.
p. cm. — (International Federation for Information Processing; 34)
Includes bibliographical references and index.
ISBN 0-7923-7731-1 (alk. paper)
1. Integrated circuits—Very large scale integration—Design and construction—Congresses. 2. Integrated circuits—Very large scale integration—Computer-aided design—Congresses. 3. Embedded computer systems—Design and construction—Congresses. 4. Computer storage devices—Design and construction—Congresses. I. Silveira, Luis Miguel. II. Devadas, Srinivas. III. Reis, Ricardo A.L. (Ricardo Augusto da Luz). IV. Title. V. International Federation for Information Processing (Series); 34.

TK7874.I328 1999
621.39'5—dc21 99-049486

Printed on acid-free paper. Printed in the United States of America.

TABLE OF CONTENTS

Preface

For over three decades now, silicon capacity has steadily been doubling every year and a half with equally staggering improvements continuously being observed in operating speeds. This increase in capacity has allowed for more complex systems to be built on a single silicon chip. Coupled with this functionality increase, speed improvements have fueled tremendous advancements in computing and have enabled new multi-media applications. Such trends, aimed at integrating higher levels of circuit functionality are tightly related to an emphasis on compactness in consumer electronic products and a widespread growth and interest in wireless communications and products. These trends are expected to persist for some time as technology and design methodologies continue to evolve and the era of Systems on a Chip has definitely come of age.

While technology improvements and spiraling silicon capacity allow designers to pack more functions onto a single piece of silicon, they also highlight a pressing challenge for system designers to keep up with such amazing complexity. To handle higher operating speeds and the constraints of portability and connectivity, new circuit techniques have appeared. Intensive research and progress in EDA tools, design methodologies and techniques is required to empower designers with the ability to make efficient use of the potential offered by this increasing silicon capacity and complexity and to enable them to design, test, verify and build such systems.

Solutions to improving designer productivity include the development of new tools and techniques, increasing the abstraction level of designs and introducing reuse of components and systems parts. This book contains a set of papers, which address various subproblems arising in VLSI design, that were presented at the tenth IFIP Very Large Scale Integrated Systems conference, VLSI'99. This conference is organized biannually by IFIP Working Group 10.5. Previous conferences in this series have taken place in Edinburgh, Trondheim, Vancouver, Munich, Grenoble, Tokyo and Gramado. This conference, at the turning point of the millenium and in an atmosphere of rapid and exciting change, took place at the Hotel Meridien Park Atlantic, in Lisbon, Portugal, from 1-4 December 1999.

The current trend towards the realization of complex and versatile Systems on a Chip requires the combined efforts and attention of experts in a wide range of areas including microsystems, embedded hardware/software systems, dedicated ASIC and programmable logic hardware, reconfigurable computing, wireless communications and RF issues, video and image processing, memory systems, low power design techniques, design, test and verification algorithms, modeling and simulation, logic synthesis, and interconnect analysis. Thus, the papers presented at VLSI'99 address a wide range of Systems on a Chip problems.

Traditionally, the conference has been organized around two parallel tracks, one dealing with VLSI Systems Design and Applications and the other discussing VLSI Design Methods and CAD. In this context the following topics were addressed:

VLSI Systems Design and Applications

- Analog Systems Design
- Analog Modeling and Design
- Image Processing
- Reconfigurable Computing
- Memory and System Design
- Low Power Design

VLSI Design Methods and CAD

- Test and Verification
- Analog CAD and Interconnect
- Fundamental CAD Algorithms
- Verification and Simulation
- CAD for Physical Design
- High-level Synthesis and Verification of Embedded Systems

Additionally a number of special sessions and embedded tutorials were organized by experts in their respective fields. These include:

- Design methodologies for Microsystems
- RF Design and Analysis
- FPGA's and reconfigurable hardware

- Architectural Synthesis and Verification
- Timing and Verification
- CAD for Microelectromemchanical systems
- Interconnect Process Parametrization
- Design of Multimedia Systems on Chip

We hereby would like to thank IFIP and more specifically IFIP TC10 and IFIP WG 10.5 for the support of this event, the members of the Organizing Committee and the reviewers that had the daunting task of carefully selecting and providing feedback for the papers submitted. Most of all however, we would like to thank all the researchers and authors that submitted papers to the conference and presented their work there, thus contributing decisively to its success!

Luis Miguel Silveira, Lisboa, Portugal
Srinivas Devadas, Cambridge, MA, U.S.A.
Ricardo Reis, Porto Alegre, RS, Brazil

December, 1999

Conference Committees

GENERAL CHAIR:

Luís Miguel Silveira, - *Systems and Computers Research Institute (INESC) and Cadence Europeans Laboratories (CEL), Dept. of Electrical and Computer Engineering, Instituto Superior Técnico (IST), Technical University of Lisbon, Lisbon, Portugal, lms@inesc.pt*

GENERAL VICE-CHAIR:

José Carlos Monteiro - *Systems and Computers Research Institute (INESC), Dept. of Informatics, Instituto Superior Técnico (IST), Technical University of Lisbon, Lisbon, Portuga, jcm@inesc.pt*

TECHNICAL PROGRAM COMMITTEE CHAIR:

Srinivas Devadas - *Laboratory for Computer Science, Dept. of Elect. Eng. and Comp. Science Massachusetts Institute of Technology, Cambridge, MA USA, devadas@lcs.mit.edu*

VICE CO-CHAIRS OF TECHNICAL PROGRAM COMMITTEE:

Systems Design Track:

Prof. Peter Ivey - *University of Sheffield, UK,p.ivey@sheffield.ac.uk*

Design Methods Track:

João Marques-Silva - *INESC/IST/CEL, Lisbon, Portugal, jpms@inesc.pt*

FINANCE CHAIR:

Arlindo Oliveira - *INESC/IST/CEL, Lisbon, Portugal, aml@inesc.pt*

TUTORIALS CHAIR:

Luciano Lavagno - *DIEGM, Universita di Udine, Udine, Italy, luciano.lavagno@diegm.uniud.it*

PUBLICATIONS CHAIR:

Ricardo Reis - *Instituto de Informática, Universidade Federal do Rio Grande do Sul, Porto Alegre, Brazil, reis@inf.ufrgs.br*

PUBLICITY CHAIR:

Bernard Courtois - *TIMA, Grenoble, France, Bernard.Courtois@imag.fr*

TECHNICAL PROGRAM COMMITTEE:

Adam Postula, *Royal Inst. of Tech., S*
Anton Chichkov, *Alcatel, BE*
Antônio Ferrari, *Univ. Aveiro, P*
Antônio Leal, *IST/INESC, P*
Arlindo Oliveira, *IST/INESC/CEL, P*
Bernard Courtois, *TIMA/IMAG, FR*
Dominique Borrione, *TIMA/IMAG, FR*
Donnatella Sciuto, *Polit. Milano, IT*
Eduard Cerny, *University of Montreal, CA*
Flávio Rech Wagner, *Univ. Rio Grande do Sul, BR*
Franz J. Rammig, *Univ. Paderborn, D*
George Milne, *Univ. of South Australia, AU*
Gerd J.E. Scheller, *Philips AG*
Grant Martin, *Cadence, USA*
Hannu Tenhunen, *KTH, S*
Hans Eveking, *Univ. Darmstadt, D*
Horácio Neto, *IST/INESC, P*
Irith Pomeranz, *Univ. Iowa, USA*
Joan Figueras, *UPC, ES*
João Marques-Silva, *IST/INESC/CEL, P*
João Paulo Teixeira, *IST/INESC, P*
João Vital, *IST/IT, P*
Jochen Jess, *Tech. Univ. Eindhoven, NL*
John Hayes, *Univ. Michigan*
José Carlos Monteiro, *IST/INESC, P*
José de Sousa, *Bell Labs, Lucent*

José Epif'nio da Franca, *IST/IT, P*
Jose Huertas, *Univ. Sevilla, ES*
Karem Sakallah, *Univ. Michigan, USA*
Karl-Heinz Diener, *Fraunhofer Inst., D*
Leonel Sousa, *IST/INESC, P*
Luciano Lavagno, *Polit. Torino, IT*
Luís Miguel Silveira, *IST/INESC/CEL, P*
M. Helena Sarmento, *IST/INESC, P*
Maciej J. Ciesielski, *Univ. Massachusetts, USA*
Manfred Glesner, *Univ. Darmstadt, D*
Manuel Medeiros Silva, *IST/INESC, P*
Martyn Edwards, *UMIST, UK*
Masahiro Fujita, *Fujitsu Labs, USA*
Mattan Kamon, *Microcosm Technologies, MA, USA*
Michel Robert, *Univ. Montpellier, FR*
Neil Weste, *Radiata Communications, AU*
Odysseas Koufopavlou, *University of Patras*
Pasupathi A. Subrahmanyam, *Bell Laboratories, Holmdel, USA*
Paul Jespers, *Univ. Catholique de Louvain, BE*
Pedro Guedes de Oliveira, *DEEC/FEUP*
Peter A. Ivey, *University of Sheffield, UK*
Pranav Ashar, *NEC, USA*
Przemyslaw Bakowski, *IRESTE Nantes, FR*
Ricardo Reis, *Univ. Rio Grande do Sul, BR*
Roger Woods, *The Queen's Univ. of Belfast, UK*
Rolf Becker, *Philips*
Rolf Drechsler, *Univ. Freiburg, D*
Stephane Donnay, *IMEC, BE*
Steve Furber, *Manchester Univ., UK*
Tiziano Villa, *Parades, IT*
Tsutomu Sasao, *Kyushu Institute of Technology, JP*
Vigyan Singhal, *Cadence Berkeley Labs, CA*
Warren A. Hunt, Jr., *IBM Corporation*
Wayne Burleson, *Univ. Massachusetts, USA*
Werner Grass, *Univ. Passau, D*
Wolfgang Kunz, *Univ. Frankfurt, D*
Wolfgang Nebel, *Univ. Oldenburg, D*
Wolfgang Rosenstiel, *Univ. Tuebingen, D*
Yusuke Matsunaga, *Fujitsu, JP*

SPECIAL SESSION CHAIRS:
Jaijeet Roychowdury, *Bell Labs, USA*
Karem Sakallah, *U. Michigan*
Klaus Mueller-Glaser, *TU Karlsruhe, Germany*
Manfred Glesner, *TU Darmstadt, Germany*
Marios Papaefthymiou, *U. Michigan*
Sérgio Bampi, *URFGS, Brazil*
Srinivas Devadas, *MIT, USA*

ORGANIZING COMMITTEE:
Luís Miguel Silveira, *INESC/IST/CEL, Lisboa, Portugal*
José Carlos Monteiro, *INESC/IST, Lisboa, Portugal*
João Marques-Silva, *INESC/IST/CEL, Lisboa, Portugal*
Arlindo Oliveira, *INESC/IST/CEL, Lisboa, Portugal*
Ana Teresa Freitas, *INESC/IST, Lisboa, Portugal*
Ricardo Reis, *UFRGS, Porto Alegre, Brazil*
Ana Cristina Medina Pinto, *UFRGS, Porto Alegre, Brazil*
Flávio Luiz de Oliveira Zimmermann, , *UFRGS, Porto Alegre, Brazil*
Luciano Volcan Agostini, *UFRGS, Porto Alegre, Brazil*

SECRETARIAT:
Ana Cardoso, *INESC, Lisboa, Portugal*
Ana de Jesus, *INESC, Lisboa, Portugal*

Optimizing Mixer Noise Performance: a 2.4 GHz Gilbert Downconversion Mixer for W-CDMA Application

Shenggao Li, Yue Wu, Chunlei Shi, Mohammed Ismail
Analog VLSI Lab, The Ohio-state University
205 Dreese Lab, 2015 Neil Ave.
Columbus, OH 43210
USA

Abstract A downconversion mixer for wideband CDMA application is designed and fabricated in a 0.5 μm CMOS technology. The mixer operates at 2.4 GHz frequency with a 3 V power supply and consumes only 4.5 mW power. Targeting in designing high performance mixers of high yield within short design cycle, this paper addresses RF mixer design issues from an optimization and statistical point of view, and derived the noise equation based on a simple noise model of the Gilbert-cell mixer. The mixer achieved a conversion gain of 14.3 dB, a SSB noise figure of 10.4 dB, and an IIP3 of -8 dBm.

Keywords: RF Mixer, Noise Optimization, WCDMA

1. INTRODUCTION

In recent years the portable communication market has been a strong driving force for IC technology development. In portable systems, low-power and low-cost are the most common requirements. Under the surface of these requirements, the full analog and mixed-signal single-chip solution is of great interest to researchers among industry and academic institutions. This in turn stimulates the evolution of new system architectures with emphasis in using submicron CMOS technology.

Along with the new architectures and new technologies are the numerous wireless standards in different regions and for different services. The current world may not be able to foresee the merge of these different standards. In fact, the standards themselves are lastingly evolving and new standards are still emerging, each with different system specifications and creating highly challenging problems for RF components design. In a portable communication system, RF components take only a very small fraction of the whole system, but the cost for RF parts is relatively much higher than that of baseband and digital parts. This is mainly due to the fact that systematic design methodologies for RF components of high performance are still limited or immature. One example to show the complexity of the problems is the design for low noise figure RF blocks. In general, a RF block can be designed to have an optimum noise figure with regard to a given source impedance. Yet optimum noise figure does not agree with optimum impedance matching, optimum power consumption, or maximal linearity. In a more complicated case, the noise performance of a mixer is even hard to be analyzed in a closed analytical form. Thus, most commonly, a RF system/block is designed in an ad-hoc fashion, with a high cost of human power and a low yield of products.

This paper presents the design of a Gilbert cell mixer which is used in a direct downconversion architecture for a wideband CDMA RF transceiver at the 2.4 GHz ISM band. The design will address the issues mentioned above, and seek to provide some half theoretical and half empirical methods to speed up the design process. Section 2 gives a description of the whole mixer. Section 3 discusses the performance optimization of the mixer with regard to transistor sizes. In Section 4 we give the design result for the mixer followed by a few comments to conclude the paper.

2. CIRCUIT DESCRIPTION

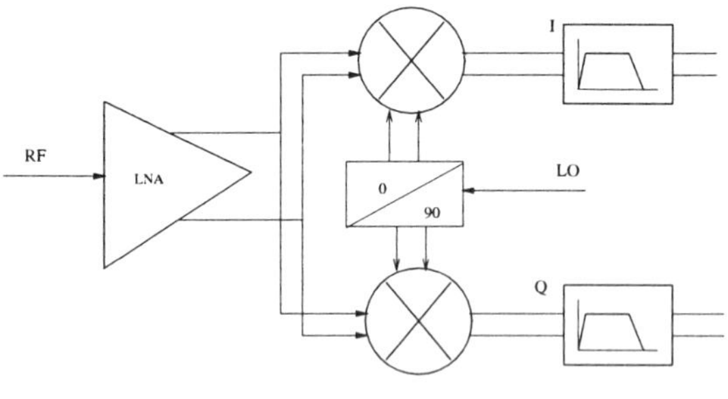

Figure 1 Direct downconversion RF front-end

The direct downconversion wideband CDMA receiver structure is shown in figure 1. The RF signal is located in the 2.4 - 2.48 GHz frequency band, being spread over a 32 MHz bandwidth. We adopted the double balanced Gilbert-cell mixer in our application mainly because the overall performance of Gilbert mixers is generally superior to that of other type of mixers at Gigahertz working frequency. Since low-voltage and low-power design is the mainstream for portable products, this work will address the LV/LP design issues with regard to Gilbert cell mixer. The power supply in this design is 3 volts. It should be noted that new mixer architectures should be sought as power supply is scaled down below 2.0 V. For RF mixers, while most design efforts are centered on designing the mixer core to achieve good noise figure, linearity, conversion gain, etc., attention should also be paid to the biasing circuits to ensure the core circuit work well and achieve good performance [7]. For instance, variation of the tail current, and variation of LO biasing, can degrade the linearity and noise figure tremendously. Figure 2 is the complete circuit diagram of the mixer,

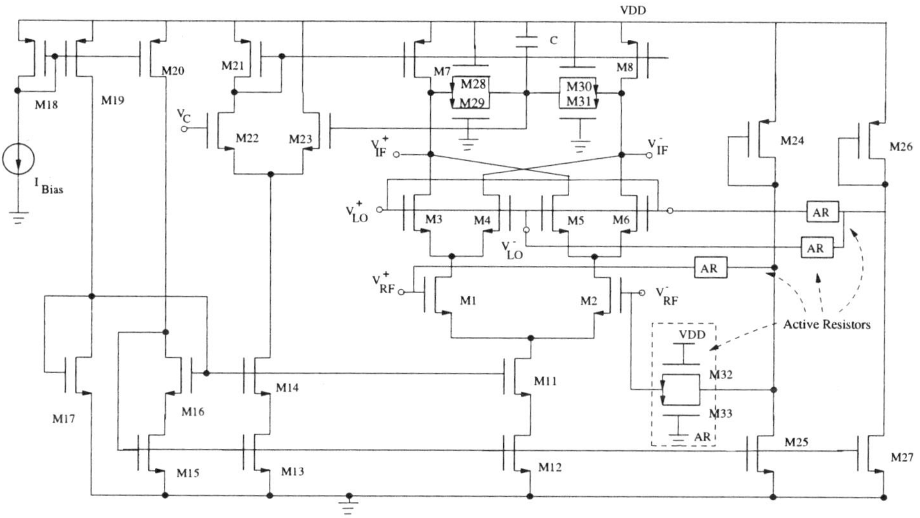

Figure 2 Schematic diagram of the mixer

similar structure can be found in [4] [5].

In this circuit, transistors M1-8 form the core of the Gilbert cell, with M1-2 the input transconductance stage, M3-6 the commuting switches, and M7-8 the PMOS active load. For certain applications, the load transistors M7-8 can be replaced by passive inductors, hence to reduce the power supply. However, inductive load is not suitable for our broadband and direct downconversion case since the inductive load essentially presents a variable gain over the frequencies and zero gain at DC.

The tail current of the Gilbert cell is supplied by a wide-swing current mirror composed of M11-12 and M15-17. The wide-swing current mirror keeps the Gilbert cell from distortion at the low end of the output voltage swing. For testing purpose, the biasing current of the whole circuit is controlled through an exterior current source of 0.1 mA(I_{Bias}). The scaling factor of the current mirror is 1:10, giving 1 mA for the nominal tail current of the Gilbert cell. Instead, the actual obtained current is 1.1 mA because device parameters such as V_T vary with regard to channel width and length. In this current mirror, M11-12 work right at the edge of the triode region, so do M15-16. To account for the parametric mismatch during fabrication that might deviate any of M11-12 and M15-16 from working in active region, the drain current of M19 is designed to be slightly larger than that of M20 [2]. With the above design, the output tail current is constant when the drain voltage of M11 is driven to as low as 0.3 V.

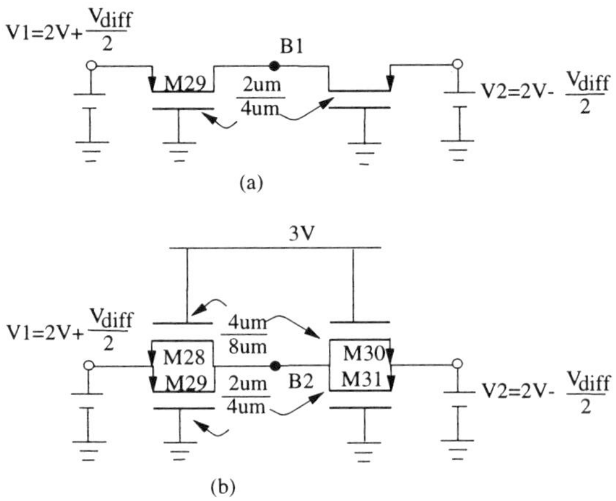

Figure 3 Common-mode voltage acquistion

Transistors M21-23 form the common-mode feedback loop. The common-mode voltage is acquired by transistors M28-31, which work in triode region and function as active resistors. Since the gate-source voltage applied to M28-29 and M30-31 are not equal, the circuit is not symmetrical, meaning the resistance given by M28-29 and M30-31 are not equal. Figure 4 shows the common-mode voltage characteristics when a differential-mode signal is

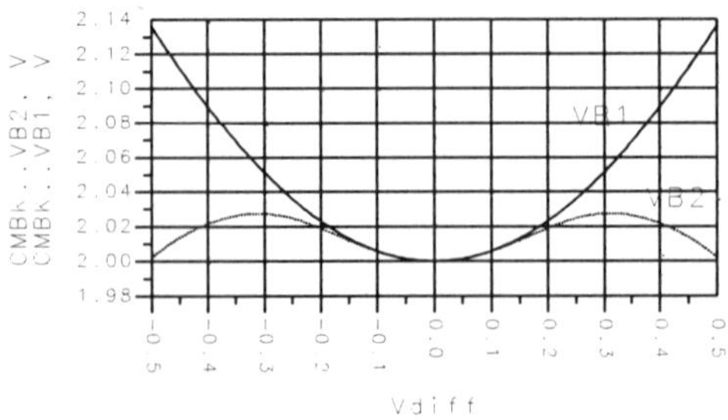

Figure 4 Common-mode voltage

applied as shown in figure 3(a)(b). Without any common-mode voltage variation, the differential-mode voltage will drive the CMFB loop, causing the common-mode output deviate from its nominal value. If the differential output has a wide swing, this deviation may cause problem for succeeding stages. Note that the NMOS transistors in this complementary structure add to the noise of the circuit. If the noise contribution from M28, M30 is significant, we should consider using only M29 and M31 [figure 3(a)]. However, the PMOS-only configuration causes even larger deviation [VB1 in figure 4(a)].

M24-27 provides DC biasing voltage to the LO and RF ports. The AR blocks, briefed for active resistors (M32-33), are used to isolate the negative and positive input signal ports, and to keep the biasing circuitry from loading the input signals.

3. PERFORMANCE OPTIMIZATION

To achieve a high quality of communication under existing technologies, the RF front ends are often required to work right at their performance limits. The design of a satisfactory RF system is a trade-off of many critical performance parameters. Generally, the design of RF components is a try-and-error procedure. It is worthwhile to investigate optimization techniques to speed up this procedure. With CMOS technologies, performance parameters of RF components are directly related to the width and length of building transistors. For a given circuit, ideally, we can represent all the performance parameters by these geometric parameters quantitatively, which gives a complete description of the problem space. As long as the problem space is completely describable, one can identify a set of optimal solutions to satisfy the targeted system specifications. It also becomes possible to investigate how each geometric parameter affects the system performance, thus to identify those most critical parameters. In reality, however, with tens of transistors in a circuit, the problem space has already been prohibitively large in terms of the variable geometric parameters, let alone that in RF frequency, one need to take high order parasitic effects into consideration. For these reasons, many works consider the optimization problem with regard to only parts of the performance parameters [6] [3], or less complicated circuits such as LNA [1]. The overall design optimization of mixers, somehow is still a myth to designers, and much effort is based on experience.

To design the Gilbert cell mixer for our direct downconversion application, we need to size transistors M1-8 carefully to approach a trade-off among the different performance parameters. The first consideration is the LO leakage and self-mixing problem. With down-scaled tail current, the aspect ratio of the M1-8 can be chosen to be small, as a result, the transistor sizes can be also small, which reduces parasitic capacitance and in turn reduces LO leakage. However, the transconductance of M1-2 should be moderately high enough in order to achieve certain conversion gain. The effective gate-source voltage of M3-6 should be small enough to ensure that the transistors can be switched on and off quickly, at the same time, to avoid pushing the source voltage of M3-6 too low such that M1-2 is driven to triode region. These two requirements demand larger aspect ratio for M1-6. Meanwhile, the aspect ratio of M7-8 is chosen to satisfy the output DC working point. To obtain proper conversion gain, the length of M7-8 can be adjusted so that the output impedance is changed.

The noise figure is affected when we scale the sizes of M1-8. We want to see how each transistor contributes to the noise of the mixer. Generally, with MOS devices, large device geometry is desirable for better noise performance. Intuitively, for the mixer, increasing the conversion gain will reduce input-referred noise level. The problem is that in our case we favor moderate gain. In other words, with moderate gain, the noise performance may not meet system specification. A fine tuning of the width and length of M1-8 is then necessary.

For the wideband direct downconversion mixer, we realize that the main noise contributors are baseband flicker and thermal noise, and thermal noise downconverted from RF frequency. The actual noise mechanism of the mixer is hard to describe analytically. We attempt to understand the noise in such a way:

- Baseband noise from M3-8 directly contributes to the output without the need to consider frequency mixing.
- Baseband noise from M1-2 also directly contributes to the output. To understand this, one just needs to notice that M3-M5, and M4-M6 always form a path from M1 or M2 to the IF ports.
- LO signal and its harmonics mix with noise from M1-2 at different noise band near LO and its harmonics down to baseband [3].

To investigate the noise behavior with regard to the transistor geometry, we simplify the problem and look at only the signal path formed by M1, M3, and M7. The equivalent noise model is illustrated in figure 5. We first assume that all three transistors working in active region (This assumption is true when LO signal are in transition).

The input-referred flicker noise is given by

$$
\begin{aligned}
v_{in,f}^2 &= v_{g1,in}^2 + v_{g3,in}^2 + v_{g7,in}^2 \\
&= v_{g1}^2 + v_{g3}^2 \frac{(\frac{I}{\lambda_n L_1})^2}{g_{m1}^2} + v_{g7}^2 \frac{g_{m7}^2}{g_{m1}^2} \\
&= \frac{K}{W_1 L_1 f} + \frac{K}{W_3 L_3 f} \frac{(\frac{I}{\lambda_n L_1})^2}{2\mu_n C_{ox} \frac{W_1}{L_1} I}
\end{aligned}
$$

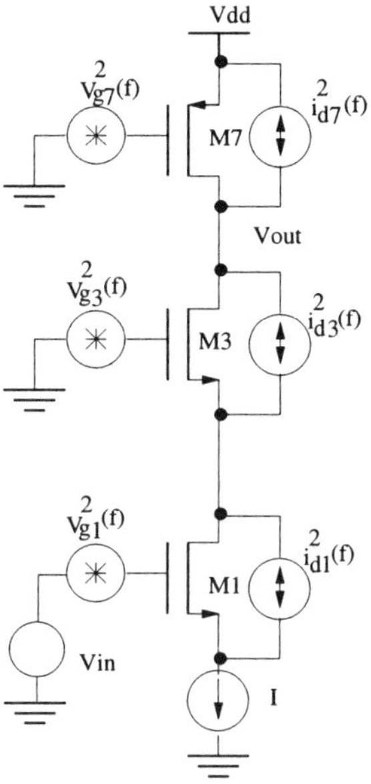

Figure 5 Noise model of the Gilbert cell

$$\begin{aligned} & + \frac{K}{W_7 L_7 f} \frac{2\mu_p C_{ox} \frac{W_7}{L_7} I}{2\mu_n C_{ox} \frac{W_1}{L_1} I} \\ = & \frac{K_1}{W_1 L_1 f} + \frac{K_2 I}{W_1 L_1 W_3 L_3 f} + \frac{K_3 L_1}{W_1 L_7^2} \end{aligned} \tag{1.1}$$

where K_1, K_2, K_3 are process dependent constants.

The input-referred thermal noise is given by

$$\begin{aligned} v_{in,t}^2 & \simeq \frac{i_{d1}^2}{g_{m1}^2} + \frac{i_{d3}^2 r_{ds3}^2}{g_{m1}^2 (r_{ds1} + r_{ds3} + r_{ds7})^2} + \frac{i_{d7}^2 r_{ds7}^2}{g_{m1}^2 r_{ds1}} \\ & \simeq \frac{4kT\frac{2}{3}g_{m1}}{g_{m1}^2} + \frac{4kT\frac{2}{3}g_{m3}}{g_{m1}^2} + \frac{4kT\frac{2}{3}g_{m7}(\frac{\lambda_p L_7}{I})^2}{g_{m1}^2(\frac{\lambda_n L_1}{I})^2} \\ & = K_4 \frac{\sqrt{L_1}}{\sqrt{W_1 I}} + K_5 \frac{L_1 \sqrt{W_3}}{W_1 \sqrt{L_3 I}} + K_6 \frac{\sqrt{W_7 L_7} L_7}{W_1 L_1 \sqrt{I}} \end{aligned} \tag{1.2}$$

where K_4, K_5, K_6 are also process dependent constants.

Now consider the situation when M3 works in triode region (M1 and M7 still in active region). The noise terms contributed by M1 and M7 are still valid in equation (1.1) and (1.2). However, in first order approximation, the flicker noise of M3 does not have much influence on $V_{in,f}^2$ and can be neglected from equation (1.1). The input referred thermal noise from M3, is now given by the following equation,

$$
\begin{aligned}
v_{in,t3}^2 &= \frac{I_{ds}^2 r_{ds3}^2}{g_{m1}^2 r_{ds1}^2} \\
&= \frac{\frac{4kT}{\mu C_{ox}\frac{W_3}{L_3}(V_{gs3}-V_t)}}{2\mu C_{ox}\frac{W_1}{L_1}I(\frac{\lambda_n L_1}{I})^2} \\
&= K5\frac{L_3 I}{W_3 W_1 L_1 (V_{gs3}-V_t)} \qquad (1.3)
\end{aligned}
$$

The obtained noise equations can be utilized in design optimization. One can find that increasing W_1 will reduce both thermal noise and flicker noise. Increasing L_1, however, does not necessarily achieve lower noise, implying that an optimal value exists for L_1. The equations also quantitatively illustrate the geometric influence of other transistors, as well as the biasing current and LO voltage. The result, in combination with those in [6] [1], is aimed to provide some insight into mixer design and performance optimization. It efficiently helped to reduce the design time in achieving appropriate conversion gain, noise figure and linearity in our design. It should be reminded that the simplified model does not take 2nd or higher order parasitic effect and short channel effect into consideration. More rigorous efforts should be taken to ensure analytical reliability and accuracy.

4. CONCLUSION

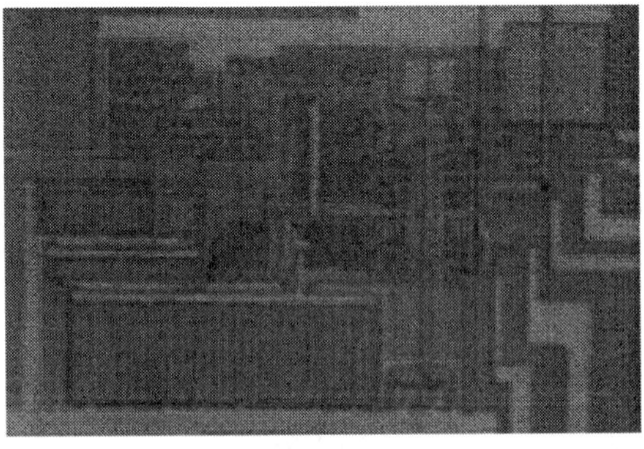

Figure 6 Die Photo

This paper presents the design of a direct downconversion mixer for wide-band CDMA application. Figure 6 is the die photo of the mixer fabricated in a 0.5 μ m CMOS technology. Table 1 summarizes the simulated performance parameters. The test is still under processing.

Specifications	Results
Power Supply	3 V
Current Consumption	1.5 mA
LO Input Power	0 dBm
RF Input Power	-60 dBm
Conversion Gain	14.26 dB
SSB Noise figure	10.66 dB
IIP3	-8.02 dBm
SFDR	65.03 dB

Table 1 Performance Summary

In this work, we discussed the design and optimization issue of mixers. We derived the noise equations according to a simplified noise model as a guideline for mixer design. For high volume and high performance RF components design, systematic design methods, along with effective statistic analysis techniques, should be pursued in the near future.

5. ACKNOWLEDGMENTS

Support for this research was provided by Texas Instrument Graduate Fellowship at The Ohio State University.

6. REFERENCES

[1] C.-H. Feng, F. Jonsson, M. Ismail and H. Olsson. "Analysis of Nonlinearities in RF CMOS Amplifiers". *6th IEEE International Conference on Electronics, Circuits and Systems (ICECS '99), Paphos, Cyprus*, Setp. 1999.

[2] David Johns, Ken Martin, Kenneth W. Martin. *Analog Integrated Circuit Design.* Jon Wiley & Sons, 1996.

[3] Keng Leong Fong. *Design and Optimization Techniques for Monolithic RF Downconversion Mixers.* University of California, Berkeley(PhD Thesis), Fall 1997.

[4] Jacques C. Rudell, Paul Gray. "A 1.9-GHz Wide-Band IF Double Conversion CMOS Receiver for Cordless Telephone Applications". *IEEE Journal of Solid-state Circuits*, 32(12):2071–2088, Dec. 1997.

[5] P.Litmanen, P.Ikalainen, K.Halonen. "A 2.0 GHz Submicron CMOS LNA and A Downconversion Mixer". *IEEE International Symposium On Circuits and Systems 1998*, pages 357–359, May 1998.

[6] Qiuting Huang, Francesco Piazza, Paolo Orsatti, and Tatsuya Ohguro. "The Impact of Scaling Down to Deep Submicron on CMOS RF Circuits". *IEEE Journal of Solid-state Circuits*, 33(7):1023–1036, July 1998.

[7] Shenggao Li, Advisor: Prof. M. Ismail. *A Low-power Submicron CMOS Downconversion Mixer for Wideband CDMA Transceiver Application.* The Ohio-state University(MS Thesis), May 1999.

An Analog Non-Volatile Storage System for Audio Signals with Signal Conditioning for Mobile Communication Devices.

Geoffrey B. Jackson, Saleel V. Awsare, Lawrence D. Engh, Mark A Hemming, Peter Holzmann, Oliver C. Kao, Chun Mai-Liu, Carl R. Palmer, Aditya Raina

Storage Devices, A Winbond Company, 2727 North.First Street, San Jose, CA 95134, USA

Key words: Mobile Communications, Multi Level Storage (MLS), Configurable System

Abstract: Presented is a system on a chip for conditioning of voiceband analog audio signals for use in mobile communication devices. The system allows for direct interface to acoustic transducer elements and provides signal conditioning to gain adjust, multiplex, filter and mix two independent signals. The system can record these processed signals as analog samples in a non-volatile flash EEPROM array for later retrieval. Control of the system is achieved via a serial interface, which is used to configure and control the device. All necessary components of the system are provided on chip including analog processing elements, non-volatile storage and high voltage and reference generation.

1. INTRODUCTION

In any mobile communication system (e.g. cellular telephony)(Fig.1), it is indispensable to have the ability to process two streams of information; namely upstream (information from the local user to the remote caller) and downstream (information from the remote caller to the local user). Other desirable features in a mobile communication environment include a voice memo function, full-duplex voice record and playback, answering machine and call screening functions. Also, minimum external components and low power consumption are vital. This paper describes a system on a chip solution capable of processing and storing voice-band signals while incorporating all of these aforementioned features. By inserting itself between the baseband module of a cell phone and the acoustic transducers, this unique system on a chip can perform the analog processes of several chips thereby enhancing system level integration.

2. CHIP ARCHITECTURE

The chip is divided into three parts (Fig. 8). The top section contains the high voltage circuits needed to program the flash cells along with the digital logic needed for the SPI interface, chip control and timing generation. The middle section consists of the array, column drivers, and row decoders. The column drivers include analog sample and hold circuits along with analog comparators to perform the analog non-volatile storage algorithm. The bottom section consists of the analog signal paths and associated signal conditioning circuits and the reference generation circuits. Three separate power buses are used for isolating noise, one for high voltage generation circuits, one to supply digital logic and a third for the analog section. The chip runs from a 2.7V-3.3V supply and incorporates programmable power down control to minimise power consumption in all modes

2.1 Memory Array and Operations

A 0.6um two poly source-side injection (SSI) cell (Fig. 2) is the basic unit of the memory array. The flash cells are arranged in an array of bit lines (Fig 5), word lines and common source lines shared by adjacent rows This memory cell consists of a select gate (SG) transistor and a floating gate (FG) transistor merged in a split-gate configuration. There are three terminals–the common source (CS), which accesses from the FG-transistor side, the drain, which accesses from the SG-transistor side and the select gate (SG). The memory array is organised in a NOR architecture, where the select gates form the word lines, the drains are strapped by first metal to form the bit lines and the common source lines, parallel to the word lines are strapped by second metal. The programming voltage is coupled to floating gate via CS diffusion to FG overlap. Hot carriers from the channel current promote impact ionisation on the source-side of the FG transistor provide efficient cell programming. Poly to poly electron tunneling erases the cell. Refer to

Table 1 for the conditions applied to the memory cell during a Read, Program and Erase operation.

2.2 Algorithm and Programming Characteristics

To write an analog sample from the sample and hold circuits to the memory cell, a writing algorithm is used. The writing algorithm is based on a closed loop iterative program and verify cycle. The cell is first erased and then subjected to a train of programming pulses applied to the common source node as illustrated in (Fig. 3a). A column is selected by sinking the appropriate programming current from the bit line as illustrated in (Fig. 3b). After each programming pulse, the cells are read back and compared to the voltage of sample and hold, capacitor. When the desired value is reached the bit line current sink is disabled barring further programming. This programming algorithm is made practical to achieve a large cell window for stored signals,. The variations of memory cells from wafer to wafer and lot to lot further reduce such window as illustrated in (Fig. 3c).

2.3 The S/H and the Writing Circuits

Once the signal has been sampled onto the sample and hold capacitors, the samples are programmed into the memory cells in parallel; hence there are multiple sample and hold (S/H) circuits on the system. This allows the actual programming of the memory cell to take much longer than the sampling time. The samples will be held and used by the writing circuit. The sample and hold circuit is shown in (Fig. 4). This S/H circuit can be connected to a unit gain operational amplifier (Op Amp), which is common to all the other S/H circuits. The 'select' signal determines which S/H will be connected. When the S/H is disconnected, the analog input sample can be retrieved from the source node of a native NMOS transistor. This voltage will then be used to program the memory cell. The signal 'bank select' connects either 'bank A' or 'bank B' of the S/H circuits. There are two banks of S/H circuits. While programming the samples of one bank the other bank can be loaded with new samples. Therefore, programming the memory array is a non-stop operation. (Fig. 5) shows how the S/H circuit including the two banks is connected to the writing circuit. During programming, a common source node and a select gate node in the memory array are selected by the 'Xdecoder'. The 'Waveshaper' and the high voltage 'Driver' supply the waveform as shown in (Fig. 3). This waveform is applied to the selected common source node. During each programming cycle a high voltage (HV) pulse is applied to the common source node, while a programming current is flowing to a selected bitline. This bitline is selected through a column multiplexer (MULTIPLEXER). After the HV pulse is applied, the source follower voltage (Vsf) of the selected cell is read and compared to the sampled voltage. If the Vsf is equal or less than the sampled voltage a latch will be reset. The latch will cause the selected bitline to be tied to an inhibit voltage 'Vxx'. This will stop further programming. There are multiple copies of the S/H circuit with comparator latch and column MULTIPLEXER on-chip. This allows the multiple cells to be programmed in parallel.

2.4 HV generation and distribution

Fig 6 illustrates a simplified block diagram of the high voltage generation and distribution. The erase and iterative programming pulses (Fig. 3a.) are generated via the block CDAC, which is a digital to analog converter. As the counter (10bit HVINC) counts up CDAC produces pulses from 6 to 12 V, which increment in 16mv steps. The pulses are applied to the CS of memory cell in the array. Two separate op amps are used during the read and the program operations. The voltage applied to the CS line is force-sensed to eliminate the drop along the decoder switches. The voltages are then passes through a predecoder (XRED) and a decoder (XDEC) according to which memory cell in the array needs to be programmed.

2.5 Analog Path

The analog path (Fig. 7) has been designed to provide maximum flexibility and ease of integration when interfacing with any mobile communication system. There are three signal inputs namely, MIC+/-, AUXIN, ANAIN and three signal outputs, ANAOUT+/-, SP+/-, and AUXOUT. Internally there are several analog processing blocks interconnected by programmable multiplexers. Fully differential signal paths are utilised on-chip to maximise signal quality and the multiplexers utilise pumped gate bias to reduce distortion and non-linearity. The processing blocks are as follows:

- Microphone Automatic Gain Control (AGC). This is designed for a 3mV to 300mV input signal with an output level fixed to maximise the array resolution. The AGC is a two-stage circuit consisting of a variable gain stage utilising NMOS transistors with a variable gate control voltage to control the gain followed by a novel switched capacitor AC coupled, fixed gain stage.

- Summation Amplifiers. These two amplifiers allow the mixing of signal paths to achieve full-duplex recording or playback functions.
- Sample Rate of the device. It has four user selectable settings to produce sample rates of 4, 5.3, 6.4 and 8kHz. The oscillator is referenced to a 0TC(0 Temperature Coefficient) current source derived from an on-chip bandgap reference.
- Low Pass Filter. This a 5th order Chebyshev filter used as an anti-aliasing filter in record mode and a smoothing filter in playback. The filter uses MOSFET resistors whose control voltage is derived from the oscillator current, forcing the cut-off frequency to track the oscillator frequency over the 4-8kHz range of sample frequencies.
- Volume Control. An 8-step volume control/attenuator is provided allowing signal adjustment in 4dB steps.
- Balanced ANAOUT amplifier. A high signal quality balanced output is provided to interface to the cellular baseband section
- Speaker Driver. A 23mW speaker driver is integrated to differentially drive an 8-Ohm load. The amplifier uses pumped voltages to allow rail-to-rail output swing.
- Variable Gain Input Amplifiers. The AUXIN and ANAIN inputs incorporate variable gain amplifiers to allow interfacing of signal levels to the array.
- Multilevel non-volatile analog memory storage array, which can be written upto one million cycles, and stores data without power consumption for 100 years.

A description of the various paths is given below (Fig.7). These paths are activated by issuing SPI commands to the system:

a) *Feed-through Mode:* In this mode the user communicates with the remote caller without the device recording or conditioning the signal. The user's signal is received at MIC+ and MIC-, goes through a 6dB gain element and is transmitted to ANAOUT+ and ANAOUT-. Also the remote user's signal can be received at ANAIN, passed through a variable gain amplifier, a multiplexer, and to a speaker driver which drives the speaker.

b) *Record Mode:* In this mode the user's signal is coupled in at MIC+, MIC- and goes through an AGC circuit which produces a signal level that fits the array window, and input multiplexer, summing amp, filer multiplexer, low pass anti-aliasing filter which smoothes this signal, another summing amp and is stored in the non-volatile array. The signal can also be recorded from AUXIN. The signal of the remote caller can be coupled to ANAIN and can be recorded with the local user talking at the same time.

c) *Play OGM Mode:* This mode is used to play an outgoing message in a mobile application or in a pure answering machine application. The signal goes from the storage array to the anaout amplifier that transmits the signal upstream through the baseband circuit, via a path consisting of the following blocks: filter multiplexer, low pass filter and summing amp2.

d) *Full duplex record mode:* In this mode both sides of a conversation (user and caller) can be recorded. The analog signal of the user is transmitted upstream to the remote caller through the signal path that includes the 6dB amp, anaout mux and anaout amp, sum2 amp, the agc amp and the input mux. The remote caller's signal is received at ANAIN and is transmitted through the anain amp, output mux, and speaker driver amplifier to the user. The remote caller's analog signal is also fed to sum1 amp, which mixes the 2 signals. This mixed signal passes through the filter to the storage array.

e) *Full duplex play mode:* This mode is used to playback a stored message to the remote caller while the user is talking to the remote caller. The signal paths involve mixing the user's analog signal at the Mic inputs with the message in the storage array and transmitting the mixed signal upstream to the remote caller. The user's analog signal is coupled to the Mic inputs and routed to sum1 amp through the AGC amplifier and the input mux. The message in the storage array goes through the filter, through the filter mux and is applied to sum1 amp which mixes the two signals. The mixed signal is routed to ANAOUT+, ANAOUT- through the anaout mux and amplifier for transmission upstream to the remote caller. The second path involves mixing the remote caller's analog signal with the message recorded in the storage array and providing the mixed signal to the user. The remote caller's analog signal is received at ANAIN input, amplified by a variable gain amplifier, through the sum2 amp which mixes the remote caller's signal with the message stored in the storage array and provides this to the volume control circuit which adjusts the level of the signal and presents it to the speaker driver through the output mux.

2.6 Command Set

For interfacing to a microcontroller, this system uses the SPI interface with a smart instruction set. The instruction set is designed to easily accomplish frequent operations such as play or a record operation or message cue operation. The analog path is configured by the user via a 32 bit configuration register which is used to set multiplexer's gains, sample rate and allow power down of unused blocks thus reducing power consumption.

3. CONCLUSION

A 2.7V to 3.3V analog signal processing and storage system has been presented (Fig. 8) for interfacing with the cellular baseband system (Table2), combining a fully integrated programmable signal interface. The system has a configurable signal path to maximize flexibility and ease system integration with all wireless and cordless chipsets. In mobile communication applications, the system allows for two-way call recording, call screening, playback or recorded message during a call, voice memo, and an answering machine/call screening function.

4. ACKNOWLEDGEMENTS

The authors would like to thank James Brennan Jr., Don Tran, Theresa Chung and Chris Chan for their support during the project.

5. REFERENCES

[1] T. Blyth, et al., *ISSCC Digest of technical Papers*, pp. 192-193, Feb. 1991.

[2] H. Tran, et al., *ISSCC Digest of technical Papers*, pp. 270-271, Feb., 1996.

[3] J. Brennan, et al., *Proc of NVSMW*, pp. 41-42, August, 1998.

[4] B. Yeh, *U.S. Patent No. 5,029,130*, July, 1991.

[5] Y. Tsividis, et al., *IEEE JSSC*, pp. 15-30, Feb., 1986

[6] Al Kordesh, et al,VLSI Technology Systems, and Applications, June 1999

[8] C.Liu, et al, VLSI Technology Systems, and Applications, June 1999

Table1. MEMORY CELL OPERATIONS

		Mode	
Parameter	Erase	Program	Read
Bit line current Ip or Id		-1uA	-1uA
Common source voltage Vcs	0V	6-12V	2.2V
Select gate voltage Vsg	~15V	2.3V	4.2V
Bit line voltage Vsf	float	~0.8V	*measure*

Table 2. CHIP CHARACTERISTICS

PARAMETER NAME	VALUE
Temperature	- 40C to 90C
No. Of equivalent bits	~ 8
VCC	3.0 V +/- 10%
SAMPLE RATES	4,5.3,6,4,8KhZ
DURATION	4-8 minutes
Analog Path SINAD	62.5dB(@ 1kHz dBm0)
AGC Input Range	3-300mV
Speaker Driver Power	24mW @ 8 Ohms
Record Cycles	100,000(typical)
Message Retention)	100 Years (typical)
ISB	< 1 μA
Die Size	4.1 x 7.6 mm
ICC	< 30 mA
Technology	0.0um Flash CMOS
SINAD	42 dB @32 mVpp, 1 kHz
THD	0.5%

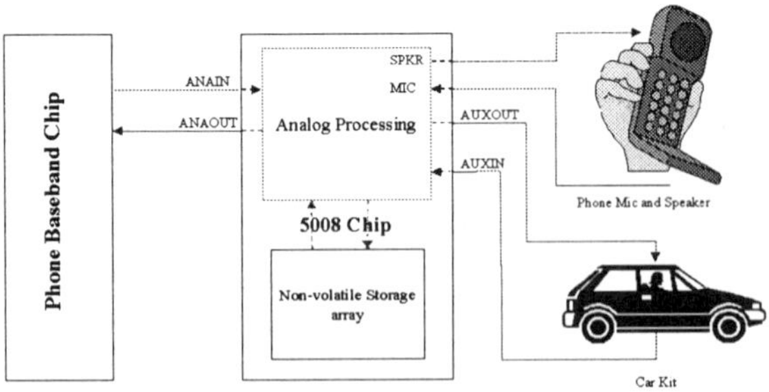

Figure 1. System Configuration for Applications

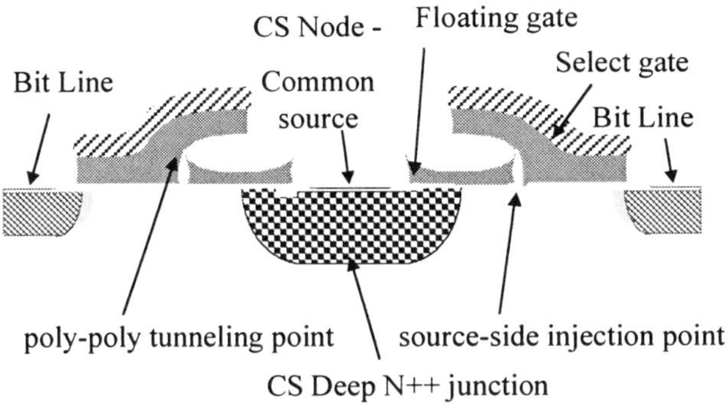

Figure 2. Cross Section of Cell

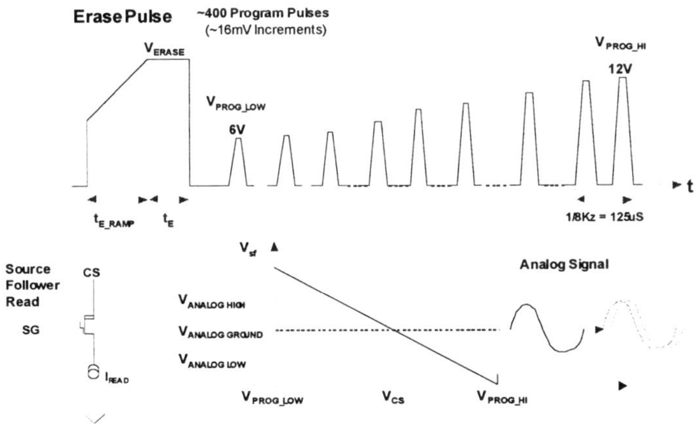

Figure 3. Cross Section of Cell a) a depiction of the erase and cumulative program sequence b) memory cell schematics c) Vsf-Vcs cumulative program characteristic curve, analog signal and analog window

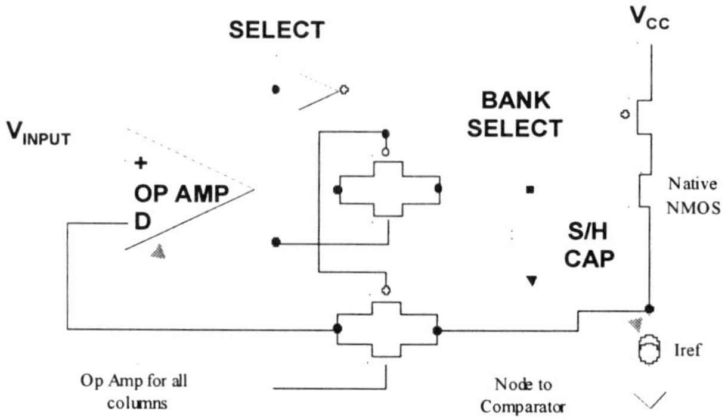

Figure 4 The S/H circuit – the Op Amp drives one S/H at a time to bring the comparator node to the same voltage as Vinput

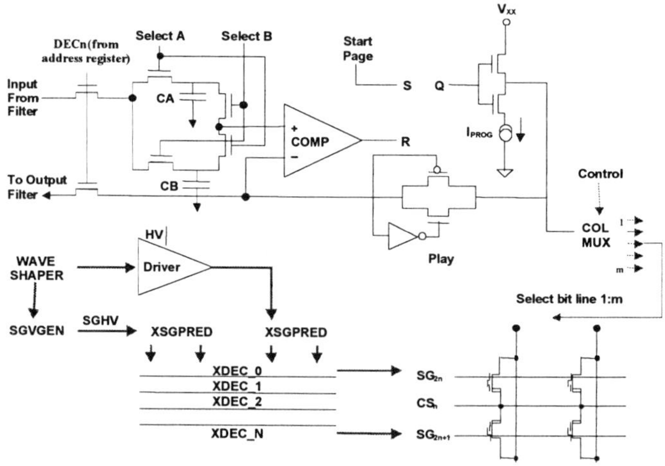

Figure 5 The writing circuits – the wave shaper, HV driver, row decoders, comparators column multiplexer and the memory array

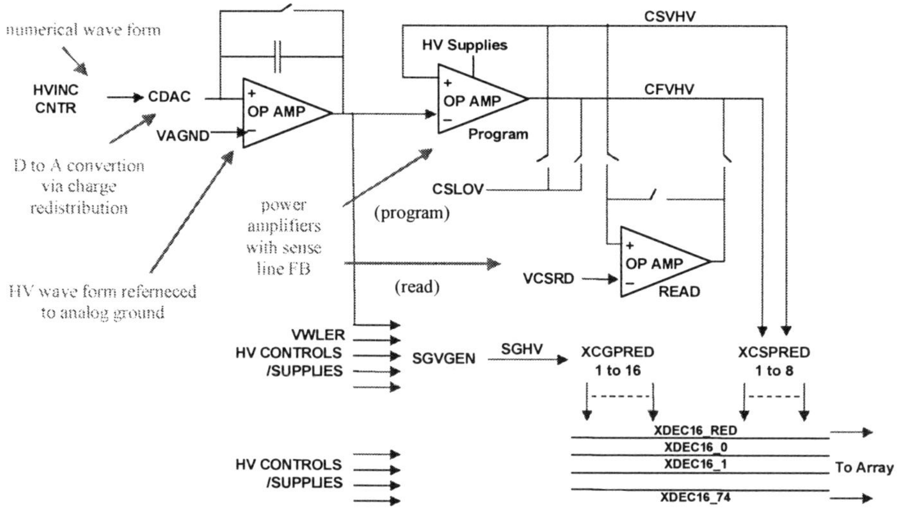

Figure 6. Data Storage Architecture

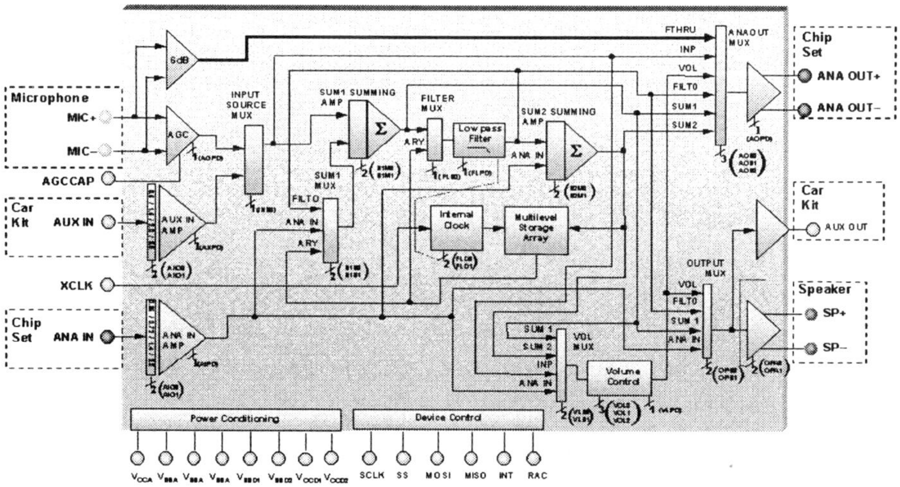

Figure 7. Analog Path Block Diagram

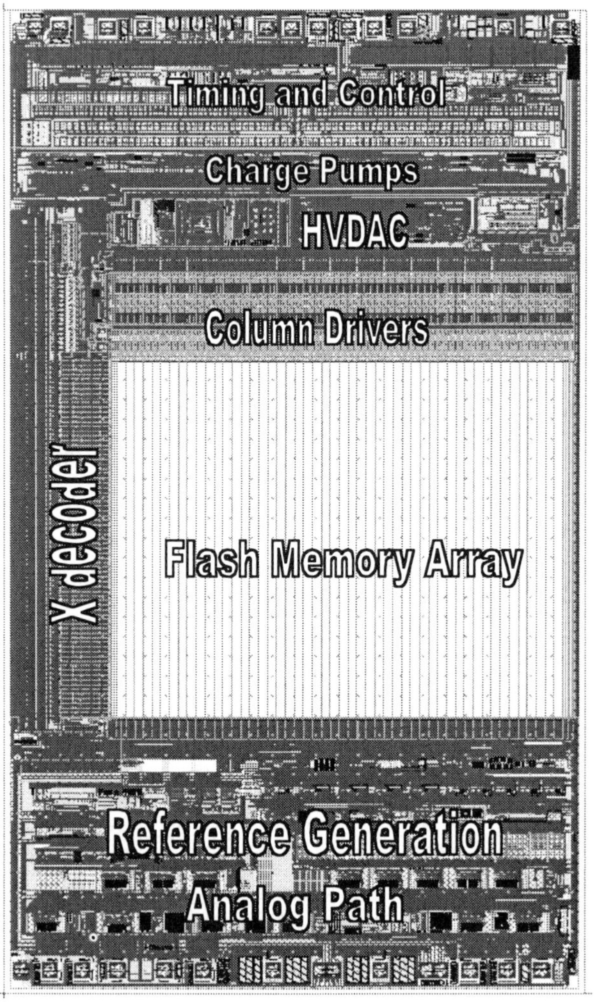

Figure 8. Die Photo

A Design of Operational Amplifiers for Sigma Delta Modulators using 0.35um CMOS Process

Bingxin Li and Hannu Tenhunen
Electronic System Design Laboratory, Royal Institute of Technology, Electrum 229, Isafjordsgatan 22, 164 40 Kista, Sweden, lbingxin, hannu@ele.kth.se

Key words: CMOS, operational amplifier, sigma delta modulator, A/D converter

Abstract: An operational amplifier designed with 0.35um CMOS technology is presented. All the transistors are realized with minimum or near-minimum channel length. As the short channel length causes performance degradation, a proper operational amplifier structure is selected to compensate the performance degradation. The op amp is designed to meet the requirement of high-speed high-resolution sigma delta modulators. It has a folded-cascode first stage and a class-A output stage. It features a DC gain of 78dB, an open-loop unity-gain frequency of 266MHZ, a slew rate of 650V/us, and consumes 10.2mW from a +/-1.5V power supply. High level simulation is used to evaluate the OTA performance in sigma delta modulators.

1. INTRODUCTION

The fast development of CMOS process technique makes it possible to integrate more and more functions into a single Digital-signal-Processing chip. However, the physical signal (which is analog) still needs an interface to be handled by DSP. A/D and D/A converters are such interfaces. In the area of high resolution A/D conversion, sigma delta converters are the best choice. They adopt oversampling and noise shaping technology to move the quantization noise out of signal band, and then digitally low-pass filter the shaped noise. From the middle of 80s they have been widely used in digital audio applications. The biggest drawback of sigma delta modulators is that they can not convert wide band signals as their counterpart, flash A/D converter can do. The resolution of a sigma delta modulator (in bit, or dB) is nearly linearly proportional to the oversampling ratio. Thus to achieve high resolution the modulator must work at a sampling frequency that is much higher than the signal band. With a signal band of 1MHz and oversampling ratio of 16 (which is already relatively low), the modulator must sample at 32MHz. Many researches have been done to increase the signal band that can be converted by sigma delta modulators [1,2], and obviously that the essential task is to design a high performance operational amplifier which is used in switch capacitor integrators. The amplifier should have a high DC gain, large bandwidth, large slew rate (driving capability) and large output swing.

Although CMOS process has already stepped into deep submicrometer, the analog transistors are still often designed with much larger channel length than in digital circuits. One of the reasons is that using small channel length transistor will degrade the performance. For example, Rds, the drain-source resistance of a NMOS transistor in small-signal model will reduce according to the left formula due to channel-length modulation. And the short-channel effects will make things even worse. It is obvious that Rds will decrease when the channel length L decreases, even though the Width/Length ratio keeps the same. When Vds is so big that short-channel effect become apparent, Rds will reduce even faster than predicted by the above formula.

$$Rds = \frac{2L}{\lambda \mu_n C_{ox} W (V_{gs}-V_{tn})^2}$$

where

$$\lambda = \frac{k_{ds}}{2L(V_{ds}-V_{eff}+\Phi_0)^{1/2}}$$

Figure 1 shows the simulated Rds of a NMOS transistor in AMS 0.35um process [3]. The transistor has a fixed Width/Length ratio (5:1) and fixed V_{eff} (V_{gs}-V_{tn}) while the value of length changes from 0.4um to 5um. The maximum Rds in active regions is plotted against the channel length. Note that in an operational amplifier Rds is directly related to the DC gain. Figure 2 shows this simulation for a simple common-source gain stage with an active load (T2). The bias current and Width/Length ratio of each transistor are fixed, while the absolute value of the channel length is varied. The gain of this stage is changed accordingly. So using a short channel length transistor will automatically degrade the DC gain unless otherwise compensated. This compensation can be done on the structure level, for example, by changing a single stage to a two-stage structure. With a proper design, analog circuit can also take the advantage of advanced CMOS process without sacrificing performance.

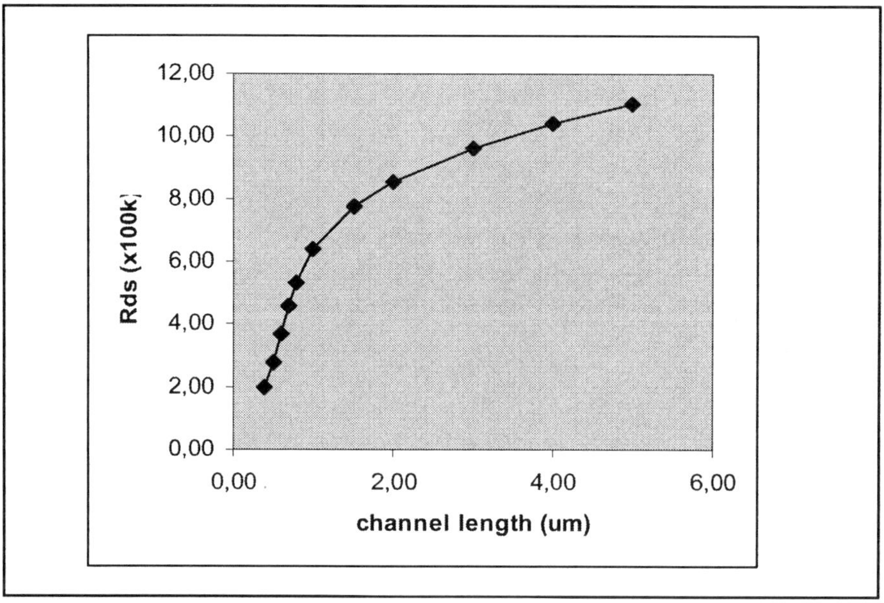

Figure 1. Rds of a NMOS with different channel lengths

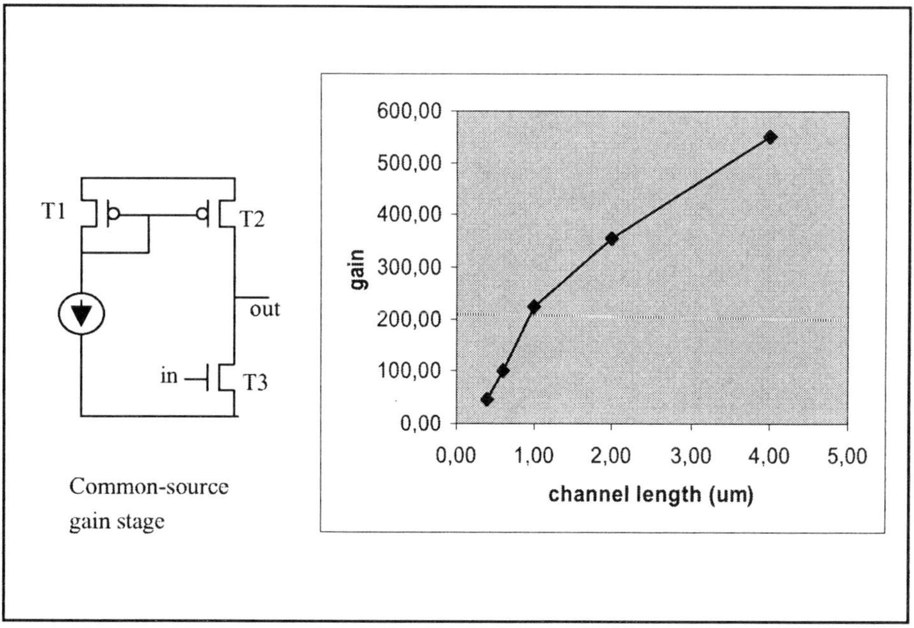

Figure 2. Gain of a common-source stage with different channel lengths

2. DESIGN TARGETS AND STRUCTURE SELECTION

The target is to design an operational amplifier for sigma delta modulators that can work at a sampling frequency above 40MHz. With high level simulation of sigma delta modulators this requirement of the modulator can be translated into several requirements of the amplifier [1], for example, the DC gain, the open loop cut-off frequency, the slew rate, the phase margin and so on.

As the amplifier will be used in switch capacitor circuits, the load is only capacitive (no resistance load). Thus the so-called Operational Transconductance Amplifier (OTA) is a good choice (output buffer stage is omitted). Folded-cascode OTA has been used extensively in switch capacitor circuits. It has the advantages of high DC gain, large output swing, large bandwidth and simple structure. Therefore it has been used successfully in many high-speed high-resolution sigma delta modulators [4]. However, to have a large bandwidth and large slew rate many of the transistors in OTA need to have very large transconductance, which also means large Width/Length ratio. This increases the die area and parasitics. If a short channel length, e.g. 0.35um, can be used to implement these transistors, we can reduce the chip size and the parasitics as well. However as mentioned before, using short channel length transistors will degrade the performance. So here a two-stage OTA [5] is selected as shown in figure 3. The inner stage is a traditional folded-cascode amplifier while the outer stage is a class-A stage. The OTA is designed using AMS 0.35um process. All the transistors have minimum or near-minimum channel length. With the addition of the second stage, the requirement of the first stage can be largely loosen, thus the transistor Width/Length ratio and bias current can be reduced. The class-A second stage provides large drive capability (slew rate) and large output swing (near rail-to-rail). As a result, it is possible to achieve higher overall performance while consumes less power and has smaller die area (compared with the single-stage folded-cascode implementation with large channel length). The other attractive advantage of this circuit is that it can work with very low operating voltage. With the careful arrangement of transistor sizes and bias currents, it can work well with 1.5V power supply [5]. This makes it a candidate circuit for future processes that require lower supply voltages.

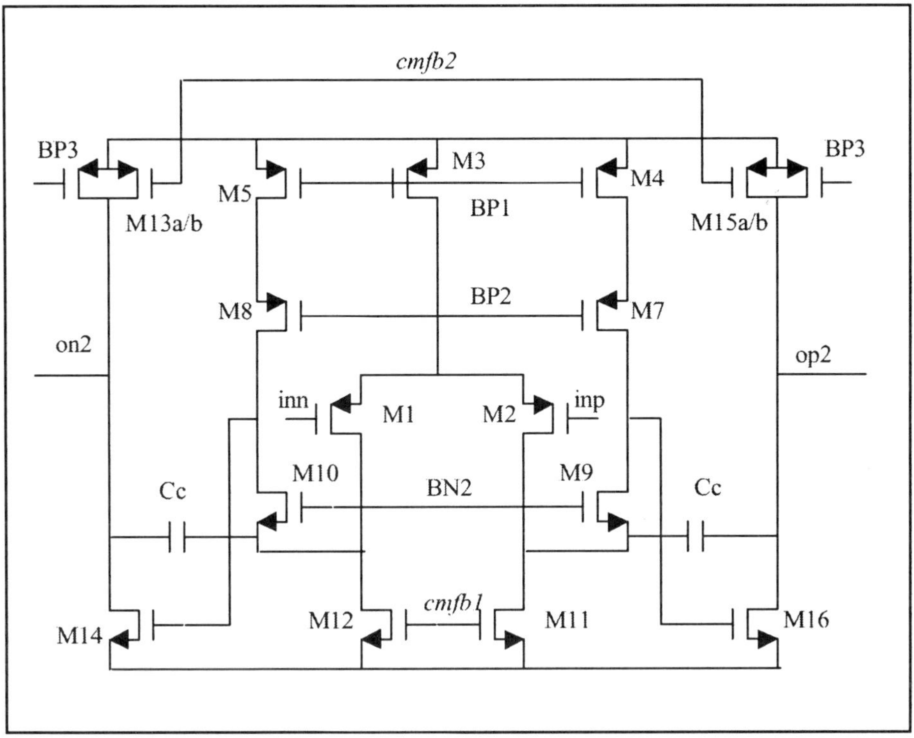

Figure 3. Two-stage OTA with folded-cascode inner stage and class-A outer stage

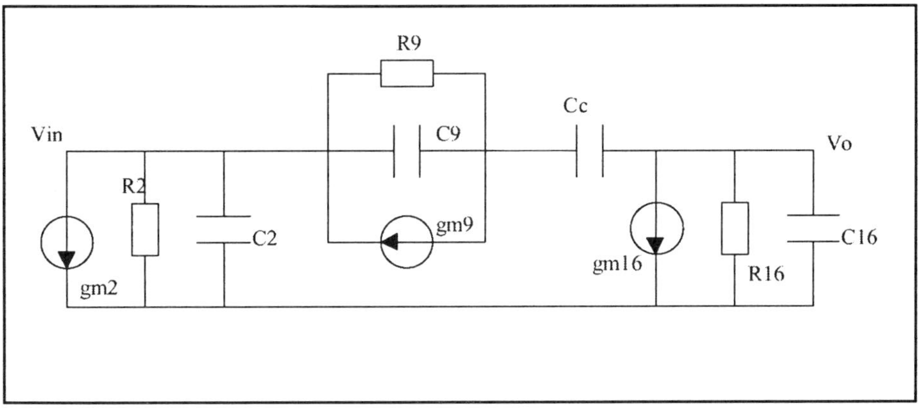

Figure 4. Simplified small signal model for differential half circuit

3. CIRCUIT ANALYSIS

3.1 DC Gain

In a sigma delta modulator the finite DC gain of the operational amplifier will cause integrator leakage [6,7], which degrades the overall performance, especially in cascaded modulators. To implement a high-resolution modulator, the DC gain should be made above 60dB. In CMOS Op amp it is conflicting to achieve both a high gain and large slew rate. A high gain requires small bias current while a large slew rate requires a large bias current. In this two-stage structure the first stage can be biased at a relatively small current to have a large gain, and the second stage be biased at large current to achieve large slew rate. The first folded-cascode stage has a DC gain in the order of $(g_m r_o)^2$, and the second class-A stage has a DC gain in the order of $g_m r_o$. So the total DC gain is in the order of $(g_m r_o)^3$. This makes it easy to achieve a DC gain of 1000 or higher.

3.2 Slew Rate

In a high-speed sigma delta modulator, the OTA slew rate must be large enough to guarantee that the output signal is accurately settled within a very short period (less than half of a clock period). Some empirical formulas can be used as a guideline to find the needed slew rate for a specific application. For example, "Slew Rate>7*Vref*fs", which is derived from high-level simulations [1]. The slew rate of the first stage is defined by the bias current in M1, M2 and their load capacitance (include compensation capacitor). The bias current of M14, M16 and the output load capacitance define the slew rate of the second stage. The overall slew rate is limited by either the first stage or the second stage, whichever is slower. If the first stage is designed to be fast enough the slew rate is defined by the bias current of the second stage and its load capacitance. The proper bias current can be found through SPICE simulation.

3.3 Frequency Response and Compensation

The first stage's frequency compensation is realised by cascode compensation capacitor Cc. Compared with standard Miller compensation, it has much better PWRR at high frequency and reduced capacitance load to the first stage. It is first introduced in [8] and then widely used in many designs [9,10]. The simplified ac small-signal model for the differential half circuit is shown in figure 4. With some reasonable assumption, the OTA can be modelled as a 3-pole 2-zero system. The first pole is formed by the input transistors M1 and M2, so their g_m (g_{m2} in figure 3) should be big enough to move the first pole to higher frequency. The distance between the second and third pole is determined by the value of Cc. With a large Cc, these two poles are moved further away from each other. This is the so-called pole-splitting effect. However Cc also introduces a zero on positive real axis which will reduce the phase margin unless it is put far away from cut-off frequency. The proper value of Cc can be found through hand calculation of the simplified ac small-signal model, and then be verified through simulation. And it is worth mentioning that to have enough phase margin, the g_m of M9, M10 (g_{m9} in figure 4) and M14, M16 (g_{m16} in figure 4) should be much larger than that of M1 or M2.

3.4 Bias Circuit

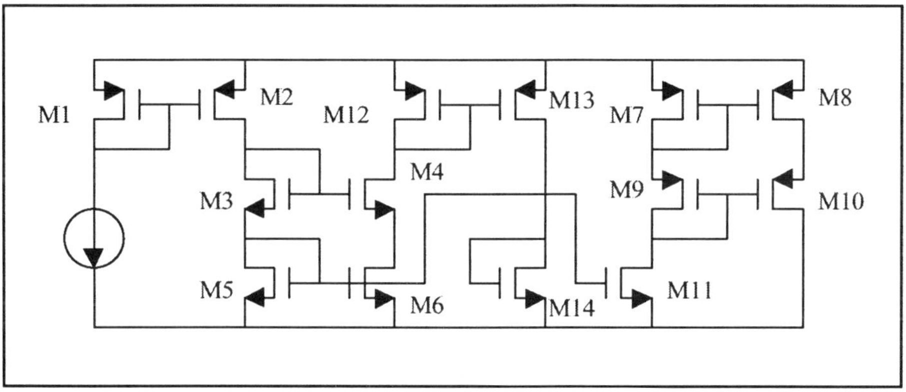

Figure 5. Bias circuit of OTA

The bias circuit of the OTA is shown in figure 5. It provides proper bias current for both PMOS and NMOS transistors. It also includes a duplicate of class-A output stage (MB13 and MB14) to establish its common-mode input voltage which is used in the first common-mode feedback circuit.

3.5 Common-mode Feedback

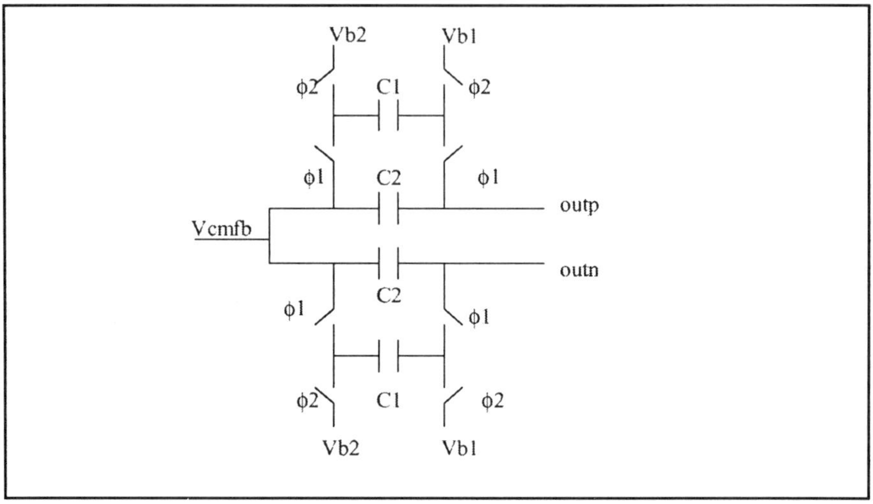

Figure 6. Switch capacitor common-mode feedback circuit

The control of common-mode output voltage is more difficult than in the folded-cascode OTA because there are two independent stages. Some designs use a single common-mode-feed-back (CMFB) circuit [2], where the common-mode voltage of the second stage is sensed and then put through a sign inversion circuit (often a current mirror, which consumes extra power) to control the bias in the first stage. In this implementation we use two separate CMFB circuits. With this scheme, we can accurately control the common-mode output voltage of the first stage to be the required input bias voltage of the class-A second stage [11]. The second stage's common-mode output voltage is set to be the middle of power suppliers. In this way the common-mode voltages of both stages can be fast and accurately controlled. The two switch-capacitor CMFB circuits are identical and shown in figure 6. They are suitable for switch-capacitor circuits and consume much less power than continuous CMFB circuits.

4. OTA PERFORMANCE SIMULATION

Figure 7 shows the simulated transient response and frequency response. In both cases each output terminal is loaded with 2pF capacitor. The slew rate is found to be 650v/us. The DC gain is 79dB, which is enough to meet the requirement of a high-resolution sigma delta modulator. The cut-off frequency is about 266MHz and the phase margin is 44 degree. Power consumption is 10.2mW (not include the bias circuit) with +/- 1.5V power supply.

5. HIGH LEVEL MODULATOR SIMULATION WITH OTA NON-IDEALITIES

Once the OTA characteristic is extracted, it can be put through high-level sigma-delta modulator simulations. Figure 8 Shows a 5th order single-stage single-bit modulator optimised for OSR=64 [12]. A full SPICE level simulation of such a modulator is too time-consuming and often leads to inaccurate results. High level simulation (e.g. based on MATLAB) is therefore often adopted to efficiently evaluate the performance. To make this kind of simulation accurate enough to be useful, we must consider non-idealities in the component model. It is known that in a sigma-delta modulator only the first integrator's non-ideality has dominant influence on the overall performance [12]. Thus in high level simulations only the first integrator needs to be modelled with non-idealities, while all other components can be modelled as ideal. The non-idealities include kT/C noise, OTA noise, clock jitter, OTA slew rate, bandwidth and limited DC gain. The modelling of these non-idealities is described in [13]. The effect of clock jitter in sigma delta modulator can be simplified as its effect on the input signal sampling. The kT/C noise and OTA internal noise can be modelled as an input-referred random noise. The effect of finite DC gain can be modelled as a modification on integrator transfer function. The finite bandwidth and slew rate can be modelled with some form of non-linear integrator gain. The simulation uses a sampling frequency of 40MHz, and the signal band is 312.5KHz (OSR=64). It achieves a SNDR of 92.3dB (figure 9). Compared with the ideal case which has a SNDR of 110 dB (figure 10), about 18 dB SNDR is lost. This means the noise power caused by integrator non-idealities is about 10 time larger than the pure quantization noise. The kT/C noise contributes most to this 18dB SNDR lost (with only kT/C noise, the SNDR will reduce to 96dB), while the OTA non-idealities contributes much less.

6. CONCLUSIONS

In analog circuits which need large Width/Length ratio transistors (e.g. Op amp) we can consider using very short channel length transistors. With a proper design, the circuit can maintain its performance while having a smaller die area and less parasitics. An OTA is designed in such a way for high-speed high-resolution sigma delta modulators. Its high performance makes it capable to work at a sampling frequency of 40MHz or more.

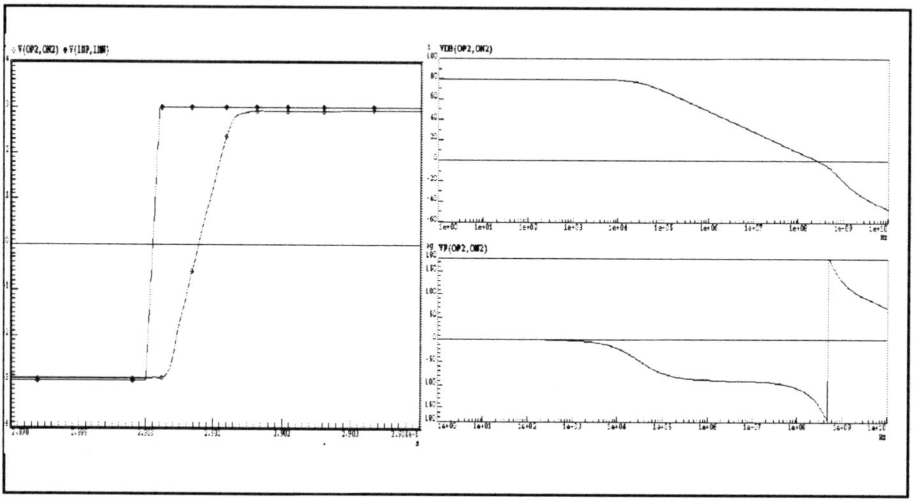

Figure 7. Transient response and frequency response of OTA

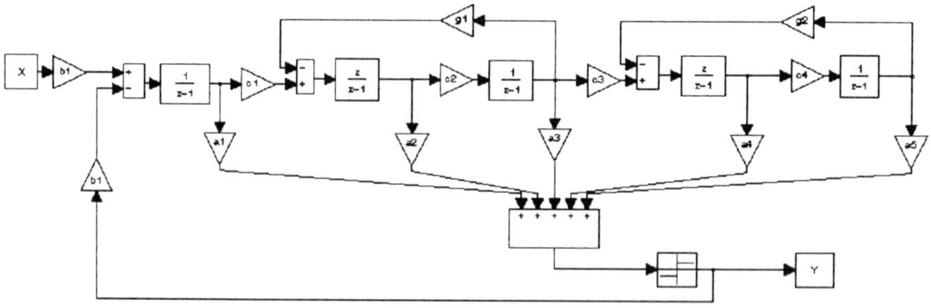

Figure 8. A 5th order single-stage single-bit sigma delta modulator (The coefficients are optimised for OSR=64. a1=0.5734; a2=0.5279; a3=0.4495; a4=0.2588; a5=0.2620; g1=0.0032; g2=0.0375; b1=0.8711; c1=0.4087; c2=0.2200; c3=0.2264; c4=0.0528)

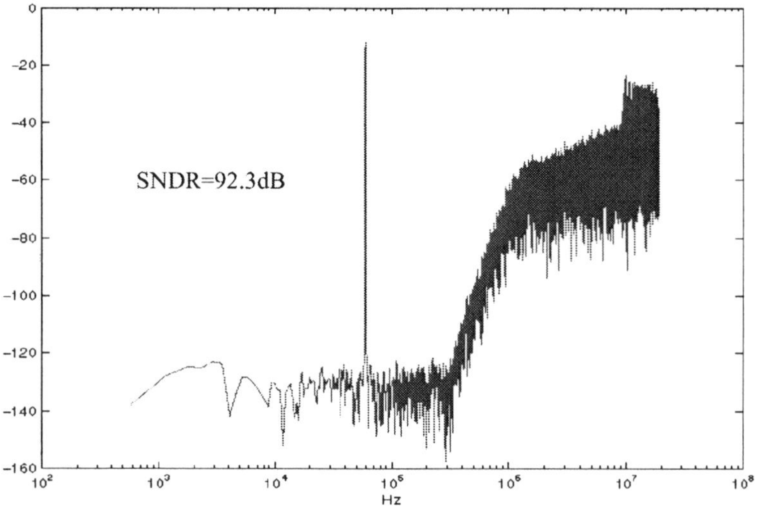

Figure 9. PSD of modulator output with first integrator's non-idealities included

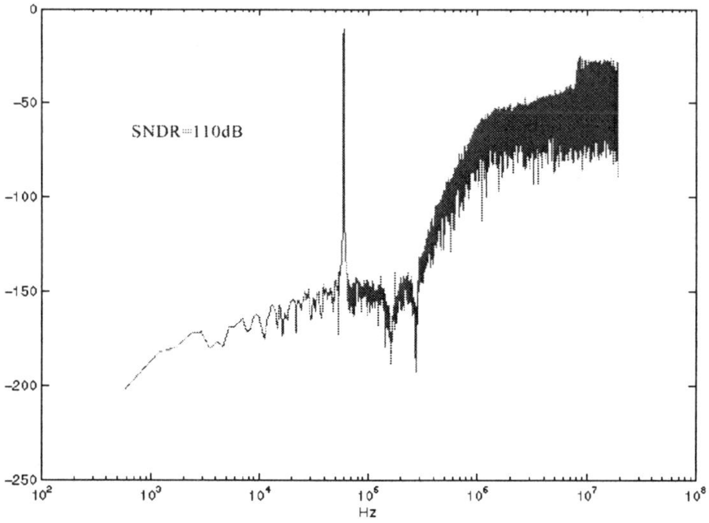

Figure 10. PSD of modulator output with all the components treated as ideal

7 REFERENCES

[1] A. M. Marques, V. Peluso, M. S. J. Steyaert and W. Sansen, "A 15-b Resolution 2-MHz Nyquist Rate DS ADC in a 1-um CMOS Technology", *IEEE J. Solid-State Circuits*, vol. 33, Jul. 1998, pp. 1065--75

[2] A. R. Feldman, B. E. Boser and P. Gray, "A 13-bit, 1.4-MS/s Sigma-Delta Modulator for RF Baseband Channel Applications", *IEEE J. Solid-State Circuits*, vol. 33, Oct. 1998, pp. 1462--69

[3] http://www.ams.co.at/corporate/profile.html

[4] S. Ingalsuo, T. Ritoniemi & H. Tenhunen, "New Differential Amplifier for High Speed Switched Capacitor Circuits", *in proc. of European Conference on Circuit Theory and Design (ECCTD-91)*, Copenhagen, Denmark, September 1991.

[5] A. M. Abo and P. Gray, "A 1.5-V, 10-bit, 14.3-MS/s CMOS Pipeline Analog-to-Digital Converter", *IEEE J. Solid-State Circuits*, vol. 34, May. 1999, pp. 599--606

[6] S. R. Northworthy, R. Schreier and G. C. Temes, *Delta -Sigma Data Converters, Theory, Design, and Simulation*, IEEE press, 1997

[7] B. E. Boser and B. A. Wooley, "The Design of Sigma-Delta Modulation Analog-to-Digital Converters", *IEEE J. Solid-State Circuits*, vol. SC-23 Dec. 1988, pp. 1298--1308

[8] B. K. Ahuja, "An improved Frequency Compensation Technique for CMOS Operational Amplifier", *IEEE J. Solid-State Circuits*, vol. SC-18, Dec. 1983, pp. 629--633

[9] D. B. Ribner and M. A. Copeland, "Design Techniques for Cascoded CMOS Op Amps with Improved PSRR and Common-Mode Input Range", *IEEE J. Solid-State Circuits*, vol. sc-19, Dec. 1984, pp. 919--625

[10] K. Nakamura, M. Hotta, L. R. Carley and D. J. Allstot, "An 85 mW, 10b, 40 Msample/s CMOS Parallel-Pipelined ADC", *IEEE J. Solid-State Circuits*, vol. 30, Mar. 1995, pp. 173--183

[11] S. Rabii and B. A. Wooley, *The Design of Low-Voltage, Low-Power Sigma-Delta Modulators*, Kluwer Academic Publishers, 1999

[12] B.X. Li and H. Tenhunen, "Hierarchical Modeling of Sigma Delta Modulators for Noise Coupling Analysis", *in proc. of 1999 Southwest Symposium on Mixed-Signal Design*, Apr., 1999, pp. 58-62

[13] S. Brigati, F. Francesconi, P. Malcovati, D. Tonietto, A. Baschirotto and F. Maloberti, "Modeling Sigma-Delta Modulator Non-idealities in Simulink", *in proc. of IEEE International Symposium on Circuits and Systems*, May, 1999, pp. II 384-387

A Low Power CMOS Micromixer for GHz Wireless Applications

Yue Wu[1,2], Shenggao Li[1], Mohammed Ismail[1,2], Håkan Olsson[2]
[1] *Analog VLSI Lab, The Ohio State University*
205 Dreese Lab, 2015 Neil Ave.
Columbus, OH 43210, USA
[2] *Radio Electronics Lab, Royal Institute of Technology,*
Electrum 229, S-164 40 Kista, Sweden
wuy, lisg, ismail@ee.eng.ohio-state.edu
hakan@ele.kth.se

Abstract This paper presents a CMOS RF mixer with low power consumption and high linearity. The low power and high linearity is achieved with a class-AB input stage. Detailed analysis of the circuit has been given. The mixer is working at 2.44GHz for WCDMA system, with a power consumption less than 3 mW, it achieved 7.5 dB conversion gain, 4.6 dBm IIP3 and 13dB NF. Simulation results performed on HPADS has been presented, HP0.5μ m CMOS process is used.

Keywords: Down-Conversion Mixer, Class-AB mode, Conversion Gain, Intermodulation, Noise Figure, WCDMA, Wireless Systems

1. INTRODUCTION

In wireless communication systems, downconversion mixer plays an important role because it converts the radio frequency (RF) signal down to intermediate frequency (IF) or baseband frequency. The reason to convert signal down to lower frequency is that it is difficult to design high Q channel filter at RF frequency with reasonable low noise and power consumption. Because of the low g_m/I ratio inherent to MOS transistors, it is difficult to design low power MOS mixer with high gain, high linearity and good noise performance([1],[2],[3]). In [4], a new mixer structure called micromixer has been proposed for BJT, this topology is also very suitable for low power MOS mixer design. In this paper, a very low power MOS mixer is designed and simulated. The circuit uses HP0.5μm technology, it can achieve 7.5 dB conversion gain, 4.6dBm IIP3 and 13dB Noise Figure with power consumpsion of only 3mW. Section 2 introduces mixer topology and performance parameters, followed by a detailed analysis of the class-AB input stage in section 3. In section 4, the design of the mixer and simulation results are presented, finally conclusion is summarized in section 5.

2. MIXER STRUCTURE AND PERFORMANCE PARAMETERS

A commutating mixer can be divided into three stages as shown in Fig.1, i.e. the input stage, the switching stage and the load stage. The input stage is a transconductance stage that converts the input voltage to a current. For the downconversion mixer case, this part is working at RF frequency, so this stage must have enough transconductance at high frequency to translate the RF voltage signal efficiently to RF current signal. The second stage is the switching stage, which really performs the frequency translation through the commuting action of a NMOS common-source differential pair. The load stage is a transimpedance stage, it converts the switched current back into voltage.

The mixer shown in Fig.1 is a double-balanced commutating mixer. In the input stage, RF input voltage has been translated into currents which are represented by current sources I_{RF1} and I_{RF2}. If we assume ideal switch at the commutating stage, which means the switching activity will not affect the work of the input stage, and there is no current loss when the switch is on, then it is easy to get the output voltage as

$$V_{out}(t) = R[I_{RF1}(t) - I_{RF2}(t)][S_1(t) - S_2(t)] \tag{1.1}$$

where R is the load impedance of the output stage. After the Fourier Series Expansion of the last item of previous equation, we have

$$V_{out}(t) = Rg_m v_{rf}(t)\frac{4}{\pi}cos(\omega_{LO}t) - Rg_m v_{rf}(t)\frac{4}{3\pi}cos(3\omega_{LO}t) + \cdot \tag{1.2}$$

The first item in the previous equation is the desired downconverted signal at the output of the mixer, higher order items will be greatly attenuated by the low pass characteristic of the circuits or by the following filters. For a single-side band downconversion mixer, the conversion gain is

$$Gain = \frac{2}{\pi}Rg_m \tag{1.3}$$

Note that in this equation, the conversion gain is not related to LO amplitute, this is because we assume ideal switch for the switching stage. In real design, when LO is too small, the transistors can not be cut off completely, there will be gain loss and high noise. LO signal can not be too large either, when it is the case, it may force the input stage into triode region and will cause gain loss too.

In the above derivation, we assume that the differential currents are linear to the input voltage, i.e. $I_{RF1}(t) - I_{RF2}(t) = g_m v_{rf}(t)$. Unfortunately, in most cases this is only an approximation which is valid when input signal is small. When input signal is strong or there exists large interference signals, we have to consider the nonlinear effects of the input stage, this is what IP3(third order intercept point) accounts for. If the input signal is a two tone signal, $v_{rf}(t) = A(cos(\omega_1 t) + cos(\omega_2 t))$, and the differential current is as following

$$I_{RF1}(t) - I_{RF2}(t) = \alpha_1 v_{rf}(t) + \alpha_2 v_{rf}^2(t) + \alpha_3 v_{rf}^3(t) + \cdots \quad (1.4)$$

we can get the amplitute of frequency component at $2\omega_1 - \omega_2$ and $2\omega_2 - \omega_1$, which are the result of third order intermodulation, to be equal to $\frac{3}{4}|\alpha_3|A^3$. If we set it equal to the fundamental frequency output, then the IIP3(3rd order Input Intercept Point) will be $A_{IP3}^2 = 4\alpha_1/3\alpha_3$[5]. Since mixer is the stage following the LNA, the input signal from antenna has been amplified, so the linearity requirement of mixer is much higher than LNA, normally it will be about -5dBm.

A third important parameter is the Noise Figure(NF). Even though the effect of mixer noise can be attenuated by the gain of LNA, it is desirable to have low noise mixer to relax the design of LNA. The calculation of mixer NF is complex because the input and output noise are in different frquency domain, a systematic analysis can be found in [6]. If the switch is fast enough, the noise will be dominated by the input stage.

From the above introduction, it is clear that the input stage plays an important role in that it can determine the gain, linearity and noise performance of a mixer. There always exist tradeoffs in RF circuit design among those parameters, the following section will introduce a new input stage which can achieve low power, high linearity and high gain simultaneously.

3. CLASS-AB INPUT STAGE

Recently, a new input stage has been proposed([4]) for bipolar mixer, and mixer with this kind of input is named as Micromixer according to the author of that paper. The key point of this structure is the class-AB behavior of the input stage which uses a common emitter transistor and a common base transistor to generate the differential input current. This idea can be used in CMOS mixer design, following is the analysis and design of the CMOS Micromixer.

Fig.2 shows the input stage of the Micromixer, it is composed of the biasing transistor Mb1, Mb2 and M2, the common gate transistor M1 and common source stage M3. If we assume that the two biasing transistors have the same size, M1-M3 also have the same size, with biasing current is I_q, then according to Fig.2, there exists a translinear loop Mb1, Mb2, M1 and M2, so we have

$$V_{gs,1} + V_{gs,2} = V_{gs,b1} + V_{gs,b2} \tag{1.5}$$

If all these four transistors are working in saturation region, and we assume that all the transistors have the same threshold voltages, V_t, then we have

$$\sqrt{I_1} + \sqrt{I_2} = 2\sqrt{\frac{k_1}{k_0}}\sqrt{I_q} \tag{1.6}$$

where $k = \frac{1}{2}\mu C_{ox}\frac{W}{L}$. Since we also have I_2-I_1=I_{in}, so it is easy to derive currents I_1 and I_2 as following

$$I_1 = \lambda(1 - \frac{I_{in}}{4\lambda})^2, I_2 = \lambda(1 + \frac{I_{in}}{4\lambda})^2 \tag{1.7}$$

where λ=$(k_1/k_0)I_q$ is a constant determined by the size ratio of the input to biasing transistors and quiescent current.

From the above equation, when $I_{in} = 0$, I_1 and I_2 will have the same value $(k_1/k_0)I_q$. When I_{in} increase, then I_1 will decrease and I_2 will rise or vice versa. The current relationship is shown in Fig.3, it is clear that this stage is working at class-AB status and will have low power consumption compared with differential pair.

Even though neither current is linear with the input current I_{in}, the difference between I_1 and I_2 is still equal to I_{in}. Note that the above equations are valid only when the input current I_{in} is smaller than $4(k_1/k_0)I_q$. Otherwise, according to the equations, one of the input transistor will cut off (i.e. I_1 or I_2 will be zero) and the derivation is no longer valid.

According to equation(3), it is important to determine the transconductance of the input stage because it determines the conversion gain of the mixer. Since we have

$$V_{RF} = V_{gs,1} = V_q + \frac{\sqrt{k_0}}{4k_1\sqrt{I_q}}I_{in} \tag{1.8}$$

where $V_q = V_t + \sqrt{I_q/k_0}$ is the DC voltage of biasing transistors, the large signal transconductance g_m will be

$$g_m = \frac{1}{R_{in}} = \frac{1}{\frac{\partial V_{RF}}{\partial I_{in}}} = \frac{4k_1\sqrt{I_q}}{\sqrt{k_0}} \tag{1.9}$$

It turns out that the large signal input impedance is a constant value which is determined by the size ratio and biasing current, or in other words, the input current difference (i.e. $I_{in} = I_1 - I_2$) is a linear function to the input voltage (i.e. V_{RF}) as long as I_{in} is less than $4(k_1/k_0)I_q$. This is very important characteristic of this input stage, compared with the result of [4], the large signal input current of bipolar Micromixer is not a really linear function to the input voltage, so we can expect higher linearity from the CMOS Micromixer.

When we compare the transconductance with the common source differencial pair, which is the input stage of Gilbert mixer, we can also find the advantage in both linearity and gain. A differential pair with biasing current I_q and input voltage V_{RF}, the differential current will be[7]:

$$I_{in} = I_1 - I_2 = kV_{RF}\sqrt{\frac{2I_q}{K} - V_{RF}^2} \tag{1.10}$$

This equation is valid if V_{RF} is smaller than $\sqrt{I_q/k}$. The transconductance of the differential pair is not linear even when the input signal is small as

$$g_m = \frac{\partial I_{in}}{\partial V_{RF}} = \frac{2I_q - 2kV_{RF}^2}{\sqrt{\frac{2I_q}{k} - V_{RF}^2}} \tag{1.11}$$

From the above equation, it is clear that the transconductance of the common source differential pair is a function of the input voltage signal, this means the degradation of linearity. Fig.4 shows the simulation result of the normalized large signal transconductance of both class-AB input stage and the differential pair, it can be seen the transconductance of class-AB input stage keep almost the same till the input signal is very large, while the transconductance of differential pair keeps on decreasing when input signal increases. The highest transconductance of differential pair is achieved at the point when input voltage is zero, which is $g_{m,max} = \sqrt{2kI_q}$. If we set $k_1 = k_0 = k$ in equation(13), then the transconductance of the class-AB input stage is $g_m = 4\sqrt{kI_q}$. So with the same biasing current and transistor sizes, the micromixer will have $2\sqrt{2}$ or about 9dB gain increase over the differential pair.

It is complex to calculate the noise figure of mixer, but in order to gain some insight of the noise performance, we can just calculate the noise of the input stage because Micromixer's other stages are the same as Gilbert mixer. The input stage's noise is determined by the three transistors M1, M2 , and M3. The noise from biasing circuit can be attenuated through large resistance as shown is Fig.5. Then we have the noise current PSD(Power Spectrum Density) of each transistor $\overline{I_n^2} = 4kT\gamma g_m$, when we mirror them back at the input stage, it should be divided by the square of the transconductance from input to that transistor. So we have the equivalent noise voltage at the input

$$\overline{V_{n,add}^2} = \overline{V_{n,1}^2} + \overline{V_{n,2}^2} + \overline{V_{n,3}^2} = \frac{12kT\gamma}{g_m} \tag{1.12}$$

where we assume three transistors have the same transconductance. The noise figure of the input stage can be

$$NF = 1 + \frac{\overline{V_{n,add}^2}}{4kTR_s} = 1 + \frac{3\gamma}{g_m R_s} \quad (1.13)$$

while for differencial pair, the noise figure can be $NF = 1+2\gamma/g_m R_s$, so there is some degradation of noise performance for the class-AB input stage. From the above equation, it can be found that large transistor transconducance will be desired for noise performance, but it will require more power or generate large parasitic capcitance.

4. IMPLEMENTATION OF A CMOS MICROMIXER

A CMOS micromixer using HP0.5μm technology has been designed, the circuit is shown in Fig.5. This mixer is designed to work at 2.44GHz in a direct downconversion system. The quiescent current of the input stage is about 450μA with 3V supply, the total power consumpsion is less than 3mW. As shown in the figure, Mb1,Mb2, and M1-M3 is the class-AB input stage, M4-M7 is the switching transistors. The biasing of switching transistors is provided by Mb2 and Mb3. Rb1-Rb3 is large resistor about 3 KOhm, the purpose of these resistors is to attenuate the noise of biasing circuit appearing at the RF and LO input. L1 is a spiral inductor, it is modeled as [8]. Lbond is bondwire inductor, it is provided by package model. Rload is the output load impedance, its value will determine the conversion gain, higher value will have higher gain. But since it also determine the DC level of the output, it can not be too large because it may force M1-M3 work in triode region. The size of M1-M3 is determined by conversion gain requirement and power consumption. From the above analysis, the higher the biasing current, the higher the gain and linearity, but once the power consumption is set, there exist some optimum point for the biasing and sizes of M1-M3. RF and LO signals are AC coupled through capacitors C1-C3. The RF input matching is very simple, taken the package model into account, we only need one series capacitor C1 to tune the matching, simulation result show the S11 can be as less than -20 dB. Differential LO signal comes from the LO buffer which can provide at least -10 dBm input power for good performance.

It is important to make the quiescent current in the two branches equal, this is because any mismatch in the quiescent currents will directly translated into LO leakage to the IF output. According to equation(1), if there exists mismatch ΔI_q, the output voltage will be

$$\begin{aligned} V_{out}(t) &= R[I_{RF1}(t) - I_{RF2}(t)][S_1(t) - S_2(t)] \\ &= R\Delta I_q[S_1(t) - S_2(t)] + Rg_m v_{rf}(t)[S_1(t) - S_2(t)] \end{aligned} \quad (1.14)$$

the first item is the LO leakage to the output of the mixer. Because of the substrate effect, we should tune the size ratio of M1,M2,M3 to eliminate the current mismatch.

Inductor L_1 is a spiral inductor with value of about 2nH, the purpose of this inductor is to adjust the phase difference of the two differential branches. Since from the input point to the two output points of the input stage, the parasitic capacitance will cause different phase shift of the two RF currents, the phase difference is more than 180 degree as shown in Fig.6(it also cause different transconductance). This will cause gain loss because of the following

$$sin(\omega t - \varphi) - sin(\omega t + \pi) = (1 + cos\varphi)sin(\omega t) - sin\varphi cos(\omega t) \quad (1.15)$$

the gain loss will be

$$Loss = 10log(\frac{2}{1 + cos\varphi}) = 3 - 10log(1 + cos\varphi)dB \quad (1.16)$$

With the inductor adjustment, we can add some phase delay to one branch so that it is possible to adjust phase difference to be 180 degree. The method is to change the size of M1 as shown in Fig.7, we find when W1 is about 135μm, the two branches will have 180 degree phase difference and equal amplitute.

According to [5], there exists an optimum size of the switching transistor for the noise performance. Sweeping the size in HPADS simulation, we get the optimum size is about 60μm which can achieve 13dB Noise Figure as shown in Fig.8. Fig.9 shows the conversion gain and IIP3 to be 7.5dB and 4.6 dBm respectively. It also shows the variation of NF with LO input power. When LO input power is low, both switching transistor will on and the noise from these transistors will dominate the noise figure. When LO input is large enough, the input stage will dominate, so the decrease of NF becomes much slower as shown in the figure. In order to achieve acceptable noise performance, the LO input must be at least -10dBm.

5. CONCLUSION

Through analysis of a class-AB input stage, it is clear that it is very suitable for high linearity low power MOS mixer design. Simulation of a CMOS micromixer can achieve 4.6dBm IIP3 and 7.5dB conversion gain consuming less than 3mW. The circuit has been layout and will be fabricated, the measurement result will come out soon.

6. REFERENCES

[1]A. Vauhkonen, K.J. Grahn"A 12-mW Wide Dynamic Range CMOS Front-End for a Portable GPS Receiver" *Ananlog Integrated Circuits and Signal Processing* , pp. 33-42, March, 1998.

Specifications	Results
Power Dissipation	<3mW
RF frequency	2.44GHz
RF input power	-50dBm
LO input frequency	2.44GHz
LO input power	-5dBm
Conversion Gain	7.5dB
IIP3	4.6 dBm
SSB NF	13dB

Table 1 Performance Summary

[2]F. Piazza, Q. Huang, "A 1.57-GHz RF Front-End for Triple Conversion GPS Receiver", *IEEE J. of Solid State Circuit,* vol. 33, no. 2, pp. 202-209, Feb., 1998.

[3]K. L. Fong and R. G. Meyer,"A Class AB Monolithic Mixer for 900Mhz Applications," *IEEE J. of Solid State Circuits* , vol. 32, no. 8, pp. 1166-1172, August, 1997.

[4]B. Gilbert,"The Micromixer: a highly linear variant of the Gilbert mixer using a bisymmetric class-AB input stage" *IEEE J. of Solid State Circuits* , vol. 32, no. 9, pp. 1412-1423, September, 1997.

[5]B. Razavi, " RF microelectronics," Prentice-Hall, New Jersey, 1998

[6] C.D. Hull and R. G. Meyer, "A systematic approach to the analysis of noise in mixer," *IEEE Trans. CAS* , vol. 40, no. 12, pp. 909-919, December, 1993.

[7] M. Ismail and T. Fiez, " Analog Integrated Circuit and Signal Processing," McGraw-Hill, New York, 1993

[8]N. Nguyen, R. Meyer"Si IC-Compatible Inductors and LC Passive Filters" *IEEE J. of Solid State Circuits* , vol. 25, no. 4, pp. 1028-1031, August, 1990.

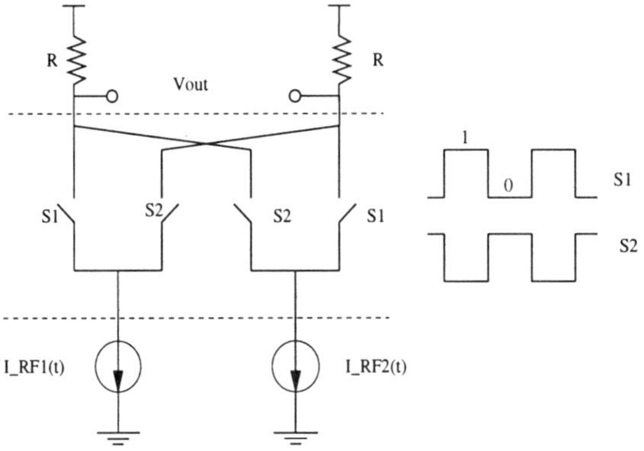

Figure 1 Commuting Mixer Structure and Switching Signal

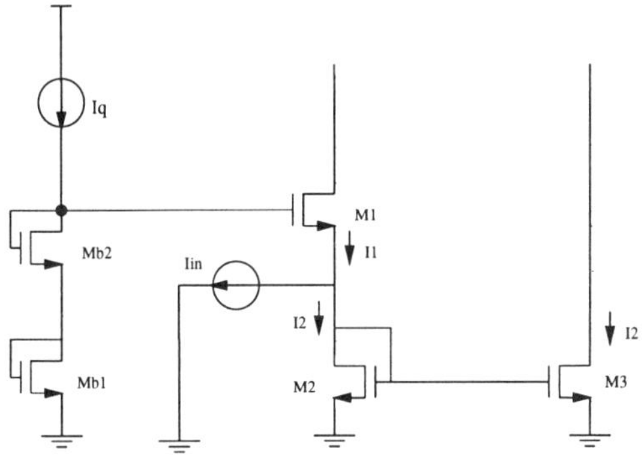

Figure 2 The Class-AB input stage

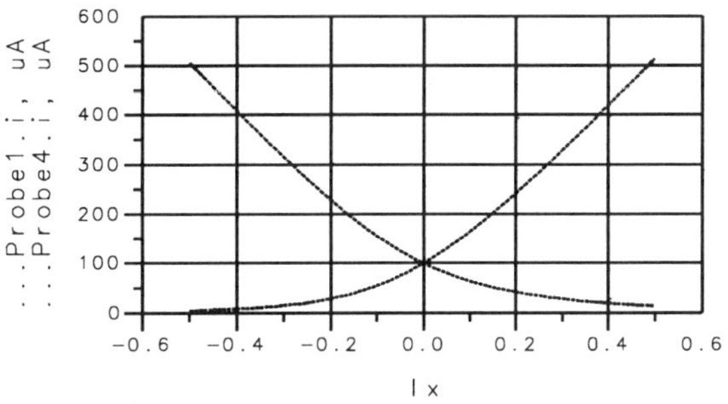

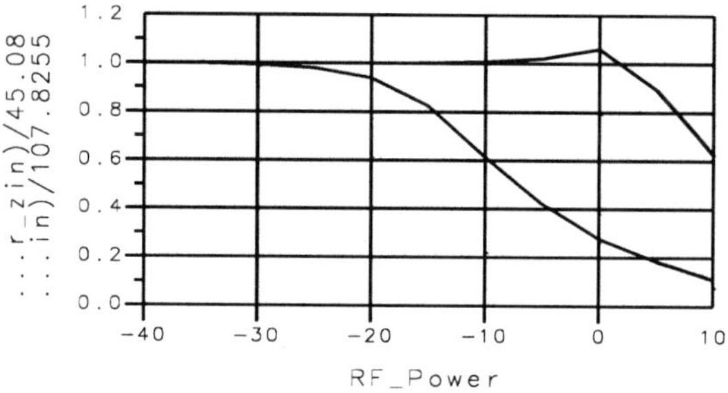

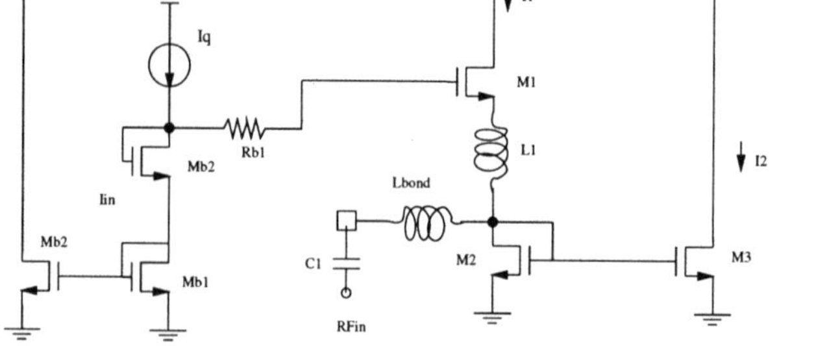

Figure 5 Low power high linearity CMOS Micromixer

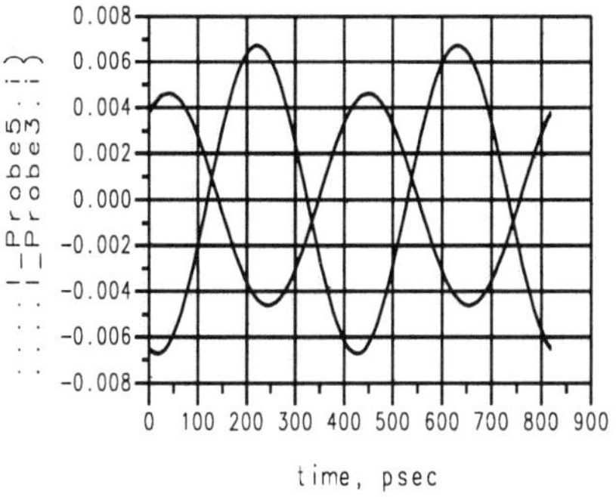
0.008
0.006
0.004
0.002
0.000
-0.002
-0.004
-0.006
-0.008
0 100 200 300 400 500 600 700 800 900
time, psec

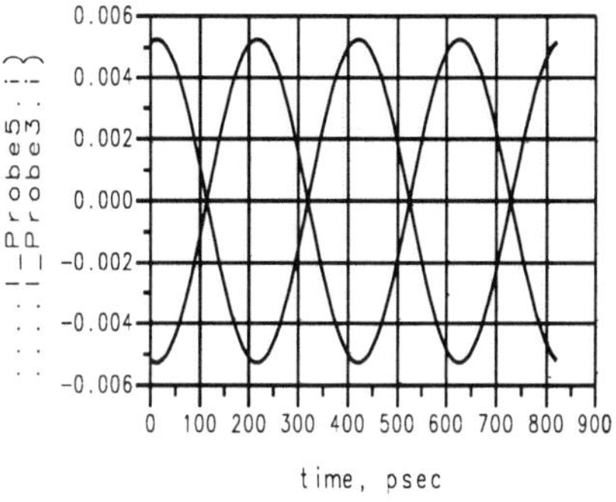
0.006
0.004
0.002
0.000
-0.002
-0.004
-0.006
0 100 200 300 400 500 600 700 800 900
time, psec

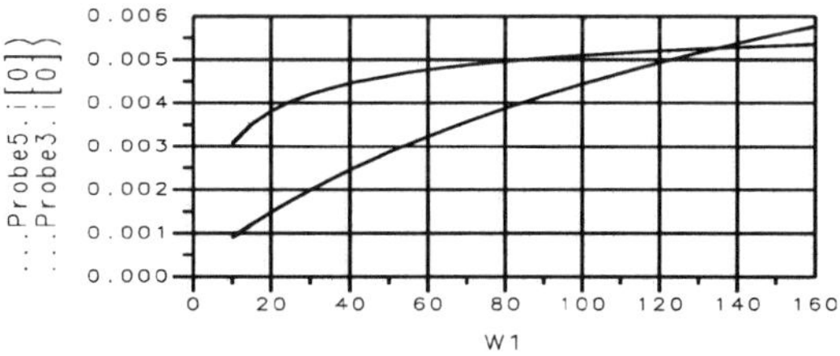
...Probe5.i[0]
...Probe3.i[0]
0.006
0.005
0.004
0.003
0.002
0.001
0.000
0 20 40 60 80 100 120 140 160
W1

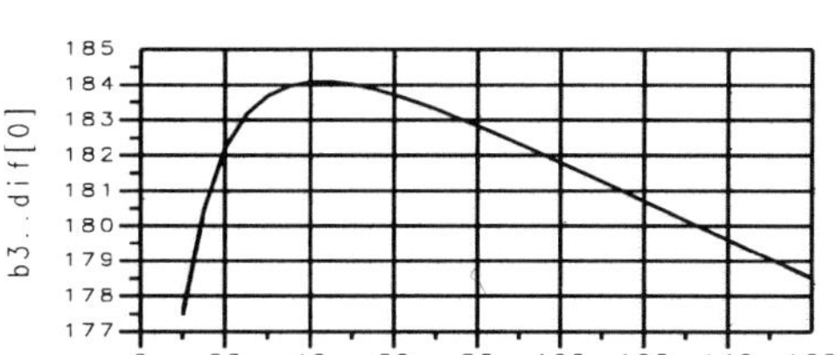
b3..dif[0]
185
184
183
182
181
180
179
178
177
0 20 40 60 80 100 120 140 160
W1

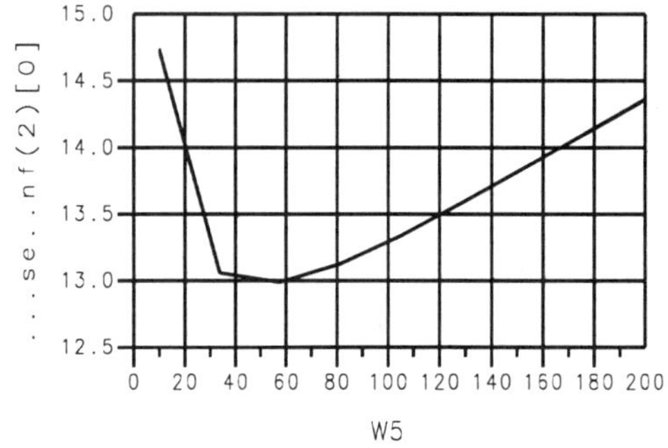

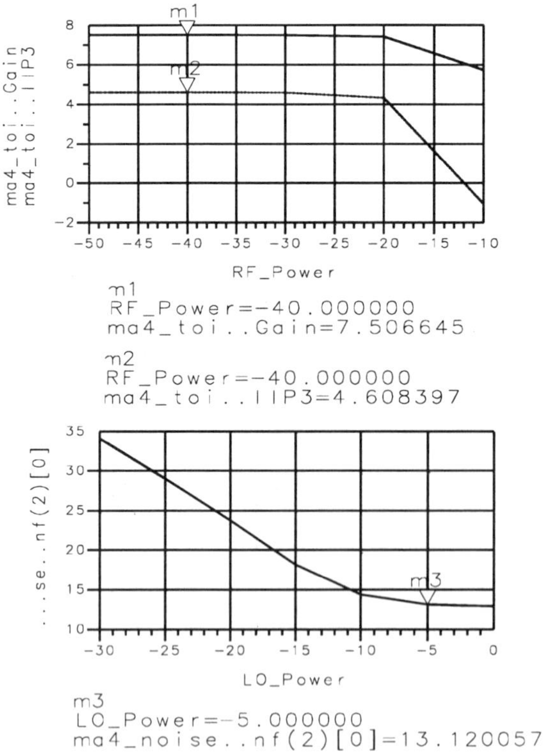

Figure 9 Conversion gain, IIP3 and Noise performance

High Current, Low Voltage Current Mirrors and Applications

S. S. RAJPUT[1] AND S. S. JAMUAR[2]

[1]*Thin Film Technology group, National Physical Laboratory, Dr. K. S. Krishnan Road, New Delhi 110012 India*

email:ssrajput@csnpl.ren.nic.in

[2]*Department of Electrical Engineering, Indian Institute of Technology, Delhi, New Delhi 110016, India*

email: ssjamuar@ee.iitd.ernet.in

Abstract. An improved current mirror structure suitable for low voltage analog applications is proposed which has high output impedance (≈5 MΩ), high bandwidth (≈400 MHz), low input (<0.4 V) and output compliance voltages (<0.5 V) and high output current capability (≈500 μA). These properties enhance the utility of this type of current mirrors. Applications of these current mirrors in analog circuits are also presented to demonstrate their superiority over their conventional counterpart. PSpice simulations confirm the suitability of the mirror for low voltage analog applications.

Key words: Analog VLSI, current mode analog signal processing, current conveyors, transconductors, current mirrors.

1. INTRODUCTION

Low voltage CMOS analog circuit structures are well established now. The requirement for such circuits stems from the requirements of high circuit density required to incorporate both analog and digital systems on a single chip. Hence, for mixed mode and complex circuit structures, smaller size devices are necessary. The device sizes are shrinking at a fast rate [1] and the threshold voltage of the MOSFET devices cannot be reduced below a certain minimum value. This necessitates investigation into new circuits using high threshold voltage MOS devices, which can operate at low voltages without sacrificing the circuit performance.

In analog circuits, a current mirror (CM) is an essential subsystem block. The overall performance of the analog circuit is dependent on the characteristics of the CM. This has generated the need for design of low voltage (LV) CM with low input and output voltage operation. Almost all-high swing CMs, which have been reported [2-4] so far, suffers due to the requirements of high voltage at the input end. Some LVCMs have been reported, which require low input voltages [5-7] for their operation. However, these CM structures possess poor frequency response (<100MHz) along with lower input current range (<100 μA). The demerits of these CMs make them unsuitable for use in low voltage high frequency analog and mixed mode signal processing applications such as current conveyors, operational floating amplifiers etc.

In this paper, we propose a LVCM circuit structure based on popular tail current enhancement technique [8,9]. This circuit has capability to operate at lower input voltages (<0.5V). The proposed LVCM circuit has been obtained through suitable modifications incorporated in the circuit of [8,9] by using a level shifter [5,7] at their input port. It is possible to obtain composite current outputs (i.e., positive and negative current simultaneously) by suitable modifications in output structure of LVCM. LVCMs, with composite current outputs are normally used in current mode analog signals processing circuits. The operating current range of the modified circuit extends from 1 μA to 500 μA with low input (<0.4 V) and output (<0.5 V) compliance voltages. The proposed LVCMs can operate up to a frequency of 400 MHz for negative output currents. However, for the positive output current the frequency response is in the range of 200 MHz. The proposed LVCM has been used to obtain class A current conveyor (CC) and voltage to current (V-I) converters.

2. PROPOSED LV CURRENT MIRRORS

2.1 CM Structure

The proposed LVCM circuit structure is shown in Fig. 1. At the input port a PMOS transistor M9 is used to shift the level of gate bias voltage required to operate the input transistor M2 in linear (triode) region. Injection of the input current I_{in} into the drain of transistor M2 causes a voltage (v_{in}), at its drain which is known as the input compliance voltage. The voltage shifter transistor M9 is designed to operate in saturation region for which an appropriate amount of the bias current is required to flow. The selection of this bias current depends over the compromise reached between the offset current and v_{in}. Input bias current is obtained from an inverted current mirror formed by M15, M16, and M17. To enhance the output impedance of the current source, a cascode structure of MOSFETs, is built using M1 and M5 at the output.

Though the cascode structure at the output port enhances the output impedance, it affects the available output signal (voltage) swing. The required gate bias for M5 is achieved through another voltage shifter PMOS (M8). A CM formed by M14, M16, and M17 provides biasing current for M8. In this configuration, the output impedance R_{out} of the transistor M1 is enhanced and is given as [8]

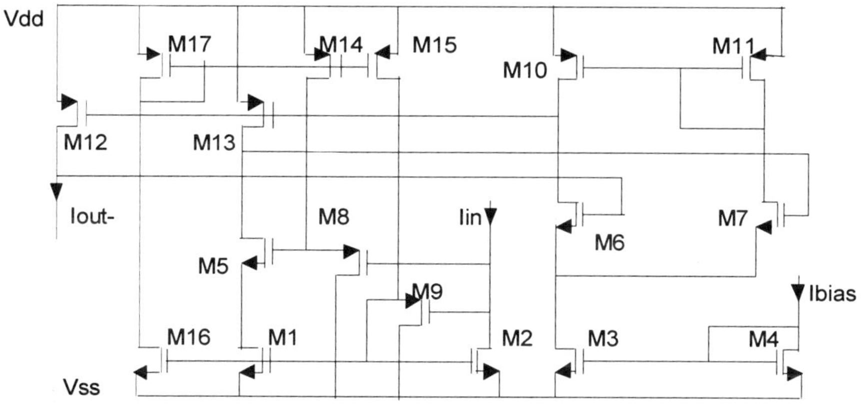

Figure 1. Proposed LV current mirror

$$R_{out} \approx - g_{m5} / g_{d1} \, g_{d5} \quad (1)$$

If g_{di} and g_{mi} are the output conductance and trans-conductance of M2 respectively, the input impedance R_{in} and v_{in} will be given by

$$R_{in} \cup \frac{1}{g_{mi} + g_{di}} \quad (2)$$

$$v_{in} \cup \Delta V_T + \sqrt{\Delta V_T^2 + \frac{2 I_{in}}{\beta_{n2}}} \quad (3)$$

where $\Delta V_T = V_{TP} - V_{TN}$, V_{TP} and V_{TN} are the threshold voltages for M9 and M2.

Due to mismatch in the threshold voltages of the NMOS and PMOS transistors which are used in the input/output circuits and level shifter; there is an output current known as offset current (I_{offset}) for zero input current. The offset current for the circuit of Fig. 1 is given by

$$I_{offset} \cup \frac{\beta_{n2}}{2}\left\{X + \Delta V_T^2 + 2\Delta V_T \sqrt{X}\right\} \quad (4)$$

where,

$$X = \sqrt{\frac{2I_{bias1}}{\beta_{p9}}}$$

If $\Delta V_T^2 >> (X+2\Delta V_T\sqrt{X})$, I_{offset} can not be decreased below a certain minimum level ($\approx\beta_{n2}(\Delta V_T)^2/2$). The offset current is highly dependent on the mismatch ΔV_T between the threshold voltages of the input/output NMOS and level shifter PMOS. Smaller the mismatch, lower will be the offset current. However, if V_{TN} and V_{TP} can be matched, then I_{offset} can be reduced quite a bit. Thus, I_{offset} sets the lower limit of operating currents available from these current mirrors.

2.2 Composite Current Mirror

The composite current mirror (CCM) is required in many analog subsystem blocks where both positive and negative currents are needed. CCM can be derived from Fig. 1 by adding transistors M18 and M19 at the output ports as is shown in Fig. 2. M19, which is used in cascode mode, enhances the required output impedance for the positive current output. The gates of M18 and M19 are connected to the sources of transistor M9 and M8 respectively. Thus the additional transistors are not required to provide the required biases to M18 and M19. The input and output resistance remains same as that of LVCM. However, R_{out} for the positive current output is given as:

$$R_{out} \cup \frac{g_{m19}}{g_{d1}g_{d19}} \quad (5)$$

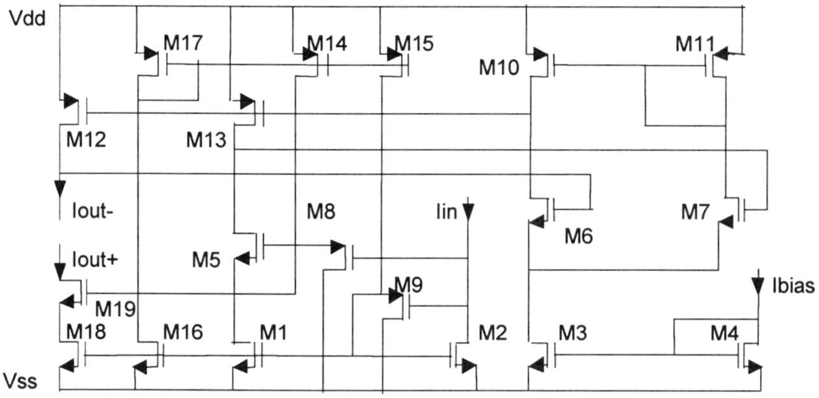

Figure 2. Proposed composite output current mirror

3. CIRCUIT ANALYSIS

3.1 CM Modeling

We have used hybrid (*h*) parameters to model the CM [10] and to analyse the circuit behaviour. Small signal analysis carried out for the circuit operation at low voltages gives the following expressions for the various low frequency hybrid parameters:

$$h_{11} \cup \frac{g_{dp9}}{g_{m2} + g_{dp9}} \tag{6}$$

$$h_{12} \cup 0.0 \tag{7}$$

$$h_{21} \cup \frac{g_{mn1}}{g_{m2}} \tag{8}$$

$$h_{22} \cup 0.0 \tag{9}$$

3.2 Frequency Response

Current transfer ratio from output to the input port is an important parameter for evaluation of the performance of LVCM. The frequency response characteristics in terms of output current I_{out} to I_{in} of the proposed CM is given as:

$$\frac{I_{out}(s)}{I_{in}(s)} \cup \frac{1+\dfrac{sC_{gdpo}}{g_{mpo}}}{1+\dfrac{s(2C_X + C_{gdpo})}{g_{mpo}}} H(s) \tag{10}$$

where C_{gdpo}, and g_{mpo}, represent the gate to drain capacitance and trans-conductance for M12 or M13 respectively. C_x and $H(s)$ are representing the stray capacitance associated at the gate of M7 and the current transfer function for the input circuit formed by M1, M2, M3, M8 and M9. Generally $H(s)$ is frequency independent and taking it to be unity, equation (10) reduces to:

$$\frac{I_{out}(s)}{I_{in}(s)} \cup \frac{1+\dfrac{sC_{gdpo}}{g_{mpo}}}{1+\dfrac{s(2C_X + C_{gdpo})}{g_{mpo}}} \tag{11}$$

As can be seen from the equation (11), the stray capacitance (C_x) severely degrades the high frequency response for the proposed LVCM.

3.3 Sensitivity Analysis

Sensitivity is one of the most important criteria used for comparing the circuit performance. We shall evaluate the sensitivity of the output current over the circuit parameters such as device dimensions and circuit structure. The sensitivity of the output current with respect to I_{bias1} and ΔV_T is given as:

$$S_{I_{bias1}}^{I_{OUT}} \cup \frac{\beta_{ni}}{2ST_{I_{bias1}}^{I_{OUT}}} \tag{12}$$

$$S_{\Delta V_T}^{I_{OUT}} \cup \frac{2\beta_{ni}\Delta V_T}{ST_{\Delta V_T}^{I_{OUT}}} \tag{13}$$

where the stability factors are represented as:

$$ST_{I_{bias}}^{I_{OUT}} \cup \frac{\beta_{ni}\left(\overline{v_{in}} + \sqrt{\dfrac{2I_{bias}}{\beta_{p5}}} + \Delta V_T\right)}{\sqrt{2I_{bias}\beta_{P5}}} \tag{14}$$

and,

$$ST^{I_{OUT}}_{\Delta V_T} \cup \beta_{ni} \overline{} v_{in} + \sqrt{\frac{2I_{bias}}{\beta_{P5}}} + \Delta V_T \tag{15}$$

As can be seen from the above expressions that the stability factors are quite high (>10^3) for the proposed circuit structures. This in turn results in low sensitivity (<10^{-5}) of the output current to the variations of the circuit (β_n and β_p) and device parameters (ΔV_T).

3.4 Simulation Results

The proposed LVCM circuits of Fig. 1 and Fig. 2 have been simulated using the PSPICE circuit simulator. The device parameters for 1.2 μm CMOS technology [11] have been assumed. The device dimensions taken for the simulation are given in Table 1.

We are primarily interested in R_{in}, R_{out}, v_{in}, and the frequency response of LVCM. For the proposed structure R_{in} is equal to 700 Ω (quite low value for any CMOS structure). R_{out} is calculated to be 5 MΩ for an input (output) current of 500 μA, which is sufficiently higher than that of a single transistor CM structure. Fig. 3 shows the I_{out} versus the voltage present at the drain of M12 characteristics for various values of I_{in} for the proposed LVCM.

Table 1. Device dimensions

Device	Device type	Aspect ratio (W/L)
M1, M2, M5, M16, M18, M19	NMOS	240 μm/2.4 μm
M3, M4, M6, M7	NMOS	60 μm/2.4 μm
M8, M9, M12, M13	PMOS	120 μm/1.2 μm
M10, M11	PMOS	240 μm/2.4 μm
M14, M17	PMOS	240 μm/1.2 μm
M15	PMOS	1.2 μm/120 μm

LVCM needs a voltage margin of less than 0.5 V at both ends of the power supply for an input current swing up to 500 μA. However, an input swing of 1.5 V for supply voltage of 2.0 V and for I_{in} < 250 μA can be achieved.

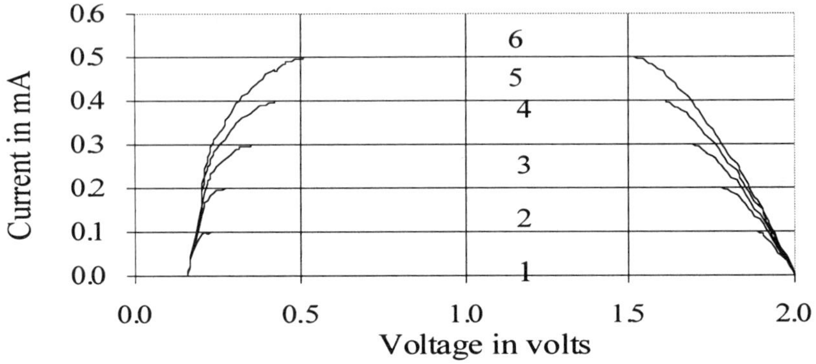

Figure 3. Output current versus the voltage present at the drain of M12 characteristics for various values of I_{in}
(1--0.0 mA, 2--0.1 mA, 3--0.2 mA, 4--0.3 mA, 5--0.4 mA, 6--0.5 mA)

The input voltage (v_{in}) characteristics are depicted in Fig. 4. The input voltage suddenly rises to a required voltage of 0.2 V as I_{in} is pumped in to the CM. Then it rises with the rise in the input current and still remains fairly low (<0.4 V) for the entire current range from 1 μA to 500 μA. The input compliance voltage for the conventional CM of [8] is also plotted in Fig. 4 for comparison. A clear gain of about 0.8 V in the input compliance voltage is visible.

Frequency response for the proposed LVCM and conventional CM of [8], are shown in Fig. 5. The proposed CM can operate up to 400 MHz while the conventional CM without the proposed modifications has a bandwidth of 270 MHz for same device parameters. The bandwidth enhancement due to the proposed modifications is explained in reference 12 and 13 in detail. The superiority of the proposed LVCM in terms of frequency response is evident from the figure. The peak present in the output is due to the pole shifting toward origin because of the stray capacitance C_x.

4. LVCM APPLICATIONS

The proposed LVCM can find use in low voltage analog signal processing circuits. We examine applications of LVCM in current conveyors and in linear trans-conductors.

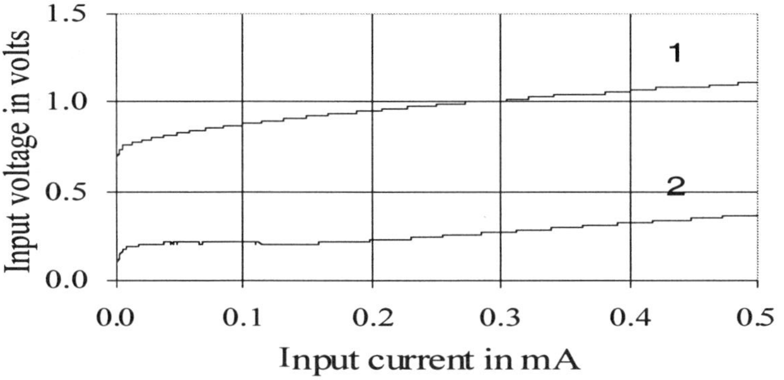

Figure 4. Input voltage characteristics
(1 - Conventional mirror [8] and 2 - Proposed LV CM)

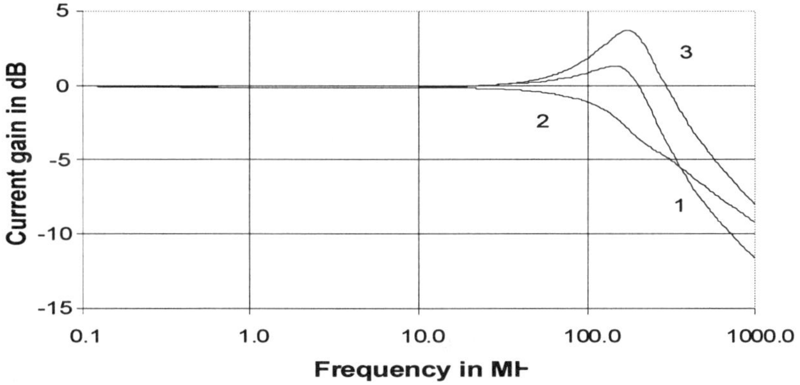

Figure 5. Simulated frequency response characteristics
(1—Conventional CM [8], 2—Proposed LVCM for positive current output and 3--- for negative current output).

4.1 Current Conveyor II

LVCM discussed in the previous section has been used for the design of a class A CMOS current conveyor III (CCIII) of the type discussed in [14, 15]. For rail to rail operation of this type of CCIIIs, low input voltage CMs are necessary.

4.1.1 Circuit Implementation

The circuit diagram of a proposed CCIII is shown in Fig. 6. The voltage transfer from port Y to port X takes place as the LVCM1 forces the equal current to flow through M1 and M2. When the aspect ratio (W/L) of the transistor are kept to ensure $\beta_{n1}=\beta_{p2}$, the drop across gate source terminals of M1 and M2 will be equal, which in turn transfer the voltage applied at the drain of M1 to the source terminal of M2. Current transfer from port X to port Y is achieved through the action of LVCM1 and LVCM2.

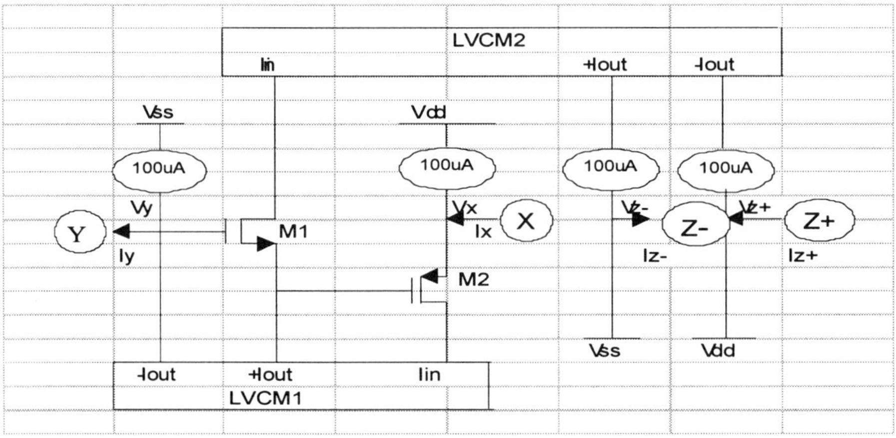

Figure 6. Proposed CCIII structure.

As evident from the figure the input signal can swing rail to rail for the positive inputs however, for negative voltages the input voltage swing is restricted to:

$$v_{in} \cup V_{SS} - V_T - V_{CM} \qquad (16)$$

When V_{CM} is very low as in the case of proposed LVCM, the swing range increases approximately by 0.8V (the difference in the input voltage of proposed and conventional CMs).

4.1.2 Simulation Results

P-Spice simulations were carried out for CCIII structure. The aspect ratio for transistor M1 and M2 were taken to be 60 μm/1.2 μm and 240 μm/1.2 μm for NMOS and PMOS transistors respectively. The bias voltages are taken to be ±1.0 V. The output characteristics for the input current swing are shown in Fig. 7. The maximum input current is ±100 μA. The voltage developed at the X terminal due to injected current is 100 mV.

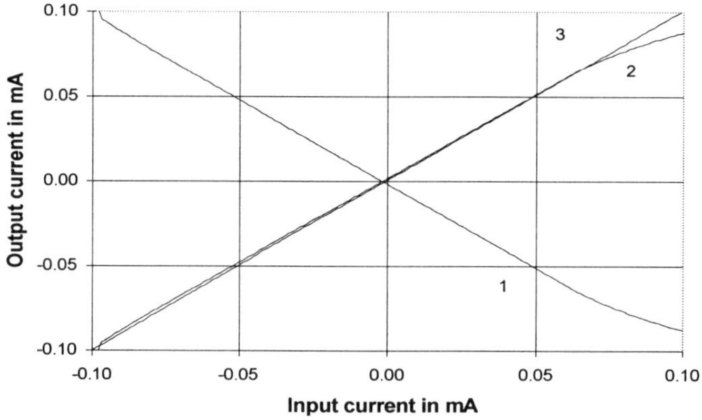

Figure 7. DC input output characteristics of the proposed CCIII
(1---I_{Z+} , 2---I_{Z-}, 3---I_Y)

For the frequency response characteristics of the proposed CCIII, an ac current is injected at the input port X, and the resultant current in the Z + and Z- terminals are recorded and presented. The frequency response characteristics are given in Fig. 8. The power consumption is 2.7 mW and the circuit has 20 MHz bandwidth.

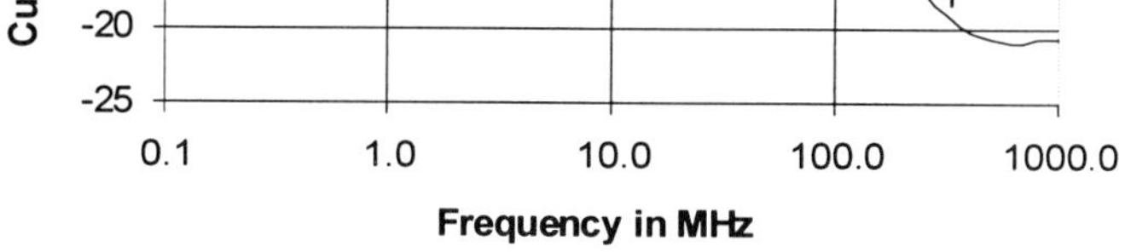

Figure 8. Frequency response of CCIII (1---Positive output current, 2---negative output current, 3---Current required at port Y).

4.2 Linear Voltage to Current Converter

Voltage to current (V-I) converters are known as trans-conductors which find many uses in signal processing circuits like analog filters. High linearity, low noise, large trans-conductance and low power dissipation are the few important parameters required for a high frequency trans-conductors The proposed LVCM is used to construct a linear trans-conductor.

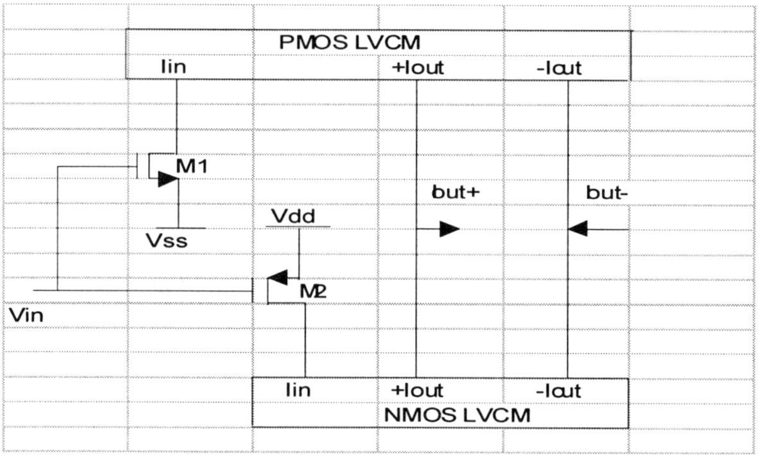

Figure 9. Proposed trans-conductor structure

4.2.1 Circuit Implementation

The circuit structure adapted for the realization of the trans-conductor is shown in Fig. 9. It is assumed that the transistor M1 and M2 operate in saturation, the currents flowing into the input transistors M1 and M2 are given by

$$i_{D1} \cup \frac{\beta_{n1}}{2}(V_{in1} - V_{TN})^2 \qquad (17)$$

$$i_{D2} \cup \frac{\beta_{p2}}{2}(V_{in2} - V_{TP})^2 \qquad (18)$$

Assuming $\beta_{n1} = \beta_{p2} = \beta$, $V_{in1} = V_{in}-V_{SS}$, $V_{in2} = V_{in}-V_{dd}$ and $V_{dd} =- V_{ss}$, the current at the output can be given as:

$$i_C = i_{D1} - i_{D2} \qquad (19)$$

$$\cup 2\beta V_{dd}(V_{in} - V_{TN}) \qquad (20)$$

Examination of the above equation yields that the trans-conductor will show a linear characteristic and the trans-conductance can be varied by suitably chosen values of β and/or supply voltages.

4.2.2 Simulation Results

For PSpice simulations of above trans-conductor, the aspect ratio for transistor M1 and M2 were taken to be 7.2 μm/2.4 μm and 30.6 μm/2.4 μm for NMOS and PMOS transistors respectively. The substrate contacts of these MOSFETs are assumed to be connected to the most positive terminal for the PMOS and the most negative terminal for the NMOS respectively and the supply voltages of ±1.0 V are taken for all simulations.

Output current versus input voltage characteristics of the circuit is depicted in Fig. 10 in which its rail to rail input voltage swing capability is evident. The trans-conductor consumes 0.6 mW power.

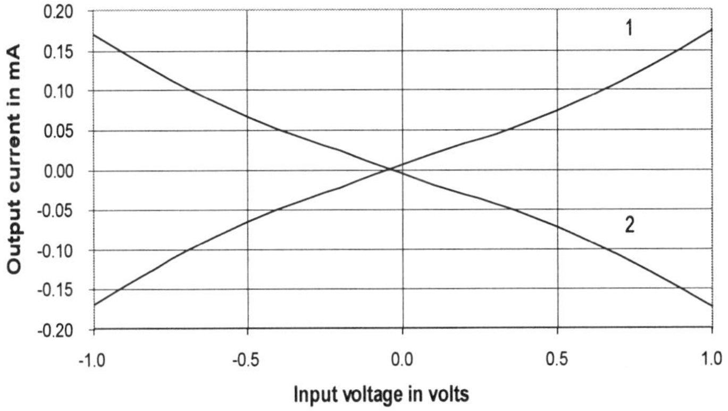

Figure 10. V-I conversion characteristics (1—I_{out+} and 2---I_{out-})

5. CONCLUSIONS

We have presented a modified low voltage, current mirror structure, capable of giving high input and output voltage swings. The operating current range of the LVCM extends from 1 μA to 500 μA with a bandwidth as high as 200 MHz. The input voltage requirement is also quite small (<0.4 V). The usefulness of the proposed LVCM over its conventional part of [8] makes it an attractive candidate for low voltage analog design.

The superiority of these LVCMs when used in the current conveyor structures and linear trans-conductor structures is also significant. The resultant CCIII and trans-conductor structures consume very low power and can operate at sufficiently low voltages present at the input port. Thus the proposed LVCM can be used for low voltage, analog circuit applications.

6. ACKNOWLEDGEMENT

One of the author (S. S. Rajput) expresses his gratitude to Dr. R. Bhattacharyya, Head, Thin Film Technology Group, NPL and to Prof. A. K. Raychodhury, Director NPL New Delhi, for their constant encouragement, and keen interest in the work.

7. REFERENCES

[1] Y. Taur, D. A. Buchanan, W. Chen, D. J. Frank, K. E. Ismail, S. Lo, G. A. Sai-halisz, R. G. Vishvanathan, H. C. Wann, S. J. Wind, and H. Wang, "CMOS scaling into the nano-meter regime", Proc. IEEE, April 1997, 486-504.

[2] J. Mulder, A. C. Woerd, W. A. Serdian, and A. H. M Roermund, "High swing cascode MOS current mirror", Electron Lett. , 4th July, 1996, 32, 1251-1252

[3] E. Sackinger, and W. Guggenbuhl, "A high swing, high impedance MOS cascode current mirror", IEEE J. Solid State Circuits, Feb. 1990, 25, 289-298

[4] A. Zeki, and H. Kuntman, "Accurate and high output impedance current mirrors suitable for CMOS current output stages", Electron Lett. , 5th June 1997, 33, 1042-1043

[5] V. I. Proddanov, and M. M. Green, "CMOS current mirrors with reduced input and output voltage requirements", Electron Lett. , 18th Jan. 1996, 32, 104-105

[6] P. Heim, and M. A. Jabri, "MOS cascode-current mirror biasing circuit operating at any current level with minimal output saturation voltage", Electron Lett, 1995, 31, 690-691

[7] B. J. Blalock, P. E. Allen, and A. R. Gariel, "Designing 1-V Op Amps using standard digital CMOS technology", IEEE Trans. Circuits and Systems-I, July 1998, 45, 769-780.

[8] F. You, S. H. K. Embabi, J. F. Duque-Carrilo, and E. Sanchez-Sinencio, "An improved tail current source for low voltage applications", IEEE J. Solid State Circuits, Aug. 1997, 37(8), 1173-1179

[9] E. Seevinck, M Plessis, T. Joubert, and A. E. Theron, "Active bootstrapped gain enhancement technique for low voltage circuits", IEEE Trans. Circuits Systems-II, Sept. 1998, 45(9), 1250-1254

[10] Jaeger, R. C., Microelectronics Circuit Design, Mc-Graw Hill, New York, 1997

[11] Silval-Martinez, J., Steyaert, M., Sansen, W., High performance CMOS continuous time filters, Kluwer Academic Publishers, Boston, 1993

[12] T. Voo, and C. Toumazau, 'High speed current mirror resistive compensation technique', *Electron Lett.*, 1995, 31 , no. 4, pp. 248-250

[13] T. Voo, and C. Toumazau, "Precision temperature stabilized tunable CMOS current mirror for filter applications", *Electron Lett.*, 1996, 32 , no. 2, pp. 105-106

[14] H. W. Cha, and K. Watanabe, "Wide band CMOS current conveyor", *Electron Lett.*, vol. 32, no. 14, pp. 1245 –1246, 4th July 1996

[15] S. S. Rajput, and S. S. Jamuar, "A new wide bandwidth current controlled CMOS current conveyor III implementation," reported to Electronics letters.

Nonlinearity Analysis of a Short Channel CMOS Circuit for RFIC Applications

Yue Wu[1,2], Hong-sun Kim[1,2], Fredrik Jonsson[2], Mohammed Ismail[1,2], Håkan Olsson[2]

[1]*Analog VLSI Lab, The Ohio State University*
205 Dreese Lab, 2015 Neil Ave.
Columbus, OH 43210, USA
[2] *Radio Electronics Lab, Royal Institute of Technology,*
Electrum 229, S-164 40 Kista, Sweden
wuy, kimho, ismail@ee.eng.ohio-state.edu
fredrikj, hakan@ele.kth.se

Abstract In this paper intermodulation phenomenon at Radio Frequency caused by the nonlinear characteristics of submicron MOS transistor has been analyzed. Common source stage with source degeneration is used for the analysis. The effects of bias current, transistor size and effective voltage on the 3rd order intermodulation(IM3) of the common source stage have been investigated. The analytical results are verified with APLAC simulations.

Keywords: Nonlinearity, Volterra Series, Low Noise Amplifier, Mixer, Intermodulation, OIP3

1. INTRODUCTION

Recently, great efforts have been spent on the integration of RF front-end circuits using CMOS/BiCMOS technologies. As the dimension of the transistors is scaling down continuously, the f_{max} of the transistors becomes large enough for typical RF applications(GSM, DECT, GPS, etc). Lots of papers [1]-[3] introduce the design methodology of Low Noise Amplifier (LNA) and downconversion mixer to achieve reasonable gain and good noise figure, but few of them mention the effect of the nonlinear characteristics of transistors in an analytical way. The reason is that it is very difficult to analyze the non-linear effects of circuits consisting of only a few transistors. Another reason is that even though transistor models nowadays are very accurate in the first order, higher order parameters, which determine the nonlinear characteristics of the transistors, are not modeled so accurately as the linear ones.

Common source stage with inductive source degeneration as shown in Fig.1 is widely used in RF front-end circuit. A detailed nonlinearity analysis of the bipolar case has been given in [4], but CMOS case has not been fully investigated yet. Since the CMOS technology has the advantage to integrate analog and digital parts together, it is very worthwhile to investigate CMOS nonlinear behavior in RF band. In this paper, nonlinearity analysis has been done on the circuit based on the short channel MOS model. In section II, the Volterra Kernals has been derived, then the effects of circuit elements' parameters on the Output Intercept Point(OIP3) have been addressed in section III. Finally sectionIV concludes this paper.

2. DERIVATION OF VOLTERRA KERNALS

Since in the case of LNA and downconversion mixer, the input signals are very small, the nonlinear behavior can be considered as weakly nonlinear, where it is enough to describe the nonlinearity with the first three orders of the input signal. In this case, Volterra Series is very suitable to predict the nonlinearity of the circuits. The circuit model we use is shown is Fig.2, C_{gs} is the gate-source capacitance, which can be assumed as a linear device as long as the transistor works in saturation region[6]. C_{gd} is neglected for simplicity. The output resistance r_o of transistor can be neglected as long as the load impedance is small compared with r_o. The above assumption is pretty reasonable because when we consider the input stage's load of a cascode LNA or a Gilbert type mixer, the output load is about $1/g_m$, which is small enough compared with r_o. Zs , Zg are the source degeneration impedance and gate impedance respectively. i_D is the drain current of the MOS transistor, which is the only nonlinear source in this analysis. For long channel devices, i_D is modeled to have a square law relationship with $V_{eff} = V_{GS} - V_t$, but for short channel devices, this relationship is no longer valid. A simple equation which models the mobility degradation effect of short channel devices is as follows[6]:

$$i_D = K\frac{(V_{eff} + v_{gs})^2}{1 + \theta(V_{eff} + v_{gs})} \tag{1.1}$$

where K=$\mu C_{ox} W/2L$, and V_{eff}=$V_{GS} - V_t$.

Using Taylor series to expand this equation and removing the DC component from i_D, we get the AC signal current i_d:

$$i_d = i_D - I_D = g_1 v_{gs} + g_2 v_{gs}^2 + g_3 v_{gs}^3 + \cdots \tag{1.2}$$

where

$$\begin{aligned} g_1 &= \frac{KV_{eff}(2+\theta V_{eff})}{(1+\theta V_{eff})^2}, \\ g_2 &= \frac{K}{(1+\theta V_{eff})^3}, \\ g_3 &= \frac{-\theta K}{(1+\theta V_{eff})^4} \end{aligned} \tag{1.3}$$

There are several ways to calculate the Volterra Series, a direct method is as follows:

First we apply the Kirchoff Voltage Law(KVL) to the circuit of Fig.2, and we get:

$$v_s = [sC_{gs}(Z_g + Zs) + 1]v_{gs} + Z_s i_d. \tag{1.4}$$

when we substitute eq.2 into eq.4, we can find the nonlinear relationship between v_{gs} and v_s:

$$\begin{aligned} v_s &= [sC_{gs}(Z_g + Zs) + 1]v_{gs} + Z_s[g_1 v_{gs} + g_2 v_{gs}^2 + g_3 v_{gs}^3] \\ &= a(s)v_{gs} + b(s)v_{gs}^2 + c(s)v_{gs}^3 \end{aligned} \tag{1.5}$$

where

$$\begin{aligned} a(s) &= sC_{gs}[Z_g(s) + Z_s(s)] + Z_s(s)g_1 + 1 \\ b(s) &= Z_s(s)g_2 \\ c(s) &= Z_s(s)g_3. \end{aligned} \tag{1.6}$$

When we write the nonlinear transfer function of i_d in terms of v_s,

$$i_d = H_1(s_1)\circ v_s + H_2(s_1, s_2)\circ v_s^2 + H_3(s_1, s_2, s_3)\circ v_s^3 \tag{1.7}$$

it is easy to find the desired Volterra Kernals are:

$$\begin{aligned} H_1(s_1) &= \frac{g_1}{a_(s_1)} \\ H_2(s_1, s_2) &= \frac{H_1(s'')H_1(s_1)H_1(s_2)[g_2 a(s'') - g_1 b(s'')]}{g_1^3} \\ H_3(s_1, s_2, s_3) &= \frac{H_1(s''')\gamma(s''')}{g_1^3}[\frac{H_1(s_1)H_1(s_2)H_1(s_3)}{g_1} \times \\ &\quad (g_3 - \frac{2g_2^2}{g_1}) + 2\overline{H_1 H_2} g_2] \end{aligned} \tag{1.8}$$

where $s'' = s_1 + s_2$, $s''' = s_1 + s_2 + s_3$,

$$\overline{H_1 H_2} = \frac{H_1(s_1)H_2(s_2, s_3) + H_1(s_2)H_2(s_1, s_3) + H_1(s_3)H_2(s_1, s_2)}{3} \tag{1.9}$$

and

$$\gamma(s) = 1 + sC_{gs}[Z_g(s) + Z_s(s)]. \tag{1.10}$$

So we have got the Volterra Kernals of i_d. From eq.8, we can calculate the first three orders of nonlinearity. In the next section, we will analyze the third order intermodulation using the above results.

3. THIRD-ORDER INTERMODULATION

Among those nonlinear effects, we are most interested in third order intermodulation components. This is because in communication systems, this component will block or desensitize the desired signal and cause significant degradation of system performance[5]. If in two-tone test with the input frequencies ω_1, ω_2 and equal amplitute V_s, the 3rd-order intermodulation at $2\omega_1 - \omega_2$ is defined as its amplitute ratio to the fundamental ones:

$$IM3 = \frac{3}{4}\left|\frac{H_3(j\omega_1, j\omega_1, -j\omega_2)}{H_1(j\omega_1)}\right| \cdot |V_s|^2 \tag{1.11}$$

Using eq.8, and since $\omega_1 \approx \omega_2 \approx \omega_c$, $\omega_1 - \omega_2 \ll \omega_c$, we have

$$\begin{aligned} IM3 &= \frac{3}{4}|V_s|^2 \frac{|\gamma(s_c)|}{|\gamma(s_c) + Z_s(s_c)g_1|^3} \times \\ & \frac{1}{g_1}\left|g_3 - \frac{2g_2^2}{3}[2B(\Delta s) + B(2s_c)]\right| \end{aligned} \tag{1.12}$$

where $s_c = j\omega_c$,$\Delta s = j\omega_2 - j\omega_1$, and $B(s) = \frac{Z_s(s)}{\gamma(s)+Z_s(s)g_1}$.

This is a very similar equation to the bipolar case. Since γ appears in the nominator, we can get similar result as [4]: Using inductive degeneration will improve the linearity because it will generate a negative real value which can reduce the real part of γ, and it means that $|\gamma|$ can be reduced at some frequency. It is better than using resistive degeneration because of noise considerations, and capacitive degeneration will only degrade the linearity of the circuit.

Since $\Delta\omega$ is very small, $B(\Delta s)$ can be neglected because of the small inductance of the inductor at low frequency. Assuming $\gamma(2s_c) \ll Z_s(2s_c)g_1$, then $B(2s_c) \approx 1/g_1$. From eq.3, g_1 will increase with V_{eff} while g_2 and $|g_3|$ will decrease. From eq.12, and note that g_3 has negative signs, increasing V_{eff} will decrease IM3, which means improvement of linearity. Fig.4 shows the OIP3 vs biasing voltage using the theoretical result from eq.12 and the result of APLAC simulation, the circuit used in simulation is shown in Fig.3 and its parameters are $L_s = 3nH, L_g = 6nH, C_g = 100pf, C_{ox} = 2.2fF/um^2$,$f_c = 2GHz$, $\Delta f = 5MHz$. We also show the relationship between OIP3 and the bias current in Fig.5, comparing with BJT[4], it can be found that given the same biasing current, MOS circuit is more linear than bipolar one as we expected.

Since g_3 is a negative number, $g_3 - \frac{2g_2^2}{3}[2B(\Delta s) + B(2s_c)]$ is actually the summation of the two items. So, decreasing any of the following items will increase the circuit's linearity: g_2,$B(2s_c)$,$B(\Delta s)$ or $|g_3|$. If θ is very small(for long channel device), g_3 will decrease dramastically while g_1 and g_2 will not change much and the OIP3 will increase. This means that long channel device will be more linear. The comparision is also shown in Fig.4, where we choose $\theta = 0.1$ and $\theta = 0.5$ respectively for typical long channel and short channel devices, it can be seen that given the same size and biasing voltage, small θ will result in higher OIP3.

Eq.3 also shows that g_1, g_2 and g_3 are all propotional to K, where $K{=}\mu C_{ox}W/2L$. According to the above assumption $B(2s_c) \approx 1/g_1$, K can be cancelled in $\frac{1}{g_1}\left|g_3 - \frac{2g_2^2}{3}[2B(\Delta s) + B(2s_c)]\right|$, so the effect of K on linearity is only determined by g_1 in $\frac{|\gamma(s_c)|}{|\gamma(s_c)+Z_s(s_c)g_1|^3}$, and IM3 will decrease when K increase. So if we increase W/L ratio, the OIP3 will increase, this can also be explained as the result of bias current increase, since we keep the bias voltage the same, then increase W/L means also the increase of bias current. Changing W/L of the transistor has another effect: tuning the gate source capacitance C_{gs}. Actually in RF front-end circuits, we always keep transistor's length at the smallest size, so we only consider the change of width. In most designs, the values of L_s and L_g are determined in order to cancel the imaginary part of the input impedance. The real part of the input impedance should be matched to the source impedance to obtain maximum power transfer. It is interesting to note that if we neglect the source impedance R_s such resonation may null out the 3rd order intermodulation. From eq. 10 and eq.12, when the width is close the resonant size, $\gamma(s)$ will decrease till to be zero at the exact resonate size, so the OIP3 will be infinite, in other words, the third-order intermodulation componant is nulled out. Keep increasing the width, the magnitude of $\gamma(s)$ will begin to increase, this will cause the drop of OIP3. When W is large enough, the OIP3 increases with W because g_1 dominates eq.12. So the OIP3 will begin to increase again with W. When we consider source impedance R_s, we can calculate by inserting it into $Z_g(s)$, $\gamma(s)$ then becomes $sC_{gs}R_s + 1 + s^2C_{gs}(L_g + L_s)$. At center frequency, $\gamma(s)$ will still have a remaining term $j\omega_c C_{gs}R_s$ which is not zero. Therefore the effect of 3rd-order nulling is not very significant. All these effects are shown in Fig.6 of APLAC simulation results. We can see from this figure how the OIP3 will change with the width of the transistor. We also notice that the smaller the R_s is, the more nulling effect exists. We can also see that when W is large enough, the slope of all those curves are almost the same, this matches with our expection because in that region, the dominating item g_1 are the same, the absolute value difference of OIP3 is caused by $\gamma(s)$ in the nominator of eq.12 of different R_s.The high peak occurs at about $W = 485\mu m$, it is a little bit smaller than the theoretical size for resonation(about $520\mu m$), this can be explained by the Miller effect of C_{gd} which we neglect in this analysis.

4. CONCLUSION

The nonlinear behavior of a special circuit has been analyzed, it is caused by the nonlinear characteristic of short channel MOS model. The Volterra Kernals have been used to derive the intermodulation expression, it is shown the linearity will increase with higher effective voltage or bias current, the size of transistor can have tuning effect on the linearity. The results have been verified by APLAC simulation with real $0.8\mu m$ CMOS process paramters.

5. REFERENCES

[1]A. R. Shahani, D. K. Shaeffer and T. H. Lee,"A 12-mW Wide Dynamic Range CMOS Front-End for a Portable GPS Receiver" it IEEE J. of Solid-state Circuits , vol. 32, no. 12, pp. 2061-2070, December, 1997.

[2]F. Piazza, Q. Huang, "A 1.57-GHz RF Front-End for Triple Conversion GPS Receiver", *IEEE J. of Solid State Circuit,* vol. 33, no. 2, pp. 202-209, Feb., 1998.

[3]K. L. Fong and R. G. Meyer,"A Class AB Monolithic Mixer for 900Mhz Applications," *IEEE J. of Solid-state Circuits* , vol. 32, no. 8, pp. 1166-1172, August, 1997.

[4]K. L. Fong and R. G. Meyer,"High Frequency Nonlinearity Analysis of Common-Emitter and Differential-Pair Transconductance Stages," *IEEE J. of Solid-state Circuits* , vol. 33, no. 4, pp. 548-555, April, 1998.

[5]R. G. Meyer and A. K. Wong,"Blocking and Desensitization in RF Amplifiers," *IEEE J. of Solid-State Circuits* , vol. 30, no. 8, pp. 944-946, August, 1995.

[6]Y. Tsividis, " Operation and modeling of the MOS transistor," McGraw-Hill, New York, 1987

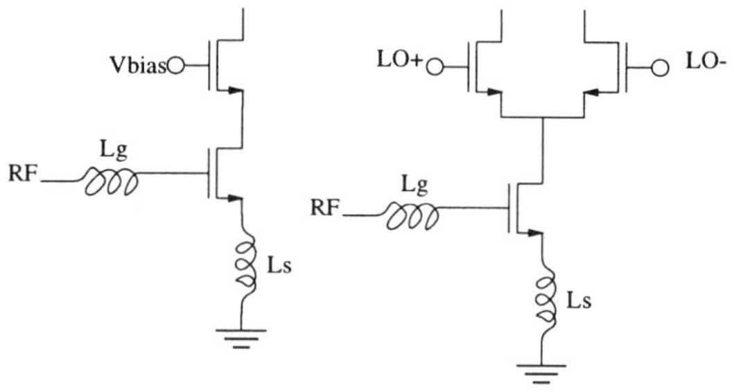

Figure 1 Common source circuit with degeneration in LNA and Mixer

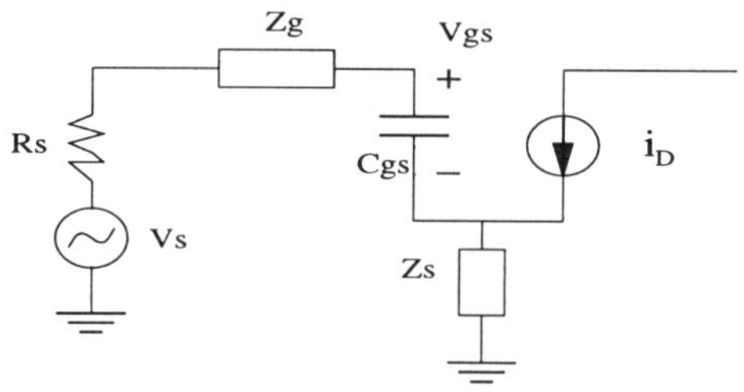

Figure 2 Circuit model used in analysis

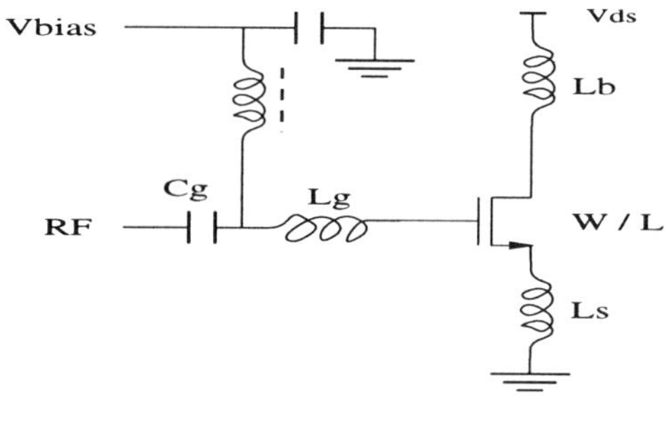

Figure 3 Circuit used in APLAC simulation

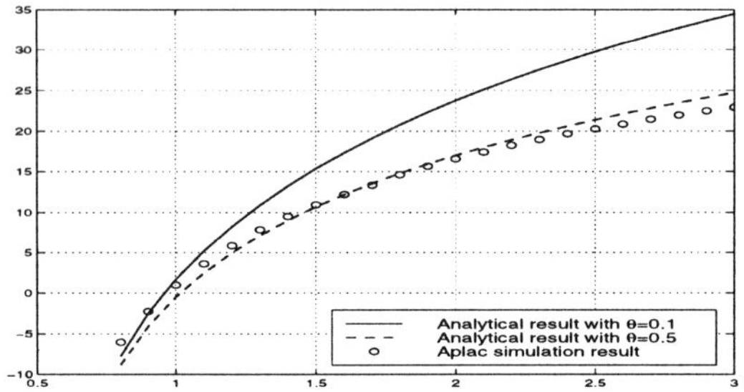

V_{gs}, analytical and APLAC results

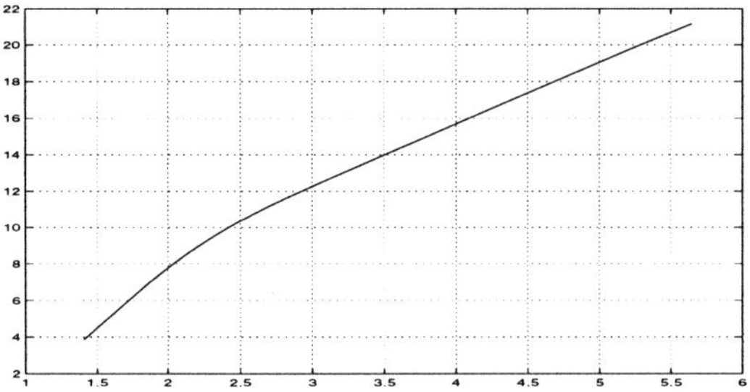

I_D

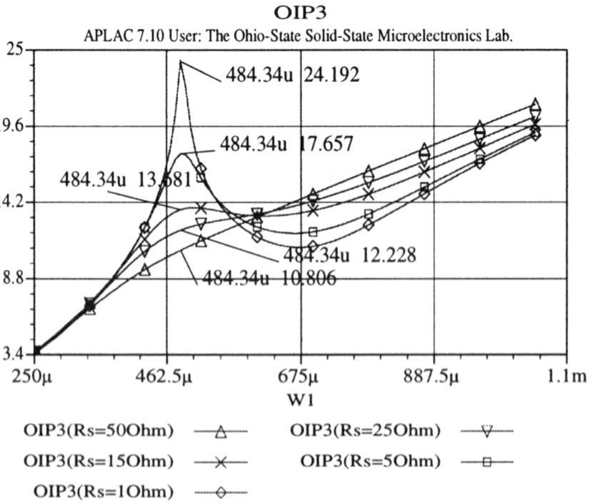

A Fast Parametric Model for Contact-Substrate Coupling

N. Masoumi M. I. Elmasry S. Safavi-Naeini

University of Waterloo, Waterloo, Canada, N2L 3G1

nmasoumi@vlsi.uwaterloo.ca elmasry@vlsi.uwaterloo.ca safavi@aftab.uwaterloo.ca*

Abstract This paper presents a fast and yet accurate modeling method for substrate coupling between a device contact and a substrate backplane. Effects of the physical parameters and geometrical characteristics of the contact and the substrate on the model are reported. We have derived model expressions for extraction of circuit model elements of the substrate. The method is very efficient for speed, memory usage, and is relevant for implementing in CAD tools. We have validated the proposed model over a wide range of frequencies up to 20 GHz.

1. INTRODUCTION

The continuous trend toward a higher IC density and a greater speed introduces the serious problem of the substrate coupling especially in mixed-signal circuits. Substrate-coupled noise introduced by fast switching digital circuits may disturb other devices in sensitive analog circuits and cause functionality failures [Schmerbeck et al., 1991].

Recently, different modeling methods for substrate coupling have been published. In finite element methods (FEM) [Johnson et al., 1984; Stanisic et al., 1994; Verghese et al., 1993] the entire bulk of the substrate is discretized to small volumes. Thus, FEMs can handle multi-layer substrates that contain different doping profiles, wells, etc. However, such methods are impractical for anything but simple problems [Stanisic et al., 1994]. This occurs because these methods encounter a huge model matrix produced by full discretization of the substrate. Boundary element methods (BEM) [Smedes et al., 1995; Gharpurey

*This work was supported in part by MICRONET (Canada), and CITO (Canada).

and Meyer, 1996] require discretization of only the surface of the contacts on the substrate and also the backplane contact, if it exists. Although the model matrix dimensions are smaller than those of the FEMs, it is still dense. Therefore, both approaches present difficulties because of long computational times.

Another problem is that in these methods the substrate is usually assumed to be purely resistive. The substrate coupling modeling approaches presented in [Su et al., 1993; Joardar, 1994] are good examples of this regard. Therefore, these techniques are valid only for low frequencies up to a few GHz.

In this work, we have investigated the details of modeling contact-to-substrate coupling and have proposed accurate and simple expressions for fast extraction of model elements. These expressions are functions of the contact geometry, as well as the thickness and physical parameters of the substrate. We are concerned with coupling from only a single contact to the substrate. As a matter of fact, the problem of substrate coupling in a multi-contact substrate with a ground backplane can be reduced to the problem of one-contact substrate when the coupling between contacts are negligible because of a large separation between them. We have used IE3D software (Zeland Software, Inc.) [Zeland Software, 1999] to verify the accuracy of our model.

In this paper, Section 2. presents general concepts of the substrate coupling. In section 3., the modeling technique for square-contact substrates is described. Section 4. addresses a general formulation method for rectangular contacts. We also expand on our modeling method, in section 5., to include 2-layer substrates.

2. GENERAL CONCEPTS

We consider a single-layer substrate as shown in Fig. 1. This is the case in most high-resistivity and low-resistivity substrates. In high-resistivity substrates, the high resistive bulk is the only common substrate for the devices. In low-resistivity substrates, the epitaxial layer is the only active medium, whereas the very low resistive bulk acts as a common node for the devices [Su et al., 1993].

In this paper, a contact between a typical device and the substrate is modeled by a conductive plate as shown in Fig. 1. *Contacts* (or ports) correspond to the areas where the designed circuit interacts with the substrate. Examples of these contacts include possible noise sources and receptors, such as contacts from a substrate or wells to supply lines, n^+ buried layer in BJTs, drain/source areas in MOS transistors, etc. The substrate circuit model consists of a parallel combination of a capacitor, C_{sub}, and a conductance, G_{sub}. These elements, in general, are nonlinear functions of a number of variables as shown below.

$$C_{sub} = f_1(a, b, h, \rho, \varepsilon_r, f) \tag{1.1}$$

$$G_{sub} = f_2(a, b, h, \rho, \varepsilon_r, f) \tag{1.2}$$

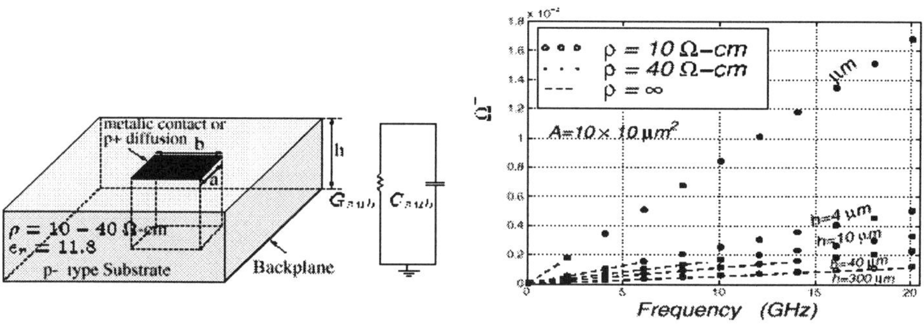

Figure 1 One-contact substrate and its equivalent circuit $B1$ for three different values of ρ_{sub}

where a and b are dimensions of the contact, and f stands for the frequency. Also, h, ρ, and ε_r represent thickness, resistivity, and relative permittivity of the substrate, respectively. In all simulations we have assumed an infinite substrate and the frequency range of 0.1 GHz to 20 GHz. Moreover, we have applied the input source along one edge of the contact.

3. MODELING APPROACH FOR SQUARE CONTACT

3.1 SUBSTRATE RESISTIVITY AND SUBSTRATE HEIGHT DEPENDENCY

The plots in Fig. 2 illustrate simulation results for the substrate with different values of the resistivity. The values of resistivity, ρ_{sub}, between 10 Ω-cm to 40 Ω-cm are commonly used for both high-resistivity and low-resistivity substrates. For each substrate with a constant resistivity, the substrate height, h, has been varied over the range of 1 μm to 300 μm. As this figure shows, the substrate susceptance, $B1$, varies linearly with the frequency. Thus, assuming $B1 = 2\pi f C_{sub}$, we consider that C_{sub} is independent of the frequency. Another notable result is that C_{sub} is a constant function of the substrate resistivity, ρ_{sub}, at least for all resistivity values greater than 10 Ω−cm, as Fig. 2 illustrates.

To extract the substrate model elements, we define the *capacitance-factor*, K_C, and the *conductance-factor*, K_G, as shown below.

$$K_{C,sim} = \frac{C_{sim}}{C_{PP}} \tag{1.3}$$

$$K_{G,sim} = \frac{G_{sim}}{G_{DC\sqcap}} \tag{1.4}$$

where C_{sim} and G_{sim} are the rigorous substrate capacitance and the substrate conductance, respectively, extracted from simulation results. Also C_{PP} stands

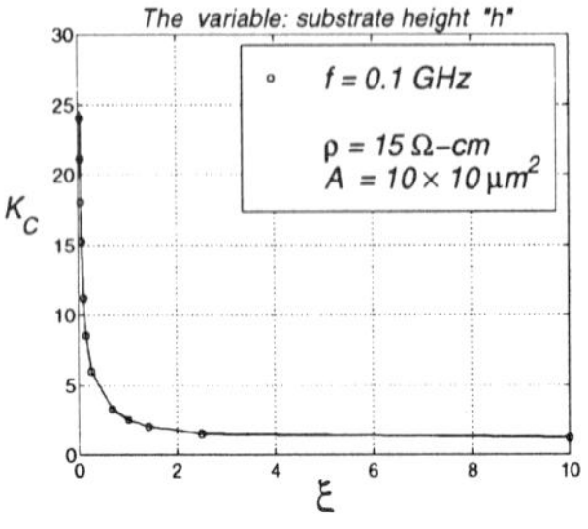

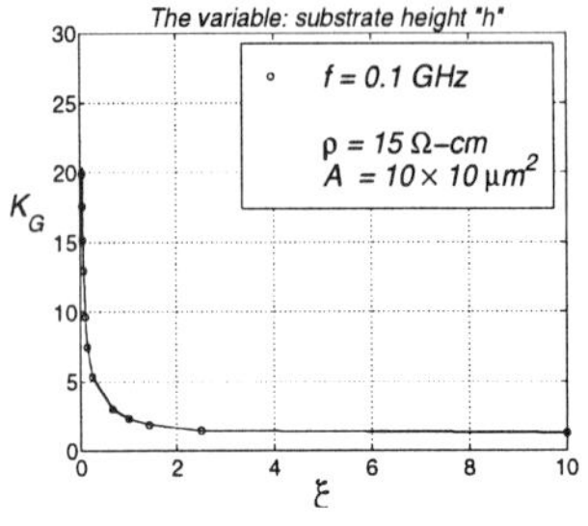

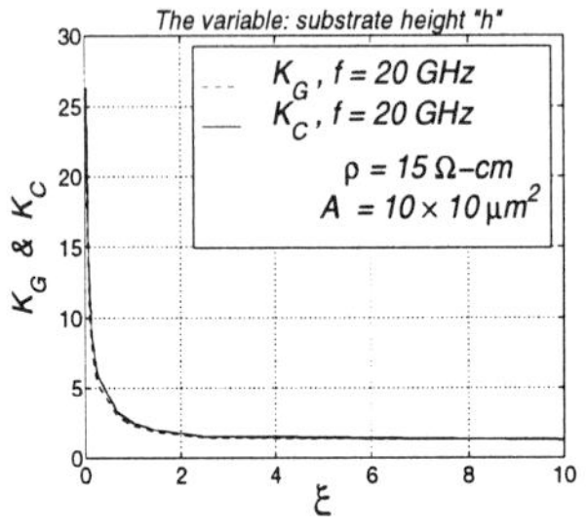

ξ ξ K_C and K_G

for the parallel plate capacitance given as, $C_{PP} = \varepsilon_0 \varepsilon_r \frac{A}{h}$, and $G_{DC\sqcap}$ denotes the conductance of the rectangular prism bulk under the contact given as, $G_{DC\sqcap} = \frac{1}{\rho} \cdot \frac{A}{h}$.

Figs. 3 and 4 show the variation of $K_{C,sim}$ and $K_{G,sim}$, respectively, versus the *shape-factor* ξ which is defined as $\xi = \frac{\sqrt{A}}{h}$. As indicated in these figures, the plots for the frequencies 0.1 GHz and 20 GHz are almost distinguishable. Finally, Fig. 2.5 illustrates a comparison between $K_{C,sim}$ and $K_{G,sim}$. As this figure indicates, we note that $K_{C,sim}$ and $K_{G,sim}$ are nearly equal (for h variable and A constant).

3.2 CONTACT SIZE DEPENDENCY

In the second stage, we investigate the silicon substrate structures with a constant height, h, while the contact area, A, is varied. Contact dimensions are varied from $10 \times 10\ \mu m^2$ to $80 \times 80\ \mu m^2$.

Figs. 2.6 and 2.7 illustrate values of the substrate capacitance C_{sim} compared with those of C_{PP}, and values of the substrate conductance G_{sim} compared with those of $G_{DC\sqcap}$, respectively. We have plotted both $K_{C,sim}$ and $K_{G,sim}$ in Fig. 2.8 for comparison. Considering this figure, we again observe that $K_{C,sim}$ and $K_{G,sim}$ are nearly equal (for h constant and A variable).

Fig. 2.9 compares values of $K_C(\xi)$ obtained from two different sets of substrate structures introduced in the previous sections. Clearly, the values of two plots are in very good agreement. Therefore, for any value of A and h (within the range we have studied) one can write: if $\xi_1 = \xi_2 \rightarrow K_C(\xi_1) \approx K_C(\xi_2)$. The effects of ε_r and ρ_{sub} are taken into account by C_{PP} and $G_{DC\sqcap}$, respectively. Finally, we have applied curve fitting technique to plots for $K_{C,sim}$. The following is the closed form relations obtained.

$$K_{C\square} = 5.1675 + \frac{0.6342}{\xi} \quad , \quad 0.01 \leq \xi < 0.15 \tag{1.5}$$

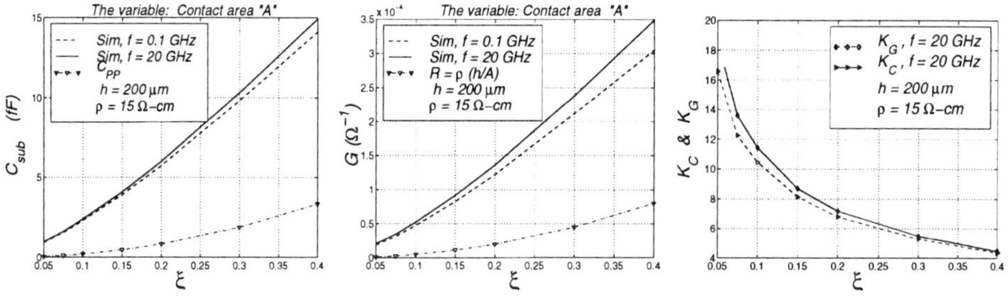

K_C and K_G

ξ ξ

$$K_{C\square} = 1.2538 + \frac{1.1783}{\xi}, \quad 0.15 \leq \xi \leq 1.5 \tag{1.6}$$

$$K_{C\square} = 1.1658 + \frac{1.0604}{\xi}, \quad 1.50 < \xi \leq 10 \tag{1.7}$$

where $\square$ implies a square contact. For simplicity, we call the above expressions $K_{C\square}$ *model expressions*. Furthermore, We have derived *single expression model*, which approximates the above model expressions, as shown below.

$$K^{s}_{C\square} = 1.7855 + \frac{0.85896}{\xi}, \quad 0.01 \leq \xi \leq 10 \tag{1.8}$$

Observing Fig. 2.10 we see that the plots for model expressions agree with those of simulations.

4. RECTANGULAR CONTACTS

Based on an extensive study of simulated data we have derived the *general parametric model expression*, F_C, which can be applied to any rectangular contact. The model expression F_C is defined as

$$F_C = \frac{C_{sub}}{C_{PP}} \tag{1.9}$$

Assuming $a \leq b$, for the contact excited from the edge, a, the capacitance C_{sub} is approximated as

$$C_{sub} = K_{C\square} C_{PPa} + \frac{2\tilde{a}}{3a}(K_{C\square} - 1) C_{PPa} \kappa_{edge} + C_{PP\tilde{a}} \tag{1.10}$$

where $\tilde{a} = b - a$, $C_{PPa} = (a/b)C_{PP}$, and $C_{PP\tilde{a}} = (\tilde{a}/b)C_{PP}$. Also κ_{edge} is the factor that accounts for the variations of the edge effects of the contact and

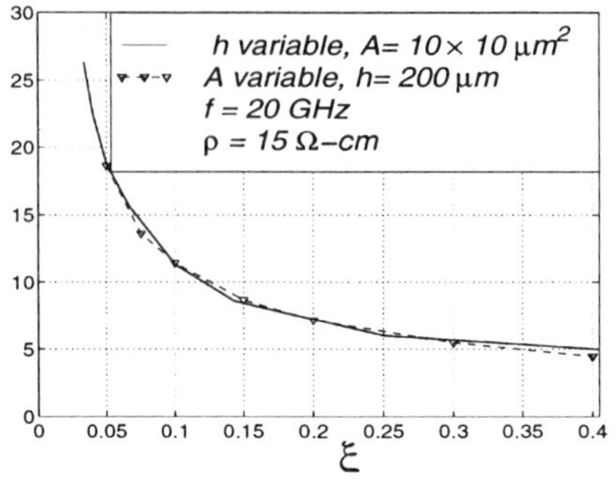

K_C for two different set of structures

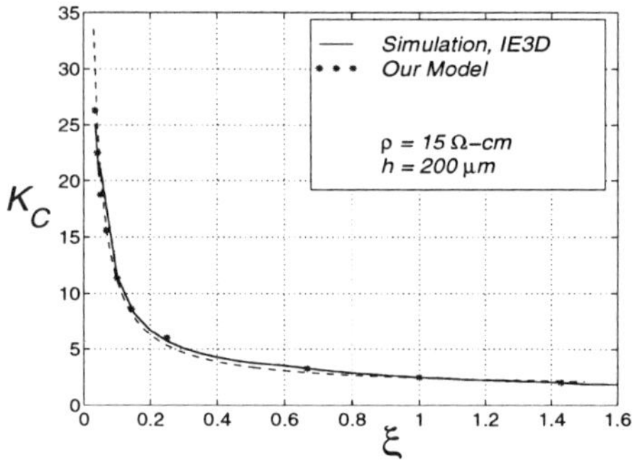

K_C extracted from IE3D simulation results, our model expressions (2.5), (2.6), (2.7), and the single model expression (2.8)

is given as

$$\kappa_{edge} = 1.2 + 0.4\, e^{-0.5\frac{\sqrt{a \cdot \tilde{a}}}{a}} \tag{1.11}$$

Eventually, the parametric model F_C is given as

$$F_C = \frac{K_{C\square}}{b}\left(a + \frac{2}{3}\tilde{a}\ \kappa_{edge}\right) + \frac{\tilde{a}}{b}\left(1 - \frac{2}{3}\kappa_{edge}\right) \tag{1.12}$$

We have compared the results and accuracy of our method with those of IE3D simulation results in Fig. 2.11. This figure indicates that our model agree with simulation results.

4.1 ADVANTAGES AND LIMITATIONS OF THE METHOD

Table 2.1 summarizes the results of the computation times, and the memory requirements for a single-contact substrate by the three methods. The major advantage of our method is its higher speed over other published methods in [Gharpurey and Meyer, 1995] and [Costa et al., 1999]. Another significant feature of the proposed model is that the memory requirements for storing the input data and computations is negligible compared to other methods.

One limitation of the proposed model is the assumption regarding the infinite substrate. However, this matter doesn't affect the adaptability of the method. The rough guideline reported in [Smedes et al., 1995] suggests that the sidewall effects may be neglected, if the contact is farther than twice the epi-layer thickness from the edges of the chip. The other limitation with the proposed model is that it doesn't extract the coupling between the contacts. As mentioned, the focus of this paper is to thoroughly examine single-contact substrates.

Table 1.1 Comparison of the computation times required for extraction of substrate coupling elements for a single-contact substrate.

Modeling Method	Comput. Time (sec)	Speedup Factor	Mem. Usage
The Model F_C (M1)	353×10^{-6}	$\frac{M1}{M3} = 17785$ $\frac{M1}{M2} = 1185$	Negligible
Eigendecomposition (M2) [Costa et al., 1999]	Not Reported	$\frac{M2}{M3} = 15$	Considerable
Green's Function (M3) [Gharpurey and Meyer, 1995]	6.28	—	263k Long Word

5. EXTENSION TO TWO-LAYER SUBSTRATES

We have extended our modeling method to a two-layer substrate, as illustrated in Fig. 12. In this approach, we consider that the total substrate capacitor is a series combination of the oxide capacitor, C_{ox}, and the substrate bulk contribution, C_{sub}. Thus, the total two-layer substrate capacitance, C_{2layer}, is given as

$$C_{2layer} = \frac{C_{ox} C_{sub}}{C_{ox} + C_{sub}} \tag{1.13}$$

The plots in Fig. 13 illustrate a comparison of the total capacitance, C_{2layer}, for several methods. The results of the model shown in this figure are accurate over a wide range of frequencies. The rigorous results obtained from the numerical solution by IE3D agree with our results, whereas the results of the other approximate method based on the uniform field assumption presented in [Valkodai and Manku, 1997] deviate substantially from the rigorous data. This significant and accurate result of our modeling method indicates that it is a very suitable application for CAD tools.

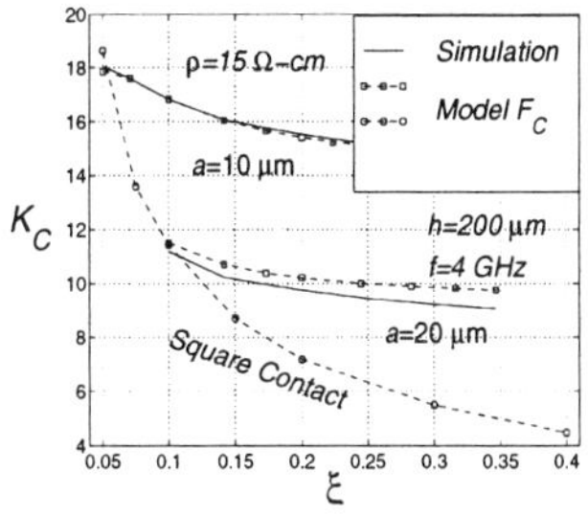

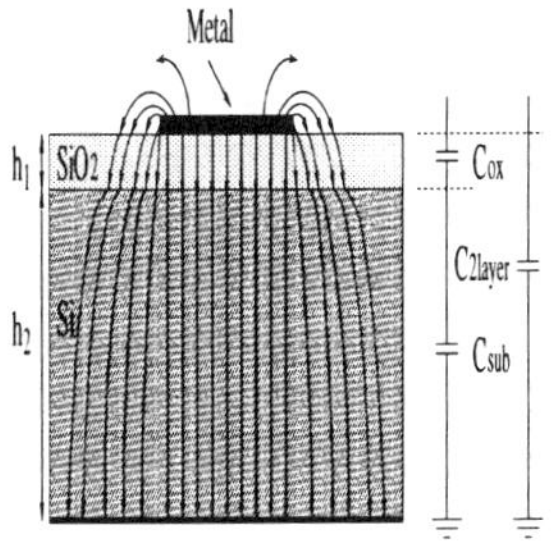

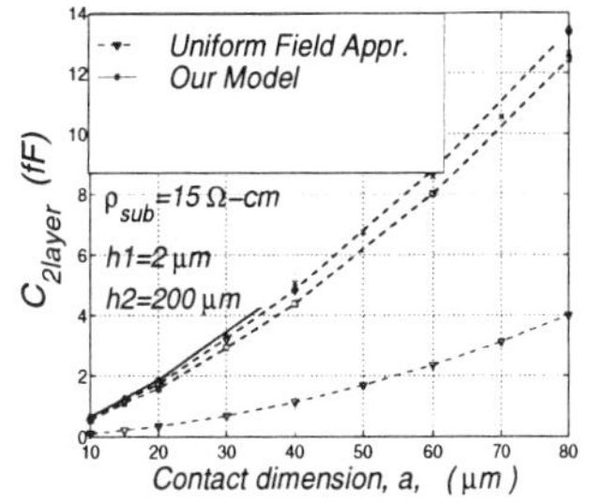

F_C with simulation results

Figure 12 A two layer substrate with fringing fields

C_{2layer} obtained by using 3 approaches

6. CONCLUSIONS

We have investigated the effects of frequency variations up to 20 GHz, as well as the effects of the substrate, and the contact physical and geometrical parameters on the model elements of the substrate coupling.

Based on extensive simulations, we have obtained the general formula, F_C, for rectangular contacts. This parametric model is accurate and applicable within a wide range of substrate resistivities, substrate permittivities, and contact sizes.

The parametric model facilitates a significant speedup factor over the published methods in [Gharpurey and Meyer, 1995; Costa et al., 1999] for single-contact substrates, and an accuracy that corresponds the simulation data of IE3D. We have applied our model to a 2-layer $Si - SiO_2$ structure.

7. REFERENCES

Costa, J. P., Chou, M., and Silveira, L. M. (1999). Efficient techniques for accurate modeling and simulation of substrate coupling in mixed-signal IC's. *IEEE Trans. on Computer-Aided Design*, 18(5):597–607.

Gharpurey, R. and Meyer, R. G. (1995). Analysis and simulation of substrate coupling in integrated circuits. *Int. Journal of Circuit Theory and Applications*, 23:381–394.

Gharpurey, R. and Meyer, R. G. (1996). Modeling and analysis of substrate coupling in integrated circuits. *IEEE Journal of Solid-State Circuits*, 31(3):344–352.

Joardar, K. (1994). A simple approach to modeling cross-talk in integrated circuits. *IEEE Journal of Solid-State Circuits*, 29(10):1212–1219.

Johnson, T. A., Knepper, R. W., Marcello, V., and Wang, W. (1984). Chip substrate resistance modeling technique for integrated circuit design. *IEEE Trans. on Computer-Aided Design*, CAD-3(2):126–134.

Schmerbeck, T. J., Richetta, R. A., and Smith, L. D. (1991). A 27 MHz mixed analog/digital magnetic recording channel DSP using partial response signalling with maximum likelihood detection. In *Dig. of Tech. Papers IEEE Int. Solid-State Circuits Conf.*, pages 136–137.

Smedes, T., van der Meijs, N. P., and van Genderen, A. J. (1995). Extraction of circuit models for substrate cross-talk. In *Proc. IEEE Int. Conf. on Computer-Aided Design (ICCAD)*, pages 199–206.

Stanisic, B. R., Verghese, N. K., Carley, L. R., and Allstot, D. J. (1994). Addressing substrate coupling in mixed-mode IC's: Simulation and power distribution synthesis. *IEEE Journal of Solid-State Circuits*, 29(3):226–238.

Su, D. K., Loinaz, M. J., Masui, S., and Wooley, B. A. (1993). Experimental results and modeling techniques for substrate noise in mixed-signal integrated circuits. *IEEE Journal of Solid-State Circuits*, 28(4):420–430.

Valkodai, A. and Manku, T. (1997). Modeling and designing thin-film inductors and transformers using HSPICE for RFIC applications. *Integration,-The-VLSI-Journal*, 24(2):159–171.

Verghese, N. K., Allstot, D. J., and Masui, S. (1993). Rapid simulation of substrate coupling effects in mixed-mode ICs. In *Proc. IEEE Custom Integrated Circuits Conf.*, pages 18.3.1–18.3.4.

Zeland Software, I. (1999). *IE3D User's Manual*. Zeland Software, Inc., release 6 edition.

A Feature Associative Processor for Image Recognition Based on a A-D Merged Architecture

Atsushi Iwata
Makoto Nagata
Hiroyuki Nakamoto
Noriaki Takeda
Mitsuru Homma
Hiroto Higashi
Takashi Morie
Faculty of Electrical Engineering, Hiroshima University
1-4-1,Kagamiyama, Higashi-Hiroshima 739-8527 Japan
iwa@dsl.hiroshima-u.ac.jp

Abstract An Image Feature Associative Processor (IFAP), which extracts local and grobal features of input image data based on bio-inspired parallel architecture, is proposed. It consists of an image sensor, a cellular automaton and pattern matching processors based on PWM analog-digital merged circuits. IFAP extracts image features at a standard video frame rate with low power dissipation.

Keywords: CMOS image sensor, pulse width modulation, vector quantization, pattern matching, cellular automata,

1. INTRODUCTION

Next generation computing systems will exhibit their capability in the application field of flexible recognition of complex objects and human faces. For implementing many kinds of image recognition systems, algorithms and architectures have been studied on conventional digital computers using a software approach. However, in order to realize real time recognition, bio-inspired architectures that feature massive parallel processing with a huge number of processing units and inputs are necessary. Implementations with binary digital systems and software are not suitable, because they operate sequentially and consume large amount of power and large chip area. Analog circuits however, are promising for realizing low-power and low-cost massively parallel systems.

In order to realize these processing circuits, neuron MOS devices and circuit architecture were proposed [1,2]. They realize arbitrary multi-input Boolean functions and various parallel analog processing circuits. Cellular automata using Neuron MOS devices were also proposed for 2D image processing[3]. Merged analog-digital circuit architecture using pulse width modulation (PWM) signals which is suitable to low-voltage, deep sub-μm CMOS devices, was also proposed.[4,5]

An image recognition system requires an on-chip image sensor. CCD imagers are dominant components in current video applications. However, They do not coordinate with deep-submicron CMOS technologies because special process technologies and high voltage clocks are required. As was reported in recent papers, CMOS imagers are appropriate for integrating various pixel level processings along with image capturing[6,7].

This paper proposes the architecture for an image feature associative processor utilizing PWM merged analog-digital circuits.

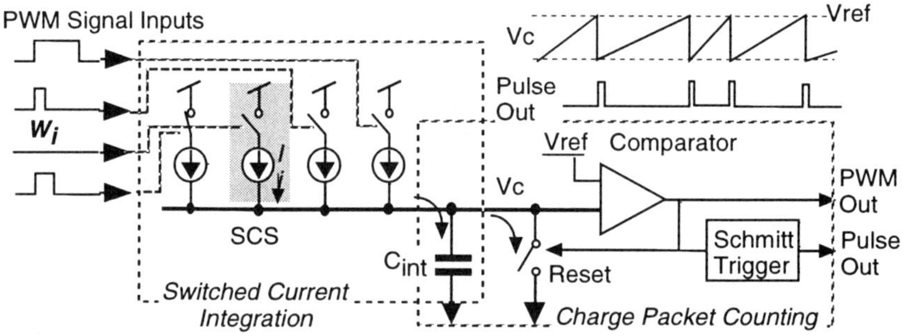

Figure 1 A switched current integration and charge packet counting for PWM signal processing.

2. SYSTEM ARCHITECTURE

2.1 PWM SIGNAL PROCESSING

We have proposed using pulse width modulation signal (PWM) for merging analog and digital processing[4,5]. A PWM signal expresses an analog value on pulse widths with binary voltage or current amplitude. This time domain expression provides immunity to dynamic range reduction caused by low voltage operation.

Figure 1 shows a switched current integration technique for the PWM arithmetic. Switched current sources (SCSs) convert voltage PWM pulses into current pulses. Current integration on C_{int} results in asynchronous and parallel additions of those pulse widths. Reduced transition activities give the advantage of low power dissipation. Charge packet counting converts a small reference charge amount $Q_{ref} = C_{int}V_{ref}$ to a pulse and removes it from Q_{int} successively during integration.[8]

2.2 ALGORITHM FOR FEATURE ASSOCIATION

The processing functions are image sensing, image pre-processing, feature extracting, and pattern matching. The input image is sensed at pixel array and read out with gray scale or binary data. To effectively extract various features from sensed natural images, pre-processing is necessary. Pre-processing includes several spatial filters for noise reduction, edge enhancement, and averaging, thresholding and thinning. To realize flexible and high-speed image recognition, two features, global features and local features, are extracted. The global features include (1) block averaging and (2) X- and Y-projections of thresholded image. They compress redundant local information and select the search area and reference vectors to be matched for efficient local feature association. The local features are associated by pattern matching of sub-block of the input data to the reference vector data. The most similar reference vector code is obtained by calculating Manhattan distance and minimum distance search that requires very high computing power. These two features are united at higher level processing for recognition of complex objects and faces.

3. SYSTEM ARCHITECTURE AND CIRCUITS

IFAP is composed of a CMOS imager, a cellular automaton (CA), and a pattern matching processor (PMP), as shown in Fig. 2. The inputs are optical image data focused on the sensor plane. The outputs are the associated codes of local features and global features. Three functional blocks communicate with each other through PWM signals through the global PWM bus, which has parallel 56-bits and a programmable pulse width in the range of 100ns-1μs. An 8bit binary data bus is used for output of the reference code and the distance value, and input of the reference data.

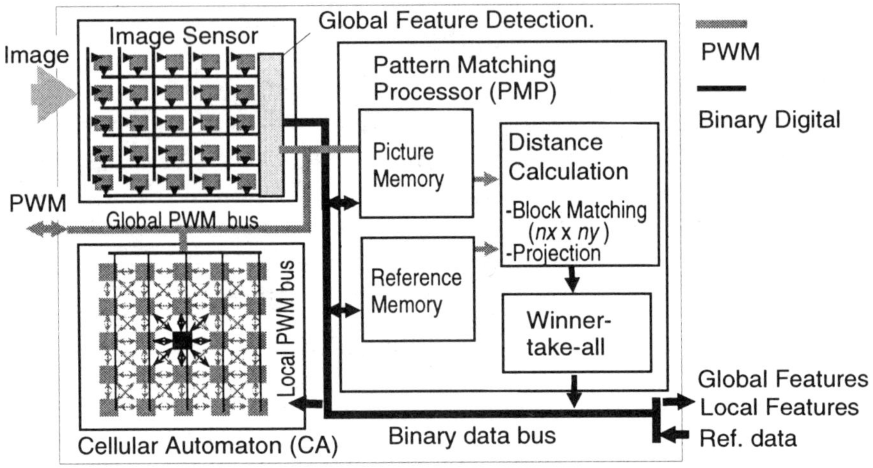

Figure 2 A system block diagram of IFAP

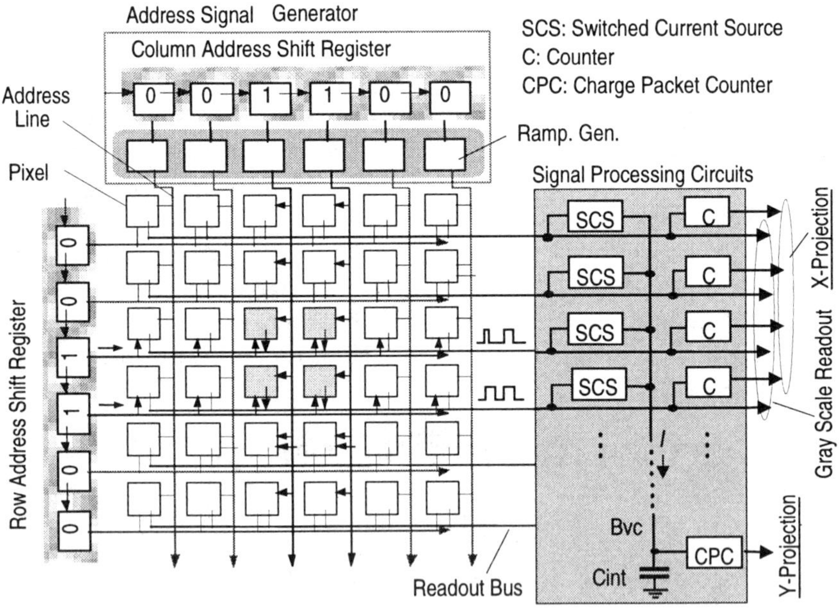

Figure 3 A block diagram of the PWM imager.

3.1 IMAGE SENSOR

The imager has four functions: (1) to read out each pixel value, (2) to threshold each pixel value, (3) to project the X- and Y-directions of threshold image, and (4) to average pixel values of the block. Figure 3 shows a block diagram of the imager. Each pixel executes non-destructive conversion of the input light intensity to PWM signals using a simple voltage comparator. [9]

The pixel has two operation modes of gray scale conversion and thresholding. Pixels asserted by the column and row address shift registers become active for readout. An address signal generator supplies linear upward ramp voltages to the selected column pixels in the gray scale conversion, or a reference voltage in the thresholding. Row pixels share a readout bus that is driven by the current pulses in parallel in the block access mode.

The signal processing circuit consists of an array of a row counter and an SCS connected to each readout bus, and a charge packet counter (CPC). In the projection calculation, pixels work in thresholding, and generate voltage pulses. X-projection values are obtained by counting by the pulses. Y-projection values are obtained by the switched current integration technique. The voltage pulses from the pixel are converted to current pulses by the SCS, and the currents are integrated on the capacitor, C_{int}. Converting the integrated charge to digital data by CPC, Y-projection values are obtained.

3.2 CELLULAR AUTOMATON

A CA cell is connected with its 8-nearest neighbors through PWM current pulses. The templates provide coefficients of connection, and control CA function. The function includes acting as spatial filters to reduce noise, enhance edges and thin. A schematic of the CA cell circuit and templates are shown in Fig. 4. The CA has two modes of operation: a multi-bit mode and a binary mode. In the multi-bit mode, some spatial filters for noise reduction and image enhancement are provided. In the binary mode, templates for dilation, erosion, edge detection and thresholding are provided. By combining these templates, effective pre-processing of sensed image data and binary image thinning can be realized. Data input/output to the cell array are executed through the column parallel local PWM bus which is connected to the global PWM bus.

The cell consists of Switched Current Sources (SCSs), two capacitors (C_1 and C_2), a latch comparator and an inverter chopper comparator. The state of each cell is represented by the charge on the capacitor C_2. The charge is converted to a PWM signal by the inverter chopper comparator, where the latch comparator detects the polarity of cell state in the multi-bit mode. The PWM output signal is transmitted to 8 neighbor cells and the self-input. The templates are determined by switching the SCSs. Multiplication of the template coefficient and addition are carried out by PWM switched current integration on the capacitor C_1. Output PWM generation and the multiplication/addition operations are carried out in parallel pipelined timing. Each template is carried out within only one cycle time because of the fully parallel operation of the cellular automaton. The cycle time is a summation of the maximum pulse width of the PWM signal and a reset timing.

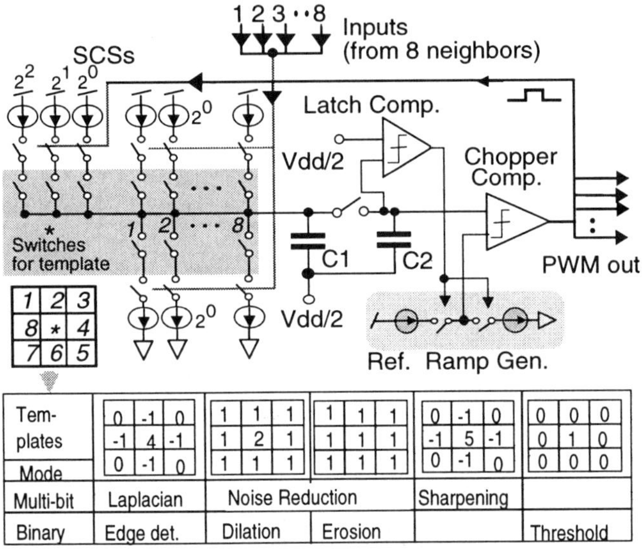

Figure 4 A schematic of the CA cell circuit and templates.

3.3 PATTERN MATCHING PROCESSOR

A block diagram of the Pattern Matching Processor (PMP) is shown in Fig. 5. The PMP is composed of a PWM-to-Digital Converter (DWC) array, picture RAM (p-RAM), reference RAM (r-RAM), a Digital-to-PWM Converter (WDC) array, processing element (PE) array, a charge-to-pulse converter (CPC) array, a winner-take-all (WTA) array and an address control. The input PWM image data is converted to binary digital data by WDC composed of binary counters, and stored in static CMOS RAM. The stored input data are converted to PWM by the DWC, and supplied to the PE array. The reference data are also converted to PWM by the DWC, and supplied to the PE array.

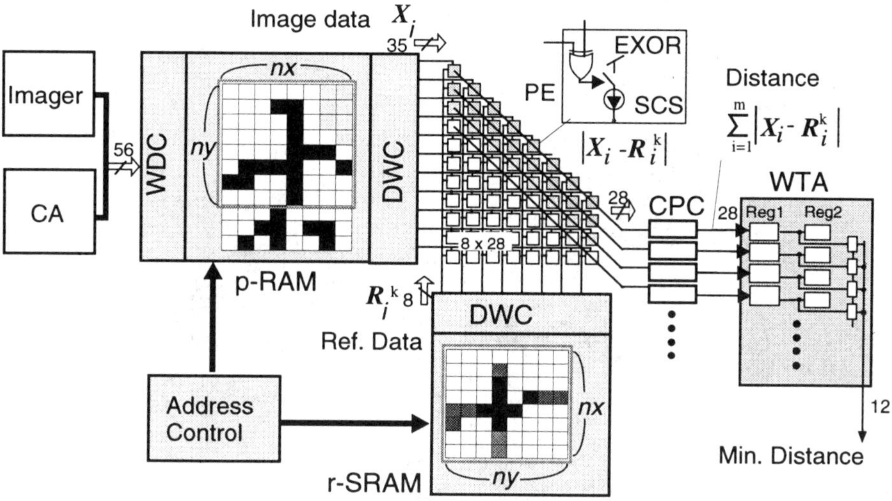

Figure 5 A block diagram of PMP.

Manhattan distance is calculated by the PE array using PWM arithmetic, and the CPC array, as shown in Fig. 5. The PE is composed of an EXOR gate, which calculates the absolute difference of PWM signals: $|X_i - R_i^k|$. All difference data are summed up in parallel by current integration of PWM pulses, as shown in Figure 1. The PWM pulse from ny PE's are added simultaneously, and the additions are sequentially carried out nx times. Thus, Manhattan distance for $nx \times ny$ block matching is calculated. Each distance value that is represented by the integrated charge is converted to binary digital data by the CPC, where nx and ny are programmable in the range from 1 to 8. The minimum distance is searched by the WTA array, which is composed of binary digital techniques based on word parallel, serial bit-by-bit comparison. A new distance value transferred from thr CPC to Reg.1 is compared with the last minimum distance value that is stored in Reg2. If the new value is smaller than that of the last value, the new value is stored in Reg1.

4. CHIP FABRICATION AND EXPERIMENTAL RESULTS

An experimental IFAP chip was designed and fabricated with $0.8\mu m$ p-well CMOS technology with double-poly and double-metal layers. A

Figure 6 A block layout of IFAP chip.

block layout of the IFAP chip is shown in Fig. 6. The chip size is 15mm×15mm. The image sensor array with 56×56 pixels, the address signal generator, and the signal processing circuit are integrated in a 6mm×6mm chip area. Each pixel circuit has a PN junction photo detector, an analog storage capacitor and PWM processing circuits in a $100\mu m \times 100\mu m$ area. The cellular automaton with 40×50 cells, the timing and the template control circuits are integrated in a 7.5mm×7.5mm area. The cell includes about 150 MOS devices and 2 capacitors in a 159μm×122μm area. PMP is integrated in an about a 50mm^2 chip area, the PE array size is 8×28 and the CPC array size is 28. The data storage for the input of PMP consumes a relatively large chip area of 8.1mm×3.2mm, because, in addition to p-RAM, the WDC and DWC arrays are necessary. In the future, analog memories will be available to store PWM data, and reduce a chip area one-half of the conventional digital approach.

4.1 CMOS FUNCTIONAL IMAGER

Figure 7(a) shows a gray scale still test image focused on the focal plane of the imager. Output PWM pulses in the parallel row gray scale readout are shown in Fig. 7(b). The exposure time (T_{sh}) is 20ms, and the readout time for 56 columns is $84\mu s$. This figure shows that large PWM pulses appear in bright area. Figures 7(c) and (d) are images reproduced from the measured pulse widths for the gray scale conversion and thresholding, respectively. Measured linearity was about 6bit for a $420ns$ maximum pulse width. It is in good agreement with SPICE simulation results. Measured analog supply current is 12.6mA and power dissipation is 41.7mW

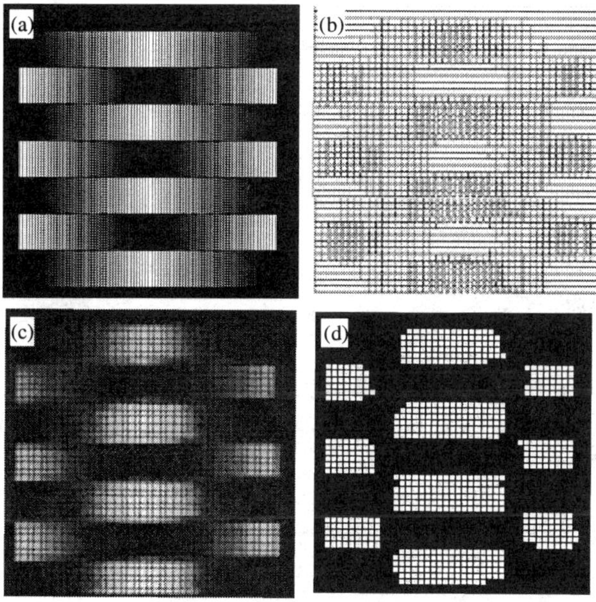

Figure 7 (a) A gray scale still test image, (b) output PWM pulses in the gray scale conversion and (c) reproduced images in the gray scale conversion and (d) in the thresholding.

4.2 PATTERN MATCHING PROCESSOR

Parallel distance calculation for the local feature association is carried out. Input and output pulses measured by HP16500B logic analyzer are shown in Fig. 8. The input image data is a part of a Chinese character with an 8×16 pixel block. The reference data is one of the cross point pattern represented by an 8×8 pixel block. The smallest distance was obtained at the 6-th output of CPC array. PMP consumes 120mW at a 3.3V power supply. Processing speed per unit power dissipation was 6.75

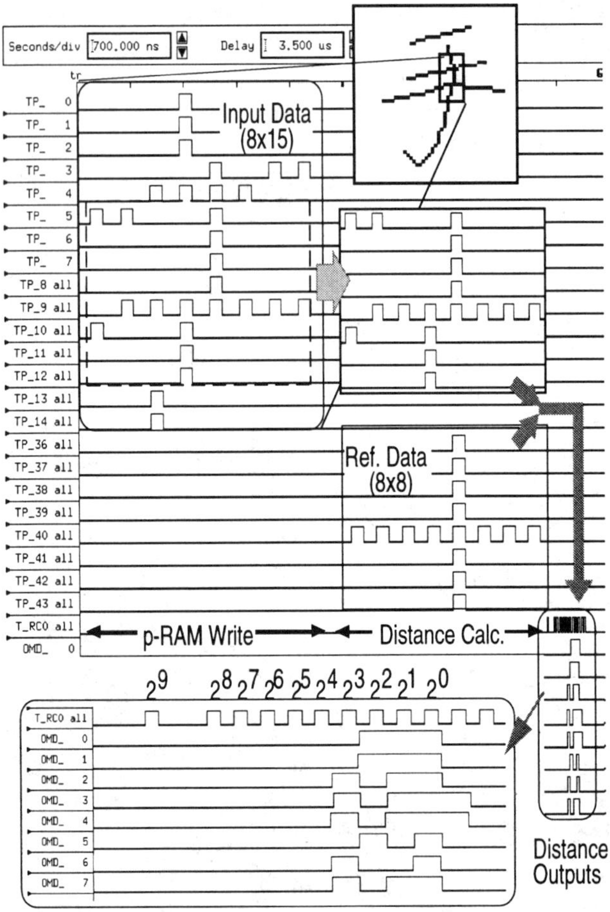

Figure 8 Measurement results of distance calculation.

GOPS/W. The power dissipation of PMP is one-fourth of the simulated value of binary digital circuits with the same CMOS technology.

5. APPLICATIONS TO HANDWRITTEN CHARACTER RECOGNITION

As a typical application of IFAP, feature association for the recognition of handwritten Chinese characters was demonstrated by simulation. An input character, thinned character, X- and Y-projections, search areas, reference vectors, associated local features and a feature map are shown in Fig. 9. Using X- and Y-projections of the binary image, searching windows and candidates of reference vectors were effectively restricted for local feature association. The local features were matched with the reference vectors composed of 8-direction terminations, 12 branches, 8 corners and 2 crosses with 3×3 pixels. The associated local features were represented by the feature map which was compacted maintaining the relative position of each local feature. Other feature maps are also shown in Fig. 1.9. These results show that deformation and size-difference of handwritten characters are compensated by this algorithm.

The features associated by IFAP are transferred to a higher level processor, and are linked with higher level symbolic information. The flexible feature association realized with IFAP will become key for intellectual recognition systems.

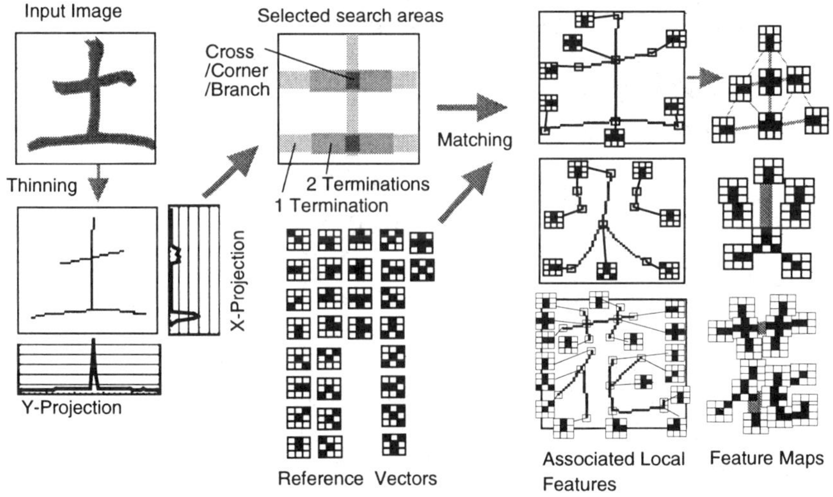

Figure 9 An example of feature association for the recognition of handwritten Chinese characters.

6. CONCLUSIONS

An Image Feature Associative Processor (IFAP) with an on-chip imager and parallel processors which utilize merged A-D circuit architecture based on pulse width modulation (PWM) method was proposed. An experimental IFAP chip was designed and fabricated in a 15mm×15mm chip with 0.8μm CMOS technology. The PWM A-D merged architecture drastically reduces power dissipation and chip area.

7. ACKNOWLEDEGMENTS

This work was partially supported by the M.E.S.S.C. of Japan under Grant-in-Aid for Scientific Research on Priority Areas (No.269), and also by NEDO and FED, Japan. The authors also thank NTT Electronics Corp. for the test chip fabrication.

8. REFERENCES

[1] T. Shibata and T. Ohmi, "A functional MOS transistor featuring gate-level weighted sum and threshold operations," IEEE Trans. Electron Devices, Vol. 39, No. 6, pp.1441-1455, 1992.

[2] T. Shibata, T. Nakai, N. M. Yu, Y. Yamashita, M. Konda and T. Ohmi, "Advances in neuron-MOS applications," ISSCC Dig. Technical Papers, pp.304-305, 1996.

[3] M. Ikebe, M. Akazawa and Y. Amemiya, "A Functional νMOS Circuit for Implementing Cellular-Automaton Picture-Processing Devices," Computers Elect. Engng, Vol.23, N0. 6, pp. 439-451, 1997.

[4] A. Iwata and M. Nagata, "A Concept of Analog-Digital merged Circuit Architecture for Future VLSI's," IEICE Trans. Fundamentals, vol. E79-A, pp. 145-157, 1996.

[5] M. Nagata and A. Iwata, "A PWM Signal Processing Core Circuit based on a Switched Current Integration Technique," IEEE J. Solid-State Circuits, Vol. 33, No. 1, pp. 53-60, 1998.

[6] E. R. Fossum, "CMOS Image Sensors: Electronic Camera-On-A-Chip," IEEE Trans. on Electron Devices, Vol. 44, No. 10, pp. 1689-1698, 1997.

[7] C. H. Aw and B. A. Wooley, "A 128 × 128–Pixel Standard-CMOS Image Sensor with Electronic Shutter," IEEE J. Solid-State Circuits, Vol. 31, No. 12, pp. 1922-1930, 1996.

[8] M. Nagata, T. Yoneda, D. Nomasaki, M. Sano, and A. Iwata, "A Minimum-Distance Search Circuit using Dual-Line PWM Signal Processing and Charge-Packet Counting Techniques," ISSCC Dig. Tech. Papers, TP2.3, pp. 42-43, 1997.

[9] M. Nagata, M. Homma, N. Takeda, and A. Iwata,"A Smart CMOS Imager with Pixel Level PWM Signal Processing", Symposium on VLSI Circits, pp.141-144, 1999.

Massively Parallel Intelligent Pixel Implementation of a Zerotree Entropy Video Codec for Multimedia Communications

A.M.Rassau*, G.Alagoda, D.Lucas, J.Austin-Crowe, K.Eshraghian
Centre for Very High Speed Microelectronic Systems,
Edith Cowan University, 100 Joondalup Drive, Joondalup, Perth WA 6027, Australia
**Department of Cybernetics, University of Reading, Whiteknights, Reading RG6 6AY, UK*
**(Currently attached to the Centre for Very High Speed Microelectronic Systems, ECU)*
Ph.: +44 (0)118 987 5123
Fax.: +44 (0)118 931 8220
E-Mail: amr@cyber.rdg.ac.uk

Key words: Parallel architectures, video processing, compression, multimedia applications.

Abstract: A recently proposed Intelligent Pixel Mobile Multimedia Communicator (IPM^3C) integrates the capture, encoding, decoding and display of optical images into a single array of Intelligent Pixel (IP) processors. The real-time video processing required both for compression of the captured images and decompression of the received data stream presents a significant challenge for the required pixel based coding. This paper presents an architecture for the implementation of a massively parallel wavelet based zerotree entropy video codec, designed for implementation on the IP array, capable of fulfilling the very low bit-rate coding requirements for M^3C.

1. INTRODUCTION

Integrated image sensor and display devices, capable of real-time image capture and processing, are of significant interest for the future development of interactive multimedia communications devices [1]. A novel single chip image processor capable of real-time simultaneous image capture, processing and display was recently proposed for application to mobile multimedia communications (M^3C) [2].

The processor consists of an array of so called Intelligent Pixel (IP) processing elements, shown schematically in Figure 2. Each element of the array incorporates three components. Firstly, a photo-detector (PD) with associated analogue to digital converter (ADC), for the conversion of the incident light level to a suitable input value. Secondly, a localised processor customised for the specific implementation of the parallel video codec algorithm and finally a liquid crystal display for the conversion of the decoded incoming video stream to an optical output [3]. To carry out the required video coding the IP array is used to emulate a mesh network of processing elements such that the inherent parallelism and interconnectivity of the architecture is fully exploited.

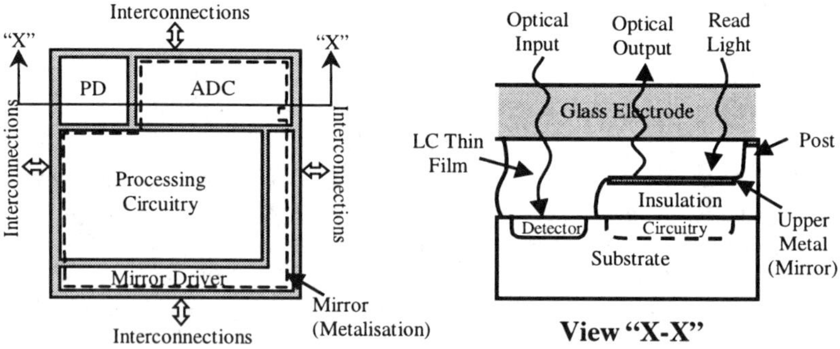

Figure 1. Intelligent Pixel processor

A massively parallel wavelet based video codec algorithm for an IP array processor was recently presented [4]. The proposed IP implementation of the algorithm is capable of real-time very-low-bit-rate video coding with performance comparable to other state-of-the-art codecs and is also highly scalable. These characteristics are crucial to the development of an M^3C device, and are compatible with the low bit-rate (~20kb/s) recommendation in MPEG-4 and the channel bandwidths of the existing GSM (23kb/s) systems. In this paper a logical architecture for the implementation of this codec within the IP array is described along with the necessary control strategy for embedded parallel array operation.

2. VERY LOW BIT-RATE VIDEO COMPRESSION FOR MULTIMEDIA COMMUNICATION

All of the current video coding standards (ITU, MPEG, etc.) are based on the same general architecture, namely motion-compensated temporal coding coupled with some form of image transform spatial coding. The requirements of the IPM^3C are for very-low-bit-rate real-time compression of video within tight hardware complexity constraints. Of all the techniques currently in development the most suitable to fulfil this requirement, in terms of performance versus complexity, is almost certainly a combination of the discrete wavelet transform (DWT) coupled with zerotree coding.

Few hardware implementations of zerotree coders have been developed to date, mainly because the structure of the standard embedded zerotree wavelet (EZW) algorithm [5] requires a complex architectural mapping. A new more efficient technique termed zerotree entropy (ZTE) coding has, however, recently been developed by Martucci and Sodagar [6] specifically for very low bit-rate coding of wavelet coefficients. ZTE coding differs from the more traditional zerotree coders in a number of ways that make it suitable for implementation on the parallel IP array. In addition the alphabet of symbols used to classify the tree nodes is changed to one that is capable of performing significantly better for very low bit-rate encoding of video. A very low bit-rate video codec algorithm based on these techniques and suitable for implementation on an IP array is presented in [4]. The system block diagram for the forward path of this algorithm is shown in Figure 3.

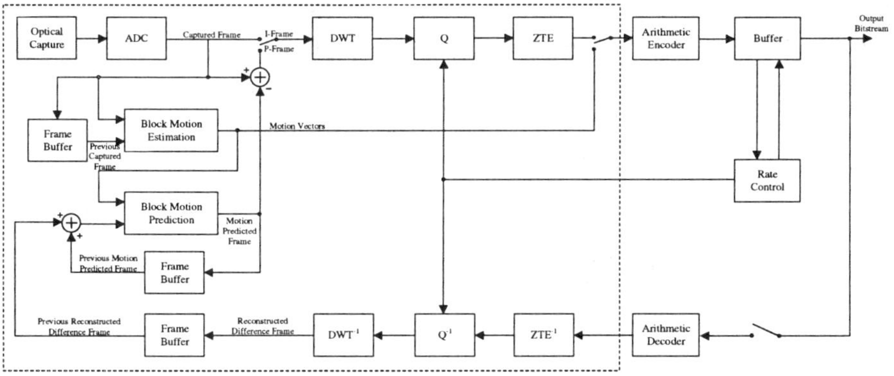

Figure 2. System block diagram of the encoder for the proposed video codec. All functionality prior to the arithmetic coder (shown enclosed within the dashed line) is embedded within the IP array.

2.1 Motion Estimation

The massively parallel IP array is well suited to cope with the large processing demands of a block matching approach to motion estimation. Due to the low resolution of the video in mobile applications, a block size of 8x8 pixels is sufficient. To achieve very low bit-rate compression the length of the motion vectors must be kept to a minimum to ensure that the motion information does not account for too large a proportion of the available bit-budget. This can be achieved by limiting the search area to a relatively small range. For videoconferencing applications it can be reasonably assumed that the majority of motion to be tracked will be slow moving, thus a small search area is acceptable. A search area of 8 pixels in all four directions from the centre of each block is adequate and by exploiting the parallel processing capabilities of the array a full search within this area can be performed. The sum of absolute differences (SAD) is used as a distortion measure, the block with the lowest SAD being selected for the estimated frame and the appropriate motion vector transmitted. The prediction error calculated from the motion compensated previous *reconstructed* frame, is further compressed using DWT and ZTE coding to remove spatial correlation.

3. IP ARRAY IMPLEMENTATION

An array of IPs is uniquely capable of performing highly parallel image processing tasks on images captured in situ within the array. This provides the potential for real-time operation at very high frame rates. The limited hardware complexity that can be realised within each pixel, however, necessitates very efficient architectural mapping of the video coding algorithms.

3.1 Motion Estimation

To perform the motion estimation a complete search is carried out in a 16x16 pixel search area around each motion block in the image to find the closest matching block. The corresponding motion vector for that block is then coded and transmitted. This search can be carried out on all motion blocks in the array in parallel by shifting the entire current frame, through the complete search area, over the previous frame and storing a motion vector in each block corresponding to the point where the distortion measure was lowest.

3.2 Discrete Wavelet Transform

A wavelet transformation is essentially a sub-band decomposition that segments an image into a number of frequency domains at different scales. A complete two-dimensional decomposition can be calculated by sequential convolution of the rows and columns of an image with a pair of suitably chosen quadrature mirror filters (QMFs) followed by half rate sub-sampling at each scale. For a VLSI implementation 1-D FIR filters can be used for these convolutions. Suitably chosen wavelet basis functions are used to derive the filter coefficients. The transform coefficients, y_i, are thus obtained by convolving the pixel input values, x_i, with the coefficients, w_s, of an FIR filter of length L:

$$y_i = \sum_{s=1}^{L} x_{i-s} w_s \qquad (1)$$

This convolution can be computed by scaling the image coefficients with each of the filter coefficients in turn and accumulating the sum of these products in each pixel. This can be carried out very efficiently on the IP array as the mesh network of processing elements allows all of the convolutions in one plane to be carried out in parallel.

3.2.1 Wavelet Transform Using Symmetrical FIR Filters

An efficient method that can be used to solve the inherent border problem that arises when performing linear convolutions, is to implement a bidirectional data shifting technique as first proposed by Lu [7]. If the FIR filters used to perform the discrete wavelet transform (DWT) are formed from a symmetric kernel of length $2P+1$ then by using the property $w_s = w_{-s}$ the convolution in equation (1) can be rewritten as:

$$y_i = \sum_{s=1}^{P} x_{i+s} w_s + \sum_{s=1}^{P} x_{i-s} w_s + x_0 w_0 \qquad (2)$$

Bidirectional shifting of the image data can then be used to calculate the complete convolution in P cycles. Figure 4 illustrates an example of the necessary data movement within one individual row. The pixel input value is replicated and stored in a pair of registers within each IP. In each cycle the data in one of these registers moves to the right while the data in the second moves to the left. After one such bidirectional shift, as shown by the arrows, each IP contains the corresponding x_{i+s}, x_{i-s} and x_0 (x_0 in a third register not shown). This provides symmetry between the first and last rows of the array and thus avoids the reconstruction error on the image border by removing the discontinuity at the edges.

In addition to solving the finite border problem this scheme lends itself particularly well to an efficient hardware implementation. Due to the symmetric nature of the filters used, only half of the filter coefficients are required to carry out the convolution. In fact the convolution sum can very easily be calculated from the bidirectionally shifted data. x_{i-s} is the right shifted data, shown at the top of each pixel in Figure 4, and x_{i+s} is the left shifted data, shown at the bottom. After each shift these two filtered data values need only be multiplied with the next positive filter coefficient and the sum of both results accumulated in each of the processors [8].

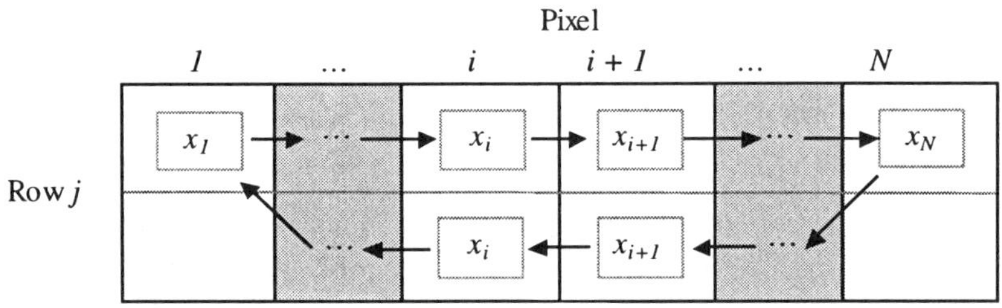

Figure 3. Schematic diagram showing data movement for the bidirectional shifting method

3.2.2 Binary Wavelet Filters and ZTE Coding

When considering low complexity hardware implementations the precision with which the filter coefficients can be represented must be considered. Integer approximations need to be made which will lead to systematic errors during reconstruction. These errors can, however, be avoided if filters with binary representable coefficients [9, 10] are used. The coefficients can then be represented by powers of 2 and the scaling required for the convolution calculations can be carried out using single-bit shifts.

Due to the limited precision available for the calculations within the IP architecture binary wavelet filters have been shown to perform as well if not better for hardware implementations of wavelet based ZTE coders than the standard real-valued filters such as Daubechies' [11]. Because the simple binary filters have inferior spatial-frequency localisation properties, they tend to produce relatively large coefficients in the high frequency sub-bands. This does not adversely affect ZTE coding to any great extent, however, due to the fact that all significant coefficients are transmitted in one pass unlike the multiple passes used with the more standard EZW coding

The use of binary representable filters allows the convolution operations necessary for the computation of the DWT to be carried out using integer arithmetic. A complete multi-scale DWT can thus be computed on the IP array using a single bit-serial adder accumulator along with two additional registers to store the current frame for left and right shifting.

3.2.3 Nucleic Coefficient Storage

After all of the convolutions for a particular scale of the decomposition have been carried out, four frequency sub-bands are obtained (LL, LH, HL and HH). For the next scale it is only the low-low (LL) filtered transform coefficients that are further decomposed. A nucleic scheme can be used to efficiently store these sub-band coefficients within the IP array. As the decomposition progresses to each subsequent scale only certain nuclei, which contain the data required at that scale remain active [2]. The rest of the pixels retain the data from the previous scales but are put into a bypass mode making them transparent to the decomposition at the current scale. The resulting arrangement is illustrated in Figure 5. $\mathbf{LL_s^{xy}}$ represents a low-low filtered coefficient at scale **s** of the decomposition from cell **x,y**.

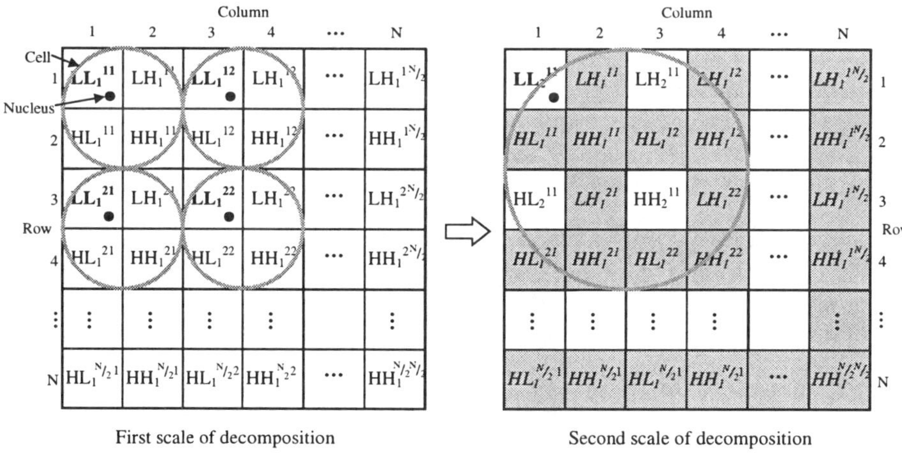

Figure 4. Two scale decomposition of an NxN image in the nucleic scheme. LL_s^{xy} represents a low-low filtered coefficient obtained at scale s of the decomposition in cell x,y

This allows the decomposition at any scale to be carried out with the same time complexity and has the additional advantage that at the end of the transform the coefficients are ordered together into wavelet blocks facilitating the implementation of the ZTE coder.

3.3 Zerotree Entropy Coder

ZTE coding is based on, but differs significantly from, the well-known EZW algorithm. Like EZW, the ZTE algorithm exploits the self-similarity inherent in the wavelet transform of image frames to predict the location of information across wavelet scales. The wavelet coefficients are quantised and zerotrees are used to reduce the number of bits required to represent the wavelet trees. In ZTE coding, the coefficients of each wavelet tree rooted in the lowest band are rearranged to form a *wavelet block*. This wavelet block provides a direct association between the wavelet coefficients and what they represent spatially in the frame. Related coefficients at all scales and orientations are included in each block. As previously discussed the transform coefficients are already arranged into these wavelet blocks after the nucleic wavelet transform so no rearrangement is necessary prior to ZTE coding.

To perform ZTE coding each wavelet block is first adaptively quantised according to scene content and frequency band as well as the desired bit-rate. The extreme quantisation required to achieve a very low bit-rate results in a high proportion of zero coefficients. The wavelet trees can thus be efficiently represented and coded by scanning each tree depth-first from the root in the low-low sub-band through the children and assigning either a *zerotree root (ZTR)*, *valued zerotree root (VZ)* or *value (V)* symbol to each appropriately. A zerotree root exists at any node where the coefficient is zero and all the node's children are zerotrees. Zerotrees do not need to be scanned further because it is known that all coefficients in these trees have amplitude zero. A valued zerotree root is a node where the coefficient has a non-zero amplitude but all four children are zerotree roots, again the scan of the tree can stop at this symbol. A value symbol identifies a coefficient that has some non-zero descendants; its amplitude can be zero or non-zero.

4. ARCHITECTURAL MAPPING

The approaches outlined above allow a hardware implementation of the proposed video codec to be achieved very efficiently. Each pixel need only contain a single bit-serial adder-accumulator, a number of shift registers for frame buffering and coefficient storage, a multiplexer to route signals between pixels and a small amount of additional control logic.

4.1 Discrete Wavelet Transform

Figure 7 outlines the logical architecture necessary to carry out the DWT using a symmetric binary representable wavelet filter. Due to the limited space available within the IPs all arithmetic operations are performed in a bit-serial fashion. The registers in each pixel need to operate as bi-directional shifters in order to perform the scaling operations. Shift-right is the inherent operation and will occur at every clock cycle, shift-left is equivalent to a no-shift-right operation for one clock cycle. The 4-way multiplexer in each of the pixels is used to select as input to the pixel the output from any one of the neighbouring pixels. In order to carry out the convolutions required for the DWT *Reg1* and *Reg2* alternately load data from the left or right neighbouring pixels respectively or pass data to the adder-accumulator. This allows the summation and the data shifting operations to be carried out in parallel.

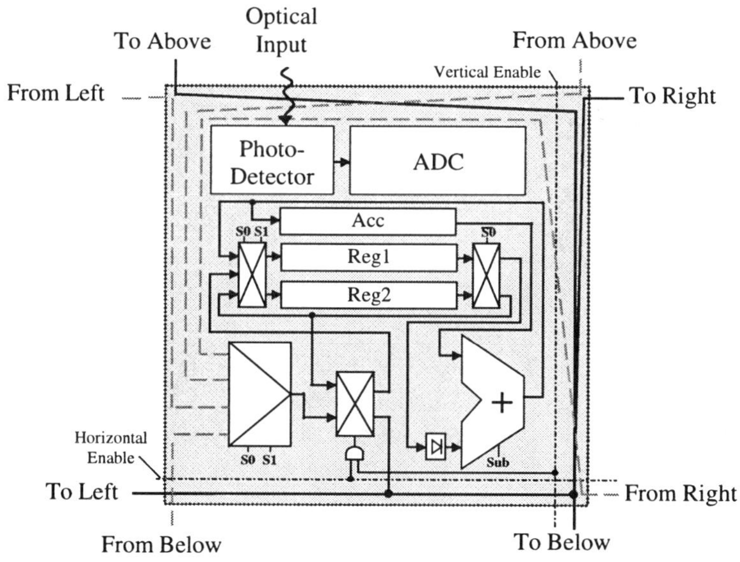

Figure 5. Logical architecture within each pixel for computation of the DWT

The decimation required after each scale of the transform is inherent in the ordering of the coefficients in the pixels. All odd numbered rows or columns of pixels perform the low pass filtering and all even numbered row or columns the high pass filtering. At the end of each scale all of the coefficients are thus arranged in the nucleic scheme introduced above. To perform the transform at the next scale the even rows and columns containing all the high frequency sub-bands need only be disabled such that the convolution operations can be performed as before on the low-low sub-band. In order to disable the high pass pixels they are placed into a bypass or transparent mode so that data can be passed directly through them. In active pixels the 4-way multiplexer is connected to one of *Reg1* or *Reg2* and the output from that register is connected to the output bus. When a pixel is switched into the bypass or transparent mode of operation the 4-way multiplexer is connected directly back to the output bus in the same pixel.

Each pixel in the array is controlled by a select line that determines whether it functions in a low pass or a high pass filter mode. These select lines run all the way through every row and column. At the start of the forward transform all even numbered rows and columns are set to operate in the high pass mode and the odd numbered rows and columns in the low pass mode. At the end of the scale all pixels that were in the high pass mode are switched into bypass mode leaving only the pixels containing the low-low filtered coefficients active. For the next scale and all subsequent scales the first row and column always remain in the low pass mode. Subsequently every alternate row or column is switched into high pass mode, noting that the bypass mode pixels are effectively transparent. The algorithm is completely symmetric so for the inverse transform this process is simply carried out in reverse.

4.2 Zerotree Entropy Coder

The arrangement produced by the nucleic DWT enables a ZTE processor to be designed that can independently process each wavelet block in parallel. To carry out ZTE coding each wavelet block is first adaptively quantised and the significance, *S*, of each pixel determined. A pixel is significant if the magnitude of its quantised coefficient is greater than zero. Child significance, *C*, is then derived from this. This signal propagates through the entire wavelet block, it is high if any descendants of a pixel are significant or have significant descendants. The pixels at the lowest level have no children hence *C* is always zero for these. Parent significance, *P*, is further derived from this, it is high if a pixels parent or any of its siblings are significant.

These three signals are used to derive two status bits, *A* and *B*, which correspond to one of four states indicating whether the pixel is a zerotree root (ZTR), a valued zerotree root (VZ), a value (V) or does not need to be transmitted as it is part of zerotree (DNT). Figure 8 shows the logical architecture required in each pixel to derive the necessary significance information.

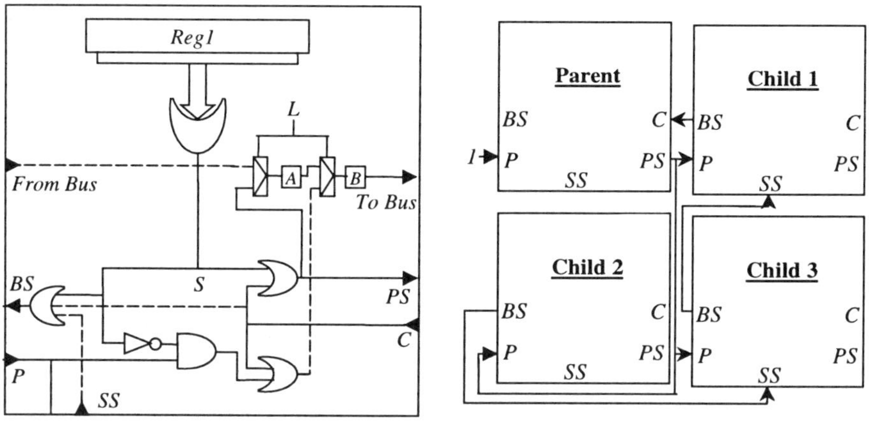

(a) (b)

Figure 6. (a) Logical architecture for the hardware implementation of the ZTE coder and (b) Parent-child interconnections for the first wavelet scale.
The SS signal routes the significance information between sibling pixels within one level of a wavelet block. The BS signal is high if either the pixel, any of its siblings or their children are significant. The PS signal is high if the pixel or any of its children are significant.

4.3 Control Strategy

The limited space available for processing within each individual pixel in the array necessitates some degree of external control. Each IP contains one additional register to store a series of control bits that define the state of the multiplexers and other logic within the IP to appropriately route data between and within the pixels. Appropriate control words generated by a state machine external to the array are thus loaded into the pixels to carry out all of the necessary stages of the video coding algorithm, i.e. motion compensation, forward WT and ZTE coding for captured frames, followed by ZTE decoding, inverse WT and motion prediction for simultaneous display.

5. CONCLUSION

The logical architecture and control strategy required for the implementation of a very low bit-rate video codec on an IP array has been described. Due to the massively parallel nature of the IP array the architecture is capable of operating in real-time at very high frame-rates. A complete three-scale forward wavelet transformation followed by ZTE coding can be performed in less than 300 clock cycles irrespective of image size. VHDL hardware simulations have been performed, which verify that the hardware implementation of the video codec performs similarly to the ZTE codec presented in the literature. Results have shown that this low complexity codec is capable of performing comparably to other state-of-the-art very low bit-rate coders for scenes with relatively low levels of motion, as would be common for a personal communication application.

6. ACKNOWLEDGEMENTS

Contributions provided by the Australian Research Council and the support of the Centre for Very High Speed Microelectronic Systems, Edith Cowan University are gratefully acknowledged.

7. REFERENCES

[1] V.O.K.Li and X.Qiu, "Personal communication systems (PCS)", Proc. of IEEE, Vol. 83, No. 9, pp. 1210-1243, September 1995.

[2] A.M.Rassau, T.C.B.Yu, H.N.Cheung, S.W.Lachowicz, K.Eshraghian, W.A.Crossland and T.D.Wilkinson, "Smart pixel implementation of a 2D parallel nucleic wavelet transform for mobile multimedia communication", Design and Test Conference, Paris, Feb. 1998.

[3] N.Collings, W.A.Crossland, P.J.Ayliffe, D.G.Vass and I.Underwood, "Evolutionary development of advanced liquid crystal spatial light modulators", Applied Optics, Vol. 28, No. 22, November 1989.

[4] A.M.Rassau, G.Alagoda and K.Eshraghian, "Massively parallel wavelet based video codec for an intelligent pixel mobile multimedia communicator", (submitted to the 5th Int. Sym. On Signal Processing and its Applications, Brisbane, Australia, August 1999).

[5] J.M.Shapiro, "Embedded image coding using zerotrees of wavelet coefficients", IEEE Trans. On Signal Processing, Vol. 41, No. 12, pp. 3445-3462, December 1993.

[6] S.A.Martucci and I.Sodagar, "Zerotree entropy encoding of wavelet coefficients for very low bit rate video", Proc. of the IEEE Int. Conf. Image Processing, Switzerland, September 1996.

[7] J.Lu, "Computation of 2-D wavelet transform on the massively parallel computer for image processing", Proc. of the 7th IEEE Multi-D Signal Processing Workshop, Lake Placid, NY, September 1991.

[8] A.M.Rassau, K.Eshraghian, H.Cheung, S.W.Lachowicz, T.C.B.Yu, W.A.Crossland and T.D.Wilkinson, "Wavelet transform algorithms for smart pixel Opto-VLSI implementation", Int. Conf. on Computational Engineering and Systems Applications, Tunisia, April 1998.

[9] E.L.Schwartz, A.Zandi and M.Boliek, "Implementation of compression with reversible embedded wavelets", Proc. of the SPIE 40th Annual Meeting, Vol. 2564-04, July 1995.

[10] G.Strang and T.Nguyen, "Wavelets and filter banks", Wellesley Cambridge Press, 1996.

[11] A.M.Rassau, Soo-Ik Chae and K.Eshraghian, "Simple binary wavelet filters for efficient VLSI implementation", Elec. Letters, Vol. 35, No. 7, pp. 555-557, April 1999.

Implementation of a Wavelet Transform Architecture for Image Processing

Camille DIOU, Lionel TORRES, Michel ROBERT
Laboratoire d'Informatique de Robotique et de Microélectronique de Montpellier, UMR CNRS - Université Montpellier II n°5506
161 Rue Ada, 34392 MONTPELLIER Cedex 5, FRANCE
Phone:(+33) 4-67-41-85-85 Fax:(+33) 4-67-41-85-00
e-mail :<author>@lirmm.fr

Key words: Image compression, integrated processor, lifting scheme, multi-resolution analysis, wavelet transform.

Abstract: The wavelet transform appears to be an efficient tool for image compression. Many works propose an implementation of the pyramid algorithm with some improvement to reduce its treatment time or to increase its performances. However, the pyramid algorithm remains silicon area costly, essentially because of its memory needs, and depending on the size of filters used. This paper proposes a new implementation of the wavelet transform using the lifting scheme. This method proposes many improvements such as in-place calculation, small memory needs, and easy inverse transform.

1. INTRODUCTION

Compression is a necessary step for data transmission or storage. Whereas there are many ways of compressing still images, the JPEG method is the most widely used. On the other hand, the video compression often remains based on the MPEG. The goal of this work is to develop a generic wavelet core that can be used as an *Integrated Processor* in integrated systems for many applications as image/video compression, edge detection, moving objects detection. This paper shows a comparison between the Mallat's pyramid algorithm (PA) [1] and the lifting scheme (LS), a new tool for creating wavelets [7,8,10].

Next, we present the pyramid algorithm's implementation based on filter banks, and our architecture based on the lifting scheme.

2. WAVELET TRANSFORM AND MULTI-RESOLUTION ANALYSIS

The multi-resolution analysis [1] uses two functions to project a signal on two spaces:

- the wavelet function extracts the details (high frequency signal);
- the scaling function keeps the approximation (low frequency signal).

By translating and dilating the wavelet, we can analyse the signal all over the time and at different resolution levels. Different wavelet techniques exist for image processing [18,19] depending on the signal to be analysed. We concentrate our efforts on the pyramid and the lifting scheme algorithms.

2.1. Filter banks and pyramid algorithm (PA)

The relationship between the wavelet multi-resolution analysis and filter banks was first shown by Mallat [1]. The image is filtered by both high-pass and low-pass filters along horizontal direction, giving, respectively, an approximation of the original image and its horizontal details. This scheme is re-applied on the two sub-images along the vertical direction, giving us the three horizontal, vertical and diagonal details sub-images, and the 2-D approximated sub-image (Figure 1, left).

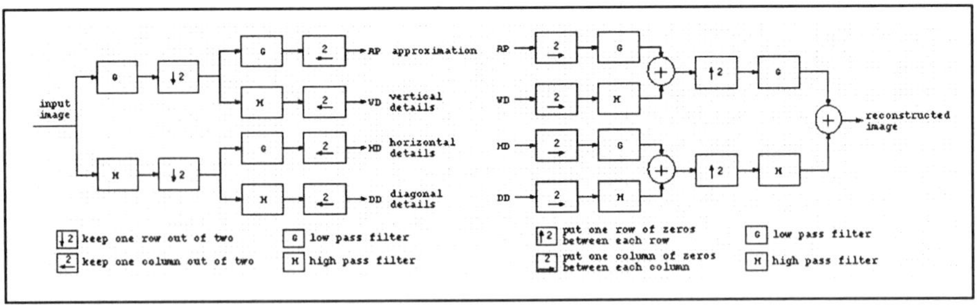

Figure 1. Wavelet transform image decomposition (left) and recomposition (right)

The inverse transform is obtained by inverse filtering with the corresponding filters (Figure 1, right).

2.2. The lifting scheme (LS)

The lifting scheme algorithm [7,8,9,10] presents many advantages compared to the pyramid algorithm [9]:

- calculations are performed in-place for an important memory saving;

- the algorithm shows an inherent SIMD parallelism at all scales;
- inverse transform is obtained easily from the direct transform;
- this scheme can be performed even when the Fourier techniques are no longer suitable (for example when the samples are not evenly placed, that causes problems with the filter banks because of the sub-sampling)

2.2.1. Transform

It is composed of the three following steps (see Figure 2 for details):

- **split**: this step separates the signal into two parts. The more correlated the sub-signals are, the better the predict; practically, the signal is divided into a set of even indexed samples (*S*), and another one of odd indexed samples (*D*);
- **predict**: the first set *S* is used to predict the second one *D*, according to a defined function; the difference between the predicted set and the original one is kept as the detail signal;
- **update**: the detail signal *D* is used to update the unmodified set *S*, in order to keep the average value of the signal.

This can be resumed by the following implementation:

$$(\text{odd}_{j-1}, \text{even}_{j-1}) := \text{Split}(s_j)$$
$$\text{odd}_{j-1} -= \text{Predict}(\text{even}_{j-1})$$
$$\text{even}_{j-1} += \text{Update}(\text{odd}_{j-1})$$

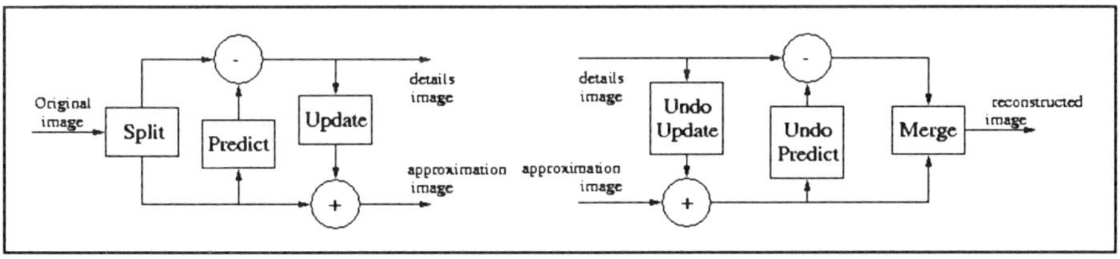

Figure 2. Signal decomposition (left) and recomposition (right) with lifting scheme

Figure 2 gives the schemes for 1-D decomposition-recomposition. For a 2-D transform, we simply apply the same method to each output, and process the scheme along the vertical direction.

The implemented wavelet depends on the prediction function. The higher the degree of the prediction function is, the smoother the wavelet. If the prediction is very close to the signal, the detail coefficients (*i.e.* the wavelet coefficients) will be very small.

Let's take an example with a linear prediction: if the original signal x is split into the detail signal d and the smooth signal s, we have:

Before the prediction:

(1) $d_k = x_{2k+1}$

(2) $s_k = x_{2k}$

After the linear prediction:

(3) $d_k = x_{2k+1} - \frac{1}{2}(x_{2k} + x_{2k+2})$

$s_k = x_{2k}$

And after the update:

$$d_k = x_{2k+1} - \frac{1}{2}(x_{2k} + x_{2k+2})$$

(4) $s_k = x_{2k} + \frac{1}{4}(d_{k-1} + d_k)$

The equation 3 gives us the corresponding high-pass filter, and when inserting equation 3 into 4, we get the corresponding low-pass filter:

$$HP = \left\lfloor -\frac{1}{2};1;-\frac{1}{2} \right\rfloor \qquad LP = \left\lfloor -\frac{1}{8};\frac{1}{4};\frac{3}{4};\frac{1}{4};-\frac{1}{8} \right\rfloor$$

2.2.2. Inverse transform

The inverse transform scheme contains three steps (undo update, undo predict and merge) which are obtained by reversing the order of the operations and by changing the signs of the operators, as shown in Figure 2.

We resume it with:

$$\text{even}_{j-1} -= \text{Update}(\text{odd}_{j-1})$$

$$\text{odd}_{j-1} += \text{Pr edict}(\text{even}_{j-1})$$

$$s_j := \text{Merge}(\text{odd}_{j-1}, \text{even}_{j-1})$$

The equivalent filter is obtained by putting one detail coefficient to 1, all others to 0, and computing the inverse transform. It gives the high-pass filter. Doing the same with a smooth coefficient gives the low-pass filter.

$$HP = \left\lfloor -\frac{1}{8};-\frac{1}{4};\frac{3}{4};-\frac{1}{4};-\frac{1}{8} \right\rfloor \qquad LP = \left\lfloor \frac{1}{2};1;\frac{1}{2} \right\rfloor$$

Note that the filters we obtain correspond to the wavelet defined by Cohen, Daubechies and Fauveau (CDF) [2].

3. ARCHITECTURES

There are many works on the implementation of the WT on FPGA [11,12] or ASIC [13,14,15,16,20]. Most of them propose improvements to the pyramid algorithm [14,17], or to the implementation (systolic or semi-systolic architecture, parallel or semi-parallel design [13,15,16]). But these methods rely on the same algorithm to perform the WT.

In this section, we present how we can implement this wavelet transform, first using filter banks, then with the lifting scheme. Then, we compare the methods to point out the number of needed operators, the silicon area cost and the memory cost of each method.

3.1. Filter banks architecture

Figure 3 shows a basic implementation of the pyramid algorithm for a 1-D wavelet transform. The number of operators is chosen in order to correspond to the CDF wavelet (the implementation with lifting scheme is shown in Section 3.2).

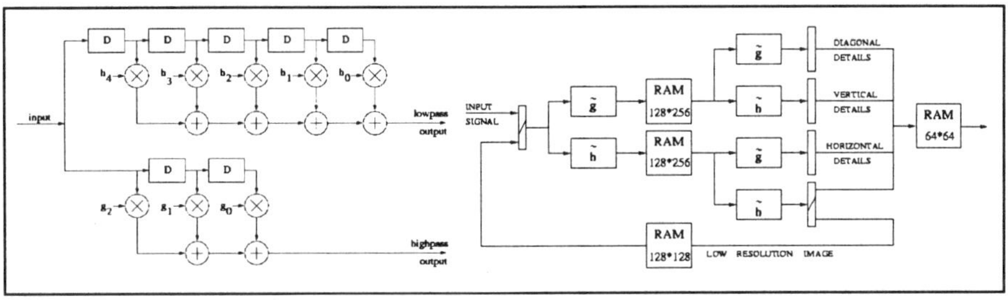

Figure 3. Pyramid algorithm implementation. Left: basic filtering cell. Right: whole structure

An implementation of the 2-D wavelet transform using the CDF wavelet is shown in [11]. We can see, in Figure 3 that the filters described above are present three times and that an important memory is necessary to perform the transform. At least the two 128×256 memories must be on-chip if we want to keep good performance during the transform. All these memories and the filters need an important silicon area. The inverse transform is not shown, but it needs a consequent memory too: 112 kB are necessary in the architecture presented in [11].

Line processing		Column processing	Total
31.7 ms		31.7 ms	64 ms
Data reading Of first cells	Data reading Of last cells	Image writing	Total
4.556 ms	8.857 ms	3.491 ms	17 ms

Table 1. Compression and decompression times of a filter banks architecture

The Table 1 shows the theoretical performances for this system for 5 levels of resolutions on a 256×256 image (see [11]), taking into account the following values: 250 *ns* per pixel for the analysis, and 40 *ns* per pixel for the synthesis. This table does not show the times of the quantization/de-quantization process, which are small compared to the transform.

The evaluated number of CLB for an implementation in a XC4005 family FPGA is about 340. Thus, the implementation of the wavelet transform needs around 5 kgates, without the memory.

3.1.1. Lifting scheme architecture

1-D structure We saw in the section 2.2 that the implementation of the lifting scheme requires few operators. We need, for each step (prediction and update), two adders in the decomposition and two adders in the recomposition. We can see in Figure 4 that there is no need for memory during the 1-D transform: the transformed coefficients overwrite the original ones during the decomposition. Thus the transform block can be seen from outside as a delay line. The transform can be performed in pseudo real-time, depending on the system clock. We get the same conclusion for the reconstruction step.

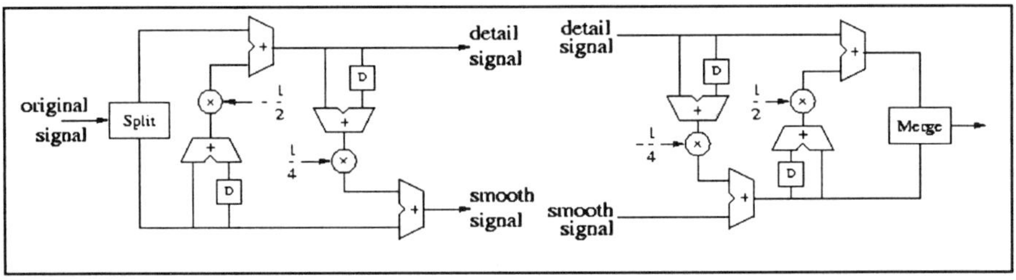

Figure 4. Structure of the lifting scheme blocks: transform (left) and inverse transform (right)

2-D structure For the 2-D implementation, we use the same blocks, organised in a cascade as shown in Figure 5.

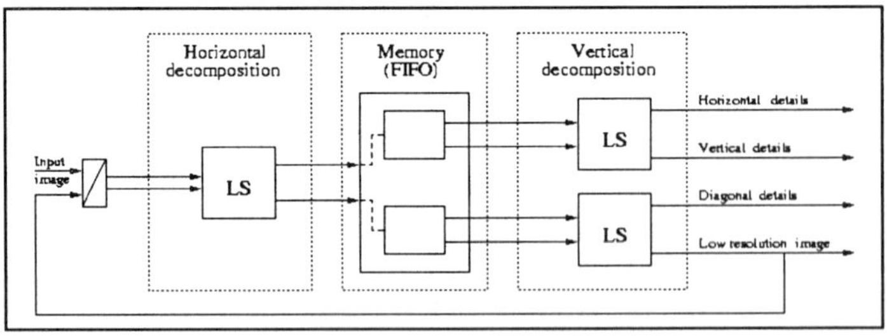

Figure 5. Structure of the lifting scheme block for a 2-D transform

Unlike the 1-D case, we need, in the 2-D case, some memory to perform the transform. Before starting the transform along the columns, we have to wait until the transform along the rows is achieved. There are different ways of doing this:

- we wait that the first 1-D decomposition is achieved, we store the image in a memory, rotate it by 90° or address the memory in a different manner, and we perform the second decomposition;

- we only wait that a few rows are treated and then start the vertical decomposition on these rows. When the following rows are achieved, we continue the vertical decomposition on them.

This last case is the most interesting from a memory point of view, but we also have to define the method to access the memory: we receive a few lines, and we want to transform them along the columns.

Figure 6 shows how we can start the vertical decomposition after the horizontal decomposition of the three or four first ones is achieved. We start the horizontal decomposition by computing the detail coefficients, and re-use them to compute the approximation. Once the 3 first lines are treated, we can start the vertical decomposition.

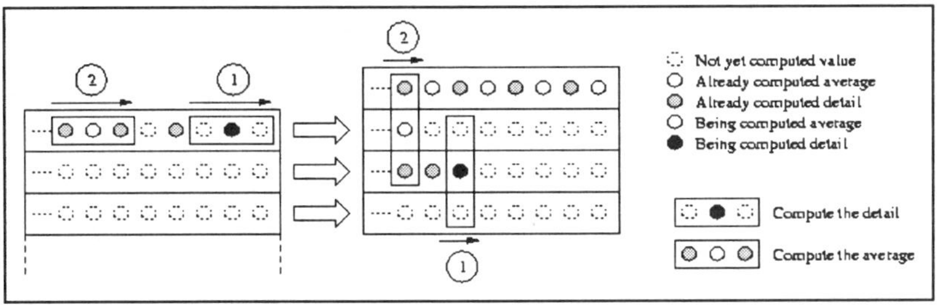

Figure 6. Memory need of the 2-D lifting scheme

When the fourth line (detail) is treated, the line of vertical detail coefficients is achieved too. Both this line and the previously treated one can then be used to compute the vertical average coefficients. While the vertical average is treated, the next vertical detail can be processed concurrently, while the horizontal average is also being calculated. Thus, in order to perform the complete 2-D transform, we only need a memory that can contain 4 lines of the image, *i.e.* for a 256×256 images at 8 bits per pixel, 4×256×8=1 kBytes.

3.2. Comparisons

Number of logic blocks. The lifting is very interesting for integrated systems because of little need for logic blocks. The table below show the difference between the lifting scheme and the filter banks.

Lifting scheme		Filter banks	
Total (2-D)	12 adders	Total (2-D)	21 adders

Table 2. Operators necessary for one step (transform or inverse transform)

We saw that the filter banks architecture needs around 340 logic cells. The lifting scheme only needs around 190 (24 adders × 8 blocks per adder).

Furthermore, to perform the multiplication, using bit shifting, the lifting scheme needs only 12 shifts whereas filter banks needs, for a 1-D transform. For a 2-D transform, this increases to 36 and 102 respectively.

Memory. One of the main advantages of the lifting scheme is the in-place calculation. We don't need any buffer memory to perform the transform: the wavelet coefficients overwrite the original ones (Figure 7). Figure 7 shows, on a real image, the Mallat and lifting scheme coefficients distribution after a 2-level decomposition.

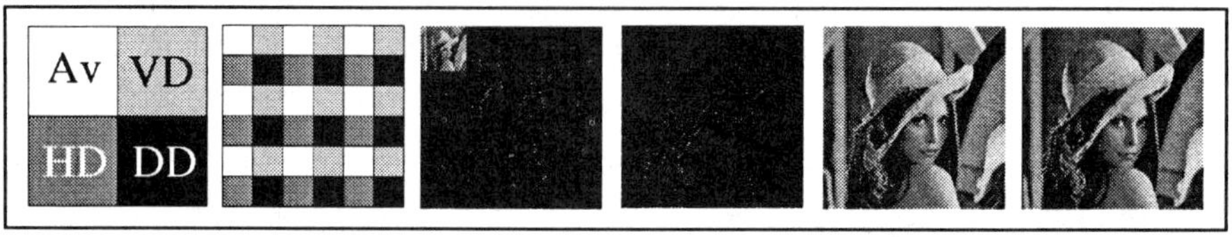

Figure 7. Distribution of the wavelet's coefficients: *a:* pyramid algorithm and *b:* lifting scheme and 2-levels decomposition of an image. *c:* Mallat and *d:* LS coefficients distribution. *e:* original and *f:* re-composed images

The on-chip memory needed for the filter banks design is approximately the size of an entire image whereas, for a lifting scheme implementation, this memory is reduced to a few lines. *Table 3* shows the memory necessary for the filter banks method as described in [11].

	Lifting scheme	Filter banks
Decomposition	1 kB	84 kB
Recomposition	1 kB	112 kB

Table 3. Memory need of the lifting scheme compare to the filter banks for a 256×256 image of 256 grey-levels

Thus, the lifting scheme implementation needs two times less logic blocks and very little memory compared to filter banks architecture. Furthermore, as the lifting scheme computes the wavelet transform in-place, we want to evaluate its performance at video rate. We see, in the next section, a first implementation of the lifting scheme.

4. HARDWARE IMPLEMENTATION

4.1. Overview

We have started to validate the lifting scheme architecture in real-time with an APTIX prototyping platform [21]. This programmable platform (Figure 8) contains Altera 10k100 FPGA, used to implement the wavelet transform with the necessary memory. A DSP core (ST D950) is used to perform the quantization and the coding of the wavelet coefficients. Figures 10 and 11 show the implementation of the 2-D lifting scheme. The input image is first decomposed horizontally. We get two sub-images at half the pixel clock frequency. Thus, we can alternatively decompose them along the vertical direction, switching between detail sub-image and smoothed sub-image. To keep the video rate, we have to interlace the two sub-images' lines - putting one data of the first sub-image in the memory, and then one data of the second sub-image - and compute the resulting image along vertical direction.

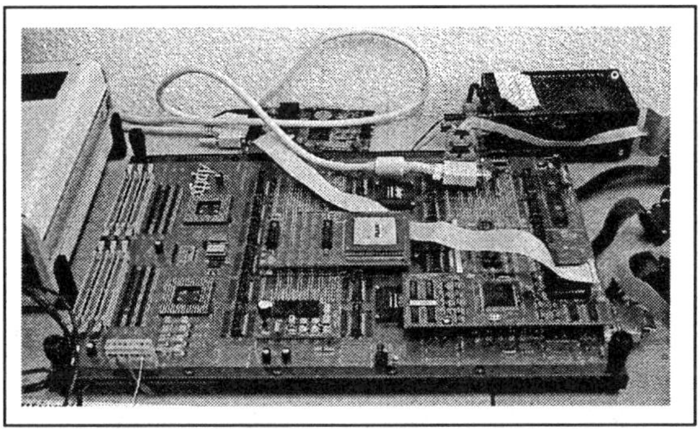

Figure 8. APTIX prototyping platform

4.2. Horizontal decomposition of the input image

Because of the in-place calculation, the horizontal decomposition of the image can be performed at the video rate. The original set of data at the frequency F is decomposed into two subsets (detail and average) at the frequency F/2. A simulation of the 1-D horizontal lifting scheme decomposition shows that this block needs 116 Altera's Flex 10k100 logic cells, that is, 2% of the chip. Thus, there is no noticeable difficulty in implementing the first 1-D horizontal decomposition block. Thus, we'll point out the vertical decomposition below.

4.3. Vertical decomposition of the sub-images

There are different ways of computing the vertical decomposition: we can use FIFO memories or RAM. The FIFO seem to be more efficient because they need no memory managing, in the strict sense of the term. But, the logic necessary to manage the four FIFO increases considerably the complexity. We describe here the two methods.

Using FIFO memories We consider that a first computed odd line DD_{n-1} is present in the first FIFO F1. A first not computed even line S_n is in the FIFO F2. A first not computed odd line D_{n+1} is in the third FIFO F3. When the second not computed even line S_{n+2} is ready, we compute the detail coefficient line from S_n, D_{n+1} and S_{n+2}. We name these coefficients DD_{n+1}.These coefficients are used to compute the average coefficients SS_n from DD_{n-1}, S_n and DD_{n+1}. The coefficients SS_n and DD_{n-1} are written to the output of the FIFO F2 and F1 respectively. The DD_{n+1} replaces the DD_{n-1} in F1 and the S_{n+2} replace the S_n in F2. Now, we have DD_{n+1} in F1 and S_{n+2} in F2; F3 is free. When D_{n+3} is ready, we stock it in F3 and wait for S_{n+4} which is necessary for calculating DD_{n+3}. Then, the scheme can be repeated from the beginning.

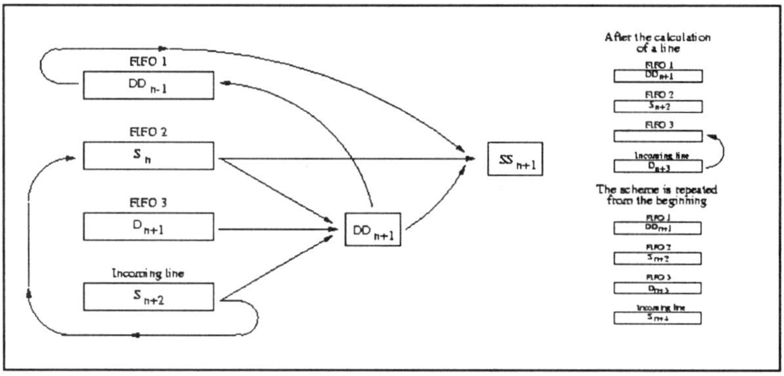

Figure 9. Three FIFO are used to perform the vertical decomposition

Using RAM The method described above shows the complexity of the FIFO memories managing. We have to use numerous multiplexers and demultiplexers for choosing between the video input (the output of the 1D transform block) or the output of the FIFO. All the logic that must be developed could be used to implement a RAM controller. Thus, instead of using FIFO memories, we could use RAM and address the data in a standard way (Figure 11).

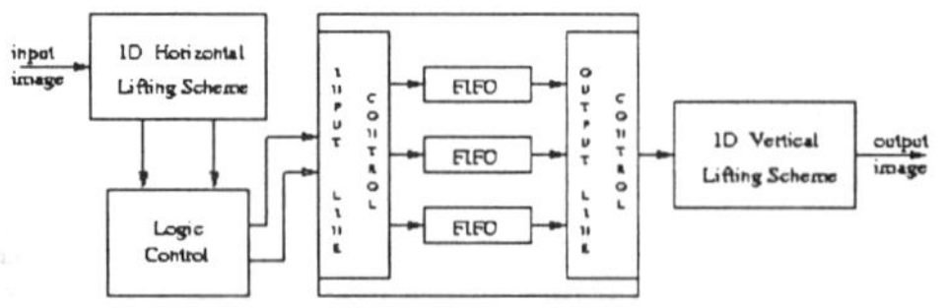

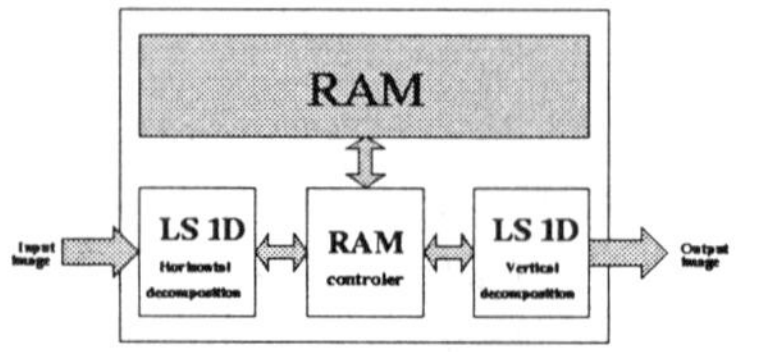

Figure 10. (left) Implementation of the lifting scheme using FIFO
Figure 11. (right) Implementation of the lifting scheme using RAM

4.4. Implementation

A first evaluation of the system presented in Figure 10 shows that the system uses 554 Altera's logic cells (*i.e.* 25% of the Flex 10k100 chip), and 6144 memory bits, that is 25% of the total memory of the Flex 10k100. All these results are given for 8 bits bus width. With a clock frequency about 30 MHz, this will allow the treatment in real-time, of images with a size of 800×800 pixels in 256 grey-level by increasing the size of the FIFO or the RAM. The needed memory can be implemented on the Altera's 10k100 FPGA. By using pipeline techniques, we estimate that we can obtain a clock frequency of 60 MHz, allowing us to process images in HDTV format.

First results of using the LS for image compression show a compression ratio of about 10 for a Peak Signal to Noise Ratio (PSNR) of 30 dB, which is the minimal admitted value for a good image quality. We are currently working on improving the compression ratio to get a value around 40, by using a well-adapted quantization and arithmetic coding [4,5,6].

5. CONCLUSION

In this paper, we have shown a new way of implementing the wavelet transform. This method combines an efficient processing rate with low area cost and memory use. It can easily be adapted to perform the inverse transform. The design re-use and IP cores concept is becoming increasingly present in the top down design flow; it is changing the way electronic engineers works. The impact on time-to-market can be considerable; for this reason, we intend to develop a complete architecture for wavelet transform which can be used as a wavelet IP core for image processing applications.

6. REFERENCE

[1] Stéphane G. Mallat, "A Theory for Multi-Resolution Signal Decomposition: The Wavelet Representation", *IEEE Transactions On Pattern Analysis And Machine Intelligence*, vol. 11, N° 7, July 1989

[2] A. Cohen, I. Daubechies and J. Feauveau, "Bi-orthogonal bases of compactly supported wavelets", *Comm. Pure Appl. Math.*, vol. 45, pp. 485-560, 1992

[3] Chok-Ki Chan and Chi-Kit Ma, "A Fast Method of Designing Better Codebooks for Image Vector Quantization", *IEEE Transactions on Communications*, vol. 42, N° 2/3/4, February-March-April 1994

[4] Marc Antonini, Michel Barlaud, Pierre Mathieu and Ingrid Daubechies, "Image Coding Using Vector Quantization in the Wavelet Transform Domain", *Proceedings IEEE ICASSP*, pp.2297-2300, April 1990

[5] Marc Antonini, Michel Barlaud, Pierre Mathieu and Ingrid Daubechies, "Image Coding Using Wavelet Transform", *IEEE Transactions on Image Processing*, vol. 1, N° 2, April 1992

[6] Pierre Mathieu, Michel Barlaud and Marc Antonini, "Compression d'image par transformée en ondelette et quantification vectorielle"

[7] Wim Sweldens, "The Lifting Scheme: A Custom-design Construction of Biorthogonal Wavelets", *Appl. Comput. Harmon. Anal.*, vol. 3, N°2, pp. 186-200, 1996

[8] Wim Sweldens, "The Lifting Scheme: A Construction of Second Generaton Wavelets", Technical Report 1995:6, Industrial Mathematics Initiative, Department of Mathematics, University of South Carolina, May 1995

[9] Wim Sweldens and Peter Schröder, "Building your Own Wavelet at Home"

[10] Gabriel Fernandez, Senthil Periaswamy and Wim Sweldens, "LIFTPACK: A Software Package for Wavelet Transforms using Lifting"

[11] Frédéric Truchetet and André Forys, "Implementation of still-image compression-decompression scheme on fpga circuits", *SPIE*, vol. 2669, pp. 66-75, January 1996

[12] Philipe Guermeur, Stéphane Guermeur and André Picaud, "Une implantation de la transformée en ondelettes discrètes à base de circuits FPGA"

[13] Jimmy C. Limqueco and Magdy A. Bayoumi, "A VLSI Architecture for Separable 2-D Discrete Wavelet Transform", *Journal of VLSI Signal Processing*, N° 18, pp. 125-140, 1998

[14] Gauthier Lafruit, Francky Catthoor, Jan P. H. Cornelis and Hugo J. De Man, "An Efficient VLSI Architecture for 2-D Wavelet Image Coding with Novel Image Scan", *IEEE Transactions on Very Large Scale Integretion (VLSI) Systems*, vol. 7, N°1, March 1999

[15] José Fridman and Elias S. Manolakos, "On the synthesis of regular VLSI architectures for the 1-D discrete wavelet transform", *SPIE Conference on Mathematical Imaging: Wavelet Applications in Signal and Image Processing II*, San Diego, July 1994

[16] Aleksander Grzeszczak, Mrinal K. Mandal, Sethuraman Panchanathan and Tet Yeap, "VLSI Implementation of Discrete Wavelet Transform", *IEEE Transactions on Very Large Scale Integretion (VLSI) Systems*, vol. 4, N°4, December 1996

[17] I. Urriza, J.I. Artigas, J.I. Garcia, L.A. Barragan, D. Navarro, "Locality Improvement of the Pyramid Algorithm to Compute the Discrete Wavelet Transform", *Proceedings of XIII Design of Circuits and Integrated Systems Conference*, pp. 30-35, November 1998

[18] B. B. Hubbard, "Ondes et Ondelettes", *Pour la Science*, dif. BELIN, Paris, 1995

[19] Frédéric Truchetet, "Ondelettes pour le signal numérique", *Collection traitement du signal*, HERMES, 1998

[20] Analog Devices, "Ultralow Cost Video Codec", *Datasheet*, 1997, http://products.analog.com

[21] APTIX, http://www.aptix.com

Scalable Run Time Reconfigurable Architecture

Abdellah Touhafi, Wouter Brissinck and Erik Dirkx
Erasmus Hogeschool Brussel nijverheidskaai 170 1070 Brussel Belgium
Vrije Universiteit Brussel Pleinlaan 2 1050 Brussel Belgium

Abstract: Currently multi-FPGA reconfigurable computing systems are still commonly used for accelerating algorithms. This technology where acceleration is achieved by spatial implementation of an algorithm in reconfigurable hardware has proven to be feasible. However, the best suiting algorithms are those who are very structured, can benefit from deep pipelining and need only local communication resources. Many algorithms can not fulfil the third requirement once the problem size grows and multi-FPGA systems become necessary. In this paper we address the emulation of a run time reconfigurable processor architecture, which scales better for this kind of computing problems.

1. INTRODUCTION

Currently multi-Field Programmable Gate Array (FPGA) reconfigurable computing systems are still commonly used for accelerating algorithms. This technology where acceleration is achieved by spatial implementation of an algorithm in reconfigurable hardware has proven to be feasible. However, research pointed out that the application must fulfil some requirements in order to achieve a high performance. The best suiting algorithms are those who are very structured, can benefit from deep pipelining and need only local communication resources. Many algorithms can not fulfil the third requirement once the problem size grows and multi-FPGA systems become necessary. This paper addresses this problem by presenting a scalable run time reconfigurable processor architecture. The presented architecture uses the fine granular nature of FPGA's to benefit from fine granular parallelism and ads the flexibility of run time reconfiguration together with instruction based programming. An emulator for this architecture is built and tested on its performance. As a case study we have implemented an ATM switch fabric simulator and show that our architecture suits better for large computing problems that require a complex interconnection scheme between successive pipeline stages.

2. MOTIVATION

Multi-FPGA systems can not simply be tiled ever larger without considering hughe latencies for wires traversing many FPGA's. The problem encountered here is comparable to interconnect problems encountered in parallel and distributed computing systems. Amdahl pointed out the existence of this problem for distributed computing systems and has shown that there is an upper bound to achievable performance gain by simply adding more computing resources. When using connected FPGA's as processing elements, the problem is even more emerging since the number of needed processing elements is a function of the size (complexity) of the problem. This implies that large computing structures with complex interconnection requirement will not always achieve a huge performance gain. Researchers in the reconfigurable computing society have remarked this effect [1] [2]. In figure 1 this effect is computed for a very simple computation model where we suppose that the communication cost scales linearly with the number of nodes.

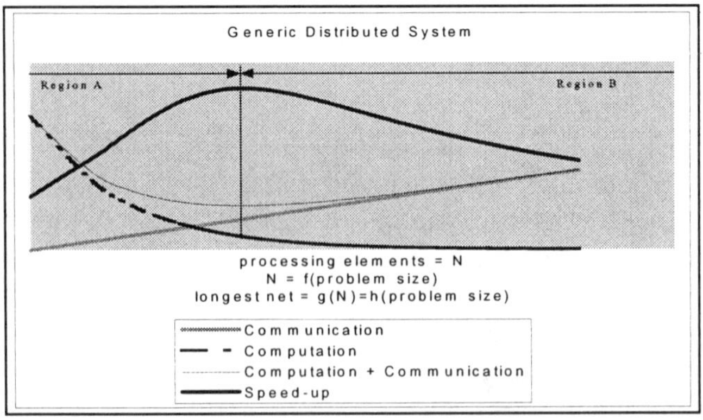

Figure 1. Generic distributed system

For pipelined systems a comparable graph will be achieved when considering the interconnected processing elements between two successive pipeline stages. If the spatial implementation of an application is located in Region A it is clear that there is potential to achieve faster implementations by exploiting more spatial resources. However, once an application enters region B it suffers from huge latency and throughput limitation resulting in larger execution times than would be possible in the ideal case. For applications located in region B a few approaches can be considered to enhance the performance. A first approach could be to provide more and faster routing resources (hierarchic)[3], a second approach is to make the basic computation cell larger [4] and a third approach is to provide a resource sharing mechanism in order to shorten virtually the communication lines. The three approaches are comparable in the sense that they all try to change the granularity of the computing system such that it fits better the application.

2.1 Computational model Targeted

A lot of research in the field of reconfigurable architectures targets solutions for general purpose computing[5][6]. Our research is more pushed by the need for scalable computing systems for high performance computing problems. As an example, we will address the simulation of large ATM switching fabrics [7] through this paper. NP complete problems like the Boolean Satisfiability problem [8] and Neural network implementations are also good candidates for the presented architecture.

The computational model targeted is based on three kinds of components. The first component is the source that provides input data to be computed. The second component is a process that executes some computation on the provided data and the third component is the sink that gathers the results. A computational problem can then be represented by a directed graph G(V,E) with vertices V and nodes E. The vertices represent communication channels and nodes are processing elements. Each processing element is characterised by its computation time and the number of input and output channels attached to it. We further suppose that each node has a "circuit" implementation, which can be mapped on runtime reconfigurable hardware. A node that represents a circuit is called a context. Independent nodes are organised in successive pipelined stages interconnected by unidirectional communication links. The nodes of a graph can then be interpreted as being contexts

and the vertices are context dependencies. A context can not be loaded for execution on a runtime reconfigurable device as long as all the other contexts on which it depends have not been processed yet.

2.2 Spatial Implementation

Given a directed graph of contexts it would be possible to implement its circuit equivalent completely in space using a multi-FPGA system. Such an approach has proven to be feasible for designs with deep pipelines. However once the interconnect between two pipeline stages becomes complex and nets between the different pipeline layers must traverse different FPGA's the well known Amdahl's bottleneck starts playing an important role. The accumulated net-delays will cause a frequency drop for the complete system. Although more gates are being evaluated in parallel for each clock cycle, the effect of the frequency drop can cause a net performance loss (i.e. less gates are being processed per time unit).

2.3 Temporal/Spatial Implementation

It is clear that other solutions must be searched for to keep gaining performance from scalable systems. The one proposed in this paper is the use of scalable (run time) dynamically reconfigurable processors with buffered communication between contexts. Runtime reconfiguration is used as a mean to make the circuit virtually less fine granular by putting some sequenciality in it. Run time reconfiguration is also used to virtually shorten the distance between computing nodes. With other words we start from a very fine granular graph specification which has a maximum amount of parallelism but at the same time can cause serious frequency drops due to accumulated net delay. From there on the fine granular nodes of the directed graph are grouped into larger nodes where the communication between the fine granular nodes is not visible anymore to the outside part of the directed graph. Further, the larger nodes are time-multiplexed on the available run time reconfigurable hardware, taking as a cost function to be minimised: Accumulated (communication time + reconfiguration time +computation time).

2.4 Buffered communication channels

Run time reconfigurable architectures are very sensitive to reconfiguration time. Some approaches proposed in [10] are based on the use of configuration prefetching, partial reconfiguration and compression techniques. Such approaches require rather difficult algorithms to be executed at runtime. Other approaches try to implement FPGA's with very fast context switch possibility [9]. For the computational model considered, a frequent reconfiguration scheme can be avoided by the use of buffered communication links. Other presented run time reconfigurable architectures

[9,12] do not provide- or provide only a very limited amount of memory cells to pass partial evaluated results from one configuration to another. This restriction has as effect that frequent reconfigurations are always necessary.

2.5 Cycle hiding

Avoiding frequent reconfigurations as explained in the previous subsection requires that communication links are not feedback. A possible solution to that restriction is to implement those contexts which contain feedback links in space as if they where one context. This requires that the reconfiguration of the computing nodes, which are used for the spatial implementation, can be synchronised with neighbour computing nodes. Whether a reconfigurable computing node is synchronised or not, with its neighbour nodes must be evaluated at each reconfiguration.

3. SUReCA: A SCALABLE ARCHITECTURE

In this paper, we introduce a processor with a scalable runtime reconfigurable computing layer and a scalable control layer, which gives a better base to deal with the computational model targeted. The SUReCA Architecture is organised as shown in figure 2.

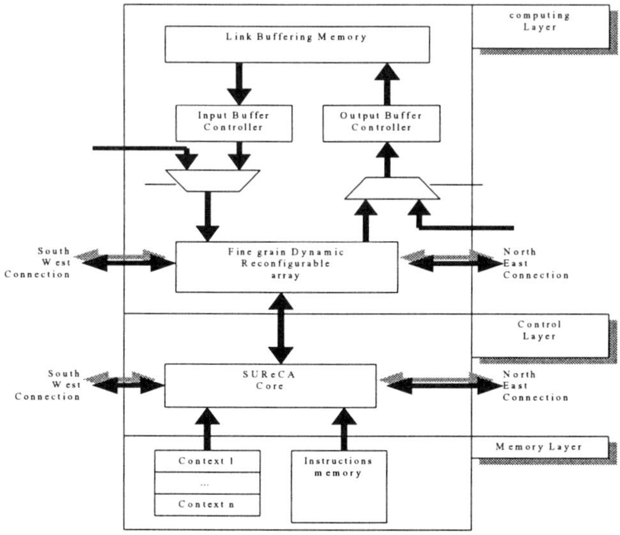

Figure 2. Scalable Uniform Reconfigurable Computing node Architecture

It consists of a computing layer, a control layer and a memory layer. A SUReCA node has communication links on the computing layer and on the control layer. Interconnected SUReCA nodes can then benefit from a large computing layer with a distributed tightly interconnected core. The virtual architecture layer contains the processor core that has full control over the processor resources.

3.1 Computing Layer

The computing layer contains a fine grain dynamic reconfigurable array like FPGA's. Further there is a fast 4 Mbytes memory and some memory control circuitry. The dynamic (run time) reconfigurable array is used as the computing medium. Different circuit contexts are being loaded sequentially onto the dynamic reconfigurable array, where they are activated for a certain number of clock cycles. The fast memory is used to pass partial results from one context to another and is organised as a set of FIFO buffers. For every circuit context their is an associated number of input and output FIFO buffers with each FIFO buffer being a communication link between two or more circuit contexts. The programmer has control over allocating buffer space and associating FIFO buffers with contexts. Memory management to handle this association is in hands of the input- and output buffer controller.

It is possible that contexts running on different SUReCA nodes share the same FIFO buffer. A multiplexer is used such that the local memory of a node can be connected to its neighbouring computing nodes. The flexibility of having such runtime routable memory can help to manage a better overlap between communication and computation.

The dynamic reconfigurable layer has near neighbour interconnections giving the ability to form a large fine granular tightly coupled computing plane. Reconfiguration of this large computing plane is synchronised and tested for I/O consistency on each reconfiguration cycle. Testing for I/O consistency is of major importance because inconsistent I/O connections can cause physical damage. When an I/O inconsistency is detected SUReCA nodes involved keep their I/O pins in high impedance state.

3.2 Control Layer

This layer forms the core of our architecture. The internal organisation of the core is given in figure 3. It is composed of an instruction decoder, a context loader, a link controller, and a condition evaluator and synchronisation circuitry. The control layer is responsible for the correct execution of the computation. The parts of the control layer are further explained in more detail.

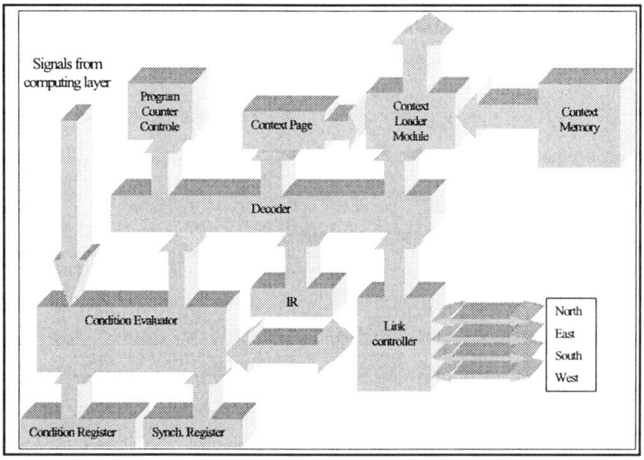

Figure 3. Sureca Core

3.2.1 Instruction Decoder

A SUReCA program is built up out of instructions and circuit contexts. The instruction set contains context-move instructions, memory management instructions, jump instructions, local resource initialisation instructions, and synchronisation and communication instructions. The instruction decoder controls the sequencing of the instructions and the circuit contexts. It also sets up the input buffer controller, and output buffer controller and link controller. The decoder is implemented as a state machine that runs in three phases. During the first phase an instruction is fetched. The second phase is used to decode the instruction and increment the program counter. The third phase is the execute phase of which the required number of clock cycles can vary between one and a few thousand depending on the instruction. In case of a context load instruction the next instructions will be already fetched decoded and eventually executed. If the instruction is a synchronisation, communication, initialisation or jump instruction it is executed. This way set up of the buffer controllers and synchronisation with the neighbour nodes can be overlapped with the ongoing context move instruction.

3.2.2 Context Loader

The context loader receives the context ID to be loaded from the instruction decoder. Before loading a new context into the dynamic reconfigurable layer all I/O pins are put into high impedance state to avoid inconsistent I/O set up. It then checks whether the context ID is the same as the one, which is already loaded. In case they are different the required context is loaded from the context memory. The next step is to initialise the state registers of the loaded context.

3.2.3 Link Controller

The link controller is a smart mesh router that provides a programming interface to the chip and at the same time gives the ability to send and receive synchronisation information from its neighbour nodes. The first byte of an incoming data-stream is an address. The address is composed of a node number and a local resource address. If the receiving node recognises its node number and its local resource, the data stream is sent to that location. The parts that can be accessed by the link controller are the memory layer and the FIFO buffers in the computing layer. This has as a nice side effect that certain SUReCA nodes can reprogram or simply change initialisation values of other SUReCA nodes. A possible application for this are online trained neural network implementations where some SUReCA nodes implement the forward pass while other nodes contain the back propagation algorithm which send the newly computed neural network node weights to the appropriate neural network nodes.

3.2.4 Condition Evaluator

The evaluation of instructions can be blocked until a certain condition or a set of conditions is met. This blocking gives the possibility to perform a conditional context switch. Hence, a context switch can be performed because of many reasons. In the best case a context switch is done because the context has finished all available input data. In other cases the reason for a context switch can be that the output FIFO buffers are full. In other cases again the active context can have reached a state in which itself asks for a context switch. The condition evaluator gives the possibility to set up the required context switch conditions. It generates a "condition-met" signal when the required conditions are met. The instruction decoder then relieves the block and starts with a new fetch cycle. The condition evaluator receives signals from the dynamically reconfigurable layer, from the instruction decoder, the context loader, and the synchronisation controller and memory buffer.

3.2.5 Synchronisation

As proposed in the motivation, for circuits with feedback channels, a spatial implementation is still the best solution because reconfiguration cost would be overwhelming. Supporting tight multi node implementations requires synchronisation with the involved nodes when working with runtime reconfigurable systems. The synchronisation control has the task to deal with this synchronisation. As can be seen in figure3, the synchronisation controller receives synchronisation information from the neighbour nodes and from the link controller. The synchronisation can be overlapped with the ongoing context loading. A synchronisation instruction is available to specify with which neighbours synchronisation is necessary.

4. SUReCA EMULATOR PLATFORM

An emulation system for the presented architecture is built and tested on performance. As shown in figure 4 our emulator is built up from of the shelf FPGA's, DPGA's, memory and connectors. Each emulation board can contain four SUReCA nodes, one master node, connectors in north, east, south and west direction and a parallel port interface to program the emulator. Two-dimensional computing arrays or a torus can be built by interconnecting emulator boards. The master controller contains a state machine that interprets incoming data streams from the hosts' parallel port; it is also the programming interface towards the four computing nodes. The master nodes of connected emulation boards can also interchange messages. Each board has a unique address, set up by external jumpers. The SUReCA node on the emulator is implemented concordant with the architectural description. For the implementation of the dynamic reconfigurable layer we've used an XC6264-2 FPGA from Xilinx. The core and buffer controllers are implemented using an XC4013E-2 FPGA also from Xilinx. We've further foreseen 4 Megabytes of 12 ns RAM organised as 1024K x 32 bits. The RAM organisation can be changed according to the requirements of the application. The board has a total of 16 Mbytes of fast static RAM divided over the four SUReCA nodes. This RAM is used as context memory, instruction memory and FIFO buffer memory.

A 16-bit mesh interconnects the computing layers of the four SUReCA nodes. To overcome the rather small bus-width we've foreseen 4 multiplex signals such that the mesh can be virtually 256 bits wide. An eight bit wide mesh interconnects the control layer. The emulation board has two clocks. A fixed 20 MHz clock for the control layer and a programmable clock for the dynamic reconfigurable layer. The clock speed can be changed at runtime on request of a SUReCA node. This is possible because each SUReCA node has access to the master node through the control layer. A program can send data to the master node requesting a certain clock period for the context that is going to be loaded. The master node has not only control over the programmable clock but also over the multiplexing signals.

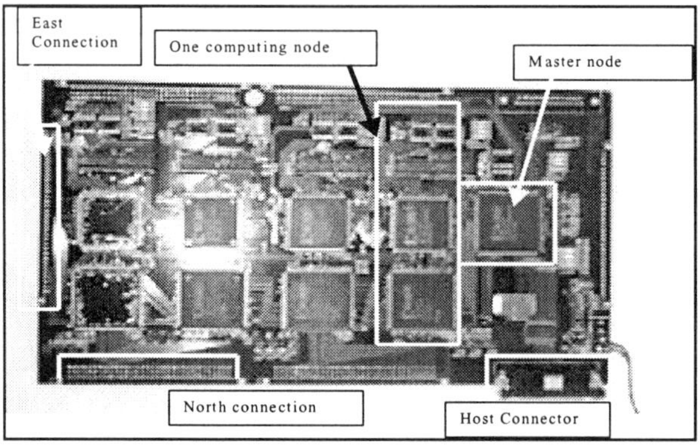

Figure 4. Emulator

Implementing large two or tree dimensional multi chip computing systems, which are modular and scalable with one global clock, such that all computing nodes can be synchronised is hard to realise. This due to clock skew appearing over connectors (Capacitive load) and clock power distribution (only limited fan-out is possible which would require buffering of the clock signal). An approach to deal with this restriction is the use of self timed modules. To enable such an approach, each computing node uses a local clock, for inter-node communication an asynchronous handshake protocol is used. Every board has one programmable clock distributed over the four SUReCA nodes. Although they have the same clock we have foreseen an asynchronous interfacing between the four computing nodes on the control layer.

5. EXAMPLES AND PERFORMANCE ISSUES

In the following example we compare the implementation of a high level simulator for a multistage ATM switching fabric on a mesh of FPGA's and an implementation on our emulator. Multistage ATM switches [7] are conceptually based on multistage interconnection networks (MIN). The switch architecture is built by using small shared memory switches as switching elements (SE) which are arranged in a MIN. The simulator built is that of an internally blocking switch of which the used MIN is classified as a Banyan network. The simulator for the ATM switch fabric is based on the hardware implementation of the queuing model of a switching element [11].

The following performance results where achieved for different problem sizes. The performance is given as the simulation time divided by the number of ATM packets handled. The simulations done handled 32.10^6 packets. We have chosen this number to make the computation time large enough, such that the set up and reconfiguration cost of the SUReCA system can be better amortised over the whole computation time.

To have comparable results we've used for the FPGA implementation the XC6264 but without reconfiguring it at runtime.

Problem size	FPGA's/ SUReCA nodes	Performance FPGA's	Performance SUReCA
16x16 switch	1 FPGA 1 SUReCA	3.75 ns/packet	3.77 ns/packet
64x64 switch	6 FPGA's 3 SUReCA	1.04 ns/packet	2.7 ns/packet
256x256 switch	40 FPGA's 20 SUReCA	0.91 ns/packet	0.85ns/packet

Interesting in these results is to see that the SUReCA system scales different from the meshed FPGA system. For the small problem size we have less performant results on the SUReCA system because only the half of the available parallelism is used. Once the size of the problem becomes large enough we remark that a better performance can be achieved on a SUReCA based system with less reconfigurable hardware. This due to the fact that the communication cost overwhelms the benefit of more fine grain parallelism in meshed FPGA systems.

6. CONCLUSIONS

We've tried to show that better performance can be achieved than the one obtained on meshed FPGA systems for large computing problems with highly complex interconnection requirements between successive pipeline stages. A processor architecture that gives a better scaling of the performance is discussed and an emulation system is built. The presented processor architecture consists of three layers and is programmed by means of instructions and circuit contexts. The first layer is a computing layer which contains an FPGA like run time reconfigurable compute medium and a FIFO buffering mechanism to exchange data between contexts. The input and output port of the FIFO buffer can be connected with the neighbour nodes, giving the possibility to have data exchanged between contexts ran at different moments on different computing nodes. The buffering mechanism is also used as a mean to avoid frequent reconfigurations. The second layer is the control layer, which next to decoding instructions and controlling all the local resources also implements a synchronisation mechanism with the neighbour computing nodes. An emulation system for the presented architecture is built and tested out on performance. The used dynamically reconfigurable FPGA needs several ms for a complete reconfiguration. This due to its size (256x256 cells) and the rather slow reconfiguration interface (approximately 5 MHz). Although, the use of a deep FIFO-memory makes it possible to have the reconfiguration cost minimal. However note that this buffering technique is not applicable for general purpose computing systems. As a case study an ATM switching fabric is simulated on a meshed FPGA system and our emulation system. For those problems that are rather small the meshed FPGA system performs far better than the Emulator does. Once the problem size is large enough our emulator starts performing better.

7. REFERENCES

[1]Jonathan Babb et al, "Solving Graph problems with dynamic computational structures",SPIE photonics East: Reconfigurable Technology for Rapid Product Development and Computing, Boston,MA, November 1996

[2]Jonathan et al, ,"The RAW Benchmark suite: Computational Structures for General Purpose Computing", IEEE symposium on field programmable Custom Computing Machines

[3] M. Sliman Kadi et al, "A fast FPGA prototyping system that uses inexpensive high performance FPIC", in proc. ACM/SIGDA Workshop Field Programmable Gate Arrays,1994

[4] Elliot Waingold et al,"Baring it all to software: The RAW machine", MIT/LCS Technical Report TR-709, March 1997

[5] Andre DeHon "Reconfigurable architectures for General Purpose Computing",MIT artificial Intelligence Laboratory, Technica Report 1586, september 1996

[6]Scott Hauck,"The Chimaera Reconfigurable Functional Unit",IEEE symposium for custom computing machines, 1997

[7] Martin De Prycker, Asynchronous Transfer Mode, Solution for Broad Band ISDN, ISBN 0-13-342171-6

[8] Zhong P. et al.,"Solving boolean satisfiability problems with Dynamic Hardware Configurations",,8th international workshop,FPL'98 Tallin Estonia, September 1998, Proceedings

[9]Steve Trimberger et al, "A Time Multiplexed FPGA",IEEE symposium on Field Programmable Custom Computing Machines, FCCM 1997, NAPA, CA, april 1997,Proceedings

[10]Scott Hauck, "Configuration Compression for the xilinx xc6200", IEEE symposium on Field Programmable Custom Computing Machines, FCCM 1998

[11]Abdellah Touhafi et al, Simulation of ATM switches using Dynamically Reconfigurable FPGA's, 8th international workshop,FPL'98 Tallin Estonia, September 1998, Proceedings

FRONTIER: A FAST PLACEMENT SYSTEM FOR FPGAS

Russell Tessier
Department of Electrical and Computer Engineering
University of Massachusetts, Amherst, Ma. 01003.
tessier@ecs.umass.edu

Abstract

In this paper we describe Frontier, an FPGA placement system that uses design macro-blocks in conjuction with a series of placement algorithms to achieve highly-routable and high-performance layouts quickly. In the first stage of design placement, a macro-based floorplanner is used to quickly identify an initial layout based on inter-macro connectivity. Next, an FPGA routability metric, previously described in [10], is used to evaluate the quality of the initial placement. Finally, if the floorplan is determined to be unroutable, a feedback-driven placement perturbation step is employed to achieve a lower cost placement. For a collection of large reconfigurable computing benchmark circuits our placement system exhibits a 4× speedup in combined place *and* route time versus commercial FPGA CAD software with improved design performance for most designs. It is shown that floorplanning, routability evaluation, and back-end optimization are all necessary to achieve efficient placement solutions.

1 INTRODUCTION

Over the past decade field-programmable gate arrays (FPGAs) have revolutionized the way digital systems are designed and built. With architectures capable of holding millions of logic gates on the horizon and planned integration of reconfigurable logic into system-on-a-chip platforms, the versatility of programmable devices is expected to increase dramatically.

When programmable logic first became available a decade ago the task of converting a high-level design into a high-performance physical implementation was frequently a time-consuming, manually-driven process requiring many days or weeks. While sizable development times are still tolerable for some applications of FPGA devices today, many uses of FPGA technology, such as reconfigurable computing and ASIC prototyping, require compilation times on the order of minutes to allow for rapid design turnaround from high-level design to physical implementation. Currently, a majority of FPGA compilation time is spent in device layout due primarily to the assumption that each collection of new design elements must be placed and routed from scratch. Given the exponential growth of FPGA logic capacity expected in the next few years, place and route times using algorithms currently employed in FPGA software systems can only be expected to get worse.

In this paper Frontier, an integrated placement system that aggressively uses macro-blocks and floorplanning to quickly converge to a high-quality placement solution, is detailed. This system can be used in place of existing placement approaches for macro-based designs targetted to devices with architectures similar to the Xilinx XC4000 [2] and Lucent Orca [1] families. Rather than using a single algorithm, the new Frontier tool set relies on a sequence of interrelated placement steps. First, in a floorplanning step, hard and soft macros are combined together into localized clusters of fixed size and shape and assigned to device regions to minimize placement cost. Following initial floorplanning, a routability evaluator, based on wire length, is used to determine if subsequent routing for a given target device is likely to complete successfully. If this evalution is pessimistic, low-temperature simulated annealing is performed on the contents of all soft macros in the design to allow for additional placement cost reduction and enhanced design routability.

2 PROBLEM STATEMENT

The target FPGA architecture used for this research is the *island-style* architecture commonly found in commercial FPGA devices such as the Xilinx XC4000 family [2] and the Lucent Orca family [1]. These architectures are characterized by a regular two-dimensional array of logic and routing *cells*. Each identical cell contains a logic block consisting of a small number of programmable lookup tables and flip flops and associated routing wires of differing segmentation lengths.

An FPGA design under placement consideration consists of N_{blocks} logic blocks grouped into M instantiated macro-blocks. Each macro-block contains an RTL component such as a datapath function or finite state machine and has a distinct logic block capacity N_{M_i}. Hard macro-blocks are assigned fixed height h_i and width w_i while soft macro-blocks have flexible shape.

The goal of our placement approach is to create a placement for N_{blocks} design logic blocks encompassed by a set of M macro-blocks onto N_{cells} array logic blocks such that subsequent routing may complete successfully. A set of $Nets_M$ inter-macro wires interconnect all macro-blocks and $Nets_{blocks}$ wires interconnect all logic blocks inclusive of $Nets_M$.

In general, placement progresses subject to the following constraints:

1. Each hard macro-block is assigned a distinct placement rectangle R_i of dimensions h_i and w_i so that no two macros overlap ($R_i \bigcap R_j = \phi$).

2. Placement is performed to maximize overall routability by minimizing overall routability-based placement cost. Initially, floorplanning considers, among other criteria, minimizing the length of all inter-macro nets $Nets_M$. Subsequently, during placement refinement, the length of all design wiring, $Nets_{blocks}$, is considered.

In addition to wire length, several supplemental cost criteria are considered in performing placement. Commercial FPGA devices, like those from the XC4000 and Orca families, have *direct* connections between logic blocks that enhance routability and long-lines that span the extent of the entire device. By building hard macros that are constructed to take advantage of these features, additional design routability can be achieved.

3 RELATED WORK

Most commercial FPGA placement packages use simulated annealing to evaluate a series of logic block swaps based on a predefined cost function. Annealing, started from an initial placement, typically achieves good placement quality at the cost of long execution times that are exponentially bounded by the number of design logic blocks [11]. In [9], recursive clustering was used to identify circuit locality prior to annealing to reduce subsequent annealing execution time. While this approach yielded a placement time speedup of four in obtaining minimized FPGA placement cost, no mechanism for supporting pre-placed macro-blocks was included.

Several floorplanning efforts for island-style FPGAs have relied on specific user design implementation styles to quickly achieve a highly-routable placement. These systems [5] [8] [7] restrict target circuits to datapaths oriented in a left-to-right linear communication pattern. Design regularity facilitates vertical bitwise abutment of macro-blocks and allows for a rapid traversal of the one-dimensional topological search space. In general, one-dimensional approaches cannot be easily modified for circuits with more irregular communication patterns and larger Rent parameters.

A floorplanning methodology based on hierarchical placement was recently described in [6]. This floorplanner clusters macros together into fixed sized bins and then optimizes bin placement using a two-step tabu search. Several of the large benchmarks targetted by the system showed considerable speedup in placement time but much more than a 100% increase in routing time. In this paper we show that this routing time increase was likely caused by the lack of a globally optimizing placement smoothing step following floorplanning to minimize localized wire length inefficiencies.

4 FRONTIER IMPLEMENTATION

The use of macro-blocks to accelerate placement for island-style FPGAs requires accommodation of the following two competing placement goals:

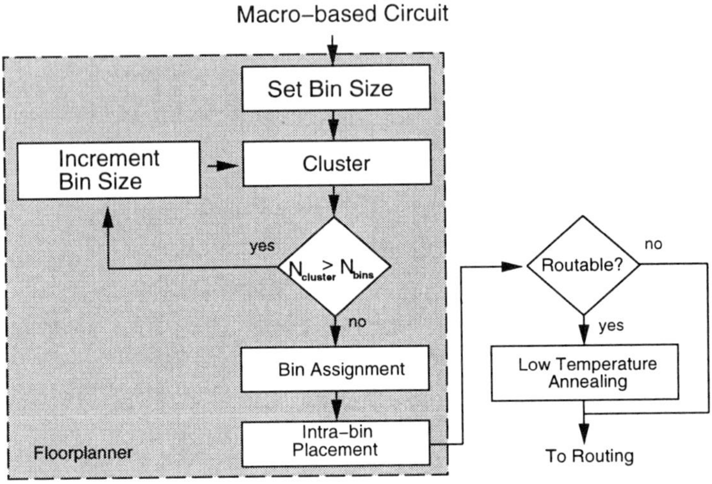

Figure 1 Frontier Placement Flowchart

- Locality information stored in pre-placed macro-block libraries should be used to avoid the need to reconstruct *local* design structure from scratch and to better take advantage of device features such as near-neighbor direct connection and long-line alignment.

- The placement system should have the flexibility to minimize *global* wire length by swapping individual logic blocks across the entire design. An approach that is insufficiently flexible will lead to high wire length placements that are likely to take additional time to route, eliminating the benefit of placement time speedup.

The placement system described in this paper is the first integrated approach that addresses both of these competing concerns in one package. The system progresses in a series of algorithmic steps by supplementing new layout techniques with recent advances in FPGA routability analysis. As illustrated in Figure 1, the layout process starts with a macro-based netlist of soft and hard macros targetted to an FPGA device containing N_{cells} logic blocks. Initially, to enhance locality, the FPGA device is decomposed into an array of placement *bins*, each of the same physical dimension, as shown in Figure 2. To determine bin contents, macros are grouped together into clusters, each of which will accommodate the volume of macro logic blocks and the physical dimensions of hard macros inside a bin. If following clustering an insufficient number of bins are available to place all clusters, bin sizes are increased and clustering is restarted. After clustering, each cluster is assigned to a physical bin location on the target device and entire bin clusters are subsequently swapped between physical bins to minimize inter-bin placement cost including connectivity to device pins. Since the number of bins allocated to a device is frequently much smaller than the number of device logic blocks, this process

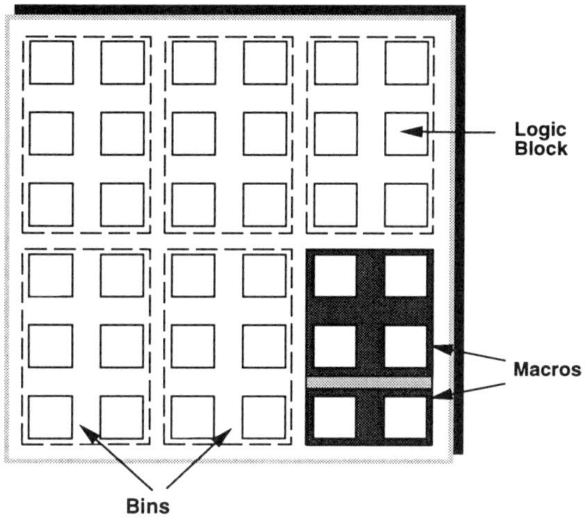

Figure 2 Bin-based Cluster Assignment

proceeds rapidly. The annealing formulation used in inter-bin swapping follows directly from logic block-level annealing used for flattened designs and is easily incorporated into the software flow. Following bin placement, hard and soft macro-blocks are placed within each bin in a space-filling fashion. All intra-bin placement is based on inter and intra-bin connectivity. Soft macros are resized at this point to meet bin shape constraints.

In Section 5, it is shown that while floorplanning alone is sufficient to provide effective placements for many designs targetted to contemporary FPGA devices, in some cases additional placement perturbation is required. In Frontier, following floorplanning, a detailed estimate of the placement wire length is determined, taking into account the special features of the FPGA device. As described previously in [10], this wire length estimate can be used to evaluate whether subsequent device routing will complete quickly, require a long period of time, or fail to route at all. For floorplans that are impossible or difficult to route, low-temperature simulated annealing is performed on soft macros to smooth wire length inefficiencies. Through a series of design examples a set of annealing parameters that lead to the best time versus performance tradeoff is determined.

4.1 PLACEMENT STEPS

Bottom-Up Clustering. In the first step of placement, macros are clustered together into placement bins of identical dimensions and CLB volume to identify inter-macro design locality. While bins must be sized to support a range of macro-block dimensions, needlessly large bins limit the number of bins available for subsequent inter-bin swapping and may have a negative impact on final floorplan quality. As previously suggested in [6], when floorplanning is started, bins are initially set to the X and Y dimensions of the largest hard macro-block or, if no hard macros exist, to the square root of the logic block volume of the largest soft macro.

Given the fixed dimensions of each bin, clustering must not only take into account connectivity, but also size feasibility of the cluster under consideration. To smooth macro-block size disparities, smaller macro-blocks should be clustered first with other like-sized blocks so that the total number of created clusters is minimized. For Frontier this is accomplished through the use of a cost function described in [12] and [6] that is weighted to take logic block counts and interconnectivity into account:

$$Cost_{ij} = feas(i,j) \times \frac{N_{blocks}}{N_{M_i} + N_{M_j}} \times \frac{min(N_{M_i}, N_{M_j})}{max(N_{M_i}, N_{M_j})} \times \sum Nets_{ij} \qquad (1.1)$$

where N_{blocks} is the total number of logic blocks in the circuit, N_{M_i} and N_{M_j} are the number of logic blocks in the macro-blocks M_i and M_j under consideration and $Nets_{ij}$ are the nets connecting M_i and M_j. The first term in Equation 1.1 determines if a candidate cluster can be feasibly shaped during intra-bin placement, using criteria described later in this section, to fit the physical dimensions of a target bin. Its value is set to 1 if a shape is feasible and 0 if it is not. The second term in the cost function prevents a specific cluster from becoming too large in relation to the rest of the circuit. The third term prevents two macros with vastly different numbers of blocks from being connected together thereby creating area inefficiencies, and the last term measures connectivity.

If, following clustering, more clusters C than bins B exist, bin dimensions are modified by increasing bin horizontal and vertical dimensions by 1 logic block and clustering is started again from scratch.

Bin Assignment. Following clustering, all macros are bound to a cluster and the number of clusters is less than or equal to the number of available device bins. The next step is to determine an assignment of clusters to specific device bins. After initial random assignment of clusters to bins, simulated annealing is used to evaluate cluster swaps based on both inter-bin and bin-to-pad wire lengths. The dynamic annealing schedule described in [4] is used to reach a good quality placement quickly. Given the small number of bins (typically less than 20) annealed swapping can typically be completed in a few seconds.

Internal Bin Placement. Once each cluster of macro-blocks is assigned to a specific bin, intra-bin placement is performed to assign macro logic blocks to specific device logic block locations. As a first step for each bin, all N_{hard} hard macro-blocks and N_{soft} soft macro-blocks in the assigned cluster are linearly ordered in the horizontal dimension using a topological sort based on intra and inter-bin connectivity. For soft macro-blocks, previously-determined library placements are used to approximate final soft macro logic block positions and wire lengths.

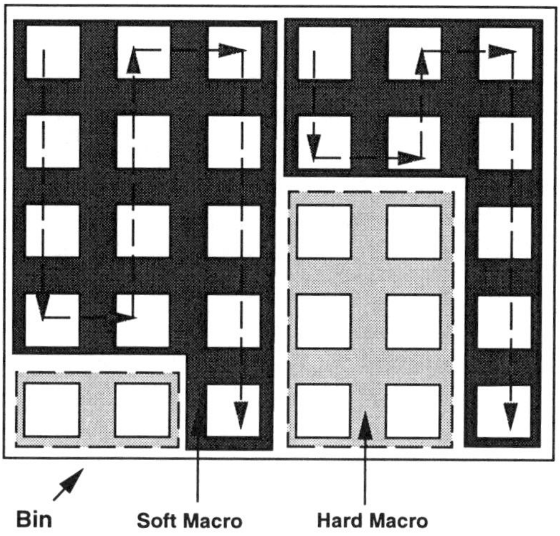

Figure 3 Internal Bin Placement

Following ordering, exact X, Y logic block positions in each bin are determined for *hard* macros by resolving inter-macro spacing. If the width of a bin is w_{bin} and the combined horizontal width of all hard macros in a bin is $\sum_i w_{M_i}$, the space between hard macros occupied by soft macros is determined to be $X_{soft} = \lfloor \frac{w_{bin} - \sum_i w_{M_i}}{N_{soft}} \rfloor$. This equation leads to an X_{soft} value of 1 for the bin shown in Figure 3. Following X_{soft} determination, hard macros are assigned X coordinates inside each bin in a left-to-right order with X_{soft} spacing inserted for each soft macro.

Subsequent to the positioning of hard macros, intra-bin X and Y locations for *soft* macro logic blocks are determined. These locations are determined by allocating bin space remaining after hard macro placement in a snake-like fashion starting in the upper left-hand corner of the bin. Individual soft macro logic blocks are assigned to specific locations within this shape by sequentially selecting logic blocks that minimize overall wire length. By following this methodology, up to 100% logic block utilization can be achieved in each bin.

Routability Prediction. Recently, a direct correlation has been formulated between the number of routing tracks in an FPGA device, the wire length of a design placement, and the amount of time needed to route a design [10]. Due to macro-block shape considerations and specific interconnection patterns of individual designs, a successful floorplanning step provides no guarantee that a placement possessing close to the global minimum cost has been achieved or that routing will subsequently succeed for a given target device. To evaluate placement fitness, a wire length-based routability metric [10] has been directly built into Frontier.

For a given placement, W_{min} may be defined as the minimum track count per FPGA routing channel required to successfully route a design. If the device track count available in a target FPGA, W_{FPGA}, exceeds $1.1 \times W_{min}$ the routing problem is defined to be *low-stress* and can be expected to complete quickly (e.g. within several minutes). If $W_{min} < W_{FPGA} < 1.1W_{min}$ the routing problem is defined as *difficult* and will likely require many minutes to complete. Generally, if $W_{FPGA} < W_{min}$ it is unlikely routing will complete successfully even after substantial routing time.

Swartz [10] noted that since a placement-only estimate of routability is required, it is necessary to use an estimated rather than an exact W_{min} value to determine routability for a design. Through experimentation it was determined that W_{min} can be estimated from placement wire length as:

$$W_{min-est} = \lceil wirelength/(2 \times N_{cells} \times U) \rceil \tag{1.2}$$

where *wirelength* is the total estimated wire length determined from placement, N_{cells} are the number of logic blocks in the device and U is a utilization factor determined to be architecture-specific. By using this equation for estimation it was possible to determine needed device routing resources following floorplanning for specific Xilinx XC4000XL devices exhibiting W_{FPGA} of 32 tracks per channel and U of 0.6.

Low-Temperature Annealing. As will be shown in Section 5, simply performing floorplanning is generally sufficient to create a placement in the low-stress routing range for many designs. In some cases, however, routability evaluation may reveal that the current placement is difficult or impossible to route given available target device routing resources. For these cases, additional placement perturbation is needed to ensure subsequent fast routing.

To overcome placement inefficiency, Frontier employs low-temperature simulated annealing of individual logic blocks to allow for smoothing of wire length across soft macros and bins without destroying the high-level placement structure achieved by the floorplanner. While detailed discussions of the simulated annealing algorithm for placement and associated controlling parameters [4] are available elsewhere, a brief description of several important parameters needed for floorplan refinement may be summarized as follows:

- Starting annealing temperature, T_{init} - Starting temperature must be set high enough so that the placement may be perturbed to a lower overall minimum, but low enough to avoid destroying the basic hierarchy determined through floorplanning. For our system, a number of experiments were performed to determine T_{init} values that lead to an effective quality versus time tradeoff.

- Inner number, β - This value varies the number of swaps made at each annealing temperature [9]. In the annealing formulation used to perturb logic block placement, the number of moves at each temperature is set to

Design	Device	CLBs	Macros	Device Utilization
bheap5	4085XL	2715	30	87%
bubble16	4044XL	1280	31	80%
bubble32	4085XL	2608	63	83%
fft16	4085XL	3032	48	97%
spm4	4036XL	1064	11	82%
spm8	4085XL	2425	23	77%
ssp16	4052XL	1491	46	77%
ssp32	4085XL	2370	79	76%

Table 1 Macro-based Design Statistics

Execution times (s)							
	Xilinx PAR-M1.4			Floorplan Only			
Design	Place	Route	Total	Fplan	Route	Total	$W_{min-est}$
bheap5	903	**343**	1246	8	**647**	655	35
bubble16	199	130	329	3	101	104	28
bubble32	1793	294	2087	15	316	331	29
fft16	710	360	1070	14	377	391	22
spm4	143	103	246	8	101	109	26
spm8	530	266	796	4	240	244	26
ssp16	180	81	261	16	72	88	26
ssp32	855	167	1022	17	136	153	22
average	664	218	882	10.6	248	259	27.5

Table 2 Design Layout Statistics - Xilinx 4000XL devices

$\beta \times N_{blocks}^{4/3}$ [4]. In Section 5, quality-time tradeoffs for a range of β values are considered and a β value of 1 is shown to exhibit the most favorable quality-time characteristics.

5 RESULTS

The placement system outlined previously was applied to eight macro-based reconfigurable computing benchmarks from the RAW Benchmark Suite [3]. Prior to experimentation, all hard and soft macros were mapped to Xilinx XC4000 logic blocks and resulting XNF netlist files were annotated with *RLOC* placement information. Macro-based netlists in XNF format were then used as input to both the Frontier system and to Xilinx PAR software, version M1.4 [2]. Design statistics for the benchmarks appear in Table 1. All run time results for both Frontier and PAR were obtained using a 140 MHz UltraSparc I with 288

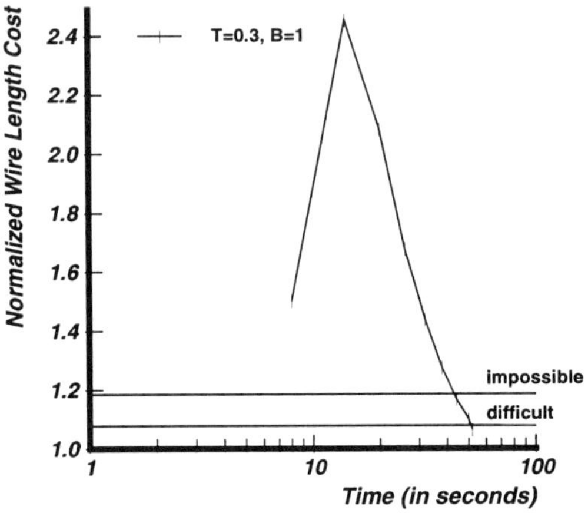

Figure 4 Placement Versus Design Quality - bheap5

Mb of memory. Routing for all designs was performed using Xilinx PAR-M1.4 software with default parameter and routing effort settings.

Execution times for PAR placement, Frontier floorplanning (without low-temperature refinement), and PAR routing appear in Table 2. For all designs except one (*bheap*5) routing times for the floorplanned designs were comparable to those found by the PAR-M1.4 placer that required 60× longer. This is not suprising since for all designs except *bheap*5 the estimated minimum track count per channel needed to route the circuit was less than the 32 tracks per channel available in Xilinx XC4000XL devices. $W_{min-est}$ values were determined by measuring the post-floorplan wire lengths of designs and then directly correlating them to required track counts via Equation 1.2. From Table 2 it can be seen that the minimum track count needed to route the floorplanned version of *bheap*5 is significantly greater than the track count available inside the XC4000XL device and route times (indicated in boldface) reflect the disparity. Following floorplanning and routability determination, it is apparent that placement refinement is needed.

To determine appropriate values for β, the annealing moves-per-iteration variable, and T_{init}, the annealing start temperature for low-temperature annealing, a series of time-quality tradeoffs were evaluated. Starting from a floorplanned placement, each design underwent low-temperature annealing for β parameters ranging between 0.2 and 5 and for T_{init} values ranging between 0.1 and 1. It was found that constraining soft macro logic blocks within the bounds of the soft macro determined during intra-macro placement or within bin boundaries resulted in worse results than allowing soft macro logic blocks to pass between bins. All results shown in the figure were collected without block movement constraints. From the data collected it was determined that the best time-quality tradeoff was achieved for $T_{init} = 0.3$ and $\beta = 1$.

Execution times (s)						Perf.
Flow	PAR	Fplan	LT Anneal	Route	Total	(MHz)
PAR-only	903	0	0	343	1246	5.08
Fplan Only	0	8	0	647	655	8.17
Fplan/Anneal	0	8	34	272	314	12.71

Table 3 Layout Execution Time/Performance Comparison - Design bheap5

Design	Xilinx PAR-M1.4	Floorplan
	MHz	MHz
bheap5	5.08	12.71
bubble16	6.93	14.79
bubble32	10.04	12.48
fft16	21.94	22.47
spm4	8.10	9.20
spm8	6.06	2.07
ssp16	7.40	13.78
ssp32	14.77	13.12

Table 4 Placement Performance Comparison (post-route)

These parameter values were used to refine the initial floorplan for *bheap*5 to a lower cost placement. A graph of relative placement cost versus time at various points during execution is shown in Figure 4 for $T_{init} = 0.3$ and $\beta = 1$. Cost points associated with impossible, difficult and low-stress routing, as defined in Section 3, are labelled. It can be seen that as low-temperature annealing is performed, placement cost is moved from the impossible-to-route range, through difficult, and into the low-stress region.

The effect of this modified placement is clear from the results shown in Table 3. Even though placement time has been extended by 34 seconds, routing time has now been significantly reduced due to the refined placement. Table 4 shows that following routing by Xilinx PAR-M1.4 software, floorplanned (and low-temperature annealed in the case of *bheap*5) circuits exhibit favorable performance characteristics compared to circuit placements created by the PAR placer.

6 CONCLUSION

In this paper a novel FPGA placement tool has been described that quickly achieves high-quality placement by leveraging design regularity in the form of pre-compiled macro-blocks. While placement achieved through initial macro-based floorplanning steps are shown to be highly routable in most cases, for some designs additional placement refinement may be necessary to achieve a routable placement. The system that has been introduced exhibits this capability by first identifying if a design is routable and then perturbing an initial floorplan with low-temperature simulated annealing.

References

[1] *Field-Programmable Gate Arrays Data Book.* Lucent Technologies, 1996.

[2] *The Programmable Logic Data Book.* Xilinx Corporation, 1996.

[3] J. Babb, M. Frank, V. Lee, E. Waingold, and R. Barua. The RAW Benchmark Suite: Computation Structures for General Purpose Computing. In *Proceedings, IEEE Workshop on FPGA-based Custom Computing Machines*, Napa, Ca, Apr. 1997.

[4] V. Betz and J. Rose. VPR: A New Packing, Placement, and Routing Tool for FPGA Research. In *Proceedings, Field Programmable Logic, Seventh International Workshop*, Oxford, UK, Sept. 1997.

[5] T. Callahan, P. Chong, A. Dehon, and J. Wawrzynek. Fast Module Mapping and Placement for Datapaths in FPGAs. In *International Symposium on Field Programmable Gate Arrays*, Monterey, Ca., Feb. 1997.

[6] J. Emmert and D. Bhatia. A Methodology for Fast FPGA Floorplanning. In *International Symposium on Field Programmable Gate Arrays*, Monterey, Ca., Feb. 1999.

[7] S. Gehring and S. Ludwig. Fast Integrated Tools for Circuit Design with FPGAs. In *International Symposium on Field Programmable Gate Arrays*, Monterey, Ca., Feb. 1997.

[8] A. Koch. Structured Design Implementation - A Strategy for Implementing Regular Datapaths on FPGAs. In *International Symposium on Field Programmable Gate Arrays*, Monterey, Ca., Feb. 1996.

[9] Y. Sankar and J. Rose. Trading Quality for Compile Time: Ultra-Fast Placement for FPGAs. In *International Symposium on Field Programmable Gate Arrays*, Monterey, Ca., Feb. 1999.

[10] J. Swartz, V. Betz, and J. Rose. A Fast Routability-Driven Router for FPGAs. In *6th International Workshop on Field-Programmable Gate Arrays*, Monterey, Ca, Feb. 1998.

[11] R. Tessier. *Fast Place and Route Approaches for FPGAs.* PhD thesis, Massachusetts Institute of Technology, Department of Electrical Engineering and Computer Science, 1998. also available as MIT LCS TR-768.

[12] T. Yamanouchi, K. Tamakashi, and T. Kambe. Hybrid Floorplanning Based on Partial Clustering and Module Rest ructuring. In *Proceedings, ACM/IEEE 33rd Design Automation Conference*, 1996.

Dynamically Reconfigurable Implementation of Control Circuits

Nuno Lau, Valery Sklyarov
Departamento de Electrónica e Telecomunicações da Universidade de Aveiro

Key words: FPGA, Reconfigurable Computing, Partial and Dynamic Configuration

Abstract: Control Circuits can be described using a top-down approach with the aid of Hierarchical Graph-Schemes (HGSs). The implementation of HGSs in a fine-grain FPGA has been done using a Hierarchical Finite State Machine structure where each sub-algorithm implementation is independent from the others. Static and dynamically reconfigurable implementations using the XC6200 FPGA have been obtained by applying a number of commercial design tools together with tools developed by the authors.

1. INTRODUCTION

FPGAs are being widely used in the design of complex digital systems involving both single and multiple FPGAs [Hauck 98], but the properties of plain, dynamic, or partial reconfigurability are not explored in most of these applications. The absence of CAD tools that support the design flow of reconfigurable circuits and the lack of published application expertise are certainly some of the main difficulties that a designer of reconfigurable circuits must handle [Hutchings95]. On the other hand several research groups have developed applications that prove the possibility (and in some cases the economic efficiency) of dynamically reconfigurable applications [Eldredge96, Wirthlin95, Robinson98, Shirazi98, Sklyarov98].

We think that one of the application areas where dynamic reconfiguration can prove its usefulness is in the implementation of control circuits. The specification of control circuits can be made using a top-down approach where the details of behaviour are at the lowest level and the top level gives an overview of the whole system. A top-down approach leads to a modular specification of the circuit that is also well adapted to a modular implementation. In our view, circuits that are modular and well-structured are good candidates for implementation using dynamic reconfiguration. A reconfiguration model has been developed that allows the management of sets of FPGA resources as if they were pages in a virtual memory system. A dynamically reconfigurable implementation of an HFSM has been used as a proof-of-concept example for the reconfiguration model. Section 2 presents the specification of control circuits using a top-down methodology known as hierarchical graph-scheme (HGS) [Rocha97]. Section 3 discusses the implementation of HGSs using a Hierarchical Finite State Machine model, the synthesis method adopted, and its static and dynamic implementation using a XC6200 FPGA.

2. HIERARCHICAL GRAPH-SCHEMES

The standard method to describe a control circuit is using a state transition table or a state transition diagram [Baranov94]. While these specification methods are very useful, the specification of control circuits can be done at a behavioural level using HGSs. An HGS specification of a control circuit gives us a visual description of the control algorithm. Any HGS must have one BEGIN node and one END node and it may have rectangular and rhomboidal nodes that link the BEGIN and END node forming a directed connected graph (*Figure 1*) [Sklyarov97]. Rectangular nodes determine which micro-operations (denoted y_i) and macro-operations (denoted z_j) have to be activated at each step. Rhomboidal nodes are used to select from two different execution flows of the HGS. The value of a conditional signal (denoted x_l) that is inserted in the rhomboidal node determines which path is executed. HGSs are powerful specification tools as they allow the specification of the control algorithm to be seen at several layers of abstraction. This is achieved by the use of macro-operations, i.e. one HGS can invoke another HGS in a way similar to procedure calls, and execution of the HGS that makes the call will only proceed when the HGS that was invoked has reached its END node.

3. HGS IMPLEMENTATION

Synthesis of a control circuit that is described by a set of HGSs is the process of transforming the HGS specification of the control algorithm to a hardware implementation. In our case we are interested in the implementation of the HGS in the Xilinx XC6200 FPGA family [Xilinx97].

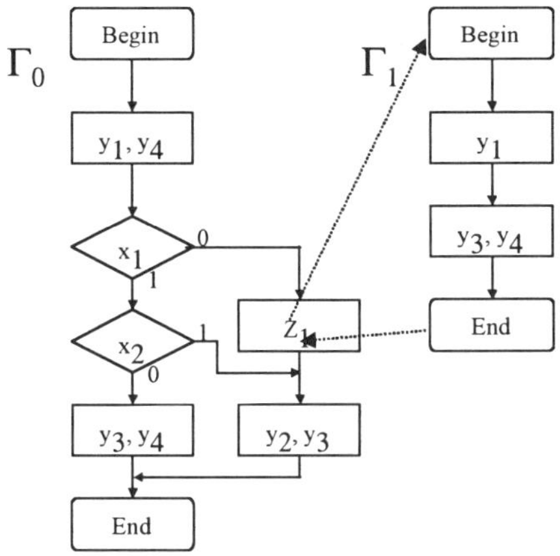

Figure 1. Example of HGS specification

This FPGA has a fine-grain sea-of-gates architecture and is dynamically and partially reconfigurable.

3.1 HFSM implementation model

The HGS specification is to be implemented in hardware using the Hierarchical Finite State Machine (HFSM) model shown in *Figure 2*. This model is an extension of the FSM model, where the state register has been replaced by a combination of a stack memory and a normal register. This modification has been introduced to accommodate hierarchical calls between HGSs. When an HGS call is executed, the current state is stored on the stack and the stack pointer is incremented. The state of the HGS that was called is recorded and managed at the new stack level (*Figure 3*). This mechanism can even be used for recursive HGS calls. The standard FSM model has also been extended with a new component called the code converter, which is responsible for the selection of the invoked HGS. This is implemented as a RAM based device. Rewriting the contents of this RAM allows an easy (and fast) change in the behaviour of the HFSM. This is an important feature as it embodies the concept of virtual HGSs, i.e., dynamic binding of sub-algorithms. The management of the HFSM model demands a multi-phase clock synchronisation scheme.

By examining the variability of the components of the HFSM model of the behaviour of the control circuit, one can separate the components into two groups: one group containing components that will be parameterised and the other containing components that will be synthesised.

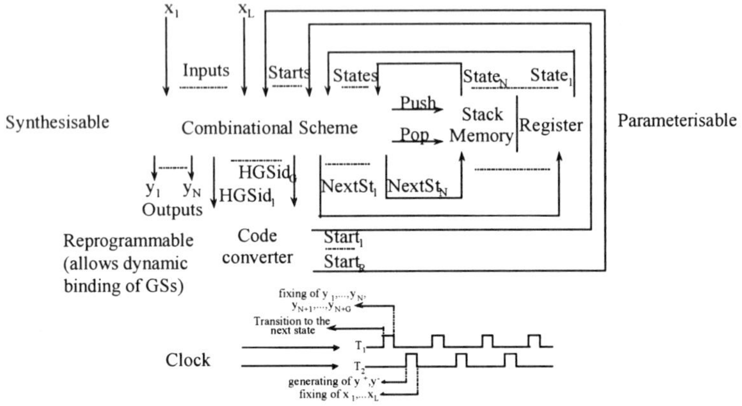

Figure 2. HFSM implementation model

The parameterised components are those that are almost independent of the behaviour of the HFSM. This group includes the stack memory, the code converter, and the synchronisation scheme. Optimised versions of structural VHDL descriptions of these components have been developed and are included in a library.

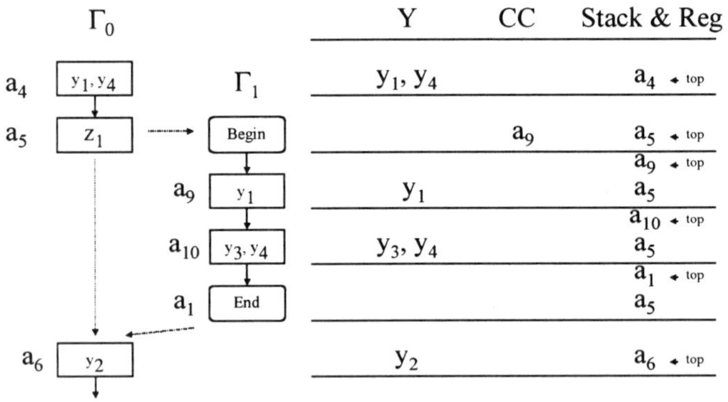

Figure 3. Execution of HGS

The synthesised group includes those components that are very dependent on the actual HFSM behaviour. These include the combinational scheme and the non-stack part of the state register. Synthesis tools have been developed to generate structural VHDL descriptions in accordance with the HGS behaviour. These tools use a textual description of the HGS (*Figure 4*). The syntax of the HGS text file was developed having in mind that it should be understandable for a human reader, its parser should be simple and it should be open to the inclusion of further HGS characteristics.

```
GRAPH 0 Z0
-- node num type next
NODE 0 0 Begin   2
NODE 1 1 End
NODE 2 3 Rect1   3 MICRO 1 4 MACRO
NODE 3 4 Rhomb   1 1 6 4
NODE 4 3 Rect2   5 MICRO MACRO 1
NODE 5 3 Rect3   1 MICRO 2 3 MACRO
NODE 6 4 Rhomb   2 2 5 7
NODE 7 3 Rect4   1 MICRO 3 4 MACRO
```

Figure 4. Textual description of Figure 1's Γ_0.

The synthesis of the combinational scheme is done considering each macro-operation as a different module, and generating a structural VHDL entity for each of them. Then the whole combinational scheme is an aggregation of the mentioned above entities [Oliveira98].

3.2 HGS synthesis

The synthesis of each macro-operation, representing a sub-algorithm, is done using a direct mapping of the macro-operation graph description onto its hardware implementation. To achieve a simple direct association between an HGS and its circuit implementation we have used one-hot state assignment This provides an implementation of a sub-algorithm that is independent of the behaviour of the others, and also allows for the possibility of internal states, as we will explain later. In the mapping for each of the rectangular nodes of an HGS there is a memory element and for each of the rhomboidal nodes there is a demultiplexer. *Figure 5* shows a simple example for which the HGS does not have macro-operations.

If one-hot state encoding is used, the mapping of the rectangular nodes of an HGS is dependent on their contents:

- Those that include macro-operations must be mapped onto stack memory. The corresponding states will be called hierarchical states.
- Those that include just micro-operations are mapped onto a D-flip-flop. The flip-flops can be instantiated as internal components of the VHDL entity that implements the macro-operation. This makes the routing of the system easier and provides a simpler interface for the entity.

The number of states that are mapped in RAM should be minimised in this implementation. If one hot state assignment is used inside the stack then the number of states that it can map is very limited, i.e., a stack of 8 bit registers would be able to map 8 states. One can minimise the problem by providing a coder and a decoder for stack contents, which enables the same stack of 8 bit registers to map 256 one-hot states, but this inevitably has significant costs in terms of area and performance of the whole system.

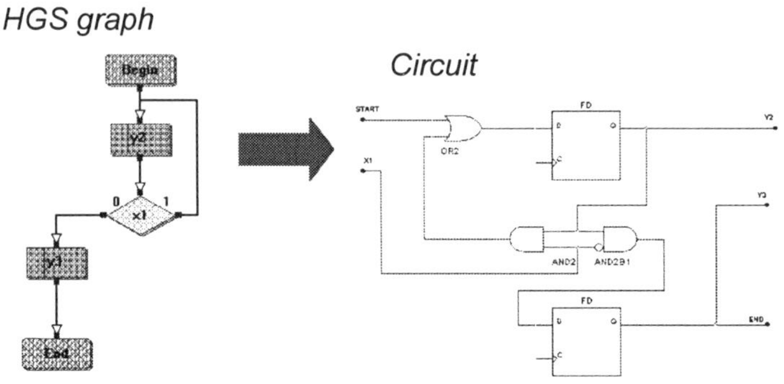

Figure 5. Mapping HGS onto the respective circuit

3.3 Static model

In this model of HGS implementation, all components of the HGS circuit are permanently mapped onto the FPGA. The combinational scheme is an aggregation of the various macro-operations optimised for each of them, i.e., if a certain sub-algorithm only uses conditional signals 1 and 2 from the four available then only these two signals are provided for that sub-algorithm. The synthesis tools generate this optimised aggregation.

The final step in the implementation of this type of circuit is the elaboration of the structural VHDL description of the circuits, which is performed by Velab and generates an EDIF file. Then the mapping, placement and routing of the circuit are executed by XACT6000, which generates the final configuration of the XC6200 FPGA.

3.4 Dynamically reconfigurable model

To explain the dynamically reconfigurable implementation of the HGS we should start by stating the general principles used in the development of our reconfiguration model.

We have defined some of the FPGA resources (functional cells and routing muxes) as "dynamically reconfigurable" (called "reconfigurable" in the rest of the paper), meaning that they are available for reconfiguration during run-time, while other ("fixed") resources must be configured only at startup. These "reconfigurable" resources have been divided into several sets where the objective is to ensure that modifying a "reconfigurable" resource of a specific set is guaranteed not to change the circuit beyond the limits of the set where it belongs. In our case these sets are non-overlapping rectangular areas inside the FPGA and in each area (set) all functional units and local routing resources have been considered "reconfigurable".

Each of the "reconfigurable" sets has the same shape and the same topology with respect to the routing resources of the FPGA, i.e., if they all use length 4 routing resources then their position relative to a 4x4 boundary has to be the same for all "reconfigurable" sets.

Each of the "reconfigurable" sets has a fixed interface, both structurally, where they all have the same signals as inputs and outputs, and spatially, where the relative position of inputs and outputs inside the set is the same for all sets.

Each partial configuration describes a circuit that uses "reconfigurable" resources of only one of the sets. There should be more partial configurations than "reconfigurable" sets, otherwise the static model implementation is preferable.

If several partial configurations use resources that belong to just one "reconfigurable" set, they are available for swapping.

These assumptions are very important because they provide each reconfigurable set with characteristics similar to those of pages in virtual memory systems:

- The configuration of a particular set is independent of the configuration of the others and does not affect the others.
- Configurations can easily be moved from one set to another.
- The maximum size of a configuration can be easily calculated. In fact we could consider configurations as being of a fixed size, but as we will see, not all of the resources of each "reconfigurable" set need to be programmed to change the configuration of that set.
- All the information about each configuration can be saved in an external device.

Another principle of our dynamically reconfigurable model is that the circuit is responsible for signalling some external device in the event that reconfiguration is needed. A reconfiguration handler implemented in the external device is responsible for loading the new partial configuration to the desired set. Additional information may be needed by the external reconfiguration handler in order to choose which configuration has to be mapped onto which of the "reconfiguration" sets.

Let's turn our attention to the dynamically reconfigurable implementation of HGSs and to the methodology and tools that have been developed to support it. We'll also discuss some of the limitations of Xilinx tools when they are applied to dynamic reconfigurable circuits and how we have overcome these limitations.

As has already been said, our "reconfigurable" sets of resources have been defined as non-overlapping rectangular areas inside the FPGA. The interface of each "reconfigurable" area can be seen in *Figure 6*. Every input is fed to every reconfiguration area and every output is also received from all areas because neither the sub-algorithms that are going to be mapped nor their input/output needs are known,. Each of these areas can accommodate one sub-algorithm of the HGS if it is needed at a certain execution time (*Figure 7*).

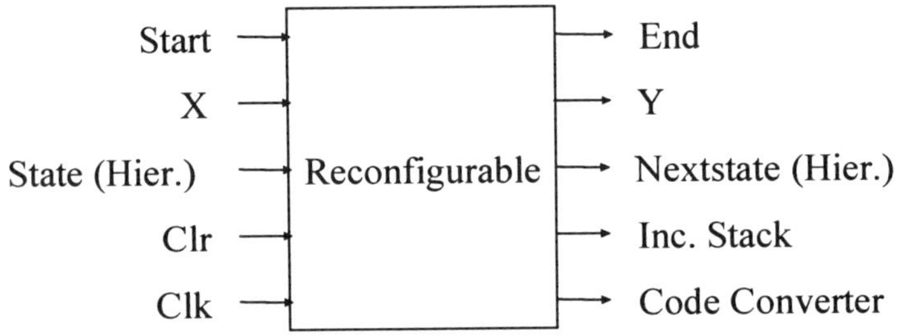

Figure 6. Template for the external interface of an HGS component

The need for reconfiguration has been easily incorporated into the HFSM model by reserving one bit of the code converter entries to specify if a certain macro-operation is configured inside the FPGA or not. If it is configured, the other code converter bits specify the mapping area, and if it is not mapped, these bits specify the macro-operation code (*Figure 8*).

When reconfiguration is required the reconfiguration handler reads the code converter register (inside the FPGA) in order to know which macro-operation has to be loaded. In this model, reconfiguration only happens

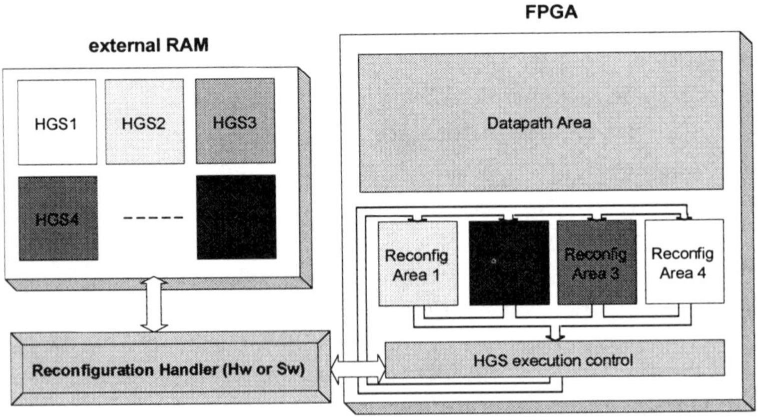

Figure 7. HGS reconfiguration model

during macro-operation invocation and not during macro-operation return, as it is a constraint of this model that all macro-operations in the path from the main graph to the present macro-operation cannot be swapped out. The reconfiguration handler maps the macro-operation onto one of the available areas, determined by reading stack contents, then changes the contents of the code converter so that it reflects the actual configuration of these areas and indicates that execution may proceed.

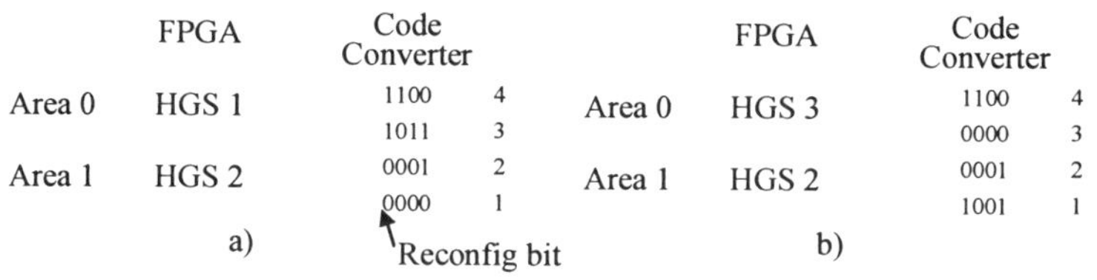

Figure 8. Code Converter contents before (a) and after (b) reconfiguration

Most of the components belonging to the parameterised group didn't have to undergo any modification to allow reconfigurability. Only the synchronisation scheme had to be changed. This is the circuit that signals the need for reconfiguration and that manages the protocol with the reconfiguration handler to resume HGS execution when reconfiguration is finished.

The synthesis tools had to be adapted to this new model as the interface of all macro-operations implementations is now the same, whether that particular implementation is using all interface signals or not (inputs that are not used are left open while unused outputs are connected to ground). The combinational scheme has a fixed structure that does not depend on the HGS specification and depends only on the specified interface and on the number of "reconfigurable" areas. The contents of each of these areas are only defined during run-time by loading the relevant partial configuration.

To respect the principles stated above, the "fixed" configuration should not use any of the "reconfigurable" resources. If they are used then partial configurations might affect the circuit implemented by the "fixed" configuration. XACT6000 provides a designer attribute to exclude all functional units within a certain area from being used, but it doesn't provide anything similar with respect to routing resources. In order to generate a safe "fixed" configuration, we have made XACT think that all "reconfigurable" resources were already being used (so that it wouldn't use them for other purposes) by generating a dummy layout (in the form of a XACT layout file [Xilinx97b]) that occupies all these resources (*Figure 9*).

In our case we have a different partial configuration for each area and for each macro-operation, so if we have two reconfiguration areas and 4 macro-operations we need 8 partial reconfigurations. Only 4 partial configurations would be required if we had a translation operator for partial configurations available, so this operator is planned for the near future.

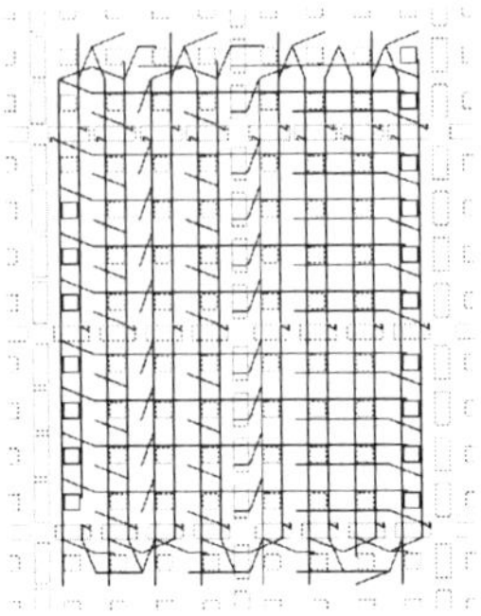

Figure 9. Dummy layout for reconfiguration area

The generation of usable partial configurations is not directly supported by XACT6000. We have created them by using XACT to generate a global configuration for the entities of each macro-operation located in each area, and then filtering these global configurations to obtain partial configurations where only "reconfigurable" resources are modified. The filtering process is based on a textual description of the "reconfigurable" resources of each area (*Figure 10*). The methodology to obtain the initial and partial reconfigurations is shown in *Figure 11.*

```
Cells (16,7-17)
  NORTH Reconfigurable
  SOUTH Reconfigurable
  EAST Reconfigurable
  WEST Fixed
  X1 Fixed
  RP Fixed
  ...
  End
```

Figure 10. Defining reconfigurable resources

The reconfigurable control circuit and design procedures have been tested in hardware using the Annapolis FireFly™ PC board. Two versions of the reconfiguration handler have been used, one implemented in software using C++ and the other implemented in hardware using an XC4010XL FPGA mounted on a XS40 board [Xess98] that interfaces to the PC through the parallel port.

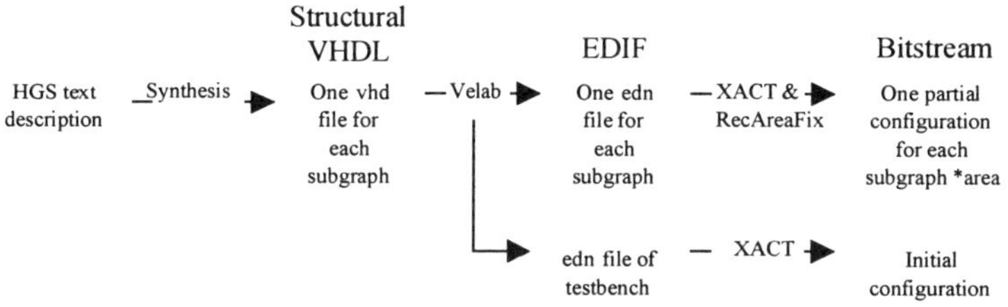

Figure 11. Creating initial and partial reconfigurations

4. CONCLUSION

Complex control circuits can be specified using HGSs by detailing the properties of the system at several layers of abstraction. These complex control circuits can be implemented in fine-grained FPGAs using the HFSM structure proposed. This structure provides a high degree of modularity for the implementation of the sub-algorithms (macro-operations) of the HGS, and so it is well adapted to a dynamically reconfigurable implementation.

General principles for a model to use reconfigurability have been presented and these principles have been applied with success to the dynamically reconfigurable implementation of HGSs using the XC6200 FPGA. Several tools had to be developed to overcome the lack of proper support for dynamic reconfigurability in standard tools.

5. ACKNOWLEDGEMENTS

We are thankful to Ricardo Monteiro, António Rui Borges, Arnaldo Oliveira, Andreia Melo and António Ferrari for fruitful discussions concerning this work. We thank Ivor Horton for reading and useful comments on this paper.

6. REFERENCES

[Baranov94] Samary Baranov, *Logic Synthesis for Control Automata*, Kluwer Academic Publishers (1994)

[Eldredge96] James Eldredge, Brad Hutchings, "Run-time Reconfiguration: A Method for Enhancing the Functional Density of SRAM-Based FPGAs", in *Journal of VLSI Signal Processing*, Volume 12, pp 67-86 (1996)

[Hauck98] Scott Hauck, "The Roles of FPGA's in Reprogrammable Systems", *Proceedings of the IEEE*, vol. 86, no. 4, pp. 615-638 April, 1998.

[Hutchings95] B.L.Hutchings, M.J. Wirthlin, "Implementation Approaches for Reconfigurable Logic Applications", In *5th International Workshop on Field Programmable Logic and Applications*, pp 419-428, Oxford, England, August (1995)

[Oliveira98] A. Oliveira, N. Lau, V. Sklyarov, "Synthesis of VHDL Code from Hierarchical Specification of Control Circuits for Dynamically Reconfigurable FPGAs", *CD Proceedings of VHDL International User Forum - VIUF*, Orlando (1998)

[Robinson98] D. Robinson, P. Lysaght and G. McGregor, "New CAD Framework Extends Simulation of Dynamically Reconfigurable Logic", in *Field Programmable Logic and Applications*, pp 1-8, Tallinn, Estonia, Sept. (1998)

[Rocha97] António Adrego da Rocha, Valery Sklyarov, António de Brito Ferrari, "Hierarchical Description and Design of Control Circuits Based on Reconfigurable and Reprogrammable Elements", *Proceeding of the Int. Workshop on Logic and Architectural Synthesis-IWLAS'97*, Grenoble, December (1997)

[Shirazi98] N. Shirazi, W. Luk and P.Y.K. Cheung, "Automating production of run-time reconfigurable designs", in *Proc. IEEE Symposium on Field-Programmable Custom Computing Machines*, IEEE Computer Society Press, (1998)

[Sklyarov97] Valery Sklyarov, António Adrego Rocha, António Ferrari, "Synthesis of Reconfigurable Control Devices Based on Object-Oriented Specifications", In A*dvanced Techniques for Embedded Systems Design and Test*, Kluwer Academic Publishers, pp. 151-177 (1997)

[Sklyarov98] V. Sklyarov, N. Lau, A. Oliveira, A. Melo, K. Kondratjuk, A. Ferrari, R. Monteiro, I. Skliarova, "Design Tools for Dynamically Reconfigurable FPGAs", *Proceedings of PACT'98 - Workshop on Reconfigurable Computing*, Paris, Oct. (1998)

[Xess98] "XS40, XSP and XS95 Board Manual", *http://www.xess.com/FPGA/manual.html*, XESS corporation (1998)

[Xilinx97] "XC6200 Field Programmable Gate Arrays", Xilinx Product Description (Version 1.10). April 24 (1997)

[Xilinx97b] "DES and CFG Format Reference", Xilinx (1997)

[Wirthlin95] Michael J. Wirthlin, Brad Hutchings, "A Dynamic Instruction Set Computer", In *Proceedings of the IEEE Workshop on FPGAs for Custom Computing Machines*, pp 99-107, Los Alamitos, California, April (1995)

An IEEE Compliant Floating Point MAF

R.V.K. Pillai, D. Al-Khalili[†]and A.J. Al-Khalili
Concordia University, Montreal, CANADA
[†]Royal Military College, Kingston, CANADA

Key words: computer-arithmetic, digital - CMOS, floating-point, low-power-design, power-consumption-model, switching-activity, vlsi

Abstract:

In this paper[1], we present a new architecture for low power floating point multiply - accumulate (MAC) fusion. The proposed architecture supports IEEE and non IEEE rounding modes. The functional partitioning of the adder segment of the MAC into three distinct, clock gated data paths allows activity reduction. The switching activity function of the adder is represented as a three state FSM. During any given operation cycle, only one of the data paths is active, during which occasion, the logic assertion status of the circuit nodes of the other data paths are maintained at their previous states. Critical path delay and latency are reduced by incorporating speculative rounding and data path simplifications. The proposed scheme offers a worst case power reduction of around 25%, in contrast to a comparable scheme reported in literature.

1. INTRODUCTION

The computation of multiply - accumulate is fundamental in many scientific and engineering applications. Since the number of computational operations envisaged by a dot product process are more than one - evaluation of a product and summation of this product with another

---

1.This work had been supported by the Natural Sciences and Engineering Research Council (NSERC) of Canada.

operand - the time complexity of dot product operations are relatively high. The execution time of such operations may, however, be reduced by incorporating concurrency. With floating point operands, the fusion of MAC operations is fairly intricate, owing to the requirements for significand alignments during addition. Though the complexity of floating point hardware units that envisage fusion of multiply - accumulate operations (MAF) is relatively high in comparison with traditional approaches, MAF architectures [1] are still the preferred choice for time critical applications. The IBM RISC/6000 reported in [1] had been the first FPU with multiply - accumulate fused architecture. While the IBM MAF demonstrates the feasibility of floating point multiply - accumulate fusion, this MAF is, however, not widely accepted owing to certain limitations as far as compliance with IEEE [2] [3] floating point standards is concerned. The IBM MAF doesn't produce results that conform to IEEE standards, though the numerical accuracy of results is probably better than that of IEEE conformal schemes. W. Kahan [4] terms this "a mixed blessing". In order that the results of multiply - accumulate operation be conformal with IEEE standards, the results of multiplication and addition need separate rounding. With the IBM scheme, rounding is performed only once, which is rather a compound rounding operation encompassing both multiplication and addition. This paper addresses the development of a low power floating point multiply accumulate fused architecture which produces results that comply with the IEEE norms. The proposed architecture also supports non IEEE rounding.

2. THE PROPOSED MAF

Fig. 1 illustrates the significand data path organization of the proposed MAF. With the proposed scheme, formation of IEEE product is rather straight forward. The partial product array compresses the partial products into two sum and carry vectors. The CP Add/round block performs the carry propagate addition/rounding operation. Pre - computation for rounding is envisaged. Once the final result taking into account the rounding/normalization decisions is arrived at, the rounding information of the product can be used for the rounding of the dot product (or sum).

With the IBM MAF scheme, since the position of the significand of the product is taken as the reference, the significand of C gets aligned all the time irrespective of the value of its exponent. For those values of exponents of C that are greater than that of the product AB, the significand of C is left shifted through an appropriate number of bit positions and vice

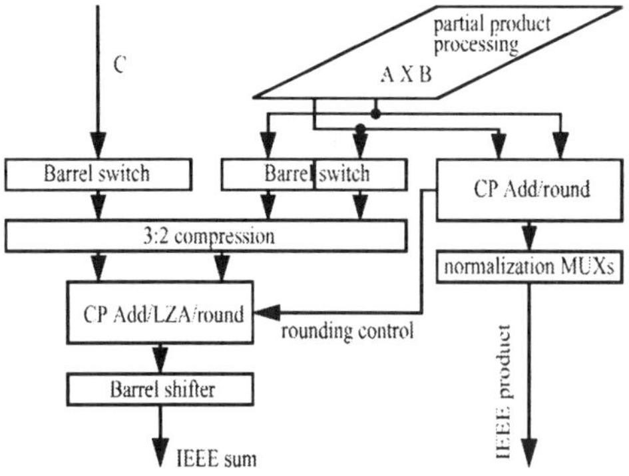

Fig. 1 - Significand data path organization of the proposed MAF

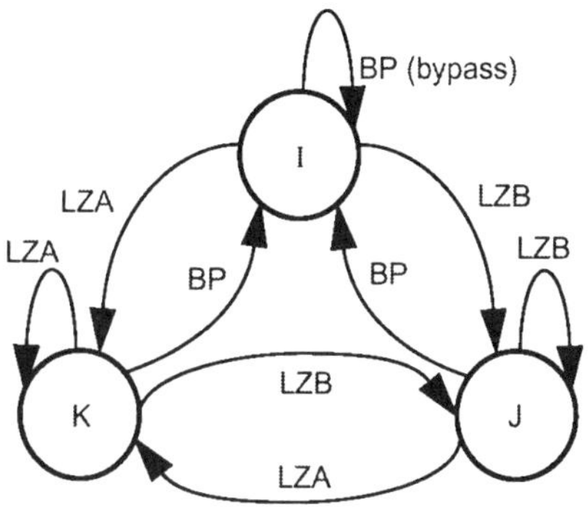

Fig. 2 - FSM representation of FADD operation

versa. With the proposed scheme, in contrast to the IBM MAF, the significands of both AB and C can be aligned in accordance with the relative magnitudes of the product and sum. During situations when the exponent of C is larger than that of AB, the output sum and carry vectors from the partial product compression circuits of the multiplier segment of the MAF are simultaneously right shifted, by using a double barrel shifter. When the exponent of AB is greater than that of C, the significand of C is right shifted by a single barrel shifter. The pre - alignment barrel shifter that shifts C can be merged with the double barrel shifter by incorporating suitable significand selection schemes. The output of the pre-alignment shifters (3 operands) is further compressed into two vectors. The significand adder accepts rounding control signals from the multiplier segment of the MAF. With the proposed scheme, support for IEEE/non IEEE rounding modes can be easily accomplished through the selection of appropriate rounding control signals from the multiplier. Pre-computation of different copies of results taking into account the various rounding/normalization requirements is envisaged. Conditional sum/carry select adders are ideal for such applications.

3. TRANSITION ACTIVITY SCALING FOR LOW POWER

In CMOS logic structures, the switching activities of functional units exhibits sensitivities towards architectural/algorithmic design decisions [5]. At the architecture level, transition activity scaling of functional units offers a viable approach for power minimization [6]. In [7], we reported the architectural design of a transition activity scaled triple data path floating point adder (TDPFADD). The approach outlined in [7] is applicable for the design of floating point MAFs as well. Significant among the observations made in [7] are: (1) The leading zero estimation circuits of FADDs that handle a variable number of leading zeros need be operational only during a limited set of additions. (2) FADDs may be bypassed during certain situations. Functional partitioning of the FADD segment of the MAF into three distinct, mutually exclusive, clock gated data paths allows activity reduction. During any computing cycle, only one of the data paths is active, during which state, the logic assertion status of the circuit nodes of the other data paths are maintained at their previous states. Fig. 2 illustrates the finite state machine representation of the transition activity scaled TDPFADD [7]. State I represents bypass conditions. State J represents the operation of the FADD during those situations when the signed magnitude addition of significands can produce at the most one leading zero while state K represents FADD operations that can produce a pre-normalized significand with a variable number of leading zeros. The time averaged power consumption of the FADD is represented by

$$P = P(I)P_I + P(J)P_J + P(K)P_K \quad (1)$$

where $P(I)$, $P(J)$ and $P(K)$ represent the probability that the FADD is operating in states I, J and K respectively. P_I, P_J and P_K represent the time averaged power consumption of the FADD when the FADD is operating in the respective state. With non activity scaled FADDs, the power consumption can be as high as $P_I + P_J + P_K$.

The proposed partitioning of floating point additions also leads to data path simplifications. Since the significand pre - alignment shifts are ≤ 2 during situations when the MAC operation produces a pre - normalized significand with a variable number of leading zeros, significand pre - alignment operations of this data path (LZA data path) can be effected by using a single level of 3X1 MUXs. With p bit significands, this data path requires a normalization barrel shifter that can handle a maximum right shift of 1 and a maximum left shift of p bits. With the leading zero bounded data path (LZB data path), significand pre - alignment shifts can be anywhere between 0 and $p + 1$. The normalization shifts for this data path are bounded, a maximum right shift of 2 bit positions and a maximum left shift of 1 bit position. For both the computing data paths, only one large barrel shifter is present, by virtue of which the power consumption, logic depths and circuit delays of the data paths are minimized.

With the proposed architecture, the FADD endures bypass conditions whenever the results are known apriori. During situations when $|C| > |AB|$

and the exponent difference is greater than $p + 1$, the result is C. When $|C| < |AB|$ and the exponent difference is greater than p, the result is AB. Apart from these situations, the FADD can also be bypassed during operations of the type $0 \pm$ operand, $\pm \infty \pm$ operand and operations that leads to NaNs.

With MAFs, the multiplier may also be activity scaled for reduced power operation. Whenever the product $|AB| \ll |C|$, the multiplier segment of MAF can be activity scaled [8]. However, for applications that need IEEE sums and products, irrespective of their relative magnitudes, the question of transition activity scaling of the multiplier segment is not very relevant. During situations when the product is known apriori, e.g., 0 X number, NaN X number, ± ∞ X number, the multiplier can be activity scaled. The activity scaling of the multiplier is best addressed from a control path perspective. In instruction driven processors, pre-computation of multiplier/MAF bypass conditions during an early stage of instruction scheduling is possible. With such schemes, transition activity scaling of multiplier doesn't slow down the speed performance of the MAF.

4. POWER MODELS

During FP additions, the transition activities of barrel shifter control lines exhibit sensitivities towards the significand alignment behavior of FADDs. In particular, the magnitudes and rate of change of alignment shifts reflect the power implications of significand alignment operations. The switching activities within the significand data paths also exhibit sensitivities towards the above parameters. The following paragraphs highlight the development of analytical models, that capture the impact of significand alignments on the power consumption of FADDs. Before we go into the specifics of this aspect of power consumption, the following definitions shall be introduced.

Definition 1 - Expected shift: The expected shift of any data alignment operation is defined by

$$E_{SH} = E[x] = \sum_{k=0}^{k_{max}} kP(x = k) = \sum_{k=0}^{k_{max}} kP_k \qquad \textbf{(2)}$$

In equation (2), x represents the shift distance, which is not necessarily the exponent difference; the precise relation between these parameters is a function of FADD data path/barrel shifter organization [8]. With equation (2), it is assumed that x is wide sense stationary (WSS). In floating

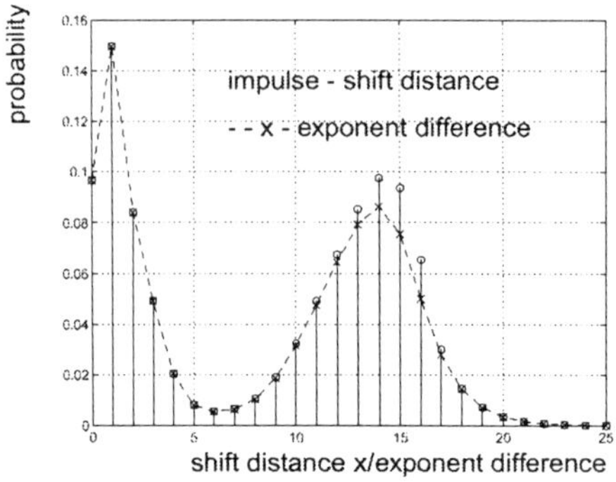

Fig. 3 - pdf of shift distances

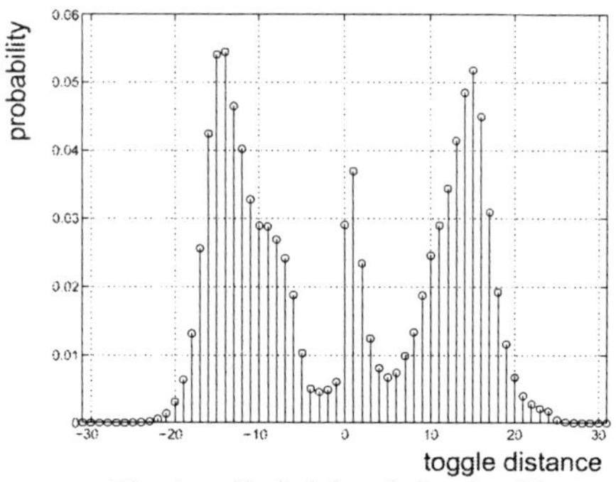

Fig. 4 - pdf of z[n] = x[n] - x[n - 1]

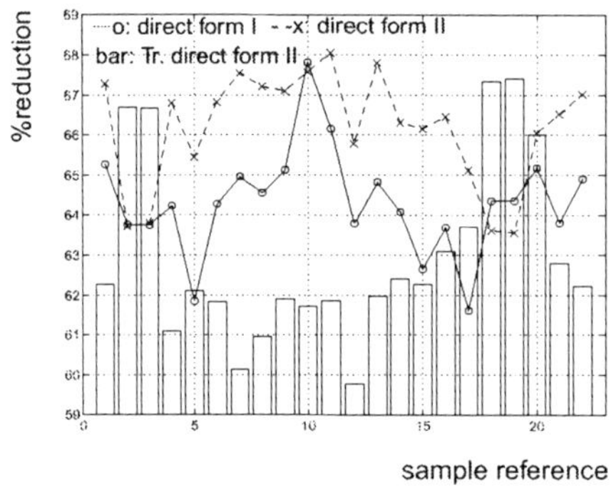

Fig. 5 - Reduction during IIR filtering

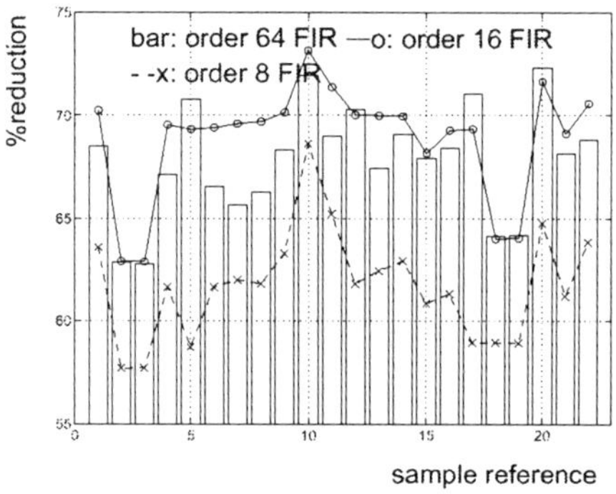

Fig. 6 - Reduction during FIR filtering

point DSP operations, strictly speaking, $x[n]$ represents a non stationary random process [9]. Though $x[n]$ is non stationary, the time averaged power consumption of FP units may still be characterized on the basis of an 'average' behavior of $x[n]$ [8]. In our simulation based study, the probability density function (pdf) of x is evaluated on the basis of the frequency distribution of $x[n]$ over the whole set of FP addition operations during the underlying experiment. Fig. 3 illustrates sample pdfs of x (significand pre alignment shift of a non activity scaled FADD) as well as the underlying exponent differences, the relevant frequency distributions of which had been observed during filtering of white noise ($N(0,1)$). A sequence of white noise samples (128K) had been low pass filtered (single precision FP operations) using an 8th order elliptical filter (transposed direct form II IIR filter - pass band ripple 0.03 dB, stop band ripple 140 dB, normalized cut-off frequency 0.2). In Fig. 3, only a truncated region of the pdf is shown, that is, the probabilities are illustrated for shift distances/exponent differences upto 25 only. The expected shift in this case is 8.5886 bits while the expected exponent difference is 14.2360 bits.

Definition 2 - Toggle distance: Toggle distance is defined as the difference between present shift and last shift, i.e., $z[n] = x[n] - x[n-1]$.

Definition 3 - Expected toggle distance: The expected toggle distance is defined as the average number of bit positions through which shift operations oscillate about the mean or expected shift. The expected toggle distance is computed as the mean value of absolute toggle distances, as given by,

$$E[|z[n]|] = \sum_{l=0}^{l_{max}} lP(|x[n] - x[n-1]| = l) \quad \textbf{(3)}$$

With conventional IEEE single precision FADDs, the index variable l can assume values between 0 and 31 (0 and 63 for double precision). With activity scaled FADDs, l is a function of the organization of barrel shifters.

With transition activity scaled FADDs, most of the parameters discussed above are scaled versions of the relevant parameters of conventional FADDs. Fig. 4 illustrates a pdf of pre-alignment toggle distances of a conventional FADD, that had been experimentally observed during filtering of random noise samples, described earlier. The expected toggle distance for this case is 10.4004 bits.

4.1 CONTROL PATH SWITCHING

In general, the control path power consumption of FADDs is dominated by the power consumption of barrel shifter control lines (both prealignment and normalization) as well as various data selection signals that facilitate the presentation of exponents, default results etc. With this, the control path power measure of FADDs can be modeled by

$$P_C \propto \sum_{i \in S} t_i Y_i \quad \textbf{(4)}$$

where S represents the set of all control signals, t_i and Y_i represent the transition activity and fanout of the ith control signal. With different architectural schemes, the number of control signals, their transition activities as well as fanouts differ.

4.2 ALIGNMENT DRIVEN DATA PATH SWITCHING

Whenever the position of the aligned significand endures oscillations about the expected shift, the significand data path bits that fall within the toggle range endure higher transitions. The power consumption due to this phenomenon is proportional to the expected toggle distance. During FP subtractions, the 1's (or 2's) complement of the aligned significand is added with the significand of the larger number. During such a scenario, the zeros appearing at the higher order bit positions (due to shift operation) of the aligned significand gets complemented into 1's. If the toggling between addition and subtraction operations is relatively significant, then the power consumption due to this activity is significant. The power consumption of significand adders of FADDs (which, by and large, reflects the alignment driven data path switching), taking into account the above effects can be represented by

$$P_{SI} = 2\hat{P}_{ADD}\left[1 + \frac{E[|z[n]|] + E[x]t_S}{p}\right] \tag{5}$$

In (5), $E[x]$ and $E[|z[n]|]$ represent the expected shift and expected toggle distance respectively while p represents the width of significand. t_s represents the probability for sign toggling. $\hat{P}_{ADD}$ represents the time averaged power consumptions of significand adders, during situations when both $E[|z[n]|]$ and $E[x]t_s$ are zeros. With FADDs that incorporate leading zero anticipatory logic, the power consumption of these units are comparable to that of significand adders. This aspect is taken into account by the scaling factor 2 in equation (5). With the proposed MAF, the power consumption of the FADD segment can be represented by

$$P_{SII} = 2\hat{P}_{ADD1}[1 + \mu_A]P(K) + \hat{P}_{ADD1}[1 + \mu_B + \gamma_B]P(J) \tag{6}$$

In the above equation, μ_A and μ_B represent the values of the parameter $E[|z[n]|]/p$ for the LZA and LZB data paths respectively of TDPFADD. γ_B represents the value of $E[x]t_s/p$ for the LZB data path. Since the LZA data path handles only subtractions, the question of alignment driven sign toggling doesn't arise in this case. With the IBM MAF scheme, the extra switching activity due to pre-alignment toggling largely affects the p MSB bit positions of the adder. With the lower order bit positions, the switching activity is, by and large, a function of the signal probabilities of the compressed partial products. With the proposed MAF scheme, the effect of pre-alignment toggle distances of the compressed partial products (sum and carry vectors) outweigh that of the significand of C, due to similar reasons.

TABLE I : DATA PATH UTILIZATION PROBABILITIES DURING IIR FILTERING

Sl. No	IIR1			IIR2			IIR3		
	P(I)	P(J)	P(K)	P(I)	P(J)	P(K)	P(I)	P(J)	P(K)
1	0.0589	0.5380	0.4031	0.0595	0.6769	0.2636	0.0588	0.5610	0.3802
2	0.0589	0.5944	0.3467	0.0588	0.5910	0.3502	0.0600	0.6671	0.2729
3	0.0588	0.5667	0.3745	0.0588	0.5666	0.3746	0.0598	0.6749	0.2653

TABLE II : DATA PATH UTILIZATION PROBABILITIES DURING FIR FILTERING

Sl. No	FIR1			FIR2			FIR3		
	P(I)	P(J)	P(K)	P(I)	P(J)	P(K)	P(I)	P(J)	P(K)
1	0.2002	0.6433	0.1565	0.1765	0.6390	0.1845	0.1112	0.6963	0.1925
2	0.2002	0.6687	0.1311	0.1765	0.7396	0.0839	0.1111	0.8364	0.0525
3	0.2003	0.6756	0.1241	0.1765	0.6937	0.1298	0.1112	0.7651	0.1237

5. RESULTS

Instrumented digital filter programs that envisage single precision FP operations, emulating the two MAF schemes had been developed. The experiments involved the filtering of an assorted collection of data samples - both synthetic and real data. The first among the synthetic signals is a sequence of white noise samples (N(0,1) IID RVs) of sample size 128K, while the second and third are auto regressive signals of the same sample size. Specifically, the AR model of the second signal is $y[n] = x[n] + 0.9*y[n-1]$ while that of the third signal is $y[n] = x[n] + 0.5*y[n-1]$. The first three filters are 8th order elliptical filters (low pass), having pass band ripple of 0.03 dB, stop band ripple of -100dB and normalized cut - off frequency 0.2. Filters I, II and III are direct form I, direct form II and transposed direct form II realizations of the same filter. The last three are low pass (normalized cut-off frequency of 0.2) FIR filters of order 64, 16 and 8 respectively. With real data, an assorted collection of bipolar audio signal samples ranging in size between 8594 and 6318040 samples had been low pass filtered using both FIR and IIR filters. During the course of filtering, frequency distributions of pre-alignment and normalization shifts, their rate of change and the relevant bit level activities had been collected.

Tables I and II present the data path utilization statistics of the FADD segment of the MAF, that had been observed during filtering of synthetic data. With the above results, the most important observation is that the probability that a variable number of leading zeros occur during the

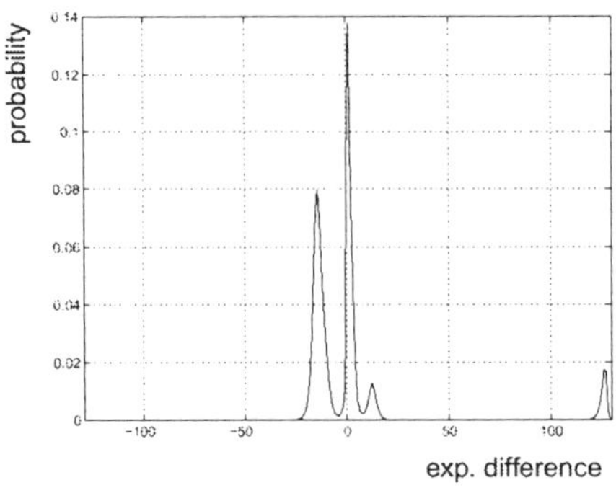

Fig. 7 - pdf of exp. diff. in IBM MAF

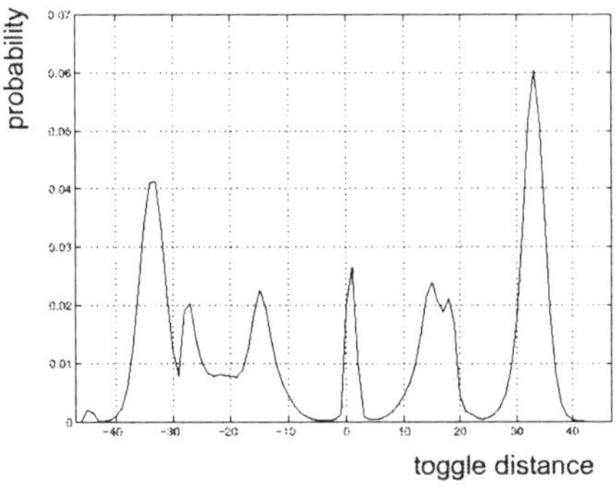

Fig. 8 - pdf of temporal shift behavior

signed magnitude addition of aligned significands is marginally low, which substantiates the efficacy of the leading zero estimation based transition activity scaling approach.

Figures 5 and 6 illustrate the percentage reduction in switching activity offered by the proposed scheme during filtering of bipolar audio signal samples, as far as significand pre-alignment control is concerned. The worst case reduction is better than 56%, which is attributed to data path simplifications and transition activity scaling. The reduction in switching activity as far as significand addition is concerned is also better than 50%. During filtering, the probability that the product is shifted is around 50%. That means, the double data path barrel shifter is not operational during 50% of the time. With the IBM MAF, the pre-alignment of the significand of *C* is handled by a non - activity scaled, bidirectional barrel shifter, the effective data path width of which is around twice that of the proposed scheme. Because of this, the fanout weighted switching activity of the barrel shifter control lines of the IBM scheme is significantly higher than that of the proposed scheme. Figures 7 and 8 illustrate the significand pre - alignment behavior as well as the rate of change of shift (evaluated as present shift - last shift) of the IBM MAF, that had been observed during IIR filtering of white noise samples. (The pdfs illustrated in Figures 3, 4, 7 and 8 represent various instances of exponent behavior observed during performance of the same experiment). In Fig. 7, negative values of shifts indicate situations during which the significand of *C* is left shifted. In contrast to the significand pre-alignment behavior depicted by Figures 3 and 4, the variances as well as entropies of the pdfs shown in Figures 7 and 8 are large. In general, the higher the variance of these pdfs, the higher the power consumption.

With normalization control, the switching activity reduction observed during our experiments is consistently better than 10X. In FADDs, normalization shifts through a large number of bit positions are required only during situations when the process of significand addition results in a large number of leading zeros. During all other situations, normalization shifts are limited. However, with the IBM MAF scheme, normalization shifts can be large even during other situations. With this scheme, with p bit significands, the leading 1 after significand addition can occur within a range of $2p + 2$ bits, and hence the normalization shifts are usually large. Because of this, the leading zero estimation logic also has to work with the $2p + 2$ bit results.

In general, the power consumption of FADDs outweigh that of FP multipliers. As discussed previously, owing to the relatively large magnitudes of switched capacitances associated with significand alignments, the power consumption of barrel shifters dominate the power consumption of FADDs. Assuming that the power consumption of the multiplier segment of the MAF is comparable that of the adder segment, it is relatively straight forward to conclude that the worst case power advantage offered by the proposed scheme is around 25%.

6. DISCUSSION

Compared to the IBM scheme, the salient features of the proposed MAF scheme that renders it an ideal choice for DSP applications as well as general purpose computing are summarized below.

(1) IEEE compatibility: The requirement for IEEE compatible floating point results is mandatory for many computing applications. The availability of IEEE product and sum is a definite advantage.

(2) Data path simplifications: With the proposed scheme, the width of the significand data path is around half of that of the IBM scheme. The removal of one barrel shifter from the critical path of the significand adder is another notable feature. Because of these simplifications, the estimated speed performance of the proposed scheme is better than that the IBM scheme. The data path simplifications also results in area reduction. The area measures of significand adders, normalization barrel shifter and leading zero anticipatory logic of the proposed MAF are less than that of the IBM scheme. The proposed scheme also envisage the handling of certain arithmetic operations by using 1's complement arithmetic units [8], which results in power/area reductions. To put it briefly, though the proposed scheme envisage a separate data path for the handling of significand additions that are likely to result in a variable number of leading zeros, the additional area implications of this data path is offset by the area reduction measures.

(3) Transition activity scaling: The transition activity scaled data path partition renders power optimal operation.

7. CONCLUSION

A proposal for floating point multiply - accumulate fusion is presented. The proposed scheme delivers IEEE compatible (as well as non IEEE) sums and products. The estimated worst case reduction in switching activity offered by the proposed scheme is around 25%. The power/ delay advantages of the proposed scheme renders it an ideal choice for floating point dot product computations.

8. REFERENCES

[1] Erdem Hokenek, Robert K. Montoye and Peter W. Cook, "Second generation RISC floating point with multiply - add fused", *IEEE Journal of Solid State Circuits*, Vol. 15, pp. 1207 - 1213, October 1990.

[2] IEEE Standard for Binary Floating - Point Arithmetic", ANSI/IEEE Std 754 - 1985, New York, *The Institute of Electrical and Electronics Engineers Inc.*, August 12, 1985.

[3] Israel Koren, "*Computer Arithmetic Algorithms*", Prentice Hall, Englewood Cliffs, 1993.

[4] W. Kahan, "Lecture Notes on the Status of IEEE Standard 754 for Binary Floating-Point Arithmetic", (http://http.cs.berkeley.edu/~wkahan/ieee754status/iee754.ps) Elect. Eng. & Computer Science, University of California, Berkeley.

[5] Kurt Keutzer and Peter Vanbekbergen, "The Impact of CAD on the Design of Low Power Digital Circuits," in *Proceedings of the 1994 IEEE Symposium on Low Power Electronics*, pp. 42 - 45.

[6] Anantha P. Chandrakasan, Randy Allmon, Anthony Stratakos, and Robert W. Brodersen, "Design of Portable Systems," in *Proceedings of the 1994 IEEE Custom Integrated Circuits Conference*, pp. 259 - 266.

[7] R. V. K. Pillai, D. Al - Khalili and A. J. Al - Khalili, "A Low Power Approach to Floating Point Adder Design", *in Proceedings of the 1997 International Conference on Computer Design*, pp. 178 - 186.

[8] R. V. K. Pillai, "On Low Power Floating Point DSP Data Path Architectures", Ph. D Thesis (under preparation) - Department of Electrical and Computer Engineering, Concordia University, Montreal, Quebec.

[9] R. W. Hamming, "On the Distribution of Numbers", *Bell System Technical Journal*, Vol. 49, No. 8, pp. 1609 - 1625, October 1970.

Design and Analysis of On-Chip CPU Pipelined Caches

C. Ninos, H. T. Vergos & D. Nikolos
Department of Computer Engineering & Informatics, University of Patras, 26 500, Greece
Computer Technology Institute, 3, Kolokotroni St., 262 21 Patras, Greece.

Key words: On-chip CPU caches, Pipelined Caches, Performance Analysis and Evaluation.

Abstract: The access time of the first level on-chip cache usually imposes the cycle time of high-performance VLSI processors. The only way to reduce the effect of cache access time on processor cycle time is the use of pipelined caches. A timing model for on-chip caches has recently been presented in [1]. In this paper the timing model given in [1] is extended so as pipelined caches can be handled. Also the possible pipelined architectures of a cache memory are investigated. The speedup of the pipelined cache against the non-pipelined one is examined as a function of the pipeline depth, the organization and the physical implementation parameters.

1. INTRODUCTION

Cache memories are used for enhancing the performance of almost every modern microprocessor. Therefore their architecture challenges have been investigated extensively in the past years [2-6]. Computer designers have the problem of building a cache that has both a low miss rate and a short access time. A low miss rate ensures that the high cost of cache misses does not dominate execution time and the short access time ensures that the cache does not slow down the rest of the CPU. A partial solution to this cache design problem is to provide more than one level of cache memory. In a two level cache hierarchy the level one (primary cache) is made small and fast to match the CPU speed and the level two (secondary) cache is made slower and much larger to keep the overall cache miss rate low [5, 6].

Although this is better than having a single large cache, the primary cache will still limit the CPU cycle time. The only way to reduce the effect of cache access time on CPU cycle time is to increase the number of cache pipeline stages. This spreads the fixed delay of the cache access time over more CPU cycles making it possible for the CPU cycle time to be reduced [7].

Several models have in the past been presented for studying different cache organizations. Examples include an area model [8], timing models [1, 9] and yield models [10, 11]. No model though has in the past been presented for studying the effect of increasing the cache pipeline stages on the cache cycle time. Although specific pipelined cache designs have been presented in the open literature (for example [12]), a general method for designing and analyzing pipelined caches has not yet been presented.

In [7] the pipeline depth for achieving CPU performance optimization using pipelined first level caches was investigated assuming that the cycle time of a pipelined cache is equal to:

{non-pipelined cache cycle time} / {pipeline depth} (1).

In this paper we present the required extensions on the analytical enhanced cache access and cycle time model presented in [1] for handling pipelined caches. By comparing the derived analytical model to an HSPICE model, it was shown to be accurate to within 6%. The derived model is used for providing the optimal layout organizational parameters of a pipelined cache as well as for studying the effect of various cache pipeline depths on its cycle time. The results indicate that relation (1) above for the cycle time of a pipelined cache is incorrect.

2. PRELIMINARIES

In [1] an analytical access and cycle time model has been presented for on-chip caches. The model was verified by comparing its results against HSPICE simulation results. In the following we give some of the terminology used in [1] for the clarity of presentation.

The array of the stored data is a rectangle area of memory cells with horizontal dimension equal to 8 x {block size} x {associativity} and vertical dimension equal to {number of sets}. This organization, in most cases, results in an array that is much larger in one direction than in the other. That means that the bit-lines or the word-lines have unequal delay times. For reducing the access time of the data memory array it is broken in N_{dwl} subarrays horizontally and N_{dbl} subarrays vertically. The corresponding parameters for the tag array are N_{twl} and N_{tbl}. Finally the N_{spd} and N_{tspd} parameters specify the number of sets that are mapped into a single word-line. The model of [1] selects the best combination of these parameters in order to achieve the minimum cycle time. Figure 1 presents the cycle and the access time of direct mapped caches with block size equal to 16 for caches of sizes of 8K up to 256K assuming the process parameters for an 0.8_m technology given in [1]. The left column is used for the data path while the right for the tag path. Figure 2, presents the contribution of each module of the design for the same caches. Figure 2 indicates that the dominant delay contributor for the data part is the decoder module. For the tag part the decoder and the comparator contribute a large percentage of the overall delay.

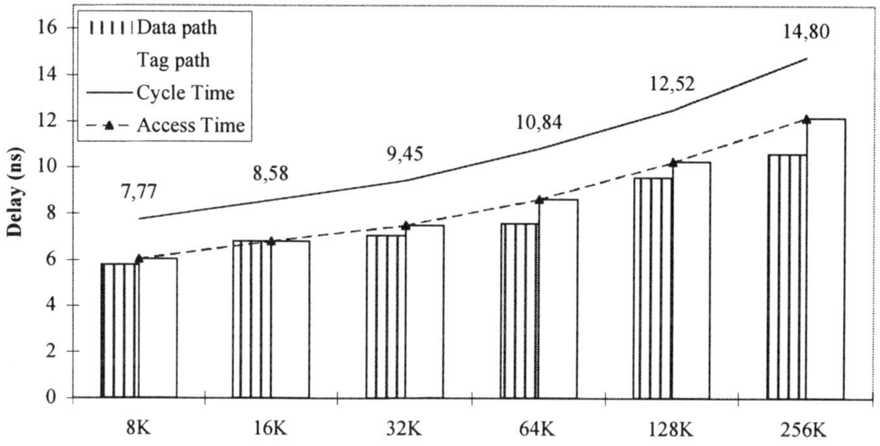

Figure 1. Access and cycle time for the non-pipelined cache design

In general, by modifying the array organization parameters, we can reduce the delay contributed by a certain module. For example, for decreasing the data decoder part delay we might decrease the N_{dwl}. Such a change though will impact the precharge time of the data bit lines increasing the cycle time overall.

3. PIPELINED-CACHE TIMING MODEL

The first level cache is usually designed as direct-mapped for having the shortest cycle time [4, 5, 13]. Due to this we focus in the following on direct mapped cache organizations. Our study is also valid for set-associative pipelined caches, but in this case we have the possibility of one more pipeline stage, the data-output stage.

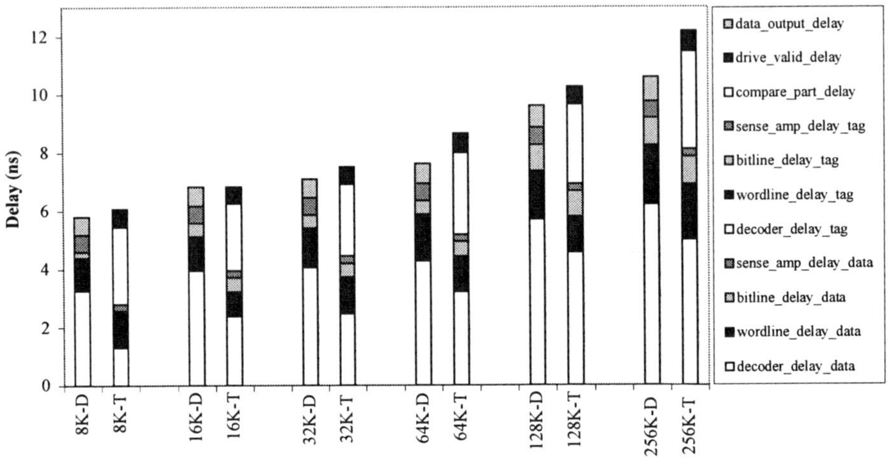

Figure 2. Distribution of the delays in single stage caches

For building a pipelined cache, some modules should be isolated with latches. Latch insertion though can not be arbitrary, since some modules should not be isolated. For example, according to [14], the data access path, formed by the word-line driving, bit-line charging, column multiplexing and sense amplifier modules can not be further split in separate stages. Thus, the modules that can form separate pipeline stages are : a) the decoder, b) the data-access path, c) the comparator, and d) the output drivers. Since the delay introduced by the output drivers in direct mapped caches is extremely small it would be unprofitable to consider them as a separate pipeline stage. Figure 3 presents a possible organization of a three stage pipelined cache. In set associative organizations, the output drivers delay may be significant; therefore it could constitute a separate pipeline stage.

The model proposed in [1] was extended as follows:

In order to hold the data that drive each stage, a latch must separate consequent stages. We utilized the inverters of each stage to form the latch by adding an extra cross-coupled inverter and a transistor, which will isolate the latch from the previous stage. This design is similar to that in [15] and targets low area overhead. Registers were utilized for holding the appropriate address bits in each stage that will later feed the comparator. The bit lines should be precharged before any memory access. This means that the sequence of operations that must be made in order to access the memory array is a) disable all word lines, b) precharge the bit lines, c) drive the appropriate word-line, and d) wait bit-lines to be evaluated and amplifiers to return the stored data. The model given in [1] covers operation (a) as being performed by the decoder and his precharge subcircuit. Since we have isolated the decoder from the data-access path, we were forced to place to the input of the word-line driver both an isolating and a pull-up transistor in order to disable all word-lines during the bit-line pre-charge period. The original (a) and the modified (b) design of the word-line driver are shown in Figure 4.

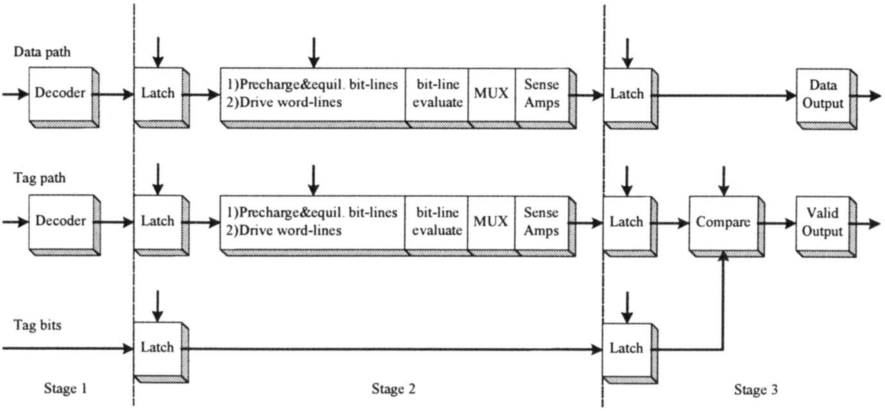

Figure 3. Design of the three-stage pipelined cache with the additional latches. The vertical arrows indicate the pre-charge or latch-enable control signals

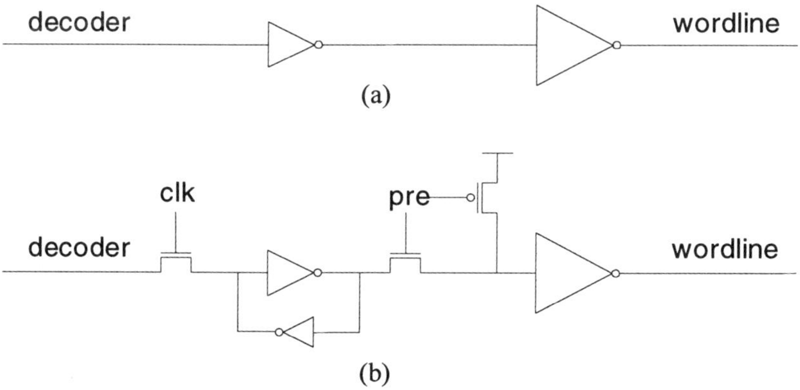

Figure 4. Implementation of the word-line driver. (a) Previous design, (b) New design

Moreover, in [1] the evaluate permission signal for the comparator was driven by a delay chain in order to be synchronized and start producing results at the right time. When the comparator is considered as a separate stage, we place that delay chain to the latch enable signal of the isolating latches in order to let the sense amplifier stabilize. That extra delay in the tag data-access path inserts more limitations to the final speedup. We left to the comparator stage only the last inverter (evaluate-INV) which works as virtual-ground to the compare-NMOS transistors. The operation sequence of the comparator-stage consists of the steps: a) drive to logic one the evaluate-INV deactivating the comparator, b) activate the pre-charge transistor in order to pre-charge the OUT line, c) stop pre-charging and drive evaluate-INV to logic zero in order to start the evaluation and d) evaluate. The equation that predicts the comparator precharge delay time is simply the charge of the RC circuit of the charger transistor and the capacitance of the OUT line. The worst delay in this case is produced when no compare-transistors are open and no extra charging path exists through the evaluate-INV and compare-transistors.

The changes that we have made to the initial model of [1] were verified by comparison against HSPICE simulations. All the results produced by our model where well within 6% of the corresponding HSPICE results. As an example in Figure 5 we present comparative results regarding the comparator precharge time as a function of the tag bits.

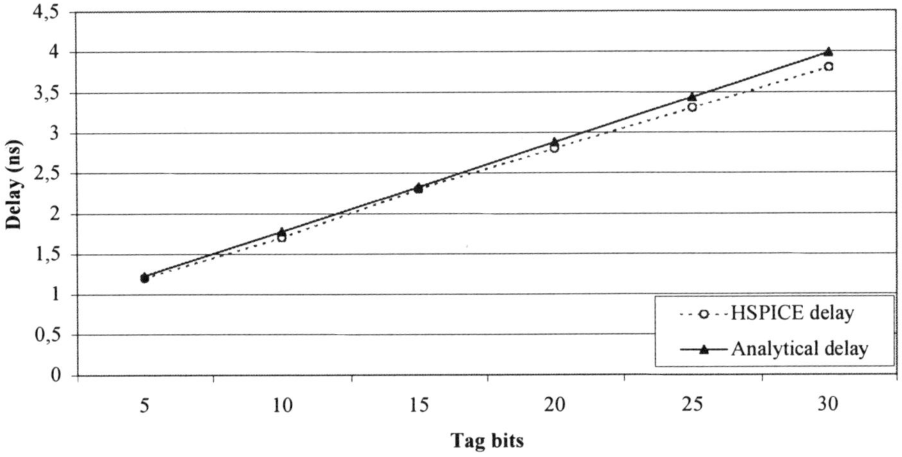

Figure 5.

In our model we assume that the decoder forms a pipeline stage by itself. If it is profitable to increase the pipeline depth, it is necessary to break the decoder in more stages. For example, in [16] a deeply pipelined architecture with a hierarchical design of the decoder is presented. Since breaking the decoder in smaller stages is implementation specific, we chose not to model any of the possibilities. We will discuss the effect of breaking the decoder in smaller stages though in the next section.

4. APPLICATION OF THE DERIVED MODEL AND DISCUSSION

For the application of the model, we used the same process parameters as [1] for an example 0.8_m process. We will present results for direct mapped caches with 16 bytes block size and with sizes ranging from 8K up to 256 K bytes. Our results for caches with smaller or larger block sizes indicate similar behavior. We assumed either a single decoding stage, case a, or a n-stage decoder, case b.

a) For a single decoding stage two different schemes for the pipelined cache architecture were investigated using the derived model:
   1. The 2 stages scheme. In the 2 stage scheme we have placed for both data and tag path the decoder as the first stage and all the rest modules as the second stage.
   2. The 3 stages scheme. In this organization the decoder forms the first stage, the data access path the second and the rest modules the third stage.

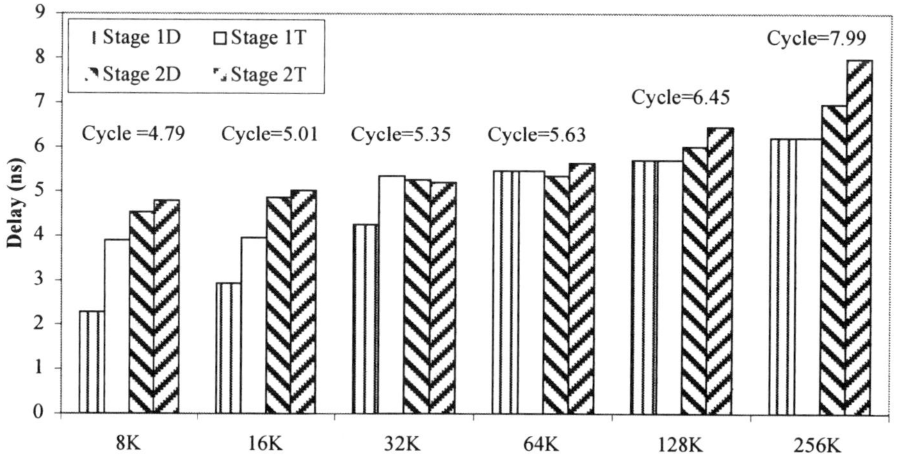

Figure 6. Delay of each stage in the 2 stage scheme

Figures 6 and 7 present the delays of each stage for the 2 and the 3 stages schemes respectively. In these Figures we set the maximum number of sub arrays to 8 for the data as well as the tag memory. We have observed that in these cases larger number of subarrays do not result into faster cycle times. For each stage, the left bar corresponds to the data part of the cache, while the right to the tag part. The resulting cycle time is also presented. Comparing Figures 6 and 7 with Figure 1, we can see that a pipelined cache can offer significantly faster cycle times for both schemes. For the two stages scheme (Figure 6) the 2nd stage of the tag path (consisting of the tag memory access and the comparison delay) is in all but one cases the stage determining the cycle time of the cache. As can be observed in Figure 7, for the largest simulated cache, the decoder stage is the stage determining the overall cycle time of the cache. In all rest cases the cycle time of the cache is determined by the data access stage when three pipelined stages are utilized.

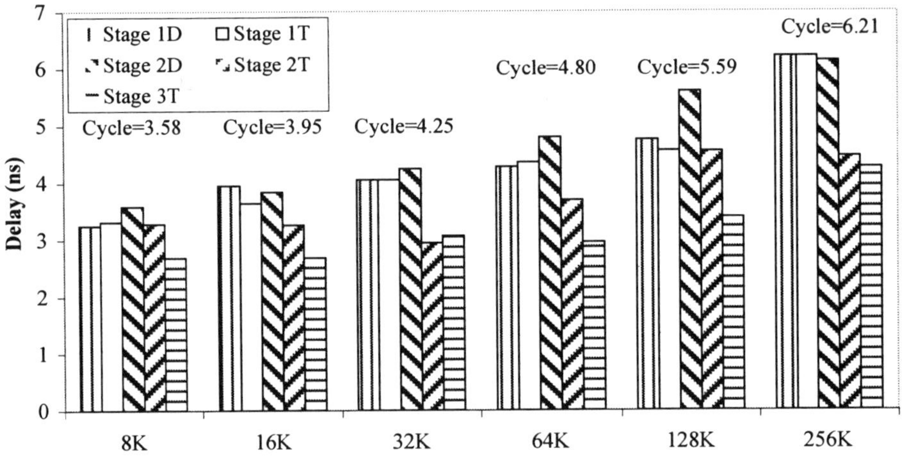

Figure 7. Delay of each stage in the 3 stage scheme. (The delay of stage 3D is always significantly smaller than that of 3T. So in this figure we give only the delay of 3T).

b) For an n-stage decoding scheme we investigated n+2 stages pipelined caches. In this organization the first n stages are decoding stages. The next stage is the data access stage while all the rest modules form the last stage. Since the value of n can be made large enough, the cycle time of the cache will be determined by the data access stage. This stage can be optimized by changing the organization parameters, for example N_{dwl}, of the cache. In Figures 8, 9 and 10 we present the n decoding stages for the tag and data arrays as one stage denoted by 1D and 1T respectively. The data access stages are represented by 2D and 2T and the last stages as 3D and 3T.

Figures 8, 9 and 10 present the optimal time of the data access stage (stage 2D) achieved among all possible combinations of the organization parameters leading to 8, 16 or 32 sub arrays respectively, for the data as well as the tag memory, ignoring the decoder stage. If we break the decoder in n stages so that the delay of each stage of the decoder is smaller than the delay of stage 2D and 2T then the maximum among the delays of stages 2D and 2T will be also the cycle time of the cache. The value of n can be determined as the maximum of the delays of stages 1D and 1T divided by the maximum of the delay of stages 2D and 2T. For 8, 16 and 32 subarrays the resulting cycle time will be denoted by opt.sa8, opt.sa16 and opt.sa32 respectively.

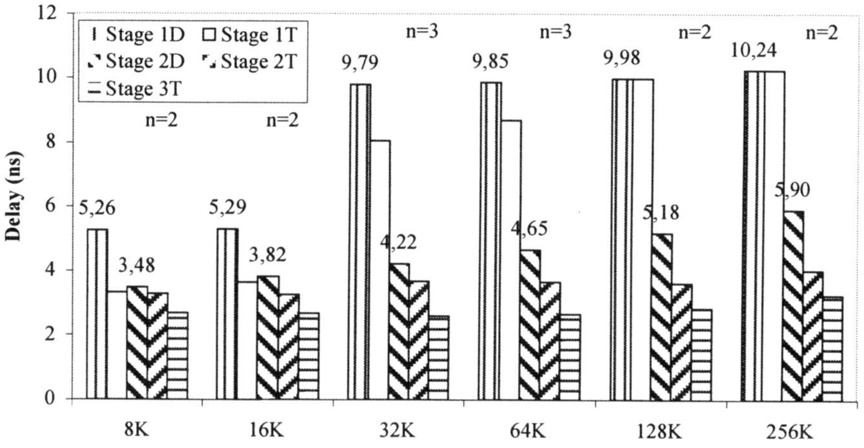

Figure 8. Data-access optimization (max subarrays: 8, stages: n+2)

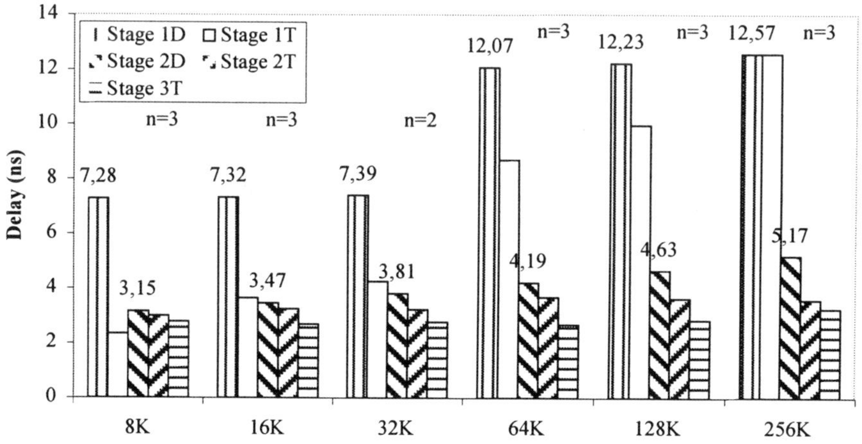

Figure 9. Data-access optimization (max subarrays: 16, stages: n+2)

Comparing Figures 7 and 8 it can be observed that breaking the decoder in two stages will provide an increasing profit in the cycle time of 2.7% up to 7.3% as the cache size increases. The profit becomes even bigger in comparing Figures 9 and 10 with 7, but then the decoder stage must be divided in 3, 4 or even 5 stages of equivalent delay.

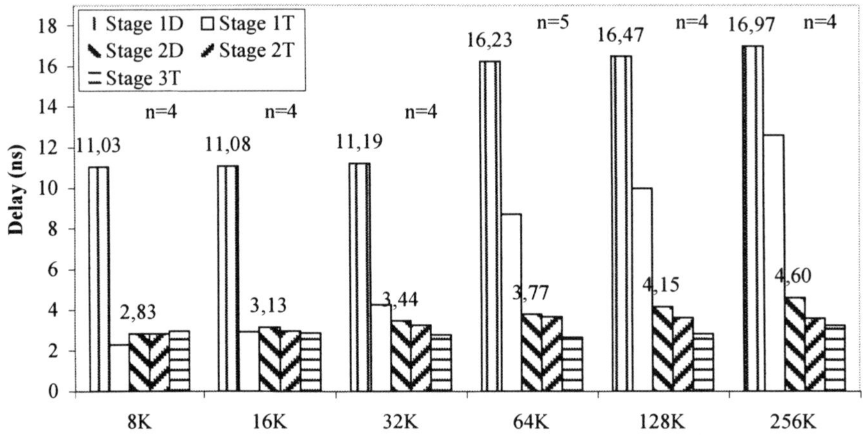

Figure 10. Data-access optimization (max subarrays: 32, stages: n+2)

We define the term speedup as {non-pipelined cache cycle time} / {pipelined cache cycle time}. Figure 11 presents the speedup that can be achieved using 2 stages, 3 stages, and n+2 stages with n=2,3 or 4. In Figure 11 we have included the graphs for opt.sa8, opt.sa16 and opt.sa32 which should be interpreted as a pipeline cache of depth 4, 5 or 6. For example in the 32Kbytes cache, when the number of sub arrays is 32, we determine from Figure 10 the value of n as $\lceil 11.19 / 3.44 \rceil = 4$ leading to a pipelined cache of 6 stages in total.

Table 1 lists the maximum speedup that can be achieved as a function of the pipeline stages. The table reveals that three pipeline stages can achieve a value close enough to the maximum speedup in most cases. The extra hardware and control complexity of more cache pipeline stages must be well justified by the nature of the application and become attractive only in the largest of the simulated caches.

As it can be observed by Table 1, the oversimplified relation (1) is far from the reality.

5. CONCLUSIONS

Caches are an essential part in every modern microprocessor chip. Although their organizational and architectural challenges have been extensively studied in the past, little effort has been devoted in the design and analysis of pipelined caches. Pipelined caches can effectively reduce the cycle time of the CPU.

In this paper the design and analysis of pipelined CPU caches was studied. A timing model given for non-pipelined caches was enhanced for handling pipelined caches. By comparing the model to an HSPICE model it was shown to be accurate within 6%. The computational complexity however, is considerably less than HSPICE. The proposed is the first, in the open literature, model handling pipelined caches.

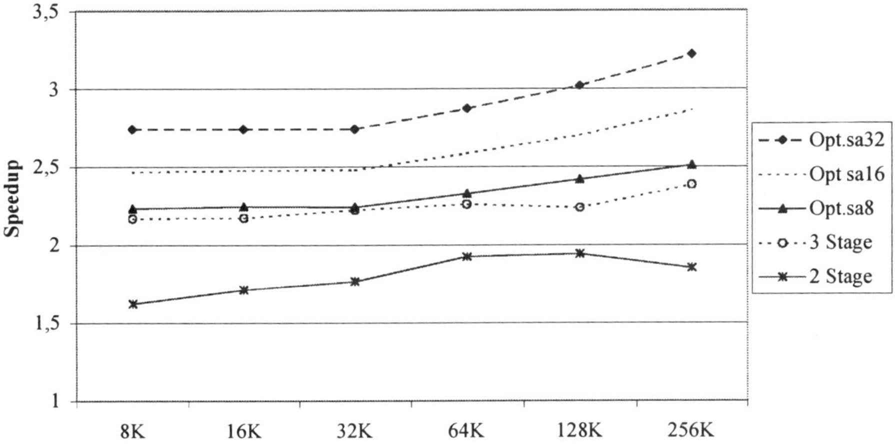

Figure 11. Speedup achieved

Table 1. Speedup vs. number of pipeline stages

Cache Size	Pipeline Stages 2	3	4	5	6
8 K	1.62	2.17	2.24	2.47	2.74
16 K	1.71	2.17	2.24	2.48	2.74
32 K	1.77	2.22	2.48	(*)	2.74
64 K	1.92	2.26	(*)	2.33	(*)
128 K	1.94	2.24	2.42	2.70	3.02
256 K	1.85	2.38	2.51	2.86	3.22

(*) This pipeline scheme does not lead to better results.

6. REFERENCES

[1] S. J. E. Wilton and N. P. Jouppi, "CACTI: An Enhanced Cache Access and Cycle Time model.", IEEE Journal of Solid - State Circuits, Vol. 31, No. 5, pp. 677- 688, May 1996.

[2] A. J. Smith, "Cache Memories", ACM Computing Surveys, Vol. 14, no.3, pp. 473-530, September 1982.

[3] H. S. Stone, *High Performance Computer Architecture*, Addison-Wesley, October 1987.

[4] Przybylski S. A., *Cache and Memory Hierarchy Design*, Morgan Kaufmann Publishers Inc., 1990.

[5] N. P. Jouppi and S. J. E. Wilton, "Tradeoffs in Two-Level On-chip Caching", Proc. of the 21st Int. Symp. on Computer Architecture, pp. 34 – 45, April 1994.

[6] J. H. Edmodson et. al., "Superscalar Instruction Execution in the 21164 Alpha Microprocessor", IEEE Micro, vol. 15, no. 2, pp. 33-43, April 1995.

[7] K. Olukotun et. al., "Performance Optimization of Pipelined Caches", Proc. of the 19th Int. Symp. on Computer Architecture, pp. 181-190, 1992.

[8] J.M. Mulder, et. al., "An Area Model for On-Chip Memories and its Application", IEEE Journal of Solid-Sate Circuits, Vol. 26, No. 2, pp. 98-106, Feb. 1991.

[9] T. Wada et. al., "An analytical access time-model for on-chip cache memories", IEEE Journal of Solid - State Circuits, Vol. 27, pp. 1147-1156, August 1992.

[10] D. Nikolos and H. T. Vergos, "On the Yield of VLSI Processors with on-chip Cache", Proc. of EDCC-2, pp. 214-229, Oct. 2-4, 1996.

[11] D. Nikolos et. al., "Yield - Performance tradeoffs for VLSI processors with partially good two level caches", Proc. of DFT-96, pp. 53-57, Nov. 6-8, 1996.

[12] A. Srivastava et. al., "190 – MHz CMOS 4-Kbyte Pipelined Caches", Proc. of Int. Symp. On Circuits and Systems, pp. 1053 – 1056, 1995.

[13] M. D. Hill, "A Case for Direct-Mapped Caches", IEEE Computer, Vol. 21, No. 12, pp. 25-41, December 1988.

[14] Hoi-Jun Yoo, "A Study of Pipeline Architectures for High-Speed Synchronous DRAM's", IEEE Journal of Solid - State Circuits, Vol. 32, No.10, pp.1597-1603, October 1997.

[15] Y.Takai, et. al., "250 Mbyte/s Synchronous DRAM Using a 3-Stage-Pipelined Architecture", IEEE Journal of Solid - State Circuits, Vol. 29, No. 4, pp.426-430, April 1994.

[16] D. Schmitt-Landsiedel, et. al., "Pipeline Architecture for Fast CMOS Buffer RAM's", IEEE Journal of Solid - State Circuits, Vol. 25, No. 3, June 1990, pp. 741-747.

Synchronous to Asynchronous Conversion

A Case Study: The Blowfish Algorithm Implementation

J. Alcântara*, S. Salomão*°, E. Granja*, V. Alves*, F. França*
**COPPE – Federal University of Rio de Janeiro, POBox 68504, 21945-970 Rio, Brazil*
°IME – Military Institute of Engineering, Praça General Tibúrcio, 80, 22290-270 Rio, Brazil

Key words: Asynchronous logic, cryptography, VLSI.

Abstract: This work introduces a novel methodology that eases the synchronous to asynchronous conversion of existing digital circuits. Synchronous single-phased circuits may have its performance improved with the use of a variable rate clock generator if the conversion is done on some key circuits. This methodology is used to improve the performance of a soft-core implementation of the Blowfish cryptographic algorithm.

1. INTRODUCTION

This work introduces the use of a distributed and deterministic mechanism on the asynchronous timing of selected digital networks. Our approach shows that any digital system can be redesigned as a synchronous system with a varying clock rate. Hardware overhead is small and will not add significantly to the real state needs of the original system.

The conversion methodology is based on Scheduling by Edge Reversal (SER), a distributed algorithm [i] [ii] that is a potentially optimal solution to Dijkstra's *Dining Philosophers* problem [iii] under heavy load. SER gives a fair and distributed solution to this problem and has been used in the implementation of many distributed processing paradigms.

All information exchanged between any two distinct neighboring digital subcircuits must then be treated in a *mutual exclusion* style. SER can be used to control the asynchronous timing among parts of a target digital system and can be seen as a generalization of the handshake protocol. [iv] SER is not topology constrained. [v] Theoretical background for the distributed mechanism behavior is introduced in section 2. Section 3 presents the conversion methodology. Section 4 describes the Blowfish algorithm. Section 5 details the implementation and the performance achieved. Concluding remarks follows in section 6.

2. SCHEDULING BY EDGE REVERSAL – SER

Relevant aspects of the SER methodology are shown in this section. For an in-depth review of the theory involved the reader is referred to [vi] [vii] [viii]. The SER methodology [ix] is composed of three basic steps.

1. Representing the target digital circuit. Consider a neighborhood-constrained system composed of a set of processes and a set of shared atomic resources represented by a connected graph $G = (N, E)$ where N is the set of processes and E, the set of edges defining the interconnection topology. Each process is represented by one node. An edge is present between any two nodes if and only if these two processes share at least one resource.

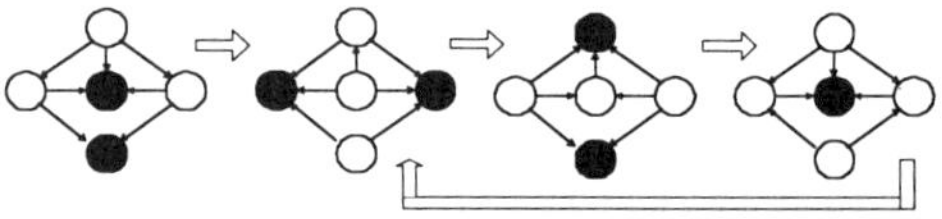

Figure 1. A graph G under SER, with m = 1 and p = 3

Figure 1 shows an example of G. SER works in the following way: starting from any acyclic orientation ω on G there is at least one *sink* node, i.e., a node that has all of its edges directed to itself. All sinks are allowed to operate while other nodes remain idle. After operation, a sink node will revert the orientation of its edges, becoming a *source*, and thus releasing resources to its neighbors. The whole process is then repeated for the new set of sinks. Note that the reversible edges in the definition graph G do not imply any data flow, they simply sign resource availability.

2. Construction of the asynchronous timing network. Let $G = (N,E)$ represent the target SER network defined to drive the timing of the target circuit C. Every node $N_i \in N$, is connected to N_{ipost}, where $N_{ipost} = N(i+1)$ mod N, if and only if the corresponding elements are neighbors.

3. Initialization of G. An appropriate acyclic orientation on G can be obtained if an edge between any two neighboring nodes N_i, $N_j \in N$ is oriented from N_i to N_j if $i > j$. Using this particular orientation node 0 is the first *sink* to happen under SER.

3. CONVERSION METHODOLOGY

In a synchronous digital system any connection topology may be defined, so that stages include but are not limited to pipelines. *Figure 2* shows an example of a possible target circuit composed of four functional stages, and its data communication. Stages having no interaction between the elements of the output vector have an algorithmic complexity on time of $O(1)$ and may include decoders, multiplexers and random logic. For circuits performing arithmetic operations, such as adders and multipliers, the mean complexity on time will be different from the worst case. An n-bit ripple carry adder, for instance, has a worst case complexity of $O(n)$ while having a mean complexity of only $O(\log_2 n)$. [x]

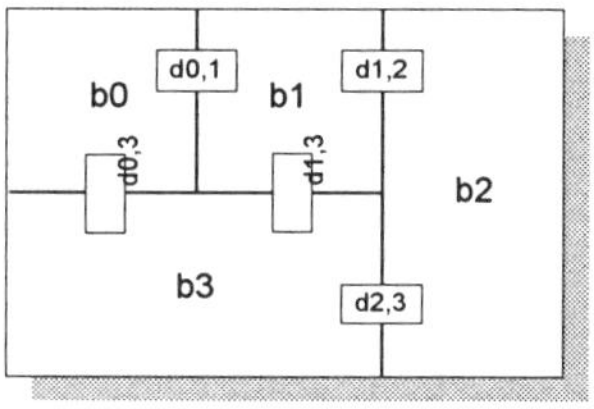

Figure 2. Synchronous stages and data communication of the target circuit

A great part of the system can then be treated just like any synchronous system and just a few stages will need to be redesigned. The result will be a faster clock for the majority of operations.

Definition. Synchronous to Asynchronous Conversion methodology.

(i) *Representation of the target digital circuit C.* The target digital circuit is represented by a graph $C=(B,D)$, where B is the set of its synchronous stages and D is the set of edges used to time their operation. A synchronizing node S is then added to the circuit as shown in *Figure 3*.

(ii) *Construction of the asynchronous timing network.* Let $G = (N,E)$ represent the target self-oscillatory SER network defined to drive the timing of the target circuit C. Nodes $N1_i$ control the length of the *high* clock level. For every stage in the circuit there is an edge between nodes $N0$ and $N1_i$. As each stage ends its operation, it reverses its edge towards $N0$. When all the stages are done, $N0$ becomes a sink and produces a clock transition from *high* to *low*. Then it operates for a very short time and prepares for a new cycle by signaling all the stages with an edge reversal.

(iii) *Reduction of the asynchronous timing network.* In order to save circuitry, and also to simplify the conversion process, only a few stages need to be asynchronously timed. Stages with a complexity $O(1)$ have a nearly constant t_d, no matter how its input vary. In this case, the stage i, for which $t_{d\min_i} > t_{d\max_j}$ for all $j ? i$, can be used to time itself and all the j stages. Other stages, according to the instantaneous values of their inputs, will have varying t_d. Those stages must be timed separately and ideally they will be the candidates for an asynchronous redesign.

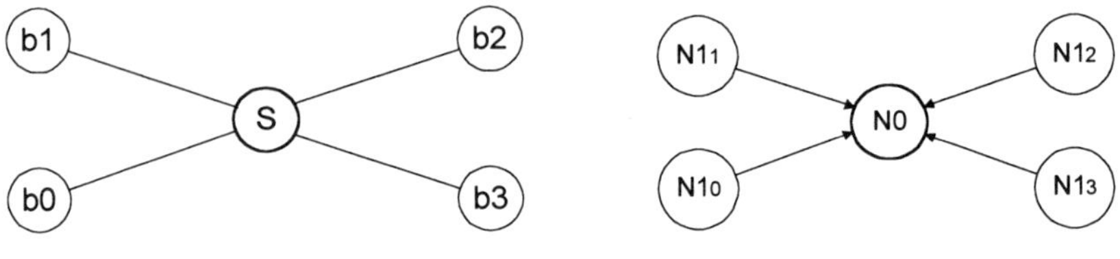

Figure 3. The target circuit as C=(B,D) and as G=(N,E)

(iv) *Initialization of G.* An appropriate acyclic orientation on *G* is easy to be obtained if we define N0 as the initial state. All the edges are initialized pointing to N0, which becomes a sink.

SER Hardware Implementation. The implementation comprises two basic circuits, namely the *Node Controller* and the *Edge Controller*, both operating on self-timed functional blocks. Note that each token represents the orientation of each edge on *G*.

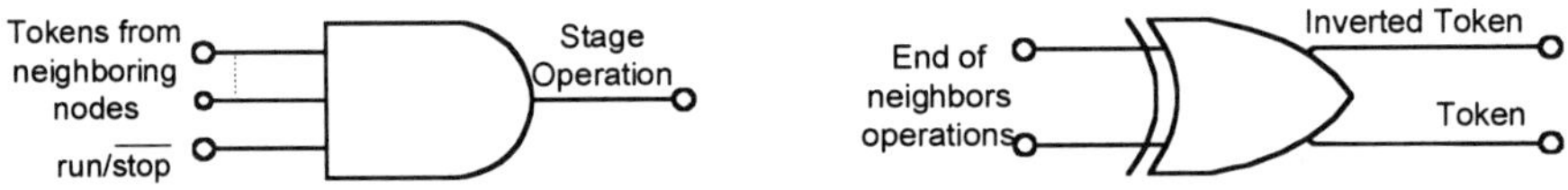

Figure 4. Schematic for the Node and Edge Controllers

Node Controllers simply evaluate the presence of tokens and, after detecting that the node has become a *sink*, they allow the target local circuit to operate. This implementation relies on a transition sensitive *end of operation* signal that is sent to all the neighboring Edge Controllers. Every time the node controller for nodes $N1_i$. is activated, all the stages start their operation. Node *N0* is simply a synchronizer. As shown in *Figure 4*, a node starts operation according to an AND gate that samples the incident tokens and as many inputs as the number of arrowheads pointing to the node.

Edge Controllers shown in *Figure 4* connect two nodes using a differential output XOR gate as a token generator. They also contain logic used to guarantee the initial acyclic orientation. Each Edge Controller will reset the token for the node that ended its operation and at the same time will send that token to a neighbor node.

4. CASE STUDY

Blowfish Cryptographic Algorithm. Blowfish[xii] [xi] is a symmetric variable-length key, 64-bit block cipher cryptographic algorithm based on Feistel Networks[xii] developed by Bruce Schneier [xii][xi]. Data encryption is performed through a 16-round Feistel network. Each round consists of a key-dependent permutation, and a key- and data-dependent substitution. All operations are done over 32-bit words.

Subkeys: These keys must be pre-computed[xii][xi]. They consist of a P-array with 18x32 subkeys and four S-boxes with 256x32 entries each.

Encryption: As stated previously, the Blowfish is a 16 round Feistel network, as seen in *Figure 5(a)*. The input is a 64-bit data element, X.

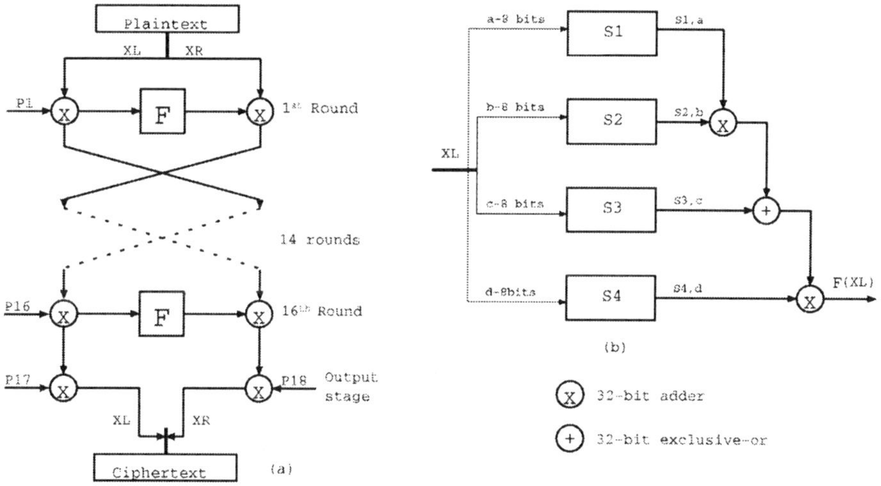

Figure 5. The structure of the Blowfish algorithm

The algorithm can be stated as:

Divide x into two 32-bit halves: X_L, X_R

For i = 1 to 16: $X_L = X_L \oplus P_i$; $X_R = F(X_L) \oplus X_R$; Swap X_L and X_R; end

$X_R = X_R$ XOR P_{17}; $X_L = X_L \oplus P_{18}$; Recombine X_L and X_R

Function F, see *Figure 5(b)*: Divide xL into four eight-bit quarters: a, b, c, and d. Then $F(X_L) = ((S_a^1 + S_b^2 \bmod 2^{32}) \oplus S_c^3) + S_d^4 \bmod 2^{32}$

Decryption: Same algorithm, with $P_1, P_2,..., P_{18}$ in the reverse order.

Implementations of Blowfish that require the fastest speeds should unroll the loop and ensure that all subkeys are stored in embedded RAMs [xii][xi]. Examining one round in the data flow shown in *Figure 5*, one can identify groups of data independent operations, pointing out temporal and spatial parallelism implemented as one round pipeline shown in *Figure 6*. FIFOs included into the pipeline hold data forwarded between non-adjacent stages.

S2A Conversion. The first step in the conversion methodology consists of circuit partitioning. An **S2A** system (*Figure* 7a), with 4 nodes corresponding to nodes $N1_i$ and the extra *N0* node was created. A single timing node N3 corresponds to the circuit blocks having a nearly constant delay. The delay corresponding to node N3 has been determined from critical path extraction. Nodes N1 and N2 correspond to asynchronous adders ADD1 and ADD2 respectively.

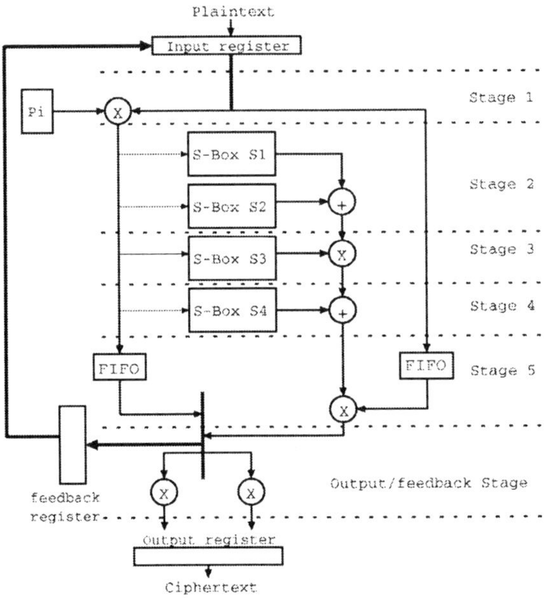

Figure 6. The pipeline structure

The clock signal is derived from node N0 and indicates that nodes N1, N2 and N3 have completed their respective task. This varying clock signal will time the pipeline described in *Figure 6.* The clock period for the converted system will vary from a minimum corresponding to the operation of N3 to the adder worst case just described. In fact, edge reversal is not instantaneous, but it imposes just a very low overhead to the cycle time.

Rules and constraints. S2A conversion methodology is very simple and powerful. However, some care must be taken regarding hardware implementation: Blocks with variable operating time have to be redesigned in order to provide an end of operation signal and end of operation signals must be free of glitches. Also, circuit outputs must be stable after end of operation until the end of the cycle. Thus, depending on the specific circuit implementation, it may be necessary to latch its outputs.

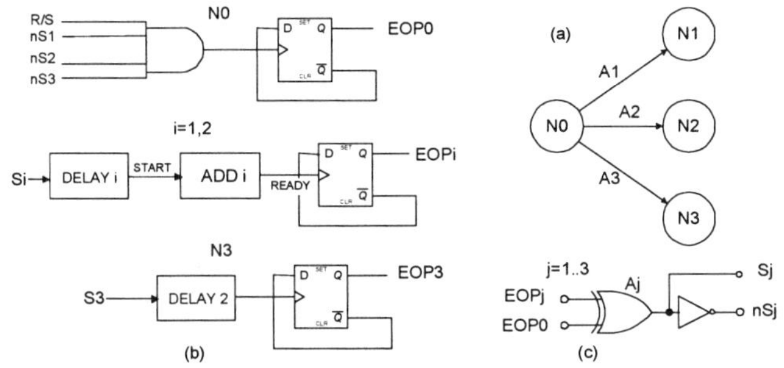

Figure 7. Asynchronous conversion

5. IMPLEMENTATION AND PERFORMANCE

Circuit description. Adder units were designed from 32_bit ripple carry adders. There are four 32_bit pipeline registers. The subkeys are stored in 6 embedded RAMs. Scheduling by Edge Reversal is used for correct circuit timing is depicted in *Figure 7*. Edges A1 to A3 are simply implemented with a XOR gate and an inverter. DELAY1, used in stages 2 and 4, represents the worst case timing for the S-boxes, while DELAY2 is the worst case obtained from critical path extraction for the remaining pipeline stages.

Table 1. Performance comparison

	Synchronous version	Asynchronous version
Number of transistors	43280	48771
Minimum cycle time	-	13ns
Average cycle time	20ns	16ns
Maximum cycle time	-	25ns
Average throughput	200Mbps	250Mbps

Performance. The asynchronous circuit obtained from S2A conversion was described in VHDL, synthesised with a 0.7μm CMOS standard cell library and simulated with Synopsis CAD tool. Variable clock generation and SER operation are shown in *Figure 8*. The circuit was validated by comparison with a C program executing the Blowfish algorithm. *Table 1* compares the performance of synchronous and asynchronous versions.

6. CONCLUSIONS

We have presented an alternative approach to the conversion of synchronous circuits into asynchronous. The use of SER's methodology simplified the conversion, as it allowed a viable implementation of a variable rate clock generator requiring very low area overhead. A pipelined implementation of the Blowfish cryptographic algorithm, its VHDL description and the resulting asynchronous circuit were simulated and synthesized, showing a reasonable performance improvement.

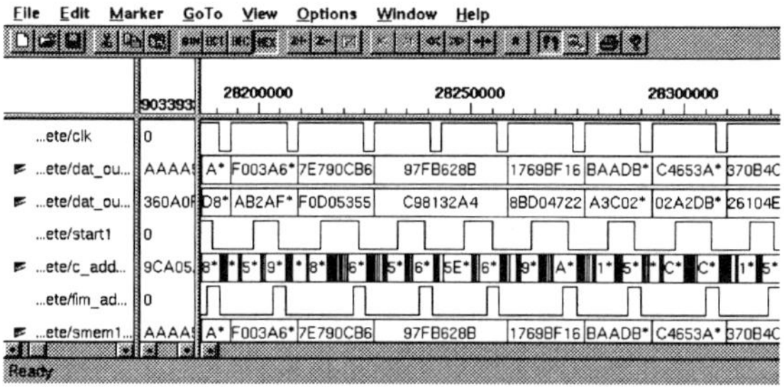

Figure 8. Simulation results

7. REFERENCES

[i] Barbosa, V., Gafni, E. "Concurrency in heavily loaded neighborhood-constrained systems". *ACM Transactions on Prog. Languages and Systems*, vol. 11, no. 4, Oct 1989.

[ii] Barbosa, V. C., *"An Introduction to Distributed Algorithms"*, MIT Press, 1996.

[iii] Dijkstra, E. "Cooperating Sequential Processes" In *Programming Languages*, F. Genuys, Academic Press, New York, 1968, pp 43-112.

[iv] Meng, T. *"Synchronization design for digital systems"* Kluwer Academic, 1991.

[v] Jou, S. and Chuang I. "Low-Power Globally Asynchronous Locally Synchronous Design Using Self-Timed Circuit Technology" *IEEE Int'l Symp. on Circuits and Systems.* Hong Kong, June 1997.

[vi] França F., Alves V., Granja E. – *Edge Reversal Based Asynchronous Timing Scheme* – International Symposyum on Circuits and Systems, Monterey, USA, 1998.

[vii] França F., Alves V., Granja E. – *A Multi-phase Asynchronous Timing Scheme* – SBCCI, Gramado, Brazil, 1997.

[viii] Alves V., França F., Granja E. – *A BIST Scheme for Asynchronous Logic* – Asian Test Symposium, Singapore, 1998.

[ix] França, F., Alves, V., Granja, E.- *Patent Req INPI BR.*

[x] Franklin, M., Pan T. "Performance Comparison of Asynchronous Adders" in *Int'l Symp. on Advanced Research in Asynchronous Circuits and Systems.* November 1994.

[xi] Schneier, B.- Description of a New Variable-Length Key, 64-bit Block Cipher (Blowfish) *Cambridge Security Workshop Proceedings,* Springer-Verlag, Dec 1993.

[xii] Schneier, B. – *Applied Cryptography* – John Wiley and Sons, New York, 1996.

Clock Distribution Strategy for IP-based Development

Rui L. Aguiar and Dinis M. Santos[1]
Universidade de Aveiro / Instituto de Telecomunicações
3810 AVEIRO, Tel:+351.34.370200, Fax:+351.34381128
email: ruilaa@ua.pt, dms@ua.pt

Key words: Clock Distribution, IP-based development, Controlled Delay Lines, Clock Jitter

Abstract: This paper presents a high-level clock distribution strategy for usage in a design-and-reuse environment. This strategy allows for controlled clock distribution across an arbitrary number of blocks through the usage of controlled delay lines. A new clock frequency multiplication structure optimised for this clock distribution strategy is finally proposed, since multifrequency clock support is highly desired.

1. INTRODUCTION

Increased integrated circuits sizes are presenting enlarged possibilities to circuit designers. Current estimates [1] indicate e.g.. microprocessor die sizes on the 4.5 cm^2 range in a three-year time frame, with nearly 100 million transistors. CMOS semiconductor technology has reached a point where the concepts of "system-in-a-chip" are becoming reality [2]. Systems previously mounted in printed circuit boards or large multi-chip modules are now to be implemented in a single chip. This trend brings increased technical benefits, as most of the input-output ports of previous circuits disappear, leading to lower power consumption with improved performance (often measured in working clock frequency).

Unfortunately, this same integration level is bringing increased problems to circuit designers, such as power and packaging issues [3]. Furthermore,

[1] This work has been sponsored partially by project Genclock, funded by Praxis XXI.

integrated system development is not possible due to increased design complexity. Circuit partitioning (and correspondent floorplanning) has been a reality for several years now: a team of designers separates the circuit in high-level blocks, and each block is developed independently. With increased system-in-a-chip developments, even this hierarchical development strategy has been showing its shortcomings. The complexity of CMOS circuits, with multiple functions, has grown in such an order that companies are finding hard to create development teams able to cover with realistic time-to-market constraints all the functions desired in the circuits.

Thus these systems are increasingly using "off-the-shelf" predesigned components, frequently coming from different sources. These components - often called Intellectual Property (IP) blocks, or Virtual Components (VC) - can be delivered in three different formats: soft IP, where a simple logical and synthesisable description is provided; firm IP, where this description is enhanced by some floorplanning and routing information; and hard IP, which provides complete physical models (e.g. layout) of the block. This approach is becoming popular even inside development teams, which are starting to handle their own blocks in a hard-IP format (and providing these blocks in this format to other groups for further design reuse) both due to company policy and to increasing needs for design reuse. Thus VCs will become progressively more common in future generations of chips.

This chip development strategy does provide fast development of complex circuits, especially as it fits naturally to current circuit partitioning methods. However, it brings a novel set of problems to clock distribution.

Even with fairly complex circuits, clock distribution is frequently handled as a problem by itself, covering the whole circuit development: it is common for a 2-million transistor circuit to be handled and analysed as a single clock distribution tree, resorting to modelling and complex simulation tools (frequently developed in-house). This clock distribution tree can present varying degrees of complexity [4], but nevertheless bounds on clock phase are usually provided [6] and maximum operating frequencies are derived for these circuits.

This approach is becoming inadequate as semiconductor technology evolves. For one-hand, as circuit sizes increase (and technology dimensions decrease), interconnect effects are becoming a dominant effect on the clock distribution tree. Furthermore, these are becoming (in percentage) more sensitive to process and environment parameter variation across the chip. Both effects increase the complexity of designing a clock distribution tree. On the other hand, no detailed control exists for the clock distribution inside reused hard-IP blocks. The system designer is effectively incapable of changing/controlling clock distribution in these blocks. This may also happen with most firm blocks, where changes in its clock distribution may affect the block performance (it is usually recommended that detailed clocking information should be delivered with firm IP [2]). Thus an increasing number of IP-blocks with its own clock distribution tree will be used as a “black-box” block by design teams of future chips.

Two questions arise naturally in this type of environment: i) how to distribute the clock to these blocks such that with minimum jitter and skew is achieved, and ii) how to handle the effects of these hard-IP blocks in terms of jitter/skew.

This paper presents a clock distribution strategy that covers these two issues. We present a high-level clock distribution architecture independent of the individual clock distribution trees in each block in Section 2. This strategy allows for precise multi-point synchronization. In Section 3, IP blocks are characterized in terms of "clock uncertainty", and this measure can also be applied to the clock distribution mechanism being used. From these values, maximum clock uncertainty bounds and simple design specifications can be achieved. This strategy can be applied in a hierarchical approach, allowing for further reuse of the circuit. Section 4 expands this approach in order to provide for local frequency multiplication, due to power consumption considerations.

2. HIGH-LEVEL CLOCK DISTRIBUTION

2.1 Clock Distribution Model

The traditional design paradigm to date is the synchronous design. This is a deterministic approach, robust and easy to implement. It is based on a state-machine concept, where states are represented by register output values, and state transitions are controlled by the results of combinatorial logic. Latching data into a new state is controlled by a set of clock signals.

Although asynchronous design has gained popularity in the last years [7], it is not the focus of this paper, as currently self-timed methods are not often used in complex circuits in a design-and-reuse environment. Self-timed approaches present some properties that seem to hold much promise for future complex chips (namely low power and heterogeneous timing capabilities) but design methodologies and supporting tools still need to evolve in order for their widespread usage in very complex systems.

Thus our synchronous reference model assumes synchronization units (I/O registers) at the input and output of every IP-block. (If synchronization points are not available, it is not possible to evaluate clock delay without detailed knowledge of circuit functions and implementation). A multi-point clock distribution tree then drives these blocks (Fig. 1).

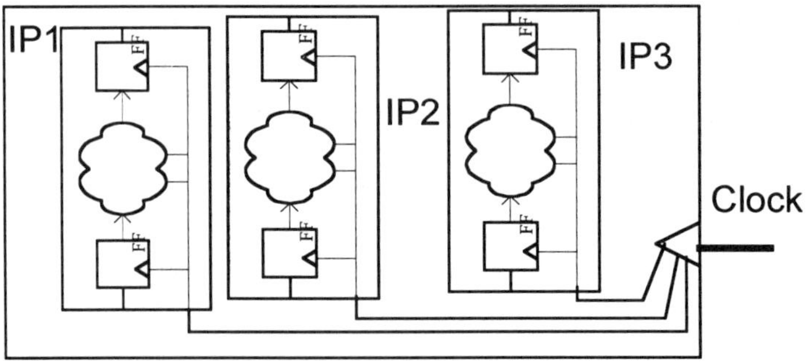

Figure 1. Reference model for clock distribution, with three hard-IP components

Synchronous design methodology forces the time reference (the clock) to be constrained by the slowest path in the whole circuit [6]. The critical path delay (T_c) can be related to the smallest clock period possible ($Tclk_{min}$) by:

$$Tclk_{min} - \delta_{max} = Tgate_{max} + Tset_{max} + Tp_{max} = T_c \qquad (1)$$

where $Tgate_{max}$ is the maximum propagation delay in the critical path gates, $Tset_{max}$ is the maximum setup and hold times of the registers, and Tp_{max} is the maximum signal propagation time in the interconnects of the critical path (which can be estimated by several propagation models); δ_{max} is the modulus of maximum clock uncertainty (due to jitter and/or skew). Although more complex clock strategies may be used [4, 6], such as e.g. cycle borrowing, two-phase clocking or pipelining, their applicability is either infeasible at the top design partitioning level we are considering, or can be readily incorporated in our discussion.

Clock phase differences between registers inside each IP block are neglected at this stage. All blocks are considered synchronization domains, with null clock phase difference between registers (The next section will remove this assumption). Then the problem can be stated as the delivery of a clock signal at the input of each block in such a way as to assure the same phase reference at all synchronization points, or, using expression (1) to minimize clock uncertainty δ_{max}. Note that as the clock uncertainty increases, clock frequency will have to be decreased.

Two different (but interrelated) issues affect clock uncertainty: skew and jitter. Skew can be defined as the deterministic delay difference between two supposedly synchronous signals, caused by such effects as different interconnect line lengths or different driver strengths. Jitter is the random variation of clock phase around its average point, caused by effects such as circuit noise or (dynamic) load variations on clock propagation lines.

Both effects are disadvantageous for clock synchronization purposes, and they represent only different aspects of clock uncertainty. For clock synchronization purposes both perfect timing between different synchronization points and perfect regularity in clock period is sought (unless complex clock cycle management is deliberately used).

Traditional clock distribution resorts to the design of a clock tree network [4, 5], or a two level tree network, with hierarchical clock buffering schemes and local skew equalization [6] with careful interconnect routing. Nevertheless, these approaches suffer from significant process and environment problems, even when they theoretically allow for perfect skew cancellation.

2.2 Clock Feedback Design Philosophy

Clock distribution across several IP-based blocks should minimize both skew and jitter, as no information on the content of each block may be available. Traditional approaches look at clock distribution essentially as a passive problem. The clock is distributed from a source towards the sinks, without any further control. Clock networks are designed in such a way that minimize skew, but no dynamic control is done over this parameter. In this section we propose a radically different way of approaching these ideas, where feedback is an important part of the clock tree behaviour. The clock is still distributed to a set of clock drivers (sinks), but this distribution now uses feedback to set remotely the clock phase at a specific point. Several synchronization domains can be set in phase, regardless both of the number or characteristics of the domains (IP-blocks) and of interconnection lengths. The concept is simple and several clock distribution implementations have already presented some sort of feedback [8, 9, 10].

This approach requires the interconnect system (which may comprise both passive interconnect lines and driving buffers) to incorporate lines with controlled Delay Elements (DEs). Several structures for feedback can be used, according to the number of blocks to control, the size of the interconnections and the clock frequency. In our clocking strategy we adhere to a variant of a generic approach, initially proposed in [8].

Figure 2 shows a three-block active clock distribution network, where three points (Φ_1, Φ_2, Φ_3) will present the same phase, regardless of the number of DE in each interconnect and on the total interconnect delay - if the control blocks are able synchronize properly. Each one of these synchronization points can now act as a clock reference point for an IP-block, as in the situation depicted in Fig. 1. The clock network uses controlled delay lines with an even number of DEs, and establishes a loop with a symmetrical path between the clock synchronization reference and each synchronization point. The synchronized points are located at the middle of these loops and will be in phase if the DEs are controlled to provide a line delay with an odd-or-even [8] number of clock periods in the total interconnection. Phase accuracy at the synchronized points depends mainly on the quality of the control blocks, on their matching, and on the matching of both clock interconnect lines for each block [8]. Control blocks should be placed together, as illustrated in Fig. 2. Both clock interconnect lines (to and from) between the clock reference point and each of the synchronization domains should be routed through the same routing channel, as the phase accuracy at the synchronization points depends on the symmetry of these two interconnections.

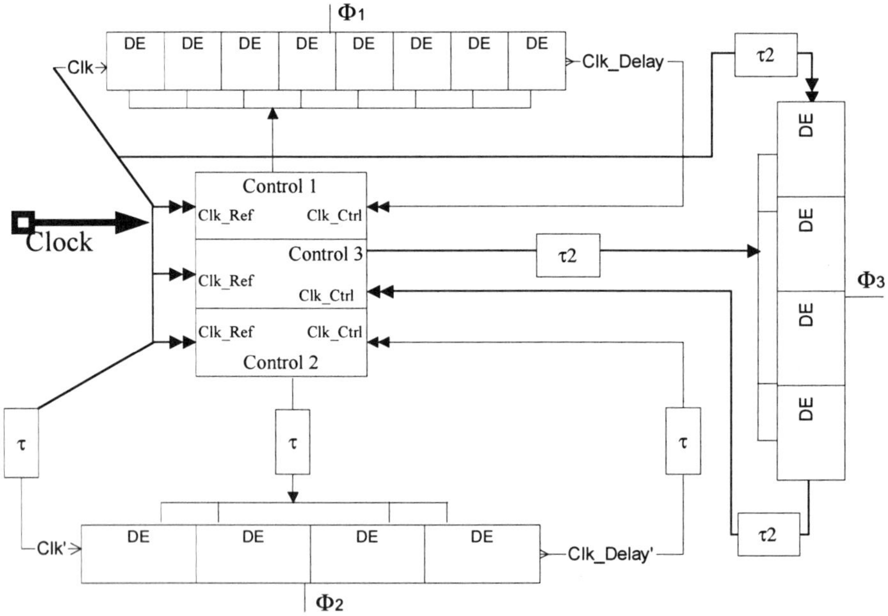

Figure 2. Three-point active clock distribution network. "Clock" is the clock reference point, and Φ_1, Φ_2, Φ_3 are the synchronised points

2.3 Controlling Delay Lines

Some sort of phase comparison and control signal is required for setting the delay of the active line. This control should be done with digital techniques, which are more resilient to the inevitable substrate and power supply noises of high frequency digital systems. Then, using an active clock distribution, total clock interconnect delay can be expressed as:

$$\Delta_{clk} = T_{DE} + T_{clk\_prop} = M.(k.\delta_{DE} + T_{DEmin}) + T_{prop} \quad (2)$$

with Δ_{clk} as the total delay in the line, T_{DE} the (controlled) total delay in each DE, and T_{prop} the delay in the connection lines; M is the number of DEs in the line, T_{DEmin} is the minimum delay of the DEs and δ_{DE} is the minimum (controlled) delay step possible; *k* is the (integer) control variable. From (1), if no clock uncertainty exists then, at maximum operating frequency we get:

$$\Delta_{clk} = n * (Tgate_{max} + Tset_{max} + Tprop_{max}) \quad (3)$$

but this would require stringent (and generically unattainable) relationships between *M, k* and δ_{DE}. (In a first approximation, the integer value of *n* is not relevant.) Thus, using any digital control in this type of system will *require* a given clock uncertainty, as the total line delay will not be an exact multiple of the *maximum possible* clock period. This clock uncertainty is not equal to this minimum delay step δ_{DE}, but depends on other design parameters.

In practice the clock uncertainty Δ_{tg} will be the design target specification. It will be chosen as the maximum value that is able to fulfil the system design specifications for all parameter (viz. process and temperature) conditions.

The following methodology can be used for the design of such an active clock distribution system (the numbers achieved by this process provide design targets to be exceeded by the synchronization system):

a) Define the target clock uncertainty $\Delta_{tg}(= M.\delta_{DE})$; this will be related with the quality of the phase comparator in the control blocks.
b) Evaluate maximum (Δ_{clkMax}) and minimum (Δ_{clkmin}) possible time delays in the interconnect system, in function of interconnect size, number of clock buffers required to maintain sharp clock transitions and parameter variation.
c) Calculate the maximum control value, the number of DEs and the minimum delay step required with equation $\Delta_{clkMax} - \Delta_{clkmin} = \Delta_{tg}.k_{max}$ (where k_{max} is the maximum control value). The structure chosen for the DE will place bounds on "reasonable" values of *M* and δ_{DE}.
d) Evaluate the synchronization period (*n*) using (2) and (3) above, and keeping in attention that all control blocks should operate on the same parity (*n* is required odd or even for all blocks) [8].

Thus we will have a control system that can assure a maximum clock uncertainty in the phase relationship of the clock delivered at each IP-block. Unfortunately, clock uncertainly will be further increased by noise in the controlled delay line (including its delay elements). Several generic analyses have been made [11, 12], mostly for voltage controlled delay lines. It is often possible for the active control circuitry to be designed such that the uncertainty introduced by noise is much smaller than the uncertainty inherent to the digital delay control used.

3. IP BLOCK CHARACTERIZATION

In the previous section we discussed how to establish a multi-point clock distribution network with a given clock uncertainty, assuming that each controlled block was a perfect synchronization domain. In reality a hard-IP block may have an arbitrary complexity (e.g. a microprocessor), and thus may have its own uncertainties related with its internal clock distribution.

Thus a hard-IP block should be characterized by two parameters: a) its critical path delay; b) the maximum clock uncertainty that may appear between the I/O registers and the clock reference point, in all possible operation parameters. (Note that in our synchronous model, both inputs and outputs of IP-blocks are directly connected to registers: only phase differences between its clock reference point and the input and output registers are then relevant.)

The critical path delay is naturally required in order to quantify the maximum operating frequency. The critical path delay will be the maximum value of the critical paths of each IP-block ($T_c(i)$) and of the interconnection logic block ($T_c(int)$ (this block includes all logic implemented for the interconnection of the several IP-blocks, and is usually specifically designed by the global system development team):

$$T_c = \max\{T_c(1), T_c(2), \ldots T_c(n), T_c(int)\} \qquad (4)$$

The critical path delay of the interconnection block has to consider the clock uncertainties in each IP block ($\delta_{max}(i)$). Thus:

$$T_c(int) = Tgate_{max} + Tset_{max} + Tprop_{max} + \max\{\delta_{max}(1),..\delta_{max}(n)\} \qquad (5)$$

Current development tools already supply the critical path of a block, and advanced clock development tools provide for bounds on clock distribution phase and on clock propagation delay across interconnect lines (e.g. [6]). These tools are able to immediately provide the values above referred.

This approach creates a hierarchical development structure for clock distribution. Each block handles its clock distribution network in accordance with the techniques that finds more appropriate. For further design usage, the IP-block displays two properties (critical path and I/O clock uncertainty) that are used for the implementation of the clock distribution network of any circuit that uses this block. These values are then recursively used in the evaluation of the clock characteristics of this new circuit - which can become a new virtual component (IP-block) by itself.

4. MULTIPLE CLOCK FREQUENCIES

Previous sections have discussed the characterization of each hard-IP block in terms of clock performance, and a clock distribution strategy able to achieve a given clock uncertainty. In those sections the idea of a single clock frequency across the whole system was present.

However, clock distribution is power expensive. This is more significant as clock frequencies and interconnection distances increase. Future clock frequency estimates present quite different values, depending on interconnection distance [1]. Trying to distribute the same high-speed frequency across increasing distances brings propagation, power consumption and noise problems.

Moreover, there is no intrinsic need for the whole system to be driven by the maximum frequency. Several sub-blocks could use higher frequencies, but for power consumption and noise considerations, each sub-block should be attacked with as small a frequency as possible to achieve global frequency targets [13]. Global system synchronization would still be maintained at the maximum required clock frequency (which could be much lower than the maximum clock frequency in some of the IP-blocks). This transposes traditional digital development frameworks (e.g. [14]) for the "systems-in-a-chip" environment.

The clock strategy discussed in the previous sections allows for the inclusion of independent clock generators in each IP-block. These clock generators could provide for clock multiplication internally, and thus allow for much smaller clock frequencies to be distributed across the whole circuit. Both clock frequency and phase adjustment has to be performed in order for minimum clock uncertainty to be achieved in the IP-block.

Our proposed clock distribution strategy uses a novel synchronization method, depicted in Fig. 3. The internal clock generator is coupled to the clock reference point (Φ_1), as a phase detector controls the output of the clock generator (Φ'_1) in function of this point. The centralized control block adjusts the overall clock phase (as discussed before), controlling several DEs in the synchronization loop. However, some of the DEs in the loop are not under the control of this unit. The local phase control block assures that the multiplied local clock is "in phase" with the global clock signal. It does this by changing the delay in some DEs of the synchronization loop, in such a way that it compensates the timing delay of the clock multiplication block. For the synchronization loop, these locally controlled DEs act as another interconnect delay. As long as this phase control affects symmetrically the synchronization loop, the middle point (Φ_1) still keeps its phase reference, and will be the same across multiple similar connections. Although frequency multiplication is done, the internal clock is "in phase" with the external distributed clock (i.e. the local high frequency clock has transitions at the same instants than the slower global clock at the reference point). All phase errors depend mainly on the performance of the phase control systems, and not directly on parameter variation. Thus this structure, associated with the clock distribution mechanism already discussed, allows the synchronization of multiple domains with multiple clock frequencies, while a precise common time reference is still assured across all domains.

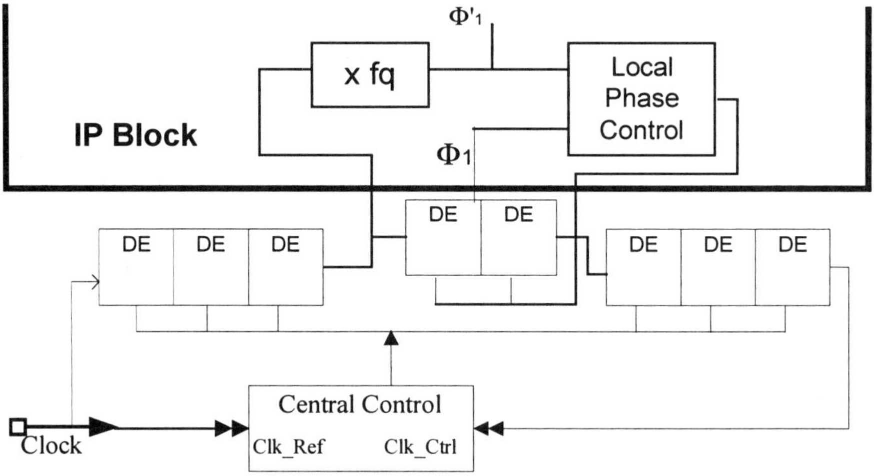

Figure 3. Local clock multiplication embedded in the synchronisation loop

In terms of expressions (4) and (5), the introduction of multiple clock frequencies does not present major changes, as simple additive processes may approximate jitter mechanisms on DLLs (for typical design parameters [15]). The major difference is that the clock uncertainty caused by any frequency multiplication has to be considered in the calculation of the clock uncertainty of that IP-block. However, this effect should only be evaluated at the global clock transition instants, when it should be small and dependent mainly on the IP-block local phase detector characteristics.

5. CONCLUSIONS

Increased usage of Virtual Components (or IP-blocks) is a consequence of the raise in complexity of current CMOS microcircuits. These components are often provided in a hard-IP format, with little information on its internal clock distribution scheme. This increased complexity is furthermore being translated as greater circuit dimensions, while technology features decrease.

All these aspects bring new problems to traditional clock distribution strategies, as large blocks are placed inside a circuit without any control of its clock distribution while, simultaneously, interconnection lengths increase.

We have presented a clock distribution strategy adequate to these new environment characteristics. This strategy uses active delay lines as top-level clock distribution, assuring bounds on phase uncertainties for each IP-block. Furthermore, two simple parameters are associated to each block (critical path and clock uncertainty). These parameters present working bounds for system engineers to incorporate these components in new systems, creating a hierarchical development strategy of arbitrary complexity.

This strategy is flexible enough for allowing multi-frequency operation in the same system, reducing power consumption. We extended this strategy with a frequency multiplication method optimised for this clocking methodology. This method adheres to a global phase reference framework while presenting an internal frequency multiple of the reference clock.

6. REFERENCES

[1] Semiconductor Industry Association, *The International Technology Roadmap for Semiconductors, 1998 Update*, 1998.

[2] VSI Alliance, *Architecture Document, version 1.0,* 1997.

[3] Dennis Herrel, "Power to the package", *IEEE Spectrum*, n. 6, July 1999, pp.46-53.

[4] Eby G. Friedman, " Clock Distribution Networks in VLSI Circuits and Systems", *Clock Distribution Networks in VLSI Circuits and Systems*, IEEE Press, 1995, pag.1-36.

[5] H. Fair, D. Bailey, "Clocking Design and Analysis of a 600MHz Alpha Microprocessor", *IEEE Int. Solid-State Circ. Conf. ISSCC'98*, pp.398-399, 1998.

[6] Jon Rabaey, "Digital Integrated Circuits, a Design Perspective", Prentice Hall, 1996.

[7] C.H.Van Berkel, M. B. Josephs, S.M. Nowick, "Scanning the Technology – Applications of Asynchronous Circuit", *Proceedings of the IEEE*, vol. 87, n. 2, Feb. 1999, pp.223-233.

[8] Rui L. Aguiar, Dinis Santos, "Wide-Area Clock Distribution Using Controlled Delay Lines", *5th IEEE Int. Conf. on Electr., Circ. and Syst., ICECS'98*, pp.63-66, Sep. 98.

[9] G. Geannopoulos, X. Dai, "An Adaptive Digital Deskewing Circuit for Clock Distribution Networks", *IEEE Int. Solid-State Circ. Conf., ISSCC'98*, pp. 400-401, 1998.

[10] H. Sutoh, and K. Yamakoshi, "A Clock Distribution Technique with an Automatic Skew Compensation Circuit", *IEICE Trans. on Electronics*, E81-C, n. 2, Feb. 98.

[11] B. Kim, T.C. Weingandt, P.R. Gray, "PLL/DLL System Noise Analysis for Low Jitter Clock Synthesizer Design", *IEEE Int. Symp. on Circ. and Syst., ISCAS'94,* 1994.

[12] Rui L. Aguiar, Dinis Santos, "Modelling Charge-pump Digital Delay Locked Loops", *6th IEEE Intl Conf. on Electr., Circ. and Syst., ICECS'99*, Cyprus, Sep. 99.

[13] A.P. Chandrakasan, S. Sheng, R. W. Brodersen, "Low-Power CMOS Digital Design", *IEEE Journal of Solid-State Circuits*, vol. 27, n. 4, pp.473-483, Apr 1992.

[14] F. Anceau, "A Synchronous Approach for Clocking VLSI Systems", *IEEE Journal of Solid-State Circuits*, vol. 17, n. 1, Feb 1982.

[15] Rui L. Aguiar, Dinis Santos, "Simulation and Modelling of Charge-pump Digital DLLs", *42nd IEEE MidWest Symp. on Circ. and Syst., MWSCAS'99*, Las Cruces, Aug. 99.

An Architectural and Circuit-Level Approach to Improving the Energy Efficiency of Microprocessor Memory Structures

David H. Albonesi

Dept. of Electrical and Computer Engineering

University of Rochester

*Rochester, NY 14627-0231 USA**

albonesi@ece.rochester.edu

Abstract We present a combined architectural and circuit technique for reducing the energy dissipation of microprocessor memory structures. This approach exploits the subarray partitioning of high speed memories and varying application requirements to dynamically disable partitions during appropriate execution periods. When applied to 4-way set associative caches, trading off a 2% performance degradation yields a combined 40% reduction in L1 Dcache and L2 cache energy dissipation.

1. INTRODUCTION

The continuing microprocessor performance gains afforded by advances in semiconductor technology have come at the cost of increased power consumption. Each new high performance microprocessor generation brings additional on-chip functionality, and thus an increase in switching capacitance, as well as increased clock speeds over the previous generation. For example, both transistor count and clock speed have roughly doubled in the three years separating the Alpha 21164 microprocessor [6, 11] and the recently introduced Alpha 21264 [14, 15]. Because the dynamic power, which is currently the dominant contributor to power consumption in high-speed CMOS circuits, is linearly related to each of these factors, it is extremely difficult for circuit-level techniques (such as voltage scaling) to singlehandedly keep power consumption from increasing under these circumstances. Indeed, the power consumption of the 21264 is 1.5-2 times that of the most recent 21164 version despite the fact that the voltage has been reduced from 3.3V to 2.2V [8, 9]. Similarly, while the UltraSparc I microprocessor [20], which was introduced in 1995, dissipated 28W at 167MHz, the forthcoming UltraSparc III design [12] is estimated to

*This research is supported by the National Science Foundation under CAREER Award CCR-9701915 and grant CCR-9811929.

dissipate 70W at 600MHz. For these reasons, in order to prevent microprocessor designers from being limited by power and energy dissipation, especially in desktop and portable environments where heat dissipation and battery life are critical constraints, it is necessary to devise architectural techniques for low power that complement circuit-level approaches.

Because of the increasing usage of microprocessor die area for on-chip caches, several architectural-level approaches to reducing *energy dissipation* in these structures have been devised. These techniques seek to reduce the amount of switching activity within the hardware for a given workload. The SA-110 embedded microprocessor [19] uses 32-way associative 16KB L1 I and Dcaches, each of which is divided into 16 fully associative subarrays. With this scheme, only one-eighth of the cache is enabled for each access which considerably reduces dynamic power. With a 160MHz target clock frequency, the SA-110 designers were able to maintain a one cycle cache latency with this degree of associativity. However, with larger on-chip caches and frequencies in the GHz range, such a solution would likely increase L1 cache latencies. The resulting increase in the branch mispredict penalty and in load latency would significantly degrade performance for many applications.

A similar approach is used to reduce power consumption in the 21164 microprocessor's 96KB, 3-way set associative L2 cache, whose data portion is split into 24 banks. The tag lookup and data access are performed in series rather than in parallel as with a conventional cache. This allows for predecoding and selection of 6 of the 24 data banks in parallel with tag access, and final access of only the 2 banks associated with the selected way[1] (as determined by the tag hit logic). Because only a small fraction of the total L2 cache is enabled on each access, the dynamic power savings is considerable, estimated at 10W [11]. However, the serial tag-data access of this technique increases cache latency as with the SA-110 approach, and thus this approach is limited to on-chip memory structures where overall performance is relatively insensitive to the latency of the structure.

Several other approaches, such as the filter cache [16] and the L-Cache [5] have been proposed for reducing the switching activity. However, each of these techniques significantly alters the on-chip memory design in order to improve energy efficiency. This ultimately results in a non-trivial performance degradation or other limitations. In addition, these schemes only address caches whereas non-trivial amounts of energy may also be dissipated in other memory structures such as Translation Lookaside Buffers (TLBs), register files, branch predictors, and instruction queues. In this paper, we introduce an alternative and more general approach: that of leveraging the subarray partitioning that is

[1] In this paper, we use the term *set* to refer to the cache block(s) pointed to by the index part of the address, and the term *way* to refer to one of the n sections in an n-way set associative cache.

often present in large on-chip memories for speed reasons in order to provide the ability to *selectively enable* particular partitions of the memory. With appropriate architectural support, these partitions can be *dynamically* enabled in an *on-demand* fashion. That is, the full unmodified memory structure is enabled when necessary to obtain good application performance, but only a subset is enabled during periods where application requirements are more modest. Such a *performance on demand* approach exploits that fact that hardware demands vary from application to application, and may also vary during the execution of an individual application [2, 23]. With the ability to enable only the precise amount of on-chip memory needed to meet performance requirements, a significant amount of energy savings can be realized in the on-chip memories, and thus in the overall microprocessor. However, unlike many other previous approaches to energy savings, this technique delivers identical performance to a conventional memory design when required by the application.

In the rest of this paper, we explore this concept in further detail. In the next section, we examine the subarray partitioning that is often necessary to minimize the access time of memory structures, and how this partitioning can be exploited to tailor the memory organization to application requirements. We then discuss in Section 3 our approach for selectively enabling partitions, including schemes for properly handling information in a disabled partition. In Section 4 we explore the application of this technique to the L1 Dcache, and evaluate the energy savings that can be realized when some performance degradation can be traded off for reduced switching activity. Finally, we conclude and present future work in Section 5.

2. THE ORGANIZATION OF ON-CHIP MEMORIES

In this section, we use a modified version of the Cacti cache cycle time model [22] to explore the partitioning of on-chip memories that is necessary to optimize access time. Cacti is an analytical delay model that evaluates in detail each component of the data and tag delay paths. In addition, Cacti includes six layout parameters that allow for partitioning of single tag and data arrays into multiple subarrays. The parameters *Ndwl* and *Ntwl* refer to the number of times the wordlines are segmented for the data and tag arrays, respectively, while *Ndbl* and *Ntbl* are the corresponding bitline parameters [21]. The parameters *Nspd* and *Ntspd* refer to the number of sets that are mapped to the same wordline for the data and tag arrays, respectively [22].

Figure 1 shows an example of a 4-way set associative cache with *Ndwl* = 4 and *Ntbl* = 2 and all other *N* parameters equal to one. The data array is partitioned into four subarrays, each with its own decoder and one-quarter the sense amps of a single data array. Here, each wordline is roughly one-quarter the length of that in a single array. Two tag subarrays are formed by segmenting

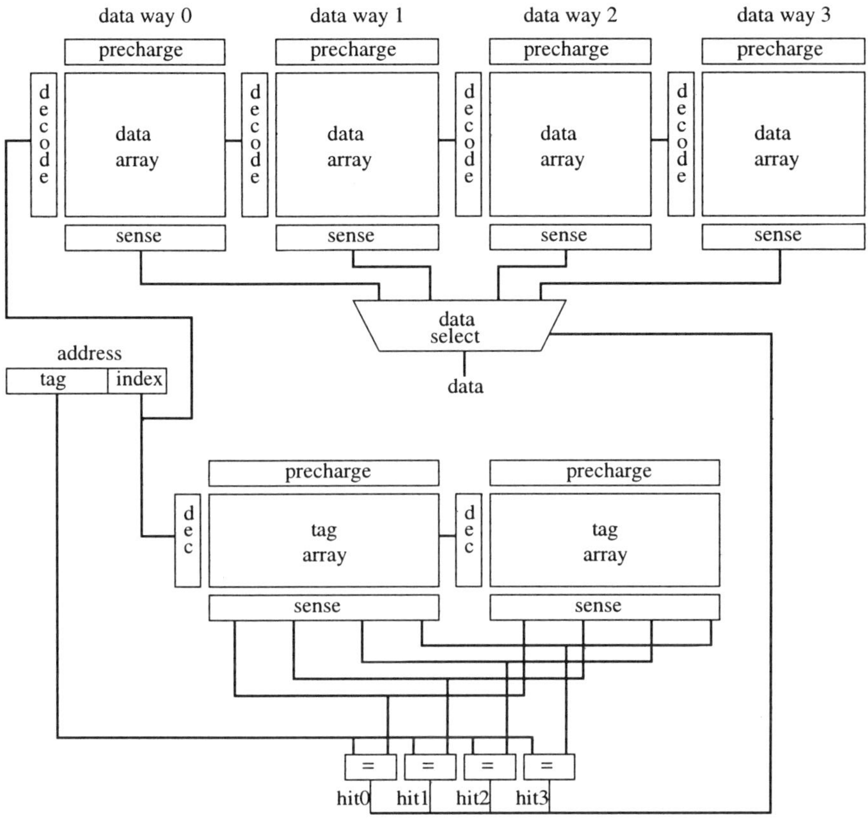

Figure 1 A 4-way set associative cache with *Ndwl* = 4, *Ntbl* = 2 and all other *N* parameters equal to one.

the bitlines, resulting in a halving of the decoder width but a doubling of the number of sense amps relative to a single tag array. Note that by segmenting the tag array, an extra output selector delay is incurred. This can be implemented by activating only one of the two sets of tag sense amps associated with the same column during each access [21].

We used Cacti to explore the partitioning necessary to minimize the access time of two different on-chip memories: TLBs and caches. The results are shown in Table 1, which displays the *N* parameters that produce the fastest access time for various organizations. Each *N* parameter is limited to a maximum value of eight to avoid unreasonable aspect ratios. Note that in all cases, a partitioning of both the data and tag arrays into multiple subarrays is necessary to minimize access time. For the TLBs, the data wordlines need to be segmented

Table 1 Optimal *N* parameters for on-chip TLB and cache structures. The block size is 16 bytes for TLBs and 32 bytes for caches.

Structure	Organization		Optimal N Parameters					
	Size	Associativity	Ndwl	Ndbl	Nspd	Ntwl	Ntbl	Ntspd
TLB	1KB	4	8	1	1	2	1	1
	2KB	4	8	1	1	2	1	1
	4KB	4	8	1	1	2	1	1
Cache	8KB	1	2	4	1	1	2	2
		2	4	2	1	1	2	1
		4	8	1	1	2	1	1
	16KB	1	2	4	1	1	2	2
		2	4	2	1	1	2	2
		4	4	2	1	1	2	1
	32KB	1	1	8	1	1	2	4
		2	8	1	1	1	2	2
		4	4	2	1	1	2	1
	64KB	1	1	8	1	1	2	4
		2	4	2	1	1	2	2
		4	4	2	1	1	2	1

eight times, while a combination of data wordline and bitline segmentation is often required for optimal cache performance.

Although there are several aspects to Cacti that limit the applicability of these results, on-chip caches are often partitioned in practice. For example, the 32KB 2-way set associative and 2-way banked L1 Dcache in the R10000 microprocessor is partitioned into four subarrays [1, 24], as is each of the two 512KB data banks of the 1MB 4-way set associative L1 Dcache of the HP PA-8500 microprocessor [18].

Note also that although we did not evaluate other on-chip memory structures, many of these need to be similarly partitioned for speed-optimality. For example, branch prediction tables are typically tagless structures whose size has increased significantly in recent microprocessors. For example, while the Alpha 21164 employed a single 2K$\times$2 branch direction predictor table, the Alpha 21264 implements four tables of sizes 1K$\times$10, 1K$\times$3, 4K$\times$2, and 4K$\times$2. The latter predictor is 3.625KB in size, over seven times that in the 21164, and may consume non-trivial amounts of energy. Each of the predictor tables when implemented as a single array would be a very long and narrow memory structure, and therefore, the bitline delay would be proportionally much longer than the other delay components. Thus, segmentation of the bitlines would likely reduce the access time of these large tables. In the next section, we discuss the mechanisms necessary to exploit this partitioning such that the enabling and disabling of memory partitions can be manipulated under software control.

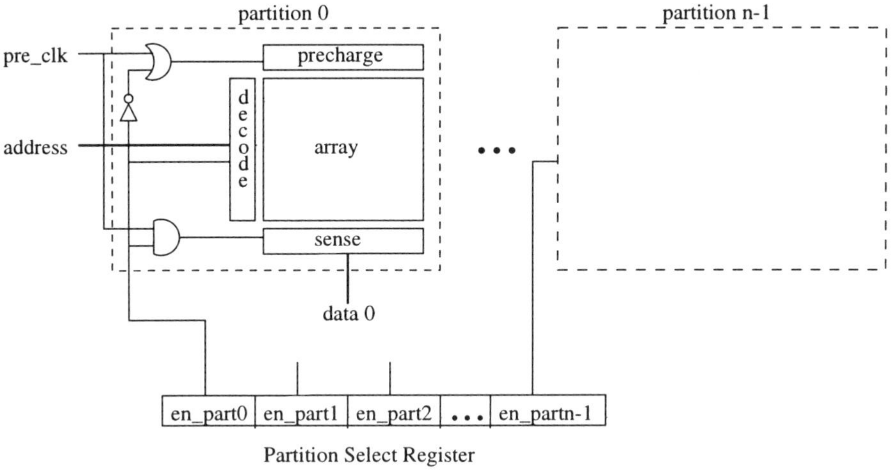

Figure 2 An on-chip organized as *n* partitions, each of which can be enabled/disabled via the Partition Select Register.

3. MECHANISMS FOR DISABLING MEMORY PARTITIONS

Figure 2 is an overall diagram showing a memory structure which has been partitioned into subarrays for speed purposes, and the small amount of additional circuitry needed to selectively disable each memory partition, which is comprised of one or more subarrays. The gating logic in this diagram is based on that used to selectively activate data subarrays in the Alpha 21164 L2 cache [6]. Each bit in the Partition Select Register (PSR) controls the enabling of one of the *n* memory partitions. If a particular bit is set to zero, then that partition is not precharged, no word lines are selected, and its sense amps are prevented from firing. Thus, no switching activity ensues and thus this partition dissipates essentially no dynamic power.

A PSR may concatenate partition control bits for several on-chip memory structures, and it may be necessary to include several PSRs in the processor. Each PSR is software readable and writable through special instructions. Several possible sources are possible for control of the PSRs including the compiler, the runtime system, the core operating system (which also needs to save the PSRs as part of the process state on context switches), special priviledged routines such as those that manage the TLB on several processors, or a continuous profiling and optimization system [4, 25]. Another approach is to have hardware that dynamically detects when partitions can be disabled. But this hardware would be complex and dissipate energy, thereby mitigating some of the benefits of our approach. These architectural issues are beyond the scope of this paper and are further discussed in [3].

Note that our ability to exploit subarray partitioning is dependent on how the partitioning divides the information stored in the memory. For example, if only the wordlines of an instruction queue are segmented, then disabling a partition would eliminate access to part of the instruction information in every queue entry, and thus the processor would not correctly operate. For memory structures that consist of a tag array in addition to a data array, different subarray partitioning may be employed. In this case, the disabling of tag and data partitions must be such that hits are masked for disabled data partitions, and that hits are detectable for enabled data partitions. In some cases, this may mean that we cannot take advantage of the tag partitioning in terms of disabling partitions and saving power. (We discuss this further in Section 4.) Thus, our scheme cannot be used for every possible partitioning of a memory structure, and in these cases, the partitioning needs to be changed from that which is speed-optimal in order to provide the ability to disable partitions. The resulting increase in access time may critically impact performance as we describe in Section 4. Therefore, in this paper we assume that we do not deviate from the speed-optimal partitioning of Table 1 in applying this approach to on-chip memory structures.

3.1 PRESERVING INFORMATION TO ENSURE CORRECT OPERATION

For some on-chip memory structures, partitions can be disabled without regard to the accessibility of the information in the disabled array. This is the case for instruction caches, branch predictors, TLBs, and register files. Both instruction caches and TLBs create no new information but rather provide faster access than main memory, from which the same information can be extracted. By ensuring that instruction TLB entries are invalidated before they are disabled, we can ensure correct operation of both the instruction TLB and instruction cache whenever entries are re-enabled. Similarly, the absence of branch prediction information may simply cause more mispredictions, but these do not result in incorrect execution. Any performance penalty resulting from having to reload instruction caches and TLBs or reconstruct branch prediction information can usually be mitigated by limiting the rate at which partitions are disabled, *e.g.*, only during context switches.

Similarly, if the register file is appropriately partitioned, no action is required by the hardware to preserve data in a disabled partition. Because the register file is compiler-managed, the compiler can precisely determine application register usage and insert instructions to disable register file partitions when requirements are modest. When more registers are required, or when the values in these disabled registers are to be used, the compiler re-enables the required partitions. The full register file must also be enabled by the operating system when saving process state to memory. The issues associated with register renaming are more complex and are not discussed in this paper.

For other structures, incorrect operation can result unless the information in a disabled partition is preserved or made accessible. For example, the partitions of instruction queues and reorder buffers cannot be disabled until the instructions associated with the disabled entries have completed execution and committed their results. Thus, some time must elapse between the execution of an instruction that writes the PSR and the actual disabling of these structures. During this period (which may last 20-40 cycles on a modern microprocessor that issues two instructions per cycle on average), no new instructions must be placed in the to-be-disabled partitions. To speed up this process, the instruction scheduling hardware can raise the priority of the instructions in these partitions.

The disabling of partitions in the L1 data cache (Dcache) and the L2 cache is not as straightforward. For these structures, modified cache blocks must be made accessible to sharing processes (on the same processor or other processors), to the same process (either on the same or another CPU), and to the I/O subsystem. In addition, the coherence state of data in a disabled partition must be properly maintained in case the partition is later re-enabled. These issues are addressed in [3] in which we discuss the performance on-demand approach of a set associative cache in which ways can be dynamically enabled to meet performance demands. In the next section, we briefly describe this approach as an example of saving energy in on-chip memory structures.

4. A PERFORMANCE ON-DEMAND L1 DCACHE

In this section, we describe the application of disabling memory partitions to on-chip caches. Specifically, we quantify the energy savings obtained, and performance degradation incurred, with set associative L1 Dcaches in which the data ways of the cache can be selectively disabled using the methods described earlier and in the rest of this section. The hardware organization of this approach, which we call *selective cache ways*, is described in the next section.

4.1 HARDWARE ORGANIZATION

Figure 3 is an overall diagram of a 4-way set associative cache using selective cache ways. The wordlines of the data array are segmented either four or eight times according to the Cacti results of Table 1, creating four separate data way partitions. The bitlines of each data way may be segmented as well, although this is not shown in the diagram. Note however, that the tag portion of the cache (which also includes the status bits) is identical to that

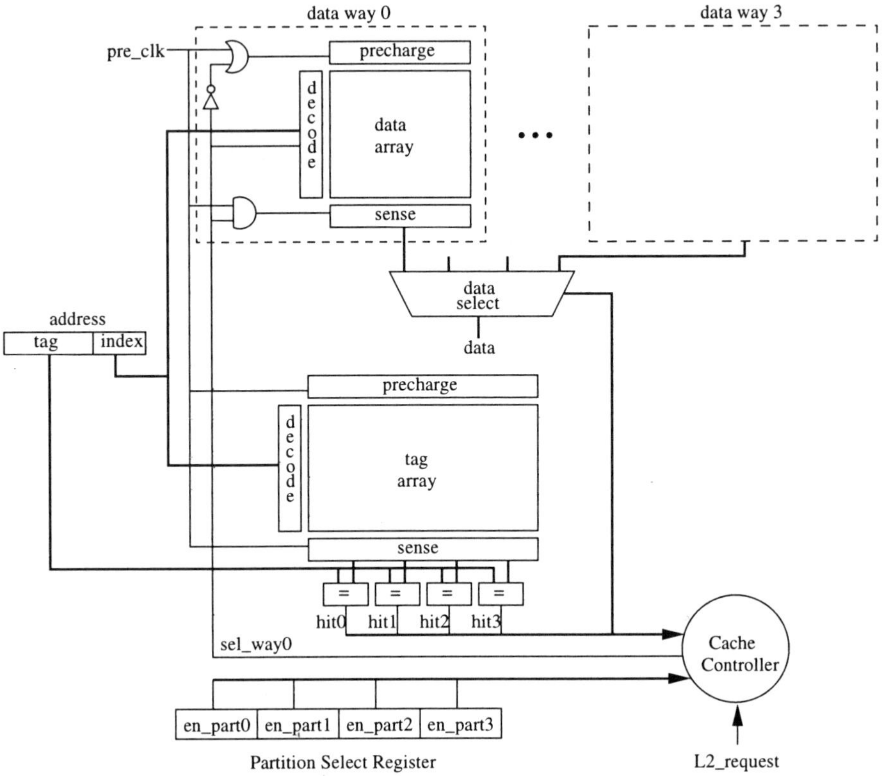

Figure 3 A 4-way set associative cache using selective cache ways. The details for data ways 1-3 are identical to way 0 but are not shown for simplicity.

Table 2 Cache cycle time degradation of tag wordline partitioning relative to the speed-optimal cache partitioning.

Cache Org	Degradation
32KB 2-way	3.7%
64KB 2-way	7.8%
32KB 4-way	4.3%
64KB 4-way	7.3%

of a conventional cache. Our Cacti-based timing estimates indicate that for the cache organizations we have studied, segmenting the tag wordlines can in some cases result in a significant cache cycle time degradation relative to the optimal tag *N* parameters of Table 1. For example, Table 2 shows that the cycle time degradation incurred when the tag wordlines are segmented is roughly 4-8% for 32KB and 64KB set associative caches. Because the L1 Dcache is

Table 3 Simulated memory hierarchy parameters.

Mem Level	Organization
L1 Icache	64KB, 4-way set assoc, 32B block, random, 1 cycle latency
L1 Dcache	64KB, 4-way set assoc, selective cache ways, 2 ports, 32B block, random, 1 cycle latency
L2 cache	512KB, 1MB, or 2MB, 4-way set assoc, 32B block, LRU, 15 cycle latency, 16 partitions
main memory	16B bus width, 75 cycle initial latency, 2 cycles thereafter

typically a critical path, especially for caches as large as those in this table, this degradation may result in an overall cycle time increase. For these reasons, we used the N parameters of Table 1 and therefore only save energy in the data portion of the cache. However, the data portion comprises roughly 90% of the total energy dissipation for the cache organizations we have studied.

Note from Figure 3 that the outputs of the PSR do not directly control the enabling of the arrays (as was the case in Figure 2), but rather are sent to the Cache Controller. This is to allow the Cache Controller the ability to access modified data in disabled ways as well as to properly handle coherence transactions as is discussed in [3].

4.2 ENERGY AND PERFORMANCE EVALUATION

In this section, we quantify the performance degradation incurred, and the energy savings obtained, with an L1 Dcache using selective cache ways. Our evaluation methodology combines detailed processor simulation for performance analysis and for gathering event counts, and analytical modeling for estimating the energy dissipation of both conventional caches and caches employing selective cache ways.

We use the SimpleScalar toolset [7] to model a modern 4-way out-of-order speculative processor with a two-level cache hierarchy that roughly corresponds to a current high-end microprocessor such as the HP PA-8000 [17] and Alpha 21264 [15]. Table 2 shows the simulator parameters for the memory hierarchy. Selective cache ways is implemented for only the L1 Dcache. The data array of the L2 cache is implemented as 16 partitions, only one of which is selected for each access. This power-saving technique is used in the Alpha 21164 on-chip L2 cache [6].

We estimate L1 Dcache and L2 cache energy dissipations using a modified version of the analytical model of Kamble and Ghose [13]. This model calculates in detail cache energy dissipation using technology and layout parameters

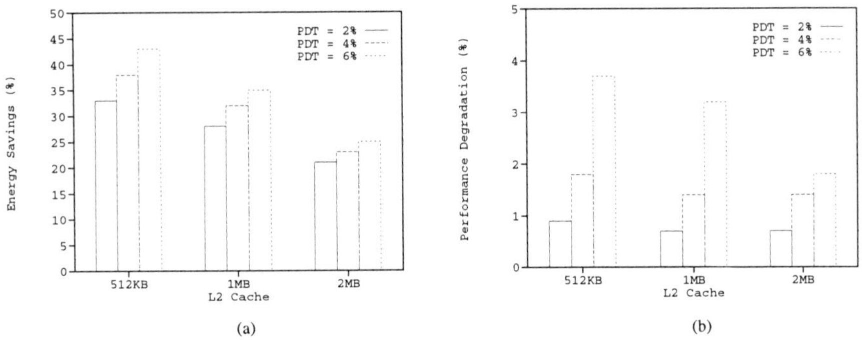

Figure 4 (a) Combined L1 Dcache and L2 cache energy savings and (b) actual performance degradation as a function of the performance degradation threshold.

from Cacti and counts of various cache events (hits, writebacks, *etc.*) as inputs. These event counts, in addition to performance results, are gathered from SimpleScalar simulations (each 400 million instructions long) of eight benchmarks: the SPEC95 benchmarks *compress*, *ijpeg*, *li*, *turb3d*, *mgrid*, *fpppp*, and *wave5*, as well as *stereo*, a multibaseline stereo benchmark from the CMU benchmark suite [10] that operates on three 256 by 240 integer arrays of image data. The number of enabled ways is determined based on overall application cache characteristics, and therefore the number of enabled cache ways is only changed during context switches. Only L1 Dcache and L2 cache energy dissipations are calculated as the L1 Icache and main memory energy dissipations do not change significantly with the number of enabled L1 Dcache ways.

The energy savings of selective cache ways depends on the amount of performance that can be traded off for energy. The *Performance Degradation Threshold (PDT)* signifies the average performance degradation relative to a cache with all ways enabled that is allowable for a given period of execution. If the PDT is 2%, and, for a given period of execution, performance is projected to degrade by 1% with three ways enabled, and 4% with two ways enabled, then three ways are enabled for that period of execution, so long as the total energy is less than that with all four ways enabled. This would not be the case if the extra misses with three ways enabled increase L2 cache energy more than the energy savings obtained with disabling one of the L1 Dcache ways. In this case, all four ways are enabled. In this study, the optimum number of enabled ways for each benchmark is determined from comparing performance and energy dissipation results. In an actual system, a runtime system such as Compaq's DCPI [4] can read cache hierarchy performance counters and make changes based on knowledge of relative L1 and L2 cache energy dissipations.

Figure 4 shows the energy savings and actual performance degradation incurred across all benchmarks as a function of the PDT. The energy savings is calculated from the average energy dissipation of all benchmarks with all ways enabled, and the average with the number of disabled ways allowable for a given PDT value. The performance degradation is similarly calculated from the corresponding Instructions Per Cycle results. The actual performance degradation incurred is significantly less than the PDT value. Overall, roughly a 40% cache hierarchy energy savings is realized with less than a 2% performance degradation for a 512KB L2 cache with a PDT of 4%. The benefits are less, yet still significant, for larger L2 caches due to the higher energy dissipated servicing an L1 Dcache miss. Even with a large 2MB on-chip L2 cache, a 25% energy savings is obtained with less than a 2% performance degradation using this technique.

5. CONCLUSIONS AND FUTURE WORK

In this paper, we have described techniques for leveraging the subarray partitioning of high-speed memory structures in order to dynamically tailor the organization of these structures to application requirements. We have discussed the mechanisms necessary to allow software to dynamically modify which partitions are disabled, as well as to ensure the proper handling of information in disabled partitions. Through detailed simulation and analytical modelling, we demonstrated that a 40% reduction in overall cache energy dissipation can be achieved for 4-way set associative L1 Dcaches with only a 1-2% overall performance degradation.

Our future work includes applying this technique in concert to multiple on-chip memory structures, as well as exploring how these structures can be dynamically tailored to changing requirements during individual application execution. Finally, we plan to combine these efforts with our previous work on dynamic speed-complexity performance tradeoffs [2] in order to dynamically optimize the energy-delay product of future high performance microprocessors.

6. ACKNOWLEDGMENTS

John Strasser coded the cache energy dissipation model, and to the author's knowledge, George Cai of Intel first coined the term "performance on demand."

7. REFERENCES

[1] A. Ahi et al. The R10000 superscalar microprocessor. *Hot Chips VII Symposium*, August 1995.

[2] D.H. Albonesi. Dynamic IPC/clock rate optimization. *Proceedings of the 25th International Symposium on Computer Architecture*, pages 282–292, June 1998.

[3] D.H. Albonesi. Selective cache ways: On-demand cache resource allocation. *Proceedings of the 32nd International Symposium on Microarchitecture*, November 1999.

[4] J. Anderson et al. Continuous profiling: Where have all the cycles gone? *Proceedings of the 16th Symposium on Operating Systems Principles*, October 1997.

[5] N. Bellas et al. Architectural and compiler support for energy reduction in the memory hierarchy of high performance microprocessors. *Proceedings of the International Symposium on Low Power Electronics and Design*, pages 70–75, August 1998.

[6] W.J. Bowhill et al. Circuit implementation of a 300-MHz 64-bit second-generation CMOS Alpha CPU. *Digital Technical Journal*, 7(1):100–118, Special Issue 1995.

[7] D. Burger and T.M. Austin. The SimpleScalar toolset, version 2.0. Technical Report TR-97-1342, University of Wisconsin-Madison, June 1997.

[8] Digital Equipment Corporation. Alpha 21164 microprocessor data sheet. August 1998.

[9] Digital Equipment Corporation. Alpha 21264 microprocessor data sheet. February 1999.

[10] P. Dinda et al. The CMU task parallel program suite. Technical Report CMU-CS-94-131, Carnegie Mellon University, March 1994.

[11] J.H. Edmondson et al. Internal organization of the Alpha 21164, a 300MHz 64-bit quad-issue CMOS RISC microprocessor. *Digital Technical Journal*, 7(1):119–135, Special Issue 1995.

[12] T. Horel and G. Lauterbach. UltraSPARC III: Designing third-generation 64-bit performance. *IEEE Micro*, 19(3):73–85, May/June 1999.

[13] M.B. Kamble and K. Ghose. Analytical energy dissipation models for low power caches. *Proceedings of the International Symposium on Low Power Electronics and Design*, pages 143–148, August 1997.

[14] R. Kessler. The Alpha 21264 microprocessor. *IEEE Micro*, 19(2):24–36, March/April 1999.

[15] R.E. Kessler, E.J. McLellan, and D.A. Webb. The Alpha 21264 microprocessor architecture. *International Conference on Computer Design*, October 1998.

[16] J. Kin, M. Gupta, and W.H. Mangione-Smith. The filter cache: An energy efficient memory structure. *Proceedings of the 29th International Symposium on Microarchitecture*, pages 184–193, December 1997.

[17] A. Kumar. The HP PA-8000 RISC CPU. *IEEE Computer*, 17(2):27–32, March 1997.

[18] G. Lesartre and D. Hunt. PA-8500: The continuing evolution of the PA-8000 family. *Proceedings of Compcon*, 1997.

[19] J. Montanaro et al. A 160-MHz, 32-b, 0.5W CMOS RISC microprocessor. *Digital Technical Journal*, 9(1):49–62, 1997.

[20] M. Tremblay and J.M. O'Connor. UltraSparc I: A four-issue processor supporting multimedia. *IEEE Micro*, 16(2):42–50, April 1996.

[21] T. Wada, S. Rajan, and S.A. Przybylski. An analytical access time model for on-chip cache memories. *IEEE Journal of Solid-State Circuits*, 27(8):1147–1156, August 1992.

[22] S.J.E. Wilton and N.P. Jouppi. An enhanced access and cycle time model for on-chip caches. Technical Report 93/5, Digital Western Research Laboratory, July 1994.

[23] B. Xu and D.H. Albonesi. A methodology for the analysis of dynamic application parallelism and its application to reconfigurable computing. *Proceedings of the SPIE International Symposium on Voice, Video, and Data Communications (Track: Reconfigurable Technology: FPGAs for Computing and Applications)*, September 1999.

[24] K.C. Yeager. The Mips R10000 superscalar microprocessor. *IEEE Micro*, 16(2):28–41, April 1996.

[25] X. Zhang et al. System support for automatic profiling and optimization. *Proceedings of the 16th Symposium on Operating Systems Principles*, October 1997.

Single Ended Pass-Transistor Logic

A comparison with CMOS and CPL

Mihai Munteanu, Peter A. Ivey, Luke Seed, Marios Psilogeorgopoulos, Neil Powell, Istvan Bogdan

University of Sheffield, E.E.E. Department, Electronic Systems Group

Key words: SPL, CPL, pass-transistor, low power

Abstract: SPL (Single-rail Pass-transistor Logic) is one of the most promising logic styles for low power circuits. This paper examines some key issues in the implementation of SPL: swing restoration, optimum number of pass-transistor stages between buffers and SPL circuits with two supply voltages. Simulation results based on netlists extracted from layout are presented to compare SPL, CPL and standard CMOS.

1. INTRODUCTION

In a survey of low power circuit design we concluded that pass transistor logic styles, and especially SPL (Single-rail Pass-transistor Logic also known as Single-ended Pass-transistor Logic or LEAP - Lean Integration with Pass-Transistors [5]), are very promising for low power circuits.

This paper presents the results that we have obtained from simulations of SPL circuits and it will examine some key issues in the implementation of SPL: swing restoration, optimum number of pass transistor stages between intermediate buffers and SPL circuits with two supply voltages. A number of simple circuits were implemented in three logic styles: SPL, CPL and CMOS using GSC200 MITEL CMOS standard library cells and simulated under a variety of conditions.

All the circuits presented in this paper were implemented in the Mitel 3.3V 0.35μm CMOS technology and all of the simulations were performed using Cadence Spectre simulator. The simulations were performed using netlists extracted from layout using supply voltages between 1.8 and 3.3Volts (3.3V is the maximum allowable supply voltage while 1.8V is the minimum supply voltage for the Mitel standard cell library).

2. SPL OVERVIEW

Our survey of contemporary low power techniques concluded that the SPL logic style is a promising logic style for low power design. One of the papers that popularised the SPL logic style is [5]. In [6], SPL is also reported to be a promising pass-transistor style: it has the advantage of efficient implementation of complex functions, especially arithmetic functions and, because it uses only NMOS transistors, the layout is very compact, simple and regular.

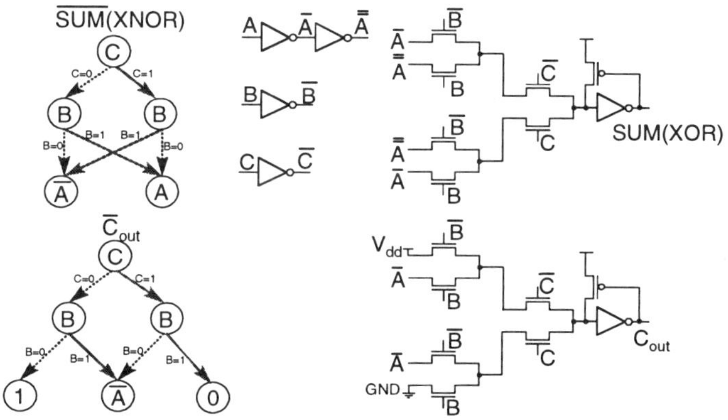

Figure 1. Full adder: BDD representation and the SPL circuit

The advantages of SPL are:

- Circuits are easy to synthesise starting from the Boolean expressions using Binary Decision Diagram (BDD) graphs. A pass-transistor network maps directly to a BDD diagram (*Figure 1*). The circuits in *Figure 1* are the SPL implementation of the BDD diagrams for a full adder (two good introductory papers about BDDs are [1])
- An SPL cell library has no more than 10 basic components [5]. These components are simple pass transistors cells, simple inverters and inverters with swing restoration
- The pass transistor network has fewer transistors than a CMOS network, especially for large functions and functions based on multiplexers and XORs

- The transistors in the pass network are all N-type transistors
- The circuit layout is very compact and regular because the pass transistor network contains only N type transistors. So the layout consist of rows of inverters alternating with pass transistor networks

The disadvantages of SPL are:

- Difficult to integrate with existing circuit synthesis tools
- The delay of the circuit is more sensitive to voltage scaling than for complementary CMOS logic. More factors have to be taken into account when designing SPL for low voltages. Simulation results presented later in this paper show that the delay of SPL circuits increases dramatically when the supply voltage approaches $V_{TN}+V_{TP}$, the theoretical lower limit of the supply voltages for SPL circuits. However [3] shows that using dynamic threshold pass-transistor logic, SPL and CPL circuits can work with a reasonable delay at very low supply voltages

3. IMPORTANT ISSUES FOR SPL

3.1 Swing Restoration

Because the SPL logic style uses only N type transistors in the pass-transistor network, the voltage swing at the end of a pass transistor network will be 0V to V_{dd}-V_{TN}, (V_{TN} is the threshold of the N type transistor). Therefore, the buffers inserted in the pass-transistor network, or at the outputs of the pass-transistor network, must have a swing restoring circuit in order to minimise the leakage currents through the inverters. In [4] a generalised circuit for converting low voltage swings to full swing is presented. Other approaches use amplifiers, but this is not good for low power design because of the permanent static current of the amplifiers. In the next section we will show that using two supply voltages it is possible to replace swing restoring inverters with simple inverters. In, *Figure 2* four versions of swing restoring inverters are presented (A, B, C and D).

Version A is the simplest and the most commonly used. B is a "faster" version of the swing restoring buffer because it loads the output with a smaller gate capacitance (M_{P2}) and the M_{P3} transistor is always on. However, simulation results show that the power consumption and the delay of this inverter are about the same as the simplest one while the area of the cell layout is bigger. By connecting the gate of the transistor M_{P3} to an enable line, it is possible to enable or to disable the swing restoring circuit.

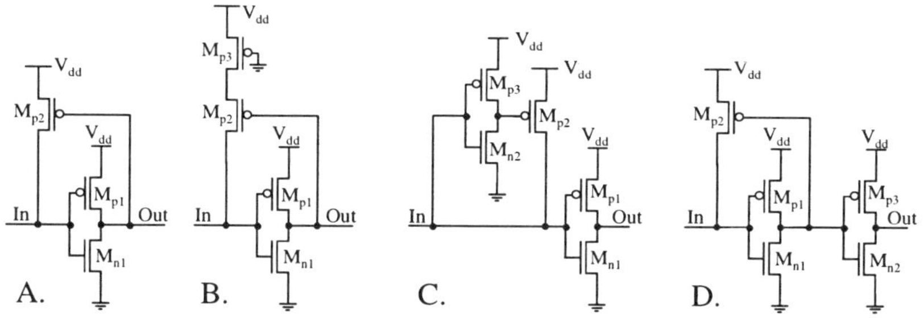

Figure 2. Swing restored inverters

The problem with the inverters A and B is that the swing restoring circuit is activated after a low to high transition is propagated through the inverter. This propagation time is affected by the load of the inverter, especially at low supply voltages. C and D avoid this problem by making the load of the inverter connected to the swing restoring transistor (M_{P3}) constant. *Figure 3* presents simulation results for swing restoring inverters. The results include the power and delay of a 4 pass transistor chain connected before the inverters and a large load. The circuits were simulated using a random input signal with an activity factor α=0.5 and frequency of 100 MHz.

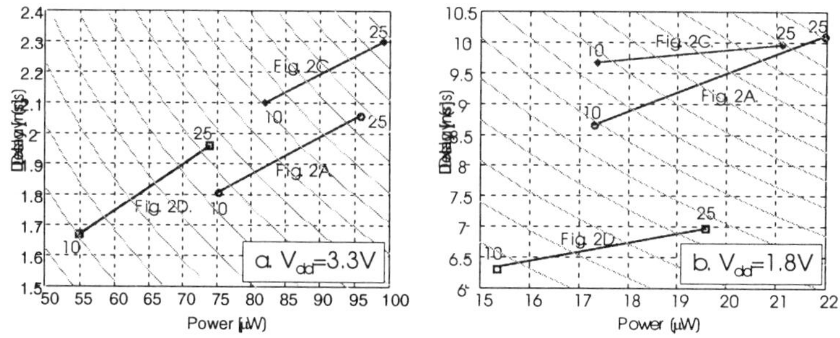

Figure 3. The Power consumption and delay of four pass transistors followed by different swing restoration inverters. Circuit load is 1mm metal track and 10 or 25 inverters

For large loads we prefer version D to version C because it uses tapered inverters. In this way the power consumption can be optimised [2]. This is confirmed by the results in *Figure 3*. However, within a pass-transistor network, a buffer with a big fan-out can be replaced by more smaller inverters within each branch.

3.2 Using Two Supply Voltages

Simulation results in the previous section show that the swing restoring inverter circuit is an important contributor to the delays of SPL circuits especially at low supply voltages or if the output is connected to a large load. By using two supply voltages it is possible to replace some of the swing restoring inverters with simple inverters. This approach reduces the power consumption of SPL circuits.

It can be noted that the only difference between the circuits in *Figure 7* and *Figure 8* is that the swing restoring inverters are replaced by normal inverters supplied at V_{dd}-V_{TN}. Given that the voltage swing at the end of a pass transistor network is 0V to V_{dd}-V_{TN}, swing restoration circuits are not required.

Not all the swing restoring inverters can be replaced in this way. If the output of an inverter is connected to a pass transistor gate, it has to have full swing on the output (0V to V_{dd}). In this case, if the inverter were to be supplied at V_{dd}-V_{TN}, another V_{TN} would be lost through the N-type pass transistor so the swing at the output would be 0V to V_{dd}-2V_{TN}.

Simulations show that the optimum power-delay product is obtained if the second supply voltage is about V_{dd}-V_{TN}. It is not practical to reduce the second supply voltage further without reducing the other supply voltage as well. In sections 4 and 6 simulation results of SPL circuits with two supply voltages are presented.

3.3 Long Pass-Transistor Chains

The delay of a pass transistor chain increases quadratically with the number of stages. To improve the delay of the pass transistor chain, intermediate buffers can be inserted. We have simulated a 60 pass-transistor chain with buffers inserted every 1,2,3,4,5 or 6 stages. *Figure 4* shows the power and the delay of the pass transistor chain at a supply voltage of 3.3V (the curves look similar for other supply voltages). The contours on the graph mark the constant power-delay product curves.

The goal of these simulations is to find the optimum power-delay product and this is obtained for a buffer every 4 or 5 stages in both cases. In real circuits it is probable that the buffers will be inserted every 3 or 4 stages. Real circuits are more complex than simple pass transistor chains, so the parasitic capacitances of the pass-transistor networks are bigger. Another reason is that this avoids high complexity of the pass-transistor networks between the buffer rows. Additionally, it must be noted that there is a significant trade-off between power and speed for varying the number of buffers and depending on the requirements of the design, circuits can be optimised differently.

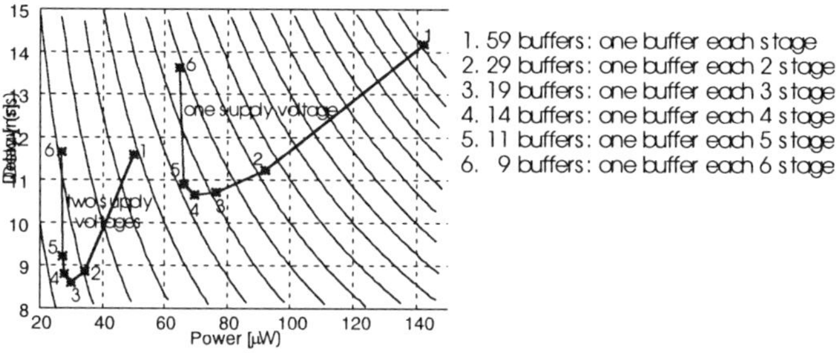

Figure 4. The power and delay of a 60 stage pass transistor chain with inserted buffers. Vcc=3.3V (3.3V and 2.5V in the two supply voltages case)

4. FULL ADDER SIMULATION RESULTS

The full adder is one of the most widely used circuits in comparing logic styles. It is implemented very efficiently in pass transistor styles, so we expect that it will perform well in SPL: simulation results confirm this fact.

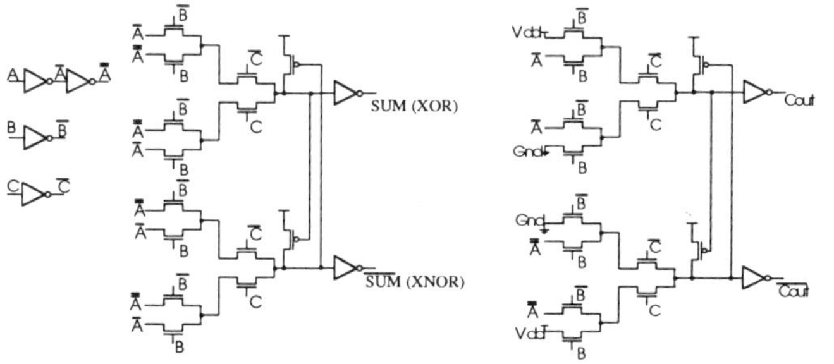

Figure 5. CPL full adder

We compared a standard cell, a SPL (*Figure 1*) and a CPL full adder (*Figure 5*). All of these adders were simulated in identical conditions:

- Stimulus: random; activity factor α=0.5; simulation interval: 0 to 2µs
- Frequency: 100MHz for all 3 inputs; rise and fal! times: 1ns
- Load circuit: RC circuit modelling a 100µm metal track followed by 1,4 or 10 parallel inverters

Figure 6 shows the power and delay for SPL, CPL and standard cell full adders. In all cases, the power consumption of the SPL full adder is about half of the standard cell library element. At 3.3V, the delay of the SPL full adder is slightly greater but is significantly greater for lower supply voltages. The power-delay product of the CPL circuit is about the same as the power-delay product of the CMOS standard cell.

At 1.8V the SPL full adder is more than twice as slow as the standard cell adder. The reason for this is the P-type swing restoring transistor on the output inverters. This inverter must be replaced with the more complex version of the swing restoration inverter. We used version D presented in section 3.1 and both cases are displayed in *Figure 6*b.

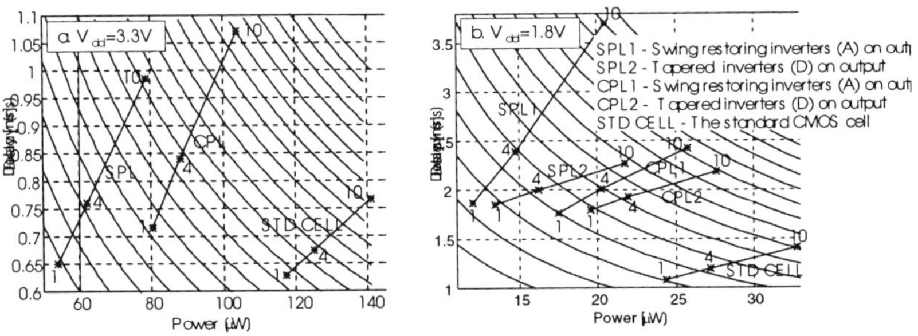

Figure 6. Power and delay for SPL, CPL and standard cell full adders. Circuit load is 100mm metal track followed by 1,4 or 10 standard inverters

CPL circuits are generally slower than SPL circuits. The reason is that the drive capability of the pass transistor network is poorer than the drive capability of an inverter in the SPL case. Only at very low supply voltages for large loads do the simulation results show that CPL circuits are faster than SPL. At 1.8V supply voltage the CPL circuits are slightly faster than SPL, but at higher supply voltages the SPL circuits become faster.

The layout area of the SPL full adder is smaller than the area of the standard cell. The SPL full adder cell is 26.8x10.3μm (276.04μm^2) comparing with 32.2x12.6μm (405.72μm^2) of the smallest full adder from the standard library.

5. 4 BIT ADDERS

5.1 SPL Carry Chain Versions

As for any adder, the critical part of an SPL adder is the carry circuit. We tested three possible versions obtained by reordering the circuit inputs. This reordering does not affect the circuit structure because the carry function is symmetrical. However reordering does affect the power consumption, the delay and, to a small degree, the area of the circuit.

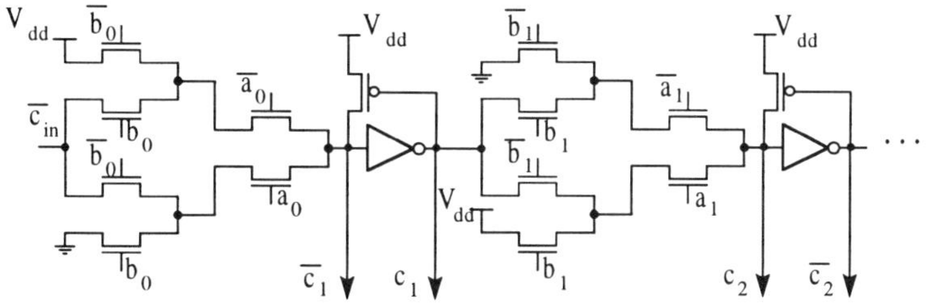

Figure 7. SPL carry chain (SPL1)

The carry chain in *Figure 7* is the simplest one. It is obtained by cascading the carry circuit of the SPL full adder (*Figure 1*). The only trick is that the inputs of the second stage are inverted, which means that the second carry will be inverted. This is a consequence of a basic BDD property [1] and it is used to avoid double inverters between two stages.

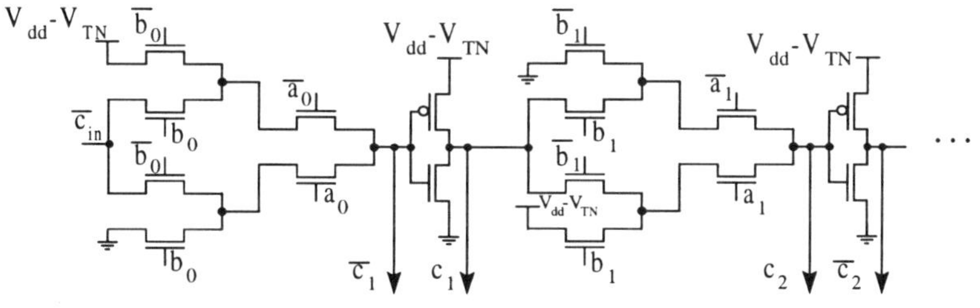

Figure 8. SPL carry chain with two supply voltages (SPL1 version)

Figure 8 shows the same circuit as *Figure 7* but in this case the two supply voltage scheme is used.

The circuit in *Figure 9* is an attempt to make the carry chain faster. Instead of passing the carry signals through all the pass transistor network, they are applied directly to the last pass transistor stage of the carry circuit.

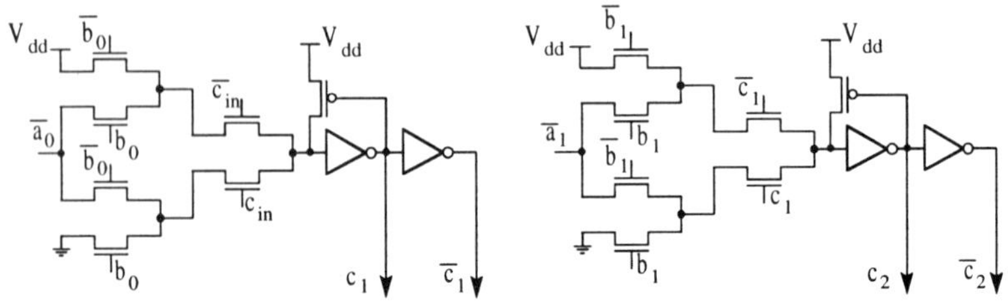

Figure 9. Faster SPL carry chain (SPL2)

Figure 10 shows the best version of a SPL carry circuit. In this case the buffers are inserted only every 4 pass-transistor stages. This also matches the optimum number of pass transistor stages between buffers, as presented in section 3.3. Simulation results in the next section show that this version is as fast as the previous one (*Figure 9*), but the power consumption is reduced.

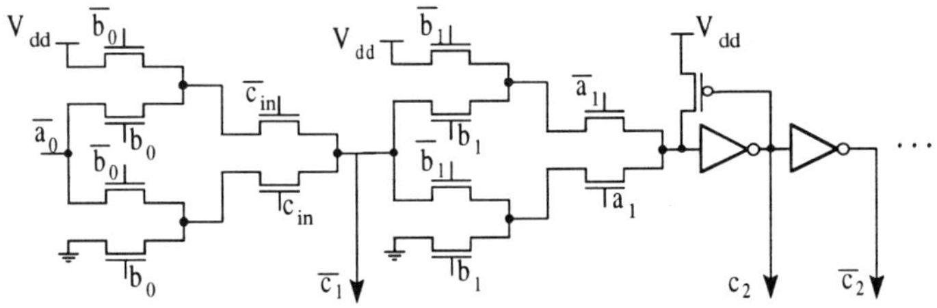

Figure 10. Low power SPL carry chain (SPL3)

For the last two versions (*Figure 9* and *Figure 10*) the swing restoring inverters cannot be replaced by a normal inverters supplied at V_{dd}-V_{TN} as the outputs of a carry stage are connected to pass transistor gates in the next stage.

5.2 Simulation Results

All of the simulated adders were implemented in the same technology and simulated with identical conditions. The simulation scenario is similar to the one presented in the previous section, except that, for simulation speed reasons, the load for each output consist of a 20fF capacitor.

Figure 11 shows the power and the delay for the four versions of SPL adders and a 4 bit ripple carry adder made of standard cell full adders. In all the cases the power is the total power of the 4 bit adder and the delay is the delay of the carry chain (measured from carry input to carry output). The graph in *Figure 11*a corresponds to a supply voltage of 3.3V and in *Figure 11*b to a supply voltage of 2V. For the case of two supplies, the voltages used are 3.3V and 2.5V in the first case and 2V and 1.2V in the second case (Note: these results cannot be compared directly with the results for the single bit adder because circuit loads vary).

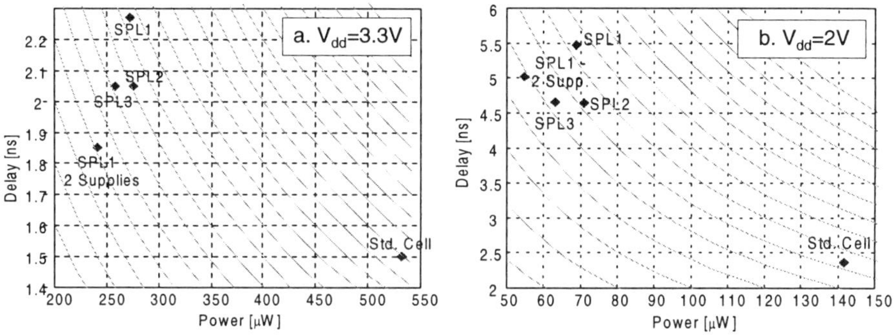

Figure 11. Power and delay of a 4bit ripple carry adder composed of full adder standard cells and 4 versions of SPL 4bit adder

At both supply voltages the power consumption of the SPL adders is half or less than standard cell adder, but the delay is worse. One of the reasons is that the driving capability of the inverters on the output of the SPL circuits is smaller. More significant is the fact that the power delay product is better for the SPL adders in almost all the cases. As we expected, the performance of the SPL adder degrades faster than the one made of standard cells when the supply voltage is decreased. For this technology V_{dd}=2V is quite close to the limit of the supply voltage for the SPL circuit, which is about 1.3V - the sum of the N-type and P-type threshold voltages.

Another important conclusion is that SPL1 with two supply voltages (*Figure 8*) has the best power-delay product in both cases and the single supply version of this circuit (*Figure 9*) is the worst among the SPL circuits.

6. SIMPLE GATE - SIMULATION RESULTS

Small gates that are very well implemented in CMOS logic style are not very efficiently implemented in SPL. This means that SPL circuits must be designed as large circuits in one step rather than breaking them into gates and implementing each gate in SPL.

We chose a simple gate from the standard library whose function is (A'B')+CD. In CMOS it can be implemented with two inverters and a one stage 4 input gate. *Figure 12* shows the SPL implementations that we simulated (the CPL circuit has two similar pass transistor networks).

It is obvious that, for such simple gates, the SPL implementation is more complex and requires more transistors than the CMOS implementation. The SPL implementation requires 17 transistors and the CMOS implementation 12 transistors (the CPL implementation uses 28 transistors).

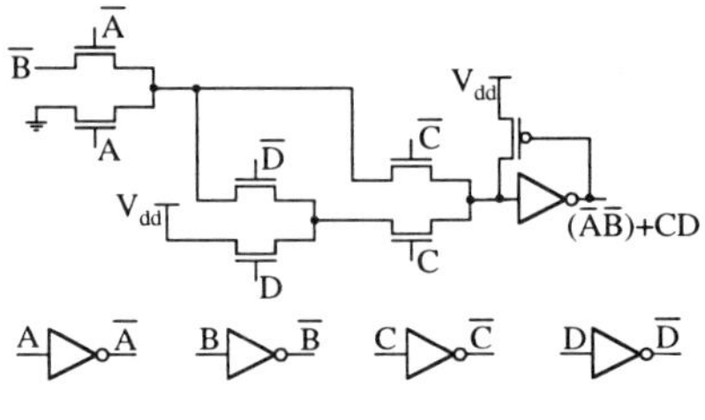

Figure 12. SPL implementation of (A'B')+CD Boolean function

Even though the SPL circuit is more complex, its performance is not much worse than that of the CMOS circuit. *Figure 13* shows the simulation results for V_{dd}=3.3V and 2V. The load of the circuit consists of a 20, 80 or 200fF capacitor. The results are consistent with the results for the adders presented in the previous sections. The CPL circuit is slower and has a greater power consumption than the SPL circuit. Again we can see that the performance of the SPL and CPL circuits is degraded at low supply voltages.

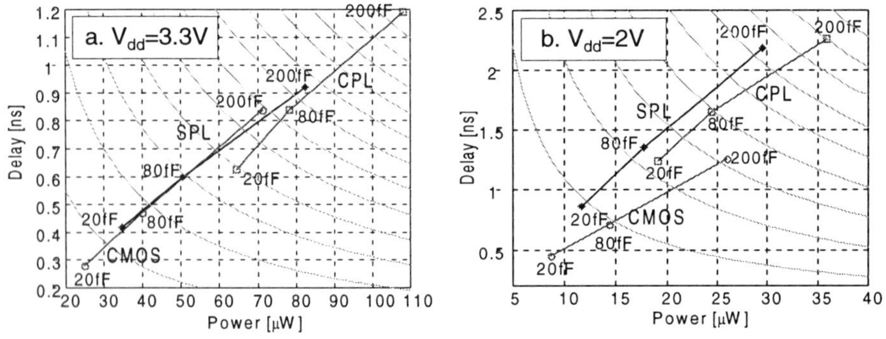

Figure 13. Power and delay for a simple gate implemented in SPL, CPL and CMOS (std. cell)

7. CONCLUSIONS

We have presented simulation results that show the type of functions that are best implemented in SPL and the type of functions that are not efficiently implemented. The conclusion is that SPL is a good low power alternative if:

- The implemented functions are arithmetic, XOR and MUX based circuits
- SPL circuits are not efficiently implemented if we try to replace the basic cells in a CMOS standard library with SPL based cells
- The supply voltage of the circuit is not too close to $V_{TN}+V_{TP}$, the theoretical lower limit of the supply voltages for SPL circuits. Simulations that we performed have shown that an optimum supply voltage is about $2(V_{TN}+V_{TP})$

We also presented some important issues about SPL and concluded that:

- For the swing restoring buffers with big loads the best approach is to use tapered inverters with the first one using a swing restoring circuit
- For the technology used, the optimum number of pass transistor stages between buffers is 4 or 5
- Two voltage supply scheme makes circuits faster and less power hungry

Further work is being undertaken to implement large complex circuit blocks in SPL logic. These larger circuits will demonstrate how efficient, from the point of view of power consumption and area, the SPL logic style is in practical cases.

8. ACKNOWLEDGEMENT

This work was partially funded by the Commission of the European Community under the ESPRIT programme (project: 25242 - PREST)

9. REFERENCES

[1] R. E. Bryant *Graph-Based Algorithms for Boolean Function Manipulation,* IEEE Trans. Computers, Vol. C-35, No. 8, August 1986, pp.677-691

[2] J. Choi, K. Lee *Design of CMOS tapered buffer for minimum power-delay product,* IEEE Journal of Solid-State Circuits, 29, No. 9, September 1994, pp. 1142-1145

[3] N. Lindert, T. Sugii et al. *Dynamic Threshold Pass-Transistor Logic for Improved Delay at Lower Power Supply Voltages,* IEEE Journal of Solid-State Circuits, Vol. 34, No. 1, January 1999, pp.85-89

[4] Y. Nakagome, K. Ytoh et al. *Sub-1-V Swing Internal Bus Architecture for Future Low-Power ULSI's,* IEEE Journal of Solid-State Circuits, Vol. 28, No. 4, April 1993, pp. 414-419

[5] K. Yano et al. *Top-Down Pass-Transistor Logic Design*, IEEE Journal of Solid-State Circuits Vol. 31, No. 6, June 1996, pp. 792-803

[6] R. Zimmermann, W. Fichtner *Low-Power Logic Styles: CMOS versus Pass-Transistor Logic*, IEEE Journal Of Solid State Circuits, Vol. 32, No 7, July 1997, pp. 1079-1090

Multithreshold Voltage Technology for Low Power Bus Architecture

A. Rjoub, O. Koufopavlou
VLSI Design Laboratory
Department of Electrical & Computer Engineering
University of Patras
6500 Patras - GREECE

Key words: Multithreshold Voltage, Low Swing, Low Power Design, Bus Architecture.

Abstract: Multithreshold voltage technology for low swing voltage bus architecture is proposed. Three new different classes of driver/repeater/receiver circuits are introduced. In the driver circuits, high threshold MOSFET transistor have been inserted in order to decrease its output swing voltage. To re-pull up the low swing to full swing voltage, innovated low delay cross-coupled latch circuit receivers are proposed. The same architecture with new repeater circuits based on multithreshold voltage are introduced in order to drive internal long interconnection lines and decreasing the total delay time. For *2*V supply voltage using *0.5*_m process technology, SPICE measurements show up to *20%* improvement in the delay time and up to *50%* saving in the total power dissipation comparing with the conventional CMOS bus architecture.

1. INTRODUCTION

Several low power design techniques have been proposed in order to decrease the total power dissipation of the circuit designs [1-7]. In [3], high area is occupied using resistor strings and buffers. In [4], careful design and reference voltage generators are required while in [5], the receiver dissipates static power, and double bus lines are used. Finally, in [6], a method based on diodes without significant decrease in power dissipation is presented. A more efficient technique with high reduction in power dissipation is used in [7] but it leads to a very complex design.

In order to achieve high performance and clock frequency, with an operating power of a few watts, the power dissipated by current designs must be reduced by 1/10. There are two main ways to meet this requirement: 1) By reducing the operating voltage below 1V, 2) By introducing a new circuit design methodology that reduces power dissipation, while keeping a high supply voltage (higher than 2V) [3]. The low swing voltage technique is an example of the second approach. Use of this technique in circuits with large fan-out or high load capacitance results in a significant reduction in the total power dissipation.

Multithreshold technology CMOS circuit designs, which have both high and low threshold voltage transistors in a single chip, can be used to deal with the leakage problem in low voltage, low power, and high performance applications [8]. It can be used in modern VLSI applications in order to increase the power savings and decrease the delay time [9].

Bus is an example where high capacitance has to be driven. It is composed from long interconnections with large fan-out and dissipates in many circuits up to 50% of the total power dissipation [10]. The purpose of this paper is to introduce new bus circuit designs that reduce power dissipation using multithreshold voltage technology while the operating supply voltage is less than 2V. In this case low area is observed also using this technique, comparing with other low swing voltage techniques [3-7].

2. CONVENTIONAL BUS ARCHITECTURE

The conventional CMOS bus architecture is shown in Fig.1. It consists of a driver, a delay element (interconnection line), and a receiver. The driver and the receiver are both conventional CMOS inverters. The input/output voltage level of the driver/receiver ranges between *0* and the power supply voltage (V_{DD}).

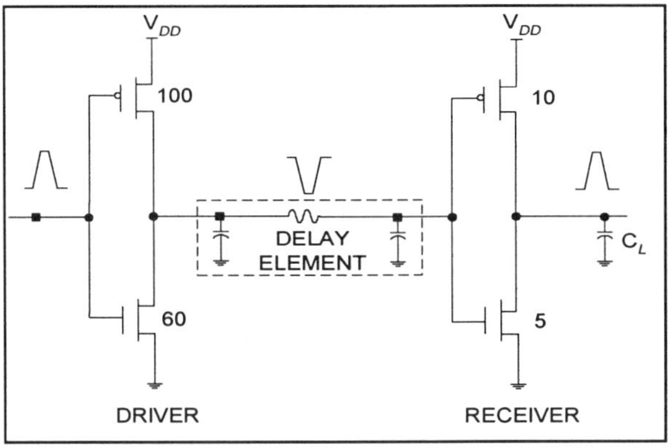

Figure 1. Conventional Bus Architecture

The decrease of the supply voltage is the most efficient way to decrease the power dissipation over the circuits, but on the other hand, it increases the delay time, which makes the new design undesirable in applications where high-speed operation is the main feature. In order to address the problem of high-speed operation in low supply voltage, new devices should be used. Therefore to keep the architecture in high speed-operation and in low supply voltage, the threshold voltage V_T of some transistors will certainly be reduced.

3. NEW BUS DRIVER CIRCUITS

Several papers have been published using the low-swing voltage technique to decrease the power dissipation over the circuits, but they were introduced the conventional way to decrease the swing voltage by inserting the same threshold voltage of MOSFET transistors. This form of design has the disadvantages of the leakage current, the high used area (inserted MOSFET transistors) and the delay time increase. Not more than 25% of power delay product reduction can be achieved. Table I shows the number of the added nMOS transistors used in previous designs in order to achieve the low swing voltage in the driver circuit.

Table 1. [Added nMOS Transistors in Previous Low Swing Designs]

Technique	# of Added Transistor
Proposed	1
[5]	3

Technique	# of Added Transistor
[11]	3
[12]	3
[13]	4
[14]	6
[15]	15

Fig. 2 shows the proposed Multithreshold Technology CMOS (MTCMOS) driver circuits. The new CMOS inverters are composed by p/nMOS transistor with low threshold voltage (Low-V_T), as the conventional CMOS driver, and by a new inserted high threshold voltage of p/nMOS transistor (High-V_T). The inserted transistor almost completely suppresses the leakage current, and decreases the output voltage by its high threshold voltage value.

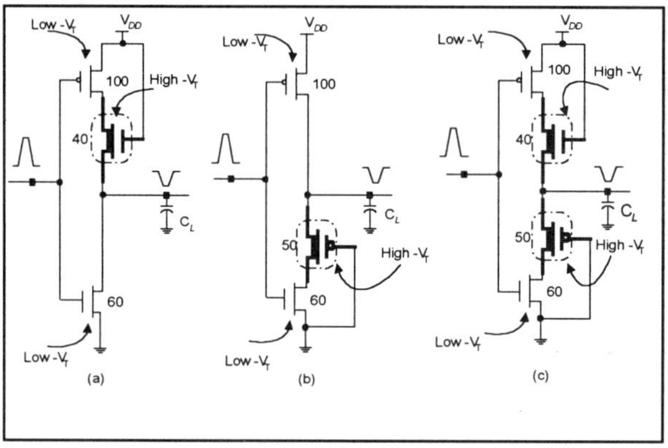

Figure 2. Proposed Multithreshold Technology Low Swing Bus Driver Circuits

According to the direction of the swing voltage reduction in drivers' output, the low swing drivers can be categorized in three different classes. The driver in Fig. 2(a) belongs to the first class called **U**p **L**ow Swing Voltage **D**river (**ULD**). In this class, the low swing output voltage (V_{LS}) ranges between *0* and (V_{DD} - V_{HTN}) where V_{HTN} is the threshold voltage of the inserted high threshold nMOS transistor. In order V_{LS} to drive successfully the succeeding CMOS gate, V_{DD} - V_{HTN} must be always greater than the low threshold of nMOS transistor voltage V_{LTN}.

The second class of the low swing voltage drivers called **D**own **L**ow Swing Voltage **D**river (**DLD**). It is shown in Fig. 2(b). In this class, the low swing output voltage (V_{LS}) ranges between V_{HTP} and V_{DD}, where V_{HTP} is the threshold voltage of the inserted high threshold pMOS transistor.

The driver in Fig. 2(c) belongs to the third class called **U**p-**D**own **L**ow Swing Voltage **D**river (**UDLD**). The output voltage of the driver ranges between V_{HTP}, for V_{DD} driver input level voltage and (V_{DD} - V_{HTN}), for *0V* driver input level voltage.

In Table 2, the normalized total power dissipation of the three proposed driver circuits for different supply voltage is compared with the conventional CMOS driver dissipation. The differences of the power dissipation between the three types of the proposed drivers are due to the different threshold voltage values of each high threshold MOSFET transistor. For example, because the threshold voltage value of the inserted high threshold voltage nMOS transistor (0.55V) in **ULD** class is less than the **DLD** (|0.65|V), it causes the **DLD** class to dissipate less power than the **ULD** [16]. The best class is the **UDLD** because of the lowest output voltage swing, it consists of both high threshold p/nMOS transistors.

Table 2. Total Power Dissipation of the Proposed Driver Circuits

Vdd	Conv.	ULD	DLD	UDLD
1	1,00	0,70	0,47	
1,5	1,01	0,74	0,47	0,16
1,8	1,06	0,80	0,52	0,20
2,1	1,23	0,93	0,63	0,24
2,4	1,49	1,09	0,79	0,33
2,7	1,90	1,40	1,06	0,57
3,0	2,50	1,93	1,50	0,95
3,3	3,50	2,81	2,26	1,55

When the output voltage swing decreases in the proposed drivers, it causes decreasing in their delay time. Table 3 shows the delay time for the proposed driver circuits compared with the conventional CMOS driver. SPICE measurements are taken with supply voltage 2V, for the same input slope and load capacitance. The delay time is calculated as the ((rise time + fall time)/2). The transistor sizes (in _m) are shown in Fig. 2.

Table 3. Propagation Delay Time

Circuit	Propagation Delay Time (nSec)
Conventional	0.173
ULD	0.170
DLD	0.163
UDLD	0.156

Comparisons of the proposed driver circuits with driver circuits proposed in other papers are presented below. The comparison simulation results prove the improvement in power dissipation reduction that achieved by the proposed circuits compared with previous designs.

The driver circuits proposed in [12] and in [20] are compared (Table 4) with the first class (**ULD**) circuit. All the SPICE measurements are normalized (with the conventional CMOS driver).

Table 4. Normalized Power Dissipation

Vdd	ULD			DLD	UDLD	
	Proposed	[12]	[20]	Proposed	Proposed	[12]
1	0,550			0,435		
1,5	0,552	0,600	0,660	0,438	0,116	0,170
1,8	0,560	0,600	0,660	0,446	0,120	0,176
2,1	0,563	0,611	0,670	0,460	0,124	0,185
2,4	0,578	0,622	0,689	0,487	0,137	0,204
2,7	0,604	0,652	0,720	0,514	0,160	0,231
3,0	0,644	0,693	0,752	0,553	0,196	0,273
3,3	0,693	0,737	0,807	0,602	0,254	0,320

For the second class (**DLD**), according to our knowledge no other driver's circuits are proposed in this class. The third class of the low swing voltage drivers **UDLD** is compared with the driver proposed in [12]. These results proved that the proposed technique using multithreshold voltage technology could operate in low supply voltage (1V) (except UDLD) while the conventional techniques operate in higher supply voltage.

4. NEW BUS RECEIVER CIRCUITS

In order to convert the driver low swing output to the full swing level, special receiver designs are required to pull-up the output low swing voltage. When a conventional CMOS inverter is used to convert the low-swing signal to a full-swing signal, the standby power could be important [3].

The proposed receivers are shown in Fig. 3. They are based on the Voltage Sense Transistor Circuit proposed in [11]. They can be implemented with multithreshold technology as the driver designs. For each category of the described drivers, an appropriate corresponding receiver is proposed.

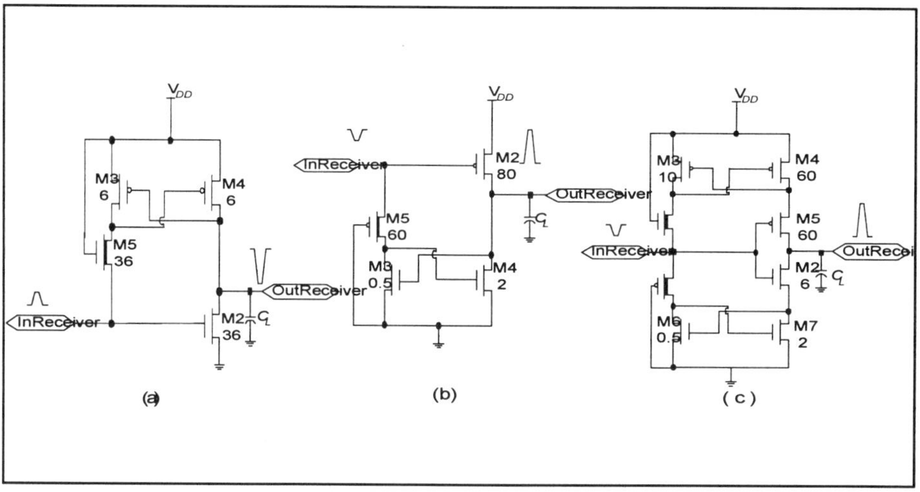

Figure 3. Proposed Receiver Circuits (a) UFR (b) DFR and (c) UDFR

For the first class of drivers, we proposed the **U**p **F**ull Swing Voltage **R**eceiver (**UFR**) as shown in Fig. 3(a). The receiver input (InReceiver) is connected to the driver output (V_{LS}), so the receiver input voltage swings between the values *0V* and (V_{DD} - V_{HTN}). When $V_{LS} = V_{DD} - V_{HTN}$, the transistor M2 turns on, discharging the receiver output node to the ground. Thus, M3 turns on, charging to V_{DD} the gate node of M4, which turns off. The high threshold voltage transistor M5 turns off. When the receiver input is *0*, the transistor M2 turns off while the transistor M5 turns on, discharging the gate of transistor M4 to *0*V. Thus, M4 turns on charging the output load to V_{DD} and ensuring a full swing operation.

The same logic of the previous receiver could be implemented in order to cooperate with the Down Low Swing Voltage Driver. A **D**own **F**ull Swing Voltage **R**eceiver (**DFR**) is shown in Fig. 3(b). The logic operation of this receiver is exactly inverse of the **UFR**. As the receiver input (InReceiver) is V_{HTP}, the transistor M2 turns on, charging the receiver output node to the V_{DD}. Thus M3 turns on, discharging to GND the gate node of M4, which turns off. The high threshold voltage M5 transistor turns off. When the receiver input is V_{DD}, the transistor M2 turns off while the transistor M5 turns on, charging the gate of transistor M4 to V_{DD}. Thus, M3 turns off discharging the output load to *0* and ensuring a full swing operation.

The combination of the above proposed receivers results in the **U**p-**D**own **F**ull Swing Voltage **R**eceiver (**UDFR**) as illustrated in Fig.3(c). It is a combination of both previous receivers (**UFR, DFR**) and a CMOS inverter. When the low swing is from up, the up half receiver converts the low swing to V_{DD}, while when the low swing is from down, the down half swing receiver converts the low swing to *0V*.

Other receiver circuits are proposed in [12], [15] and [17]. They have the common target to pull-up the low swing to full swing voltage. The normalized delay time ((rise + fall)/2) for the different classes are shown in Table 5. The measurements have been taken by using the same load capacitance and input voltage ramp.

Table 5. Normalized Propagation Delay Time

Vdd	UFR				DFR		UDFR		
	Prop.	[12]	[15]	[17]	Prop.	[17]	Prop.	[12]	[17]
1	1,48				1,20				
1,5	1,42				1,21	1,34	1,69		
1,8	1,36	3,10	2,38	2,80	1,23	1,36	1,67	3,50	2,84
2,1	1,32	2,80	2,16	2,55	1,24	1,39	1,65	3,40	2,65
2,4	1,31	2,60	2,03	2,35	1,25	1,40	1,64	3,36	2,43
2,7	1,30	2,50	1,94	2,22	1,29	1,45	1,61	3,31	2,32
3,0	1,29	2,40	1,87	2,13	1,33	1,52	1,60	3,28	2,28
3,3	1,27	2,30	1,83	2,10	1,39	1,61	1,61	3,26	2,26

In table 6, the total power dissipation of each proposed receiver comparing with other receivers proposed in previous papers are shown. For the power dissipation measurements, the power meter circuit proposed in [18] has been used.

Table 6. Normalized Power Dissipation

Vdd	UFR				DFR		UDFR		
	Prop.	[12]	[15]	[17]	Prop	[17]	Prop.	[12]	[17]
1	1,23	1,96	2,43	2,36	1,25			3,5	2,95
1,5	1,21	1,94	2,41	2,34	1,24	1,45	1,68	3,4	2,92
1,8	1,18	1,86	2,38	2,31	1,22	1,42	1,66	3,33	2,89
2,1	1,16	1,83	2,34	2,29	1,21	1,40	1,64	3,25	2,87
2,4	1,13	1,81	2,31	2,26	1,20	1,36	1,63	3,21	2,84
2,7	1,1	1,78	2,26	2,22	1,18	1,32	1,61	3,18	2,81
3,0	0,98	1,76	2,25	2,18	1,160	1,293	1,59	3,12	2,76
3,3	0,95	1,72	2,22	2,16	1,14	1,28	1,57	3,00	2,74

The data of the above tables show that using the proposed receivers could pull up the low swing to full swing in low supply voltage (1Volt) (except third class). This advantage is important for future VLSI designs where low supply voltage is required. This advantage could not be achieved using the conventional CMOS technology. We should mention also that the receiver in Fig. 3(c) reduces less power and operates in higher speed than the other receivers mentioned in [12][17], because the output in this class, is the output of the CMOS inverter. This class, the output has sharp edges ranges between *0* and V_{DD}, so less dynamic current dissipation occurred.

Simulation results of the three proposed bus technique classes (driver/receiver circuits) are shown in Fig. 4. The power delay product of each bus design is normalized with the one of the conventional bus. In all the measurements the same input ramps and load capacitance are used. The results prove, for 1V supply voltage, up to 50% in power delay product reduction could be achieved.

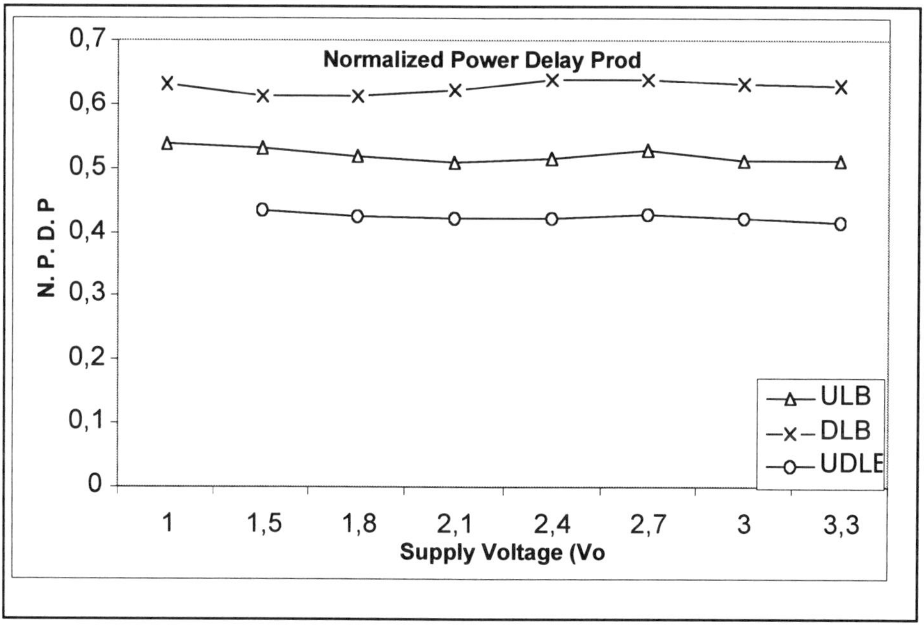

Figure 4. Normalized Power Delay Product for the Three Bus Architecture Classes

5. NEW BUS REPEATER CIRCUITS

The delay of a long line with distributed resistive and capacitive components grows as the square of its length. This results in a propagation speed that decreases with line length. To avoid this dependence, a common solution is to separate regularly the interconnection line in segments with the same length, which are driven by repeaters [4].

The line segments could be modelled in different ways. A good approximation of the delay line is obtained with the _ model, achieving accuracy better than *3%* in the delay calculation [4]. Several configurations [19] have been proposed for this architecture. However, these configurations operate well in terms of area and delay, but they do not take into account the power dissipation. In order to achieve lower power dissipation on the interconnection lines, low voltage swing repeater designs are proposed. The drivers and receivers, introduced in the previous sections, are used.

Three types of repeaters are presented, corresponding to the driver's classes. For **ULD**, the Up Low Swing Repeater (**ULR**) and for **DLD**, the Down Low Swing Repeater (**DLR**) are shown in Fig. 5(a) and Fig. 5(b). Merging these two designs a new Up-Down Low Swing Repeater (**UDLR**) is derived (Fig. 5(c)).

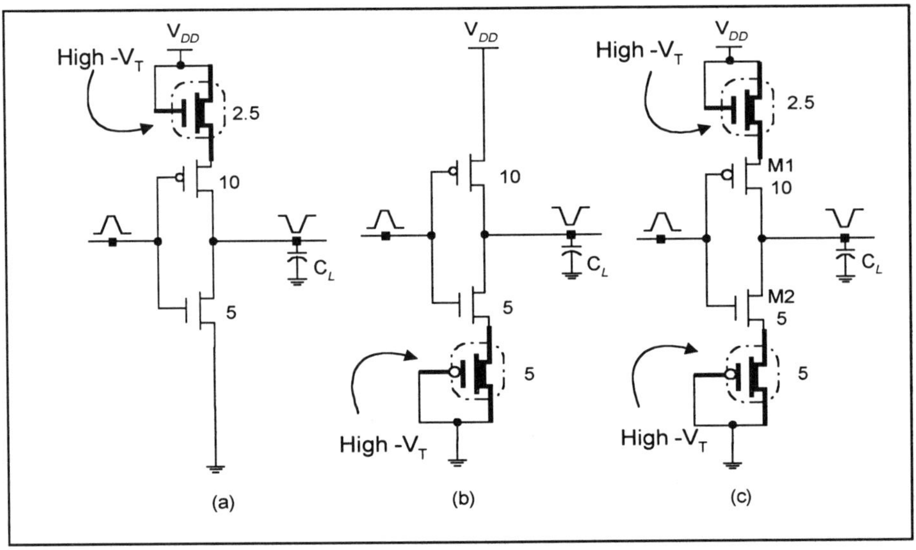

Figure 5. Proposed Repeater Circuits (a) ULR (b) DLR and (c) UDLR

The input of the **UDLR** (output of the driver) ranges between values V_{HTP} and (V_{DD} - V_{HTN}). When the input value is V_{HTP}, the pMOS transistor (M1) turns off and the nMOS transistor (M2) turns on (its source voltage value is V_{HTP}). The voltage value in the source of transistor M1 could not exceed the value of (V_{DD} - V_{HTN}), which is the output voltage of the driver. When the input voltage is (V_{DD} - V_{HTN}), the transistor M2 turns off and the transistor M1 turns on (its source voltage value is V_{HTN}). The output of the repeater in this case could not exceed the voltage value V_{HTP}.

In order to show the improvement of the different types of the repeater circuits in the delay time as in the power-delay product, we compared each proposed bus type using different number of repeaters and the conventional bus architecture. In the case of conventional bus, it is used conventional driver, *2p*F capacitor, *0.5_* resistance, *2p*F capacitor, conventional receiver, and *2p*F load capacitance. In the proposed bus, we distributed these values according to the number of the repeaters. Table 7 shows the decrement in the normalized delay time, and the decrement in the power delay product for each proposed bus (driver-repeaters-receiver) comparing with the conventional bus (driver-receiver). From Table 7, it is clear that an increase in the number of repeaters causes a decrease in the normalized delay time (as it is expected).

Table 7. Measurments for Different Number of Repeater Circuits

BUS TYPE	# REPEATERS	NORM. DELAY TIME	NORM. POWER DISSIPATION	NORM. POWER DELAY PRODUCT
UP	5	0.94	1.034	0.985
	10	0.84	1.056	0.945
	15	0.63	1.083	0.855
	20	0.54	1.118	0.825
DOWN	5	0.95	1.016	0.980
	10	0.84	1.042	0.940
	15	0.62	1.089	0.850
	20	0.53	1.111	0.820
UP-DOWN	5	0.94	1.022	0.983
	10	0.83	1.056	0.942
	15	0.64	1.073	0.854
	20	0.53	1.118	0.825

The simulation results have been derived by using *0.5-_*m double metal MTCMOS process. The device parameters are summarized in Table 8.

Table 8. SPICE Parameters multithreshold technology

Parameter	High Threshold Voltage Transistor	Low Threshold Voltage Transistor
Gate Lengh	0.55 _m	0.65 _m
Gate Oxide Thickness	110 Å	110 Å
N-Channel: Vth	0.55V	0.25
P-Channel: Vth	-0.65 V	-0.35 V

6. NOISE MARGIN AND MILLER EFFECT

Noise margin is an important parameter in VLSI design, especially in low swing voltage applications. It appears as problem in long line interconnections and bus designs. The acceptable case of noise margin in either low noise margin or high noise margin is greater than $0.1V_{DD}$[10]. In this paper we refrain from the noise margin problems by using low swing voltage to be $0.35V_{DD}$ and separate the bus by number of repeaters.

The gate-to-drain capacitance effect is called Miller effect, a phenomenon by which a feedback path between the input and the output of an electronic device is provided by the interelectrode capacitance. At high frequencies where the gate-to-drain capacitance is not negligible, the circuit is not open but involves a capacitance that is a function of the voltage gain. In the proposed design, for different values of supply voltages in different values of speed operation, Miller effect did not appeared as obstacle in the correct operation of the circuits.

7. CONCLUSIONS

Three classes of drivers/receivers for low power design applications are proposed in this paper. They are based on low swing technique using multithreshold voltage technology. In the new driver designs the usage of high threshold voltage transistors causes a significant decrease in the their output swing level. For each proposed driver design, a corresponding receiver circuit is introduced in order to recover the low swing levels. Simulation results show that, by using the proposed architecture, up to 50% power dissipation savings and 40% of power delay product reduction could be achieved also. Noise margin and Miller phenomenon have not any effect using the proposed technique.

8. REFERENCES

[1] J. M. Rabaey, M. Pedram, "Low Power Design Methodologies," *Kluwer Academic Publishers*, 1996.

[2] A. P. Chandrakasan, R. W. Brodersen, "Low Power Digital CMOS Design," *Kluwer Academic Publishers*, 1995.

[3] Y. Nakagome, K. Itoh, M. Isoda, K. Takeuchi, M. Aoki, "Sub-1-V Swing Internal Bus Architecture for Future Low-Power ULSI's," *IEEE Journal of Solid-State Circuits*, vol. 28, no. 4, pp. 414-419, April 1993.

[4] M. Nekili and Y. Savaria, "Optimal Methods for Driving Long Interconnections in VLSI Circuits," in Proc. *IEEE International. Symposium on Circuits and Systems*, vol. 4, pp. 343-346, May 1994.

[5] G. Carlo Cardarilli, M. Salmeri, A. Salsano, O. Simonelli, "Bus Architecture for low-power VLSI digital circuits," in Proc. *IEEE International. Symposium on Circuits and Systems*, vol. 4, pp. 21-24, May 1996.

[6] Z. Zhu, B. S. Carlson,"Dynamic circuits for CMOS BICMOS low power VLSI design," in Proc. *IEEE International. Symposium on Circuits and Systems*, vol.??? pp.197-200,ISCAS 1996.

[7] H. Yamauchi, H. Akamatsu, T. Fujita, "An Asymptotically zero power charge-recycling bus architecture for battery-operated Ultrahigh Data Rate ULSI's", IEEE journal of solid-state circuits, vol. 30,pp. 423-431,April 1995.

[8] S. Mutoh, S. Shigematsu, Y. Matsuya, H. Fukuda, T. Kaneko, J. Yamada, "A 1-V Multithreshold-Voltage CMOS Digital Signal Processor for Mobile Phone Application," *IEEE Journal of Solid-State Circuits*, vol. 31, no. 11, pp. 1795-1801, November 1996.

[9] S. Shigematsu, S. Mutoh, Y. Matsuya, Y. Tanabe, J. Yamada, "A 1-V High-Speed MTCMOS Circuit Scheme for Power-Down Application Circuits, " *IEEE Journal of Solid-State Circuits*, vol. 32, no. 6, pp. 861-869, June 1997.

[10] N. H. E. West, K. Eshraghian, "Principles of CMOS VLSI Design: A Systems Perspective," *Addison-Wesley*, 2nd Edition, Reading, 1993.

[11] Rjoub, S. Nikolaidis, O. Koufopavlou, T. Stouraitis, "A New Efficient Low Power Bus Architecture," in Proc. *IEEE International. Symposium on Circuits and Systems*, vol. 3, pp. 1864-1867, June 1997.

[12] H. Zhang, J. Rabaey, "Low-Swing Interconnect Interface Circuits," in the Proc. *IEEE International Symposium on Low Power Electronics and Design*, pp. 161-166, June 1998.

[13] M. Hiraki, H. Kojima, H. Misawa, T. Akazawa, Y. Hatano," Data-dependent logic swing internal bus architecture for ultralow-power LSI's," *IEEE Journal of Solid-State Circuits*, vol. 30, no. 4, pp. 397-402, April 1995.

[14] H. Kojima, S. Tanaka, K. Sasaki, "Half-Swing Clocking Scheme for 75% Power Saving in Clocking Circuitry," *IEEE Journal of Solid-State Circuits*, vol. 30, no. 4, pp. 432-435, April 1995.

[15] R. Golshan, B. Haroun, " A Novel Reduced Swing CMOS BUS Interface Circuit for High Speed Low Power VLSI Systems," in Proc. *IEEE International Symposium on Circuits and Systems*, vol. 4, pp. 351-354, May 1994.

[16] A. Rjoub, L. Bisdounis, O. Koufopavlou, "Influence of the nMOS and pMOS Transistor Threshold Voltage on CMOS Circuits Power Dissipation," in proc. *IEEE International Conference on Electronics Circuits and Systems*, vol. 2, pp. 545-549, 1997.

[17] A. Bellaouar, I. S. Abu-Khader, and M. I. Elmasry, " Low-Power CMOS/BiCMOS Drivers and Receivers for on-Chip Interconnects, " *IEEE Journal of Solid-State Circuits*, vol. 30, no. 6, pp. 696-700, June 1995.

[18] S. M. Kang, "Accurate Simulation of Power Dissipation in VLSI Circuits," *IEEE Journal Solid State Circuits*, vol. 21, no. 5, pp.889-891, October 1986.

[19] Sakurai, "Approximation of Wiring Delay in MOSFET LSI," *IEEE Journal of Solid-State Cicuits,* vol. SC-18, pp. 418-426, August 1983.

[20] H. Kawaguchi, T. Sakurai, " A Reduced Clock-Swing Flip-Flop (RCSFF) for 63% Power Reduction," *IEEE Journal of Solid-State Circuits*, vol. 33, no. 5, pp. 807-811,May1998.

Integrating Dynamic Power Management in the Design Flow

António Mota, Nuno Ferreira, Arlindo Oliveira, José Monteiro
IST/INESC
{mota,nmlf,aml,jcm}@inesc.pt

Abstract

Power dissipation has recently emerged as one of the most critical design constraints. A wide range of techniques has already been proposed for the optimization of logic circuits for low power. Power management methods are among the most effective techniques for power reduction. These methods detect periods of time during which parts of the circuit are not doing useful work and shut them down by either turning off the power supply or the clock signal.

In this work, we describe the integration of dynamic power management tools in a design flow. The designer can thus easily apply these techniques to the design, evaluate the optimization achieved and decide if the changes are worthwhile to include in the final design.

We have used this design flow with a real project. An HDLC controller has been designed and these power management techniques have been applied.

Keywords: Power optimization, power management, design flow

INTRODUCTION

Power consumption has become a primary concern in the design of integrated circuits. Two independent factors have contributed for this. On one hand, low power consumption is essential to achieve longer autonomy for portable devices. On the other hand, increasingly higher circuit density and higher clock frequencies are creating heat dissipation problems, which in turn raise reliability concerns and lead to more expensive packaging.

In the last few years, research on techniques for low power at various levels of design has intensified. Techniques based on disabling the input/state registers when some input conditions are met have been proposed and shown to be among the most effective in reducing the overall switching activity in sequential circuits. The disabling of the input/state registers is decided on a clock-cycle basis and can be done either by using a register load-enable signal or by gating the clock. A common feature in these methods is the addition of extra circuitry that is able to identify input conditions for which some or all of the input/state registers can be disabled. This class of techniques is sometimes referred to as *logic level* or *dynamic power management.*

In this work, the objective was to test these power management techniques on a real project. First, some of the techniques were integrated in the design flow: from a high-level VHDL description the circuit is synthesized to logic level; tools can be applied at this level and an evaluation of the power reduction can be obtained; the power managed design can then be mapped to an FPGA and tested under normal operating conditions.

Using this design flow, a power managed HDLC controller has been designed. This circuit is part of an expansion board for a PC that implements the interface to an ISDN (Integrated Services Digital Network) line.

In Section 1, the power management techniques that have been integrated in the design flow are presented. The complete design flow is described in Section 2. A brief description of the circuit that has been designed is given in Section 3. Section 4 presents the experiments that were carried out and the power reductions obtained. A discussion about these results is given in Section 5.

1. DYNAMIC POWER MANAGEMENT

During normal operation of well designed CMOS circuits, power consumption is determined by the switching activity in the circuit (Chandrakasan et al., 1992). Under a generally accepted simplified model, the power dissipation at the output of a gate g in a logic circuit is given by:

$$P_g = \frac{1}{2} \cdot C_g \cdot V_{DD}^2 \cdot f \cdot N_g \tag{1.1}$$

where P_g denotes power, V_{DD} the supply voltage, and f the clock frequency. C_g represents the capacitance gate g is driving and N_g is the switching activity at the output of gate g, *i.e.*, the average number of gate output transitions per clock cycle. The product $C_g \cdot N_g$ is called *switched capacitance.*

Most power optimization techniques at different levels of abstraction target the minimization of the switched capacitance in the circuit (Devadas and Malik, 1995). So called power management techniques shutdown blocks of hardware for periods of time in which they are not producing useful data. Shutdown can be accomplished by either turning off the power supply or by disabling the clock signal. A system-level approach is to identify idle periods for entire modules and turn off the clock lines for these modules for the duration of the idle periods (Chandrakasan et al., 1992, Chapter 10). However, there still is no real methodology for system level power management. It is up to the designer to devise a strategy for power management for a particular project.

In contrast, a few techniques have been proposed at the gate level, e.g., Alidina et al., 1994; Benini et al., 1996a; Benini et al., 1996b; Chow et al., 1996; Monteiro and Oliveira, 1998; Tiwari et al., 1995. These techniques are based on disabling the input/state registers when some input conditions are met, either by using a register load-enable signal or by gating the clock. In this situation there will be zero switching activity in the logic driven by input signals coming from the disabled registers. The main difference from system-level power management is that the shutdown of hardware is decided on every clock cycle, hence the name *dynamic power management.*

In order to decide whether to load or not new values into the registers, some extra logic has to be added to the original circuit. Naturally, this is redundant circuitry, increasing both area and power dissipation. In fact, the basic tradeoff in dynamic power management techniques is area for lower power consumption. The argument is that power is becoming the main constraint and that area is no longer critical.

Even if area is not a concern, this extra logic, which is active all the time, translates to additional power consumption. The power savings obtained by shutting down the registers must compensate for this overhead. The more complex this logic is, the larger the power overhead. Thus, on one hand, the more input conditions that are targeted for register shutdown, the larger the period of time during which the original circuit is being powered down. On the other hand, the larger the power penalty from the extra logic. In general, there is an optimum size for the extra logic and the goal of the different power management techniques at the logic level is to find this optimum.

The motto of logic-level power management is to have a small amount of logic that is active most of the time, but that is able to shutdown a much larger circuit during that time.

We have integrated two of these power management techniques in our design flow, which we briefly describe next.

1.1 PRECOMPUTATION

Precomputation, as proposed by Alidina et al., 1994 was one of the first logic-level shutdown methods proposed. In this method a simple combinational circuit (the precomputation logic) is added to the original circuit. Under certain input conditions, the precomputation logic disables the loading of all or a subset of the input registers. Under these input conditions, no power is dissipated in the portions of the original circuit with only disabled registers as inputs.

The basic architecture of this method is shown in Figure 1. A is the original combinational logic. Blocks g_1 and g_2 constitute the precomputation logic and are designed such that they are a function of a subset of the inputs to A. Power

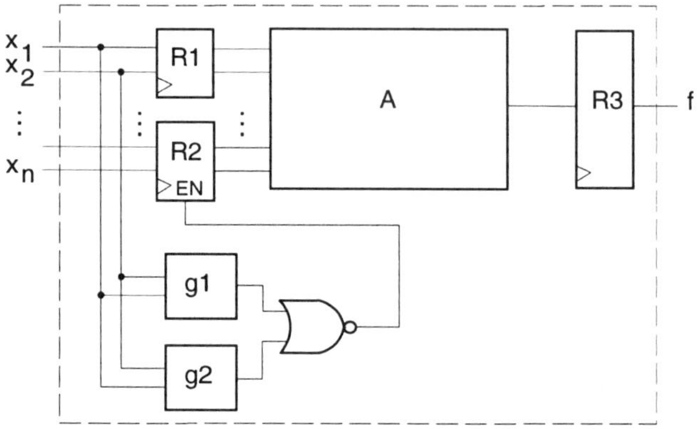

Figure 1 The precomputation architecture.

dissipation in the original circuit A is reduced when the outputs of either g_1 or g_2 evaluate to 1.

The choice and the number of inputs to use for the g_1 and g_2 functions is critical. The more inputs used, the highest the probability the precomputation logic will be active, thus disabling logic in block A. However, the size of the precomputation logic, a circuitry overhead that is active all the time, also increases, thus offsetting the gains obtained by disabling A a larger fraction of the time.

Once the number of inputs to the precomputation logic is fixed, the input selection is based on the probability that the outputs can be computed without the knowledge of a specific input, *i.e.*, the size of the observability don't-care set (ODC):

$$\text{ODC}_i = f_{x_i} \cdot f_{\overline{x_i}} + \overline{f}_{x_i} \cdot \overline{f}_{\overline{x_i}} \tag{1.2}$$

Inputs with lowest $prob(\text{ODC}_i)$ are selected to be in the precomputation logic.

1.2 FINITE STATE MACHINE DECOMPOSITION

Decomposition of finite state machines (FSMs) targeted for low power has been recently proposed by Chow et al., 1996 and Monteiro and Oliveira, 1998. The basic idea is to decompose the STG of the original finite state machine into two coupled STGs that together have the same functionality as the original machine. Except for transitions that involve going from one state in one sub-FSM to a state in the other, only one of the sub-FSMs needs to be clocked. The techniques described in Chow et al., 1996 and Monteiro and Oliveira, 1998 differ both in the way the partitioning of the states is performed and in the structure of the final circuit.

We have selected the technique of Monteiro and Oliveira, 1998 to integrate in our design flow. This technique follows the standard general decomposition

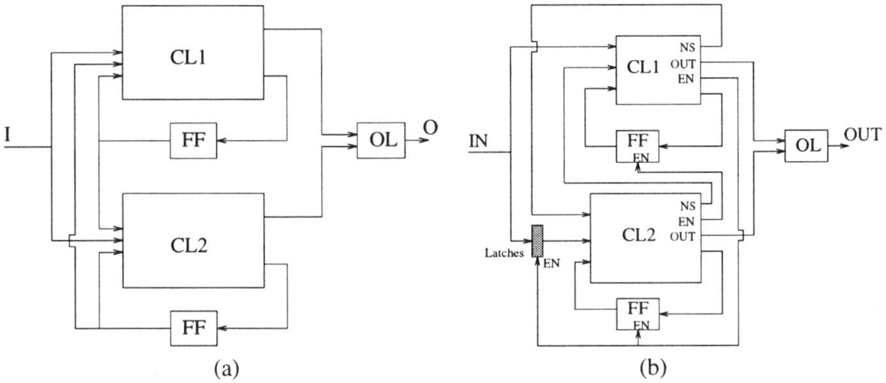

Figure 2 FSM general decomposition: (a) traditional, (b) with power management.

structure, which is shown in Figure 2(a). The selection of the states is such that only a small number is selected for one of the sub-FSMs. This selection consists in searching for a small cluster of states such that summation of the probability of transitions between states in the cluster is high and with a very low probability of transition to and from states outside of the cluster. The aim is to have a small sub-FSM that is active most of the time, disabling the larger sub-FSM. The reason for requiring a small number of transitions to/from the other sub-FSM is that this corresponds to the worst situation when both sub-FSMs are active.

The power optimized structure is shown is Figure 2(b). Each sub-FSM has an extra output that disables the state registers of the other sub-FSM. This extra output is also used to stop transitions at the inputs of the large sub-FSM. To avoid the area/power overhead incurred by adding latches, and since when this technique is effective the small sub-FSM is in operation most of the time, the inputs to the small sub-FSM are not filtered.

2. DESIGN FLOW

For the designers to be able to use dynamic power management tools, they have to be completely integrated in the design flow. We have done this for the two techniques described in the previous section. The design flow graph is depicted in Figure 3.

We start with a RTL description of the design in VHDL. We are using Synopsys to synthesize this VHDL description to gate-level. The power management techniques, precomputation and FSM decomposition, are applied at this level. However, these tools are available inside Berkeley's SIS package (Sentovich et al., 1992) and there is no common circuit description language accepted by

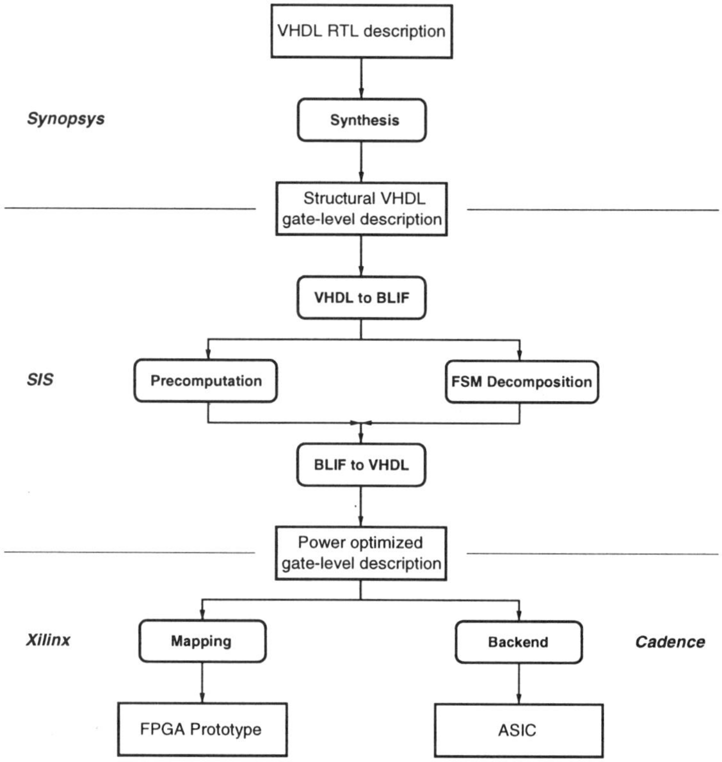

Figure 3 Design flow integrating the dynamic power management tools.

both frameworks. We have developed a translator between a mapped gate-level VHDL description and a mapped BLIF description. A translator in the opposite direction was also developed. The translation process involves mapping the gate-level description to a generic library that is common to both Synopsys and SIS.

Given the mapped BLIF description, the design is read into SIS and the power management tools can be applied. These tools are still not fully automatic in the sense that there are some parameters that the user has to specify. For precomputation, the number of inputs to use for the precomputation logic has to be specified. Thus, the designer may have to try different values in the search for the best possible power savings. Similarly, in the case of FSM decomposition, the number of states for the small machine has to be given. The typical behavior of the power estimates in this search is, as each of these values increase, the power reduces due to the increase of the fraction of time that the registers are stopped. After a certain value, the power starts to go up as the complexity of the extra circuitry required to generate the disabling signal increases significantly, offsetting the gains from the blocking of the registers.

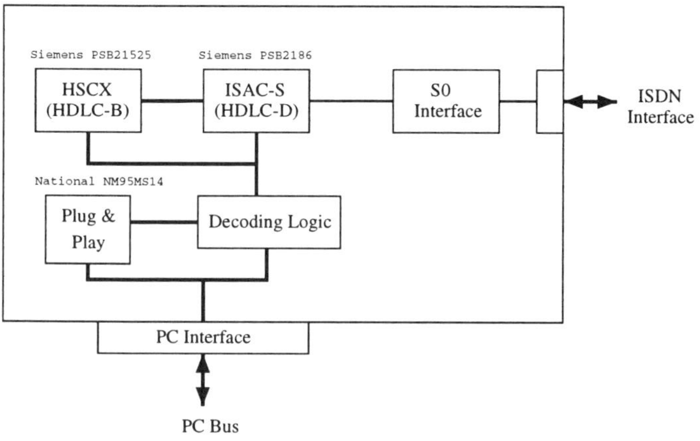

Figure 4 Schematic of the PCBIT board.

The power estimation tool in SIS permits the evaluation of the solution found by the power optimization tool. Translators from BLIF to SLS (van Genderen, 1989) and Spice formats have also been developed to allow more precise power estimates, at switch and electrical levels.

Once the designer has found the optimum in terms of power optimization, this circuit is again mapped to the common generic library and the VHDL description is obtained using the BLIF to VHDL translator. Again inside Synopsys, the circuit can be targeted for an ASIC or FPGA implementation. Depending on the target, the circuit is mapped to the corresponding library and the backend part of the process is performed using Cadence or Xilinx, respectively.

3. HDLC CONTROLLER

In this section is described the design that was used to demonstrate the integration of the dynamic power optimization tools in the design flow and their effectiveness when applied to a real design.

The PCBIT board is a PC expansion board currently being commercialized that implements an ISDN interface, allowing for a PC to be connected to an ISDN line. This board is being redesigned to a PC-Card format. One of the concerns is the power consumption of the new board. To this end, the plan is to integrate all the functionality of the board in a single ASIC and explore to the full extent the use of all power optimization techniques.

The results reported in this paper refer only to the part that implements the functionality of the Siemens PSB21525 chip. This circuit handles the layer 2 of the ISDN protocol (ITU, 1993), namely the HDLC protocol (High-level Data Link Control) for channels B1 and B2. On one side it interfaces using IOM-2 frames (Siemens, 1991) with the circuit that is responsible for layer 1.

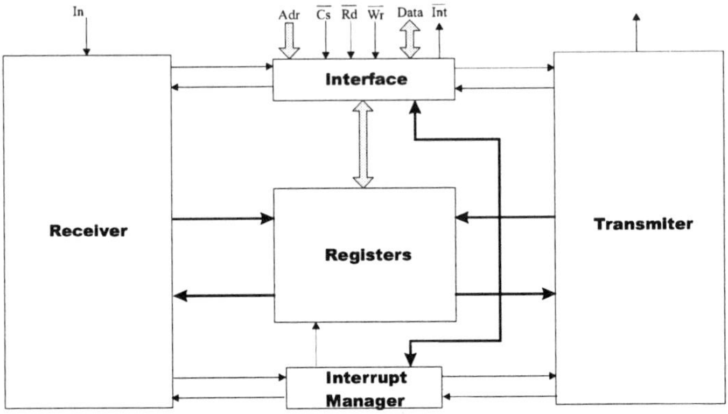

Figure 5 Block diagram of the demonstrator.

On the other side it is connected to the PC ISA bus. The block diagram of the circuit is depicted in Figure 5.

The two main modules in the circuit are the transmitter and the receiver. The transmitter receives data from the PC through a FIFO and constructs a IOM-2 frame to communicate with the layer 1 circuit. This process includes generating flags, bit-stuffing, and the frame check sequence (FCS). The receiver performs the reverse operation: decodes the information from the IOM-2 frame, writes the data to a FIFO and sends an interrupt to the PC to notify it that there is data available. The block diagrams for each of these modules are shown in Figure 6. Each is basically made of finite state machines that, with the help of counters, tracks the field of the IOM-2 frame that is being read or written. The FIFOs hold the data that has been received from the PC and is waiting to be processed by the circuit and vice-versa, data that has been read from the ISDN line and is waiting to be read by the PC. The power management techniques were applied mainly to the finite state machines in these circuits.

There are three other modules in the circuit of Figure 5. The interface module includes a decoder to generate the chip-select for the board and some latches for the interface with the PC bus. All the transfers to the PC bus are interrupt driven and this is handled by the interrupt-manager module. Finally, the registers shown in the figure are used to configure the board during system reset and to communicate with the ISA bus, namely by holding pending interrupts.

4. RESULTS

The circuit of the previous section was first described in VHDL and verified using logic simulation. It has been synthesized using Synopsys and mapped to a Xilinx FPGA, following the design flow presented in Section 2. This

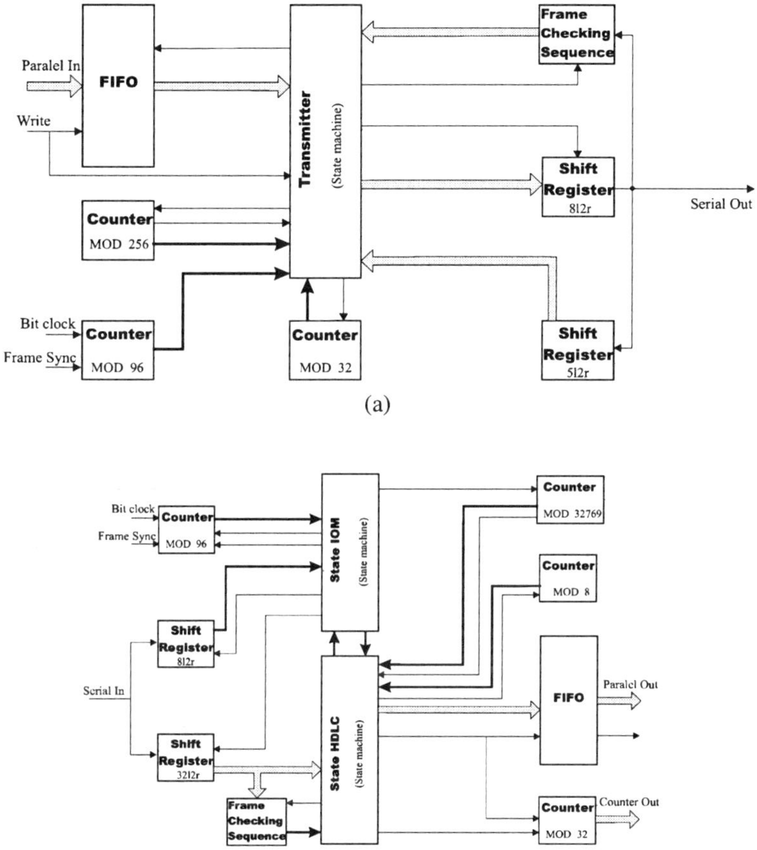

Figure 6 Block diagram: (a) transmitter, (b) receiver.

circuit was then tested under real operation by replacing the original Siemens PSB21525 circuit in the PCBIT board (see Figure 4) with the FPGA.

The power management techniques described in Section 1, precomputation and finite state machine decomposition, were applied separately to the finite state machines inside the transmitter and receiver modules. The FIFOs and counters shown in Figure 6 were not included in the FSMs since they make the state space much larger and increase significantly the number of transitions between states, two factors that necessarily limit the effectiveness of the power management techniques under test.

Table 1 Statistics for the circuits for which power management was applied.

FSM	PIs	POs	Registers	States
Trm	67	35	10	432
Rec	12	94	14	7,680

The statistics for the two FSMs are given in Table 1, namely, the number of primary inputs (PIs), number of primary outputs (POs), the number of registers and the number of states of the FSM.

Experiments with the power management technique based on FSM decomposition were carried out using a range of 1 to 10 for the number of states in the small FSM. All the solutions found by the tool led to an increase of the power consumption. Therefore this technique was not effective on these two test cases.

Precomputation was tested with 1 to 15 inputs in the precomputation logic. This technique was also not effective for the receiver. The results were not very good for the transmitter either: power reduction was only achieved when using 1 input for the precomputation logic and the reduction was a mere 0.1%.

Although these automatic tools found no good solution, still some power management for the circuit can be accomplished by inspection. The IOM-2 frame is composed of 12 windows of 1 byte each, as shown in Figure 7. The first two windows correspond to channels B1 and B2, respectively. Since this circuit only handles these channels, clock-gating can be used outside these windows to stop most counters in the circuit. Instead of adding additional logic, the `FSC` signal is being used as the gating signal. Figure 8 shows a switch-level simulation of the circuit using the gated-clock signal, `bcl_aux`, instead of the regular clock, `bcl`.

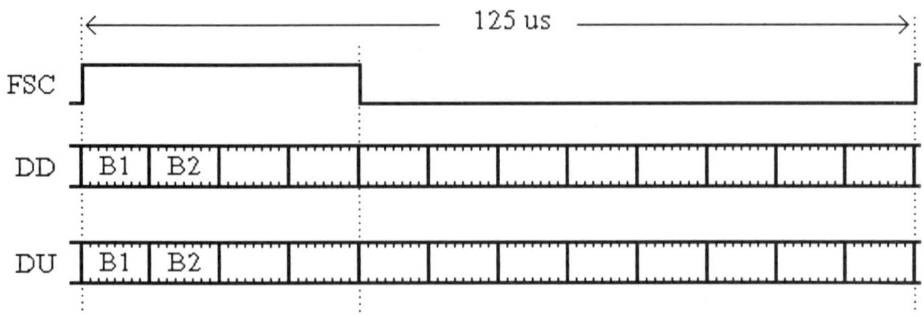

Figure 7 One frame of the IOM-2 protocol.

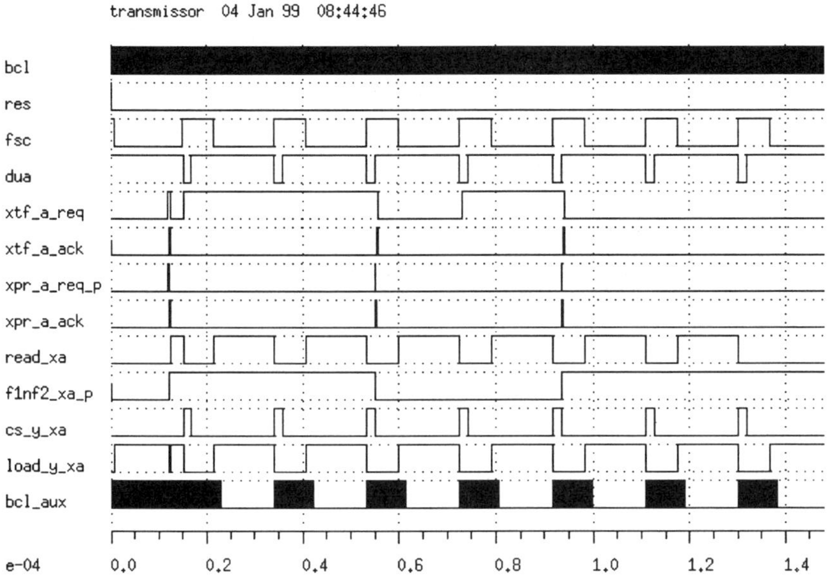

Figure 8 Simulation of the power managed version of the Transmitter.

Table 2 Statistics for the circuits for which power management was applied.

FSM	Original		FSM	Precomputation				Clock-gating			
	Tran.	Pwr	Decomp.	Tran.	% T	Pwr	% P	Tran.	% T	Pwr	% P
Trm	1,804	0.91	-	1,832	1.5	0.91	0.1	1,821	0.90	0.62	31.7
Rec	4,629	4.77	-		-			4,655	0.6	3.12	34.6

A summary of the results obtained is presented in Table 2. The power estimates in mW for the original and optimized circuit, together with the percentage variation are shown. These estimates were obtained using the switch-level simulator SLS (van Genderen, 1989). Similar statistics are given for the number of transistors in the circuit.

5. CONCLUSIONS

In this work, we have described a framework that integrates tools for dynamic power management in the design flow. The designer can thus easily experiment different techniques, evaluate their effectiveness and decide whether or not to include them in the final design.

The tools currently available in the design flow are precomputation and finite state machine decomposition. We have applied them on a real project, an HDLC controller. The typical circuits for which these techniques are applicable are controller-type sequential logic circuits. Hence, the finite state machines inside the HDLC circuit were selected as modules to be power managed. Regrettably, for this circuit, the technique based on finite state machine decomposition was not effective at all. No solution was found that reduced the power dissipation of the circuit. Moreover, even precomputation only achieved negligible power gains. Therefore, none of these techniques will be incorporated in the final design. Instead, a simple clock-gating mechanism will be implemented, with expected power savings around 30%.

Despite the negative results obtained with the automatic power management tools for this particular circuit, the integration of these techniques in the design flow makes them easy to use and, in general, worth the effort of applying them to power critical designs.

6. REFERENCES

Alidina, M., Monteiro, J., Devadas, S., Ghosh, A., and Papaefthymiou, M. (1994). Precomputation-Based Sequential Logic Optimization for Low Power. *IEEE Transactions on VLSI Systems*, 2(4):426–436.

Benini, L., Siegel, P., and Micheli, G. D. (1996a). Automatic Synthesis of Low-Power Gated-Clock Finite-State Machines. *IEEE Transactions on Computer-Aided Design*, 15(6):630–643.

Benini, L., Vuillod, P., Coelho, C., and Micheli, G. D. (1996b). Synthesis of Low-Power Partially-Clocked Systems from High-Level Specifications. In *9th International Symposium on System Synthesis*.

Chandrakasan, A., Sheng, T., and Brodersen, R. (1992). Low-Power CMOS Digital Design. *IEEE Journal of Solid-State Circuits*, 27(4):473–484.

Chow, S.-H., Ho, Y.-C., and Hwang, T. (1996). Low Power Realization of Finite State Machines – A Decomposition Approach. *ACM Transactions on Design Automation of Electronic Systems*, 1(3):315–340.

Devadas, S. and Malik, S. (1995). A Survey of Optimization Techniques Targeting Low Power VLSI Circuits. In *Proceedings of the* 32nd *Design Automation Conference*, pages 242–247.

ITU (1993). *Digital Subscriber Signalling, System No. 1*. ITU-T Recommendation Q.921.

Monteiro, J. and Oliveira, A. (1998). Finite State Machine Decomposition for Low Power. In *Proceedings of the* 35th *Design Automation Conference*, pages 758–763.

Sentovich, E., Singh, K., Moon, C., Savoj, H., Brayton, R., and Sangiovanni-Vincentelli, A. (1992). Sequential Circuit Design Using Synthesis and Optimization. In *Proceedings of the International Conference on Computer Design: VLSI in Computers and Processors*, pages 328–333.

Siemens (1991). *IOM-2 Interface Reference Guide*.

Tiwari, V., Ashar, P., and Malik, S. (1995). Guarded Evaluation: Pushing Power Management to Logic Synthesis/Design. In *Proceedings of the International Symposium on Low Power Electronics and Design*, pages 221–226.

van Genderen, A. (1989). SLS: An Efficient Switch-Level Timing Simulator Using Min-Max Voltage Waveforms. In *Proceedings of the VLSI'89 Conference*, pages 79–88.

Self-Timed Techniques for Low-Power Digital Arithmetic in GaAs VLSI

S. Lachowicz[†], K. Eshraghian[†], H-J. Pfleiderer[‡]
[†] *Edith Cowan University, Australia,* [‡] *Ulm University, Germany*

Key words: Gallium Arsenide, Self-Timed Systems, Very High Speed Systems

Abstract: This article presents a self-timed approach to digital Gallium Arsenide logic applicable to high performance VLSI circuits and systems. The design techniques are based on GaAs Latch-Coupled FET Logic (LCFL) in order to achieve reasonable power-delay-area trade-off. The complexities due to clock skew are avoided and power savings achieved through the pipelined architecture. A range of arithmetic circuits is presented and their performance evaluated.

1. INTRODUCTION

Clock skew is the speed limiting factor in digital synchronous systems. Moreover, the clock distribution system can dissipate a considerable amount of power, reaching up to 40% of the total power dissipation of the system [1]. On the other hand, self-timed digital systems do not suffer from the problem of the clock skew. However, the penalty is the increased complexity of the system and the requirement to incorporate a handshaking circuitry that permits reliable communications between asynchronous modules. Each module generates an event (in the form of a signal transition) when it is ready to accept data, and another event on completion of its computation. The use of transition signalling is common in self-timed applications due to the achievable time and power savings [2]. The handshaking modules can be implemented using Event Driven Logic described in [3]. Several standard circuit elements commonly required to process transition signals have been developed. These include: Muller-C elements (AND for events), Exclusive-OR gates (EDLXOR), Inclusive-OR gates (EDLINCOR), and inverters. The implementations of theses circuits in Gallium Arsenide technology are readily available [4][5]. Especially important is the Muller-C element which is the core building block used in the two- and four-phase handshake protocols [6]. This paper presents a design methodology for Gallium Arsenide self-timed integrated systems [7] using LCFL latched logic primitives [8].

2. GaAs LATCH-COUPLED FET LOGIC (LCFL)

Latch-Coupled FET Logic (LCFL) [8] together with Pseudo-Dynamic Latched Logic are MESFET [9], logic families overcoming the inability of the Direct Coupled FET Logic (DCFL) to support AND connection of the Enhancement type transistors in the pull-down section of the GaAs logic gate as the latch refreshing the output voltage compensates for the leakage current. The basic structure of the LCFL gate is presented in Figure 1.

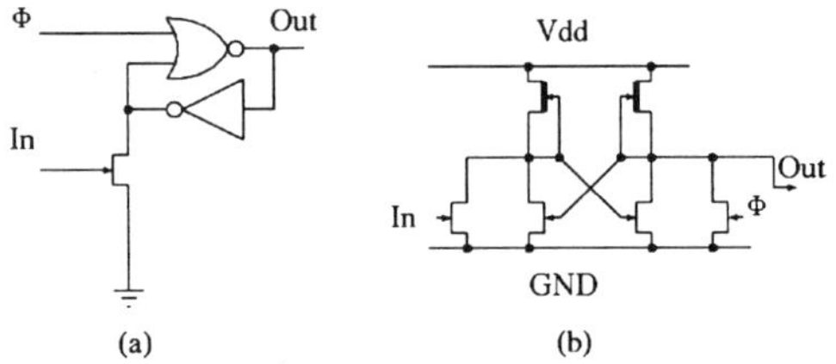

Figure 1. LCFL gate.(a) gate level schematic, (b) transistor level schematic

In order to explain the basic operation of the LCFL cell, a two-cell part of a two-phase shift register, shown in Figure 2. together with the LCFL AND gate from Figure 3(a), is considered. The cells are clocked by Φ and $\overline{\Phi}$ respectively. During the first half of the clock period the signal Φ is high and, therefore, V_{out} of the first stage is low and there is no interaction with the next stage. Since J5 is "on" and the output is low, J2 is "off". With J0 being "off" the internal node will be high and its voltage will be limited to approximately 0.7 V because of the J4 gate conduction. For the other case of J0 being "on", the internal node will be low. In other words, the transistor stage consisting of J0 and J1 acts as a DCFL inverter connected to the input J4 of the next inverter stage J4/J3. Depending on the input signal *in* two combinations are possible at the end of the first half of the clock period:

1. *in* = H *internal* = L, *out* = L (L/L)
2. *in* = L *internal* = H, *out* = L (H/L)

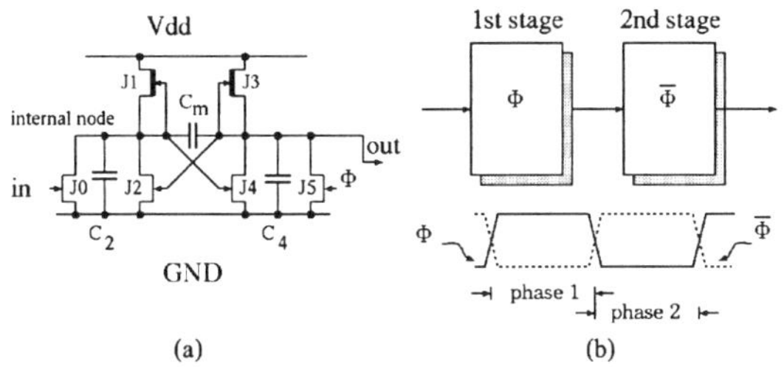

Figure 2. An LCFL shift register. Cell circuit diagram (a), two consecutive cells (b)

These are the initial conditions for the latch consisting of the two inverters, J2/J1 and J4/J3, which operates during the second half of the clock period. To discuss this behaviour the state diagram for $V_{internal}$ and V_{out}, shown in Fig. 3, will be used. It shows the static transfer curves of the latch J1/J2, J3/J4 where the DC load presented by J2 and J4 is included. The separatrix (thick line) is superimposed on the same graph. The separatrix determines which state, (L/H) or (H/L), will be reached at the end of the second half of the clock period. For example, with the initial condition set to (L/L) in phase 1 which corresponds to V_{in} = H, the latch reaches the final state (L/H) in phase 2, as shown by the trajectory in Figure 3. The initial condition set to (H/L) is in the vicinity of the final endpoint (H/L) and, therefore, poses no problem. The minimum distance between the starting point and the separatrix may be thought of as a noise margin for this type of latched logic because if (L/L) is shifted to the right of the separatrix by noise sources, this will result in an incorrect state. By proper dimensioning the sizes of transistors J0 ... J5 as well as considering the three capacitances, C_2, C_m and C_4, this distance can be made approximately 150 mV as shown in Fig. 3 for the case of the AND gate. The method of the calculation of the separatrix can be found in [10]. The slope of the separatrix will not change due to the temperature variations since it is determined by transistor parameter and capacitance ratios resulting in a rather robust circuit performance.

In terms of power dissipation, LCFL is very efficient because it uses the currents of the pull-up transistors in the latch twice: during phase 1 for the logic evaluation and in phase 2 for the latch function. Moreover, there is an inherent latch property which decouples adjacent stages. This is a major reason that the AND function in the logic is possible as opposed to DCFL.

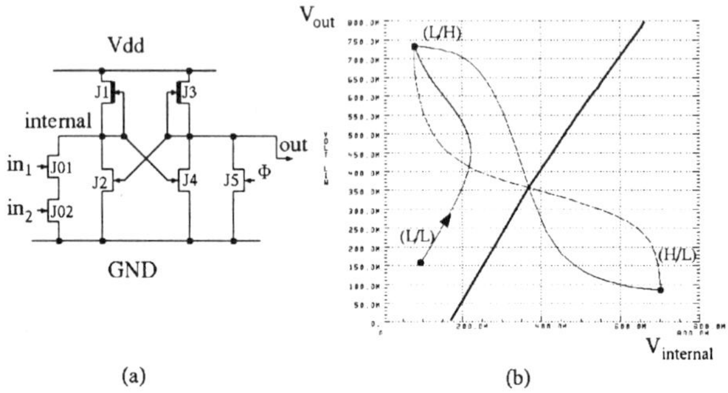

Figure 3. State transition diagram of the LCFL latch used in an AND gate. Circuit schematic (a) and the transition diagram (b)

3. SELF-TIMED SYSTEMS IN MESFET GAAS

Self-timed systems require that logic cells have several control inputs and that they generate at least one control signal for handshaking. For the typical 4-phase handshaking protocol the input signals are *Enable* and *Start* and the required generated signal is *Done* (*Complete*) as shown in Figure 4. The *Done* signal triggers the *Request* input in the next stage's handshaking block.

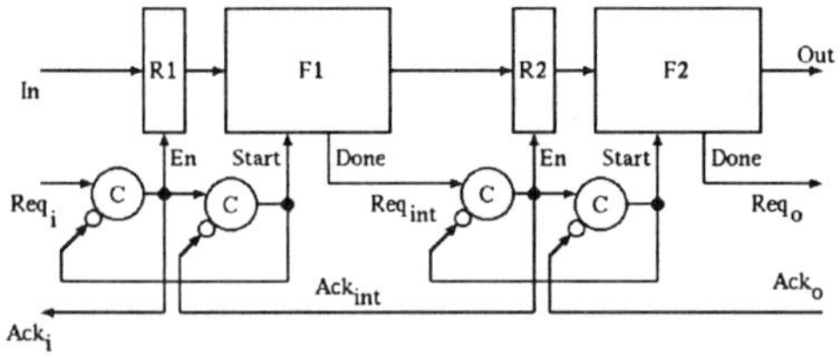

Figure 4. A classical four-phase pipeline.

The logic path consists of the register latching the input signals and the functional block implementing the logic function. The detailed operation of this type of a self-timed pipeline can be found in [6]. The GaAs LCFL logic family can be used efficiently to implement self-timed systems. The clock input can readily be used as the *Request* line, and the logic cell contains a latch which, if proper hanshaking protocol is applied, should allow elimination of the separate latches from the pipeline structure. The *Done* signal, indicating when the logic evaluation is completed, needs to be generated by extra hardware, as does the *Enable* line.

3.1 The Muller-C element

As can be seen in Figure 4, the Muller-C element is the fundamental component of the handshaking path of the self-timed pipeline. In terms of logic operation, it implements the AND function for events, such that if a specific transition takes place at one input and it is coincident with, or followed by, a similar transition of the other input(s), then that transition will be presented at the output [3]. In conventional logic terms its function can be described as:

$$Y(i+1) = Y(i)(A+B) + AB$$

Using an LCFL gate this equation can be implemented in the structure presented in Figure 4.

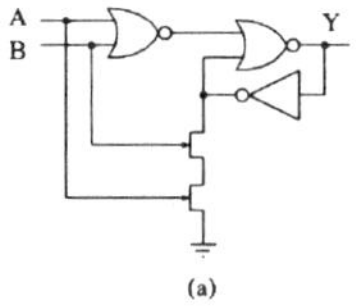

A	B	Y(i+1)
0	0	0
0	1	Y(i)
1	0	Y(i)
1	1	1

(b)

Figure 5. LCFL implementation of the Muller-C gate

3.2 Self-timed GaAs LCFL pipeline

Figure 6 shows an LCFL cell for self-timed applications. As can be seen the latch is an inherent component of the cell. This property together with the appropriate modifications to the handshake path to suit the GaAs latched logic design style can be utilised to eliminate the separate latch blocks from the pipeline. The modified pipeline is shown in Figure 7. In the self-timed LCFL cell from Figure 6, the $\overline{Complete}$ signal is generated by first producing the complement of the output with a NOR gate which is also controlled by the $\overline{Request}$ signal. This NOR gate is sized appropriately to achieve equal signal delay at the input of the following NOR gate producing the $\overline{Complete}$ signal.

The logic cell operates as follows: When the $\overline{Request}$ line is high the cell is in the reset state, both lines *Out* and $\overline{Out}$ are low and the $\overline{Complete}$ line is high. When $\overline{Request}$ goes low and *Enable* is high the cell evaluates the output and one of the lines *Out* or $\overline{Out}$ conditionally goes high which causes the $\overline{Complete}$ line to go low and this indicates that logic evaluation has been completed.

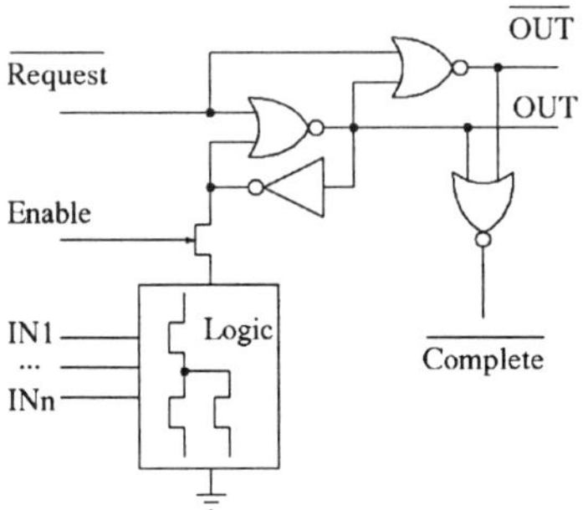

Figure 6. LCFL cell for the self-timed pipeline

The *Enable* line is not always needed and in the pipeline structure from Figure 7(a) the logic cells do not contain the *Enable* line. There are cases, however, when *Enable* line cannot be omitted as in the bubble-shift register presented in Section 4.2.

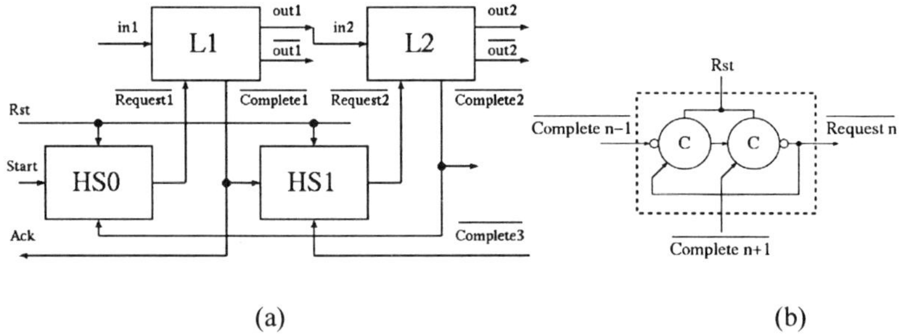

Figure 7. LCFL pipeline (a), details of the hanshaking block (b).

The modification of the handshake path comparing to the standard four-phase handshake protocol is such that a particular LCFL logic cell is not allowed to enter into the reset state until the following cell completes its evaluation, and also it cannot perform the next evaluation until the cell following it enters the reset state. This handshake protocol operation removes the need for the separate latches between the logic stages.

4. ARITHMETIC BUILDING BLOCKS

The following sections demonstrate the design of several arithmetic building blocks using the self-timed, pipelined approach. The operation of standard shift-register and bubble shift-register as well as the adder and the accumulator are presented. Especially interesting is the accumulator design because it requires the implementation of a special memory cell needed in the feedback path, as the value stored in the feedback path may have to wait for the asynchronous, input data for arbitrary periods of time.

4.1 Shift register

The pipeline from Figure 7(a), with the logic cells like the one from Figure 6, with only one transistor in the logic block and no *Enable* transistor, forms a self-timed shift-register.

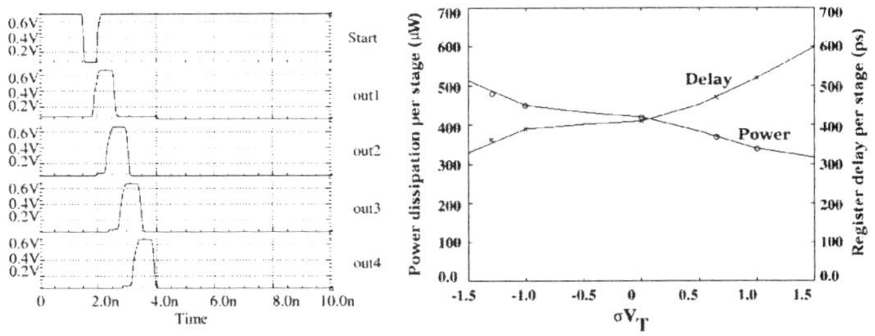

(a) (b)

Figure 8. Self-timed GaAs LCFL shift register operation

Figure 8(a) presents the HSPICE simulation of the operation of the chain of 4 cells and Figure 8(b) the performance as the function of the spread of the threshold voltage in the range $(-1.5 \div 1.5)\sigma V_T$ for the 0.6 μm MESFET technology.

4.2 Bubble shift register

Sometimes the area occupied by the register may become the main constraint, and, if the speed margin provided by the Gallium Arsenide technology is sufficient, the register may be implemented as a bubble-shift register reducing the area requirement by almost 50%. The bubble shift operation is shown in Figure 9(a). There is one more cell, than is needed to store all the data word bits. This cell is used as a bubble and is shifted left. Shifting the bubble from the rightmost to the leftmost position in the register chain is equivalent to shifting the data word one position right.

The implementation of the bubble shift register as a GaAs self-timed pipeline for the simplified case of two data bits and three logic cells is shown in Figure 9(c). The feedback path connecting the output of the register to its input is necessary if the data word needs to be preserved. The logic cells in the bubble shift register have to utilise the *Enable* line as in this case adjacent cells store the logic value and the only interaction between the logic cells is allowed when a cell holding no logic value reads the contents of is left neighbouring cell which is equivalent to the bubble being passed from a cell currently holding it to its immediate neighbour on the left.

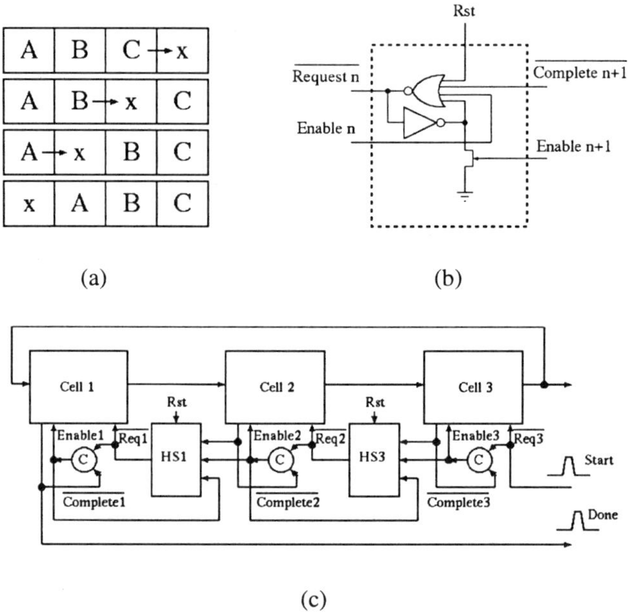

Figure 9. Bubble shift operation

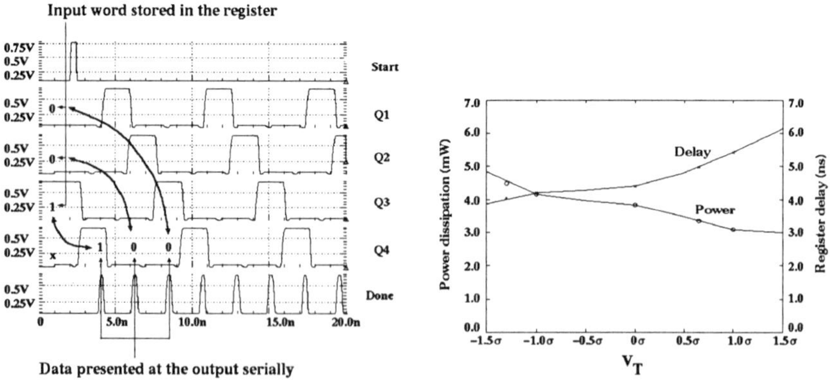

(a) (b)

Figure 10. Operation and performance of the bubble-shift register

The handshake path direction is opposite to the direction of the logic path as the handshake protocol controls passing of the bubble from right to left. The special handshake block is a slight modification of an LCFL cell where two additional reset lines to accommodate *Rst* and *Enable n* control signals have been incorporated.

TABLE 1 - Power dissipation comparison between synchronous and self-timed designs

Design	Power excluding clock	Power in the clock drivers	Total power
Synchronous	4 mW	3.5 mW	7.5 mW
Self-timed	4 mW	0	4 mW

As in the previous case, the performance of the circuit was assessed using HSPICE. Figure 10(a) shows the waveforms at the output of the register for the simplified case of four logic cells (3-bit data word) and Figure 10(b) shows the relationship between the spread of 0.6μm MESFET process parameters and the delay and power dissipation of the circuit.

The comparison of the power dissipation of the 12-bit synchronous LCFL register and the self-timed 12-bit bubble shift register is shown in Table 1.

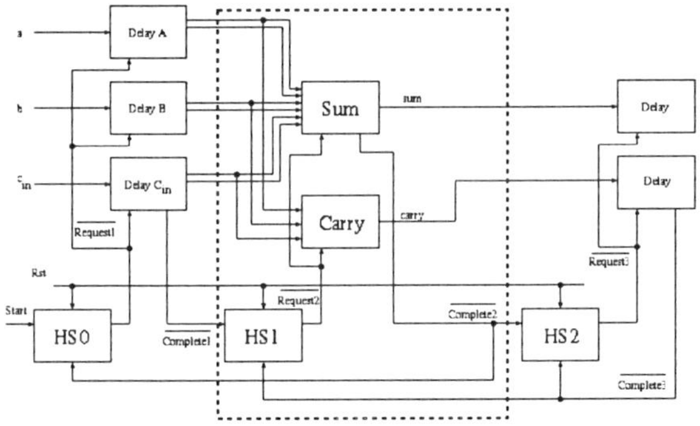

Figure 11. A one-bit, self-timed LCFL GaAs adder

4.3 The Adder

The adder is the centre point of any arithmetic unit. In bit serial calculations a one-bit adder is everything that is required. A one-bit adder is a building block for larger, pipelined stages as is demonstrated in the next section. The appropriate self-timed architecture in GaAs MESFET is depicted in Figure 11. The details of the sum and carry blocks are shown in Figure 12. The adder circuit has been simulated using the models of the 0.6 μm E/D GaAs MESFET technology. Fig 13(a) shows the operation of the adder and Fig 13(b) presents the performance. The delay has been defined as the time between the falling slopes of the $\overline{Request}$ and $\overline{Complete}$ signals. The adder exhibits the delay of 190 ps and power dissipation of 200 μW for the typical value of V_T of the MESFET process.

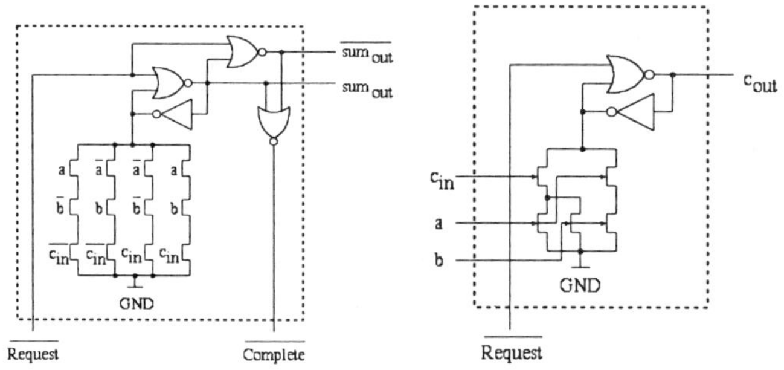

(a) (b)

Figure 12. Details of the sum (a), and carry cell (b).

4.4 The Accumulator

The accumulator structure for the case of 4 bits is shown in Figure 14(a). It can be observed that the handshaking hardware overhead is not significant, especially for higher numbers of bits, although because of the limited fan-out of the GaAs gates some buffering might be required.

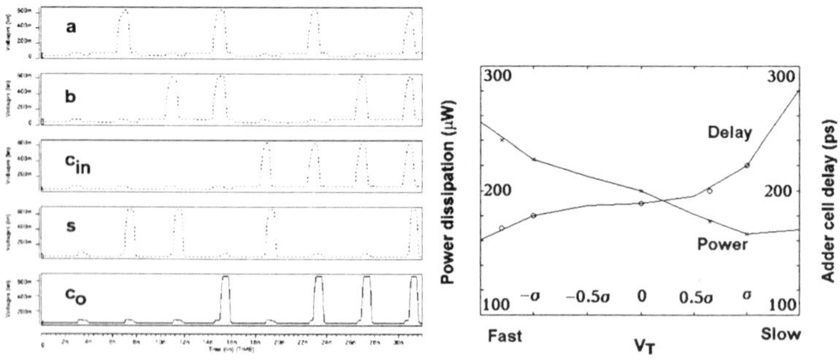

(a) (b)

Figure 13. Operation and performance of the adder

The accumulator contains one-bit adder cells, and pre-skew and de-skew sections consisting of simple delay cells. However, because of the feedback present in the accumulator, special memory cells have to be employed. The memory cells ensure that regardless of the delay in the input data (which can be asynchronous) the accumulator adds correctly the new set of data to the current contents. The memory cell, shown in Fig. 14(b), uses the basic cell from Figure 6 and one Muller-C cell from Figure 5 to read the output of the adder and is triggered by the $\overline{Complete}$ signal from the adder cell.

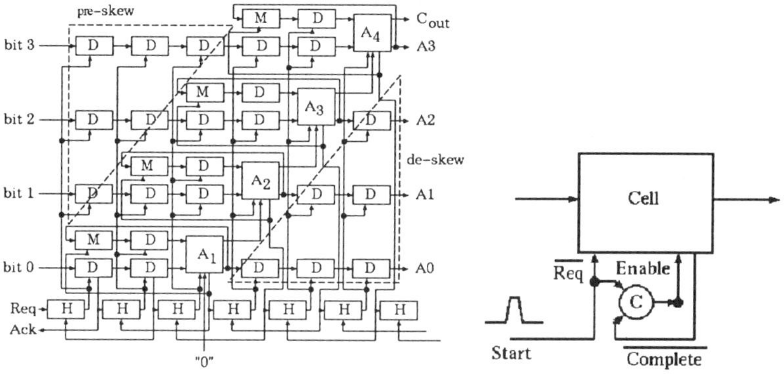

(a) (b)

Figure 14. A 4 bit accumulator (a) and a special memory cell (b).

Figure 15(a) shows the waveforms at the output of the accumulator for a 4-bit data word 0011, and Figure 15(b) shows the relationship between the spread of 0.6µm GaAs MESFET process parameters and the power dissipation of the circuit. The circuit throughput is 0.5 Gsps and does not depend on the data word width. It is expected that this value will further increase for the 0.4µm process.

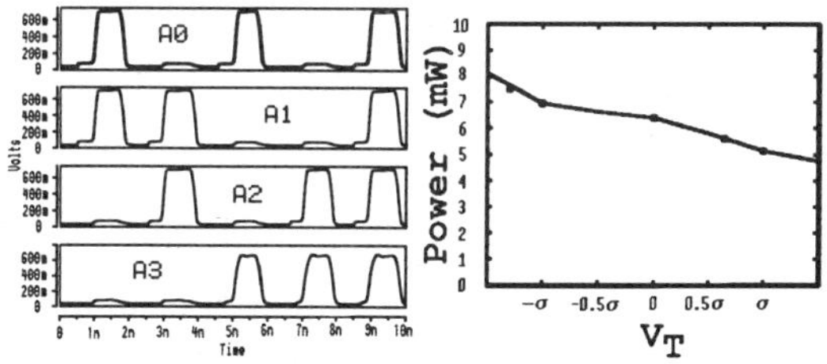

(a) (b)

Figure 15. Accumulator simulation results: (a) output waveforms for 0011 data word, (b) power dissipation as a function of V_T spread

5. CONCLUSIONS

The paper demonstrates a unified design methodology for Gallium Arsenide MESFET self-timed integrated systems useful for high performance computing. The four-phase handshake protocol has been modified specifically to suit the requirements of GaAs latched logic circuits. The resulting circuits are inherently delay insensitive and power efficient as the clock signal has been entirely eliminated. The latches present in between logic stages of the classic micropipeline have been eliminated using the inherent latching property of the LCFL GaAs logic family leading to further savings in power dissipation and die area. A range of design examples of various arithmetic circuits has been included.

6. ACKNOWLEDGMENT

The support of the Australian Research Council, German Academic Exchange Office (DAAD) and Centre for Very High Speed Microelectronic Systems at Edith Cowan University is gratefully acknowledged.

7. REFERENCES

[1] W. Bowhill, *A 300 MHz Quad-Issue CMOS RISC Microprocessor*, Technical Digest of the 1995 ISSCC Conference, San Francisco, February 1995.

[2] N.R. Poole, *Self-timed logic circuits*, Electronics & Communication Engineering Journal, pp 261-270, December 1994.

[3] D.A. Pucknell, *Event-driven logic (EDL) approach to digital system representation and related system design processes*, IEE Proceedings E, Vol. 140(2), pp. 119-126, 1993.

[4] K. Eshraghian, S.W. Lachowicz, J.F. López, and R. Sarmiento, *Efficient design of gallium arsenide Muller-C element*, Electronics Letters, Vol. 33(9), pp. 757-759, April 1997.

[5] S.W. Lachowicz, K. Eshraghian, J.F. López, R. Sarmiento, and H-J. Pfleiderer, *Low Power GaAs Methodologies for Asynchronous Systems*, Chapter in: "Mixed Design of Integrated Circuits and Systems", Kluwer Academic Publishers, London, pp. 157-162, 1998.

[6] I. E. Sutherland, *Micropipelines*, Communications of the ACM, Vol. 32(6) pp. 720-738, June 1989.

[7] S.W. Lachowicz, K. Eshraghian, J. F. López, and H-J. Pfleiderer, *Design of Self-Timed Gallium Arsenide Integrated Systems for Multimedia Computing*, Proceedings of the MIXDES'98 Conference, June 1998, Lodz (Poland), pp 93-98.

[8] S.W. Lachowicz, K. Eshraghian, and H-J. Pfleiderer, *Efficient low power circuit design with Gallium Arsenide latched logic*, Proceedings of the ISIC-97 Conference, pp 158-161, Singapore, September 1997.

[9] J.F. López, K. Eshraghian, R. Sarmiento, A. Núñez and D. Abbott, *Gallium Arsenide pseudo-dynamic latched logic for high performance processor cores*, IEEE Journal of Solid State Circuits, Vol. 32(8), pp. 1297-1303, August 1997.

[10] K.O. Jeppson, *Comments on the Metastable Behavior of Mismatched CMOS latches*, *IEEE J. Solid-State Circ.*, 1996, **31**, pp, 275-277

On Defect Level Estimation and the Clustering Effect

José T. de Sousa
INESC/IST
Lisbon, Portugal *
jts@hobbes.inesc.pt

Abstract This paper presents a review of existing defect level models and introduces a new defect level model that accounts for the fault clustering effect. The model uses generalized negative binomial statistics to model the probability distribution of the number of faults in a chip. This analysis shows that clustering, in addition to naturally increasing the yield, also raises the detection probability and therefore lowers the defect level. By accounting for clustering, the new model predicts a less stringent fault coverage requirement than other models.

Keywords: Defect clustering, defect level, fault clustering, fault coverage, reject ratio

Nomenclature

α: fault clustering parameter.

α_d: defect clustering parameter.

λ: average number of faults per chip.

λ_d: average number of defects per chip.

Ω: fault coverage.

Ω_{max}: maximum attainable fault coverage.

$\bar{D}$: fault average area density.

* Work done while at Bell Labs, Murray Hill, NJ

DL: defect level.

R: realistic to stuck-at fault detectability ratio.

T: stuck-at fault coverage.

Y: true yield.

Y_m: measured yield.

n_0: average number of faults in a faulty circuit.

r: number of faults in a circuit.

1. INTRODUCTION

Defect level (DL) is the fraction of faulty chips among the chips that passed production test. These chips are taken as good devices and shipped as so. Later they are likely to fail in the field, causing the manufacturer to incur in significant expenses. There are also important invisible costs such as customer satisfaction, company prestige, etc. The economical importance of defect level hardly needs to be highlighted. The problem is how to predict and control its value.

The cause of field rejects are faults caused by manufacturing defects; the same defects that are responsible for yield loss. To cope with the complexities of physical defect phenomena, sophisticated yield models have been developed. These models take into account the non-equiprobability of physical defects, by considering a weighted variety of possible causes for yield loss. They also account for the defect clustering phenomenon, which produces significantly more accurate yield estimates than theories that assume defects that are probabilistically independent.

Despite the relevance of the clustering effect, many defect level models do not account for it [Williams and Brown, 1981, Agrawal et al., 1982, Sousa et al., 1996], but there are a few models that do [Seth and Agrawal, 1984, Singh and Krishna, 1996]. In [Seth and Agrawal, 1984] the clustering effect is implicitly taken into account by using the negative binomial distribution for the number of defects in a chip. In [Singh and Krishna, 1996] defect clustering is exploited to identify dice with different DL values in the same wafer. These dice are then placed in different bins, according to their quality level (the inverse of defect level). However, in none of these theories the overall effect of clustering on defect level has been investigated. Since it is well known how defect clustering affects the yield of integrated circuits, we now ask the question of how it affects DL. This is the problem addressed in this paper.

The assumptions of this work are the same as in current yield theories — defects of random size and location, governed by specific probability distributions. However, we directly model the probability that a chip will contain r

faults, rather than the probability that it will contain r defects. We observed that if one defect can produce multiple faults, this simply corresponds to a higher degree of fault clustering. In this way, there is no need to introduce a relationship between faults and defects as in [Seth and Agrawal, 1984]. Consequently, the new DL model is simpler, using one parameter less than the models in [Seth and Agrawal, 1984].

This paper is organized as follows. In Section 2. we discuss existing DL models, and provide the motivation for the new model. In Section 3. the new model is introduced and analyzed. Section 4. concludes the paper and gives directions for future developments.

2. BACKGROUND

Assuming that faults in a circuit are probabilistically *independent* and have the same occurrence probability, Williams and Brown [Williams and Brown, 1981], in their seminal work, derived the following DL model:

$$DL = 1 - Y^{(1-\Omega)}. \tag{1.1}$$

This model constituted the first attempt to show how DL depends on the yield Y and the fault coverage Ω. Equating Ω to the stuck-at fault coverage T, causes the Williams-Brown model to spread generalized panic: unless 100% or very high fault coverage is obtained, the defect level will be unacceptable. For example, suppose $Y = 80\%$ and we want $DL = 100$ parts per million (p.p.m.). According to the Williams-Brown formula the fault coverage requirement is $\Omega = 99.55\%$. The belief that the stuck-at fault coverage T actually represents the real fault coverage Ω led test engineers to demand 100% stuck-at fault coverage no matter the cost. Paradoxically, the very same engineers would rest completely assured if 100% stuck-at fault coverage had in fact been achieved. In fact, reality is somewhat different: 100% stuck-at fault coverage may not be needed in practice; on the other hand, 100% stuck-at fault coverage does not prevent some other faults from escaping the test.

Despite its illustrative power, the Williams-Brown model has very low accuracy when used with the stuck-at fault coverage T. Some studies of defect level data from real manufacturing processes have demonstrated this fact [Maxwell and Aitken, 1991]. The most striking difference between the $DL(T)$ curves obtained with real data and the $DL(T)$ curves obtained with the Williams-Brown model is the type of curvature exhibited. Real $DL(T)$ plots show concave curvature (positive second derivative), whereas the Williams-Brown model shows convex curvature (negative second derivative).

It is also striking that Y, the yield parameter that appears in Equation (1.1) is usually computed under completely different assumptions. Y is obtained assuming that the number of faults in a chip has a generalized negative binomial distribution [Stapper et al., 1983]. This distribution is given by the following equation:

$$P(r) = \frac{\Gamma(\alpha + r)}{r!\,\Gamma(\alpha)} \frac{(\frac{\lambda}{\alpha})^r}{(1 + \frac{\lambda}{\alpha})^{r+\alpha}}. \tag{1.2}$$

where λ is the average number of faults per chip, and α is the fault clustering parameter. The parameter λ is given by

$$\lambda = A\bar{D}, \tag{1.3}$$

where A is the chip area and $\bar{D}$ the average fault density. The clustering effect varies inversely with α: α close to zero indicates strong clustering, whereas α large indicates weak clustering. Equation (1.2) is defined for any positive real values of α and λ and any integer $r = 0,\ 1,\ 2,\ ...,\ \infty$. In reality, the number of faults in a chip is large but finite. Nevertheless, $P(r)$ being defined in the range $r = 0,\ 1,\ 2,\ ...,\ \infty$ is accurate, since $P(r)$ decreases fast with r. The yield Y is obviously given by $Y = P(0)$, which produces the well known negative binomial yield formula:

$$Y = \left(1 + \frac{\lambda}{\alpha}\right)^{-\alpha}. \tag{1.4}$$

The formula above accounts for fault interdependence due to clustering. In contrast, the Williams-Brown DL model assumes fault independence.

The assumptions of the Williams-Brown model imply that the number of faults r in a chip follows a Poisson distribution rather than a negative binomial distribution:

$$P(r) = \frac{\lambda^r}{r!} e^{-\lambda}. \tag{1.5}$$

For this distribution the yield $Y = P(0)$ is given by

$$Y = e^{-\lambda}. \tag{1.6}$$

The result above can also be obtained using Equation (1.4) and supposing weak clustering (large α). In fact, using Stirling's formula, the limit of Equation (1.4) when $\alpha \to \infty$ results in Equation (1.6):

$$\lim_{\alpha \to \infty} (1 + \frac{\lambda}{\alpha})^{-\alpha} = e^{-\lambda}. \tag{1.7}$$

Substituting Equation (1.6) in Equation (1.1), we obtain an expression for the Williams-Brown model which is preferred in this work:

$$DL = 1 - e^{-\lambda(1-\Omega)} \tag{1.8}$$

In [Sousa et al., 1996] it is suggested that the reason why the Williams-Brown formula combined with the stuck-at fault coverage T cannot track experimental fallout data is the fact that, unlike T, in the real fault coverage Ω each fault should be weighted with its probability of occurrence. In [Sousa et al., 1994] it is shown by extensive simulation of more accurate fault models that different weighting of the faults produce different non-linear relationships between Ω and T, which would explain the real shape of $DL(T)$ curves. The clustering effect is briefly mentioned in [Sousa et al., 1996] but its fundamental importance is not realized.

Agrawal et al. [Agrawal et al., 1982] supposed that the number of faults *in a faulty circuit* is Poisson distributed, with average $n_0 \geq 1$. This assumption produced the following model:

$$DL = \frac{(1-\Omega)(1-Y)e^{-(n_0-1)\Omega}}{Y+(1-\Omega)(1-Y)e^{-(n_0-1)\Omega}}. \tag{1.9}$$

This model provided a good fit to experimental DL data [Maxwell and Aitken, 1991] using the stuck-at fault coverage T as the fault coverage Ω. The value of n_0 can be determined using, for example, a least squares fitting method. The model realistically reproduces the concave curvature of the $DL(T)$ curve.

We regard the assumption that faults in a faulty circuit are Poisson distributed as a first attempt to incorporate the clustering effect on DL models. In fact, a higher value for n_0 merely indicates a higher propensity for multiple faults, which is basically what clustering is. As discussed for the Williams-Brown model, the assumptions underlying the Agrawal et al. model are not consistent with the assumptions for deriving Y, a parameter used in the model.

Seth and Agrawal proposed a defect level model based on negative binomial statistics [Seth and Agrawal, 1984]. This model is derived from a formulation that also enables characterizing the yield equation using wafer test data. The resulting model is the following:

$$DL = 1 - \left[\frac{\alpha_d + \lambda_d(1-e^{-c})}{\alpha_d + \lambda_d(1-e^{-c\Omega})}\right]^{-\alpha_d}, \tag{1.10}$$

where λ_d and α_d are the defect density and defect clustering parameter, respectively, and c is the average number of faults per defect, assumed to be Poisson distributed.

This model also reproduces the concave curvature of the $DL(T)$ curve in a realistic manner. Defect clustering is incorporated in the Seth-Agrawal model by means of the parameter α_d. However, the need to model the occurrence of defects and, separately, the relation between logical faults and defects is questionable. Note that the existence of more parameters than needed in a

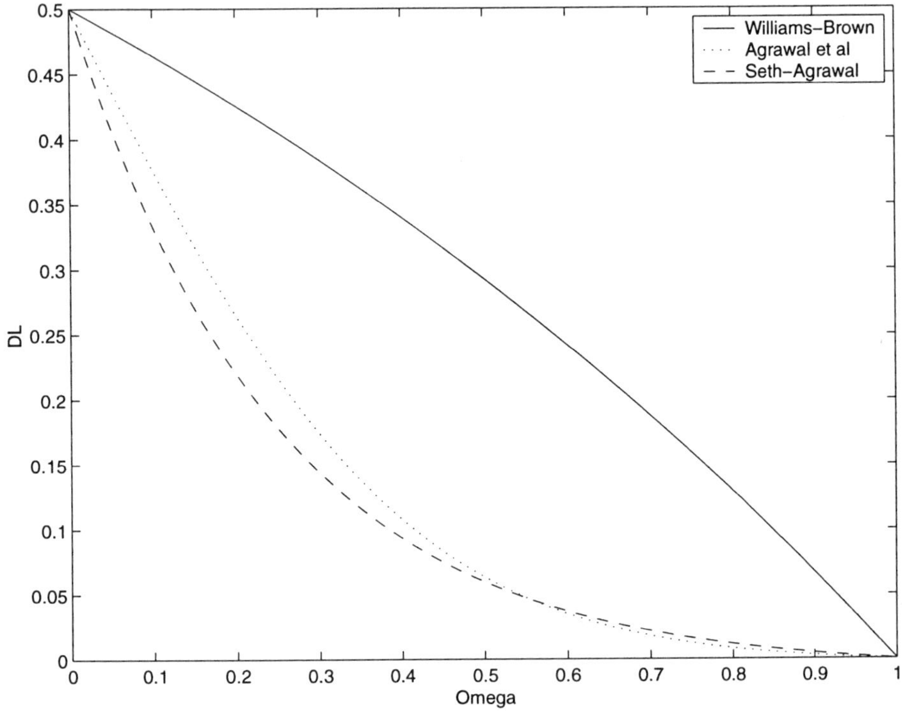

Figure 1 Comparison of DL models.

model leads to lack of decisiveness. That is, a good fit occurs for a family of values of the parameters, instead of for just one combination of values. This means that the parameters are related in a particular way; any combination of values that respects this relation will do. Also, albeit the clustering effect is accounted for, the effect of varying its intensity has not been studied.

The three models discussed are compared in Figure 1, which shows DL as a function of the fault coverage Ω, for a hypothetical circuit for which $Y = 0.5$ (note that $Y = 1 - DL$ at $\Omega = 0$). Since this paper presents a theoretical study, we have no need to relate Ω to a practical measure of fault coverage, such as the stuck-at fault coverage. All models are analyzed assuming the availability of a realistic fault coverage figure Ω.

For the Agrawal et al. model we used $n_0 = 5$. For the Agrawal-Seth model we chose the parameters $\alpha = 1$, $\lambda = 1.0187$ and $c = 4$. The values of the parameters were exaggerated to clarify the points made before, and illustrate what the models are capable of. The parameters are not chosen to track any particular data set.

While the other two models have second derivatives which can be tuned by their parameters, the Williams-Brown model exhibits negative second derivative for all values of Ω:

$$\frac{d^2 DL}{d\Omega^2} = -\lambda^2 e^{-\lambda(1-\Omega)}. \tag{1.11}$$

Another interesting study is that of the first derivative at $\Omega = 1$. The following three equations give the first derivative at $\Omega = 1$ for the three models, in the same order they were presented:

$$\left.\frac{dDL}{d\Omega}\right|_{\Omega=1} = -\lambda \tag{1.12}$$

$$\left.\frac{dDL}{d\Omega}\right|_{\Omega=1} = -\frac{1-Y}{Y} e^{-(n_0-1)} \tag{1.13}$$

$$\left.\frac{dDL}{d\Omega}\right|_{\Omega=1} = -\frac{\alpha\lambda c e^{-c}}{\alpha + \lambda(1-e^{-c})} \tag{1.14}$$

The Williams-Brown model has only one parameter (λ) to match the slope at $\Omega = 1$, as well as the whole data set. The slope at $\Omega = 1$ is very important for computing DL at high fault coverage using a linear approximation. The other two models have more flexibility, since they have two and three parameters, respectively.

3. THE NEW DEFECT LEVEL MODEL

This section introduces the new DL model. The question of whether the clustering effect significantly affects DL is examined thoroughly. We start by deriving the model from a definition of DL, and then we present an analysis of the new model.

DL is the probability that a chip is faulty given it passed the test. This can be written

$$DL = P(\text{chip faulty} \mid \text{chip passed the test}), \tag{1.15}$$

which is equivalent to

$$DL = 1 - P(\text{chip good} \mid \text{chip passed the test}). \tag{1.16}$$

Applying Bayes' formula we can write

$$DL = 1 - \frac{P(\text{chip good AND chip passed the test})}{P(\text{chip passed the test})}. \tag{1.17}$$

Since all good chips pass the test

$$DL = 1 - \frac{P(\text{chip good})}{P(\text{chip passed the test})}. \tag{1.18}$$

The probability that a chip is good is the yield Y. The probability that a chip passes the test can be measured by counting such chips and dividing that number by the total number of chips. This quantity is obviously equivalent to the measured yield. Moreover, if the chips are counted after the application of each test vector, and if the fault coverage after application of each test vector is known, we can estimate the measured yield as a function of the fault coverage and denote this quantity $Y_m(\Omega)$. Therefore, the definition of DL as a function of the fault coverage Ω is

$$DL = 1 - \frac{Y}{Y_m(\Omega)}. \tag{1.19}$$

At this point we need to assume a distribution for the number of faults in a chip. In [Seth and Agrawal, 1984], a distribution for the number of defects (not faults) is considered first, and then another distribution for the number of faults per defect is postulated. In our method we directly consider faults, and assume that the number of faults has a distribution $P(r)$. Thus, we can obtain Y and $Y_m(\Omega)$ respectively by

$$Y = P(0), \tag{1.20}$$

$$Y_m(\Omega) = \sum_{r=0}^{\infty}(1-\Omega)^r P(r). \tag{1.21}$$

Note that the $(1-\Omega)^r$ is the probability that none of the r faults is detected.

Like in [Seth and Agrawal, 1984], we make use of the probability generating function (p.g.f.) method. The p.g.f. $G(s)$ of a probability distribution $P(r)$ is defined as

$$G(s) = \sum_{r=0}^{\infty} s^r P(r). \tag{1.22}$$

Since our objective is to study the effect of fault clustering, we will assume that $P(r)$ is a negative binomial distribution with parameters λ and α. The p.g.f. of the negative binomial distribution is known to be

$$G(s) = \left[1 + \frac{\lambda}{\alpha}(1-s)\right]^{-\alpha} \tag{1.23}$$

Thus, Equation (1.21) can be rewritten

$$Y_m(\Omega) = G(1-\Omega) = \left(1 + \frac{\lambda}{\alpha}\Omega\right)^{-\alpha}. \tag{1.24}$$

Comparing the equation above with the yield expression given by Equation (1.4), and bearing in mind the relationship between λ and the chip area A

as given by Equation (1.3), we conclude that Equation (1.24) gives the yield $Y_m(\Omega)$ of a chip whose area is $A\Omega$. This physical meaning agrees with that of expected measured yield as given before: a chip passes the test if it does not contain any faults in its tested area.

Replacing Equations (1.4) and (1.24) in Equation (1.19) results in the new defect level model, which can be written as

$$DL = 1 - \left(\frac{\alpha + \lambda}{\alpha + \lambda\Omega}\right)^{-\alpha}. \tag{1.25}$$

As a DL model, the new model gives $DL = 1 - Y$ at $\Omega = 0$ and $DL = 0$ at $\Omega = 1$. Moreover, it is interesting to note that

$$\lim_{\alpha \to \infty} 1 - \left(\frac{\alpha + \lambda}{\alpha + \lambda\Omega}\right)^{-\alpha} = 1 - e^{-\lambda(1-\Omega)}. \tag{1.26}$$

That is, as clustering weakens, the new DL model given by Equation (1.25) becomes equivalent to Williams-Brown model as given by Equation (1.8).

The new model resembles the Seth-Agrawal model. However, by directly considering faults, it does not need to model the number of logical faults caused by each defect. It is assumed that the fact that some defects may cause multiple faults is another form of fault clustering, which is subsumed by the clustering parameter α. In fact, it is possible to show that increasing parameter c (average number of faults per defect) in the Seth-Agrawal model has a similar effect on the DL curve as that of increasing the defect clustering parameter α_d. In our model, to study the effect of clustering we just need to vary the parameter α. That is done is Figure 2 for three values of α. It can be seen that as clustering increases (α decreases) the yield $Y = 1 - DL(0)$ increases and DL decreases for any fault coverage. It can also be seen that a low enough α can realistically reproduce the concave curvature of the DL curve. The curve for $\alpha = 0.1$ has positive second derivative.

The most important benefit of modeling clustering is the fact that we obtain a much more accurate fault coverage requirement, which is much easier to meet compared to the fault coverage required by the Williams-Brown model. To study the effect of clustering on fault coverage requirement we will use *test transparency* [McCluskey and Buelow, 1988] instead of fault coverage. The test transparency TT is defined by

$$TT = 1 - \Omega. \tag{1.27}$$

The maximum allowable test transparency TT_{max} is a better measure for the test effort because it tells us what is the fraction of the chip that we may afford to leave untested. Using a linear approximation in the neighborhood of $\Omega = 1$,

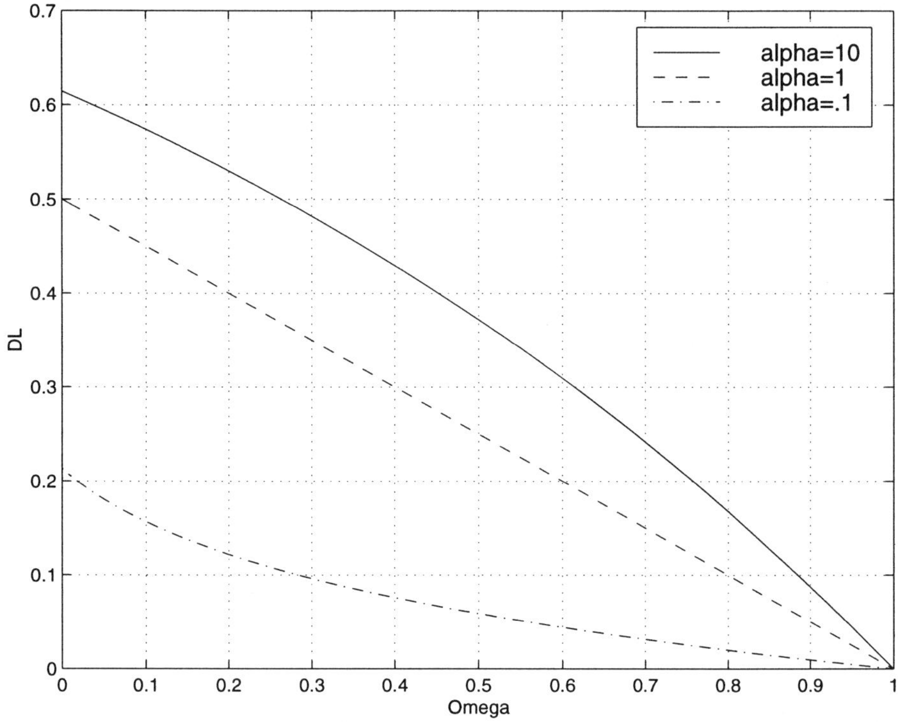

Figure 2 DL as a function of Ω with the new model for three values of α.

for a required DL_{max}, we need a TT_{max} given by

$$TT_{max} = \frac{DL_{max}}{\left|\frac{dDL}{d\Omega}\right|_{\Omega=1}}. \tag{1.28}$$

The first derivative at $\Omega = 1$ for the new DL is given by

$$\left.\frac{dDL}{d\Omega}\right|_{\Omega=1} = -\frac{\alpha\lambda}{\alpha+\lambda}. \tag{1.29}$$

The first derivative at $\Omega = 1$ for the Williams-Brown model is given by Equation (1.13). Then, comparing TT_{max} for the new model and TT_{max} for the Williams-Brown model, we obtain

$$\frac{TT_{max}(\text{new model})}{TT_{max}(\text{Williams-Brown})} = 1 + \frac{\lambda}{\alpha} \tag{1.30}$$

Typical values of α can be easily 10 times smaller than typical values of λ. Thence, the expression above can be further simplified to

$$\frac{TT_{max}(\text{new model})}{TT_{max}(\text{Williams-Brown})} \approx \frac{\lambda}{\alpha} \tag{1.31}$$

The equation above shows that the maximum allowable test transparency is inversely proportional to the clustering parameter α. This explains why the fault coverage requirement as predicted by the new model can be radically lower than that predicted by the Williams-Brown formula. For the example given before, if $DL = 100$ p.p.m. is required and $Y = 80\%$, the Williams-Brown model requires $\Omega = 99.55\%$, i.e., a maximum allowable test transparency $TT_{max} = .45\%$. If α is 10 times smaller than λ, $TT_{max} = 4.5\%$ for the new DL model. That is, instead of 99.55% fault coverage the new model requires only 95.5%.

4. CONCLUSION

In this paper a critical review of existing defect level theories has been presented, and a model that accounts for the clustering effect has been proposed. The original feature of the new model is that it assumes that the distribution of the number faults per chip is given by the generalized negative binomial distribution.

Other models in the literature either do not account for clustering or assume one distribution for the number of defects, and another distribution for the number of faults per defect. Our method of directly considering faults, eliminates possible overlapping between the roles of the parameters in the methods that consider two distributions. In this way, we were able to study the effect of clustering by varying a single clustering parameter.

Analysis of the new method revealed that the clustering effect is a very significant one, which cannot be ignored. Models such as the Williams-Brown model, that do not account for clustering, can easily underestimate the maximum allowable test transparency in one order of magnitude. In the case study presented in this paper, the Williams-Brown model required 99.5% fault coverage, while the new model required about 95.55%. This is much closer to what test engineers usually observe, and raises the optimism and confidence of manufacturers.

Directions for continuing this work are various. The newly derived model needs to be validated with actual DL data. The question of which fault models to use in order to represent the real faults remains open. Another question is whether the experimental DL curves contain or not any information about unmodeled faults. Should the *jumps* that appear on the measured yield versus fault coverage curve be modeled? In [Das et al., 1993], these jumps are not ignored and are treated as an integral part of the $Y_m(T)$ curve, but it is not known whether this is important.

5. ACKNOWLEDGMENTS

The author is grateful to Vishwani Agrawal for long and fruitful discussions on this theme.

6. REFERENCES

[Agrawal et al., 1982] Agrawal, V. D., Seth, S. C., and Agrawal, P. (1982). "Fault Coverage Requirement in Production Testing of LSI Circuits". *IEEE Journal of Solid State Circuits*, SC-17(1):57–61.

[Das et al., 1993] Das, D., Seth, S. C., and Agrawal, V. D. (1993). "Accurate Computation of Field Reject Ratio Based on Fault Latency". *IEEE Trans. on VLSI*, 1(4).

[Maxwell and Aitken, 1991] Maxwell, P. C. and Aitken, R. C. (1991). "The Effect of Different Tests Sets on Quality Level Prediction: When is 80% Better than 90%?". In *Proc. Int. Test Conference (ITC)*, pages 358–364.

[McCluskey and Buelow, 1988] McCluskey, E. J. and Buelow, F. (1988). "IC Quality and Test Transparency". In *Proc. Int. Test Conference (ITC)*, pages 295–301.

[Seth and Agrawal, 1984] Seth, S. C. and Agrawal, V. D. (1984). "Characterizing the LSI Yield Equation from Wafer Test Data". *IEEE Trans. on CAD*, CAD-3(2):123–126.

[Singh and Krishna, 1996] Singh, A. D. and Krishna, C. M. (1996). "On the effect of Defect Clustering on Test Transparency and IC Test Optimisation". *IEEE Transactions on Computers*, C-45(6):753–757.

[Sousa et al., 1996] Sousa, J. J. T., Goncalves, F. M., Teixeira, J. P., Marzocca, C., Corsi, F., and Williams, T. W. (1996). "Defect Level Evaluation in an IC Design Environment". *IEEE Trans. on CAD*, 15:1286–1293.

[Sousa et al., 1994] Sousa, J. T., Goncalves, F. M., Teixeira, J. P., and Williams, T. (1994). "Fault Modeling and Defect Level Projections in Digital IC's". In *Proc. European Design and Test Conf. (EDTC)*, pages 436–442.

[Stapper et al., 1983] Stapper, C. H., Armstrong, F., and Saji, K. (1983). "Integrated Circuit Yield Statistics". *Proc. IEEE*, 71:453–470.

[Williams and Brown, 1981] Williams, T. W. and Brown, N. C. (1981). "Defect Level as a Function of Fault Coverage". *IEEE Transactions on Computers*, C-30(12):987–988.

FASTNR: An Efficient Fault Simulator for Linear and Nonlinear DC Circuits

J. Soares Augusto and C. F. Beltrán Almeida
INESC/IST, Aptd. 13069, 1000 Lisboa, Portugal.
E-mail: jasa@dalton.inesc.pt and cfb@dalton.inesc.pt

Abstract

In this paper we describe the simulator FASTNR (*FAST Newton-Raphson*) where an efficient methodology for solving the faulty circuit equations, called *FAult RUBber Stamps* (FARUBS), is implemented. Its application to single fault simulation in linear and nonlinear circuits is reported. The efficient fault simulation in nonlinear DC circuits is due both to the development of original linearized Newton-Raphson models for electronic devices and to the simulation of fault values in a "continuation" stream. Fault simulation in linear cascades with up to 5000 nodes has shown an improvement of four orders of magnitude in simulation time, when compared to that of the nominal circuit. In nonlinear circuits, the time efficiency is sometimes better than two orders of magnitude.

Keywords: Efficient Analog Fault Simulation; Solution of Linear and Nonlinear Circuits; Analog Test and Diagnosis.

1. INTRODUCTION

Efficient fault simulation is an important issue when Simulation-Before-Test (SBT) dictionary-based techniques are used in fault diagnosis of electronic circuits. A survey of efficient fault simulation in analog circuits is given in [1, 2, 3]. In this paper, we describe some results of single fault simulation with the FASTNR simulator [4], built around the modified Newton-Raphson (NR) algorithm, that implements a methodology called *fault rubber stamps* for inserting faults in the circuit equations [3]. After a very concise (due to space limitations) survey of the literature on efficient fault simulation, we present the FARUBS methodology applied to linear circuits, and then we extend it to efficient fault simulation in nonlinear circuits. Several examples, including large circuits, are given in order to validate the approach. It isn't made any comparison with previous results since the works on the subject previously reported are focused on the methodology instead of on simulation times.

The Householder or Sherman-Morrison formula [5, 6] has been used for simulating efficiently faults in linear circuits by Temes and other authors [1, 7]. This rank one modification formula states that if $\mathbf{M}^{-1}$ is known, then the inverse of $\mathbf{M}+s\mathbf{v}\mathbf{w}^T$, where $\mathbf{v}$ and $\mathbf{w}$ are vectors and s is a scalar, is $(\mathbf{M}+s\,\mathbf{v}\mathbf{w}^T)^{-1} = \mathbf{M}^{-1}(\mathbf{I}-\alpha^{-1}\mathbf{v}\mathbf{w}^T\mathbf{M}^{-1})$, with $\alpha^{-1} = (s^{-1}+\mathbf{w}^T\mathbf{M}^{-1}\mathbf{v})^{-1}$. Since $\mathbf{v}$ and $\mathbf{w}$ are sparse, the amount of operations involved in the application of this formula is rather small when compared with the full inversion of the modified matrix. There are variants of that formula that apply to the $\mathbf{L}$ and $\mathbf{U}$ factors of $\mathbf{M}$ instead of to the inverse (see [8]), and other methods for updating triangular factors are given in [9] and its references.

Efficient fault simulation in nonlinear DC circuits was reported by Lin [10] and by Prasad [11]. In the former, it is used the piecewise linear modeling of the nonlinear devices and then ideal diodes are the only nonlinear elements. These elements load the ports of a multiport comprising the linear elements, and the circuit equations are formulated as a linear complementarity problem. In the latter work, the overall circuit is modeled as a linear multiport terminated by nonlinear elements and the equations of the faulty circuit are written by applying the modification formula to a matrix depending only on linear elements. The solution of the reduced nonlinear circuit equations is not discussed, as well as the potential efficiency that can be achieved.

Recently, other approaches [3] have been tried in this area: the simulation of faults in parallel on several distributed processors (or computers); the hierarchization of circuits in several description levels and the use of simulation tools capable of simulating these levels; and the preliminary development of simulators in the time domain capable of simulating faults in parallel.

In the present work, faults are simulated in linear circuits by *increasing* the nominal triangular LU factors by one line and one column: the nominal factors remain unchanged. In nonlinear circuits this same strategy is applied jointly with specially developed Newton-Raphson companion models for the electronic devices [5], what allows for a special structure in the NR matrix where only a small matrix block is updated in each NR iteration.

2. FAULT RUBBER STAMPS IN LINEAR CIRCUITS

The modified nodal analysis (MNA) equations can be assembled with "rubber stamps" on an element by element basis, by inserting the respective contributions in the circuit matrix [5, 12]. We extended that concept to *rubber stamps in faulty elements* and called this fault insertion and simulation methodology *fault rubber stamps*. This methodology is based on *reusing* the $\mathbf{L}$ and $\mathbf{U}$ factors of the nominal circuit matrix, obtained with the Crout algorithm [5, 6], to simulate a fault. We described this methodology for linear circuits in [4] and here we extend it to nonlinear circuits.

Consider the MNA equations of a linear circuit

$$\mathbf{Mx} = \mathbf{b} \tag{1.1}$$

where $\mathbf{M}$ is the $n \times n$ circuit matrix, $\mathbf{x}$ is the vector of circuit variables and $\mathbf{b}$ is the vector of independent sources. In circuit simulators, 1.1 is usually solved through the LU factorization of $\mathbf{M}$, that is, by calculating a lower triangular matrix $\mathbf{L}$ and an upper triangular matrix $\mathbf{U}$ with 1's in the diagonal such that $\mathbf{M} = \mathbf{LU}$. The solution of (1.1) is obtained by solving for $\mathbf{z}$ the lower triangular system $\mathbf{Lz} = \mathbf{b}$ and by solving for $\mathbf{x}$ the upper triangular system $\mathbf{Ux} = \mathbf{z}$

The LU factorization with the Crout algorithm calculates the elements l_{ij} and u_{ij} from left to right and from up to down. This means that if an element in $\mathbf{M}$ changes — say m_{ab} —, the l_{ij} and u_{ij} values with $i < a$ or $j < b$ remain unchanged and don't need to be calculated when obtaining the new $\mathbf{L}$ and $\mathbf{U}$ factors of the modified $\mathbf{M}$ matrix. If the modification in the system, due to the existence of a fault, occurs in the lowest rightmost part of the matrix, then most of the nominal $\mathbf{L}$ and $\mathbf{U}$ factors can be reused.

The method developed in order to *push the modifications in the equations due to the fault to the lowest rightmost part of* $\mathbf{M}$, consists in joining to (1.1) one more variable ϕ and one more equation related to the fault, what leads to the following system of equations for the faulty circuit

$$\underbrace{\begin{bmatrix} \mathbf{M} & \mathbf{v}_c \\ \mathbf{v}_\ell^T & -\delta^{-1} \end{bmatrix}}_{\mathbf{M}_a} \begin{bmatrix} \mathbf{y} \\ \phi \end{bmatrix} = \begin{bmatrix} \mathbf{b} \\ 0 \end{bmatrix} \tag{1.2}$$

where $\mathbf{v}_\ell$ and $\mathbf{v}_c$ are $n-$vectors filled with zeros, except in 2 positions at most, and $\mathbf{y}$ is the faulty circuit solution. The factors $\mathbf{L}_a$ and $\mathbf{U}_a$ of the matrix $\mathbf{M}_a$ lead to two triangular systems that are solved in sequence to simulate the fault

$$\underbrace{\begin{bmatrix} \mathbf{L} & \mathbf{0} \\ \mathbf{l}^T & \tau \end{bmatrix}}_{\mathbf{L}_a} \begin{bmatrix} \mathbf{z} \\ \phi \end{bmatrix} = \begin{bmatrix} \mathbf{b} \\ 0 \end{bmatrix}, \qquad \underbrace{\begin{bmatrix} \mathbf{U} & \mathbf{u} \\ \mathbf{0}^T & 1 \end{bmatrix}}_{\mathbf{U}_a} \begin{bmatrix} \mathbf{y} \\ \phi \end{bmatrix} = \begin{bmatrix} \mathbf{z} \\ \phi \end{bmatrix} \tag{1.3}$$

and comparing these triangular factors with $\mathbf{M}_a$ we conclude that

$$\mathbf{M} = \mathbf{LU}, \quad \mathbf{v}_c = \mathbf{Lu}, \quad \mathbf{v}_\ell = \mathbf{U}^T \mathbf{l}, \quad -\delta^{-1} = \mathbf{l}^T \mathbf{u} + \tau. \tag{1.4}$$

$\mathbf{L}$ and $\mathbf{U}$ and are calculated only once when solving the nominal circuit. To calculate the vectors $\mathbf{l}$ and $\mathbf{u}$ it is necessary to solve two $n \times n$ triangular systems of linear equations and to calculate τ the dot product $\mathbf{l}^T\mathbf{u}$ must be carried out.

The intermediate vector $\mathbf{z}$ is the nominal one, already calculated. After doing some algebra with (1.4) and with the lower triangular system in (1.3) we also calculate

$$\phi = \frac{\delta\, \mathbf{1}^T \mathbf{z}}{1 + \delta\, \mathbf{1}^T \mathbf{u}} \tag{1.5}$$

and the faulty circuit solution $\mathbf{y}$ is obtained from the system

$$\mathbf{U}\mathbf{y} = \mathbf{z} - \phi \mathbf{u}. \tag{1.6}$$

To exemplify a *fault rubber stamp* we consider an admittance between nodes j and j' with nominal value Y. The MNA stamp of this element in the circuit matrix is [12, 5]

$$\begin{array}{c} \\ j \\ j' \end{array} \begin{array}{c} \begin{array}{cc} v_j & v_{j'} \end{array} \\ \left[\begin{array}{cc} +Y & -Y \\ -Y & +Y \end{array} \right] \end{array}$$

where the left labels j and j' indicate the correspondence with the Kirchhoff's Current Law (KCL) in those nodes.

Let's suppose Y is faulty and its value changes to $Y + \delta$. We introduce a *fault variable* ϕ, the current from node j to node j' in the faulty admittance δ in parallel with Y (see figure 1). The *fault rubber stamp* of Y *plus* δ is

$$\begin{array}{c} \\ j \\ j' \\ \\ \end{array} \begin{array}{c} \begin{array}{ccc} v_j & v_{j'} & \phi \end{array} \\ \left[\begin{array}{cc|c} +Y & -Y & +1 \\ -Y & +Y & -1 \\ \hline +1 & -1 & -\delta^{-1} \end{array} \right] \end{array} \tag{1.7}$$

where the equation $v_j - v_{j'} - \phi/\delta = 0$ of the faulty admittance, called the *fault element equation,* was joined to the nominal rubber stamp increasing by 1 the dimension of the system. We remark that *the fault element equation (variable) is always located in the last row (column) of the faulty circuit matrix* and that the lowest rightmost element is always $-\delta^{-1}$.

We summarize the fault rubber stamps of admittances, impedances and VCCSs. Other stamps of linear elements are given in [3]. The fault variable is ϕ and δ is the *fault value,* that is, the deviation from nominal of the parameter that characterizes the element.

In figure 1 a) is shown a faulty admittance whose stamp was given in (1.7). In figure 1 b) and c) are shown a faulty impedance $Z \quad + \delta$ and a faulty

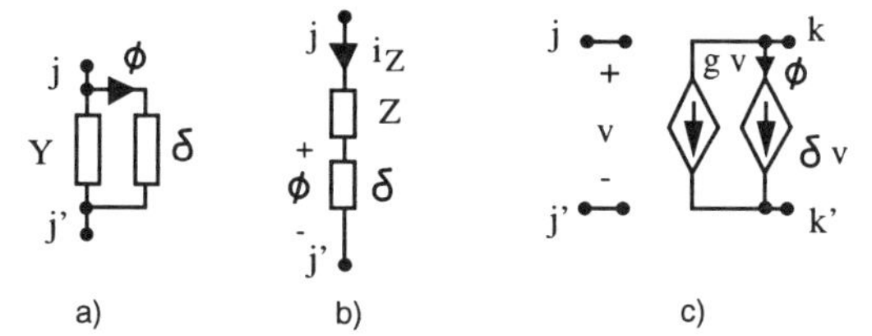

Figure 1 Faulty a) admittance (Y), b) impedance (Z), c) VCCS (g).

transconductance $g + \delta$ whose stamps are, respectively,

$$\begin{array}{c} \\ j \\ j' \\ Z\text{ eq.} \\ \\ \end{array} \begin{array}{c} \begin{array}{cccc} v_j & v_{j'} & i_Z & \phi \end{array} \\ \left[\begin{array}{cc|c|c} & & +1 & \\ & & -1 & \\ \hline +1 & -1 & -Z & -1 \\ \hline & & +1 & -\delta^{-1} \end{array}\right] \end{array} \qquad \begin{array}{c} \\ k \\ k' \\ f \end{array} \begin{array}{c} \begin{array}{ccc} v_j & v_{j'} & \phi \end{array} \\ \left[\begin{array}{cc|c} +g & -g & +1 \\ -g & +g & -1 \\ \hline +1 & -1 & -\delta^{-1} \end{array}\right] \end{array} .$$

Suppose we are dealing with full matrices and vectors. The computational cost of LU factorization is $\sim n^3$ while the computational cost of solving a triangular system is only $\sim n^2$. Thus, *we can simulate a fault with a complexity of $\sim n^2$ after the nominal circuit has been solved.* When, as is usually the case in circuit simulation, the matrices and vectors are sparse, the real asymptotes are lower than those above, but there are still substantial savings in fault simulation with FARUBS. The most dramatic efficiency is achieved *when simulating several deviations δ for the same fault parameter*: in this case we only need to calculate the new τ, ϕ, and then solve the upper triangular system in (1.3).

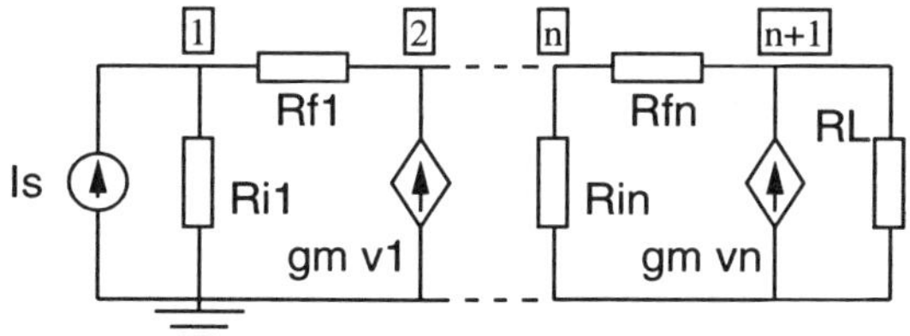

Ri*=1 Ohm; Rf*=100 KOhm; gm*=1 S

Figure 2 Cascade of n amplifiers.

The performance of the simulator was evaluated in the cascade of n amplifiers shown in figure 2 for values of n between 200 and 5000. The CPU simulation times are collected in figure 3.

This figure displays the experimental evidence of the efficiency of fault simulation in the above cascade of amplifiers. We compare the nominal circuit simulation time (□), the factorization time of the last row and column for each faulty element (+), and the estimated time of fault simulation for each fault value (i. e., the computation of τ and solution of the upper triangular system (1.3) with backward substitutions) (◇). This one *grows linearly with n,* what

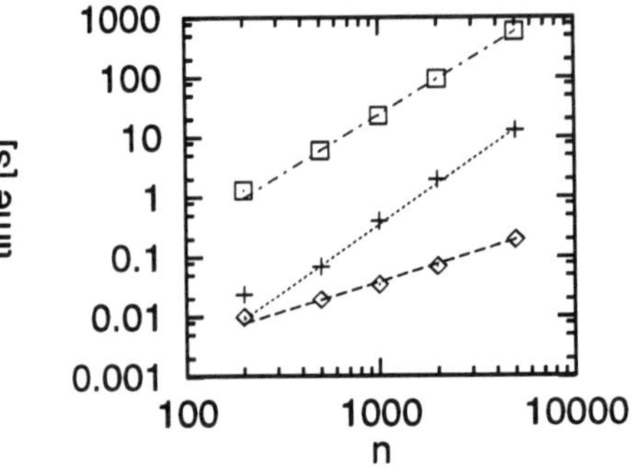

Figure 3 CPU simulation times of faults in the cascade of n amplifiers.

means that the simulation of several values of a given faulty parameter is very cheap, even in large circuits.

The simulation time of the nominal cascade (□) and the factorization time of the last row and column for each faulty element (+) follow approximately an n^2 law. Although both present the same polynomial complexity, *simulating one faulty element is about two orders of magnitude faster than simulating the nominal circuit*, as is seen by comparing the (□) with the (+) points in figure 3. We must say, however, that the nominal matrix is assembled and factorized with dynamic data structures that allow the exploitation of sparsity, while the fault line and column are written in memory and factorized as full vectors what doesn't allow us to take advantage of their sparsity.

3. SIMULATION OF FAULTS IN NONLINEAR CIRCUITS

Consider the simulation of a nonlinear circuit with the Newton-Raphson (NR) algorithm [5]. The electrical characteristics of nonlinear elements are linearized and in each iteration ν it must be solved a linear system of equations (departing from an initial guess $\mathbf{x}^0$)

$$\mathbf{M}^\nu \mathbf{x}^{\nu+1} = \mathbf{b}^\nu \qquad \nu = 0, 1, \ldots \tag{1.8}$$

To calculate $\mathbf{x}^{\nu+1}$ with LU factorization, the factors $\mathbf{L}^\nu$ and $\mathbf{U}^\nu$ such that $\mathbf{M}^\nu = \mathbf{L}^\nu \mathbf{U}^\nu$ must be obtained and two triangular systems must be solved. Since $\mathbf{M}^\nu \equiv \mathbf{M}(\mathbf{x}^\nu)$ changes from iteration to iteration, the triangular factors must be re-calculated in each iteration, and there is no advantage in simulating a fault $p + \delta$ by solving the augmented system

$$\begin{bmatrix} \mathbf{M}^\nu & \mathbf{v}_c \\ \mathbf{v}_\ell^T & -\delta^{-1} \end{bmatrix} \begin{bmatrix} \mathbf{y}^{\nu+1} \\ \phi^{\nu+1} \end{bmatrix} = \begin{bmatrix} \mathbf{b}^\nu \\ 0 \end{bmatrix}. \tag{1.9}$$

This picture changes if we can find a way of bounding the iteration dependent part of $\mathbf{M}^\nu$ to a lower-right block. To simulate efficiently faults in nonlinear circuits we developed new Newton-Raphson models for the electronic devices. We expose the motivation behind the approach using the diode as an example.

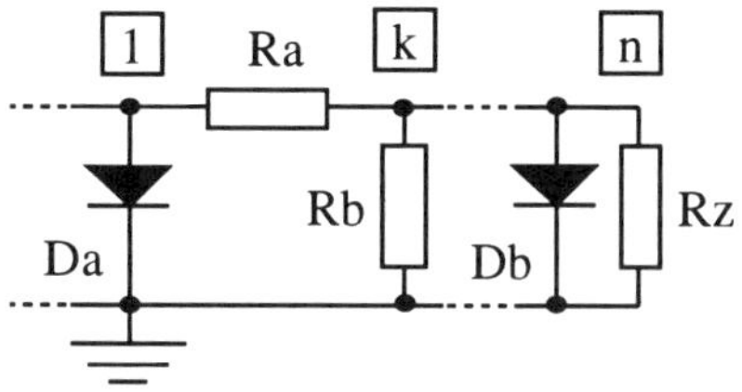

Figure 4 Nonlinear circuit with two nonlinear devices (diodes) between nodes 1 and n and ground.

The circuit in figure 4 has *only two diodes,* D_a between node 1 and ground and D_b between node n and ground. Their $v - i$ characteristic is $i = f(v)$, whose derivative is $g(v) = \frac{df(v)}{dv}$. In each NR iteration the diode current is approximated by a first order Taylor series $i^{\nu+1} \approx i^\nu + g(v^\nu)(v^{\nu+1} - v^\nu)$. Thus, we have the following NR system of equations for this circuit in the presence of a fault, if no reordering of the matrix is performed

$$\left[\begin{array}{ccc|c} g_a^\nu & \cdots & \cdots & \\ \vdots & & & \mathbf{v}_c \\ \vdots & & g_b^\nu & \\ \hline & \mathbf{v}_\ell^T & & -\delta^{-1} \end{array}\right] \left[\begin{array}{c} v_1^{\nu+1} \\ \vdots \\ v_n^{\nu+1} \\ \hline \phi^{\nu+1} \end{array}\right] = \left[\begin{array}{c} b_1^\nu \\ \vdots \\ b_n^\nu \\ \hline 0 \end{array}\right]$$

where $g_a^\nu = g(v_1^\nu)$, $b_1^\nu = g_a^\nu v_1^\nu - i_a^\nu$, $g_b^\nu = g(v_n^\nu)$ and $b_1^\nu = g_b^\nu v_n^\nu - i_b^\nu$. Since both g_a^ν changes in each iteration, all the matrix must be re-factorized and the advantage of the FARUBS methodology is lost.

If, however, D_a didn't exist in the circuit, the only matrix element changing in each NR iteration would be g_b^ν. In this case, the change in $\mathbf{L}^\nu$ and $\mathbf{U}^\nu$ from iteration to iteration would be located on their lower-right 2×2 blocks.

The obvious conclusion is: if the NR iteration-dependent elements of the matrix are pushed to the lowest rightmost block, only this block is re-factorized in each iteration. This allows for achieving some efficiency in simulating faults with the FARUBS methodology, even in nonlinear circuits.

Our approach to implementing the above reasoning consists in *modeling the diode as a current-controlled element* and introducing its current i as a new circuit variable. The diode (for instance D_a) is described by $v = h(i)$, where $h() \equiv f^{-1}()$ is the inverse function of $f()$, with derivative $r(i) = \frac{dh(i)}{di}$. The Taylor series approximation is now $v_1^{\nu+1} \approx v_1^\nu + r(i_a^\nu)(i_a^{\nu+1} - i_a^\nu)$ and we have the NR iteration

$$\left[\begin{array}{ccccc|c} & & & 1 & & \\ & & & \vdots & & \mathbf{v}_c \\ & & & \vdots & & \\ 1 & \cdots & \cdots & -r_a^\nu & & 0 \\ \hline & \mathbf{v}_\ell^T & & 0 & & -\delta^{-1} \end{array}\right] \left[\begin{array}{c} v_1^{\nu+1} \\ \vdots \\ \vdots \\ i_a^{\nu+1} \\ \hline \phi^{\nu+1} \end{array}\right] = \left[\begin{array}{c} \vdots \\ \vdots \\ \vdots \\ e_a^\nu \\ \hline 0 \end{array}\right]$$

where $r_a^\nu = r(i_a^\nu)$ and $e_a^\nu = v_1^\nu - r_a^\nu i_a^\nu$. It is obvious that in the matrix only r_a^ν changes in each iteration and the only part of the **L** and **U** factors that must be updated corresponds to the lower right 2×2 block of **M**.

Generalizing the approach to any nonlinear circuit, the NR system will be written as

$$\left[\begin{array}{cc|c} \mathbf{M}_{11} & \mathbf{M}_{12} & \mathbf{v}_c \\ \mathbf{M}_{21} & \mathbf{M}_{22}^\nu & \mathbf{0} \\ \hline \mathbf{v}_\ell^T & \mathbf{0}^T & -\delta^{-1} \end{array}\right] \left[\begin{array}{c} \mathbf{y}_1^{\nu+1} \\ \mathbf{y}_2^{\nu+1} \\ \hline \phi^{\nu+1} \end{array}\right] = \left[\begin{array}{c} \mathbf{b}_1 \\ \mathbf{e}_2^\nu \\ \hline 0 \end{array}\right]$$

where the sub-matrix $\mathbf{M}_{22}^\nu$ and the RHS vector $\mathbf{e}_2^\nu$ are the only that change in each iteration, and the **L** and **U** factors of the above matrix will be

$$\left[\begin{array}{cc|c} \mathbf{L}_{11} & \mathbf{0} & \mathbf{0} \\ \mathbf{L}_{21} & \mathbf{L}_{22}^\nu & \mathbf{0} \\ \hline \mathbf{1}^T & \mathbf{p}^{T\nu} & \tau^\nu \end{array}\right] \qquad \left[\begin{array}{cc|c} \mathbf{U}_{11} & \mathbf{U}_{12} & \mathbf{u} \\ \mathbf{0} & \mathbf{U}_{22}^\nu & \mathbf{q}^\nu \\ \hline \mathbf{0}^T & \mathbf{0}^T & 1 \end{array}\right]$$

where is clear that only the lower-right sub-matrix of both triangular factors changes in each iteration.

The smaller the number of nonlinear devices in the circuit, the smaller the dimension of the iteration-dependent blocks of **L** and **U**: thus, *this approach is extremely suitable for circuits with only a few nonlinear devices.*

The aforementioned approach was applied also to *bipolar transistors* described by the Ebers-Moll model, to *MOSFETs* described by the quadratic model and to *OPAMPs* described with a "tanh" equation that models saturation at the output. The models are less precise than those in standard simulation tools, but *its purpose is fault simulation not detailed design verification.* Precise models would increase the dimension of the iteration-dependent block. Those models are detailed in [3].

SIMULATION WITH AN ORDERED FAULT LIST

A further efficiency improvement in FASTNR is achieved by simulating the fault values of each faulty element within a "continuation" philosophy. Suppose *the fault values in a parameter p* to be simulated are ordered by increasing values as $p_{f1}, p_{f2}, p_{f3}, \ldots$ The NR system of equations when simulating the first value p_{f1} can be written as:

$$\mathbf{M}(\mathbf{x}^{\nu}, p_{f1})\, \mathbf{x}^{\nu+1} = \mathbf{b}(\mathbf{x}^{\nu}) \tag{1.10}$$

and its solution $\mathbf{x}_{f1}$ strictly satisfies the equality

$$\mathbf{M}(\mathbf{x}_{f1}, p_{f1})\, \mathbf{x}_{f1} \equiv \mathbf{b}(\mathbf{x}_{f1}) \tag{1.11}$$

(that is, $\mathbf{x}^{\nu+1} = \mathbf{x}^{\nu} + \epsilon$ after convergence is achieved, with $\epsilon \approx 0$.) To simulate the next fault value p_{f2}, which is just after p_{f1} in the ordered list, we use for the initial guess $\mathbf{x}^0$ the solution just obtained: we make $\mathbf{x}^0 \leftarrow \mathbf{x}_{f1}$ and begin the NR iteration for fault p_{f2} with the solution of

$$\mathbf{M}(\mathbf{x}_{f1}, p_{f2})\, \mathbf{x}^{1} = \mathbf{b}(\mathbf{x}_{f1}) \tag{1.12}$$

after the actualization and factorization of the matrix and RHS vector. The limit of this iteration will be the solution $\mathbf{x}_{f2}$ of the circuit with the parameter p presenting a fault p_{f2}. The reason behind the success of this continuation procedure is that, often, the solution $\mathbf{x}_{f1}$ of the previous fault value lies inside the domain of quadratic convergence to the solution $\mathbf{x}_{f2}$ of the the NR algorithm and, thus, is a good initial guess. When this happens, usually not more than 3 or 4 iterations are needed to get a solution. This depends, of course, on the "distance" between $\mathbf{x}_{f1}$ and $\mathbf{x}_{f2}$: when, for example, one or more devices change their operating region with the new fault value (e.g. a bipolar transistor goes from the active region into saturation) the convergence is not so fast.

When switching from one faulty element to another, the solution of the nominal circuit is reloaded since the first p_f simulated fault value is small (remember that $p_f \equiv \delta$ is the *deviation from the nominal value*) and thus the nominal solution is a good guess for starting the NR iterations.

EXAMPLES WITH NONLINEAR CIRCUITS

We present some results from fault simulations in nonlinear circuits. The video amplifier in figure 5 a) has tw o bipolar transistors and is an example where most of the elements are linear. The dimension of the circuit matrix is $n = 23$, including the extra line and column to handle faults. The sub-matrix $\mathbf{M}_{22}^{\nu}$ is only 4×4, which means that a 5×5 matrix must be updated and factorized in each NR iteration when simulating faults. 40 and 80 fault values were simulated for each of the 9 resistances in the amplifier and the results are tabulated in figure 7. The speed-up achieved is around 18 when considering total time; however, *when estimating the time of simulation of a single fault value, the speed-up is 192.*

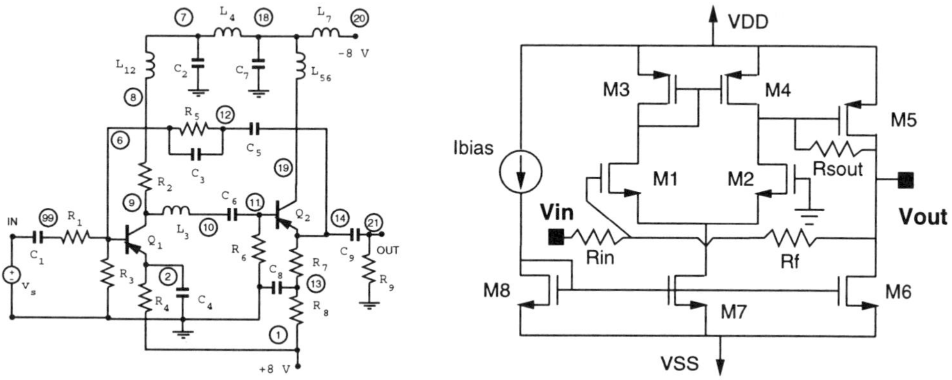

Figure 5 a) Video amplifier. b)Two-stage CMOS OPAMP.

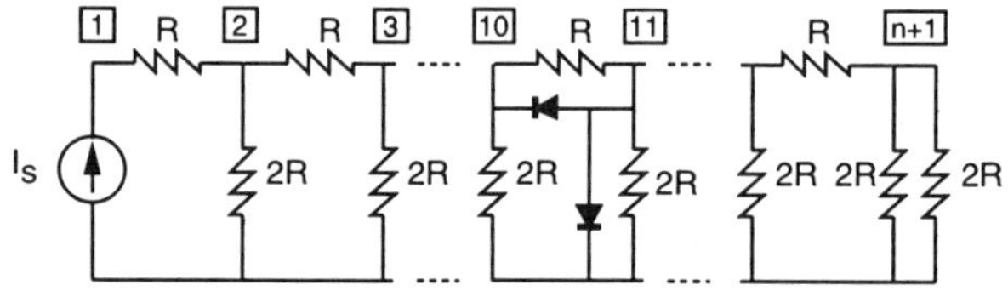

b)

Figure 6 a) Nonlinear ladder with two diodes periodically inserted in the stages.

The circuit in figure 6 is a n-stage nonlinear $RR2$ ladder with $2n+1$ resistors, where $R = 1\ \Omega$, and has, periodically, two diodes in parallel with the resistors in the stage. We simulated 500-stage, 1000-stage and 2000-stage ladders. The *speed-up achieved in the larger ladder is 44.* The simulation results are summarized in the table in figure 7.

We finish this section with the presentation of results from a circuit consisting almost entirely of nonlinear devices, the CMOS two-stage OPAMP in figure 5 b), mounted as an inverting amplifier with voltage gain $G_v = -10$. The fault simulated in this circuit is quite common in CMOS processes: it is a Gate-Oxide-Short (GOS) fault in transistor `M5`. The simulation command in the circuit file ordered the simulation of 100 values of `RSOUT` below the nominal ($1\ T\Omega$). This resistance was introduced for fault simulation purposes only. The fault simulation results in this example are also tabulated in figure 7.

The dimension of the NR equations was 29 (including the fault line and column) and the dimension of the iteration dependent block was 16×16. Despite being an almost nonlinear circuit, a speed-up of 31 was observed when simulating each value of `RSOUT`.

It must be stressed that the time overhead resulting from reading the circuit file and assembling the internal data structures only penalizes the simulation of the nominal circuit.

CIRCUIT	fel	#f	Dim	%fi	its.	spu.	nom./fa.
CMOS	-	-	28	42	40	-	**31**
OPAMP	3	40	29	46	431	**14.7**	-
	3	200	"	"	1266	**25.3**	-
video	-	-	22	25	9	-	**192**
amp.	9	40	23	32	820	**17.5**	-
	9	80	"	"	1591	**18.3**	-
nonl. lad.	-	-	1201	0.8	10	-	**49**
(1000/10)	8	20	1202	1.0	488	**42**	-
	8	40	"	"	1178	**45**	-
nonl. lad.	-	-	2801	0.8	7	-	**44**
(2000/5)	2	10	2802	0.8	101	**9**	-
	8	10	"	"	408	**14**	-

Figure 7 Results of fault simulation in nonlinear circuits. LEGEND: fel is the number of faulty elements, $\#f$ is the number of fault values for each element, Dim is the dimension of the system, $\%fi$ is the percentage of fills in the LU factors, $its.$ is the total number of NR iterations, $spu.$ is the overall speed-up and $nom./fa.$ is the ratio between the simulation times of the nominal circuit and of each *fault value*. The number of faults is $fel \times \#f$.

4. CONCLUSIONS

In this paper we described the FARUBS methodology dedicated to the efficient simulation of faults in linear and in nonlinear circuits, and we also described the FASTNR simulator where it was implemented.

The strategies developed in order to achieve the above goal, consisted in reusing the nominal circuit equations and solution, and in simulating several fault values in the same circuit element in an ordered sequence. This allows for good NR starting points in the NR iteration when simulating those faults.

The observed efficiency reached four orders of magnitude in large linear cascades with 5000 nodes. In nonlinear circuits it was between one and two orders of magnitude. Even in an almost completely nonlinear circuit, a CMOS OPAMP, it was observed a speed-up of 31 when simulating a GOS fault.

It was shown that FARUBS is well suited for efficient fault simulation in linear DC circuits and in nonlinear DC circuits with a small number of nonlinear elements. The simulation of faults in nonlinear devices, as well as fault simulation in the AC domain, are still not implemented.

5. ACKNOWLEDGMENTS

This work has been partially supported by the Portuguese Government funded PRAXIS XXI 2/2.1/TIT/1661/95 project entitled *MIXCHIP - Advanced Techniques for the Design, Simulation and Test of Mixed Analogue-Digital Circuits.*

6. REFERENCES

[1] J. W. Bandler and A. E. Salama, "Fault diagnosis of analog circuits," *Proceedings of the IEEE, vol. 73, 8, Aug.*, 1985.

[2] J. L. Huertas, *Test and Design for Testability of Analog and Mixed-Signal Integrated Circuits: Theoretical Basis and Pragmatical Approaches*, in H. Dedieu (Ed.), *Selected Topics in Circuits and Systems,* Elsevier, EC-CTD'93, Davos, Switzerland, 1993.

[3] J. S. Augusto, *Fault Simulation and Diagnosis in Analog Circuits*, Ph. D. Thesis (in Portuguese), Instituto Superior Técnico, Lisbon, 1999.

[4] J. S. Augusto and C. B. Almeida, Efficient Fault Simulation, Diagnosis and Testability Evaluation in Linear and Nonlinear Circuits with Fault Rubber Stamps. *ETW97 Proc., Cagliari, Italy*, 1997.

[5] J. Vlach and K. Singhal, *Computer Methods for Circuit Analysis and Design.* Van Nostrand Reinhold, second edition, 1994.

[6] W. Press, S. Teukolsky, W. Vetterling, and B. Flannery, *Numerical Recipes in C.* Cambridge University Press, 1992.

[7] A. Pahwa and R. Rohrer, Band Faults: Efficient Approximations to Fault Bands for the Simulation Before Fault Diagnosis of Linear Circuits. *IEEE Tr. CAS-29, no. 2, pp. 81-8, Feb.* 1982.

[8] I. S. Duff, A. M. Erisman, and J. K. Reid, *Direct Methods for Sparse Matrices.* Oxford, 1986.

[9] I. N. Hajj, Updating method for LU factorization. *Electronic Letters, vol. 8, no. 7*, 1972.

[10] Lin P-M. and Elcherif Y., "Analogue Circuits Fault Dictionary - New Approaches and Implementation," *Cir. Th. and Appl., 12, pp. 149-172,* 1985.

[11] V. Prasad, Equations of Nonlinear Analogue Circuit for Fault Dictionary Generation. *Electronics Letters, vol. 26, no. 1, pp. 24-5, Jan.*, 1990.

[12] P. A. Brennan, C.-W. Ho, A. E. Ruehli and P. A. Brennan, "The modified nodal approach to network analysis," *IEEE Transactions on Circuits and Systems, CAS-22, pp. 504-509*, 1975.

Design Error Diagnosis in Digital Circuits without Error Model

Raimund Ubar[1], Dominique Borrione
TIMA Laboratory, Grenoble

raiub@pld.ttu.ee, Dominique.Borrione@imag.fr

Key words: combinational circuits, design errors, stuck-at fault simulation, diagnostic analysis, redesign based error correcting

Abstract: We describe a new method for design error diagnosis in digital circuits that doesn't use any error model. For representing the information about erroneous signal paths in the circuit, a stuck-at fault model is used. This allows to adopt the methods and tools of fault diagnosis used in hardware testing for the use in design error diagnosis. A diagnostic specific pre-analysis of the circuit based on stuck-at fault model extracts iteratively subcircuits suspected to be erroneous. Contrary to other published works, here the necessary re-synthesis of the extracted subcircuit need not be applied to the whole function of an internal signal in terms of primary inputs, but may stop at arbitrary nodes inside the circuit. As the subcircuits to be redesigned are kept as small as possible, the redesign procedure is simple and fast. The search for subcircuits to be rectified is carried out by a systematic procedure. The procedure is guided by a heuristic priority function derived from the results of diagnostic pre-analysis. Experimental data show the high speed of diagnostic pre-analysis.

[1] On the leave from Tallinn Technical University, Estonia

1. INTRODUCTION

Design error diagnosis plays an essential role in providing correct VLSI products. Despite the use of CAD tools to produce *correct by construction* circuits, experience shows that a phase of design correction is necessary [1]. Designs made by CAD tools are manually modified to improve some aspects such as performance or area overhead. During this phase, errors are likely to be inserted. Automatic error diagnosis saves a lot of design debugging time. Existing logic rectification approaches can be classified: error-model based [2-5], structure based [6-8], and re-synthesis based approaches [9-11].

In error model based approaches, the implementation is rectified by matching the error with an error type in the model. The method is restricted because it may fail in error cases not covered by the model. The multiple error case has not been investigated at all because of complexity.

In [6], a structural approach was proposed for *engineering change* [12]. In order to re-use the engineering effort spent on the old implementation, logic rectification is performed to realize the new specification by modifying the old design. Verification techniques are used to narrow down the potential error region. Then a heuristic called *back-substitution* is employed in hopes of fixing the error incrementally. This approach requires that a structural correspondence between the specification and the implementation be provided. If this requirement is not fulfilled the method cannot be used.

Re-synthesis approaches are more general; they rely on the symbolic error-diagnosis techniques to find an internal signal in the implementation that satisfies the *single fix condition*, i.e. the condition of fixing the entire implementation by changing the function of an internal signal. Once such a signal is found, a new function is realized to replace the old function of this signal to fix the error. In the worst case, it may completely re-synthesize every primary output function. The major drawback is that it cannot handle larger designs, because it uses Binary Decision Diagrams (BDD) [13].

In this paper, the re-synthesis approach is applied not to the whole function of an internal signal given in terms of input signals, but to internal subfunctions for smaller subcircuits. By diagnostic preanalysis, a subcircuit suspected to be erroneous is extracted and redesigned to match the verification results. If the redesign does not solve the problem (cannot correct the circuit), the initial extracted subcircuit is extended either towards the inputs or towards the outputs, and the redesign procedure is repeated. As the subcircuits to be redesigned are as small as possible, the redesign procedure is simple and fast. The size of the subcircuit depends substantially on the quality of diagnostic pre-analysis. In the worst case of many design errors, or if the errors are spread all over the circuit it may be needed to redesign all the circuit, as in the case of known methods [9, 11].

2. PRELIMINARY

Consider a circuit specification, and implementation, both at the Boolean level. The specification output is given by a set of variables $W = \{w_i\}$, and the implementation output is given by a set of variables $Y = \{y_j\}$. Let $X = \{x_k\}$ be the set of input variables. The implementation is a network of components (gates), and Z is the set of connections (edges) between components, labelled by signal variables. Let S be the set of all variables in the implementation $S = Y$ _ Z _ X. The components are described by Boolean functions $s = f(s_1, s_2, \ldots, s_h)$ where s _ Y _ Z, s_i _ Z _ X. If s_k is a fanout variable then all the branches of the fanout are denoted by the second index: $s_{k,1}, s_{k,2}, \ldots, s_{k,p}$, where p is the number of branches. Denote the subset of variables in Z which represent the branches of fanout signals as Z^B.

Example: In Fig.1 a combinational circuit is represented with $X=\{s_1,s_2, s_3,s_4,s_5,s_6,s_7\}$, $Z=\{s_{3,1},s_{3,2},s_8,s_{8,1},s_{8,2},s_9,s_{10},s_{11},s_{12},s_{13},s_{14}\}$, $Z^B=\{s_{3,1},s_{3,2},s_{8,1},s_{8,2}\}$, and $Y=\{s_{15}\}$. We assume that there exists a double design error in the circuit: AND_{12} should be XOR gate, and OR_{14} should be AND gate. _

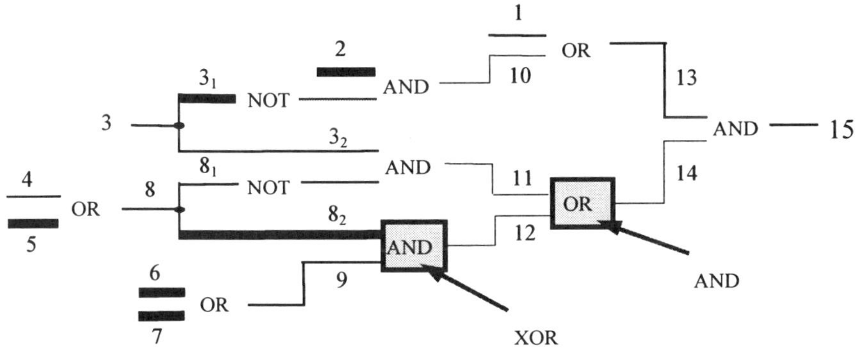

Figure 1. Combinational circuit with two design errors

Definition 1. The *cone* $C(s_k)$, s_k _ S, is the subset of all variables s _ S from which there exists a path from s to s_k (in the direction of signal flow).

A cone $C(s_k)$ is a function $s_k = f(s_1, s_2, \ldots, s_p)$ with a set of arguments $S_k = \{s_1, s_2, \ldots, s_p\} = C(s_k)$ _ X. It is a subnetwork $N(C(s_k)$ _ $X, s_k)$ with a set of inputs $C(s_k)$ _ X, and with output s_k. A subcone $C'(s_k)$ _ $C(s_k)$ may have arguments which are not primary inputs. A gate is a smallest subcone.

Definition 2. *Test patterns.* For a circuit with n inputs, a test pattern T_i is a n-bit ternary vector TR^n, $TR=\{0,1,u\}$, where u - is a don't care. A set of test patterns $T = \{T_i\}$ is a *test*.

Definition 3. *Stuck-at fault set.* Let F be the set of stuck-at-1(0) faults s/1(0) for the variables s _ Z^B _ X of the circuit. F is a representative set of faults: to test all the stuck-at faults in the circuit, it is sufficient to test only the faults in F [14]. On the other hand, when testing F, we test all the signal paths in all the tree-like subcircuits for transmitting both signals 1 and 0.

Example: The representative set of faults in the circuit of Fig.1 consists of 22 faults and is related to the variables: $s_1, s_2, s_3, s_4, s_5, s_6, s_7, s_{3,1}, s_{3,2}, s_{8,1}, s_{8,2}$. _

Note, the stuck-at fault model is not used here as a model for design error diagnosis. Its role is to provide only a measure for stating that one or another signal path is found to be erroneous. Detecting a fault in a path means that the path is going through an erroneous area (wrong gate, connection error etc.) in the circuit. We consider here the paths inside tree-like subcircuits of the design. There is a one-to-one correspondence between the paths and the variables s _ Z^B _ X used for defining F. A variable s represents the path from the edge labelled by s up to the closest edge labelled by s_k _ $(Z - Z^B)$ _ Y. For example, $s_{3,1}$ represents the path through edges $s_{3,1}, s_{10}, s_{13}, s_{15}$. Detecting the fault $s_{3,1}/0$ means that an error has been detected by testing this path for the value $s_{3,1}=1$. The information about faulty paths in the form of detected faults in F is used later for carrying out diagnostic pre-analysis in order to identify as exact as possible the suspected erroneous areas in the circuit. The second advantage of using the stuck-at fault model is to have a possibility to use for design error diagnosis common ATPGs and fault simulators which have been developed initially for hardware testing purposes.

Consider a test $T = \{T_i\}$ that detects the representative set of faults F = {s/0 and s/1 for s _ X _ Z^B}. This test may be obtained by any standard test generation technique for digital circuits [14]. The test is then simulated on the description of the implementation and on the specification. To improve the resolution of diagnostic pre-analysis or to detect the design errors not covered by the stuck-at fault test, additional test patterns may be needed.

Definition 4. *Failed test patterns*. If the result $y_j(T_i)$ _ $w_j(T_i)$ is observed when applying the test pattern T_i to the implementation and the specification, we say that the test pattern T_i *fails*. Let T^* _ T be the subset of test patterns which fail during the design verification test.

Definition 5. *Fault table*. The fault table is a matrix _$a_{i,j}$_, where $a_{i,j}$ = 1 (0) if the test pattern T_i is able to detect the fault s_j/1(0), and is undetermined otherwise. Let $F_i \subset F$ be the set of faults which may be detected by T_i.

Definition 6. *Suspected faults*. A fault s/*e*, *e* _ {0,1}, is called suspected if there is a pattern T_i _ T^* which is able to detect this fault, and there exists no other pattern T_j _ $T - T^*$ which also is able to detect the same fault. Denote the whole set of suspected faults for the test T at failed patterns T^* by F^*.

Example: In Table 1, the first 7 rows form a set of 7 test patterns which cover all the representative faults of F in Fig.1. In the left part of Table 1 the values of input and output variables for test patterns are given. In the right part detectable faults for test patterns are depicted. For example, the set of faults detectable by the pattern T_1 is $F_1 = \{s_1/0, s_3/0, s_{3,2}/0, s_4/1, s_5/1, s_{8,1}/1\}$. _

Table 1. A test and its fault table for the circuit in Fig. 1 at given test

i	Test Patterns (T_i)								Fault Table										
	Inputs j:s_j X							Output	j: s_j Z^B X										
	1	2	3	4	5	6	7	s_{15}	1	2	3_1	3	3_2	4	5	6	7	8_1	8_2
1	1	0	1	0	0	0	0	1	0			0	0	1	1			1	
2	0	1	0	1	0	1	0	1		0	1	0		0		0			0
3	1	0	0	0	1	1	0	1	0						0		0		0
4	0	0	0	1	1	0	1	0	1	1									
5	0	1	1	0	0	0	1	0	1		0								
6	1	0	0	0	0	1	0	0				1	1	1	1				1
7	1	0	1	1	0	0	0	0						0		1	1	0	
8	1	0	1	0	0	1	1	1	0			0	0	1	1			1	

Definition 7. *Error level of the variable*. Let $F(s_k)$ _ F be the set of possible faults in the cone $C(s_k)$, and $F^*(s_k)$ _ $F(s_k)$ the set of suspected faults in $C(s_k)$. $E(s_k)$=_$F^*(s_k)$_/_$F(s_k)$_ is called *error level* of the variable s_k.

3. GENERAL DESCRIPTION OF THE METHOD

The procedure for error diagnosis and for circuit correction proposed in the present paper consists of the following steps.

1. Verification test with the goal to get a knowledge about faulty signal paths. This knowledge helps finding suspected erroneous areas in the circuit.
2. Diagnostic pre-analysis based on computing error levels $E(s)$ for all variables s _ Y _ $(Z - Z^B)$.
3. Defining a suspected erroneous subcircuit (subcone) $C'(s)$.
4. Rectification of the function of the subcircuit $C'(s)$ based on the results of the verification test, generating and executing new test patterns if needed.
5. If the rectification procedure corrects the design the problem is solved. Otherwise, steps 3 and 4 should be iteratively repeated. _

In this procedure, for verification and for checking if the circuit is corrected by rectification (Step 4), the given set of test patterns T is used. The test T may be extended by additional patterns either during Step 1 for increasing diagnostic resolution, or during Step 4. It may happen that Step 5 gives a positive result, although the verification with an equivalence prover still shows that the circuit rectified in Step 4 is not correct. When this happens, the test T should be extended by additional patterns.

4. DIAGNOSTIC PRE-ANALYSIS

The result of the verification test is a subset T^* _ T of failed test patterns. If no errors are found during verification test (T^* = _), no diagnosis and no error correction should be made. On the basis of the test results and the fault table, the numbers of suspected faults $|F^*(s)|$ and error levels $E(s)$ for all variables s _ Y _ $(Z - Z^B)$ are computed by the following algorithm.

Algorithm 1.

1. From the list T^* of failed test patterns, the preliminary set of suspected faults is constructed: $F^1 = \cup_{T_i} {}_{-}{}_{T^*} F_i$.

2. From the list $T - T^*$ of passed test patterns, the set of faults which definitely cannot be present is calculated: $F^2 = \cup_{T_i} {}_{-}{}_{T - T^*} F(T_i)$.

3. The subset of suspected faults is now calculated as: $F^* = F^1 - F^2$.

4. The subsets of suspected faults observable (detectable) at edges s _ Y $\cup (Z - Z^B)$ are calculated as: $F^*(s) = F(s)$ _ F^*.

5. Error levels for s _ Y _ $(Z - Z^B)$ are calculated: $E(s) =$ _$F^*(s)$_ / _$F(s)$_.

Example: Consider the circuit in Fig.1 with two design errors. The test given in Table 1 is good for error detection, as it covers all the possible stuck-at faults. On the other hand, the diagnostic resolution for the given test is very low. The failure of patterns T_1, T_2, and T_3 and the non-failure of the other patterns implies that all the representative edges (paths) of the circuit except s_4 remain suspected as faulty. The set of suspected faults is $F^* = \{s_1/0, s_2/0, s_{3,1}/1, s_{3,2}/0, s_5/0, s_6/0, s_7/0, s_{8,1}/1, s_{8,2}/0\}$. Such information helps very little for locating the erroneous area in the circuit. Adding an additional pattern $T_8 = (1010011)$ that does not fail, significally improves the diagnostic resolution. The set of suspected faults now consists of 6 faults: $F^* = \{s_2/0, s_{3,1}/1, s_5/0, s_6/0, s_7/0, s_{8,2}/0\}$ (located at bold lines in Fig.1). The results of calculating error levels for the circuit are shown in Table 2. _

Table 2. Error levels for variables of the circuit in Fig.1 for the given test experiment

Variables (s)	s_8	s_9	s_{10}	s_{11}	s_{12}	s_{13}	s_{14}	s_{15}
Number of faults in the cone C(s)	4	4	4	8	10	6	14	22
Number of susp. faults in C(s)	1	2	2	1	4	2	4	6
Error level of the variable E(s)	0,25	0,5	0,5	0,13	0,4	0,33	0,29	0,27

On the basis of the results of Algorithm 1, the suspected erroneous area may now be predicted. This is done by two procedures: elimination of pseudofaults, and suppressing the faulty area by using error level function.

4.1 Elimination of not distinguishable pseudofaults

It is rather realistic to suppose that there is in F^* a subset of suspected faults $F' \subset F^*$ which belongs to the paths not crossing the actual erroneous area in the circuit. Let us call these suspected faults in F' as *pseudofaults.* The existence of pseudofaults in F^* may mislead us in making decisions about suspected erroneous areas. The reason of existing of pseudofaults is the absence of a nonfailing test pattern in T which would be able to distinguish the faults in F' from the faults in F^*, i.e. which would be able to detect a fault in F', and not to detect faults in $F^* - F'$.

Consider the circuit in Fig.2 with an erroneous gate OR_5 (instead of OR_5 there should be AND gate). The test T_1 = 110 will fail because of the erroneous OR_5. On the other hand, T_1 detects faults $s_1/0$ and $s_2/0$, and there is no other test pattern which can distinguish these faults. Hence, from failing of the test T_1 we have to conclude that both OR_4 and OR_5 may be erroneous.

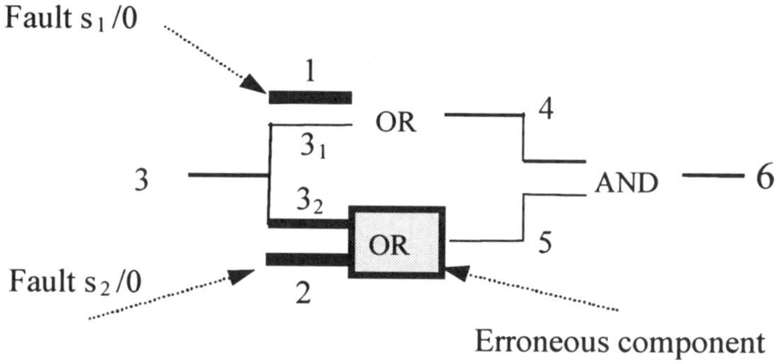

Figure 2. Diagnosis with bad resolution (faults s_1 /0 and s_2 /0 are not distinguishable

Consider now the full test for the circuit. The set of failed test patterns will be T^* = {001, 101, 110} which gives a set of suspected faults F^* = {$s_1/0$, $s_2/0$, $s_{3,2}/0$}. Only the cone $C(s_6)$ covers all the faults in F^*, which leads to the conclusion that rectification should be carried out for the whole circuit. This would be needed for a multiple error when both $C(s_4)$ and $C(s_5)$ were erroneous. Considering that the probability of multiple errors is low, we assume that only one of the cones $C(s_4)$ and $C(s_5)$ can be erroneous, and that the other one contains pseudofaults. It is also reasonable to expect that the number of pseudofaults is always less than the number of other suspected faults in F^*. This results from the expectation that not all suspected faults in the erroneous area have a counterpart in F'. Here, $F' = F^*(s_4)$ contains the pseudofault $s_1/0$ which is not distinguishable from $s_2/0$ in $F^*(s_5)$. However, $s_{3,2}/0$ in $F^*(s_5)$ has no distinguishable counterpart in $F^*(s_4)$.

Rule 1. If two subcones $C(s_i)$ and $C(s_j)$ with $F^*(s_i)$ and $F^*(s_j)$ are feeding the same component, and if $|C(s_i)| > |C(s_j)|$ then only the cone $C(s_i)$ should be suspected as erroneous, and the redesign should start with this cone.

All the cones qualified to contain pseudofaults like $C(s_j)$, are put into STACK for possible backtracking in the case of wrong predictions.

Example: In Fig.1 only the full cone $C(s_{15})$ covers all suspected faults. However, for the inputs of AND_{15} we have: $|C(s_{13})|$=2, and $|C(s_{14})|$=4, which along Rule 1 makes us to predict that the faults in $C(s_{13})$ are pseudo. Further, for the inputs of OR_{14} we have: $|C(s_{11})|$=1, and $|C(s_{12})|$=4. However, the fault in $C(s_{11})$ contains also in $C(s_{12})$, and therefore, the Rule 1 is not usable. _

4.2 Diagnosis by error level function.

After choosing by Rule 1 the cone $C(s)$ suspected to be erroneous, we may try to suppress even more the suspected area, instead of rectifying the whole network defined by $C(s)$. As the error level $E(s)$ refers to the number of suspected faults in relation to the whole number of faults in $C(s)$, we are justified to expect that the higher the error level of s, the higher is the probability that $C(s)$ is erroneous. This leads to the second rule.

Rule 2. Among all the possible cones in $C(s)$ chosen by Rule 1, a cone $C(s_k)$ where $E(s_k)$ = max over all s_i _ $\{C(s) - X, s\}$ should be rectified.

Assuming again that the probability of multiple design errors is less than the probability of single errors, we may expect that all the faulty paths in the suspected cone $C(s_k)$ are caused by an error related to the top component of the cone. This leads to the third rule for choosing the target for redesign.

Rule 3. In the cone $C(s_k)$ chosen by Rule 2, at first, its top component (the smallest subcone $C'(s_k)$ in $C(s_k)$) should be taken for rectification.

If the rectification of $C'(s_k)$ does not correct the circuit, we have to either increase the suspected area in $C(s_k)$ towards inputs or to choose another cone from $C(s)$ (or from STACK). The decision is made again on the basis of error level values. The procedure for correcting the circuit is as follows.

Algorithm 2.

1. Select by Rule 1 a cone $C(s)$, and create STACK of eliminated cones.
2. Select by Rule 2 a cone $C(s_k)$ in $C(s)$.
3. Select by Rule 3 a subcone $C'(s_k)$ for redesigning to correct the error.
4. If redesign explains the error, algorithm is ended. Otherwise, go to 8.
5. Choose the variable $s_l \in X$ with $E(s_l)$=max over all the inputs of $C'(s_k)$ and over all the variables in $C(s)$ outside $C'(s_k)$.
6. If s_l is an input of $C'(s_k)$, update the subcone $C'(s_k)$ by including the top component of the cone $C(s_l)$ into it. Go to 4.
7. If s_l is a variable outside $C'(s_k)$, take $s_k = s_l$. Go to 3.
8. If $C'(s_k)=C(s)$, take $C(s)$ from STACK, and go to 2. Otherwise, go to 5.

Example: According to Step 1 in Algorithm 2 (see previous Example), we eliminate $C(s_{13})$, and accept $C(s_{14})$ as suspected erroneous area. Steps 2, 3 suggest to restrict this area to the cone $C(s_9)=OR_9$ where $E(s_9)$=0,5=max. However, in Step 4 we find that redesign of OR_9 cannot explain the error. Steps 5,7,3 suggest now to try with AND_{12} in $C(s_{12})$ where $E(s_{12})$=0,4=max. The next choice will be $C'(s_9)=OR_9$, where $E(s_9)$=0,5, and, according to Step 6, we try to explain the design error by redesigning the network of OR_9 and AND_{12}. After the rectification again fails, we try with $C'(s_{14}) = OR_{14}$, where $E(s_{14})$ = 0,29, and then include to this subcone the gate $C'(s_{12}) = AND_{12}$ since $E(s_{12}) > E(s_{11})$. The last attempt to correct the circuit by redesigning the network of OR_{14} and AND_{12} has success (see Example in Section 5) . _

5. RECTIFICATION OF THE CIRCUIT

For rectifying the function of a subnetwork $N(S_k,s_k)$ with output s_k and input set S_k, we have to choose test patterns from the initial test set T (or create new patterns if the needed ones are missing in T), to put together a new set of patterns T' so that $T^* \subseteq T'$, and which includes all possible value combinations of the variables s _ S_k, so that for each T_i _ T' at least for one y_j _ Y the following holds: _y_j / _$s_k = 1$. The last condition makes the variable s_k observable at the output(s) and the behaviour of the subnetwork $N(S_k,s_k)$ can be corrected, if $y_j(T_i)$ _ $w_j(T_i)$ is observed. The number of patterns in T' is $|T'| \geq 2^p$, where p is the number of variables in S_k. Correction of the chosen subnetwork $N(S_k,s_k)$ means that when extracting the behaviour of s_k from the simulation results of T', the value of s_k should be changed for all the failed test patterns in T'. For the new function derived in such a way, a new network $N'(S_k,s_k)$ is designed.

<u>*Rule 4.*</u> If there is at least one pattern T_i _ T^* such that _y_j / _$s_k = 0$ for all y_j _ Y, then the circuit cannot be corrected by rectifying only $N(S_k,s_k)$.

Indeed, because of _y_j / _$s_k = 0$ for all y_j _ Y, the cause of failing T_i should be somewhere else than the changed value of s_k.

For the circuit with new $N'(S_k,s_k)$, test verification experiment is repeated with T'. If all the patterns in T' pass, the new design is correct in relation to the verification test. If at least one pattern from T' fails, the rectification procedure is repeated with another suspected erroneous area.

Example: In the previous example, corresponding to Algorithm 2 we appointed for rectification several subnetworks in the following order: OR_9, AND_{12}, $AND_{12} + OR_9$, OR_{14}, and $OR_{14} + AND_{12}$. In the following the rectification trials are shown for AND_{12}, OR_{14}, and $OR_{14} + AND_{12}$. The results of rectifying AND_{12} are given in Table 3 in the column $s_{12}= f(s_8,s_9)$ where $T'=\{T_1, T_2, T_3, T_6, T_7\}$. The results of testing are given in the column Y/W. Already the first pattern T_1 shows that along Rule 4, by rectifying AND_{12} the circuit cannot be corrected: for T_1 we find that _y_{15} / _$s_{12} = 0$. Hence, the error in AND_{12} cannot explain the erroneous behaviour of the circuit. The results of rectifying OR_{14} are given in the column $s_{14}=f(s_{11},s_{12})$. As the pattern $(s_{11}=s_{12}=1)$ is not possible we can substitute this function by constant 0. However, the repeated simulation of the test T shows that the circuit is still erroneous. By extending now the suspected area towards inputs, we try to rectify the network $OR_{14} + AND_{12}$ with function $s_{14} = f(s_8,s_9,s_{11})$. The results are in the last column. To create a test T', we have to generate new patterns T_{10}, T_{11}, T_{12} to get the full set of possible values at the inputs of the network under redesign, and to satisfy the condition _y_{15} / _$s_{14} =$ 1. It comes out that two input combinations are inconsistent for the given circuit, which means that $s_{14}= u$ *(don`t care)* for the inputs 101 and 111.

After substituting $u=0$, we find the rectified function as: $s_{14}=s_3\&\_\ s_8\&s_9$. It is easy to see that this is equivalent to $s_{14}=s_{11}\&(s_8\_\ s_9)$. The circuit is corrected.

Table 3. Rectification of suspected erroneous functions

No	Test patterns			Rectification actions					
	Inputs	Internal signals	Y/W	s_{12}=f(s_8,s_9)		s_{14}=f(s_{11},s_{12})		s_{14}=f(s_8,s_9,s_{11})	
	1 2 3 4 5 6 7	8 9 11 12 13 14	15	8 9	12	11 12	14	8 9 11	14
1	1 0 1 0 0 0 0	0 0 1 0 1 1	1/0	0 0	-	1 0	0	0 0 1	0
2	0 1 0 1 0 1 0	1 1 0 1 1 1	1/0	1 1	0	0 1	0	1 1 0	0
3	1 0 0 0 0 1 0	1 1 0 1 1 1	1/0	1 1	0	0 1	0	1 1 0	0
4	0 0 0 1 1 0 1	1 1 0 1 0 1	0/0	1 1	0				
5	0 1 1 0 0 0 1	0 1 1 0 0 1	0/0						
6	1 0 0 0 0 1 0	0 1 0 0 1 0	0/0	0 1	0	0 0	0	0 1 0	0
7	1 0 1 1 0 0 0	1 0 0 0 1 0	0/0	1 0	0	0 0	0	1 0 0	0
8	1 0 1 0 0 1 1	0 1 1 0 1 1	1/1					0 1 1	1
9	Not possible					1 1	-		
10	1 0 0 0 0 0 0	0 0 0 0 1 0	0/0					0 0 0	0
11	Not possible							1 0 1	-
12	Not possible							1 1 1	-

6. EXPERIMENTAL DATA

The main goal of the experiments was to evaluate the design error diagnostic properties of test patterns generated by traditional ATPGs for only stuck-at fault detecting purposes. The computer platform was: Sun SparcServer 20 (2 SuperSparc II microprocessors, 75MHz), Solaris 2.5.1. operating system.

In the experiments, ISCAS'85 benchmarks were used (columns 1,2,3,4 in Table 4). Test patterns for detecting stuck-at faults were created by the test generator Turbo Tester described in [15]. Fault coverage (6) and test generation time in seconds (11) are presented in Table 4. For all circuits, randomly selected single gate errors were inserted. The numbers of experiments carried out for each benchmark circuit are shown in the column 5. The total diagnosis time (13) consists of: test generation time (11) and fault diagnosis time (12). The time for rectification is not taken into account in these experiments.

The columns 7-9 show, correspondingly, the minimum, maximum and average diagnostic resolutions (numbers of suspected gates) reached by the test patterns generated only for fault detecting purposes and not for fault localization. The numbers in the column 10 illustrate the average sizes of the suspected erroneous areas in the circuits for the given test patterns.

Table 4. Experimental data about design error diagnosis with ISCAS'85 benchmarks

Circuit	Number of				Fault	Suspected gates				Time, s		
	inputs	outputs	gates	experi-ments	cover %	Min	Max	Av.	Av. %	Test gener.	Fault analysis	Total
1	2	3	4	5	6	7	8	9	10	11	12	13
c432	36	7	232	116	91,07	1	107	9	3,8	0,8	0,1	0,9
c499	41	32	618	106	99,33	1	307	77	12,3	1,0	1,4	2,4
c880	60	26	357	177	100	1	33	6	1,7	0,19	0,5	0,69
c1355	41	32	514	418	99,51	1	248	58	11,3	1,35	1,5	2,85
c1908	33	25	718	378	99,31	1	76	11	1,6	0,93	1,6	2,53
c2670	233	140	997	343	94,97	1	161	25	2,6	3,55	14,1	17,7
c3540	50	22	1446	458	95,27	1	86	10	0,7	3,08	3,7	6,78
c5315	178	123	1994	695	98,69	1	239	1	0,6	2,38	29,4	31,8
c6288	32	32	2416	2128	99,34	1	138	8	0,4	2,17	2,7	4,87
c7552	207	108	2978	1326	95,95	1	269	6	0,5	12,06	44,8	79,6

7. CONCLUSIONS

In this paper, a new approach to design error diagnosis in combinational circuits without error model is proposed. The procedure combines a diagnostic pre-analysis for predicting the suspected erroneous area, and re-synthesis for correcting the design. For representing the information about erroneous signal paths in the circuit, a stuck-at fault model is used. This allows to adopt the methods and tools of fault diagnosis used in hardware testing [14] for the use in design error diagnostic pre-analysis.

Differently from the known design error diagnosis methods which apply the re-design technique for randomly chosen cones of the circuit, in our paper, diagnostic pre-analysis is used to systematically compress the suspected erroneous area. Contrary to other published works, here the necessary re-synthesis of the extracted subcircuit need not be applied to the whole function of an internal signal in terms of primary inputs, but may stop at arbitrary nodes inside the circuit to rectify embedded subfunctions.

Our approach has two advantages compared to other re-design based methods: we use diagnostic pre-analysis to concentrate the rectification to small suspected erroneous areas, and by rectifying embedded subfunctions we can avoid the combinatorial explosion of OBDD based methods. The size of the circuits to be rectified depends on the quality of diagnostic pre-analysis. The method can be applied also to sequential circuits if scan-path technique is used.

The shortcomings of the proposed method are the lack of exact deterministic technique for diagnostic pre-analysis. Some heuristics is used for selecting the suspected erroneous areas; this is rather natural, because a very general case is considered: (1) all errors including multiple ones are allowed, and therefore no error model is used; (2) no structural similarity is assumed between the specification and the implementation.

Future work will be devoted to the development of deterministic approaches for more exact prediction of the erroneous area and to the development of methods for generating diagnostic test patterns with better resolution.

Acknowledgement

The authors appreciate the work of Jaan Raik and Artur Jutman from Tallinn Technical University for carrying out the experiments. They thank also the reviewers for valuable comments and suggestions.

8. REFERENCES

[1] A.M. Wahba, D. Borrione. Connection Error Location in Combinational Circuits. Proc. European Design & Test Conference, Paris, March 17-20, 1997, pp.235-241.

[2] P.Y. Chung, Y.M. Wang, I.N. Hajj. Diagnosis and Correction of Logic Design Errors in Digital Circuits. ACM/IEEE Design Automation Conference, pp. 503-508, 1993.

[3] I. Pomeranz, S.M. Reddy. On Error Correcting in Macro-based Circuits. Proceedings. of ICCAD, pp.568-575, 1994.

[4] A.M. Wahba, D. Borrione. A Method for Automatic Design Error Location and Correction in Combinational Logic Circuits. J. of Electronic Testing: Theory and Applications 8, 1996, pp. 113-127.

[5] R. Ubar, D. Borrione. Single Gate Design Error Diagnosis in Combinational Circuits. Proc. Estonian Acad. Sci. Engng., 1999, 5, 1, 3-21.

[6] D. Brand, A. Drumm, S. Kundu, P. Narrain. Incremental Synthesis. Proceedings of ICCAD, pp.14-18, 1994.

[7] S.-Y. Huang, K.-C. Chen, K.-T. Cheng. Error Correction based on Verification Techniques. ACM/IEEE Design Automation Conference, pp. 258-261, 1996.

[8] S.-Y. Huang, K.-C. Chen, K.-T. Cheng. Incremental Logic Rectification. Proc. of VLSI Test Symposium, 1997, pp. 134-139.

[9] J.C. Madre, O.Coudert, J.P. Billon. Automating the Diagnosis and the Rectification of the Design Errors with PRIAM. Proceedings of ICCAD, pp. 30-33, 1989.

[10] C.-C. Lin, K.-C. Chen, S.-C. Chang, M. Marek-Sadowska, K.-T. Cheng. Logic Synthesis for Engineering Change. ACM/IEEE DAC, pp. 647-652, 1995.

[11] D.W. Hoffmann, T. Kropf. Automatic Error Correction of Large Circuits Using Boolean Decomposition and Abstraction. Proc. of CHARME, to be published.

[12] U. Watanabe, R.K. Brayton. Incremental Synthesis for Engineering Change. Proceedings of ICCAD, pp. 40-43, 1991.

[13] R.E. Bryant. Graph-based Algorithm for Boolean Function Manipulation. Proceedings of ICCAD, pp. 316-319, 1989.

[14] M. Abramovici, M.A. Breuer, A.D. Friedman. Digital Systems Testing and Testable Designs. 1998.

[15] J.Raik, R.Ubar. Feasibility of Structurally Synthesized BDD Models for Test Generation. Proc. of the IEEE European Test Workshop. Barcelona, May 27-29, 1998, pp.145-146.

Efficient RLC Macromodels for Digital IC Interconnect

Bogdan Tutuianu
Motorola Inc., Austin, Texas
bogdan@ibmoto.com

Daksh Lehther
Motorola Inc., Austin, Texas
daksh@ibmoto.com

Madhulima Pandey
Motorola Inc., Austin, Texas
madhu@ibmoto.com

Ross Baldick
The University of Texas at Austin
baldick@ece.utexas.edu

Abstract The paper presents a new method to synthesize macromodels for very large on-chip interconnection networks which can be simulated very efficiently with traditional SPICE-like simulators. The method is taking advantage of the fact that in CMOS VLSI circuits, the receiving ports of the clock distribution network can be accurately modeled by lumped passive impedances. Our method simplifies the task of simulating the interconnect by building a reduced order macromodel only for the subset of driving ports of the net. The signals at the passive ports can be determined, in a later stage, as linear combinations of a reduced set of primary signals, obtained during the simulation of the driving ports macromodel. The simulation time for the macromodels generated in this way is greatly reduced, the size of the macromodels is kept small and the accuracy is preserved.

Keywords: simulation, macromodels, clock distribution networks

INTRODUCTION

In modern microprocessors, due to higher operating frequencies, tight design constraints are imposed on critical nets such as the clock distribution networks. As a consequence, the designers need to perform a more detailed circuit analysis with an accurate SPICE-like circuit simulator, as opposed to relying on the traditional static timing analysis. At the same time, the size of the clock circuits that must be simulated has increased dramatically due to two factors: increased transistor and wiring density, and high operating frequencies. Increased transistor and wiring densities allow for more transistors to be packed on the same chip and, naturally, this results in much larger clock distribution networks. In order to capture accurately the large bandwidth of the signals, the distributed RLC representation of the interconnect also needs to be more complex. So, the designers are confronted with the fact that they have to design much larger structures and, for that, they have to use more and more accurate simulators, which are very slow. This makes the design of clock networks a very time consuming task and, for a faster design cycle time, new ways to improve things are needed. One such way to reduce the simulation time of very large circuit structures is to use reduced order macromodels for the RLC interconnect sections.

The problem of building macromodels for linear circuits is very well understood and not new. However, the problem of building macromodels is usually left at the point where a reduced model is obtained, usually in the frequency-domain form, and most of the effort is put into quantifying the accuracy and the stability of the reduced order model. One of the first attempts to generate macromodels that could be integrated with general purpose simulators is presented in [Kim 94]. The macromodels generated in [Kim 94] are based on AWE [Pillage 90] and, as a consequence, they are not guaranteed to be stable. Also based on AWE is the method proposed in [Dartu 96], which takes advantage of the fact that most of the ports (the receiving ports) can be modeled as passive loads and can be included in the macromodel, but it has the draw-back that can be used only for RC interconnect circuits. Another method to generate macromodels for simulation purposes is [Feldman 95], which uses the Padé-Via-Lanczos algorithm, thus resulting in more stable reduced order models, as compared to AWE. Other notable methods for generating reduced order approximations as well as lumped circuit models are [Kerns 96, Mangold 98]. A powerful new method is the coordinate-transformed Arnoldi algorithm [Silveira 96, Elfadel 97], which guarantees stable reduced order models. In our approach, we use an implementation of the Arnoldi algorithm, PRIMA [Odabas 97], which guarantees the passivity of the reduced order models.

In an industrial environment, several other problems, beside passivity and accuracy, can appear: how to use the macromodels with an existing simulator, how to build models which are reduced in size (preferably smaller than the initial circuit), how to simulate the macromodels in the most efficient way, how to store and post-process the simulation results, how to integrate the macromodels into the general design flow. Some of these "implementation issues" can have a direct negative impact on the entire macromodel generation algorithm and the best theoretical approach may not be the most feasible one in practice.

The use of frequency-domain reduced order models normally requires access to the core of the simulator, because inputting such models in file format to a SPICE-like simulator is prohibitive. One solution to this problem is to use synthesized lumped circuit macromodels based on the frequency-domain reduced order representation because lumped circuit elements are part of the input format of any commercial circuit simulator.

The benefit of "linearizing" receiving (sink) ports and including their passive loads into the interconnect [Dartu 96] is that the signal activity is fully characterized only by the driving port signals. So, with a macromodel built only for the subset of driving ports, we can fully capture the behavior of the interconnect, if the sink port loads are known a priori. But, in order to generate the signals at the sink ports, we have to post-process the simulation results for the macromodel corresponding to the driving ports. An efficient way to do that is to generate a reduced set of "primary signals" during the simulation of the macromodel, signals which are later re-used as construction blocks for any other signal of interest (sink port response) through simple linear combinations. As a consequence, the macromodel size is small (basically only the subset of driving ports) and the simulation time is greatly improved while the accuracy is preserved.

In Section 1 we describe the modified macromodel equations based on the admittance formalism. In Section 2 we describe in detail the actual macromodel building process and the problems associated with it. In Section 3, considerations on the simulation speed are discussed, while in Section 4 we present some performance statistics of a tool that has been developed based on our macromodeling method and is used at Motorola's PowerPC^{TM} [1] microprocessor design center, Somerset Design Center in Austin, Texas.

1. THE ADMITTANCE FORMALISM OF MULTIPORT STRUCTURES

In the admittance (Y) formalism [Valken 60] a sink port is "active" even when no current is flowing through the port because its effect at other ports is seen as a function of the voltage across the port pins. For a circuit with p ports,

[1] PowerPC is a trademark of International Business Machines Corp. and is used under license by Motorola Inc.

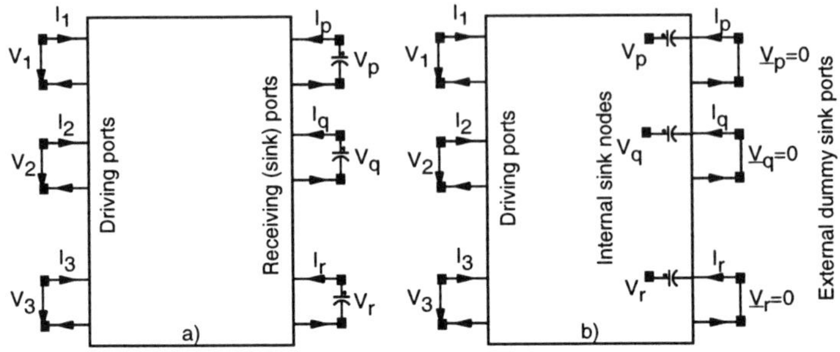

Figure 1 The multi-port representation: a) typical Y formalism; b) modified Y formalism.

d drivers and $p-d$ sinks, the equation of port j is given in (1.1):

$$I_j(s) = V_1 \cdot Y_{1j} + V_2 \cdot Y_{2j} + \cdots + V_p \cdot Y_{pj}, \quad \text{for} \quad 1 \le j \le p. \tag{1.1}$$

A multi-port structure is generally represented as in figure 1a, where the sink ports are "linearized" by capacitive loads. The voltage across sink ports is in reality determined only by the active port voltages. In figure 1b, the electrical structure of the netlist is modified such that when a port is passive, the voltage across that port is zero. This is effectively done by including the port load into the interconnect and by defining the sink port as a short circuit. In this case, the port equations will be different at the sink ports from the driver ports. For a driving port a and a sink port m, the port equations are:

$$I_a(s) = V_1 \cdot Y_{1a} + V_2 \cdot Y_{2a} + \ldots + V_d \cdot Y_{da}, \quad \text{for} \quad 1 \le a \le d \quad \text{drivers only,} \tag{1.2}$$

$$V_m(s) = V_1 \cdot H_{1m} + V_2 \cdot H_{2m} + \ldots + V_d \cdot H_{dm}, \quad \text{for} \quad d \le m \le p \quad \text{sinks only.} \tag{1.3}$$

2. BUILDING THE MULTIPORT MACROMODEL

Each transfer admittance (or voltage transfer) term is described in terms of circuit poles and residues:

$$Y_{jk} = \sum_{i=1}^{N} \frac{f_{jk}^i s + g_{jk}^i}{h^i s^2 + k^j s + 1}. \tag{1.4}$$

A very important observation is that the poles apply to the linear network as a whole and only the residues are specific to each transfer admittance (voltage transfer) term. Using the expression of the admittance from (1.4), and knowing

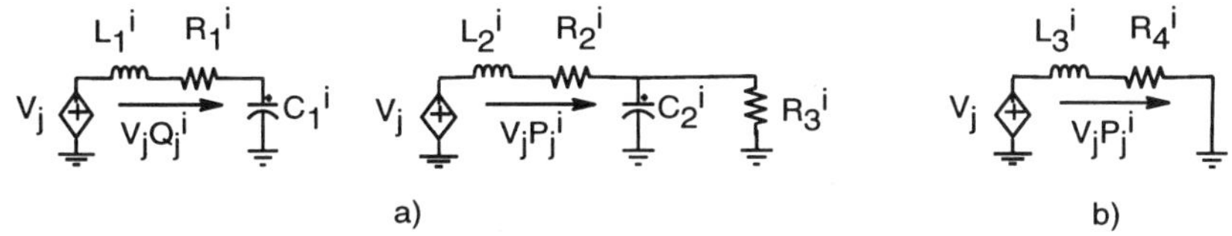

Figure 2 The synthesis of the pole component: a) the two admittances for a pair of complex conjugate poles b) the admittance for a real pole.

that only d ports are driving ports and that all the sink port loads are known and present in the circuit, we can rewrite the a-th port equation as:

$$I_a(s) \;=\; \sum_{j=1}^{d} V_j \cdot \sum_{i=1}^{N} \frac{f_{ja}^i s + g_{ja}^i}{h^i s^2 + k^i s + 1}. \tag{1.5}$$

Equation (1.5) can be rewritten as:

$$\begin{aligned} I_a(s) &= \sum_{j=1}^{d} \left(V_j \cdot \sum_{i=1}^{N} q_{ja}^i \cdot \frac{m^i s + n^i}{h^i s^2 + k^i s + 1} + r_{ja}^i \cdot \frac{p^i s}{h^i s^2 + k^i s + 1} \right), \\ &= \sum_{j=1}^{d} \left(\sum_{i=1}^{N} q_{ja}^i \cdot \frac{(m^i s + n^i) \cdot V_j}{h^i s^2 + k^i s + 1} + r_{ja}^i \cdot \frac{p^i s \cdot V_j}{h^i s^2 + k^i s + 1} \right), \\ &= \sum_{j=1}^{d} \left(\sum_{i=1}^{N} V_j \cdot P_j^i \cdot q_{ja}^i + V_j \cdot Q_j^i \cdot r_{ja}^i \right). \end{aligned} \tag{1.6}$$

$$P_j^i = \frac{m^i s + n^i}{h^i s^2 + k^i s + 1} \text{ and } Q_j^i = \frac{p^i s}{h^i s^2 + k^i s + 1} \tag{1.7}$$

In the case of a driving port, we need to determine the effect of other drivers on our port. From (1.5), the contribution of another driving port j to the total current of port a is a current $I_{ja} = V_j \cdot Y_{ja}$. Inspecting (1.6) we can see that each current contribution can be split by poles and each pole current contribution can be described using a current controlled current source. The control current is $V_j \cdot P_j^i$ or $V_j \cdot Q_j^i$ and the scalar multiplier is q_{ja}^i or r_{ja}^i, respectively. The reason why the poles are split into two components, see (1.7), is that each sub-component can be synthesized in the form of an input admittance. The synthesized RLC structures used for the two sub-components of a complex pole pair and the component of a real pole are shown in figure 2. The control currents $V_j \cdot P_j^i$ and $V_j \cdot Q_j^i$ are the currents flowing through these simple RLC synthesized structures driven by a voltage source V_j. The values for the synthesized elements are determined using only the pole values (1.8). The α coefficient in (1.8) is any number greater than 1 such that C_2^i has a real positive value.

$$
\begin{aligned}
&\text{choose} \quad \alpha > 1.0, \\
&R_1^i = 1 \quad R_2^i = 1 \quad R_3^i = \max\left\{1, \alpha \cdot \left(\frac{4h^i}{(k^i)^2} - 1\right)\right\} \quad R_4^i = 1, \\
&C_1^i = k^i \quad C_2^i = \frac{k^i \cdot (R_2^i + R_3^i) + \sqrt{(k^i)^2 \cdot (R_2^i + R_3^i)^2 - 4h^i \cdot (R_2^i + R_3^i)}}{2R_3^i}, \\
&L_1^i = \frac{h^i}{k^i} \quad L_2^i = \frac{h^i \cdot (R_2^i + R_3^i)}{C_2^i \cdot R_3^i} \quad L_3^i = k^i.
\end{aligned} \tag{1.8}
$$

The m, n and p coefficients are determined using the R, L, and C elements computed in (1.8):

$$
\begin{aligned}
\frac{m^i s + n^i}{h^i s^2 + k^i s + 1} &= \frac{\left(\frac{C_2^i R_3^i}{R_2^i + R_3^i}\right) s + \left(\frac{1}{R_2^i + R_3^i}\right)}{\left(\frac{L_2^i C_2^i R_3^i}{R_2^i + R_3^i}\right) s^2 + \left(\frac{L_2^i + C_2^i R_3^i R_2^i}{R_2^i + R_3^i}\right) s + 1}, \\
\frac{p^i s}{h^i s^2 + k^i s + 1} &= \frac{C_1^i s}{L_1^i C_1^i s^2 + R_1^i C_1^i s + 1}, \\
\frac{n^i}{k^i s + 1} &= \frac{\frac{1}{R_4^i}}{\frac{L_3^i}{R_4^i} s + 1}.
\end{aligned}
$$

The q and r coefficients are determined such that the original residues are matched:

$$
q_{ja}^i = \frac{f_{ja}^i}{m^i} \quad \text{and} \quad r_{ja}^i = \frac{g_{ja}^i - f_{ja}^i \cdot \frac{n^i}{m^i}}{p^i}.
$$

The block level circuitry for a two-driver circuit is shown in figure 3. The controlling currents (primary signals) are determined only by the drivers. This means that it will be more efficient to synthesize and derive these primary signals $V_j \cdot P_j^i$ and $V_j \cdot Q_j^i$ in a driver auxiliary circuitry rather than into the sink circuitry. These signals are used to model the transfer admittances at the driving ports in figure 3 as well as the transfer functions at the sink ports. The reduction in circuit size is achieved because the number of drivers is usually much smaller than the number of sinks, especially in large nets like clock distribution networks.

For each sink voltage given by (1.3) we have a set of voltage transfer functions, each corresponding to a driver. As opposed to the case of a driver port, the primary signals ($V_j \cdot P_j^i \cdot q_j^i$ and $V_j \cdot Q_j^i \cdot r_j^i$) are now voltages and the sink voltage is obtained using a chain of current controlled voltage sources. Note that the components $V_j \cdot P_j^i$ and $V_j \cdot Q_j^i$ are the same as the components of

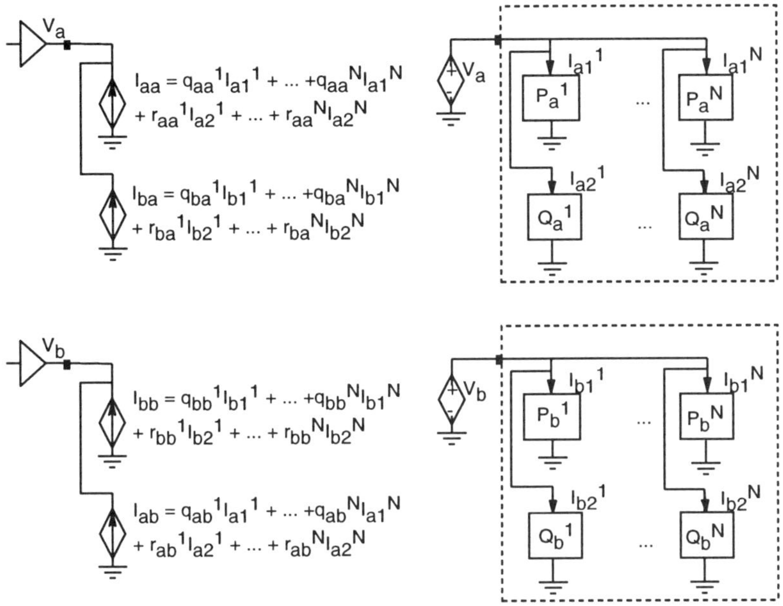

Figure 3 The block level circuitry for a circuit with two-drivers (a and b). Each driver has a driver auxiliary circuit where the primary signals are generated.

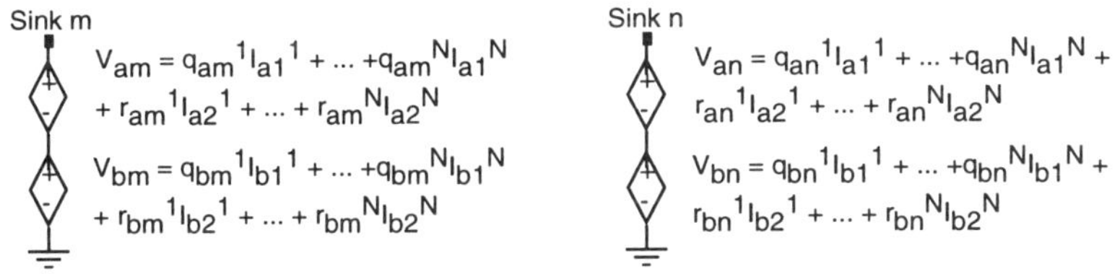

Figure 4 The structure of the sink circuitry in the case of a two-driver (a and b) two-sink (m and n) circuit.

(1.6) so the sink ports use the same partial responses from the driver auxiliary circuits. The difference is that the scalar coefficients q and r are, in the sink case, acting as resistances. As outlined in the previous paragraph, the sink response becomes just a linear composition of driver dependent partial responses and this eliminates any simulation that has to be done in the sink circuitry. For a circuit with two drivers (a and b) and two sinks (m and n), the circuitry of the sink ports is shown in figure 4.

The size of the macromodel is a function of the number of drivers (d), the accuracy (number of poles required ($2N$)) and the number of sinks ($p - d$):

1. resistors: $\leq 3dN$ (equality when all poles are complex),
2. capacitors: $2dN + p - d$,

3. inductors: $\leq 2dN$,
4. voltage controlled voltage sources: d,
5. voltage controlled current sources: d,
6. general format current sources: d^2 (each source is a linear combination of voltages - at drivers) [MCSpice 95],
7. general format voltage sources: $dp - d^2$ each source is a linear combination of voltages - at sinks) [MCSpice 95].

Maximum total number of elements: $d \cdot (7N + p + 1)$.

PRIMA also provides general macromodels for linear circuits and these are the most compact synthesized circuits with a size of O(NP). But the disadvantage of PRIMA's approach is that all the nodes of interest must be considered as ports in the multiport structure. In our case, the number of elements is O(dP) when we are building a full macromodel with all the sinks present in the circuit. The most practical approach is to build a macromodel only for the active drivers (in which case the size of our macromodel is O(dN)), simulate this circuit and save the partial responses ($V_j \cdot P_j^i$ and $V_j \cdot Q_j^i$) on disk and post-process them later in order to obtain the sink responses. Note that, in this approach, the macromodels that we are building become more advantageous than PRIMA's macromodels because, normally, $d \ll P$. Our macromodels are currently synthesizing the same set of N pairs of poles for each driving port. In reality the synthesized circuit can be much more compact, with N pairs of synthesized poles for all the driving ports. But, from an implementation point of view, this requires much more pre-processing work and makes the macromodel structure much more complicated and less intuitive.

3. THE SIMULATION OF A MACROMODEL STRUCTURE

In this section we present some considerations on the simulation speed of the macromodels built using our approach. Because the synthesized macromodel uses decoupled pole representations, it has the great advantage that the structure of the circuit admittance matrix is very sparse. All the matrix entries that correspond to the synthesized auxiliary circuits will be ordered in a block diagonal sub-matrix as in figure 5. All the constant coefficients used for controlled sources will be present in two off-diagonal blocks. All the driver ports will generate a sub-matrix that is an identity matrix because there are no direct connections between the active ports. From the general structure of the matrix, it is apparent that such a structure is very convenient for matrix LU factorization which is the most time-consuming part of a simulation. This is the reason why, sometimes, macromodels may be easier to analyze than the initial

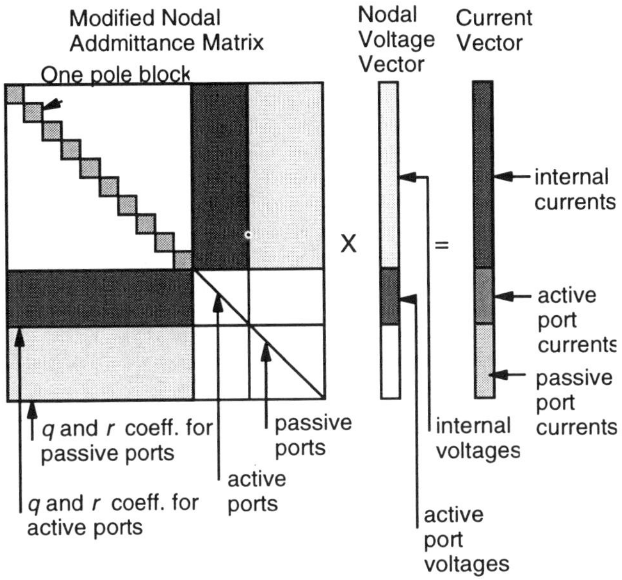

Figure 5 The macromodel admittance matrix has a very well organized structure which is convenient for matrix factorization.

circuit even if they have more components. In normal interconnect circuitry, because of the tight spacing between on-chip wires, there is a lot of capacitive and mutual inductive coupling and that results in less sparse admittance matrices. As a consequence, the MNA matrices corresponding to these nets are more dense and the use of macromodels becomes very advantageous from the simulation time point of view.

As described in the previous section, in a practical implementation, the coefficients corresponding to the passive ports are not present in the MNA matrix. These coefficients are used later in the post-processing stage in order to determine the sink responses.

4. RESULTS

In this section we present some results which demonstrate the accuracy and the efficiency of our approach. This macromodeling technique has been implemented in a program called Macrosim, which is primarily used for the design of very large clock networks at the Somerset Design Center. The tests performed on the tool have shown excellent accuracy (50% delay comparison and 10% to 90% rise time comparison) with an average of less than 1% error (2% maximum). The simulation time speed-up achieved varies from $3\times$ to $30\times$ over the full netlist.

In order to illustrate the accuracy and the speed-up of our method, we have chosen a section of the clock distribution network of one of our microprocessors. The clock section has one driver port and 370 sink ports and has an electrical model of 27484 elements (15148 capacitors, 7413 resistors and 4923 inductors). In table 1 we present the run times for the full netlist and for the macromodels. The macromodels are run in two different cases: in the first case, all the sinks are present in the macromodel and no post-processing is needed, and the second case, when no sinks are present in the macromodel and post-processing is needed. For the second case, the two numbers indicate the run time of the SPICE simulation and the run time of the post-processor stage. The driver of the clock section has 40 NMOS and 70 PMOS transistors, and 112 capacitors.

Table 1 Run-times for the full clock section netlist and its corresponding macromodels

full netlist	*macromodel with all sinks*	*macromodel with no sinks*
434.88s	66.11s	8.94s/7.32s

In figure 6 the signals at the output of the clock section driver and at one of the sink ports is shown. The macromodels used in this example (and in the normal tool flow) have 10 poles which are enough for our accuracy requirements. Note that the signal at the output node of the driver (*a drv* on the plot) is very accurately captured by the macromodel (*b drv*) despite the fact that, due to the inductance of the lines, there is a reflected wave that produces a voltage spike.

The use of Macrosim proved to be very appealing to the clock network designers because, although it is not as fast as the static timing analysis, it is much faster than the full SPICE simulation and gives the user the flexibility to analyze the drivers separately or to monitor only some sink nodes of interest. In table 2 we show the accuracy of our macromodeling technique in comparison to the typical static analysis run that is performed on clock nets. All the errors are with respect to the full netlist SPICE simulation.

Table 2 Accuracy comparison between the static timing analysis and the macromodels

Error type	*Static timing*	*Macromodels*
Avg. delay error	1.6%	0.6%
Avg. rise time error	11.5%	0.8%
Max. delay error	3.7%	0.7%
Max. rise time error	13.5%	2.0%

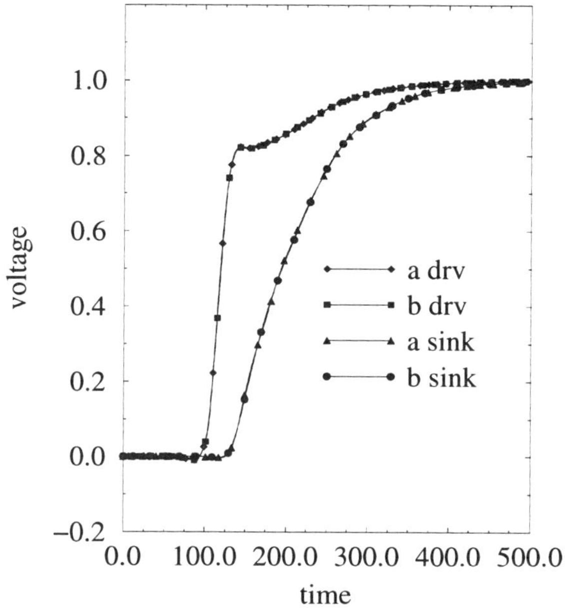

Figure 6 Simulation results for the output node of the driver (*drv*) and one sink node (*sink*). The *a* signals are from the full network, while the *b* signals are from the macromodel. The time and voltage units on the x and y axis have been scaled.

5. REFERENCES

[Kim 94] S.Y. Kim, N. Gopal and L.T. Pillage. (1994). *Time Domain Macromodels for VLSI Interconnect Analysis,* IEEE Transactions on CAD, vol.10, pp. 1257-1270.

[Pillage 90] L.T. Pillage, R.A. Rohrer. (1990). *Asymptotic Waveform Evaluation for Timing Analysis,* IEEE Transactions on CAD, vol.4, pp. 352-366.

[Dartu 96] F. Dartu, B. Tutuianu and L.T. Pileggi. (1996). *RC-Interconnect Macromodels for Timing Simulation,* IEEE/ACM Proceedings of DAC, pp. 544-547.

[Feldman 95] P. Feldman and R.W. Freund. (1995). *Efficient linear circuit analysis by Padé approximation via the Lanczos process,* IEEE Transactions on CAD, vol.5, pp. 639-649.

[Kerns 96] K.J. Kerns and A.T. Yang. (1996). *Stable and Efficient Reduction of Large Multiport RC Networks by Pole Analysis via Congruence Transformations,* IEEE/ACM Proceedings of DAC, pp. 280-285.

[Mangold 98] T. Mangold and P.Russer. (1998) *A systematic approach to the problem of equivalent circuit model generation,* IEEE MTT-S International Microwave Symposium Digest, Vol. 3, pp. 1409-1412.

[Silveira 96] L.M. Silveira, M. Kamon, I. Elfadel and J. White. (1994) *A coordinate-transformed Arnoldi algorithm for generating guaranteed stable reduced-order models of arbitrary RLC circuits,* IEEE/ACM Proceedings of ICCAD, pp. 288-294.

[Elfadel 97] M. Elfadel and D.D. Ling. (1997). *Zeros and passivity of Arnoldi reduced-order models for interconnect networks,* IEEE/ACM Proceedings of DAC.

[Odabas 97] A. Odabasioglu, M. Celik and L.T. Pileggi. (1997). *PRIMA: Passive Reduced-order Interconnect Macromodeling Algorithm,* IEEE/ACM Proceedings of ICCAD.

[MCSpice 95] B. Mulvaney, et. all. (1995). *MCSpice - User's Manual,* SSDT Circuit Design Technology, Motorola Inc.

[Valken 60] M.E. Van Valkenburg. (1960). *Introduction to Modern Network Synthesis,* John Wiley and Sons Inc.

[Arun 97] R. Arunachalam, F. Dartu and L.T.Pileggi. (1997). *CMOS Gate Delay Models for General RLC Loading,* Proceedings of ICCD, pp. 224-229.

A Decomposition-Based Symbolic Analysis Method for Analog Synthesis from Behavioral Specifications

Alex Doboli and Ranga Vemuri
Digital Design Environments Laboratory, Department of ECECS
University of Cincinnati, Cincinnati, OH 45221-0030
Email: { adoboli,ranga } ececs.uc.edu

Abstract This paper discusses a technique for symbolic analysis of large analog systems. The method exploits the hierarchical structure and uniformity of a system for producing **compact** symbolic expressions. The technique is based on decomposing a system through the method of **tearing** [5] as opposed to traditional symbolic methods that use nodal analysis. After producing the symbolic model of a system, a set of reduction rules is applied to the model. Reductions attempt to decrease the number of arithmetic operations performed for numerically evaluating the symbolic model. The discussed technique is useful for synthesis, inside an exploration-loop, as it avoids repeatedly computing the symbolic models.

1. INTRODUCTION

Symbolic analysis is the task of automatically deducing relationships between overall parameters of a system and parameters of the composing elements of the system [6] [3]. For example, a symbolic expression can describe the transfer-function of a filter in terms of parameters of its building elements, i.e. op amps, resistors, and capacitors. Analysis techniques include determinant-based methods and signal-flow graph methods [6]. Determinant-based methods use Cramer's rule for solving the set of linear equations implied by symbolic analysis [3]. Signal-flow graph methods represent a set of linear equations as a weighted graph, and use Mason's rule for solving the equations [6]. Symbolic analysis has a wide range of applications i.e. frequency response calculation, sensitivity analysis, hardware synthesis etc. This paper proposes on a new symbolic analysis technique for analog synthesis.

The main challenge for symbolic analysis is the exponential growth of the produced symbolic expressions (10^{11} terms for an op amp [4]). Current research considers two ways of handling this aspect: approximation and hierar-

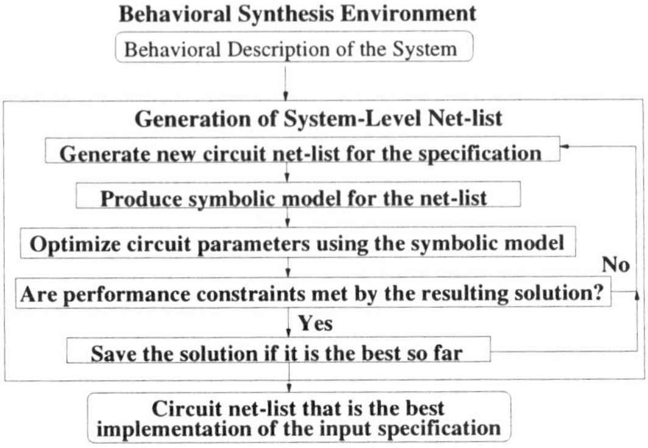

Figure 1 **Behavioral synthesis environment**

chical methods. Approximation methods [12] retain only the significant terms of the symbolic expressions and eliminate the insignificant ones. The difficulty, however, lies in identifying what terms to eliminate, and what the resulting approximation error could be. Hierarchical methods [11] [4] decompose the global analysis problem into smaller sub-problems. The global expressions are described as sequences of sub-expressions for the sub-problems. An exponential number of operations is still needed for their numerical evaluation.

Our approach addresses the exponential complexity of symbolic analysis by considering specific aspects of the analyzed system. A system has a *structural hierarchy* and *uniformity* that can be exploited for more effective symbolic analysis. For example, a filter is typically composed of multiple stages of similar blocks connected in identical patterns. Such knowledge of hierarchy and uniformity can be used for (1) simplifying symbolic calculations, (2) reducing their memory size and (3) diminishing their derivation time.

This paper discusses a method for symbolic analysis of large analog systems. For an overall parameter of a system, the algorithm produces a *computational tree* that describes how its values depend on the parameters of the blocks composing the system. A computational tree (referred in this paper as *Analog Performance Tree*) is an uninterpreted variant of the closed-form, symbolic expressions that are produced by traditional methods [6] [3]. In our approach, computational trees are produced in a top-down manner by traversing the system hierarchy and without explicitly solving equations. The algorithm exploits the hierarchical structure and uniformity of a system for producing compact symbolic expressions and for diminishing the generation time of the symbolic trees. Once a tree is produced, a set of reduction rules is applied for simplifying the computational tree and decreasing the number of arithmetic operations performed for numerically evaluating the tree.

We use this method for behavioral synthesis of analog systems. The synthesis flow is depicted in Figure 1. The application to be synthesized is specified behaviorally by expressing its signal flows and processing. During synthesis, different structural net-lists of interconnected circuits are produced for the specification. For each net-list, a *cost function* (symbolic model) is derived by the described symbolic technique. The cost function relates system parameters, i.e. overall transfer function etc. to the composing block parameters, i.e. gains, bandwidth, impedances etc. It is used for performing optimization based trade-off exploration for the circuit parameters. However, all net-lists

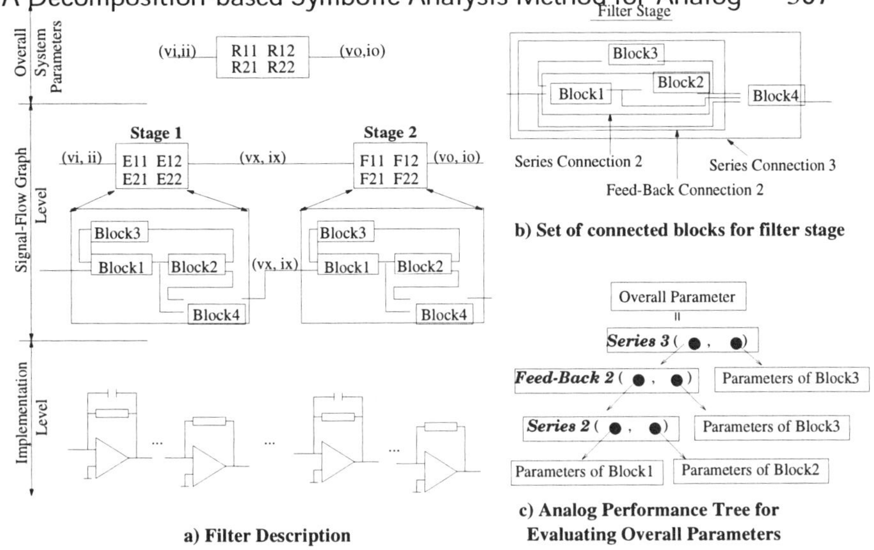

Figure 2 **Hierarchical description and Analog Performance Tree for a filter**

share a common factor in that they realize the same signal-processing flow. This can be exploited in reducing both the memory size and set-up time for the symbolic expressions, and is the motivation behind this paper. The two gains are important as symbolic analysis is repeatedly performed during synthesis.

The rest of the paper is organized as follows. Section 2 describes the main features and advantages of our symbolic method by using a motivational example. In Section 3, we present the model of analog systems that is used for symbolic analysis. Section 4 focuses on our symbolic method. Experimental results are discussed in Section 5, followed by conclusions in Section 6.

2. MOTIVATING EXAMPLE

To give a brief insight on the addressed problem and the solution we propose, we present a fourth-order filter [11] as a motivating example. The middle part of Figure 2a sketches the filter specification represented at the *Signal-Flow Graph Level*. The fourth-order function of the filter is obtained by cascading two second-order stages. Each of the second-order blocks implements the same close-loop flow, as depicted in the middle part of Figur.2a. This way of describing the signal processing and flows of a system is called *Signal-Flow Graph* (SFG) [7]. We assume that the behavior of an analog system is described as an SFG and is the input for behavioral synthesis.

Behavioral synthesis now derives the best implementation for an SFG specification, so that, the described signal-path is accomplished, certain design/performances are met, and a cost function is optimized. During synthesis, different implementations are explored, and hence, the symbolic analysis part is repeatedly done. An observation is that all implementations will actually realize the same SFG described in the specification. We exploit this similarity to reduce the required time for finding the symbolic expressions of an implementation. Secondly, there are same common factors between parts of an implementation (i.e. *Stage 1* and *Stage 2* in Figur.2a). We intend to take advantage of these similarities by analyzing only one of the stages, and re-using its symbolic model for the second stage. This shortens the set-up time of global symbolic expressions and also compacts the required memory.

For exemplifying the analysis technique, let's assume that we intend to calculate for different filter implementations their transfer-functions defined as $TF = \frac{v_o}{v_i}$, when $i_o = 0$. Values v_o and v_i are voltages at the input and output ports, and values i_i and i_o are the currents. This analysis is useful to check whether a real implementation is within an acceptable error range from the input specification.

For analysis, the overall system can be modeled as a two-port described by the symbolic matrix $[R_{ij}]_{2\times 2}$ (Figure 1.2a), where:

$$\begin{aligned} v_o &= R_{11}v_i + R_{12}i_o \\ i_i &= R_{21}v_i + R_{22}i_o \end{aligned}$$

The filter transfer function is given by $TF = R_{11}$, and thus the symbolic formula of R_{11} is calculated. Note that there is no need for also computing R_{12}, R_{21} or R_{22}.

The goal is to exploit the invariance of the SFG "algorithm" for different implementations. We perform a *top-down* symbolic analysis by considering *generic* parameters for the blocks (matrix E for *Stage 1* and matrix F for *Stage 2*, where the nature of the matrix elements is not fixed). This differs from traditional methods that are bottom-up approaches, and use physical parameters like impedances, admittances, etc. [3] [4] [6]. However, if physical parameters are used in the analysis, then the obtained formulas are specific to an implementation.

The elements of the overall matrix $[R_{ij}]_{2\times 2}$ relate to elements of matrices $[E_{ij}]_{2\times 2}$ and $[F_{ij}]_{2\times 2}$ according to:

$$\begin{aligned} R_{11} &= \frac{E_{11}+F_{11}(E_{12}E_{21}-E_{11}E_{22})}{1-E_{22}F_{11}}, \quad R_{12} = \frac{E_{12}F_{12}}{1-E_{22}F_{11}} \\ R_{21} &= \frac{E_{21}F_{21}}{1-E_{22}F_{11}}, \quad R_{22} = \frac{F_{22}+E_{22}(F_{12}F_{21}-F_{11}F_{22})}{1-E_{22}F_{11}} \end{aligned}$$

These relations were obtained by setting up the equations for *Stage 1* and *Stage 2* and eliminating the unknowns v_x and i_x.

The four expressions depend on the parameters of the composing blocks and their connection style that is *series connection* for this example. However, they are not influenced by the nature of the block parameters. As matrices E_{ij}, and F_{ij} are generic, the four expressions can also be seen as functions that link for *any* series connection the overall symbolic parameters R_{ij} of the connection, to the parameters E_{ij} and F_{ij} of the composing blocks. Similar solving functions can be found for other popular connection styles, i.e. feed-back, feed-forward, divergent, etc. and which are discussed in more detail in Section 4. Solving functions are re-used in our technique for describing all connections of the same type. This reduces both the size of the overall expressions, and the set-up time as similar connections are handled only once.

Assume that solving functions *series, feed-back, feed-forward*, etc. relate the overall parameters of their connections to the parameters of the composing blocks. Then for a set of connected blocks, a *computational tree* named *Analog Performance Tree* (APT) can be built, describing how parameters relate to each other. Figure 2b depicts a set of connected blocks and Figure 2c shows the

APT for the transfer-function of the connection. *Internal nodes* describe the solving functions of the connection styles for the blocks. Various connections are also described in the first part of Section 4 and in Table 1. The parameters for each solving function relate to the composing blocks. The returned values are the parameters of the overall block. *Leaf nodes* describe the basic building blocks of an implementation, i.e. gain stages, integrators, RC stages, etc. Their parameters have a physical meaning, i.e. impedances, admittances, etc.

An APT is a symbolic expression in an unevaluated form. This is not a disadvantage as we are not interested in getting insight of a system, and where closed-form expressions would help. Besides, the tree structure of an APT allows easy simplifications or approximations.

In our technique, the computational tree is set-up by composing functions for the basic connection patterns, and not by solving equation sets like in traditional methods. This reduces the overall set-up time for the symbolic expressions. If basic connection styles cannot be recognized at one step then a decomposition of the system is performed. Decomposition is achieved through the principle of *tearing* [5]. Tearing separates a set of equations into sub-sets with less number of unknowns. Also, whenever similar block connections are identified (i.e *Stage 1* and *Stage 2* in Figure 2a) only one of them is analyzed.

During synthesis, different implementations are explored for the blocks in an SFG. However, the entire APT tree remains unchanged, except the leaf nodes that are updated according to the chosen implementations for the blocks. This avoids repeated symbolic analysis inside the synthesis cycle.

3. SYSTEM MODEL FOR SYMBOLIC ANALYSIS

As seen from Section 1, the synthesis task bridges consecutive levels of abstraction: it starts with an abstract specification of a system, and goes down to hardware implementations by selecting circuit net-lists, building their symbolic model and performing the necessary parameter optimizations. For describing these synthesis tasks an appropriate representation model for an analog system is needed. The model must easily relate to the input specification. Also, it should allow the description of all relevant aspects for net-list generation, symbolic analysis, and parameter optimization.

Input specifications are described as SFGs (see Figure 2a). The net-list generation algorithm [1] also considers net-lists at the block-level. Thus, it is important that the model for symbolic analysis easily describes a hierarchy of connected blocks and how the overall performance parameters of the connected blocks relate to those of the composing blocks.

Our mathematical model for symbolic analysis captures (1) the signal processing and structural aspects of a system, and (2) the description of the computational path for calculating linear (linearized) performance values:

$$\boxed{System =< SFG, APT >} \qquad (1.1)$$

SFG is the *Signal-Flow Graph* used for describing the processing/structural attributes of a system, and *APT* is the *Analog Performance Tree*, that symbolically captures the computing path for performance attributes used in synthesis. This paper presents a method for setting-up the symbolic *APT* of a system when its *SFG* is known.

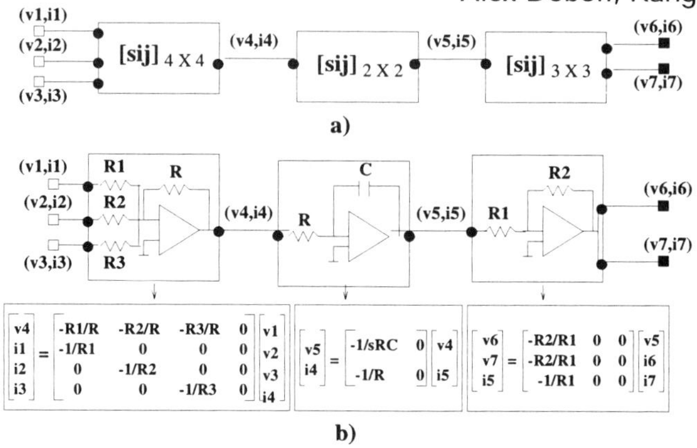

Figure 3 **Representation of the model for symbolic analysis**

SFG is the quint-tuple:

$$\boxed{SFG =< Wires, Blocks, Edges, Inputs, Outputs >} \quad (1.2)$$

Each $wire \in Wires$ represents an ordered pair of symbolic values *(v,i)*, meaning that the two values are interdependent and they always appear in conjunction. *Wires* models continuous-time signals, *v* being a voltage and *i* a current.

Blocks is the set of all operational blocks that realize the signal processing of a system. Each block has a set of ports for connecting it to other blocks by using wires. Also, each block has a behavior defined in terms of the values v and i of the wires connected at its ports.

The behavior of a block with m ports can be modeled as a set of m linear equations with m knowns and m unknowns. Equations for composed blocks are similar to the input-output equations of a linear system [7]. For basic blocks (blocks that correspond to electric circuits), equations are derived from the nodal equations of the block. Knowns and unknowns are the i and v values of the m wires at the block ports. Depending on how the sets of knowns and unknowns are selected, the symbolic coefficients in the equations have different meanings. For example, if the v values of wires are considered to be knowns, and the i values unknowns then a block with m ports is described by the equation set:

$$\begin{aligned} i_1 &= s_{11}v_1 + s_{12}v_2 + \cdots + s_{1n}v_m \\ i_2 &= s_{21}v_1 + s_{22}v_2 + \cdots + s_{2n}v_m \\ & \cdots \quad \cdots \\ i_m &= s_{m1}v_1 + s_{m2}v_2 + \cdots + s_{mn}v_m \end{aligned}$$

Coefficients s_{ij} are symbolic expressions and they describe admittances. The matrix $[s_{ij}]_{m \times m}$ is called *block matrix.*

A basic block describes either an active circuit or an RC network. Thus, coefficients for the basic blocks have a physical meaning as they are transfer-functions, impedances, admittances. They are expressed as Laplace transforms for capturing frequency dependencies. Also, these elements depend only on the type of the circuit and can be stored in a library. Kirchhoff's laws are satisfied by block matrices.

The block model is similar to the *port network model* in network theory [8]. It has the advantage that the block entities for symbolic analysis are easily identifiable with the specification entities and those of the net-list generation step, as they also consider a block level. Other representations, i.e. Mason's signal-flow graphs [7] do not offer any structural information on how their elements relate to the composing blocks of a design. However, such knowledge is important for structural synthesis i.e. net-list generation.

Edges is the function:

$$\boxed{Edges : (Blocks, N) \times (Blocks, N) \rightarrow Wires} \tag{1.3}$$

$Edges((block_i, I), (block_j, J))$ is either the wire that interconnects port I of $block_i$ to port J of $block_j$, or $\emptyset$ if the two ports are not linked together.

Inputs and *Outputs* are the sets of all input and output ports of the SFG.

A representation of the model is showed in Figure 3a and Figure 3b. For the building block it presents its block matrix. Input ports are small white squares and output ports are small black squares.

An *Analog Performance Tree* (APT) describes the computational path for calculating linear (linearized) attributes of an *SFG*. It indicates how attributes for higher-level blocks are calculated using parameters of its composing blocks and composition functions that depend on the connection style. APT can be used to describe linear attributes, i.e AC small-signal parameters: overall transfer functions for implementations, input/output impedances etc.

An APT has two types of nodes. (1) *Leaf nodes* that describe basic blocks of an SFG (blocks that are not further decomposed). Leaf nodes contain equations for relating block parameters to their design parameters. These equations are not necessarily linear and describe elements, i.e. DC biasing, manufacturablity aspects. (2) *Internal nodes* that describe the connections of leaf nodes or internal nodes. They correspond to series, feed-back, feed-forward etc. connections. Depending on the connection, an internal node is annotated with a composition (solving) function that links overall parameters to the parameters of composing blocks. Figure 2c depicts an APT.

System Decomposition and Modeling of Block Interactions by Tearing

Assume that two blocks are interconnected as in Figure 4. Knowns for the block matrices are bordered by a square.

Tearing Rule [5]: Wire (v_x, i_x) can be torn apart so that v_x is a known and i_x an unknown for the first block. For the second block, i_x is a known and v_x an unknown.

The other way of assigning the values v_x and i_x as knowns-unknowns is also possible. The only restriction is that, when selecting the block matrices for the basic blocks, the known- unknown character assigned by tearing to the values v and i of the torn wires, must be reflected.

Tearing decomposes a system into smaller sub-systems. It is a key concept for our technique of building APTs, as it allows identification of connection styles i.e. series, feed-back, etc. After handling the sub-systems their APTs have to be re-composed for finding APTs for the system parameters. Assume that v_i and v_j are the knowns in Figure 1.4. After we wrote equations for unknowns i_i ... i_k and i_x for block 1 and unknowns i_j ... i_k and v_x for

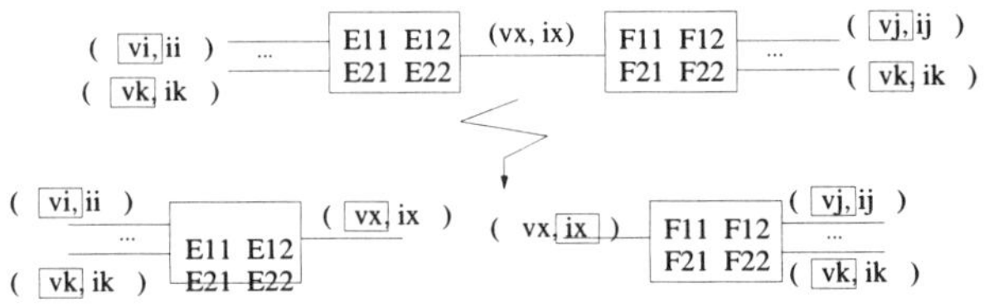

Figure 4 **Decomposition through wire tearing**

block 2, we eliminated unknowns v_x and i_x from the equation set. Following expressions resulted for unknowns i_k of block 1 and block2 depending only on knowns v_i and v_j. The parameters for the re-composed system are:

$$\boxed{i_k = \Sigma(t_{ki} + \frac{\beta\gamma_k r_i}{1-\alpha\beta})v_i + \Sigma(\frac{p_j\gamma_k}{1-\alpha\beta})v_j, \text{ for block 1.}} \tag{1.4}$$

$$\boxed{i_k = \Sigma(\frac{r_i\theta_k}{1-\alpha\beta})v_i + \Sigma(q_{kj} + \frac{\alpha\theta_k p_j}{1-\alpha\beta})v_j, \text{ for block 2.}} \tag{1.5}$$

The equations show how the overall parameters of the re-composed system depend on those of the two sub-systems. They can be easily coded into a computer program for hierarchical analysis of a system.

4. TOP-DOWN SYMBOLIC ANALYSIS METHOD FOR BEHAVIORAL SYNTHESIS

4.1 SOLVING FUNCTIONS FOR LINEAR SYSTEMS

As already seen in Section 2, our symbolic analysis algorithm constructs an APT that relates a linear parameter of the overall system to the parameters of its basic blocks. The structure of the APT remains constant during the entire synthesis algorithm as an APT reflects the invariant control (SFG) algorithm of a system. The synthesis flow instantiates different physical blocks for the leaf blocks of an APT. Thus, the block matrices of the leaves have to be correspondingly updated so that the overall-system parameters relate to the physical parameters of the leaves.

It is worth mentioning that the APT could be set-up by using only tearings and re-compositions. However, we defined a library of popular connection styles and their solving functions so that it avoids repeated tearing and re-composition for the same connections. Solving functions relate parameters of the overall connection to the parameters of its composing blocks. The algorithm for APT construction repeatedly decomposes a system through tearing so that *basic* connection styles (patterns) i.e. series, feed-back, feed-forward, divergent, etc. are identified. After the entire system is decomposed into basic connection patterns, the symbolic expressions for the sub-systems are re-composed to eliminate the tearing effects.

Selecting a library of solving functions should consider not only how often their related connection patterns are used, but also the size of their symbolic expressions. For a set of connected blocks, the size of the overall symbolic expressions increases as the number of unknowns and symbolic parameters is bigger. For example, for a connection type characterized by 9 equations with 9 unknowns and 24 symbolic parameters, the overall symbolic expressions in

	# of blocks	# of unknowns	# of symb. param.	# of knowns	# of operations Addit.	Multipl.	Total	Description of connection
1	2	4	8	2	7	17	24	Series connection
2	2	6	13	4	20	29	49	Series connection 2
3	2	7	25	3	105	233	338	Series connection 3
4	3	7	12	3	23	53	76	Convergent connection
5	2	6	8	2	30	55	85	Feed-back connection 1
6	2	5	13	2	43	131	174	Feed-back connection 2
7	2	5	13	2	14	23	37	Connection of teared wires

Table 1 **Set of connection patterns and their symbolic expressions**

unoptimized form implied 264 additions and 714 additions. This result shows that for connected blocks with 24 symbolic parameters the size of the symbolic expressions tends to be very large.

The previous discussion motivates two aspects: (1) The library of solving functions has to be designed so that the number of symbolic parameters is not very high. If a connection pattern involves more parameters, then a decomposition through tearing is prefered. (2) As expression sizes grow rapidly, optimizations i.e. finding common sub-expressions, factorization are very important. This also motivates the feasibility of using a library of solving functions, where each function is carefully optimized.

In our experiments, we used the solving functions for the connection patterns in Table 1. Functions have similar forms to those in Section 2, and were to big to be introduced in the paper. Different tearing situations are handled by patterns 1, 2 and 7, while the rest of the patterns correspond to connection styles that we often found for systems. Connection patterns 2, 3 and 6 were previously exemplified in Figure 2b. Table columns have the following meaning: Column 2 indicates the number of blocks in a connection. Columns 3, 4 and 5 show the number of unknowns, symbolic parameters and knowns of the equation sets for a connection. Columns 6, 7 and 8 describe the size of the solving functions expressed as number of additions and multiplications.

4.2 ALGORITHM FOR BUILDING THE APT OF A SYSTEM

The algorithm for building the APT traverses the hierarchical specification in a top-down fashion and infers symbolic expressions, without explicitly solving a system of equations as in traditional symbolic methods. If required, wires are torn so that already existing patterns are reused. By this strategy, both the required memory and the set-up time of an APT are reduced.

The algorithm for building the APT of a system is depicted in Figure 5. It starts with the top-most abstraction level, and traverses top-down the overall hierarchy. First, it verifies if blocks are connected in one of the known connection styles (line 1 in Figure 5). These styles are either a library pattern or a previously handled connection pattern. If the connection style is known then a new node in the APT is produced (line 2 in Figure 5). Its *function field*

```
    procedure built_APT ( current_hierarchy_level in a system) is
(1)     if blocks of current_hierarchy_level are connected in a
            library pattern or a previously handled connection pattern then
(2)         Build a new APT node, and label it with the function of the
              already handled pattern. Actual function parameters are
              parameters of the blocks that form current_hierarchy_level;
        else
(3)         Identify a set of already handled patterns so that each block
              in current_hierarchy_level belongs to only one pattern;
(4)         Tear the wires that do not belong to any identified pattern;
(5)         Produce APT nodes for the teared wires. Label them with the
              functions that describe the re-connection of torn wires;
        end if;
(6)     for each block ∈ current_hierarchy_level do
(7)       if ( block has a hierarchical structure) then
(8)         call built_APT ( block) for finding the parameters of block;
          end if;
        end for;
    end procedure built_APT.
```

Figure 5 **Algorithm for building the APT**

points to the solving function of the already encountered pattern. Actual parameters for the function are parameters of the blocks forming the connection. Thus, uniformity inside a system is considered by re-using solving functions for similar patterns. As shown by experimental results in Section 5, compact APT with less nodes result. If the connection style was not previously encountered then a number of wires are torn, so that each block belongs to a known connection pattern and no block appears in two patterns (lines 3,4,5 in Figure 5). APT nodes are produced to reflect the tearing process (their functions correspond to pattern 7 in Table 1). After processing the current hierarchy level and checking if any of the blocks has a hierarchical structure, the algorithm continues by recursively creating APTs for these blocks (line 8 in Figure 5).

It is worth mentioning that the algorithm builds APT nodes only for the parameters required in the symbolic analysis. For example, assume that only parameter R_{11} of the overall block matrix is needed and not parameter R_{12}. Obviously, no computations for parameter R_{12} should be performed by an analysis method. This requirement can be easily accomplished in a top-down method, where at each step the required parameters can be identified. Only the required parameters are expanded during the traversal of the hierarchy. In a bottom-up method, knowledge about the usefulness of certain computations is hard to establish. This is also an advantage of our method over traditional bottom-up approaches.

4.3 ANALOG PERFORMANCE TREE REDUCTION

Reduction rules attempt to decrease the number of arithmetic operations performed for numerically evaluating a symbolic APT by executing simplifications i.e. addition of value 0, subtraction of two identical expressions, multiplication with values 0 and 1, etc. Reducing the number of operations is important for a practical synthesis tool, as numerical evaluations of an APT are repeatedly performed inside the synthesis loop. Besides, the algorithm for

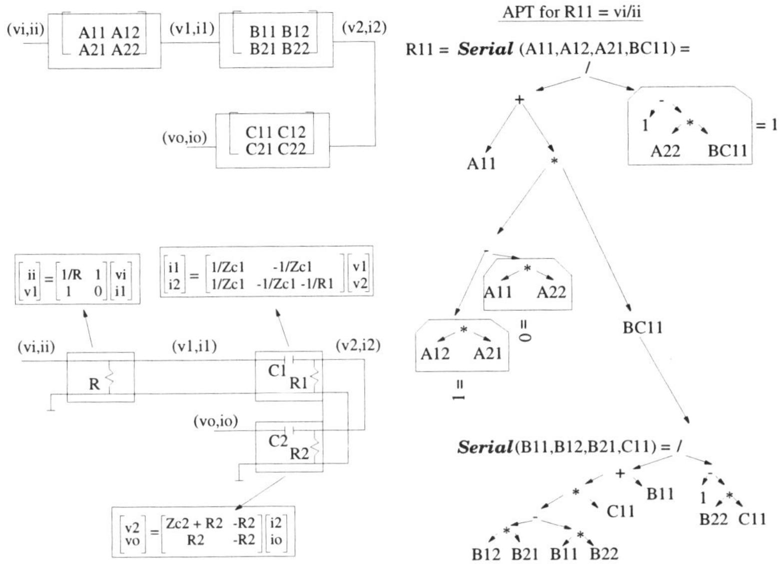

Figure 6 **Reduction of the APT for ladder structure**

APT building does not consider what the actual block parameters represent (i.e. impedances, admittances, etc.). This information is not available at the time the APT is constructed. During synthesis, blocks are instantiated with physical components. Simplifications of an APT can result as a consequence of block parameters having similar values.

We refer to the symbolic expression of the input resistance of the RC ladder in Figure 6 to motivate how reduction rules can be effective in APT simplifications. The input resistance is defined as $R_{in} = \frac{v_i}{i_i} = \frac{1}{R_{11}}$, where R_{11} pertains to the equivalent block matrix of the ladder. Figure 1.6 depicts the block matrices after mapping each block to electronic components. The selected knowns/unknowns for each block are also indicated in the figure (unknowns appear on the left hand side of the matrix equations). The APT for R_{11} is represented in the right part of the figure. Due to specific values i.e. $A_{12} = A_{21} = 1, A_{22} = 0$ the APT can be reduced, the simplified sub-trees being indicated as bordered.

5. EXPERIMENTAL RESULTS AND DISCUSSION

Experiments of our symbolic method were carried out on a set of five examples. The observed elements were: the memory size for the symbolic APTs, the number and types of the solving functions that were shared among different connections and the CPU time for setting-up the APTs. Experiments were run on a SUN Sparc 5 workstation. The five examples are: *ladder7* an RC ladder with 7 stages [10], *ladder21* an RC ladder with 21 stages [10], *filter4* a 4th order low-pass filter network [11] [4], *filter8* an 8th order band-pass filter [11], and *cascode op amp* the small-signal model for a cascode op amp [2] for which we manually built a hierarchical description.

The results of our experiments are summarized in Table 2 Columns 2 and 3 indicate the sizes of the analyzed systems (circuits) in terms of their

Example	*# of unknowns*	*# of symbols*	*Required memory*		*Function sharings*			*CPU Time*
			# of nodes if no sharing	*# of nodes with sharing*	*Total*	*Same param.*	*Distinct param.*	*(ms)*
ladder7	7	17	108	52	4	2	1	9.26
ladder21	21	59	503	62	25	11	14	14.35
filter4	42	74	1115	115	66	65	1	106.15
filter8	72	136	910	102	44	15	29	50.86
cascode op amp	36	64	120	65	4	4	-	16.45

Table 2 **Experimental results for the symbolic analysis method**

number of unknowns and the number of symbolic parameters in their block matrices. The sizes of the resulting APTs without any function sharings are indicated in Column 4. Column 5 shows the APT sizes if solving functions were shared among similar connections. Columns 6, 7, and 8 show the nature of sharings. Column 6 describes how many solving function were re-used. Column 7 indicates the number of cases where the same block matrix coefficient of a complex connection was re-used. Column 8 presents the number of situations where solving functions were re-used for connections involving different blocks. Finally, column 9 shows the CPU execution time.

Comparing Columns 4 and 5, we see significant reductions of APT sizes resulting from re-using solving functions for the connections. Column 5 shows that APTs tend to be small even for systems with a larger number of unknowns. The reason is that for any encountered connection type only one replica of its solving function is produced. For structures built by cascading similar stages, the method is very effective, as additional APT nodes are required only to indicate how the actual block parameters are passed to the solving functions for the connection of a stage. The growth of APT sizes is linear with the number of parameters. Columns 6, 7 and 8 indicate that for small systems, i.e *ladder7* and *cascode op amp*, sharings of solving functions correspond to the block matrix parameters of a connection that is re-used in several places. For bigger systems i.e. *ladder21* and *filter8*, a significant number of solving functions were shared among connections with distinct blocks. Column 9 shows that execution times were small even for larger examples.

By experimenting with our reduction rules, we observed that they are effective for connections between physical blocks. These blocks have similar symbolic terms, and hence simplification, i.e. term cancelations were more probable. At higher levels, expressions for block parameters are more abstract and tend to have less common elements.

As opposed to other decomposition methods [4] that require a partitioning of the overall model, our technique uses information on hierarchy available in a specification. The technique can be adapted to work in conjunction with a partitioning algorithm that finds connection patterns in a system. Similar to BDD-like representations [10], APTs are a compact representation. BDD-like representations avoid creating any arithmetic operators because of their semantics. APTs require extra memory as they explicitly indicate the arithmetic operations. However, they gain by using higher-order functions for describing complex connections. Similar to [9], our technique is *lazy* as it computes only sub-expressions that are required for the overall parameters. Lazy computation is natural in a top-down approach, where the elements to be calculated are known and propagated to the lower levels.

6. CONCLUSION

This paper describes a novel symbolic analysis technique that exploits the hierarchy and uniformity of a system for producing compact symbolic expressions for system parameters. Symbolic expressions are built in a top-down fashion by applying the *tearing* decomposition method. The algorithm inspects the hierarchical structure of a system, and tears apart wires so that predefined block connection patterns are recognized. The method is useful for synthesis as it avoids repeated recomputing of symbolic models inside an exploration-loop.

We identified more directions for future work. (1) In the current approach, if more wires have to be torn then their tearing order is arbitrarily selected. We plan to analyze the impact of tearing order on produced symbolic expressions. (2) The approach we used for expression reduction was not very effective for connections at the higher levels of the hierarchy. This limitation is critical for strongly connected blocks as it requires many tearing and re-connection steps.

9. ACKNOWLEDGMENT

This work was sponsored by the USAF, Air Force Research Laboratories, Wright Patterson Air Force Base under contract number F33615-96-C-1911.

8. REFERENCES

[1] A. Doboli *et al*, "Behavioral Synthesis of Analog Systems using Two-Layered Design Space Exploration", *Proc. of DAC*, 1999, pp.951-957.

[2] R. Geiger, P. Allen, N. Strader, "VLSI Design Techniques for Analog and Digital Circuits", McGraw-Hill, 1990.

[3] G. Gielen, W. Sansen, "Symbolic Analysis for Automated Design of Analog Integrated Circuits", Kluwer Academic Publishers, 1991.

[4] M. Hassoun, P.-M. Lin, "Hierarchical Network Approach to Symbolic Analysis", *IEEE Trans. on Circuits and Systems*, April 1995,pp.200-211.

[5] G. Kron, "Diakoptics - The Piecewise Solution of Large-scale Systems ", Mac Donald & Co., London, 1963.

[6] P.-M. Lin, "Symbolic Network Analysis", Elsevier, 1991.

[7] K. Ogata, "Modern Control Engineering", Prentice-Hall, 1990.

[8] T. Schubert, E. Kim, "Active and non-linear electronics", J. Wiley, 1996.

[9] S. Seda *et al*, "Lazy-Expansion Symbolic Expression Approximation in SYNAP", *Proc. of ICCAD*, 1992, pp.310-317.

[10] R. Shi, X. Tan, "Symbolic Analysis of Large Analog Circuits with Determinant Decision Diagrams", *Proc. of ICCAD*, 1997, pp.366-373.

[11] J. Starzyk, A. Konczykowska, "Flowgraph Analysis of Large Electronic Networks", *IEEE Trans. on Circuits and Systems*, March 1986, pp.302-315.

[12] Q. Yu, C. Sechen, "A Unified Approach to the Approximate Symbolic Analysis of Large Analog Integrated Circuits", *IEEE Trans. on Circuits and Systems*, August 1996, pp.656-669.

A Linear Programming Approach for Synthesis of Mixed-Signal Interface Elements

Adrián Núñez-Aldana and Ranga Vemuri
Electrical and Computer Engineering and Computer Science
University of Cincinnat; Cincinnati, OH 45221-0030, USA
anunez@ececs.uc.edu, ranga.vemuri@uc.edu

Abstract This paper presents a novel methodology to perform component selection and constraint transformation of analog-digital interface elements of mixed-signal systems. Sharing techniques are used to minimize area while meeting system constraints. The methodology selects components from a pre-defined library using a linear programming approach. The selection process is guided by a knowledge-based performance estimator. With the estimated parameters, linear models are generated helping the constraint transformation process and speeding up the component selection process. Experimental results show the effectiveness of the methodology in relatively short execution times.

Introduction

The synthesis of mixed-signal designs consists on three parts: (1) synthesis of digital section, (2) synthesis of analog section, and (3) synthesis of analog-digital interface section. The synthesis of digital systems is in a relative mature phase. CAD tools to synthesize digital circuits from behavioral description level have been developed for several years and relatively stable implementations of various high-level synthesis algorithms have started to emerge [10]. The synthesis of analog circuits has been relegated to hand-craft designs. Lately, some CAD tools have been developed to accelerate the analog circuit design process [1]. Tools for the design of analog components and physical layouts have been implemented in recent years. Currently, new methodologies are emerging to automatically synthesize analog systems from behavioral description level [4]. Finally, the synthesis of the analog-digital interface section has been relegated to a separated design process; however, with the ability to integrated analog and digital circuits in a single chip, it has gained more attention.

To completely automate the design process, it is critical to embed the synthesis of analog-digital interface modules to the mixed-signal design methodologies. The interface elements establish the communication channels between the two different domains (continuous and discrete) providing means of signal conversion and synchronization. Usually the interface consists of analog-to-digital (A/D) and digital-to-analog converters (D/A), control logic, and synchronization modules. A/D and D/A converters are used to convert back and forth between the two types of signals. The control logic and synchronization modules coordinate the communication protocol between the digital and analog signals, such that the timing constraints are met.

In this paper, we present a methodology to automate the high-level synthesize of the analog-digital interface section of mixed-signal systems. The methodology is driven by the digital and analog communication requirements and system constraints. We use sharing techniques to minimize silicon area while meeting the design constraints.

1. RELATED WORK

Some effort has been done to automate the synthesis of mixed-signal systems. Most of the work has been dedicated to layout synthesis. Costa et al. [3] developed a technique to extract substrate coupling parameters in mixed-signal designs and to generate models that reflect these parameters. Miliozzi et al. [6] presented a methodology for automatic layout generation for a class of mixed-signal circuits in presence of substrate induced noise. KOAN/ANAGRAM [2] is a layout tool from CMU, which automatically generate layout for CMOS mixed-signal designs.

Different topologies of analog-to-digital and digital-to-analog converters have been published in the literature. Each topology is designed to meet different constraints. A/D and D/A converters are designed to reduce noise, speed up conversion time, be linear, consume low power, work in high range of temperatures, etc. Besides, synthesis methodologies have been developed to automate the design process of A/D and D/A converters. Sarraj [9] presents a technique to design high speed pipeline A/D converters. Johnston [5] describes a procedure to calibrate features in Delta-Sigma A/D converters. Morling et al [7] presents the design of a sigma delta CODEC for telecommunication applications. However, A/D and D/A converters have not been integrated to the synthesis of mixed-signal designs.

2. SYNTHESIS OF A-D INTERFACE SYSTEMS

The analog-digital interface elements are constrained by the analog and digital communication requirements. The sampling theorem states that, to avoid aliasing, the sampling frequency f_s must be higher that two times the bandwidth of the input signal. Therefore, the conversion time should be less than $T_s = 1/f_s$. Besides, specifications of acceptable quantization error determine the number of bits required for the converters. Topologies and transistor sizes determine the area and performance of the interface elements. The selection process should select interface elements from a component library such that those and other constraints are met while minimizing silicon area.

Digital-to-analog converters. Several topologies of D/A converters have been published in the literature. Each topology is characterized by static and dynamic properties. Static properties establish linearity, resolution, zero and full-scale error, and monotonicity of the converters. Dynamic properties describe the behavior of the converters when the input word is changing. The settling time is a dynamic property which determines the response time.

Table 1 shows characteristics of four different D/A converters topologies. Style 1 and style 2 are voltage-scaling approaches. They have the advantages of being fast and insensitive to parasitic capacitances, but they are non-linear and require large silicon area. Style 3 is a charge-scaling converter. As it is based in switched-capacitor technology, it is very accurate, but it is limited to medium speed applications. Style 4 uses an algorithmic approach. It uses small area but it is very slow.

Analog-to-digital converters. Table 2 shows four different A/D converter topologies. Style 1 is a serial converter, therefore it is the slowest, but it uses small silicon area. Using a successive approximation approach, style 2 speeds up the conversion time of style 1. The fastest A/D converter is the style 3. It makes the conversion in parallel reducing the conversion time to one operational amplifier propagation delay. However, it has the disadvantage to use large silicon area. Style 4 uses a $\Sigma\Delta$ modulator to perform the conversion. It has the ability to be easily integrated on a chip that is predominantly made up of complex digital circuitry.

Estimation and model geneartion of A/D - D/A. The area and speed of the converters depend on the topology, number of bits and sizing of the analog components. To determine the performance of each converter, the components should be sized first. Then, a wide range of converters with different performance parameters can be generated based on the topologies presented in previous sections by changing the circuit sizing. However, circuit sizing process is computational expensive, and the possible combinations increase

Table 1 Characteristics of digital to analog converters

	Style 1	*Style 2*	*Style 3*	*Style 4*
Name	Weighted R	R-2R	Weighted C	Serial
Adv.	Fast Insensitive	Fast Small elements	Best Accuracy	Min area
Disadv.	Large elements Precision R's	Precision R's Non-monotonic	Large elements Non-monotonic	Slow Complex
Area (A)	$\sum_{i=0}^{N-1} 2^i A(R)$ $+\frac{1}{2}A(R)$ $+N * A(Sw)$ $+A(OA)$	$3N * A(R)$ $+\frac{1}{2}A(R)$ $+N * A(Sw)$ $+A(OA)$	$\sum_{i=0}^{N-1} A(\frac{C}{2^i})$ $+A(\frac{C}{2^{N-1}})$ $+(N+2)A(Sw)$ $+A(OA)$	$2 * A(C)$ $+4 * A(Sw)$ $+A(OA)$ $+A(logic)$
Power (P)	$V_{ref}^2 \sum_{i=0}^{N-1} \frac{1}{2^i R}$ $+P(OA)$	$\frac{Vref^2}{R}$ $+P(OA)$	$P(OA)$	$P(OA)$
Speed	$Delay(Sw)$ $+\Delta T(OA)$	$Delay(Sw)$ $+\Delta T(OA)$	$Delay(Sw)$ $+4.6RC$ $+\Delta T(OA)$	$2NDelay(Sw)$ $+4.6NRC$ $+\Delta T(OA)$

Note: N specifies the number of bits of the converters

exponentially as thea number of possible sizing elements increase. Therefore, critical to the success of our methodology is the accuracy and speed of an analog performance estimator. We use an analog performance estimator (APE) [8] to evaluate the performance of the interface modules.

To automate and speed up the design process, we generate linear performance models of the constituent components of the A/D and D/A converters presented in previous sections. Sweeping a wide range of the design parameters, we estimate the performance of the components using APE. Then, using an interpolation method, we generate linear component performance models of the following format:

$$perf = \alpha_1 * designP_0 + \alpha_2 * designP_2 + \ldots + \alpha_r * designP_r + \alpha_{r+1} \quad (1.1)$$

Formulation of the problem. We assume that the number of analog-to-digital and digital-to-analog conversions, sampling frequency (f_s) and number of bits of the converters (N) are defined by the user. Let's start by formulating the problem for the A/D converters first, then the D/A converters.

Table 2 Characteristics of analog to digital converters

	Style 1	*Style 2*	*Style 3*	*Style 4*
Name	Serial	Succ. Approx.	Flash	Over Sampling
Notes	Low Complex High resolution Slow	Low Complex Medium speed Medium area	Fast Large area Low resolution	Complicate Medium speed high SNR
Elem	Ramp Generator 2 N-bit Count's opamp	N-bit DAC opamp N-bit Register N-bit Shift Reg.	$(2^N - 1)$ opamps 2^N resistors $(2^N - 1)$ decoder	Modulator Decimator LP Filter
Area (A)	*(Small)* $A(ramp)$ $+2A(count[N])$ $+A(opamp)$	*(Medium)* $A(DAC[N])$ $+A(opamp)$ $+A(Reg[N])$ $+A(Shift[N])$	*(Large)* $\sum_{i=1}^{2^N-1} A(opamp)$ $+2^N A(R)$ $+A(dec[2^N - 1])$	*(Small)* $A(Mod)$ $+A(FF)$ $+A(LPF)$
Speed	$2^N *$ T	$N *$ T	$1 *$ T	$4 * N *$ T
Clk	$Delay(DAC)$ $+\Delta T(opamp)$ $+Delay(logic)$	$Delay(DAC)$ $+\Delta T(opamp)$ $+Delay(logic)$	$\Delta T(opamp)$ $+Delay(dec)$	$Delay(\Sigma\Delta)$ $+Delay(FF)$

Analog-to-digital converters. Let's define S to be the set containing all analog signals which are read by the digital section in a mixed-signal system. Then, $p = ||S||$ is the number of signals that require an A-to-D conversion.

Several frequencies are used in mixed-signal systems. First, the sampling frequency $f_s = \frac{1}{T_s}$ is the rate of conversion defined by the sampling theorem. T_s constrains the conversion time CT between analog and digital signals, $CT < T_s$. Second, Let $f_d = \frac{1}{T_d}$ be the clock frequency of the digital logic. The digital logic requires to read all analog-to-digital ports at the sampling frequency; therefore, $p * T_d < T_s$. Finally, the A/D converters operating frequency $f_c = \frac{1}{T_c}$ determines the speed ratio of the converters. T_c is a function of the A/D converter topology, sizing, and fabrication process parameters.

The propagation delay T_i of the ith converter can be estimated as follows:

$$\begin{aligned} T_{c_i} &= \textstyle\sum_{k \in ADC_i} DelayOf(k) \\ &= \textstyle\sum_{k \in ADC_i} (\alpha_{1_k} * designP_0 + \ldots \alpha_{r_k} * designP_r + \alpha_{r+1_k}) \\ &= \textstyle\sum_{j=1}^{n} (a_{ij} * P_{ij}) + b_i \end{aligned} \quad (1.2)$$

Where P_{ij} is the low-level design parameter j of the converter type i, and a_{ij} and b_i are constants defined as follows:

$$\begin{aligned} a_{ij} &= \textstyle\sum_{k \in ADC_i} \alpha_{jk} \\ b_i &= \textstyle\sum_{k \in ADC_i} \alpha_{r+1_k} \end{aligned} \tag{1.3}$$

Similarly, we can estimate the $Area_i$ of the ith converter (with c_{ij} and d_i being the constants introduced by the performance models) as follows:

$$Area_i = \sum_{j=1}^{n} (c_{ij} * P_{ij}) + d_i \tag{1.4}$$

Let's define C to be a function which maps the A/D converter type and the number of bits N to number of cycles T_c required to perform the conversion: $C : Type \times NoBits \rightarrow NoOfCycles$:

$$C(i, N) = \begin{cases} 2^N & \text{for i=1} \quad \text{(Serial)} \\ N & \text{for i=2} \quad \text{(Successive)} \\ 1 & \text{for i=3} \quad \text{(Parallel)} \\ 4 * N & \text{for i=4} \quad (\Sigma\Delta) \end{cases} \tag{1.5}$$

CT is a function from the converter type to the conversion time:

$$CT(i, N) = C(i, N) * T_{c_i} \tag{1.6}$$

We define the maximum load of a converter as follows:

$$Load_{i \times N} = \left\lfloor \frac{T_s}{CT(i, N)} \right\rfloor \tag{1.7}$$

Finally, we model the synthesis of mixed-signal interface elements as an optimization problem. The component selection process selects elements from the analog component library such that the area is minimized while all the constraints are met. Let X_i be the number of converters type i selected (X_i is defined to be a positive integer number). Assuming the library contains u A/D converters, the optimization and constraint functions are as follows:

$$\begin{array}{ll} minimize & Area = \sum_{i=1}^{u} Area_i * X_i \\ ST & \sum_{i=1}^{u} Load_{i \times N} * X_i \geq p \end{array} \tag{1.8}$$

Digital-to-analog converters. Similarly, we formulate the problem of selecting D/A converters. Let's define Q to be the set containing all digital output signals communicating with the analog section, and $q = ||Q||$ the number of digital signals which require a digital-to-analog conversion. The digital logic requires to write to all digital-to-analog ports at the sampling frequency; then, $q * T_d < T_s$. The conversion time T_c, area $Area_i$ and load $Load_{i \times N}$ of the D/A converters are defined similarly to the A/D converters. Assuming the library contains v D/A converters, the optimization and constraint functions are as follows:

$$\begin{array}{ll} minimize & Area = \sum_{i=1}^{v} Area_i * X_i \\ ST & \sum_{i=1}^{v} Load_{i \times N} * X_i \geq q \end{array} \tag{1.9}$$

Methodology. To solve an integer linear problem, the object function, constraint and right-hand coefficients should be known. However, in equations (1.8) and (1.9), $Area_i$ and $Load_i$ are functions of the topology, sizing and process parameters. Evaluating all solutions are infeasible; therefore, we need to determine how changes in the parametric values affect the optimal solution, such that only parametric ranges are evaluated. Sensitivity analysis determines ranges for the coefficients of a linear model such that the set of basic varibles in the optimal solution is unchanged, although their values may change.

3. EXPERIMENTAL RESULTS

In this section, we present experimental results involving circuit synthesis of mixed-signal interface elements using the methodology presented in this paper. Table 1.3 shows the synthesis results of several real life examples. Circuit 1 is a T1 channel service unit with 24 full duplex voice channels. Circuit 2 calculates the real, RMS and reactive power of a three phases electrical line. Circuit 3 is a bank of 16 digital filters. Circuit 4 is a programmable logic controller which has 8 analog outputs and 4 analog inputs. Circuit 5 is an example of a high-speed communication channel which uses the asynchronous transfer mode. Finally, circuit 6 is an equation solver.

Table 3 shows the execution time required to generate the models, and perform the selection process. The model generation is a one-time process. The last column shows the number of converters selected for each example. In all cases, the total number of converters is less than the total number of A/D and D/A signals, which indicates that converters are being shared.

4. CONCLUSIONS

We present a methodology to perform the component selection and constraint transformation process for analog-digital interface elements of mixed-signal systems. The problem was formulated as an integer linear programming model. Using sensitivity analysis, we speed up the selection process avoiding the evaluation at every single solution point. The methodology is guided by an analog performance estimator, which is used to generate performance models. In future work, we plan to extend the methodology to work with non-linear

Table 3 Results of synthesis of mixed-signal interface elements

ckt	D/A conv	A/D conv	T_s ($\mu sec.$)	model gen ($sec.$)	exec time ($sec.$)	area (μ^2)	no. conv.
T1	24	24	125	9.62	40.01	51200	12
Power	3	6	4000	9.62	28.87	10988	2
Filter	16	16	5	9.62	37.13	151848	15
PLC	8	4	100	9.62	29.39	24188	5
ATM	4	4	2	9.62	35.02	101825	6
Eq	32	32	10	9.62	33.89	99125	18

models, and use interleaving techniques to meet fast conversion time application requirements.

5. REFERENCES

[1] L. R. Carley, G. Gielen, R. A. Rutenbar, and W. Sansen. Synthesis tools for mixed-signal ics: Progress on frontend and backend strategies. *proceedings of Design Automation Conference*, pages 298–303, June 1996.

[2] J. M. Cohn, D. J. Garrod, R. A. Rutenbar, and L. R. Carley. Koan/anagram ii: New tools for device-level analog placement and routing. *IEEE JSSC*, 26(3), March 1991.

[3] J. P. Costa, M. Chou, and L. M. Silveira. Efficient techniques for accurate extraction and modeling of substrate coupling in mixed-signal ic's. *Conference proceedings of Design Automation and Test in Europe*, pages 396–400, March 1999.

[4] A. Doboli, A. Nunez-Aldana, N. Dhanwada, S. Ganesan, and R. Vemuri. Behavioral synthesis of analog systems using two-layered design space exploration. *36th Design Automation Conference*, June 1999.

[5] J. E. Johnston. Calibration features in delta-sigma a/d converters. *Proceedings of Analog & Mixed-Signal Applications Conference*, pages 209–213, July 1998.

[6] P. Miliozzi, I. Vassiliou, E. Charbon, and E. Malavasi. Use of sensitivities and generalized substrate models in mixed-signal ic design. *Proceedings of Design Automation Conference*, pages 227–232, June 1996.

[7] R. C. Morling, I. Kale, C. W. Tsang, S. J. Morris, G. Hague, and C. Foran. The design of a sigma delta codec for mobil telephone applications. *Second International Conference onf Advanced A-D and D-A conversion techiques and their applications*, July 1994.

[8] A. Nunez-Aldana and R. Vemuri. An analog performance estimator for improving the effectiveness of cmos analog systems circuit synthesis. *Proceedings of Design, Automation and Test in Europe*, pages 406–411, March 1999.

[9] M. Sarraj. New high speed technique for pipeline adc design. *Proceedings of Analog & Mixed-Signal Applications Conference*, pages 205–208, July 1998.

[10] R. Walker and R. Camposano. A survey of high level synthesis systems. *Kluwer Academic Publishers*, 1991.

RF Interface Design Using Mixed Mode Methodology

Augusto Gallegos[(1)], Philippe Silvestre[(1)], Michel Robert[(2)] and Daniel Auvergne[(2)]
[(1)] *VLSI Technology, a subsidiary of Philips Semiconductors CPG - 505, Route des lucioles, 06560 Sophia Antipolis France* [(2)] *LIRMM, UMR CNRS – Université de Montpellier II – 161, rue Ada, 34392 Monpellier Cedex 5, France*

Key words: mixed-signal modeling, RF circuit, DECT system

Abstract: A VHDL-SPICE mixed-signal modeling methodology is applied to the RF interface of a DECT ASIC designed in CMOS 0.35 μm technology. An overview of actual mixed design flows is presented followed by the DECT system description. Then a mixed-signal methodology is introduced to validate the whole circuit behavior. We illustrate the approach with detailed models, simulation results and experimental data. The chip was successfully tested and produced by VLSI Technology.

1. INTRODUCTION

Telecommunication system integration level in ASIC design needs a consistent verification mechanism for complete analog-digital circuits. Thus, efforts to improve IC modeling and verification are essential. The trend of increasing circuit functionalities gives rise to more complex digital-analog interfaces, higher operation frequencies and sophisticated circuit behaviors. Unifying system knowledge with CAD methodologies results in correct realization of these critical designs.

Mixed-Mode design state-of-the-art: Prototyping is a common but expensive verification method used nowadays for complex mixed-signal ASICs. Nevertheless, various CAD tools present faster and less expensive design methodologies:

i. Full-digital design process (*Figure 1*). With CAD advanced development in digital technology, designers choose generally to simulate the full circuit with a digital HDL simulator [1]. Design flows start by the system specification. During this phase, the preliminary analog and digital architectures are defined. The next step is system simulation. Digital parts are then modeled using a standard HDL. Instead, in order to model the analog parts with a digital HDL, the user must define a time domain function, which is a very time consuming task [2]. Besides, the resulting analog model cannot be directly compared to the transistor level description of the circuit using only a logic simulator. Two simulation phases are generally necessary: The behavioral validation and the structural verification. During the first phase, digital models are represented by RTL descriptions and analog parts are modeled with HDL behavioral models. The next steps are digital synthesis, analog layout design, place and route. When the full ASIC layout is available, a back-annotated description for each digital part can be extracted. However, analog circuits have no post-layout model. Consequently, for the structural simulation phase only the behavioral analog descriptions can be used. When this last phase is done, the ASIC is ready to be produced.

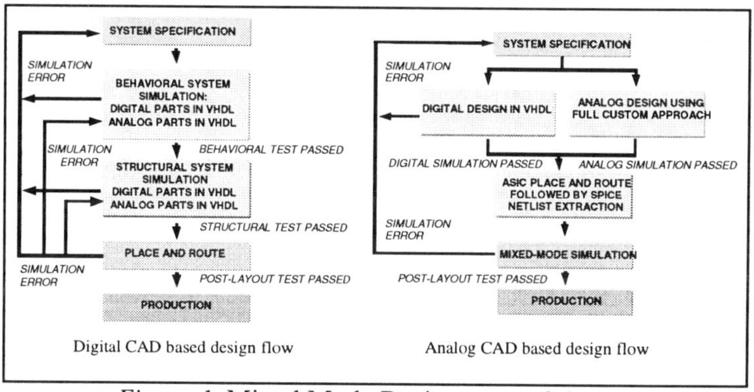

Figure 1. Mixed Mode Design state-of-the-art

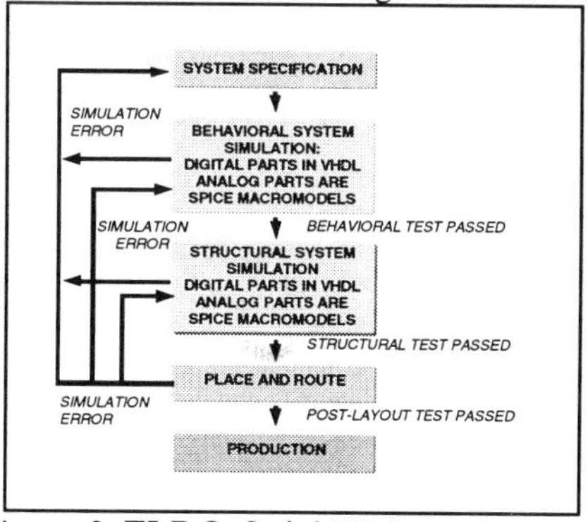

Figure 2. ELDO-QuickHDL Design Flow

ii. Event-driven transistor-level simulation (*Figure 1*). A uniform description of the IC can be obtained at transistor level. Designers verify the SPICE netlist with a simulator based on an advanced electrical simulation approach: The event-driven transistor-level simulation (EDTLS) [3]. The transistor model simulation speed is improved with look-up table methods, spare matrix processing and simplified transistor models. The design flow starts by system specification too. Then the digital and analog elements are separately implemented and simulated. The first integration process takes place during layout. The mixed signal system simulation is carried out only at the post-layout phase, when all transistor level descriptions are available. System debug needs complete design cycle iterations. Consequently, the methodology is not suitable for short time-to-market ASICs.

In the last few years, CAD tools were introduced to support both, SPICE and digital HDLs [4]. One approach uses a backplane to couple an analog and a digital simulator [5]. This paper presents a design flow (*Figure 2*) and modeling methodology using Mentor Graphics® ELDO-QuickHDL environment. To model digital parts, VHDL behavioral and structural descriptions are built. To represent analog parts, SPICE macromodels are extracted from the transistor level description. Both kinds of models allow accurate and fast system evaluation; thus, the whole architecture is precisely defined after the first simulation phase. The remaining possible defaults are those related to clock-skew for digital circuits and to parasitic resistance and capacitance for analog parts. Only when the system has been fully evaluated, the structural description implementation may start. After final ASIC place and route, the digital part is back annotated. The SPICE macromodels for this phase are extracted from post-layout transistor level netlists. In this way, the final structural simulation targets only timing problems. Usually, this kind of problems doesn't need significant architecture modifications to be solved. As a result, it's not necessary to repeat the full design cycle.

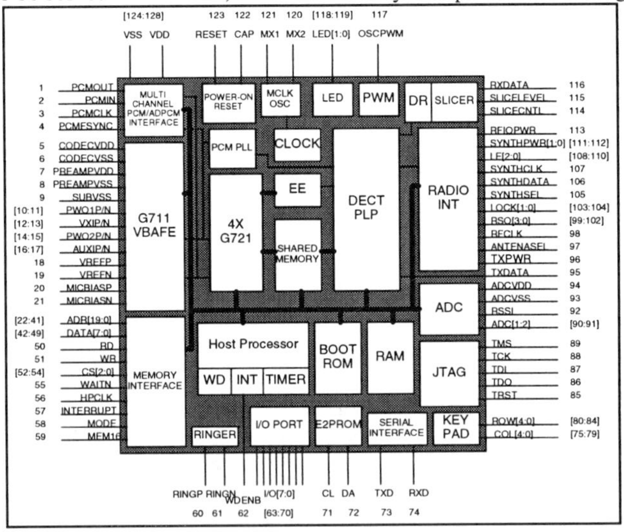

Figure 3. DECT base band IC architecture

2. THE DECT SYSTEM

The DECT (Digital Enhanced Cordless Telecommunications) system allows a 12 full-duplex channel digital wireless communication. TDMA frames are used to set 24 communication slots. Digital data is sent and received at a base band frequency of 1.152 kbps. In Europe, 10 radio carriers are allocated to modulate digital information in the 1800-1900 MHz band. Typical DECT system implementations require a base band processing ASIC and a RF module. *Figure 3* describes the developed base band ASIC architecture, which has been implemented by VLSI Technology using 0.35 μm CMOS technology. The base band circuit operation (*Figure 4*) is summarized as follows: Audio signal is filtered and converted to PCM data by a Voice Band Analog Front-End circuit (VBAFE). A microprocessor may process this data e.g. for echo cancellation. The G721-coder circuit compresses PCM to ADPCM and stores information in a shared memory. The DECT Physical Layer Processor (DECTPLP) and Encryption Engine (EE) build the TDMA encrypted frame and forward it to an external RF module via a radio interface. The latter is used to illustrate the mixed-signal approach of this paper.

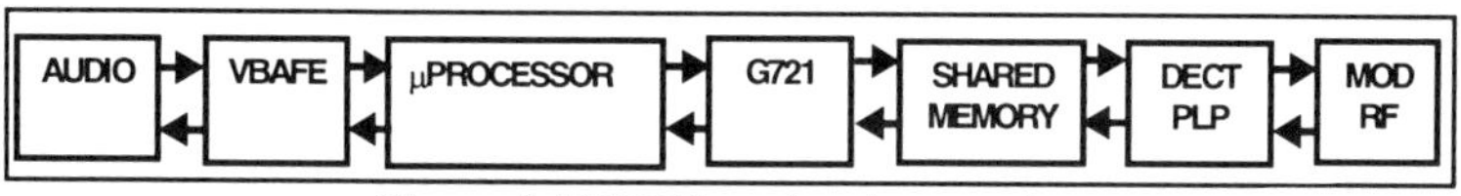

Figure 4. DECT base band processing

3. DECT RADIO INTERFACE

Our approach has been used for the whole chip design. However, this contribution illustrates only how we apply the mixed-mode methodology to the most challenging part, the RF interface. The RF interface architecture (*Figure 5*) is based on a sequencer, the Radio Signal Controller (RSC). A sequence RAM contains the control signal values and the DECT time to set them. DECTPLP provides the current frame timing signals: Slot start, slot stop, DECT bit number, A-Field start, B-Field start, etc. The sequencer sets the control signals following the RAM values, which correspond, to the frame timing. These signals are sent to several digital blocks: the clock recovery circuit, synthesizer programmer and RSSI controller. The analog circuits are the pseudo-gaussian filter, the data slicer and the RSSI ADC, which are also controlled by the RSC. *Figure 5* shaded areas show the analog blocks.

A typical RF IC block diagram is represented in *Figure 6*. GMSK modulation is usually achieved sending the baseband digital data to a gaussian LPF which output drives directly the IF VCO. This signal is then upconverted by a mixer. The Gaussian LPF may also be implemented in the baseband ASIC with look-up tables or any other digital processing method. During demodulation, the RF signal passes through a low-noise amplifier to the IF mixer. The IF signal is then demodulated by a PLL. The demodulated signal must be sliced to recover digital data. Furthermore, data slicing is crucial to clock recovery. The Data Slicer is usually integrated in the base band ASIC. So, mixed-mode modeling must be suitable to correctly analyze the full reception path. Indeed, during reception, several factors need to be considered in addition to the input/output impedance models. The demodulated RF signal contains high-frequency noise and it's affected by multi-path interference. In addition, this data slicer has to be turned on only during the receive slot to save power. The mixed-signal designer may need to test several data slicer architectures. Furthermore, he would rather prefer to use a circuit description close to the real implementation and not just a behavioral model, which performances need to be accurately measured with the original transistor level circuit.

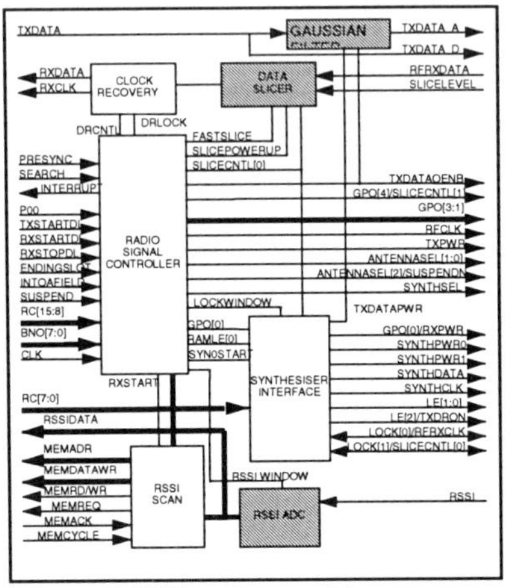

Figure 5. RF interface architecture

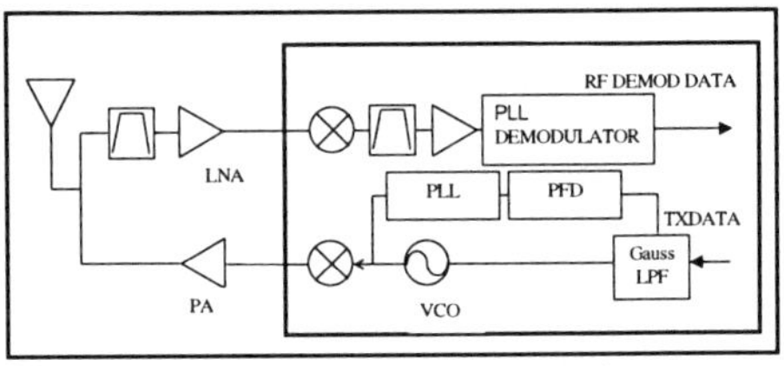

Figure 6. RF IC block diagram

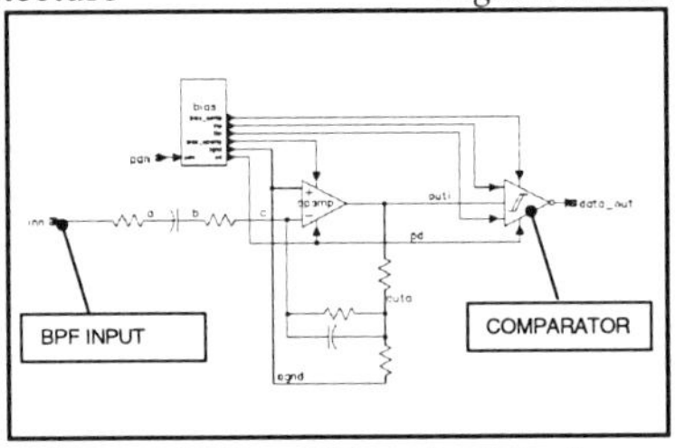

Figure 7. Data Slicer Architecture

4. MODELING APPROACH

The radio interface digital parts at RTL and structural levels are described in VHDL. The analog modeling methodology will be illustrated with the Data Slicer circuit (DS). *Figure 7* shows the preliminary DS architecture, which transistor level description will not be put in layout until the system verification has been correctly achieved. The demodulated RF signal model will be implemented in VHDL using the following equation:

$$g(t) = \frac{K}{2}[erf\{\sqrt{\frac{2}{\ln 2}}\pi B(t - \frac{T}{2})\} + erf\{\sqrt{\frac{2}{\ln 2}}\pi B(t + \frac{T}{2})\}]$$

Where B is the gaussian filter 3-dB bandwidth, T is the DECT pulse width and *erf(x)* is the gaussian error function. The g(t) function is sent to the input of a band pass filter (BPF), which midband is located at 1 MHz is used to remove RF noise. The BPF output is forwarded to a 2-threshold (Thn, Thp) comparator to extract DECT data. This data is forwarded to the clock recovery circuit and both, data and clock signals are provided to DECTPLP circuit that processes the TDMA frame. The BPF transfer function is:

$$F(s) = \frac{a_2 s^2 + a_1 s^2 + a_0}{b_5 s^5 + b_4 s^4 + b_3 s^3 + b_2 s^2 + b_1 s^2 + b_0}$$

Where: $a_0 = -0.31022 \times 10^{10}$, $a_1 = -0.13665 \times 10^{14}$, $a_2 = -0.49055 \times 10^{5}$, $b_0 = 0.62986 \times 10^{19}$, $b_1 = 0.30249 \times 10^{13}$, $b_2 = 0.297201 \times 10^{6}$, $b_3 = 0.927115 \times 10^{-2}$, $b_4 = 0.354149 \times 10^{-10}$, $b_5 = 0.1 \times 10^{-18}$. Such function would represent a difficult modeling task using a digital HDL. The macromodel in ELDO, for the whole data slicer (BPF and the comparator) is implemented as shown in

Table 1.

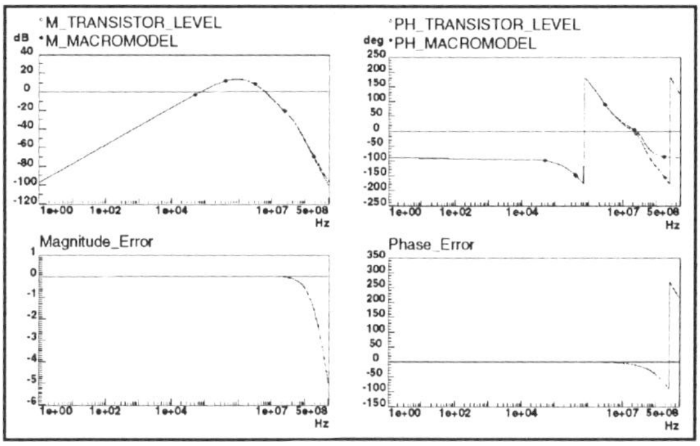

Figure 8. AC analysis

ELDO Pole-Zero Post-Processor extracts the FNSfilter model from the BPF architecture. This model represents the correct behavior within the DECT bandwidth as shown by the error curves (*Figure 8*). An ELDO macromodel (ytrig) is used to model the 2-threshold comparator architecture. Note that the slicer output and input, *a_out* and *inn* respectively, are mixed-signal nodes that exchange data between ELDO and the digital simulator QuickHDL. We assume that the RF demodulator output has a very low impedance compared to *inn*, so we didn't implement a more complex A-to-A interface description, like a voltage controlled source. If this were required, it would be a very simple task using the SPICE language. The whole model was implemented and evaluated in few hours. In addition, it's not a transistor level description neither a complex mathematical formula, which are difficult to analyze and debug. The simulation speed has been 50-fold improved.

Figure 9 shows the mixed-mode system testbench, where only the main blocks have been represented. For the first mixed-signal simulation phase, all digital parts (DECTPLP, CR, Shared RAM, RISC processor etc.) are available in VHDL RTL descriptions. The analog SPICE macromodels are those extracted before the analog layout implementation. The microprocessor initializes the system and the DECT reception starts running as shown in *Figure 4*. As the test module sends the RF signal of demodulated data, the DS and clock recovery (CR) circuits restore digital data for the DECTPLP. The latter detects the synchronization field for each slot to freeze CR clock phase and also recovers B-Field data storing it in the shared memory. *Figure 10* displays the simulation results computed with the preliminary DS architecture. The test module data sent in digital and analog forms, **TESTBENCHDATA** and **SLICE_IN** respectively, is compared to the recovered data **SLICE_OUT**. For the preliminary DS architecture, the synchronization field (E89Ah) after the preamble pattern (AAAAh) is not correctly recovered. After the system behavior analysis, the DS has been completely redefined. At this design phase the DS circuit can be modified and validated without involving layout design. The final DS circuit includes a DC offset latch circuit. *Figure 11* shows the correctly recovered **SP_MACR_RXDATA** signal, where **V(SLICEIN)** is the RF demodulated data and **SP_MACR_VC** is the latched offset value. Results are directly compared to experimental data. A mixed-mode test done for one DECT frame needs 20 min. The same simulation bypassing analog circuits, thus using only QuickHDL logic simulator, takes 5 min. This proves that simulation speed is affordable for complex system evaluations.

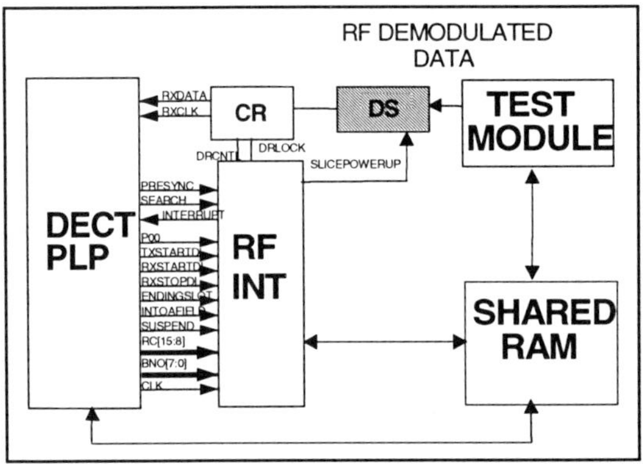

Figure 9. Mixed-Signal Testbench

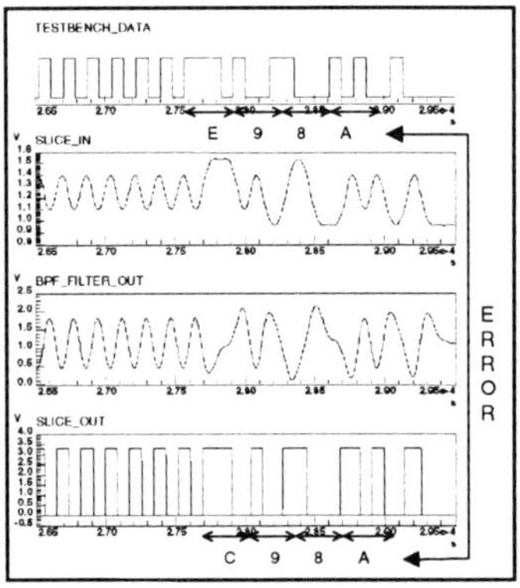

Figure 10. Simulation result: Error detection

Table 1. Data Slicer macromodel

```
FNS_MACROSPICE data slicer
Vrefere     refv       0    3.3
Voffs       outioff    0    1.1314E+00
Roffs       outioff    0    1
FNSfilter  inn outiac
+ -0.310225E+10 -0.136656E+14
+ -0.490556E+05
+ ,
+  0.629860E+19  0.302495E+13
+  0.297201E+06  0.927115E-02
+  0.354149E-10
+  0.100000E-18
yoffs add pin: outiac outioff outi
ytrig lev_d outi 0 outp outn refv
+ param: tr=9.54E-11 tf=6.5E-11
+ tpd=33.3E-09 v0=0 v1=3.3 voff=0
+ vrl=1.2309E+00 vru=1.0302E+00
.connect outn a_out
.model model_rd2a dtoa MODE=REAL
.D2A SIM=BP inn    MOD=model_rd2a
.model model_a2d atod MODE=MVL9
+ VTH1=1.0 VTH2=2.0
.A2D SIM=BP a_out MOD=model_a2d
.end
```

5. CONCLUSIONS.

The presented hierarchical modeling approach was used to design a DECT ASIC in CMOS 0.35 μm technology, ensuring circuit quality with full system simulations and improving dramatically the time-to-market, due to an early validation procedure. We compared simulation results to experimental data to prove the methodology validity as the presented circuit has been successfully tested and produced by VLSI Technology Inc. (part no. VP40553). Future works may extend models to include multi-path interference effects, noise and other RF module characteristics.

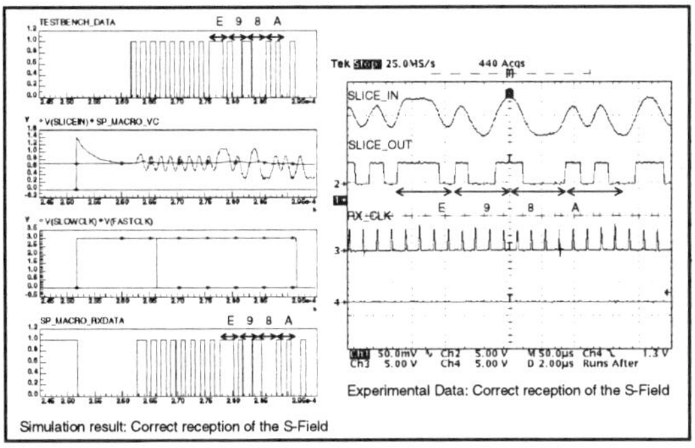

Figure 11. Mixed-Mode simulation results comparison to experimental data

6. REFERENCES

[1] M. K. Mayes, S.W. Chin "All Verilog Mixed-Signal Simulator with Analog Behavioral and Noise Models" IEEE Symposium on VLSI Circuits Digest of Technical Papers, 1996.

[2] J. Dabrowski, A. Pulka. "Discrete approach to PWL analog modeling in VHDL environment". Analog Integrated Circuits and Signal Processing International Journal. Vol. 16, Num. 2, June 1998.

[3] An-Chang Deng, "Transistor-level, event-driven simulation proves effective for mixed-signal design", Computer Design, suppl. Guide to analog and Mixed-Signal Design, pp. 32-33, December 1997.

[4] D. Overhauser, R. Saleh "Evaluating Mixed-Signal Simulators". Proc. IEEE Custom Integrated Circuits Conference, pp. 7.1.1-7.1.8. 1995

[5] M. Zwolinski, C. Garagate, Z. Mrcarica, T. J. Kazmierski, A. D. Brown. "Anatomy of a simulation backplane" IEE Proceedings on Computer and Digital Techniques. Vol. 1142, iss. 6. Pp. 377-385. November 1995.

History-Based Dynamic Minimization During BDD Construction

Rolf Drechsler
Wolfgang Günther
Institute of Computer Science
Albert-Ludwigs-University
79110 Freiburg im Breisgau, Germany
{drechsle,guenther}@informatik.uni-freiburg.de

Abstract Binary Decision Diagrams (BDDs) are the state-of-the-art data structure in VLSI CAD. Since their size largely depends on the chosen variable ordering, dynamic variable reordering methods, like sifting, often have to be applied while the BDD for a given circuit is constructed. Usually sifting is called each time a given node limit is reached and it is therefore called frequently during the construction of large BDDs. Often most of the runtime is spent for sifting while the BDD is built.

In this paper we propose an approach to reduce runtime (and space requirement) during BDD construction by using history-based decision procedures. Dependent on the history of the construction process different types of sifting are called. We propose two methods that consider the quality of the hash table and the size reduction of previous sifting runs, respectively. Experimental results show that both approaches reduce the runtime significantly, i.e. by more than 40% on average.

Keywords: BDDs, Verification, Dynamic Reordering.

1. INTRODUCTION

Decision Diagrams (DDs) are often used in VLSI CAD systems for efficient representation and manipulation of Boolean functions. The most popular data structure are *ordered Binary Decision Diagrams* (BDDs) [Bryant, 1986]. Since they provide a canonical representation, for example functional equivalence of two circuits can be checked easily by building the BDDs for each circuit and then checking whether the two BDDs are isomorphic.

However, as well known BDDs are very sensitive to the variable ordering, i.e. the size of a BDD (measured in the number of nodes) may vary from linear to exponential. Finding the optimal variable ordering is an NP-hard problem [Bollig and Wegener, 1996] and the best known algorithms have exponential worst case runtime [Friedman and Supowit, 1987, Drechsler et al., 1998].

This is the reason why many authors presented heuristics for finding good variable orderings from circuit descriptions in the last few years (see e.g. [Fujii et al., 1993]). The most promising methods for BDD minimization are based on *Dynamic Variable Ordering* (DVO) [Fujita et al., 1991], i.e. improving graph size using exchanges of neighboring variables. The best results measured in the number of nodes of the resulting BDD were obtained using *sifting* [Rudell, 1993, Panda and Somenzi, 1995], but unfortunately sifting is very time consuming for large functions. Furthermore, during BDD construction it is often necessary to start sifting several times.

For this, recently several techniques for speeding up sifting have been proposed. In [Meinel and Slobodová, 1997] an algorithm has been used that partitions the search space to improve sifting runtimes. But this algorithm is largely dependent on the initial variable ordering. Another approach based on "sampling" has been suggested in [Slobodová and Meinel, 1998, Jain et al., 1998], but dependent on the chosen candidates the quality of the result varies widely, i.e. the results can be up to a factor of two worse than "classical" sifting. Lower bound sifting [Drechsler and Günther, 1999] is an approach to speed up sifting without loss of quality by computing lower bounds during variable reordering. Also a relaxed version has been proposed where a parameter has to be set by the user allowing to trade off runtime versus quality.

The focus of this paper is to show that it is not clever to use the same sifting technique all the time during BDD construction. Usually, dynamic minimization is called if a node limit is reached. Dependent on the previous results of dynamic minimization or the time elapsed since the last call of sifting, respectively, our history-driven minimization approach automatically chooses one minimization strategy. Experiments show that speed-ups of more than 40% can be observed on average.

The paper is structured as follows: In Section 2. we review basic definitions of BDDs. The sifting algorithm is discussed in Section 3. In Section 4. speed-up techniques are proposed. The new algorithms for choosing the sifting technique are introduced in Section 5. Section 6. gives experimental results. Finally, the results are summarized.

2. BINARY DECISION DIAGRAMS

As is well-known, each Boolean function $f : \mathbf{B}^n \rightarrow \mathbf{B}$ can be represented by a *Binary Decision Diagram* (BDD) [Bryant, 1986], i.e. a directed acyclic

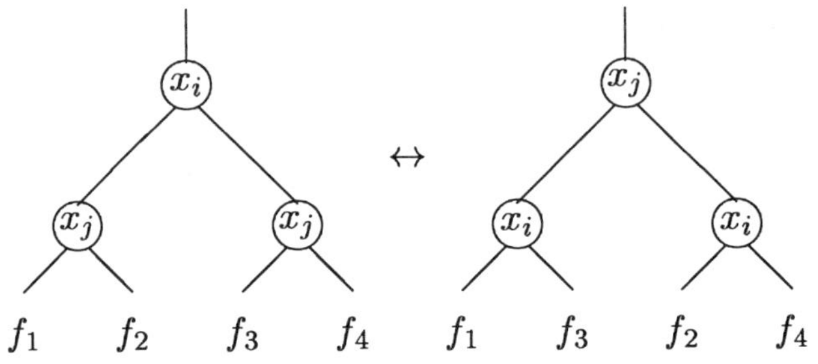

Figure 1 Exchange of i-th and adjacent variable

graph where a Shannon decomposition

$$f = \overline{x}_i f_{x_i=0} + x_i f_{x_i=1} \quad (1 \leq i \leq n)$$

is carried out in each node.

A BDD is called *ordered* if each variable is encountered at most once on each path from the root to a terminal node and if the variables are encountered in the same order on all such paths. A BDD is called *reduced* if it does neither contain vertices with isomorphic sub-graphs nor with both edges pointing to the same node. Reduced and ordered BDDs are canonical, i.e. for each Boolean function the BDD can be uniquely determined. Furthermore, for functions represented by reduced ordered BDDs efficient manipulations are possible [Bryant, 1986, Brace et al., 1990, Drechsler and Becker, 1998]. In the following, only reduced, ordered BDDs are considered and for briefness these graphs are called BDDs.

We briefly review an example from [Bryant, 1986] to show the importance of the variable ordering:

Example 1 Let $f = x_1x_2 + \ldots + x_{2n-1}x_{2n}$. If the variable ordering is given by $(x_1, x_2, \ldots, x_{2n})$ the size of the resulting BDD is $2n$. On the other hand if the variable ordering is chosen as $(x_1, x_{n+1}, x_2, x_{n+2}, \ldots, x_{2n})$ the size of the BDD is $\Omega(2^{n-1})$. Thus the number of nodes in the graph varies from linear to exponential depending on the variable ordering.

3. SIFTING

The basic operation of dynamic variable ordering is the exchange of adjacent variables [Fujita et al., 1991, Rudell, 1993]. The general case of exchanging a variable x_i and an adjacent variable x_j is shown in Figure 1. The exchange is performed very quickly since only edges must be redirected within these levels. Thus, the size is optimized without a complete reconstruction of the BDD, i.e. only local transformations for the two levels are performed, since BDDs are a canonical form.

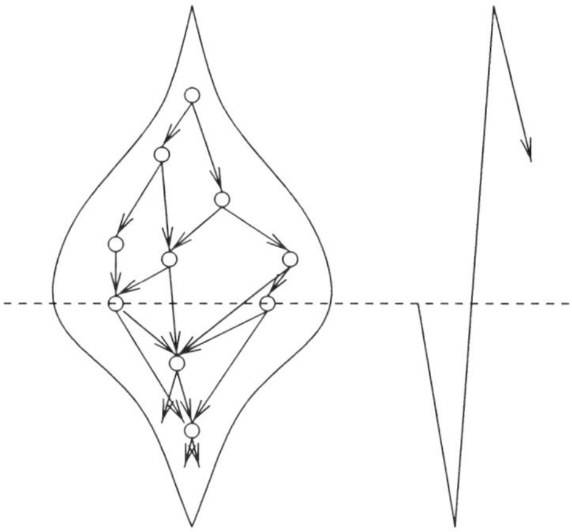

Figure 2 Sifting one variable

The sifting algorithm [Rudell, 1993] successively considers all variables of a given BDD. When a variable is chosen, the goal is to find the best position of the variable, assuming that the relative order of all other variables remains the same. In a first step, the order in which the variables are considered is determined. This is done by sorting the levels according to their size with largest level first. To find the best position, the variable is moved across the whole BDD. In [Rudell, 1993], this is done in three steps (see Figure 2):

1. The variable is exchanged with its successor variable until it is the last variable in the ordering.
2. The variable is exchanged with its predecessor until it is the topmost variable.
3. The variable is moved back to the closest position which has led to the minimal size of the BDD.

4. SPEEDING UP SIFTING

Some improvements to the original sifting algorithm have already been proposed:

Upper limit: As the size of the BDD can grow much during the movement of one variable, it is possible to set an upper limit to the growth of the BDD. If this limit is exceeded, moving into this direction is aborted. This avoids large intermediate BDDs. Using the notation from [Somenzi, 1998] in the following this "growing factor" is denoted as $maxgrowth$, i.e. $maxgrowth = 1.2$ means that the BDD may grow by at most 20% of its original size.

```
unique _find_or_add(node) {
    if (node exists) return node;
    /* a new node has to be created */
    if (next_reorder node limit is reached) {
        call sifting;
        increase next_reorder node limit;
    }
    /* some other test may follow */
    create and return new node;
}
```

Figure 3 Sketch of unique _find_or_add

Closest end: The considered variable is not always moved downwards first. Instead it is moved to the closest end and then to the opposite one.

Several other techniques have been proposed like the use of symmetries [Panda and Somenzi, 1995] and of interaction matrix[1] [Somenzi, 1998]. In [Drechsler and Günther, 1999] computations of lower bounds have been used to speed up sifting (so-called *lb-sifting*), i.e. often it is not necessary to move a variable to all positions. Also a *relaxed* lb-sifting has been considered. There not only a tight lower bound of $\frac{1}{2}$ is used, but a straightforward extension to $\frac{1}{b}$ ($b \geq 2$). (For more details see [Drechsler and Günther, 1999].) This extension allows to trade off runtime versus quality, i.e. the larger b, the faster the algorithm, but the resulting sizes get larger on average.

In CUDD [Somenzi, 1998], a state-of-the-art BDD package, dynamic reordering is applied if a given node limit is reached[2]. This is tested each time a new node is created. If this happens while e.g. ITE is computed [Brace et al., 1990], ITE is started again after sifting has been carried out. A sketch of the look-up technique used in ITE is shown in Figure 3.

Finally, for two examples it is shown how a typical BDD construction works.

Example 2 *Some data observed during BDD construction of circuits* c2670 *and* c7552 *using the CUDD package [Somenzi, 1998] is given in Figures 4 and 5, respectively. The solid line gives the relative number of nodes which are not found in the hash table during an ITE operation (see also Figure 3), i.e. the percentage of newly created nodes (left y-axis). The x-axis shows the number of calls of* **unique_table_find_or_add()** *during symbolic simulation. The vertical lines show where sifting is called during construction. The total memory consumption counted in number of nodes is given by the dashed line (right y-axis).*

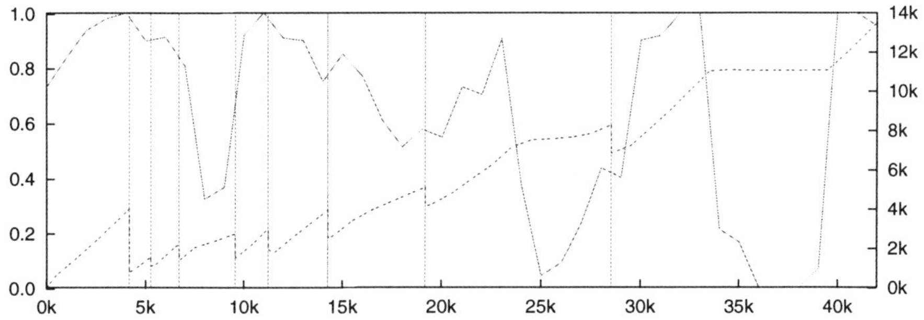

Figure 4 BDD construction of c2670

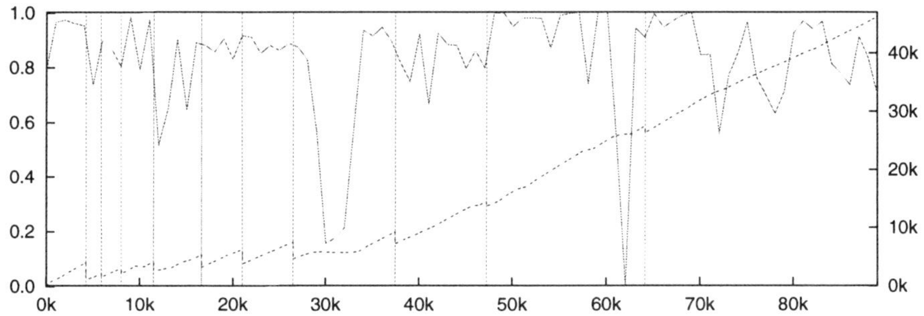

Figure 5 BDD construction of c7552

The following can be observed:

1. The percentage of nodes found in the unique table varies a lot and it is (more or less) randomly distributed and independent of the starting points of sifting.

2. Nevertheless, the behavior is very robust, i.e. there are phases where nodes are found all the time and phases where nodes are found very rarely. Oscillation cannot be observed.

3. At the beginning of the construction sifting gives very good reductions, i.e. often more than a factor of two, while at the end (where the BDDs are large and sifting becomes even more time consuming) the profit is marginal in many cases.

For this, calling the original sifting algorithm each time is not clever, since the algorithm is very slow, especially for large BDDs. We present in the following criteria to chose faster variants of sifting dependent on the *history* of the minimization process.

5. HISTORY-BASED DYNAMIC MINIMIZATION

In this section we present two methods for selecting a reordering heuristic that is called during BDD construction if an upper node limit is reached. For both methods we make use of relaxed lower bound sifting presented in [Drechsler and Günther, 1999]. Relaxed lb-sifting allows to dynamically trade off runtime versus quality by choosing the parameter b (see Section 4.). For sifting, we used the lower bound improvement which reduces runtime without loss of quality ("lb-sifting"). In order to distinguish lb-sifting and relaxed lb-sifting more clearly, in the following lb-sifting is also called "exact lb-sifting". Since early calls of sifting often reduce the BDD sizes much, the first call of sifting of both approaches is always exact lb-sifting.

5.1 LAST REDUCTION APPROACH

If the last call of sifting has not been very successful, it is very unlikely that a large reduction can be obtained this time. Therefore, relaxed lb-sifting can be used in that case. If, on the other side, the last reordering has reduced the BDD size much, then it is likely that this call of sifting can also reduce the BDD size, and exact lb-sifting is used. This approach prevents that too much time is spent by always applying sifting if the BDD blows-up during the construction. (For some functions the BDD size is very large independent of the variable ordering. In these cases sifting would be called frequently, vasting much runtime.)

More precisely, this can be seen as follows: if the reduction caused by sifting is small, then it is expected that also moving a single variable in the variable ordering only leads to small reductions. Therefore, the lower bound used in lb-sifting can be relaxed without much loss of quality.

To compute value b of relaxed lb-sifting, the reduction factor r of the last call of sifting is used (independently of the parameter b used in this last call). If $r \geq 0.5$, i.e. a reduction of more than 50% was observed, exact lb-sifting is used. Otherwise, b is computed by

$$1 + (maximum\_b - 1)(2r - 1).$$

Thus, the smaller the reduction by the previous call, the smaller the amount of time invested in the current run. ($maximum\_b$ denotes an upper limit for value b set by the user. In experiments we found that $b = 16$ is a good choice. This method is called *methodR* (reduction) in the following.

5.2 ELAPSED TIME APPROACH

In the beginning of BDD construction, sifting is called very frequently and large size reductions can be observed. On the other hand, improvements are smaller in the long run, and sifting is called less frequently. This observation leads to the following algorithm: the parameter b is chosen dependent on the time which has elapsed since the last call of sifting:

$$b = (\text{time since last sifting}) \times const,$$

i.e. if sifting has been called long ago, then relaxed lb-sifting is used with a large value of b, otherwise it is assumed that BDD construction has just started, and exact lb-sifting is used. For this method, "time" is measured in number of calls to **unique_find_or_add**(), i.e. the number of calls to the function which returns a node for a given function and creates a new node if necessary.

In our experiments it turned out that $const = \frac{1}{2000}$ is a good choice. We refer to this technique as $methodE$ (elapsed time) in the following.

6. EXPERIMENTAL RESULTS

In this section we describe experimental results that have been carried out on a *SUN Ultra 1* with 128 MBytes. All runtimes are given in CPU seconds. The algorithms have been integrated in the CUDD package [Somenzi, 1998]. We set a memory limit of 100 MBytes and all node counts are given in 1000 nodes. We used several benchmarks from LGSynth91 to build the BDD for and measured runtime and BDD peak node count[3]. Since results largely depend on the initial variable ordering [Harlow and Brglez, 1998], we report average values for 64 randomly chosen initial orderings. The starting value for performing sifting was 4000 nodes (which is the default value in CUDD). Then sifting was called each time the size doubled compared to the size after the last call of dynamic minimization.

To give some more insight in the experiments carried out, we first discuss the $maxgrowth$ parameter. This parameter limits the increase in size of the BDD during sifting and it turns out to be very important. For some choices of $maxgrowth$ the results are reported in Table 1 and 2. For each value the peak node count and the runtime is given, respectively. The smaller the factor is chosen the faster sifting runs, but if the value becomes too small the quality decreases significantly, and for some benchmarks the BDD cannot be built at all within the given memory limit. (Numbers in parentheses denote that the BDD could not be built for all initial variable orders.) Surprisingly, the quality did not reach its optimum for ∞.

We compared our approach to the best results obtained using lb-sifting each time, i.e. using a value of 1.2 for $maxgrowth$. For both methods we also set $maxgrowth$ to 1.2. Results are given in Table 3. Additionally, the average number of calls of sifting is given in column *calls*. One might expect that these numbers increase by speeding up sifting; however, this has not been observed.

As can be seen both algorithms clearly outperform sifting. Both methods significantly reduce runtime, while the peak node count is in the same range. $methodR$ even gives better results for both runtime and memory on average compared to sifting, while $methodE$ increases the peak node count by 10%,

Table 1 Effect of *maxgrowt* on peak node count

circuit	#in	∞	2.0	1.3	1.2	1.1	1.0
bigtest	328	307.0	307.0	270.3	255.1	262.5	(—)
c432	36	7.9	7.9	7.6	7.5	7.3	110.6
c1355	41	153.0	153.0	146.3	141.1	136.4	328.1
c1908	33	39.4	39.4	39.3	39.3	39.1	74.1
c2670	233	32.3	32.3	28.6	27.4	28.9	(464.0)
c3540	50	292.4	292.4	233.9	225.9	208.0	1157.3
c5315	178	15.8	15.8	15.1	14.6	14.4	20.0
c7552	207	73.8	73.8	63.7	60.2	57.2	(822.8)
dalu	75	12.3	12.3	12.3	12.3	12.2	24.0
des	256	18.5	18.5	18.5	18.5	18.5	25.0
frg2	143	10.3	10.3	10.3	10.3	10.3	10.3
i8	133	16.4	16.4	16.4	16.4	16.5	31.6
i10	257	149.0	149.0	132.9	126.2	121.1	556.8
pair	173	16.1	16.1	16.1	16.1	16.0	45.5
rot	135	12.9	12.9	12.9	12.9	12.9	49.4
s1423	91	5.4	5.4	5.4	5.3	5.3	8.3
s5378	199	8.3	8.3	8.3	8.3	8.3	15.5
s9234.1	247	12.1	12.1	12.1	12.0	11.8	58.9
s13207.1	700	17.1	17.1	17.1	17.1	17.0	46.0
s15850.1	611	53.6	53.6	53.2	52.1	47.8	116.7
s38584.1	1464	51.3	51.3	50.9	49.1	46.6	(1341.4)
sum		1305.1	1305.1	1171.2	1127.7	1098.0	(5306.2)

but further reduces runtime. Compared to lb-sifting, *methodR* is 45% faster and *methodE* is 59% faster on average. In some cases the reduction is even more than a factor of five (see e.g. *bigtest*).

All in all, history-based selection of the reordering technique significantly speeds up the computation and often also reduces the memory needed.

7. CONCLUSIONS

In this paper history-based dynamic minimization techniques for BDD construction have been proposed. Instead of applying the same reordering algorithm all the time, dependent on the history a parameter for the reordering algorithm is determined and a faster heuristic may be used in some cases.

Table 2 Effect of *maxgrowth* on CPU time

circuit	#in	∞	2.0	1.3	1.2	1.1	1.0
bigtest	328	617.0	618.1	518.2	433.7	322.2	(—)
c432	36	1.3	1.3	1.2	1.0	0.7	4.8
c1355	41	131.6	131.5	122.2	110.3	85.5	34.9
c1908	33	10.1	10.1	9.6	8.9	7.1	3.4
c2670	233	19.8	19.7	18.7	17.6	16.6	(39.8)
c3540	50	125.9	126.4	115.6	105.5	73.9	125.0
c5315	178	4.8	4.8	4.7	4.7	4.5	1.7
c7552	207	63.6	62.2	59.1	55.4	44.1	(80.5)
dalu	75	2.1	2.1	2.0	2.0	1.8	1.0
des	256	4.6	4.6	4.6	4.6	4.5	2.4
frg2	143	1.2	1.2	1.2	1.2	1.1	0.5
i8	133	2.3	2.3	2.2	2.1	1.8	1.3
i10	257	203.7	203.9	167.6	140.3	97.8	52.8
pair	173	8.9	8.9	8.9	8.8	8.2	2.1
rot	135	3.7	3.7	3.6	3.3	2.6	1.7
s1423	91	2.0	2.1	2.0	1.9	1.6	0.7
s5378	199	2.5	2.5	2.5	2.6	2.5	1.2
s9234.1	247	8.3	8.4	8.3	8.3	7.8	4.5
s13207.1	700	44.3	44.3	44.4	44.4	41.4	26.0
s15850.1	611	74.2	74.3	72.7	71.1	64.7	18.8
s38584.1	1464	213.2	212.8	212.9	212.2	208.4	(272.0)
sum		1545.2	1545.3	1382.1	1240.0	998.9	(675.0)

Experiments have shown that significant reductions of 50% in runtime on average can be observed, while the memory requirement is in the same range (±10%).

Notes

1. For each two variables the interaction matrix provides the information whether there is an output in the BDD which essentially depends on both variables. If two levels do not interact, the level exchange can be computed in constant time.
2. Other packages, like CAL [Ranjan and Sanghavi, 1997] from Berkeley, use different criteria. All algorithms presented in the following are also applicable in this case.
3. For sequential circuits, the transition function was used, i.e. latches were treated as additional inputs and outputs.

Table 3 History-based dynamic minimization for $maxgrowth = 1\ 2$

circuit	exact lb-sifting			*methodR*			*methodE*		
	nodes	time	calls	nodes	time	calls	nodes	time	calls
bigtest	255.1	433.7	13.3	210.4	171.3	11.8	247.9	72.4	12.4
c432	7.5	1.0	6.4	7.3	0.7	6.4	7.4	0.9	6.2
c1355	141.1	110.3	13.0	128.7	59.4	13.1	133.9	35.5	13.1
c1908	39.3	8.9	11.0	39.5	6.3	10.9	39.0	6.4	11.2
c2670	27.4	17.6	9.3	26.3	11.4	9.2	29.3	12.8	9.1
c3540	225.9	105.5	10.4	217.2	44.3	10.3	229.5	30.1	10.4
c5315	14.6	4.7	6.2	14.0	3.7	6.2	15.1	4.7	6.1
c7552	60.2	55.4	11.8	64.3	42.2	11.1	71.9	45.1	11.8
dalu	12.3	2.0	3.3	11.8	1.7	3.3	12.0	2.3	4.0
des	18.5	4.6	6.1	21.4	3.0	5.2	21.4	2.2	4.1
frg2	10.3	1.2	2.1	10.3	1.2	2.1	10.3	1.1	2.1
i8	16.4	2.1	3.9	15.9	1.7	3.9	16.0	1.8	3.5
i10	126.2	140.3	10.9	123.2	83.2	10.8	212.8	48.2	10.7
pair	16.1	8.8	12.0	15.2	6.9	11.5	14.7	7.2	11.2
rot	12.9	3.3	6.8	13.7	2.6	6.9	13.3	3.2	7.2
s1423	5.3	1.9	5.4	5.2	1.3	5.8	5.2	1.9	6.2
s5378	8.3	2.6	3.1	8.1	2.2	3.0	8.1	2.3	3.0
s9234.1	12.0	8.3	8.2	13.9	7.6	8.4	12.9	7.8	8.4
s13207.1	17.1	44.4	14.9	15.6	18.7	6.5	15.7	33.8	12.0
s15850.1	52.1	71.1	9.3	70.5	97.5	10.0	58.6	68.9	8.4
s38584.1	49.1	212.2	19.1	54.1	115.5	11.1	63.4	100.0	10.5
sum	1127.7	1240.0	186.7	1086.5	682.3	167.5	1238.3	488.5	171.5

8. REFERENCES

[Bollig and Wegener, 1996] Bollig, B. and Wegener, I. (1996). Improving the variable ordering of OBDDs is NP-complete. *IEEE Trans. on Comp.*, 45(9):993–1002.

[Brace et al., 1990] Brace, K., Rudell, R., and Bryant, R. (1990). Efficient implementation of a BDD package. In *Design Automation Conf.*, pages 40–45.

[Bryant, 1986] Bryant, R. (1986). Graph - based algorithms for Boolean function manipulation. *IEEE Trans. on Comp.*, 35(8):677–691.

[Drechsler and Becker, 1998] Drechsler, R. and Becker, B. (1998). *Binary Decision Diagrams – Theory and Implementation.* Kluwer Academic Publishers.

[Drechsler et al., 1998] Drechsler, R., Drechsler, N., and Günther, W. (1998). Fast exact minimization of BDDs. In *Design Automation Conf.*, pages 200–205.

[Drechsler and Günther, 1999] Drechsler, R. and Günther, W. (1999). Using lower bounds during dynamic BDD minimization. In *Design Automation Conf.*, pages 29–32.

[Friedman and Supowit, 1987] Friedman, S. and Supowit, K. (1987). Finding the optimal variable ordering for binary decision diagrams. In *Design Automation Conf.*, pages 348–356.

[Fujii et al., 1993] Fujii, H., Ootomo, G., and Hori, C. (1993). Interleaving based variable ordering methods for ordered binary decision diagrams. In *Int'l Conf. on CAD*, pages 38–41.

[Fujita et al., 1991] Fujita, M., Matsunaga, Y., and Kakuda, T. (1991). On variable ordering of binary decision diagrams for the application of multi-level synthesis. In *European Conf. on Design Automation*, pages 50–54.

[Harlow and Brglez, 1998] Harlow, J. and Brglez, F. (1998). Design of experiments in BDD variable ordering: Lessons learned. In *Int'l Conf. on CAD*, pages 646–652.

[Jain et al., 1998] Jain, J., Adams, W., and Fujita, M. (1998). Sampling schemes for computing OBDD variable orderings. In *Int'l Conf. on CAD*, pages 631–638.

[Meinel and Slobodová, 1997] Meinel, C. and Slobodová, A. (1997). Speeding up variable reordering of OBDD. In *Int'l Conf. on Comp. Design*, pages 338–343.

[Panda and Somenzi, 1995] Panda, S. and Somenzi, F. (1995). Who are the variables in your neighborhood. In *Int'l Conf. on CAD*, pages 74–77.

[Ranjan and Sanghavi, 1997] Ranjan, R. and Sanghavi, J. (1997). *CAL-1.2: Breadth-first manipulation based BDD library.* University of California at Berkeley.

[Rudell, 1993] Rudell, R. (1993). Dynamic variable ordering for ordered binary decision diagrams. In *Int'l Conf. on CAD*, pages 42–47.

[Slobodová and Meinel, 1998] Slobodová, A. and Meinel, C. (1998). Sample method for minimization of OBDD. In *Int'l Workshop on Logic Synth.*, pages 311–316.

[Somenzi, 1998] Somenzi, F. (1998). *CUDD: CU Decision Diagram Package Release 2.3.0.* University of Colorado at Boulder.

AURA II: Combining Negative Thinking and Branch-and -Bound in Unate Covering Problems

Luca P. Carloni
EECS Department, University of California at Berkeley, Berkeley, CA 94720

Evguenii I. Goldberg
Cadence Berkeley Laboratories, Berkeley, CA 94704

Tiziano Villa
PARADES, 00186 Roma

Robert K. Brayton
EECS Department, University of California at Berkeley, Berkeley, CA 94720

Alberto L. Sangiovanni-Vincentelli
EECS Department, University of California at Berkeley, Berkeley, CA 94720

Abstract Recently a novel technique has been published to augment traditional Branch-and-Bound (B&B) while solving exactly a discrete optimization problem [Goldberg et al., 1997]. This technique is based on the *negative thinking* paradigm and has been applied to develop AURA, a Unate Covering Problem (UCP) solver which reportedly was able to deal efficiently with some time-consuming benchmark problems. However, on average AURA was not able to compete with SCHERZO, a classical UCP solver based on several new bounding techniques proposed by O. Coudert in his breakthrough paper [Coudert, 1996]. This fact left open the question on the practical impact of the negative thinking paradigm. The present work is meant to settle this question. The paper discusses the details of AURA II, a new implementation of the negative thinking paradigm for UCP which combines the best of SCHERZO and AURA. Experimental results show the dramatic impact of the negative thinking paradigm in searching the solution space and propose AURA II as the most efficient available tool for unate covering.

Keywords: Combinatioral optimization, branch-and-bound, covering problem.

1. INTRODUCTION

The Unate Covering Problem (UCP) [Kam et al., 1997] occurs often in logic synthesis and operations research and is defined as:

- Given a Boolean matrix A (all entries are 0 or 1), with m rows, denoted as $Row(A)$, and n columns, denoted as $Col(A)$, and a cost vector c of the columns of A (c_i is the cost of the i-th column), minimize the cost $x^T c = \sum_{j=1}^{n} x_j c_j$, over all $x \in \{0,1\}^n$, subject to $A\,x \geq (1, 1, \cdots, 1)^T$.

Informally the minimum unate covering problem requires to find a set of columns of minimum cost, such that each row intersects - "is covered by" - at least once a column in the set (i.e., the entry at the intersection is a 1). For simplicity assume that all columns have the same cost. An instance of UCP with matrix A is denoted $UCP(A)$.

In [Goldberg et al., 1997] the authors applied to UCP a novel technique to augment Branch-and-Bound (B&B) by a new way of exploring solutions, inspired by a paradigm called negative thinking. An algorithm named *raiser* realizing negative thinking by means of incremental problem solving was implemented in a computer program called AURA. This paper discusses the details of the *raiser* algorithm and reports the results obtained with AURA II, a new UCP solver which combines the best techniques of the traditional B&B with the negative thinking paradigm.

An exact solution of UCP may be obtained by a B&B recursive algorithm, variants of which have been implemented in successful computer programs [Coudert, 1994, Coudert, 1996, Coudert and Madre, 1995, Rudell and Sangiovanni-Vincentelli, 1987]. Branching is done by columns, i.e., subproblems are generated by considering whether a chosen branching column is or is not in the solution. A run of the algorithm, say *mincov*, can be described by a computation tree, where the root is the input of the problem, an edge represents a call to *mincov* and an internal node is a reduced input. A leaf is reached when a complete solution is found or the search is bounded away. From the root to any internal node there is a unique path, which is the current path for that node. The path leading to the node gives a partial solution and a submatrix A_N obtained from A by removing some rows and columns. On the path some columns are included in the partial solution and they are denoted by $path(A_N)$.

Suppose that we know that any minimal cover of A_N is greater or equal to a value $L(A_N)$. The value is called a lower bound of the solutions of $UCP(A_N)$; e.g., a Maximal Set of Independent Rows (MSIR) is a lower bound (independent means that they have at most 1 one per column). So the size of any solution of $UCP(A)$ including the columns in $path(A_N)$ is greater or equal to $L(A_N)$ + $|path(A_N)|$. Hence, if we found before a solution $best$ with the same or a smaller number of columns, i.e., $|best| \leq L(A_N) + path(A_N)$ we can stop the recursion and backtrack to the parent node of A_N. Denote by $K(A_N)$ the value $|best| - L(A_N) - |path(A_N)|$. The condition to stop the recursion is given by $K(A_N) \leq 0$. On the other hand, if $K(A_N)$ has a large positive value, usually it means that $L(A_N)$ is far from the size of a minimal solution to $UCP(A_N)$ and so a lot of branching is expected from A_N before a leaf can be reached.

Suppose that there is no way of improving the solution $best$ in the search tree rooted at A_N, yet $K(A_N)$ is positive. Usually a B&B algorithm must continue branching. However, there is another way of making $K(A_N)$ negative or zero: it is to improve the lower bound $L(A_N)$. The first way is "*positive*", in the sense that the algorithm tries to construct a better solution, and branching columns are chosen in the hope of improving the current best solution. The second way is "*negative*", in the sense that the algorithm tries to prove that there is no better solution in the tree rooted at A_N. Often in the first leaf a solution very close to a minimum one is found, so only few improvements are required to get a minimum solution. Therefore "positive" search will succeed and yield a new better solution only in a few of the potential 2^n subproblems at the n-th level of the computation tree. In the overwhelming majority of the subproblems "negative" search is more natural. The less frequently the best current solution is improved during the search, the more the "negative" search is justified.

To exploit both "positive" and "negative" search, B&B was modified in [Goldberg et al., 1997] as follows: start solving the initial problem with "positive thinking" in the ordinary column branching mode, called PT-mode. Then, when the number of subproblems generated in the column branching mode becomes large "enough", solve each subproblem in the "negative thinking" mode, called NT-mode. Modes are switched depending on the ratio of the expected number of improvements to the number of subproblems generated at this level of the search tree. The smaller the ratio, the more appropriate it is to switch to the NT-mode.

In [Goldberg et al., 1997] the results of comparing AURA against ESPRESSO [Rudell and Sangiovanni-Vincentelli, 1987] and SCHERZO [Coudert, 1994, Coudert, 1996, Coudert and Madre, 1995] were reported. AURA could outperform ESPRESSO on every benchmark, but was not always able to beat the performance of SCHERZO, due to its improvements in the computation of the lower bounds; partition-based pruning and further modifications in the organization of the B&B scheme. In principle, these features are orthogonal to the introduction of the negative thinking paradigm.

To assess further the strength of *raiser*, an approach (only partially explored in [Goldberg et al., 1997]) would have been to reproduce systematically all the features of SCHERZO within AURA. This paper reports the results of the alternative choice to re-implement *raiser* on top of SCHERZO, yielding the program AURA II, in order to exploit the algorithmic and programming virtues of SCHERZO together with the power of negative thinking available through *raiser*. In Section 3. the results of this comparison are reported, showing that AURA II is faster than SCHERZO, especially in the most time-consuming examples. As far as we know AURA II is currently the most efficient available tool for unate covering. AURA II combines the best of both worlds, and settles some experimental questions left open in [Goldberg et al., 1997].

2. THE RAISING ALGORITHM

Figure 1 shows how the traditional branch-and-bound algorithm *mincov* [Villa et al., 1997a] is modified to incorporate the technique of incrementally raising the lower bound. After the computation of the lower bound, if the gap *difference* between the upper and lower bound is small, i.e., less than a global parameter $maxRaiser$, a new procedure $raiser$ is invoked with a parameter n set to the value of *difference*. The parameter $maxRaiser$ currently is decided a-priori, but ideally it should be adapted dynamically. Intuitively if the gap is small, we conjecture that a search in this subtree will not improve the best solution and so we trigger the procedure $raiser$ that may either confirm the conjecture and prove that no better solution can be found here or disprove the conjecture and improve the best solution, updating the current one.

2.1 RAISING ALGORITHM: OVERVIEW

As discussed in [Goldberg et al., 1997] we developed an n-*raiser* procedure, based on row branching. Given a covering matrix A, let A' be a submatrix of A and A_p a row from $Row(A) \setminus Row(A')$. Let S be a solution of $UCP(A')$. Denote by $O(A_p)$ the set $\{j \mid A_{pj} = 1\}$, i.e., the set of all columns covering A_p and by $Rec(A' + A_p, S)$ a set of solutions of $UCP(A' + A_p)$ obtained according to the following rules:

1. if S is a solution of $UCP(A' + A_p)$, then $Rec(A' + A_p, S) = \{S\}$;

2. if S is not a solution of $UCP(A' + A_p)$, i.e., no column of S covers A_p then $Rec(A' + A_p, S) = \{S \cup \{j\} \mid j \in O(A_p)\}$.

So $Rec(A' + A_p, S)$ gives the solutions of $UCP(A' + A_p)$ that can be obtained from the solution S of $UCP(A')$. According to 2., if S is not a solution of $UCP(A' + A_p)$, then we obtain $|O(A_p)|$ solutions of $UCP(A' + A_p)$ by adding to S the columns covering A_p.

As discussed in [Goldberg et al., 1997], we represent the solutions of $UCP(A)$ by sets with a structure of multi-valued cubes [Rudell and Sangiovanni-Vincentelli, 1987]. We define a *cube* to be the set $C = D_1 \times \cdots \times D_d$ where

```
AuraMincov(A, path, weight, lbound, ubound) {
    /* Apply row dominance, column dominance, and select essentials */                    (1)
    if (not reduce(A, path, weight, ubound)) return empty_solution
    /* See if Gimpel's reduction technique applies */                                      (2)
    if (gimpel_reduce(A, path, weight, lbound, ubound, best)) return best
    /* Find lower bound from here to final solution by independent set */                  (3)
    MSIR = maximal_independent_set(A, weight)
    /* Make sure the lower bound is monotonically increasing */                            (4)
    lbound_new = max(cost(path) + cost(MSIR), lbound)
    difference = ubound − lbound_new
    /* Bounding based on no better solution possible */                                    (5)
    if (difference ≤ 0) best = empty_solution
    else if (difference ≤ maxRaiser) { /* Apply raiser with n = difference */              (16)
        SolCube = cover_MSIR(MSIR)                                                         (17)
        lowerBound = |SolCube|                                                             (18)
        a = raiser(SolCube, difference, A, lowerBound, bestSol, ubound)                    (19)
        if (a = 1) best = empty_solution                                                   (20)
        else best = path ∪ bestSol /* (answer = 0) */                                      (21)
    }
    else if (A is empty) { /* New best solution at current level */                        (6)
        best = solution_dup(path)
    } else if (block_partition(A, A1, A2) gives non-trivial bi-partitions) {               (7)
        path1 = empty_solution
        best1 = mincov(A1, path1, weight, 0, ubound − cost(path))                          (8)
        /* Add best solution to the selected set */                                        (9)
        if (best1 = empty_solution) best = empty_solution
        else {                                                                             (10)
            path = path ∪ best1
            best = mincov(A2, path, weight, lbound_new, ubound)
        }
    } else { /* Branch on cyclic core and recur */                                         (11)
        branch = select_column(A, weight, MSIR)
        path1 = solution_dup(path) ∪ branch
        let A_branch be the reduced table assuming branch in solution                      (12)
        best1 = mincov(A_branch, path1, weight, lbound_new, ubound)
        /* Update the upper bound if we found a better solution */                         (13)
        if (best1 ≠ empty_solution) /* It implies (ubound > cost(best1)) */
            ubound = cost(best1)
        /* Do not branch if lower bound matched */                                         (14)
        if (best1 ≠ empty_solution) and (cost(best1) = lbound_new) return best1
        let A_(branch-bar) be the reduced table assuming branch not in solution            (15)
        best2 = mincov(A_(branch-bar), path, weight, lbound_new, ubound)
        best = best_solution(best1, best2)
    }
    return best
}
```

Figure 1 AuraMincov: Traditional *mincov* algorithm enhanced by incremental raising.

$D_i \cap D_j = \emptyset$, $i \neq j$ and $D_i \subset Col(A)$, $1 \leq i, j \leq d$. The subsets D_i are the *domains* of cube C. So cube C denotes a set of sets consisting of d columns.

Let A' be a submatrix of A. The set of all irredundant (and minimum) solutions of $UCP(A')$ can be represented as the cube $O(A_{i_1}) \times \cdots \times O(A_{i_d})$, where $A_{i_1}, \cdots, A_{i_d}$ are the rows forming A'. Let $C = D_1 \times \cdots \times D_d$ be a cube of solutions of $UCP(A')$. Then, choose a "good" row A_p from from

$Row(A) \setminus Row(A')$. From the definition of the Rec operator [1] it follows that

$$Rec(A' + A_p, C) = part1(C) \cup part2(C) \times O(A_p) \qquad (1.1)$$

where $part1(C)$ is the set of solutions contained in C which cover A_p and $part2(C)$ is the set of solutions contained in C which do not cover A_p. Hence, $Rec(A' + A_p, C)$ can be represented by $r + 1$ cubes where r is the number of rows of the $MSIR(A)$ intersecting A_p. Then, perform recursively the process for each of the $r + 1$ cubes, i.e., choose a new row from those not yet selected for each of the $r + 1$ cubes of solutions and split each cube according to the rule explained in [Goldberg et al., 1997].

The entire process can be described by a search tree, called **cube branching tree**. The initial cube of solutions C corresponds to the root node, to which we associate also a pair of matrices $MSIR(A)$ and $A\ MSIR(A)$. In each node a choice of an unselected row from the second matrix of the node is made. The chosen row is removed from the second matrix of the pair and added to the first matrix of the pair. The number of branches leaving a node is equal to the number of cubes in which the cube corresponding to the node is partitioned by the Rec operation, and each child of a node gets one of the cubes obtained after splitting. So the cube corresponding to a node represents a set of solutions covering the first matrix of the pair (that is a "lower bound submatrix" for the node).

Some useful facts are:

- When applying an *n-raiser*, the branches corresponding to cubes of more than $|MSIR(A)| + n$ domains are pruned.
- If at a node, a row A_p is chosen such that no solution from the cube C of the node covers A_p, then there is no splitting of the cube, since Rec yields only one cube $C \times [O(A_p) \setminus (D_1 \cup \cdots \cup D_d)]$.
- At each node, the following reduction rule can be applied to the second matrix of the pair: if a row of the second matrix is covered by every solution of the cube C corresponding to the node, then the row can be removed from the matrix since, if we add it to the lower bound submatrix of the pair, then the recomputed cube will be equal to C.

The recursion terminates if one of the two following conditions hold:

1. There is a node such that there are no rows left in the second matrix of the pair and the corresponding cube has k domains, where $k < |MSIR| + n$. This means that the lower bound $|MSIR|$ cannot be improved by n.

[1] With the natural extension that $Rec(A, C) = \bigcup_{c \in C} Rec(A, c)$.

Any solution from the cube can be taken as the best current solution of $UCP(A)$.

2. From all branches, nodes are reached corresponding to cubes with a number of domains greater than $|MSIR| + n$. In this case the lower bound has been raised to $|MSIR| + n$, since no solution S of $UCP(A)$ exists such that $|S| \leq |MSIR| + n$.

The correctness of the n-raiser procedure, applied to matrix A with lower bound $|MSIR(A)|$, has been proved in [Goldberg et al., 1997].

2.2 RAISING ALGORITHM: IMPLEMENTATION

The procedure *raiser* returns 1 if the lower bound can be raised by n, otherwise it returns 0, which means that the current best solution has been improved at least once by *raiser*. The following parameters are needed:

- A is the matrix of rows not yet considered. Initially $A = A' \setminus MSIR$, where A' is the covering matrix at the node (of the column branching tree) that called *raiser*, and $MSIR$ is the maximal independent set of rows, found at the node (of the column branching tree) that called *raiser*. Hence, A' is the covering matrix related to the subproblem obtained by choosing the columns in the path from the root to the node that called *raiser*. The set of chosen columns is denoted by $path$.

- $SolCube$ is a cube which encodes a set of partial solutions of the covering matrix A'. Initially $SolCube$ is equal to the set of solutions covering the $MSIR$.

- n is number by which the lower bound $lbound$ must be raised. n is an input-output parameter initially equal to $ubound - |MSIR| - |path|$, which is decreased if *raiser* decreases the best current solution.

- $lbound$ is an input parameter for $raiser$ equal to $|MSIR|$. Notice that $lbound$ differs from the original lower bound [2] by a quantity equal to $|path|$, for consistency with the previous definition of n.

- $ubound$ is the cardinality of the best solution known at the time of the current call of *raiser*.

- $bestSolution$ is an output parameter which contains the new best solution found by $raiser$, if the lower bound could not be raised by n.

[2] $lbound\_new = |MSIR| + |path|$.

```
raiser(SolCube, n, A, lbound, bestSolution, ubound) {
    /* returns 1 if solutions in SolCube raise lower bound of A by n */
    stillToRaise = lbound + n − number_domains(SolCube)
    if (stillToRaise ≤ 0) return 1
    /* If A = ∅ then path + solutions of A in SolCube beats upper bound */
    if (A = ∅)  return found_solution(SolCube, n, bestSolution, ubound)
    /* consider rows of A not covered by any solution from SolCube */
    BSONIR = find_best_set_of_non_intersecting_rows(A, SolCube)
    foreach row r_i ∈ BSONIR {
        /* add a new domain for the columns covering r_i ∈ A */
        SolCube = add_domain(SolCube, A, r_i)
        stillToRaise = stillToRaise − 1
        if (stillToRaise ≤ 0) return 1
    }
    /* Remove the covered rows from A and check again if A is empty */
    A = A \ BSONIR
    if (A = ∅)  return found_solution(SolCube, n, bestSolution, ubound)
    if (stillToRaise = 1) {
        /* Cover with SolCube and remove from A the 1-intersecting rows */
        /* If 2 rows intersect 2 different cols in the same domain, prune the branch */
        if (add_set_of_1intersecting_rows(A, SolCube) = 1) return 1
        if (A = ∅)
            return found_solution(SolCube, n, bestSolution, ubound)
    }
    /* select next "best" row to be covered with SolCube and remove it from A */
    r_i = select_best_uncovered_row(A, SolCube)
    A = A \ {r_i}
    /* Splitting: part1 = {SolCube_1, ···, SolCube_k}; part2 = {SolCube_k+1} */
    split_cubes(SolCube, A, r_i, part1, part2)
    /* add to SolCube_2 ∈ part2 new domain of the columns covering r_i */
    SolCube_k+1 = add_domain(SolCube_k+1, A, r_i)
    /* branching on cubes of part1 and part2 */
    returnValue = 1
    while (part1 ∪ part2 ≠ ∅) {
        /* select first cubes from part1, then cube from part2 */
        SolCube_j = get_next_cube(part1 ∪ part2)
        /* if a better global solution has been found set returnValue to 0 */
        if (raiser(SolCube_j, n, A, lbound, bestSolution, ubound) = 0)
            returnValue = 0
    }
    return returnValue
}

found_solution(SolCube, n, bestSolution, ubound) {
    /* extract any solution from SolCube by picking a column from each domain */
    bestSolution = get_solution(SolCube)
    newUbound = cost(bestSolution)
    newN = n − (ubound − newUbound)
    n = newN
    ubound = newUbound
    return 0
}
```

Figure 2 Algorithm to raise the lower bound.

Fig. 2 shows the flow of $raiser$, the procedure that attempts to raise the lower bound of A. Notice that it requires a routine *split_cubes* which, for a selection of a row r_i covered by k of the d domains of $SolCube$, partitions $SolCube$ in $k+1$ disjoint cubes, each of d domains; so $part1$ has k cubes of solutions from $SolCube$ covering r_i, whereas $part2$ has one cube of solutions from $SolCube$ not covering r_i. The number of domains of $SolCube$ is computed by *number_domains*.

$raiser$ is a recursive procedure which starts by handling two terminal cases. The first one occurs when the variable $stillToRaise$ [3], which measures the gap between the upper bound and the current lower bound, is less or equal to zero. If so, we know that the solutions in $SolCube$ raise the lower bound of A by at least n, so that no solutions of A can beat the current upper bound. The second terminal case occurs when, after some recursive calls, A has become empty, and so any solution obtained as the union of a solution of A in $SolCube$ together with the columns in the current $path$ is the new best solution.

After these preliminary checks, $find\_best\_set\_of\_non\_intersecting\_rows$ is called. This routine, reported in Figure 3, implements a fast heuristic to find a good subset of rows of A which do not intersect any domain of $SolCube$ and which do not intersect each other. Ideally, we would like to get the best $\mathcal{BSONIR}$ set, which is a sort of "maximum set of independent rows" related to $SolCube$, but this would require the solution of another NP-complete problem. We implemented instead the heuristic to insert first in the set $\mathcal{BSONIR}$ the largest row that intersects neither a domain of $SolCube$ nor a row previously inserted into $\mathcal{BSONIR}$.

Thereafter, since each row r_i in $\mathcal{BSONIR}$ is not covered by any solution encoded in $SolCube$, we must add a new domain to $SolCube$ made by the columns which cover r_i. While we are adding these new domains, we keep decreasing the variable $stillToRaise$ and checking if its value becomes equal to zero. Finally, we can remove the set $\mathcal{BSONIR}$ from A because the rows have been covered by the new added domains. Notice that during the first call of $raiser$ the set $\mathcal{BSONIR}$ is empty because $SolCube$ encodes the $MSIR$ and, by definition, every row not in the $MSIR$ must intersect at least one row in the $MSIR$. However, during the following recursive calls of $raiser$ the original domains of $SolCube$ may change, namely decrease in cardinality due to $split\_cubes$ and $add\_set\_of\_1intersecting\_rows$. Hence, at some node of

[3] By definition,

$$\begin{aligned} stillToRaise &= lbound + n - numberDomains(SolCube) \\ &= |MSIR| + ubound - |MSIR| - |path| - numberDomains(SolCube) \\ &= ubound - |path| - numberDomains(SolCube) \end{aligned}$$

```
find_best_set_of_non_intersecting_rows(A, SolCube) {
    /* Heuristic to find best set of rows non intersecting SolCube domains. */
    emptyInterRows = ∅
    bestRow = ∅
    foreach row r ∈ A {
        /* D is the set of SolCube domains intersected by r */
        D = compute_set_of_intersected_domains(SolCube, r)
        if (D = ∅) {
            emptyInterRows = emptyInterRows ∪ r
            if (length(bestRow) < length(r))
                bestRow = r
        }
    }
    /* If every row intersects domains of SolCube then return the empty set */
    if (emptyInterRows = ∅)
        return ∅
    else {
        /* Build BSONIR starting from bestRow */
            BSONIR = ∅
        do {
            BSONIR = BSONIR ∪ bestRow
            emptyInterRows = emptyInterRows \ bestRow
            /* Find the new bestRow within emptyInterRows*/
            foreach row r ∈ emptyInterRows {
                if ((r ∩ BSONIR) ≠ ∅)
                    emptyInterRows = emptyInterRows \ r
                else if (length(bestRow) < length(r))
                    bestRow = r
            }
        } while (emptyInterRows ≠ ∅)
    }
    return BSONIR
}
```

Figure 3 Algorithm to find the best set of rows not intersecting $SolCube$.

the recursion tree, it may happen that a row of A is not covered anymore by any domain of $SolCube$.

After having removed the rows belonging to $\mathcal{BSONIR}$, another optimization step can be applied successively before splitting $SolCube$. If at this point $stillToRaise$ is equal to 1, it means that we have already raised the lower bound by $n-1$. Therefore, if we are forced to add one more domain to $SolCube$, then we can prune the current branch. Hence, a simple condition which leads immediately to pruning is the following: consider two rows r_1 and r_2 of A which intersect $SolCube$ only in one domain $d = \{c^1, c^2, \cdots, c^l\}$, and suppose that r_1 intersects only the column c^i, while r_2 intersects only the column c^j. This fact allows us to prune the current branch because to cover one of the rows we may choose either one of the two distinct columns of the domain. Say w.l.o.g that we cover r_1 with c^i, then to cover r_2 we must use a column which does not belong to any domain of $SolCube$ and so we are forced to add one more domain to $SolCube$, thereby raising the lower bound by n.

Figure 1.4 illustrates the procedure $add\_set\_of\_1intersecting\_rows$, which exploits the previous situation and, in practice, is invoked often because the condition $stillToRaise = 1$ happens very commonly in hard problems. Basically, the routine is based on two nested cycles. The external cycle is repeated until the internal cycle does not modify $SolCube$ anymore. The internal cycle computes, for each row r of A, the set D of the domains of $SolCube$ intersected by r. If the cardinality of D is equal to 1, e.g., $D = \{d\}$, we remove from d all the columns which are not intersected by r and then we remove r from A, since r has been covered.

Notice that $add\_set\_of\_1intersecting\_rows$ is called just after having removed from A the set of non-intersecting rows $\mathcal{BSONIR}$ and therefore when all the remaining rows of A intersect at least one domain of $SolCube$. However, after cycling inside this routine and removing some columns (thereby making "leaner" some domains), it is possible that a row of A is not covered anymore, i.e., $|D| = 0$. As discussed above, this happens, e.g., when two 1-intersecting rows intersect two different columns in the same domain D. In this case the routine returns 1 in order to inform the caller to prune the current branch. If this fact does not happen before the end of both cycles, a 0 is returned but, at least a certain number of rows have been removed from A and the corresponding intersected domains of $SolCube$ have been made "leaner". After calling $add\_set\_of\_1intersecting\_rows$ and removing 1-intersecting rows, it is possible that A has become empty. If so, $raiser$ calls $found\_solution$ to update the variables $bestSolution$, $ubound$ and n.

After all these special cases have been addressed, we must select a new row r_i to be covered with $SolCube$. The row r_i is removed from A and drives the splitting of $SolCube$. The strategy to select the best row in order to split the current $SolCube$, before calling recursively $raiser$, is to look for the row of A which intersects the minimum number of domains of $SolCube$. The reason is to reduce the number of branches from the node [4]. Notice that at this stage each row of A intersects at least 2 domains of $SolCube$. In case of ties between different rows, the row having the highest weight is chosen. The weight of a row A_p is defined as $\prod_{k=1}^{m} \frac{|D'_{i_k}|}{|D_{i_k}|}$, where m is the number of domains of $SolCube$ intersecting A_p, D_{i_k} is a domain intersected by A_p and $D'_{i_k} = D_{i_k} \setminus O(A_p)$. So the weight of A_p is just the fraction of solutions from $SolCube$ that do not cover A_p, that is the quantity that we want to maximize when selecting a new

[4]Recall that there is a branch for each domain intersecting the row plus one more branch for the non-intersecting domains.

```
add_set_of_1intersecting_rows(A, SolCube) {
    /* This routine is called only if stillToRaise = 1. It covers */
    /* with SolCube and removes from A the 1-intersecting rows, */
    /* i.e., the rows intersecting only one domain of SolCube. */
    /* If 2 rows intersect 2 different columns in the same domain, */
    /* return 1 to the caller to prune the current branch */
    do {
        reducingDomains = FALSE
        foreach row r ∈ A {
            /* D is the set of SolCube domains intersected by r */
            D = compute_set_of_intersected_domains(SolCube, r)
            if (| D |= 1) {
                reducingDomains = TRUE
                /* Get the domain d of SolCube covering r and */
                /* remove from d all the cols which do not cover r */
                d = get_covering_domain(SolCube, r)
                simplify_domain(d, r)
                /* Remove the covered row r from A */
                A = A \ {r}
            }
            else if (| D |= 0) {
                /* After removing some columns, a row may not be */
                /* covered anymore, so current branch must be pruned. */
            }
            /* else (| D |> 1): do nothing */
            /* because r is not a 1-intersecting row */
        }
    } while (reducingDomains)
    return 0
}
```

Figure 4 Algorithm to handle the 1-intersecting rows.

row. If $D'_{i_k} = \emptyset$, for some k, this means that A_p is covered by any solution from $SolCube$. Such a row is simply removed from A'' and added to A'.

After performing the splitting of $SolCube$ as explained in [Goldberg et al., 1997], $raiser$ is called recursively on the disjoint cubes of the recomputed solution. If the current best solution is not improved in any of the calls, then raiser returns 1, meaning that the lower bound has been raised by n. If instead the current best solution has been improved once or more times, $raiser$ returns 0 after having updated the current best solution and upper bound.

3. EXPERIMENTAL RESULTS

In [Goldberg et al., 1997], AURA was compared against the routine *mincov* available in ESPRESSO, and against the results of SCHERZO [Coudert, 1994, Coudert, 1996, Coudert and Madre, 1995], the most effective UCP solver available then. Compared to ESPRESSO, SCHERZO features a collection of new lower bounds (easy lower bound, logarithmic lower bound, left hand side lower bound, limit lower bound), and partition-based pruning. In this paper we

			SCHERZO		AURA II			time
matrix	$R \times C(S\%)$	Sol.	nodes	time	nodes/A-nodes	time	r	ratio
ex5	831×2428 (2)	37	614631	11397.1	614510/156	11066.5	**1**	0.97
ex5	831×2428 (2)	37	614631	11397.1	31185/243184	1346.67	**2**	0.12
ex5	831×2428 (2)	37	614631	11397.1	1905/195190	746.85	**3**	0.06
max1024	1090×1264 (0.5)	245	533635	5535.67	533632/52	5244.54	**1**	0.95
max1024	1090×1264 (0.5)	245	533635	5535.67	91345/667471	2994.88	**2**	0.54
max1024	1090×1264 (0.5)	245	533635	5535.67	15353/1624827	5967.92	**3**	1.10
prom2	1924×2611 (0.3)	278	26143	1506.75	26143/16	1454.81	**1**	0.97
prom2	1924×2611 (0.3)	278	26143	1506.75	6115/115460	1685.36	**2**	1.10
prom2	1924×2611 (0.3)	278	26143	1506.75	1389/754564	10162	**3**	6.70
saucier	171×6207 (47)	6	187089	11876.1	7/36	24.0	**1**	0.002
saucier	171×6207 (47)	6	187089	11876.1	7/36	24.0	**2**	0.002
saucier	171×6207 (47)	6	187089	11876.1	7/36	24.0	**3**	0.002

Table 1 Results on *Espresso* benchmarks (SCHERZO VS. AURA II).

compare AURA II, that is *raiser* implemented in SCHERZO, against SCHERZO. The benchmarks used belong to three classes: Table 1 contains difficult cases from the collection of ESPRESSO (we start from the matrix obtained by ESPRESSO after removing the essential primes) and some matrix encoding constraints satisfaction problems from [Villa et al., 1997b]; Table 2 contains random generated matrices with varying row/column ratios and densities (e.g., $m200\_100\_30\_70$ means a matrix with 200 rows, 100 columns, and each column having a number of ones between 30 and 70). For each of these matrices, their size ($R \times C$ in the tables) and sparsity (S expressed as a percentage in the tables) are reported. The experiments were performed with a 1GB 625Mhz Alpha with timeout set to 4 hours of cpu time. Tables 1 and 2 report two types of data for comparison: the number of nodes of the column branching computation tree and the running time. Concerning the number of nodes we clarify the following points:

1. AURA II has two types of nodes: those of the column branching computation tree and those of the cube branching computation tree (called A-nodes in the tables). Indeed AURA II follows a dual strategy: it builds the column branching computation tree, but when at a node the difference between the upper bound and the lower bound is less than or equal to the raising parameter r (or $maxRaiser$), AURA II calls the procedure $raiser$ which builds a cube branching computation tree (appended at the node where $raiser$ was called). So we need to report both numbers of nodes to measure a run of AURA II.

2. Nodes of the cube branching computation tree usually take much less computing time than those of the column branching computation tree, even though a time ratio between the two types of nodes is not known a-priori. The reason is that expensive procedures for finding dominance relations and $MSIR$ are applied in each node of the column branching tree.

matrix	$R \times C(S\%)$	Sol.	SCHERZO nodes	SCHERZO time	AURA II nodes/A-nodes	AURA II time	r	time ratio
m100_100_10_10	100×100 (10)	12	95086	36.87	3180/121892	20.33	3	0.55
m100_100_10_15	100×100 (12)	10	10335	6.12	269/11071	2.41	3	0.39
m100_100_10_30	100×100 (20)	8	4618	4.05	84/2726	0.78	3	0.19
m100_100_30_30	100×100 (30)	5	1752	2.44	49/1288	0.64	3	0.26
m100_100_50_50	100×100 (50)	4	4015	6.1	5/857	0.69	3	0.11
m100_100_70_70	100×100 (70)	3	171	2.21	3/112	0.19	3	0.09
m100_100_90_90	100×100 (90)	2	2	0.02	2/0	0.02	3	1
m100_300_10_10	100×293 (3)	21	351183	235.16	10144/612753	175.37	3	0.75
m100_300_10_14	100×297 (4)	19	1906835	1257.62	70998/3453419	993.83	3	0.79
m100_300_10_15	100×297 (4)	19	11596849	7066.57	329794/16381322	4385.16	3	0.62
m100_300_10_20	100×299 (5)	17	5240615	3641.41	138572/6904928	2036.72	3	0.56
m100_50_10_10	100×50 (20)	8	2079	0.92	85/2411	0.42	3	0.46
m100_50_20_20	100×50 (40)	5	1825	1.02	23/889	0.27	3	0.26
m100_50_30_30	100×50 (60)	3	63	0.34	3/24	0.03	3	0.09
m100_50_40_40	100×50 (80)	2	2	0.01	2/0	0.01	3	1
m50_100_10_10	50×99 (10)	8	92	0.02	12/133	0.02	3	1
m50_100_30_30	50×100 (30)	4	65	0.06	5/61	0.02	3	0.33
m50_100_50_50	50×100 (50)	3	107	0.22	3/32	0.02	3	0.09
m50_100_70_70	50×100 (70)	2	2	0.01	2/0	0.01	3	1
m50_100_90_90	50×100 (90)	2	2	0.01	2/0	0.01	3	1
m100_200_10_30	100×200 (10)	12	281845	242.65	2915/161571	45.61	3	0.19
m100_200_10_50	100×200 (10)	12	281845	241.06	2915/161571	45.36	3	0.19
m100_200_10_70	100×200 (20)	8	19135	22.8	82/6538	2.36	3	0.10
m100_200_30_30	100×200 (15)	8	154475	117.5	31499/775717	220.05	3	1.90
m100_200_30_50	100×200 (19)	7	50613	78.03	4019/136979	59.58	3	0.76
m100_200_30_70	100×200 (25)	6	30577	61.55	707/15289	10.43	3	0.17
m100_200_50_50	100×200 (25)	6	32214	63.84	3753/78023	44.67	3	0.70
m100_200_50_70	100×200 (29)	5	4867	17.19	163/5581	4.94	3	0.29
m100_200_70_70	100×200 (35)	5	26588	63.73	245/22860	16.47	3	0.26
m200_100_10_10	200×100 (10)	16	13889095	10776.6	464553/16098542	3830.34	3	0.36
m200_100_10_100	200×100 (54)	6	317	1.79	9/250	0.21	3	0.12
m200_100_10_30	200×100 (19)	11	564302	584.54	9156/371430	115.52	3	0.20
m200_100_10_50	200×100 (28)	8	29803	46.64	528/17689	8.91	3	0.19
m200_100_10_70	200×100 (40)	7	1735	4.87	37/1046	1.01	3	0.21
m200_100_30_100	200×100 (64)	4	1725	11.09	5/185	0.38	3	0.03
m200_100_30_30	200×100 (30)	6	65468	115.44	883/31293	18	3	0.16
m200_100_30_50	200×100 (39)	6	123621	170.09	1177/51624	33.41	3	0.20
m200_100_30_70	200×100 (51)	4	2036	17.07	7/190	0.39	3	0.02
m200_100_50_100	200×100 (74)	3	145	7.08	3/52	0.33	3	0.05
m200_100_50_50	200×100 (50)	4	8076	35.4	9/1607	1.79	3	0.05
m200_100_50_70	200×100 (60)	4	5413	32.48	5/1302	2.31	3	0.07
m200_100_70_100	200×100 (84)	2	2	0.03	2/0	0.03	3	1
m200_100_70_70	200×100 (70)	3	169	10.89	3/90	0.46	3	0.04
m200_200_100_100	200×200 (50)	4	16313	259.45	5/2642	7.11	3	0.03

Table 2 Results on random benchmarks (SCHERZO VS. AURA II).

3. The raising parameter r is an input to AURA II. The higher the raising parameter, the fewer column branching nodes compared to cube branching nodes there will be. With a value that is high enough, there will be a single column node and the rest will be all row nodes.

The experiments show that AURA II is faster than SCHERZO, especially in the most time-consuming examples. For each of the difficult cases of Table 1, we have run AURA II with $r = 1, 2, 3$. There is always a value of r which allows AURA II to solve the problem faster than SCHERZO and in general this value is either 2 or 3. However, for the problem *prom2* the higher is the value of r the lower is the performance of AURA II: in fact, since this problem presents an highly diversified solution space, the raising procedure often terminates only after it has found a better solution (and, therefore, without having been able to prune rapidly the current branch). On the other hand, in the case of the problem *saucier*, whose solution space is poorly diversified, AURA II finds the solution in 24 second with any possible value of r while SCHERZO takes 11876 seconds. These results are in concord with the philosophy of "negative thinking" as discussed in Section 1.: the less frequently the best current solution is improved during the search, the more the "negative" search is justified. Now, when we are running a very time-consuming problem, the overwhelming majority of the subproblems do not lead to a solution improvement and, therefore, "negative" search is more natural and, if applied, leads to spectacular savings in total time. This is confirmed by the experiments with the random generated matrices of Table 2, for which we have kept the raising parameter r constantly equal to 3. In the most time-consuming of these examples AURA II takes between 36% and 75% of the time of SCHERZO.

3.1 OTHER COMPARISONS

We do not have a systematic comparison with the results by BCU, a very efficient recently-developed ILP-based covering solver [Liao and Devadas, 1997]. However, the intuition is that an algorithm based on linear programming is better suited for problems with a solution space diversified in the costs, i.e., for problems which are "closer" to numerical ones. To test the conjecture we asked the authors of [Liao and Devadas, 1997] to run BCU on *saucier.t*, whose solution space is poorly diversified (a minimum solution has 6 columns, while most of the irredundant solutions cost in the range from 6 to 8). BCU ran out of memory after 20000 seconds of computations (the information was kindly provided by S.Liao), while AURA II completes the example in 24 seconds. It would be of interest to study if the virtues of an ILP-based solver and of *raiser* could be combined in a single algorithm.

4. CONCLUSIONS

In [Goldberg et al., 1997] the authors applied to UCP a novel technique to augment Branch-and-Bound (B&B) using a new way of exploring solutions, inspired by a paradigm called negative thinking. Traditional UCP solvers are based on the mincov algorithm [Rudell and Sangiovanni-Vincentelli, 1987] which keep searching the solution space in the hope of finding a better solution (positive thinking mode) The new paradigm led to the development of the *raiser* algorithm which can be coupled with *mincov* to better guide the exploration of the binary tree representing the solution space: in fact, the search for a better solution can be appropriately interleaved with the attempt to prove that no better solution can be found in the current branching node (negative thinking mode). This paper discusses the details of the *raiser* algorithm. Moreover, by reporting experimental results obtained with AURA II, a new state-of-the-art UCP solver which combines the best of both worlds, we settle some experimental questions left open in [Goldberg et al., 1997]. Future work includes the extension of AURA II to solve the binate covering problem.

Acknowledgments

We gratefully thank Dr. Olivier Coudert (Monterey Design Systems) who kindly provided us a version of Scherzo and was always available for technical discussions.

References

[Coudert, 1994] Coudert, O. (1994). Two-level logic minimization: an overview. *Integration*, 17-2:97–140.

[Coudert, 1996] Coudert, O. (1996). On solving binate covering problems. In *The Proceedings of the Design Automation Conference*, pages 197–202.

[Coudert and Madre, 1995] Coudert, O. and Madre, J. (1995). New ideas for solving covering problems. In *The Proceedings of the Design Automation Conference*, pages 641–646.

[Goldberg et al., 1997] Goldberg, E., Carloni, L. P., Villa, T., Brayton, R. K., and Sangiovanni-Vincentelli, A. (1997). Negative thinking by incremental problem solving: application to unate covering. In *The Proceedings of the International Conference on Computer-Aided Design*, pages 91–99.

[Kam et al., 1997] Kam, T., Villa, T., Brayton, R. K., and Sangiovanni-Vincentelli, A. (1997). *Synthesis of FSMs: functional optimization*. Kluwer Academic Publishers.

[Liao and Devadas, 1997] Liao, S. and Devadas, S. (1997). Solving covering problems using LPR-based lower bounds. In *The Proceedings of the Design Automation Conference*.

[Rudell and Sangiovanni-Vincentelli, 1987] Rudell, R. and Sangiovanni-Vincentelli, A. (1987). Multiple-valued minimization for PLA optimization. *IEEE Transactions on Computer-Aided Design*, CAD-6:727–750.

[Villa et al., 1997a] Villa, T., Kam, T., Brayton, R. K., and Sangiovanni-Vincentelli, A. (1997a). Explicit and implicit algorithms for binate covering problems. *IEEE Transactions on Computer-Aided Design*, 16(7):677–691.

[Villa et al., 1997b] Villa, T., Kam, T., Brayton, R. K., and Sangiovanni-Vincentelli, A. (1997b). *Synthesis of FSMs: logic optimization*. Kluwer Academic Publishers.

Satisfiability-Based Functional Delay Fault Testing

Joonyoung Kim[†], *João Marques Silva*[††] *and Karem A. Sakallah*[†]
[†]*Advanced Computer Architecture Laboratory, Department of Electrical Engineering and Computer Science, University of Michigan 1301 Beal Ave. Ann Arbor, MI 48109, U.S.A*
[††]*IST/INESC, Cadence European Labs Lisbon, Portugal*

Key words: testing, delay fault testing, boolean satisfiability

Functional delay fault testing is concerned with propagating a transition from a primary input to a primary output of a combinational circuit. Since it does not consider individual paths in the circuit, it can overcome the biggest limitation of path delay fault testing: the explosion in the size of fault lists. Functional delay fault testing can also be used to derive test sets for IP (Intellectual Property) circuits whose implementation details are not provided. Boolean Satisfiability (SAT) and BDDs have been widely used for a variety of EDA (Electronic Design Automation) applications. Even though there have been few experimental studies to conclude the superiority of one to the other, they have been compared for a number of specific tasks in the EDA field. In this paper we show that SAT-based functional delay fault testing can yield very competitive results with careful construction of the CNF formulas for the target faults. In particular, using simple structural analysis of the circuit formulas of minimum size can be easily generated. CNF formula construction based on the circuit consistency function is presented and experimental results for ISCAS 85 and 89 circuits are reported.

1. INTRODUCTION

Delay fault testing, which addresses manufacturing defects that affect temporal behavior, is usually performed after a fabricated circuit has been tested for stuck-at faults. There are two fault models that have been widely used in delay fault testing: the *gate delay fault model* [3] which ascribes faulty behavior to individual gates having excessive delay, and the *path delay fault model* [16] which attempts to capture the distributed effects of defects over entire circuit paths leading to excessive path delays. In either case, the existence of a delay fault causes the circuit to fail to operate at the expected clock frequency.

There have been numerous research efforts on delay fault testing based on the path delay fault model [5, 10]. The major drawback of this model is that the size of the fault list (number of paths) for circuits with a large amount of reconvergence can become exponentially large. Attempts to overcome this limitation included the use of incremental path sensitization [6, 8] and targeting a group of paths using the primitive path delay fault model [7]. Such attempts, however, still fail to handle circuits with billions of paths (e.g., C6288 of the ISCAS 85 benchmark suites [2]).

An alternative to these two fault models, called the *functional delay fault model*, was first proposed in [12] and also investigated in [17]. In [13] two approaches for functional delay fault test generation, one based on a Binary Decision Diagram (BDDs) formulation and the other on a Boolean Satisfiability (SAT) formulation, were described and compared.[1] The experimental results indicated that several hours of run time for the ISCAS 85 benchmarks circuits were required in both approaches, with the BDD method resulting in a slightly better performance.

The Boolean satisfiability problem (SAT) has received a lot of attention recently, resulting in a number of robust and efficient heuristics and implementations [1, 14, 18]. It has also been used in many CAD applications such as test pattern generation for both stuck-at and delay fault models, logic verification and timing analysis [4, 8, 9, 11, 15]. In this paper, we present a SAT-based test pattern generation method for the functional delay fault model. We show that with careful construction of the target CNF formulas, this method can yield very competitive results. Promising results for both ISCAS 85 and 89 circuits are presented.

The remainder of the paper is organized as follows. In the next section, definitions that are used throughout the paper are presented. In Section 3, a method to generate the target CNF formula for functional delay fault testing based on the circuit consistency functions is presented with an example. In Section 4, experimental results for ISCAS 85 and 89 circuits are presented and Section 5 concludes the paper.

2. DEFINITIONS

Most of the definitions in Section 2.1 and Section 2.2 are taken directly from [15]. They are repeated here for completeness.

[1] The details of CNF formula generation for the SAT approach were not described.

2.1 Combinational Circuits

A combinational circuit C is represented as a directed acyclic graph $C = (V, E)$ where V denotes the circuit nodes and $E \subseteq V \times V$ denotes the connections between nodes. The following definitions also apply:

- $O(x)$ denotes the **fanout** nodes of node x, i.e., $\{y \in V \mid (x, y) \in E\}$.
- $O^*(x)$ denotes the **transitive fanout** nodes of node x, i.e., the set of all nodes y such that there is a path from x to y.
- $I(x)$ denotes the **fanin** nodes of node x, i.e., $\{y \in V \mid (y, x) \in E\}$.
- $I^*(x)$ denotes the **transitive fanin** nodes of node x, i.e., the set of all nodes y such that there is a path from y to x.
- $SI(\Psi)$ denotes the **side inputs** of a set of nodes $\Psi \subseteq V$ and is defined as follows:

$$SI(\Psi) = \{x \mid x \in I(w) \wedge w \in \Psi \wedge x \notin \Psi\}$$

The set of primary input nodes are referred to as PI, and the set of primary output nodes as PO. Figure 1 illustrates the definitions given in this section for a small benchmark circuit from [2].

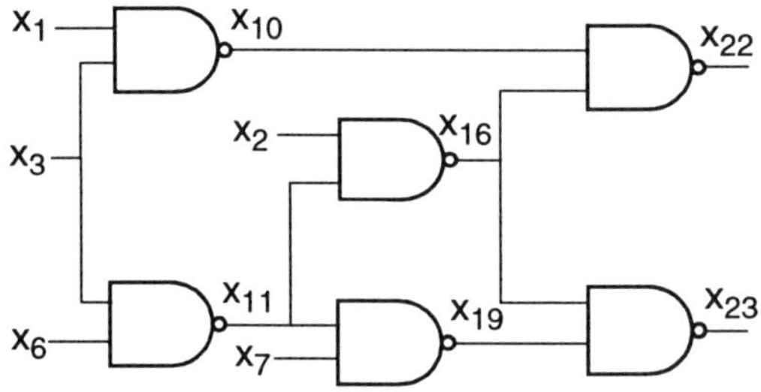

$O(x_{11})$	x_{16}, x_{19}
$O^*(x_{11})$	$x_{16}, x_{19},$
$I(x_{16})$	x_2, x_{11}
$I^*(x_{16})$	$x_2, x_3, x_6,$
$SI(x_3, x_{11}, x_{16},$	x_2, x_6, x_{10}

Figure 1. Example ISCAS 85 circuit C17

2.2 Boolean Satisfiability

We consider Boolean functions represented in Conjunctive Normal Form (CNF). A *literal* is an occurrence of a variable x or its complement x'. A *clause* is the disjunction (OR) of literals. A CNF formula φ on n binary variables $x_1, x_2, \ldots, x_n$ is the conjunction (AND) of m clauses $w_1, w_2, \ldots, w_m$. A CNF formula is said to be *satisfiable* when there is at least one truth assignment to its variables that makes all clauses equal to 1. A CNF formula is said to be *unsatisfiable* when no such assignment exists.

For each gate g in a circuit, the *gate consistency function* φ_g is a Boolean function that denotes the valid input-output assignments admissible by the gate's logic function. A detailed description of the gate consistency function and its CNF representation for primitive gates is given in [15]. The *circuit consistency function* of is defined as the conjunction of the gate consistency functions for each node in the circuit. If we view a CNF formula as a set of clauses, the circuit consistency function can be defined using a set union operator as follows:

$$\varphi = \bigcup_{x \in V} \varphi_x$$

2.3 Functional Delay Fault Testing

The objective of functional delay testing is to propagate a transition $tI \in \{rising, falling\}$ from a primary input node to a primary output node as a transition $tO \in \{rising, falling\}$. By analogy to the path delay fault model, robust propagation for functional delay fault testing can be defined as follows:

Definition *1: For a given fault* (I, O, tI, tO), *a two-pattern input combination* (v_1, v_2) *is said to function-robustly propagate a transition tI from input I to a transition tO on output O if the value on O changes if and only if the value on I changes.*

In generating a test vector pair, two modes can be considered. In the single-input-transition mode, only one primary input is allowed to change, all other inputs being set to fixed values. In the multi-input- transition mode, any number of primary inputs are allowed to change In this paper, we consider only the single-input-transition mode; it must be noted, though, that the proposed test pattern generation procedure can be readily extended to the multi-input-transition mode.

Definition *2: A two-pattern input combination* (v_1, v_2) *is a function robust test for a given fault* (I, O, tI, tO) *under the single-input-transition mode if and only if I is the only input that changes its values between* v_1 *and* v_2 *as a transition tI and the transition tO is observed at the output O.*

Consider a primary output O of a circuit that implements the function f and a primary input I of the circuit. The detection of the faults $(I, O, rising, rising)$ and $(I, O, falling, falling)$ can be accomplished by checking the function $f_I\ ?(f_{I'})'$ for satisfiability. To detect the faults $(I, O, rising, falling)$ and $(I, O, falling, rising)$ the satisfiability of the function $f_{I'}\ ?(f_I)'$ must be checked. Note that these two formulas represent the two terms of df / dI, the Boolean derivative of f with respect to I.

3. FUNCTIONAL DELAY FAULT GENERATION

3.1 Fault List Construction

Before the test pattern generation procedure can begin, a fault list needs to be created. The method used in [12] is used in constructing fault lists. The upper bound on the size of the fault list for a circuit with $|PI|$ inputs and $|PO|$ outputs is $4|PI||PO|$ since for each input/output pair, there are four (I,O,tI,tO) tuples.

There are two techniques to reduce the size of the fault list. First, for a given output, the fan-in cone for the output node can be identified and only the primary input nodes that appear in the fan-in cone are considered when creating the fault list. Second, a check can be performed for each input/output pair to determine whether there is at least one path between them with an even or an odd number of inversions. If there are no paths with an even number of inversions (I,O,tI,tO) tuples with identical (tI,tO) values can be dropped from the fault list. Similarly, (I,O,tI,tO) tuples with opposite (tI,tO) values can be dropped if there are no path between I and O with an odd number of inversions.

Figure 2 shows the fault list created for the C17 circuit. E denotes that there is at least one path between the given primary input and primary output nodes with an even number of inversions and O denotes that there is at least one path with an odd number of inversion. "-" denotes that there are no paths between the two given nodes.

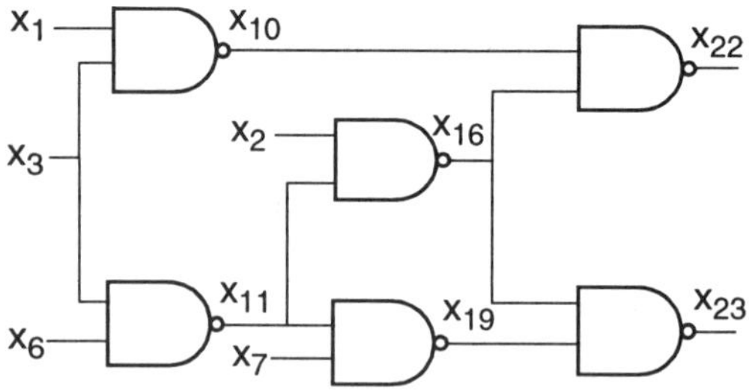

PI	PO	Parity
x_1	x_{22}	E
x_2	x_{22}	E
x_3	x_{22}	E, O
x_6	x_{22}	O
x_7	x_{22}	-
x_1	x_{23}	-
x_2	x_{23}	E
x_3	x_{23}	O
x_6	x_{23}	O
x_7	x_{23}	E

Figure 2. Fault list for C17

3.2 CNF Formula Construction

Once a fault list is set up, test generation is performed for each fault in the fault list. The overall flow of test pattern generation for delay fault testing is shown in Figure 3.

Since the formulas that must be checked involve both the positive and negative cofactors of each output with respect to each input in its transitive fan-in, the circuit is duplicated and consistency CNF formulas for both "co-factor" circuits are generated. A set of clauses from the CNF formulas thus created are then selectively chosen to create a target SAT instance for a specific functional delay fault. The clauses from the duplicate circuit are added to the target formula only when necessary. After generating the circuit consistency function, outputs are chosen one at a time. For each PO, the fan-in cone is identified. Then for each PI which has at least one path to the chosen output node, the fanout cone of the node is identified. For each (PI, PO) pair (i, o), the set of nodes that can potentially propagate a transition can be identified by $(I^*(o) | \ o) \quad (O^*(i) | \ i)$. The function duplicate_nodes() will extract the gate consistency functions from the duplicated circuit for those nodes.

Figure 4 presents an example of the test pattern generation procedure with a target fault $(x_6, x_{22}, \text{rising}, \text{rising})$. The fan-in cone for the target output node x_{22} is shown in the upper half. Since $(I^*(x_{22}) | \ x_{22}) \quad (O^*(x_6) | \ x_6) = \{x_6, x_{11}, x_{16}, x_{22}\}$, the duplicated nodes for these four nodes are identified and their gate consistency functions are selected by the function duplicate_nodes(). The function connect_side_inputs() identifies the side inputs of the given nodes in the duplicated circuit and connects them to the corresponding nodes in the original circuit. In this example, $SI(x_6, x_{11}, x_{16}, x_{22}) = \{x_2, x_3, x_{10}\}$, and the clauses that enforce the condition that each node in this set has the same value as its duplicate equivalent are added, which has the same effect as connecting a node in the original circuit to its duplicate node. Finally, for each fault between the target primary input and output pair, constraints are added based on the direction of the input and output transitions. For the example in Figure 4, the fault $(x_6, x_{22}, \text{rising}, \text{rising})$ is detected if the function

$$f_{x_6} \ ? \overline{f_{\overline{x_6}}}$$

is satisfiable, where f is the logic function at output node x_{22}. This is accomplished by setting x_6 and x_{22} to logic 1 and x_6' and x_{22}' to logic value 0.

```
begin functional delay fault generation
foreach o of PO
  φ_in = Identify I*(o);
  foreach i of PI // consider only iL I*(o)
    Identify O*(i);
    φ_dup = duplicate_nodes((I*(o)| o)   (O*(i)| i));
    φ_conn = connect_side_inputs(SI((I*(o)| o)   (O*(i)| i)));
    foreach (tI,tO)L {(r,r),(r,f)}
      φ_cons = generate_constraints((tI,tO));
      φ = φ_in | φ_dup | φ_conn | φ_cons;
      solve(φ);
end
```

Figure 3. Overall flow of functional delay fault test generation

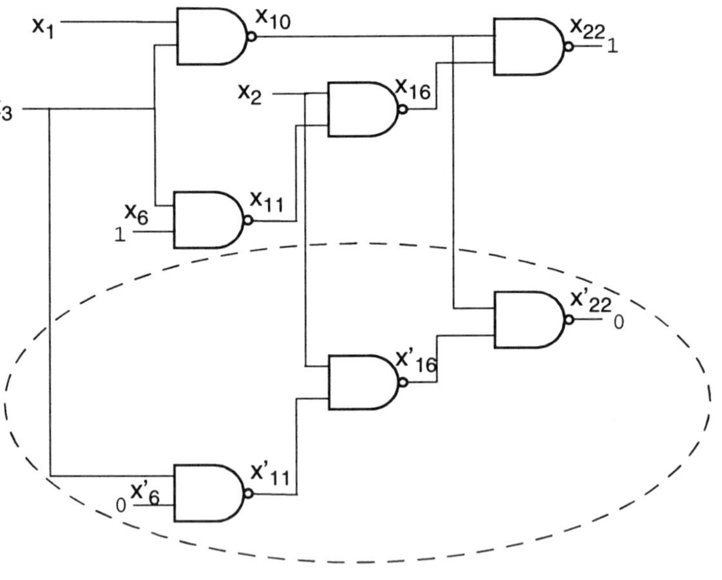

Figure 4. Target circuit for a fault $(x_6, x_{22}, \text{rising}, \text{rising})$

4. EXPERIMENTAL RESULTS

The functional delay fault test pattern generator described in this paper was implemented in C++ and integrated with the GRASP SAT solver [14]. The experiment was run on a PC running Linux with a Pentium II 300 MHz CPU and 256 Mbyte of memory. The results for ISCAS 85 circuits are presented in Table 1.

The first two columns list the name and number of faults of each circuit. From column 3 to column 5, D denotes detectable faults, R redundant faults and A aborted faults. %D in column 6 denotes the percentage of detectable faults with respect to the total number of faults. Columns 7 and 8 give the average size of the generated CNF formulas for each circuit. Although the complexity of SAT problems is not necessarily proportional to the size of the target SAT formula, it is observed that large CNF formulas usually take longer to solve. Column 7 gives average number of variables and column 8 gives the average number of clauses in each formula. Finally column 9 shows the run time in seconds. The GRASP SAT solver is called with the "+dDLCS" switch, one of the decision making heuristics that turned out to be most effective for this application. The time limit for each fault was set to 100 seconds.

Functional delay test pattern can be generated for all the circuits in Table 1 with no aborted faults except for C6288. For this circuit, we applied an approximate method introduced in [13] (see Table 2). With this approximation method, a certain percentage of primary inputs are assigned random fixed values (0 or 1) before the target SAT instance is solved. The first column in Table 2 denotes the percentage of primary inputs that are assigned fixed values. This method can cause a detectable fault to be declared as redundant, but will not make a redundant fault detectable; it provides a lower bound on the number of detectable faults.

The same experiment was performed with the combinational portions of the ISCAS 89 benchmark circuits. The results are reported in Table 3. None of the faults were aborted in this case, with very reasonable run times.

Table 1. Results for ISCAS 85 circuits

Circuit	# faults	#D	#R	#A	%D	Avg # vars	Avg # cls	Time(s)
C432	794	538	256	0	67.8	249.0	667.6	84.1
C499	5248	5184	64	0	98.8	190.5	518.7	525.8
C880	1004	1004	0	0	100.0	162.2	385.4	79.8
C1355	5248	5184	64	0	98.8	442.6	1210.3	2628.4
C1908	1870	1758	112	0	94.0	619.4	1522.4	322.7
C2670	2440	2160	280	0	88.5	456.2	1125.5	487.8
C3540	2284	2242	42	0	98.2	1033.4	2817.4	5308.2
C5315	7440	7328	112	0	98.5	368.5	906.3	901.3
C6288	3036	2862	44	130	94.3	2299.3	6826.7	30490.0

Circuit	# faults	#D	#R	#A	%D	Avg # vars	Avg # cls	Time(s)
C7552	7144	6228	916	0	87.2	581.0	1374.5	1215.3

Table 2. Results for C6288 using approximation

Approximation factor	#D	#R	#A	%D	time (s)
10%	2950	46	40	97.2	35950.9
25%	2972	54	10	97.9	32456.7
50%	2972	62	2	97.9	31162.7
75%	2964	72	0	97.6	30876.7

Table 3. Results for ISCAS 89 circuits

Circuit	# faults	#D	#R	#A	%D	Avg # vars	Avg # cls	Time(s)
s208.1	152	152	0	0	100.0	48.8	108.2	6.8
s298	230	218	12	0	94.8	39.3	96.7	9.9
s344	282	278	4	0	98.6	47.4	108.2	12.8
s349	284	280	4	0	98.6	47.6	108.8	12.3
s382	408	400	8	0	98.0	41.0	91.4	17.5
s386	308	308	0	0	100.0	49.7	111.7	13.5
s400	388	380	8	0	97.9	41.3	93.1	16.8
s420	390	390	0	0	100.0	72.5	157.0	19.0
s420.1	424	424	0	0	100.0	72.3	156.8	20.9
s444	544	528	16	0	97.1	46.5	105.9	24.0
s510	310	310	0	0	100.0	77.1	189.8	16.0
s526	492	480	12	0	97.6	47.4	113.6	21.8
s526n	492	480	12	0	97.6	47.4	113.6	22.0
s641	1064	1010	54	0	94.9	160.1	349.6	71.4
s713	1148	994	154	0	86.6	165.3	370.4	78.5
s820	514	514	0	0	100.0	64.3	165.3	25.8
s832	514	514	0	0	100.0	64.4	167.8	25.7
s838	790	790	0	0	100.0	112.4	242.7	45.3
s838.1	1352	1352	0	0	100.0	107.2	222.8	75.0
s953	934	934	0	0	100.0	99.1	232.7	52.3
s1196	1208	1158	50	0	95.9	183.3	469.8	94.5
s1238	1224	1160	64	0	94.8	188.2	497.8	98.4
s1423	5836	5458	378	0	93.5	209.7	477.0	469.4
s1488	794	792	2	0	99.7	95.9	240.5	44.6
s1494	794	792	2	0	99.7	95.7	241.3	45.1
s5378	5760	5404	356	0	93.8	197.1	436.4	442.6
s9234.1	9474	8070	1404	0	85.2	327.0	738.7	999.9
s13207.1	11448	10286	1162	0	89.8	445.8	1135.4	1676.9
s15850.1	57160	44326	12834	0	77.5	973.6	2369.7	15440.1
s38417	87394	83236	4158	0	95.2	403.4	922.6	11117.4
s38584.1	47272	41420	5852	0	87.6	194.1	449.5	3697.8
s35932	22274	20674	1600	0	92.8	92.5	223.6	1228.4

5. CONCLUSIONS

In this paper, a satisfiability-based functional delay fault test generation method which uses the circuit consistency function in generating the target CNF formula was presented. Promising results for ISCAS benchmark circuits are reported.

[12] suggests generating more than one test per each fault in the fault list in order to achieve high path delay fault coverage. Incremental Satisfiability (ISAT) has been proven successful in solving multiple instances of closely related SAT problems [8]. We are currently looking at using ISAT to generate multiple tests per each functional delay fault.

6. REFERENCES

[1] R. Bayardo Jr. and R. Schrag, "Using CSP Look-Back Techniques to Solve Real-World SAT Instances," *National Conference on Artificial Intelligence*, pp. 203-208, 1997.

[2] F. Brglez and H. Fujiwara, "A Neutral List of 10 Combinational Benchmark Circuits and a Target Translator in FORTRAN," *International Symposium on Circuits and Systems*, 1985.

[3] J. L. Carter, V. S. Iyengar and B. K. Rosen, "Efficient Test Coverage Determination for Delay Faults," *International Test Conference*, pp. 418-427, September, 1987.

[4] C. -A. Chen and S. K. Gupta, "A Satisfiability-Based Test Generator for Path Delay Faults in Combinational Circuits," *Design Automation Conference*, pp. 209-214, 1996.

[5] K. -T. Cheng and H. -C. Chen, "Delay Testing For Non-Robust Untestable Circuits," *International Test Conference*, pp. 954-961, 1993.

[6] K. Fuchs, F. Fink and M. H. Schulz, "DYNAMITE: An Efficient Automatic Test Pattern Generation System for Path Delay Faults," *IEEE Trans. on Computer-Aided Design*, vol. 10, pp. 1323-1335, October 1991.

[7] W. Ke and P. R. Menon, "Synthesis of Delay-Verifiable Combinational Circuits," *IEEE Trans. on Computers*, vol. 44, pp. 213-222, February 1995.

[8] J. Kim, J. Whittemore, J. P. M. Silva and K. A. Sakallah, "Incremental Satisfiability and its Application to Delay Fault Testing," *International Workshop on Logic Synthesis*, 1999.

[9] T. Larrabee, "Test Pattern Generation Using Boolean Satisfiability," *IEEE Trans. on Computer-Aided Design*, vol. 11, pp. 4-15, January 1992.

[10] C. J. Lin and S. M. Reddy, "On Delay Fault Testing in Logic Circuits," *IEEE Trans. on Computer- Aided Design*, vol. 6, pp. 694-703, 1987.

[11] P. C. McGeer, A. Saldanha, P. R. Stephan, R. K. Brayton and A. L. Sangiovanni-Vincentelli, "Timing Analysis and Delay-Fault Test Generation Using Path-Recursive Functions," *International Conference on Computer Aided Design*, pp. 180-183, November, 1991.

[12] I. Pomeranz and S. M. Reddy, "Functional Test Generation for Delay Faults in Combinational Circuits," *International Conference on Computer Aided Design*, pp. 687-694, 1995.

[13] S. Tragoudas and M. Michael, "ATPG tools for Delay Faults at the Functional Level," *Design, Automation and Test in Europe*, 1999.

[14] J. P. M. Silva and K. A. Sakallah, "GRASP- A New Algorithm for Satisfiability," *International Conference on Computer Aided Design*, pp. 220-227, 1996.
[15] J. P. M. Silva and K. A. Sakallah, "Robust Search Algorithms for Test Pattern Generation," *Fault Tolerant Computing Symposium*, 1997.
[16] G. L. Smith, "Model for Delay Faults Based upon Paths," *International Test Conference*, pp. 342- 349, 1985.
[17] B. Underwood, W. -O. Law, S. Kang and H. Konuk, "Fastpath: A Path-Delay Test Generator for Standard Scan Designs," *International Test Conference*, pp. 154-163, 1994.
[18] H. Zhang, "SATO: An Efficient Propositional Prover," *International Conference on Automatic Deduction*, pp. 272-275, 1997.

Verification of Abstracted Instruction Cache of Titac2: A Case Study

Tomohiro Yoneda
Department of Computer Science
Tokyo Institute of Technology
yoneda@cs.titech.ac.jp

Abstract In this paper, we demonstrate the formal verification of a practical timed asynchronous circuit. The target circuit is obtained by abstracting the instruction cache subsystem of a real asynchronous processor, TITAC 2. We also show several techniques to improve our verification method. The improved verifier could verify the target circuit in approximately 15 minutes, using less than 20 MBytes of memory.

Keywords: Formal verification, timed asynchronous circuits, partial order reduction, time Petri nets, instruction cache.

1. INTRODUCTION

In order to avoid the various difficulties which arise in designing large synchronous circuits, such as clock skews, high power consumption, and so on, designing asynchronous circuits without any clock systems has been attracting notice. In fact, several research groups demonstrated that an entire microprocessor could be designed and fabricated in this manner[Martin et al., 1989; Furber et al., 1994; Furber et al., 1996; Nanya et al., 1994; Takamura et al., 1997]. One significant problem that asynchronous circuit designers face is a lack of CAD systems. From a verification point of view, the cost of verifying asynchronous circuits is considerably higher than that for synchronous circuits because each wire of asynchronous circuits has states, and as a result, the state spaces of asynchronous circuits are often very large. Furthermore, in recent asynchronous circuit design, designers have preferred to use timed circuits for implementing fast and compact circuits. This also makes verification difficult, and prevents us from applying the latest verification techniques for untimed systems, such as symbolic model checking[Burch et al., 1992] or partial order reduction[Valmari, 1990; Katz and Peled, 1990; Godefroid, 1990]. Thus, our recent interest has been in developing efficient verification tools for timed asynchronous circuits. VINAS-P (VerIfier based on time petri Nets for timed Asynchronous Systems using Partial order reduction) is our newest formal verification tool for timed asynchronous circuits using techniques proposed in [Yoneda and Ryu, 1999]. The main idea in these techniques is partial order reduction based on the timed version[Yoneda and Schlingloff, 1997] of the Stubborn set method[Valmari, 1990]. The most closely related work is probably the ATACS which was developed at the University of Utah[Belluomini and Myers, 1998], however, these two tools employ different treatments for the trade-off between efficiency and expressibility in specifications.

The purpose of this paper is to demonstrate the verification of a practical-sized timed asynchronous circuit using VINAS-P. The target circuit is obtained from TITAC 2[Takamura et al., 1997]. TITAC 2 is a real 32-bit fully asynchronous processor which was developed at the University of Tokyo and the Tokyo Institute of Technology in 1997. It accepts almost all MIPS R2000 instructions, and contains half a million CMOS transistors. It was designed under the scalable-delay-insensitive (SDI) model, where its verification problem can be reduced to that of bounded delay circuits. Thus, it contains numerous subcircuits which are suitable as benchmark circuits for timed asynchronous circuit verification. We focus on the instruction cache subsystem of TITAC 2 because it is one of the most complicated subsystems in TITAC 2, and verifying it formally is a challenging task. However, for the formal verification, 32-bit data/address buses are too large to be handled. Therefore, we need to obtain an abstracted and simplified version of the subsystem in which many interesting properties can still be verified. This abstracted circuit contains approximately 200 gates, and its time Petri net model includes 1527 places and 1697 transitions. The original VINAS-P could handle this circuit, but it was very slow. This was mainly due to the on-line analysis needed for the partial order reduction. This paper proposes some techniques to improve the performance of VINAS-P. The improved VINAS-P could finally verify the target circuit in about 15 minutes, using less than 20 MBytes of memory.

The rest of this paper is organized as follows. Section 2 shows the overview of the TITAC 2 instruction cache subsystem, the abstracted circuit, and its specification. In Section 3, after briefly explaining the verification method of VINAS-P, we propose several techniques for the improvement of VINAS-P, and show some experimental results. Finally, we summarize the discussion.

2. TITAC 2 INSTRUCTION CACHE

2.1 OVERVIEW

Figure 1 shows the block diagram of the TITAC 2 instruction cache subsystem.[1] The cache memory contains 256 line (or block) frames, and each line frame contains eight words. Thus, the size of the cache memory is 8KBytes (256 line frames $\times$ 8 words $\times$ 32 bits). The lines are direct-mapped and fetched in the early restart manner with the critical word first[Hennessy and Patterson, 1996]. Thus, when a word with address *adr* is read and it is not in the cache memory, the line containing the word is fetched in the order $[adr], [adr+1], \cdots, [adr+n], [adr-m], \cdots, [adr-1]$, where $m = adr \bmod 8$ and $n = 7 - m$. Furthermore, access for other words within the same line is responded to as soon as the words are fetched, while all access for words not in the line is suspended until the line fetch is completed, even if the access is on a hit (see Figure 2).

A brief explanation of the operation of the instruction cache subsystem is as follows:

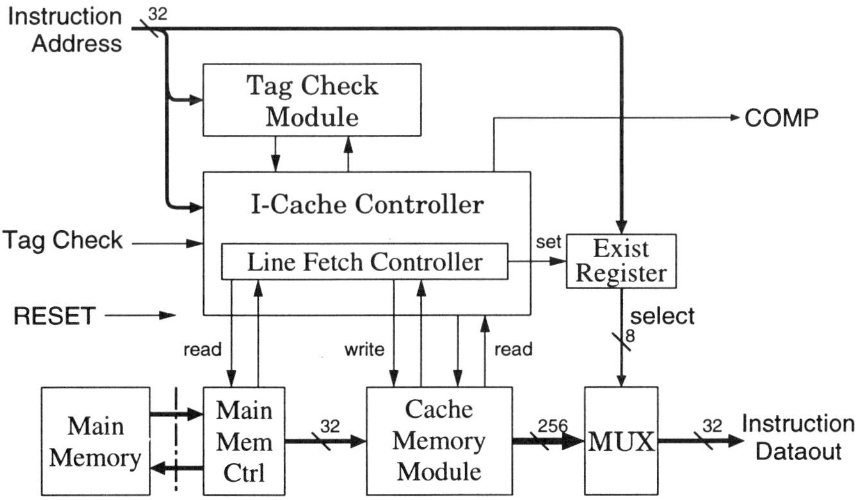

Figure 1 Block diagram of the TITAC 2 instruction cache subsystem.

[1]All information regarding the original circuit of the TITAC 2 instruction cache subsystem and several figures are from the Bachelor's thesis of Makoto Ishikawa[Ishikawa, 1997].

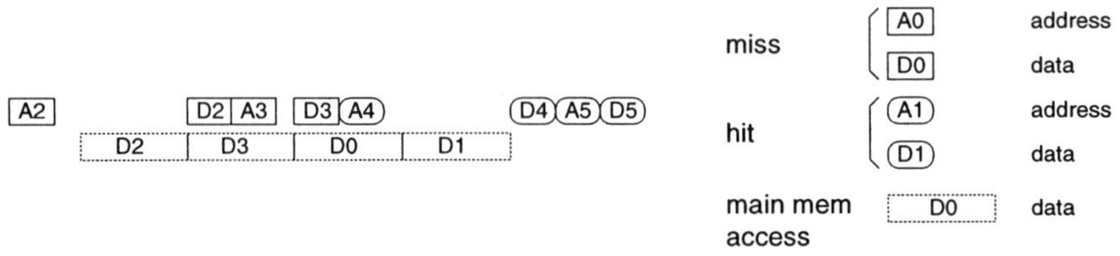

Figure 2 Early restart with critical word first (line size = 4).

1. Instruction Address *adr* is given, and Tag Check signal is activated.
2. If a line which does not contain *adr* is being fetched, access is suspended until the line fetch is completed.
3. I-Cache Controller starts both the tag check operation with *adr*[31:13] and the cache memory read operation with *adr*[12:5]. The Cache Memory Module consists of eight banks. Thus, all words in the specified line frame will be read simultaneously.
4. On a hit, when the cache memory read operation is completed, the corresponding word is selected by MUX according to *adr*[4:2].
5. On a miss, Line Fetch Controller starts the line fetch. When the first word is read from Main Memory, and is written into Cache Memory Module, the word is selected by MUX. Line Fetch Controller also sets the corresponding bit in Exist Register, which indicates the available word in the line currently being fetched. When other words within this line are read, they will be selected or the read operation is suspended according to the corresponding bits in the Exist Register.

The address bus and data bus are two-rail coded, that is, each bit is represented by two wires such that (01) corresponds to 0 and (10) to 1.

The completion of the instruction cache read operation is indicated by both COMP and Instruction Dataout. The operation is completed only when both, COMP is 1 and Instruction Dataout is a code word of two-rail code (i.e., every pair of wires in Instruction Dataout is (10) or (01)). Note that it is not known which of these two events occurs first.

After the completion of every read operation, a resetting phase is necessary. This is started by setting 0 to each bit of Instruction Address and Tag Check, and is completed when COMP and each bit of Instruction Dataout become 0.

2.2 ABSTRACTED CIRCUIT

We aim to verify that the TITAC 2 instruction cache subsystem works correctly. However, it is difficult to obtain an abstracted model for such a general property. Therefore, we focus on the LSB of the instruction and the main memory location 0 and 1. That is, in this case study, we verify the following property of the instruction cache subsystem :

> The LSB of the instruction read from the cache subsystem is the one stored in the location of the main memory with the given address 0 or 1 independent of the result (hit or miss) of the tag check operation.

Although this property is rather restricted, we believe that it verifies the control circuit of the TITAC 2 instruction cache subsystem almost completely, and that its data path circuit is also fairly well verified. For this property, we can easily obtain an abstracted instruction cache subsystem, denoted by AIC, which has a 1-bit address bus and a 1-bit data bus.

Furthermore, in order to check all possible sequences of misses and hits, we separate the tag check module from AIC, and give new external lines HIT and MISS to AIC, where those inputs are controlled by the specification. This means that the actual tag check module will not be verified due to this simplification.

The gate level circuit of AIC is shown in our technical report [Yoneda, 1999]. In AIC, every gate delay is assumed as [4,5], and every delay element has [100,100] delay. The text files which contain Verilog-like descriptions for this circuit as well as the above document can be obtained from http://yoneda–www.cs.titech.ac.jp/~yoneda/pub.html.

2.3 SPECIFICATION

Since VINAS-P is based on the trace theoretic verification[Dill, 1988], the specification needs to express the expected input and output relation. The following is an outline of our specification for the above property.

RESET is kept at 1 for a sufficient time period, and is then set to 0. Either (01) or (10) is given to Instruction Address. Tag Check is then activated. Either hit or miss operation is selected. On a hit, HIT is activated. On a miss, the main memory mode is selected. Table 1 shows the relation between the main memory mode and "address"/"data value". Then, MISS is activated. If the main memory address changes, the corresponding data is set according to the main memory mode. At this state, either COMP or Instruction Dataout can change. When COMP changes to 1, Instruction Dataout must be either

(00) or the code word (i.e., (01) or (10)) which is correct with respect to Instruction Address and the main memory mode. In the former case, the latter must follow eventually. If Instruction Dataout becomes an incorrect code word, it will be detected as a failure state as discussed in the next section. Once Instruction Dataout becomes a code word, it must not change until Tag Check becomes 0. Instruction Dataout can change in any way (except (11)) as long as COMP = 0. After COMP becomes 1 and Instruction Dataout becomes a code word, Tag Check is set to 0 and Instruction Address is set to (00). Then, either MISS or HIT is set to 0. If COMP becomes 0, then go back to the point where Instruction Address is set. Note that the choices for the instruction address, hit/miss operation, and the main memory mode are nondeterministic.

The formal expression of this specification is also shown in [Yoneda, 1999].

Table 1 Main memory mode.

main memory mode = "normal"		main memory mode = "opposite"	
address	data value	address	data value
0	0	0	1
1	1	1	0

3. VERIFICATION

3.1 METHOD

In VINAS-P, the gates in a circuit are translated by using a gate library into time Petri net modules which model the behavior of the gates, and a specification is expressed by a time Petri net module. A time Petri net[Merlin and Faber, 1976] consists of transitions (thick bars), places (circles), and arcs between transitions and places. A token (large dot) can occupy a place, and when every input place of a transition is occupied, the transition becomes enabled. Each transition has two times, the earliest firing time and the latest firing time. An enabled transition becomes ready to fire when it has been continuously enabled for its earliest firing time, and cannot be continuously enabled for more than the latest firing time, i.e., it must fire unless it is disabled. The firing of a transition consumes tokens in its input places and produces tokens in its output places. If transitions have one or more common input places, then we say that those transitions are in conflict. Usually, the firing of such one transition disables the remaining conflicting transitions.

Verification is performed by traversing the state spaces of the set of time Petri net modules with simultaneously firing every transition associated with the same wire. The circuit is considered to be correct with respect to the specification, if no failure state is reached in the state-space enumeration process. A failure state is a state where a module wants to change an output wire but the corresponding input wire is not ready to change in some module. More precisely, if a transition associated with an output wire can fire without an advance of time, but there exists a module in which no transitions associated with the corresponding input wire are enabled, then it is a failure state. Typically, a failure state is reached when a circuit produces a bad output which the specification does not expect. In this case, the corresponding input transition is not enabled in the specification.

The current version of VINAS-P only supports the checking of this kind of safety properties and simple deadlock checking. Although checking other properties, such as liveness, may occasionally be needed, what VINAS-P can verify covers many interesting and important properties. Other restrictions of the current version are that only $[0, \infty]$ bounds are allowed for transitions associated with input wires, and that multiple transitions associated with the same output wire cannot exist in a module. These restrictions are important for reducing the computational cost of the algorithm.

The idea of the partial order reduction is to prune some successor states in the state-space traversal as long as the correctness of the verification results is not affected. For example, when two transitions are ready to fire, as shown in Figure 3(a), generating (untimed) states such as $\{p_1, p_4\}$, $\{p_2, p_3\}$, and $\{p_3, p_4\}$ are often too much for checking the reachability of failure states. In such cases, one firing sequence generating $\{p_1, p_4\}$ and $\{p_3, p_4\}$ is sufficient. Even if a failure state is generated

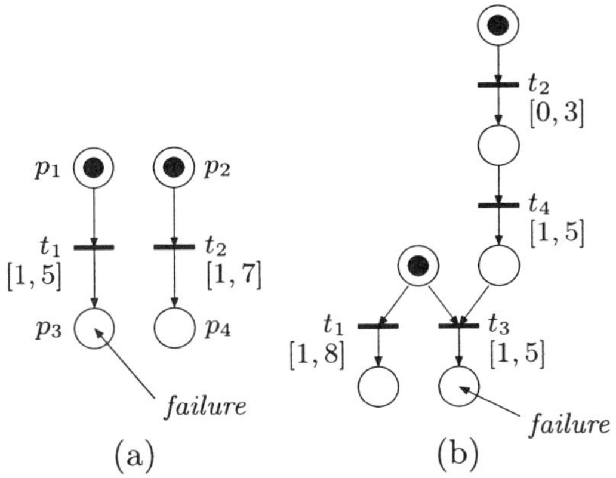

Figure 3 Concurrent and conflicting transitions.

by the firing of t_1, it is reached anyway by the above firing sequence. On the other hand, in the time Petri net shown in Figure 3(b), the firing sequence starting from t_1 may miss a failure state caused by the firing of t_3, because the firing of t_1 eliminates the possibility of the firing of t_3. Actually, the firing sequence starting from t_2 can make t_3 ready to fire if the firing of t_1 is postponed. Therefore, we cannot prune the successor state by t_2 in this case. In order to handle general cases, if we want to fire a transition t at a state s, we compute $\mathsf{dependent}(s, t)$ which is a set of enabled output transitions such that the interleavings of the firings of those transitions should be generated for the correct results (for example, $\mathsf{dependent}(s, t_1) = \{t_1\}$ for Figure 1.3(a), and $\mathsf{dependent}(s, t_1) = \{t_1, t_2\}$ for Figure 3(b)). In VINAS-P, the computation of $\mathsf{dependent}(s, t)$ is implemented such that the transitions which enable the transitions in conflict with t (e.g., t_3 in Figure 3(b)) in the future are searched backwards until some enabled output transition (e.g., t_2) is found.[2] If the given time Petri nets are large and contain a lots of conflicts, the cost to compute the dependent sets becomes very high. The verification algorithm of VINAS-P is formally described in [Yoneda and Ryu, 1999].

3.2 IMPROVEMENT

For experimental reasons, we prepare three variants of the specification of AIC, denoted by *spec*1, *spec*2, *spec*3, where

- *spec*1: Instruction Address = (01), MISS = 1 and the main memory mode "normal" are always chosen.
- *spec*2: MISS = 1 and the main memory mode "normal" are always chosen.
- *spec*3: the main memory mode "normal" is always chosen.

For example, in *spec*2, the access to the instruction address 0 and 1 are verified, but only the miss case with the main memory mode "normal" is examined. Let *spec*4 denote the complete specification mentioned in Section 2.3. The costs of verification for the four different specifications vary significantly, even if the same circuit is used. Let AIC1, AIC2, AIC3, and AIC4 denote the AICs with *spec*1, *spec*2, *spec*3, and *spec*4, respectively. In this subsection, we propose several techniques to improve VINAS-P, and evaluate them by using these different verification examples. All measurement in this paper was carried out on a UNIX workstation (Pentium II 450MHz, 512MB main memory).

[2] t_4 is not included because it is not enabled.

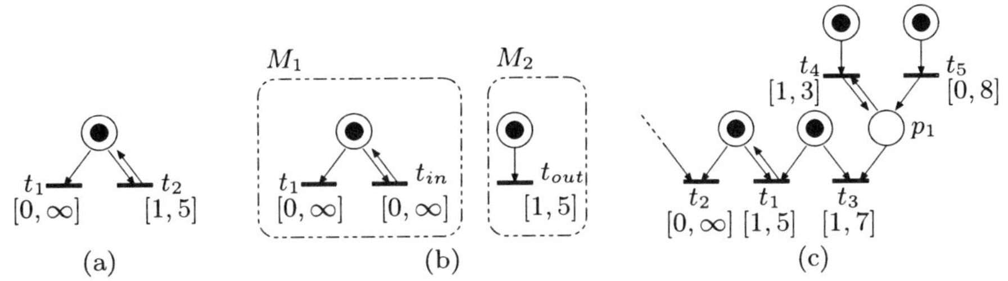

Figure 4 Ineffective conflicts.

In Figure 4(a), transitions t_1 and t_2 are in conflict. In fact, if t_1 fires, then t_2 can no longer fire. However, even if t_2 fires, t_1 can still fire. Actually, if the earliest and latest firing times of t_1 are $[0, 0]$ or $[0, \infty]$, the firing of t_2 does not influence the firing of t_1.[3] Thus, the first technique for the improvement is to ignore such ineffective conflicting relations during the dependent set computation. There is, however, one exceptional case. In the case shown in Figure 4(b), suppose that t_{in} is associated with an input wire, and that t_{out} is associated with the corresponding output wire. Then, the firing of t_1 leads to a failure state, because in the state obtained by firing t_1, an output can change but the corresponding input is not ready to change (i.e., t_{out} can fire, but t_{in} is not enabled). On the other hand, the firing of t_{out} eliminates the chance to reach the above failure state, because t_{out} is no longer enabled after firing. Hence, $\mathsf{dependent}(s, t_{out})$ must include t_1 in order to correctly detect failure states. The current algorithm of VINAS-P achieves this by using the conflicting relation between t_{in} and t_1 (see [Yoneda and Ryu, 1999] for details). For this reason, conflicting relations involving transitions which are associated with the input wires must not be ignored, even if the firings of those transitions never affect the firings of other transitions. For example, in Figure 4(c), if t_1 is associated with an output wire, $\mathsf{dependent}(s, t_1)$ searches the transitions backwards from t_3 but not from t_2, while the backward searches from both are necessary in the case where t_1 is associated with an input wire.

Similarly, transitions, such as t_4 in the above example, play no role in enabling t_3 in the future, because t_4 cannot generate a new token into p_1. Thus, we also ignore these transitions in the backward search process.

Secondly, because the same transitions are reached many times via different paths during the backward search process, we introduced a

[3] If t_1 has other firing times, then the firing of t_2 resets the clock attached to t_1 and the firing time of t_1 is affected by the firing of t_2.

Table 2 Comparison of performance of improved VINAS-P.

		AIC1	AIC2
original	(sec)	1190	56494
technique 1	(sec)	145	3715
technique 2	(sec)	99	1626
both techniques	(sec)	20	214

caching mechanism for the dependent set computation. We will cache the results of backward searches at a fixed state, i.e., the cache is flushed when a new state is reached, and we do not include states s in the keys for the cache. This is because numerous backward searches are activated in each state, and it is not reasonable to prepare such a large cache area that the cached results are used when the same states (or markings) are revisited. Thus, in each cache entry, we keep a transition and the result of the backward search from the transition. It is expected that many re-backward searches from the same transition will be omitted using this caching mechanism.

Table 2 summarizes how the CPU times are reduced using the above techniques. The row for "technique n" shows the performance values when only "technique n" is applied where $n = 1, 2$. Since these techniques are independent, they can be applied simultaneously. The row for "both techniques" shows the performance values when both techniques are applied. Because combining these techniques reduces the cost of the dependent set computation significantly, AIC3 and AIC4 can also be

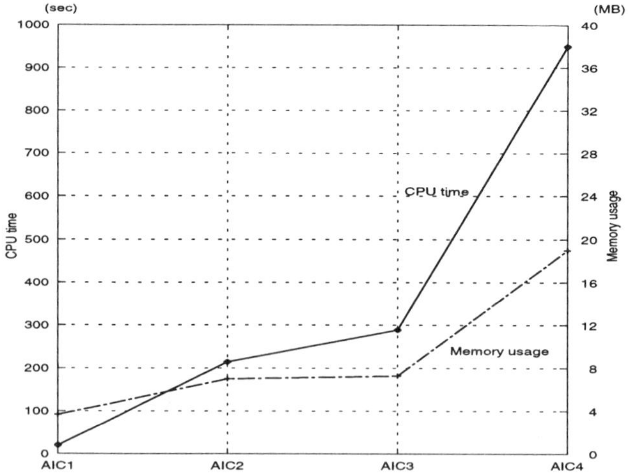

Figure 5 Performance of improved VINAS-P.

verified rather easily. Figure 5 shows the CPU times for the verification of AIC1 through AIC4 using the final version of VINAS-P. It also shows the memory usage required for each verification. The number of generated states for AIC4 was 25976.

Since TITAC 2 was designed under the SDI model, it may not work correctly in cases where some components have delays which are too large. In order to demonstrate the timing aspect of the verification, we modified the delay of one C-element in AIC4 and set it to [500,500]. Then, VINAS-P found a failure state after generating 6381 states in 196 seconds.

4. CONCLUSION

This paper demonstrated the formal verification of a practical timed asynchronous circuit. The target circuit was obtained by abstracting and simplifying the instruction cache subsystem of a real asynchronous processor, TITAC 2. We could verify an interesting property of the abstracted model. We believe that this is one of the largest benchmark examples for the formal verification of timed asynchronous circuits.

Furthermore, in order to improve our verification tool VINAS-P, we implemented several techniques, and the improved VINAS-P could efficiently verify the target circuit.

On the other hand, we faced some difficulties in dealing with large specification time Petri nets. We need to develop some formal specification language which allows us to easily create and modify large specifications.

5. ACKNOWLEDGMENTS

We would like to thank Professor Takashi Nanya, Dr. Masashi Kuwako, Rafael Morizawa, and Hiroshi Ryu for valuable discussions.

6. REFERENCES

Belluomini, W. and Myers, C. (1998). Verification of timed systems using POSETs. *LNCS 1427 Computer Aided Verification*, pages 403–415.

Burch, J. R., Clarke, E. M., Dill, D. L., Hwang, L. J., and McMillan, K. L. (1992). Symbolic model checking: 10^{20} states and beyond. *Academic Press*, 98(2):142–170.

Dill, D. L. (1988). *Trace Theory for Automatic Hierarchical Verification of Speed-Independent Circuits.* MIT press.

Furber, S. B., Day, P., Garside, J. D., Paver, N. C., and Temple, S. (1996). AMULET2e. In Muller-Schloer, C., Geerinckx, F., Stanford-

Smith, B., and van Riet, R., editors, *Embedded Microprocessor Systems*. Proceedings of EMSYS'96 - OMI Sixth Annual Conference.

Furber, S. B., Day, P., Garside, J. D., Paver, N. C., Temple, S., and Woods, J. V. (1994). The design and evaluation of an asynchronous microprocessor. In *Proc. International Conf. Computer Design (ICCD)*. IEEE Computer Society Press.

Godefroid, P. (1990). Using partial orders to improve automatic verification methods. *Proc. of Computer Aided Verification Workshop.*

Hennessy, J. and Patterson, D. (1996). *Computer Architecture: A Quantitative Approach, Second Edition.* Morgan Kaufmann Publishers.

Ishikawa, M. (1997). Study on instruction cache system for asynchronous processors. *Bachelor's thesis (in Japanese), Tokyo Institute of Technology.*

Katz, S. and Peled, D. (1990). Defining conditional independence using collapses. *Semantics for concurrency, BCS-FACS Workshop, M. Kwiatkowska (ed.), Springer.*

Martin, A. J., Burns, S. M., Lee, T. K., Borkovic, D., and Hazewindus, P. J. (1989). The first asynchronous microprocessor: the test results. *Computer Architecture News*, 17(4):95–110.

Merlin, P. and Faber, D. J. (1976). Recoverability of communication protocols. *IEEE Trans. on Communication*, COM-24(9).

Nanya, T., Ueno, Y., Kagotani, H., Kuwako, M., and Takamura, A. (1994). TITAC : Design of a quasi–delay–insensitive microprocessor. *IEEE Design & Test of Computers*, 11(2):50–63.

Takamura, A., Kuwako, M., Imai, M., Fujii, T., Ozawa, M., Fukasaku, I., Ueno, Y., and Nanya, T. (1997). TITAC-2: An asynchronous 32-bit microprocessor based on scalable-delay-insensitive model. In *Proc. International Conf. Computer Design (ICCD)*, pages 288–294.

Valmari, A. (1990). A stubborn attack on state explosion. *Proc. of Workshop on Computer Aided Verification.*

Yoneda, T. (1999). Abstracted instruction cache of TITAC2 — as a benchmark circuit for timed asynchronous circuit verification. *TIT Technical Report*, 99TR-0002.

Yoneda, T. and Ryu, H. (1999). Timed trace theoretic verification using partial order reduction. *Proc. of Fifth International Symposium on Advanced Research in Asynchronous Circuits and Systems*, pages 108–121.

Yoneda, T. and Schlingloff, H. (1997). Efficient verification of parallel real–time systems. *Formal Method in System Design*, pages 187–215.

Speeding up Look-up-Table Driven Logic Simulation

Rajeev Murgai and Masahiro Fujita
Fujitsu Laboratories of America, Inc.,
Sunnyvale, CA, USA
murgai@fla.fujitsu.com, fujita@fla.fujitsu.com

Fumiyasu Hirose
Cadence Design Systems, Inc.
Yokohama, Japan
hirose@cadence.com

Abstract Logic simulation is one of the most important steps in the design of a digital circuit. Due to the growing complexity of the designs, a large number of test vectors is needed, making simulation a big bottleneck. One way to speed up logic simulation is by designing distributed or parallel hardware architectures that are optimized for simulation. One such hardware accelerator is Fujitsu's TP5000, which was shown to be over 300 times faster than a state-of-the-art software simulator on large circuits. TP5000 uses a memory-based, event-driven simulator [9]. In this paper, we propose logic restructuring techniques to further speed up functional simulation on TP5000. The key idea is to generate a perfectly balanced network logically equivalent to the original network that fits in the TP5000 memory. A perfectly balanced network gets rid of useless evaluations. We use logic decomposition, partial collapsing, and buffer insertion to generate such a network. Experimental results indicate that our techniques reduce the number of events in TP5000 by 33% as compared to a commonly-used technique that optimizes the network and does a straightforward mapping on the simulator, and by 22% as compared to the technique of [7]. On some benchmarks, the reduction is by a factor of 3.

Keywords: Logic simulation, events, logic synthesis.

1. INTRODUCTION

Logic simulation is one of the most important steps in the design of a digital circuit, and is used to verify the correctness of the designed circuit. Test inputs are applied to see if the circuit responds to these inputs as it was supposed to. For large circuits, typically millions of such inputs are applied. Simulation can be done at either behavioral, logic, or circuit level. Also, it can be performed either in **software** or **hardware**. In software simulation, models for the basic circuit components or **primitives** (such as inverter, two-input AND, two-input OR gates in logic level simulation) are used. An AND gate, for instance, is modeled either with its truth table or using the bit-wise AND instruction of the processor on which the software runs. The circuit is converted into these basic primitives by the simulator. These primitives are evaluated either in a statically determined, topological order starting from the inputs (**compiled code simulation**) or dynamically in an **event-driven** manner. Since these primitives are two-input gates, their number is very high. This, along with a large input-vector set and the software nature of testing, makes the simulation process rather slow. **Hardware accelerators** have been proposed to speed up the simulation process [1, 3]. The circuit to be simulated is mapped on to the **hardware primitives** of the accelerator that are designed specifically for simulation and have short evaluation times. In addition, accelerators such as Fujitsu's Thread Processor TP5000 [9] divide the algorithm into fragments that can store and access data independently, and can be pipelined, thus achieving a higher throughput.

In this paper, we focus on an event-driven, logic simulator implemented on TP5000. This simulator was shown to be 300 times faster than the state-of-the-art software simulator (Cadence's Leapfrog^{TM}) [9] on circuits having more than half a million gates. Our goal in this work is to further speed up the TP5000 simulator. The paper is organized as follows. Section 2 describes the working of the TP5000 simulator. The problem statement and related work are in Section 3. Section 4 establishes a connection between useless evaluations in the simulator and glitches in a circuit under unit delay model. This leads us to propose synthesis techniques for removing glitches, specifically for generating a perfectly balanced network under the TP5000 memory constraint; these are described in Section 5. In Section 6, we present experimental results that demonstrate effectiveness of the proposed techniques in reducing the number of events during simulation. The conclusions and possible extensions are presented in Section 7.

2. USING TP5000 FOR SIMULATION

Given a design (network, circuit) to be simulated on TP5000, the first step is to map it onto the logic primitives provided by TP5000. One such logic primitive is a truth table L, which is actually the memory associated with one of the TP5000 processors. L has an 18-bit wide address bus and hence a **memory capacity** $M = 2^{18}$. L can be partitioned arbitrarily into smaller chunks, each chunk implementing a logic function. For example, L can implement either one function of 18 inputs, or one function of 17 inputs and two functions of 16 inputs, etc. As illustrated in Figure 1 (a), the truth tables of two functions $f(x_1, x_2, x_3)$ and $g(y_1, y_2)$ are stored together in the memory. Given an input combination, the value of the function (f or g) can be read from the appropriate location of the memory. Mapping of a design on TP5000 involves generating a circuit, all of whose functions can be packed together in L. This typically entails some restructuring of the design.

After the design is mapped, TP5000 starts simulating it on a set of test vectors. The event-driven simulation algorithm implemented on top of TP5000 [9] evaluates one logic function at a time. Its operation is shown in Figure 1 (b). When a new test vector is applied, an **event** (a transition from 0 to 1 or from 1 to 0) is recorded at each primary input that changed its value from the last vector. The logic functions to which such primary inputs fan out are put into the

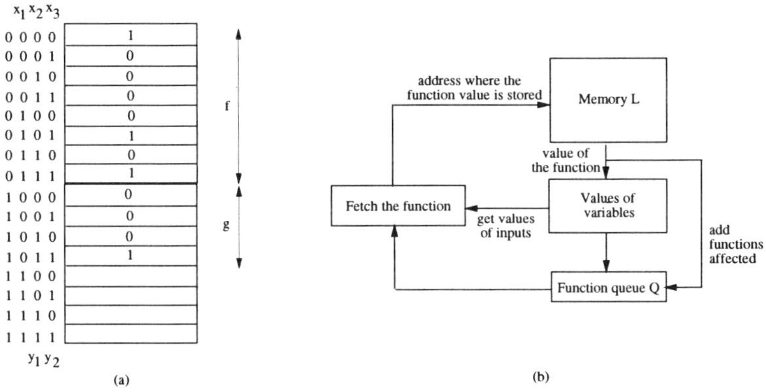

Figure 1 (a) Storing two functions f & g in an SRAM with $M = 16$; (b) Working principle of the event-driven simulator of TP5000

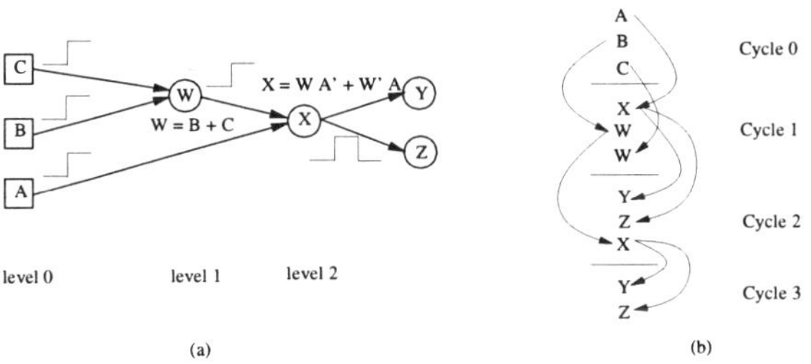

Figure 2 (a) Event-driven simulation of TP5000 in action; (b) Causality among evaluations

function queue Q for **evaluation**. *Evaluating a function f means determining the value of f, given the values of its fanins and its representation (i.e., the truth table).* The first function, say f, is fetched from the head of the queue Q for evaluation. The values of fanins of f determine the memory address that should be accessed for evaluating f. If the value stored at this location is different from the previous value of f, the value of f is updated and an event is recorded at f, in which case the fanouts of f are inserted at the end of Q for future evaluation. Evaluation of a circuit output does not cause further evaluations. This process continues until Q is empty. The following example illustrates the event-driven simulation of TP5000.

Example 1.1 *Figure 2 (a) shows a part of a network η with A, B, and C as primary inputs. The truth tables of W, X, Y, and Z are stored in the memory L of* TP5000. *Let $(A, B, C) = (0, 0, 0)$ initially, resulting in $W = X = 0$. Now, $(A, B, C) = (1, 1, 1)$ is applied. First, all the primary inputs A, B, and C are placed on the queue Q. A being at the head of Q is evaluated first. An event occurs since A goes from 0 to 1. Hence the fanout X of A is placed at the end of Q. Figure 2 (b) shows the causality between various events and evaluations (ignore the cycle numbers). For instance, the arrow from A to X means that an event took place on A, causing an evaluation of X. Now $Q = (B, C, X)$. Next B and then C are evaluated, causing W, the fanout of B and C, to be placed on Q, which becomes (X, W, W). Note that one input vector can cause a function (in this case W) to be evaluated more than once. Next X is fetched from Q. Its evaluation uses the updated value of A (i.e., 1) and $W = 0$. As a result, $X = WA' + W'A$ switches from 0 to 1. The new value of X is* *recorded and an event is generated at X. X's fanouts, Y and Z, are placed at the end of Q, which becomes (W, W, Y, Z). W is fetched next. Using $B = C = 1$, $W = B + C = 1$. This is an event at W, causing X to be put on Q. Now $Q = (W, Y, Z, X)$. The next evaluation of W does not result in any events, since W's fanins have not changed since W was evaluated last.*

As mentioned earlier, not all evaluations cause further evaluations. This happens when the value of a node does not change. For example, the second evaluation of W.
Two-valued vs. four-valued simulation: In a two-valued simulation, each signal can be either 0 or 1. Then an n-input function takes 2^n bits in the memory. TP5000 actually implements a four-valued simulation, where each signal can take values 0, 1, X, or Z. X means unknown and Z means high impedance. Two bits are needed to encode a signal. An n-input function then uses $2 \times 4^n = 2^{(2n+1)}$ bits of storage. The factor 2 refers to the two-bit output value, and 4^n is the total number of input combinations. In our experiments, the memory capacity $M = 2^{18}$. So no mapped logic function can have more than 8 inputs.

3. SYNTHESIS FOR SIMULATION

Assume we are given a network η that is to be simulated on TP5000 for functional correctness. Since we are interested only in functional correctness (and not in timing correctness or glitch analysis), we can restructure η (i.e., change its intermediate logic functions) as long as its global logic functionality does not change. Since different structures can result in different simulation times on a given set of simulation vectors, **our goal is to transform η into $\tilde{\eta}$ such that $\tilde{\eta}$ is logically equivalent to η, $\tilde{\eta}$ fits in L using the available memory M, and the total simulation time for $\tilde{\eta}$ on a given set of vectors is minimized.** One constraint was that we could not change the simulator. Examine the requirement of minimizing the total simulation time. The simulation time depends on the number of function evaluations and the time taken for each evaluation. Since a function evaluation involves a memory read, the time taken for an evaluation is essentially independent of the complexity of the function. So the requirement reduces to minimizing the number of evaluations. Since each function evaluation corresponds to an event at an input of the function, we wish to minimize the number of events in the network. It is not easy to model the number of events. Since simulation is serial in TP5000 (i.e., only one function is evaluated at a time), the following approximation was made in [7]: *model the total number of events by the number of functions (primitives) in the network mapped on L.* Thus, the goal was to minimize the number of functions in the network subject to the memory constraint M. Decomposition and partial collapsing techniques were investigated. The following example illustrates partial collapsing.

Example 1.2 *In Figure 2 (a), the total space used by W and X is 8 bits (assuming 2-valued simulation). Collapsing W into X, we obtain a new three-input function at X, which also uses 8 bits. This transformation does not increase the memory usage and also reduces the number of network primitives by one. So it will be applied in [7].*

In [7], it was observed empirically that a 2-input decomposition followed by partial collapsing targeted for minimum primitive-count subject to the memory constraint yielded the minimum number of primitives. This technique (which we shall call min-fn) reduced the number of primitives of many benchmarks by a factor of 2 to 4. Although the number of events also decreased significantly, it did not go down in the same proportion as the primitives. *In this paper, we will describe techniques that target minimizing the number of events directly instead of the number of primitives.*

To speed up cycle-based simulation, a decision-diagram based method was proposed by McGeer *et al.* [4]. Although their technique is not for event-driven simulation, our problem and

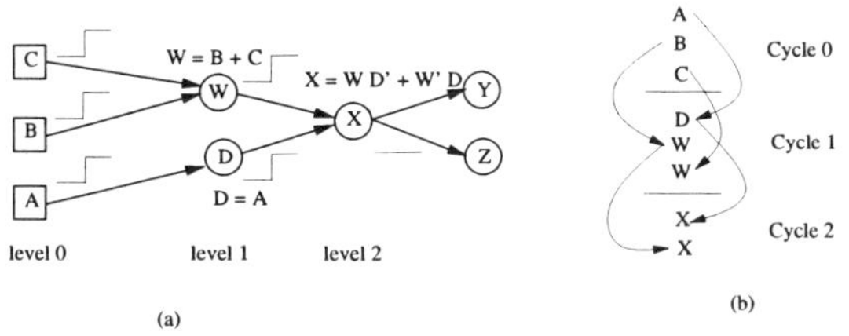

Figure 3 (a) Adding a buffer to reduce events, (b) Evaluations after buffer insertion

their problem are similar in essence. The solution techniques are, however, different. Whereas McGeer *et al.* use partitioned decision diagrams to represent logic functions and circuits, the architecture of TP5000 forces us to store logic functions as truth tables in the memory. Also, while their method is based on evaluating a node of the decision diagram, in our technique evaluating a complex logic function is an atomic operation – a memory read, which can lead to potential speed-ups in simulation. A direct comparison is, however, not possible, since the simulation targets are completely different.

4. USELESS SIMULATOR EVALUATIONS AND GLITCHES IN THE CIRCUIT

In Figure 2 (a), each node N is assigned a **level** $\ell(N)$ as follows. $\ell(N) = 0$ if N is a primary input. Otherwise, $\ell(N) = 1 + \max_{i \text{ fanin of } N}\{\ell(i)\}$. $\ell(A) = \ell(B) = \ell(C) = 0$ and $\ell(W) = 1, \ell(X) = 2$. Similarly, we can divide evaluations into **cycles**. All primary input evaluations happen in cycle 0. An evaluation directly caused by an evaluation (event) in cycle i is placed in cycle $(i + 1)$. In Figure 2 (b), the evaluations of A, B, and C are placed in cycle 0. The evaluation of A causes an evaluation of X, which is placed in cycle 1. Note that *the latest cycle in which a node can be evaluated is the same as the level of the node.*

In Figure 2 (b) X was evaluated twice, in cycles 1 and 2. The cycle 1 evaluation was caused by the event at A in cycle 0, and the cycle 2 evaluation by the event at W in cycle 1. Both evaluations resulted in events at X and caused the fanouts Y and Z to be evaluated, once in cycle 2 and then in cycle 3. The cycle 2 evaluations of Y and Z are **useless** for functional simulation, since Y and Z will be evaluated in cycle 3 in any case. If we could reduce the number of events at X from two to one, Y and Z will not be evaluated twice, thus reducing the total number of evaluations and the simulation time. This can be done, for instance, by *delaying the arrival of the signal A at the input of X by one cycle* by inserting a buffer D between A and X, as shown in Figure 3 (a). The resulting evaluations and events are shown in Figure 3 (b). The event at A reaches X in cycle 1. So X is evaluated only in cycle 2, although twice (which is same as before): once due to the event at D and then due to the event at W. However, when X is evaluated the first time in cycle 2, both D and W have already switched from 0 to 1. The new value of $X = WD' + W'D = 0$, the same as before. The second evaluation also does not change the value of X, since the fanins of X did not change values since the first evaluation. Thus, Y and Z are not evaluated at all, thus saving 4 evaluations. Recall that in Figure 2 (b), each of them was evaluated twice. However, we did incur an extra transition at the buffer D. Thus, the net reduction in the number of evaluations is 3.

In this analysis, we did not consider transitive fanouts of Y and Z. Had we, the savings in evaluations would have been potentially higher. Also note that even if X were not an XOR, at most one event could have occurred at X, still one less than Figure 2.

The above situation can be viewed in terms of **glitches** in the circuit. *Glitches are spurious transitions (events) at the outputs of gates before they settle down to their final values.* Since a glitch at a gate g can propagate to its transitive fanout gates, it causes useless evaluations of g and its transitive fanouts. If we can get rid of the glitches, we will get rid of most of the useless evaluations in an event-driven simulator, thus speeding up simulation. This is what we did with X, by removing the glitch in cycle 1. *We would like to synthesize a network that does not have glitches.*

One way to get rid of glitches is to obtain a **perfectly balanced network**. *A perfectly balanced network is one in which all the fanins of each node (or gate) are at the same level.* Thus all the paths to a node are of the same length (we call such a node a **perfectly balanced node**). This ensures that each node N is evaluated at most only in cycle $\ell(N)$, where $\ell(N)$ is the level of node N. Although N may be evaluated as many times as the number of fanins it has, only at most one evaluation can lead to an event. Next, we describe synthesis techniques for obtaining a perfectly balanced network.

5. SYNTHESIZING A PERFECTLY BALANCED NETWORK

Logic synthesis consists of two phases: **optimization** followed by **mapping**. In optimization, a minimal representation of the logic network is sought for area, delay, or power. In mapping, the optimized circuit is mapped on to a target technology (e.g., standard-cell library, FPGAs) minimizing some cost function. In this paper, we focus only on the mapping problem. In the mapping problem for the TP5000 processor, *the target technology is the processor memory and the cost function is the simulation time* (approximated by the number of evaluations or the number of events). Thus the goal is to convert a given network into a logically equivalent network that fits in the processor memory and has minimum number of evaluations or events. Intuitively, since perfectly balanced networks are free of useless events, we will restrict our problem to that of generating a perfectly balanced network under the memory constraint.

5.1 BUFFER INSERTION

Given an arbitrary network η, a simple and effective way to make it perfectly balanced is by inserting buffers appropriately at the primary inputs and gate outputs of η. The following example illustrates the idea.

Example 1.3 *Figure 4 (a) shows part of a network, with node N, at level 10, fanning out to nodes P and Q at levels 12 and 14 respectively. Then, add a series chain of 3 buffers at the output of N, where the final buffer feeds Q. This is shown in Figure 4 (b). Each buffer corresponds to one level. This ensures that if there is a path from a primary input to the output of Q that goes through N, it is of length 14. If each fanin of Q is buffered appropriately in the same manner, all paths to the output of Q will be of length 14. Then Q will be perfectly balanced. Note that the connection (N, P) is replaced by the connection from the output of the first buffer after N, thus saving buffers. The corresponding input of P arrives at time unit 11.*

The algorithm to make a network η perfectly balanced by buffering is straightforward and follows from Example 1.3. First levelize η, i.e., compute $\ell(N)$ for each node N of η. Traverse η such that a node is visited only after all its fanins. Let the current node be N. Let $FO(N)$ denote the set of fanouts of N. Compute $\ell_{max} = \max_{f \in FO(N)}\{\ell(f)\}$. Add $\ell_{max} - 1 - \ell(N)$ buffers at the output of N. This makes all the paths to the output of each fanout f of N that go through N of length $\ell(f)$. Add these buffers in a series structure to the output of N. For each

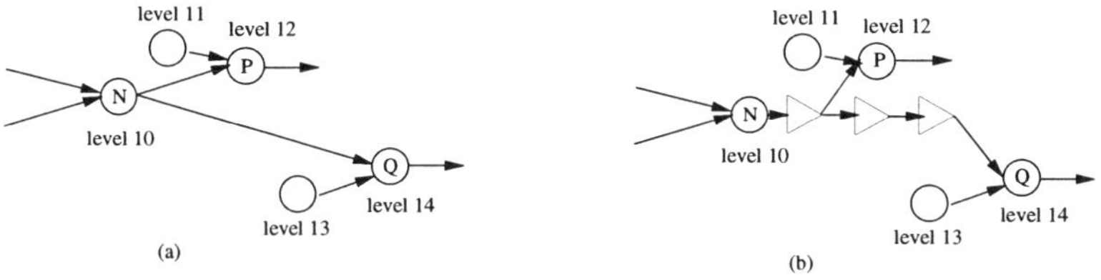

Figure 4 Adding buffers at N to balance path lengths for P and Q

$f \in FO(N)$, replace the (N, f) edge appropriately by an edge from a suitable buffer (as was illustrated in Example 1.3).

We must mention that buffering is a standard technique used in logic synthesis for reducing the delay due to fanout loading and for wave pipelining [11].

As we will see in Section 6, buffering is a powerful technique and works extremely well. However, each buffer comes with a cost, which is the extra transitions at its output. Therefore care must be taken before adding buffers, especially if the initial network is extremely unbalanced. In our approach, we use other synthesis techniques (such as decomposition and partial collapse) to balance the network as much as possible, and then use buffering as the last step to obtain a perfectly balanced network.

5.2 DECOMPOSITION

Decomposition is the process of expressing a given logic function in terms of simpler functions. If the goal is to have the fewest literals, a simple function should have fewer literals.

Example 1.4 *Let* $f = abc + abd + a'c'd' + b'c'd'$. *One way to decompose* f *is as follows.* $f = xy + x'y'$, $x = ab$, $y = c + d$. *An alternate way of decomposing* f *is:* $f = w + x + y + z$, $w = abc$, $x = abd$, $y = a'c'd'$, $z = b'c'd'$. *If fewest literals is the goal, clearly the first decomposition is preferable.*

We need decomposition in our application, because for instance, if there is a 9-input function f in the network, and the simulation is 4-valued, f will not fit in the memory of size $M = 2^{18}$. In our decomposition, we restrict each node function to have no more than k fanins, where k is a fixed constant. It was empirically found that $k = 6$ yields the best results (Section 6). If the resulting network is still too large to fit in the memory L, smaller values of k are used.

It follows then that in the context of TP5000, simplicity is determined by the total space the function takes up in the memory L. So one goal is to *decompose a function into sub-functions whose total space requirement is the minimum.* In Example 1.4, for a 2-valued simulation, the original f takes up 16 locations, the first decomposition (using XNOR) 4 + 4 + 4 = 12, and the second one 16 + 8 + 8 + 8 = 40 locations. Clearly, the first decomposition is the desired one for minimum memory.

We now come to the second goal: *the decomposition of a node should yield a perfectly balanced sub-network.* We first propose an algorithm that decomposes a multi-input AND node into a balanced network. It sorts the fanins of the node with the minimum-level fanin first. It *combines* inputs that have identical level until either k inputs have been combined or an input is reached whose level is strictly greater than the last input's. In either case, an intermediate node is added. In the second case, appropriate number of buffers may also be added to balance the fanin. We continue until all the fanins have been processed. The following example explains it.

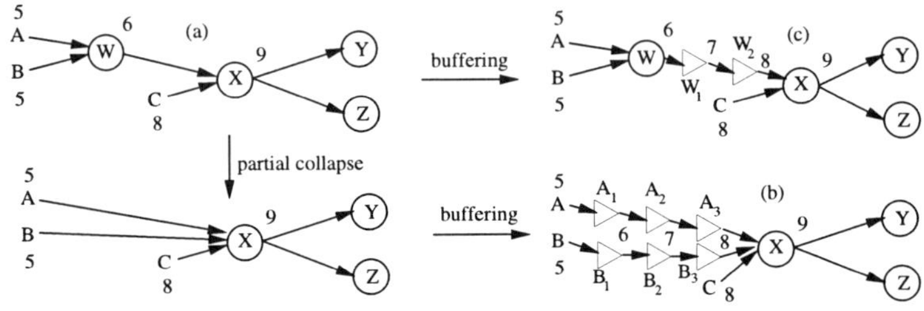

Figure 5 Collapsing vs. buffering: case 1

Example 1.5 *Let k be 4. Let $f = abcde$. Let $\ell(a) = \ell(b) = 1$, $\ell(c) = \ell(d) = \ell(e) = 2$. We first combine a and b (but not c, since $\ell(c) > \ell(a)$). This results in $x = ab$, $f = xcde$, with $\ell(x) = 2$. f need not be decomposed any further, since it satisfies the k constraint and the levels of all the fanins are identical. Both f and x are perfectly balanced and have at most k inputs.*

If $k = 3$ instead, still $x = ab$, $f = xcde$. However, since f has more than 3 fanins, it is decomposed further. Continuing the same procedure and combining inputs x, c, and d (all with level 2), we obtain node $y = xcd$, with $\ell(y) = 3$. Now $f = ye$. Node e has the minimum level. Since $\ell(y) \neq \ell(e)$, e is separated as $z = e$, and then $f = zy$. Note that z is a buffer, with $\ell(z) = 3 = \ell(y)$. Thus f is balanced. So the final decomposition is $x = ab$; $y = xcd$; $z = e$; $f = yz$. Each function has at most 3 inputs and is perfectly balanced.

Note that the procedure may use buffers ($z = e$ for $k = 3$) to ensure a perfectly balanced decomposition. However, it attempts to minimize the number of buffers. An OR node is decomposed similarly. An arbitrary node $f = c_1 + c_2 + \ldots c_p$ (where c_is are product terms) is first decomposed into p AND nodes, and an OR node at the top (f itself). The above procedure is first applied to the AND nodes one by one, the levels of the AND nodes are updated, and finally the procedure is applied to the root OR node. To apply this decomposition on a network, traverse the network topologically from primary inputs to outputs. This ensures that the levels of the fanins are correct when a node is visited and decomposed. This yields a perfectly balanced decomposition of the network.

5.3 PARTIAL COLLAPSE (ELIMINATION)

This step corresponds to tree-covering of conventional technology mappers. We start with the network obtained from the decomposition step. The goal is to transform the network by combining nodes so that the resulting network has potentially fewer evaluations.

Example 1.6 *Consider the network of Figure 2. We proposed earlier that one way to get rid of the glitch at X is to add a buffer at A. Another way is to collapse W into X. Then all the three fanins to X will have the same level 0 and X will be perfectly balanced. However, if W were fanning out to other nodes besides X, collapsing W into X alone may not help. Any events on B and C will still cause W to be evaluated. At the same time, B and C will cause their new fanout X to be evaluated as well. This means a possible increase in the number of evaluations and events.*

The last example underscores the need for deciding if we should collapse or wait until the end to add the required number of buffers, in order to balance the node. We explain how we resolve this by means of an example.

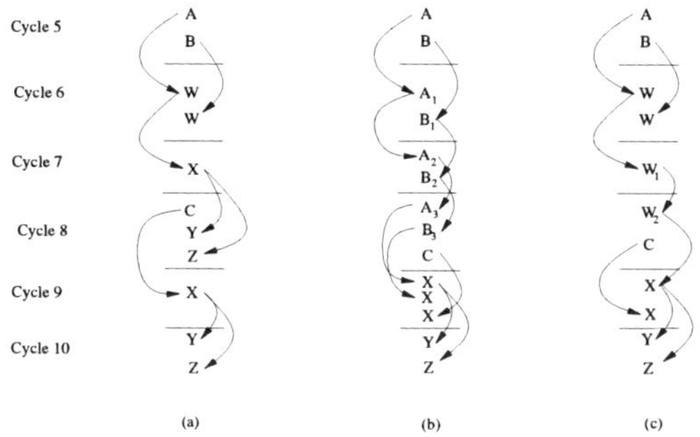

Figure 6 Collapsing vs. buffering: causality

Figure 7 Collapsing vs. buffering: case 2

Example 1.7 *Consider a perfectly balanced node W in Figure 5 (a); the numbers represent the node levels. Let W be collapsed into X. X is not perfectly balanced; 3 buffers need to be added to A and B to balance X: A_1, A_2, A_3 and B_1, B_2, B_3 respectively (Figure 5 (b)). The corresponding event causality diagram is shown in Figure 6 (b). Assuming the worst case scenario that every evaluation leads to an event, the number of evaluations is 12 + $|TFO(X)|$, where $TFO(X)$ denotes the transitive fanout of X. $TFO(X) = \{Y, Z\}$.*

Had we not collapsed W into X, we would need two buffers W_1 and W_2 to make X balanced (Figure 5 (c)). The corresponding event causality diagram is shown in Figure 6 (c). The worst case evaluations are 9 + $|TFO(X)|$.

Finally consider the case when we neither collapse W nor add buffers, i.e., implement the sub-network of Figure 5 (a) as such on the simulator. The corresponding event causality diagram is in Figure 6 (a). The worst case evaluations are 7 + $2|TFO(X)|$. There is a factor 2 with $TFO(X)$, since each of the two evaluations of X can lead to evaluation of its TFO.

We conclude that (c) is strictly better than (b). Also, (c) is better than (a) as long as $|TFO(X)| > 2$. So the best decision (almost always) is not to collapse W into X, but instead to add two buffers at the output of W.

In the last example, our decision was not to collapse W into X. Figure 7 shows a case where it is beneficial to collapse W into X.

For each node, the *partial collapse* algorithm considers which case is applicable (there are more cases than the two described here), and if collapsing would be appropriate in that case. Of course, the memory increase resulting from the collapse should not be more than the leftover memory capacity. The algorithm then evaluates all the node collapses, assigns a cost to each collapse (in the current implementation, the cost is the memory increase resulting from the collapse; in future, it may also reflect change in the number of evaluations), and greedily selects the collapse with the minimum cost. This step is repeated until either no collapse is feasible or each feasible collapse has a memory overhead higher than the remaining memory.

example	pi	po	nodes	edges	lits (fac)	vec.
5xp1	7	10	18	77	114	128
9sym	9	1	12	46	145	512
alu2	10	6	54	273	347	1000
b9	41	21	30	115	124	512
bw	5	28	35	155	160	32
clip	9	5	16	82	117	512
duke2	22	29	81	392	428	1000
e64	65	65	116	253	253	1000
f51m	8	8	16	50	80	256
misex2	25	18	25	103	104	1000
rd84	8	4	18	71	148	256
sao2	10	4	18	93	131	1000
vg2	25	8	10	65	88	1000
rot	135	107	167	597	664	1000
C1355	41	32	162	344	552	1000
apex2	39	3	43	219	268	1000
C1908	33	25	146	383	535	1000

Table 1 Statistics of benchmark circuits

6. EXPERIMENTAL RESULTS

We selected a set of standard MCNC combinational benchmarks and optimized each of them by running twice the optimization script *script.rugged* of sis [2]. We will denote an optimized benchmark by $\widehat{\eta}$. Table 1 provides some information about these benchmarks. The columns *pi* and *po* denote the numbers of primary inputs and outputs respectively. The last four columns refer to numbers of nodes, edges, literals in the factored form in the area-optimized networks $\widehat{\eta}$, and the number of vectors used for simulation. We generated pseudo-random 0-1 vectors to simulate the networks. The probability of each input being a 0 was set to 0.5. For each benchmark, one thousand vectors were generated, unless the number of primary inputs, *pi*, is too small, in which case 2^{pi} vectors were generated. The same set of vectors is used to simulate a benchmark in all the experiments.

The experimental set-up used is shown in Figure 8. Given the optimized network $\widehat{\eta}$, a synthesis technique or script converts it into a network that is more suitable for the TP5000 simulator. For each synthesis technique, the memory constraint M was set to $2^{18} = 262,144$. We do not report the memory used by the final network implementation, since it is always at most M. Once the network is stored in the TP5000 memory, input vectors are applied and the simulation results are compiled. The TP5000 simulator keeps track of the number of events over the entire input vector set, which is what we report in Table 2.

We compare various synthesis techniques discussed in the paper by reporting in Table 2 the number of events generated in the resulting networks during their simulation by TP5000. For each benchmark, the bold entry represents the minimum event count.

The column 2ip-ao corresponds to applying on $\widehat{\eta}$ a 2-input AND-OR decomposition, which decomposes the network into 2-input AND, OR, and inverter gates. The motivation for this was that 2-input gates are the primitives that a software simulator typically uses. So 2ip-ao attempts to mimic the performance of software simulators.

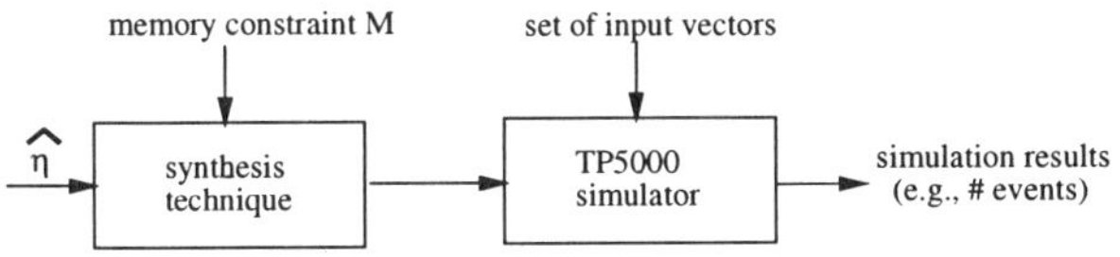

Figure 8 The experimental set-up

The column min-fn corresponds to the technique of [7], which minimizes the total number of functions in the network subject to the memory constraint M (see Section 3). In [7], no simulation results were presented for min-fn. Comparing 2ip-ao with min-fn, we find that in almost all examples, minimizing the number of functions helped reduce the number of events, sometimes by more than a factor of 2. However, in two benchmarks, the number of events went up. For *C1355*, it more than doubled! Although min-fn incorporates a good cost function, it is not always accurate.

To see the effect of buffer insertion, on each optimized network $\hat{\eta}$ we applied min-fn followed by the buffer insertion algorithm of Section 5.1. This generates a perfectly balanced network. The corresponding simulation results are shown in the column min-fn + buf. Although min-fn + buf can add a lot of buffers (on average, 3.4 buffers per function), 11 out of 17 benchmarks had fewer events as a result of buffer addition as compared to min-fn. In some examples such as *9sym*, the reduction in events is by more than a factor of 3, although the number of functions in the buffered network is 3 times! Over the entire benchmark suite, buffering reduced the average number of events by 12.6% over min-fn. However, not all benchmarks benefited from buffer insertion. For instance, *e64, rot,* and *C1908*. This, as mentioned in Section 5.1, is because there are transitions at the buffer outputs and adding too many buffers can backfire. On these three benchmarks, the buffers added were 4 to 5.5 times the number of functions!

Since our goal of reducing the number of events and glitches in a circuit is covered by the power minimization research, we decided to borrow and evaluate techniques that generate low-power circuits [8, 10]. Unfortunately, we could not obtain a tool that minimizes glitches. The power minimization tools currently available in the public-domain use a zero delay model, which precludes consideration of glitching activity. However, we could get our hands on a power estimation tool that considers unit delay model and glitching-activity as well [5]. We used it as follows.

1. Given the area-optimized network $\hat{\eta}$, use the technique proposed in [6] to generate a network $\tilde{\eta}$, which is in terms of 2-input AND and OR gates and has possibly much fewer transitions. The technique of [6] specifically considers glitch minimization.

2. Using power-estimation of [5], identify the node of $\tilde{\eta}$ with the highest switching-activity that can be collapsed into all its fanouts without violating the memory constraint. The idea is to get rid of nodes with high switching activity. Collapse this node into all its fanouts. Repeat this step until memory constraint can no longer be satisfied.

3. Use buffer insertion to balance the network.

We call this technique min-pwr; the corresponding simulation results are shown in the column min-pwr of Table 2. The results are presented only on a subset of the benchmarks, since the power estimation tool could not finish on others. Comparing with min-fn + buf, it is evident that min-pwr is not effective except on *f51m*. We believe one reason for its poor showing is that collapsing a node into its fanouts changes the switching activity of the transitive fanout nodes. This change is difficult to model and is ignored in our technique.

The last two columns 2ip + bal and bal-dec + bal use the techniques of Sections 5.2 and 5.3. Both have three steps: decomposition, partial collapse (of Section 5.3) and buffer insertion.

ex	2ip-ao	min-fn	min-fn + buf	min-pwr	2ip + bal	bal-dec + bal
5xp1	14.2 K	6.8 K	3.2 K	5.2 K	**2.3 K**	2.4 K
9sym	77.2 K	34.9 K	**10.0 K**	36.1 K	12.7 K	45.3 K
alu2	177.8 K	125.4 K	92.9 K	196.3 K	**73.6 K**	87.1 K
b9	64.4 K	36.7 K	37.8 K	51.6 K	**35.8 K**	40.6 K
bw	5.7 K	2.5 K	1.2 K	1.1 K	**0.8 K**	1.1 K
clip	58.4 K	29.0 K	14.6 K	44.6 K	**13.5 K**	24.8 K
duke2	262.9 K	218.6 K	223.6 K	247.8 K	234.0 K	**180.3 K**
e64	**121.8 K**	155.8 K	164.7 K		160.1 K	176.6 K
f51m	20.7 K	14.5 K	15.9 K	**4.4 K**	15.9 K	12.1 K
misex2	74.8 K	60.3 K	**49.7 K**	59.2 K	50.6 K	67.6 K
rd84	47.8 K	15.3 K	9.0 K	19.3 K	**5.1 K**	10.2 K
sao2	110.7 K	61.9 K	**41.2 K**	109.7 K	53.7 K	46.7 K
vg2	78.5 K	61.1 K	60.1 K	51.7 K	56.2 K	**37.0 K**
rot	73.0 K	**59.0 K**	74.7 K		71.4 K	68.5 K
C1355	266.5 K	567.2 K	394.5 K		424.1 K	**189.4 K**
apex2	205.0 K	121.4 K	**114.7 K**		121.4 K	156.6 K
C1908	450.4 K	253.0 K	311.2 K		**161.9 K**	276.2 K
total	2109.8 K	1823.4 K	1619.0 K		1493.1 K	1422.5 K

Table 2 Comparing # events for all synthesis techniques

The first technique, 2ip + bal, uses two-input FPGA decomposition, the one used in min-fn. The second technique, bal-dec + bal, instead, employs the balanced decomposition of Section 5.2. The default is a 6-input decomposition, i.e., $k = 6$. We experimented with several values of k; $k = 6$ yielded the best results. If the resulting network does not satisfy the memory constraint, a 4-input balanced decomposition is done. For no benchmark in our suite did we need to decompose any more. Partial collapse and buffer insertion steps are identical for 2ip + bal and bal-dec + bal.

In terms of the total number of events, bal-dec + bal is the best among all the six techniques of Table 2, closely followed by 2ip + bal. bal-dec + bal is 5.0% better than 2ip + bal, 22.0% better than min-fn, and 32.6% better than 2ip-ao. Out of the 17 benchmarks, 2ip + bal has the minimum events in 7, min-fn + buf in 4, bal-dec + bal in 3, and min-fn, 2ip-ao, and min-pwr in 1 each. Finally, we observed that all the synthesis techniques except min-pwr are really fast; none took more than 50 seconds on a Sparcstation 20 for any benchmark. min-pwr is slow because of repeated invocations of the power estimation tool.

We conclude that *generating a perfectly balanced network by* bal-dec + bal *and* 2ip + bal *is good for minimizing the event count.* Since no technique wins all the time, we believe that *predicting switching activity accurately in a logic network is difficult.*

7. CONCLUSIONS

In this paper, we investigated the problem of speeding up an event-driven simulator implemented on a lookup-table based hardware accelerator TP5000. However, the techniques we proposed are general and can be applied to other domains such as software functional simulation and synthesis of glitch-free, low-power circuits.

Our methodology has a few drawbacks. First, it only considers the worst case behavior, that each evaluation leads to an event. It will be better to take into account the transition probability of each node when considering the propagation of evaluations and make transformation decisions accordingly. Second, it only addresses the mapping problem. We should also solve the optimization problem for simulation speed-up. E.g., we will like to address the *simplification* problem for TP5000: *Given a node N with the associated logic function f and a corresponding sum-of-products representation (*SOP*), derive another* SOP *for f that makes N more balanced.* Finally, we wish to push large industrial designs through our techniques.

Acknowledgments

The authors thank Satoshi Kowatari for helping with the TP5000 interface and Ankur Srivastava for proof-reading the paper.

References

[1] M. M. Denneau. The Yorktown Simulation Engine. In *Proceedings of the Design Automation Conference*, June 1982.

[2] Ellen M. Sentovich et. al. *SIS: A System for Sequential Circuit Synthesis*. Memorandum No. UCB/ERL M92/41, Electronics Research Laboratory, College of Engineering, University of California, Berkeley, CA 94720, May 1992.

[3] F. Hirose. Simulation Processor SP. In *Proceedings of the International Conference on Computer-Aided Design*, November 1987.

[4] P. C. McGeer, K. L. McMillan, A. Saldanha, A. L. Sangiovanni-Vincentelli, and P. Scaglia. Fast discrete function evaluation using decision diagrams. In *Proceedings of the International Workshop on Logic Synthesis*, May 1995.

[5] J. Monteiro, S. Devadas, and B. Lin. A Methodology for Efficient Estimation of Switching Activity in Sequential Logic Circuits. In *Proceedings of the Design Automation Conference*, pages 12–17, June 1994.

[6] R. Murgai, R. K Brayton, and A. S. Vincentelli. Decomposition of Logic Functions for Minimum Transition Activity. In *European Design and Test Conference*, 1995.

[7] R. Murgai, M. Fujita, and F. Hirose. Logic Synthesis for a Single Large Look-up Table. In *Proceedings of the International Conference on Computer Design*, 1995.

[8] A. Shen, S. Devadas, J. White, A. Ghosh, and K. Keutzer. A Combinational Logic Design Methodology Targeting Low Power Applications. In *MIT Technical Report (available from the authors)*, February 1993.

[9] K. Takayama, M. Shoji, S. Shimogori, and F. Hirose. VHDL Simulator Using TP5000. In *Fujitsu Scientific and Technical Journal*, volume 31, December 1995.

[10] V. Tiwari, P. Ashar, and S. Malik. Technology Mapping for Low Power. In *Proceedings of the 30th Design Automation Conference*, pages 74–79, June 1993.

[11] D. Wong, G. De Micheli, and M Flynn. Inserting active delay elements to achieve wave pipelining. In *Proceedings of the International Conference on Computer-Aided Design*, pages 270–273, November 1989.

Efficient Verification of Behavioral Models Using the Sequential Sampling Technique

Tom Chen[1]
Isabelle Munn[2]
Anneliese von Mayrhauser[2]
Amjad Hajjar[1]
[1] *Department of Electrical and Computer Engineering*
[2] *Department of Computer Science*
Colorado State University, Fort Collins, CO 80523

Abstract HDL model validation can involve billions of cycles of simulations. To improve validation efficiency we propose a stopping rule to determine when a validation phase using a specific type of patterns has reached a point of diminishing return.

Keywords: Behavioral Model Verification, Statistical Stopping Rules, VHDL.

1. INTRODUCTION

With rapid advances in VLSI technology, the complexity of VLSI chips has reached millions of gates per chip. Modeling of these complex chips often resulted in complex behavioral models with thousands of lines of HDL (Hardware Description Language) code. Verification of HDL models has, therefore, become a critical and a time consuming task. The conventional approach to verifying HDL models is through extensive functional simulations. Design engineers who wrote the HDL models are often responsible for generating test cases to verify the models. This approach works reasonably well for small to medium size HDL models. For large and complex HDL models, this approach is less effective because of the exponentially increasing input space that has to be explored during verification. Large and complex designs that employed the conventional approach often resulted in a huge amount of test patterns applied to the models. For example, in verifying the PowerPC-601 chip's behavioral model, the design team at IBM generated more than a billion instruction-level test cases [13]. Yet often, at the end of the verification phase, the designers do not know the extent and quality of their verification efforts.

Applying verification coverage to HDL models is a relatively new concept that is rapidly gaining popularity in the VLSI design community. Using a set of coverage metrics to guide the verification process can help designers monitor simulation quality and, therefore, reduce time to market by providing a quantitative measure of simulation completeness. There are many commercial tools available to assist HDL code coverage measurement during simulation. The widely used coverage metrics include conventional coverages such as statement coverage and branch (decision) coverage, as well as state machine coverages related to state visitation and state transition. Although these tools provide instant feedback about the quality of a simulation, they do not guide designers to achieve the highest coverage in the shortest amount of time, i.e. design and verification productivity. For example, in the process of verifying UltraSPARC I, various coverage measures were taken including statement coverage, finitestate-machine trans ition coverage, and gate-level nodal toggle count coverage, and yet, 5 billion instruction simulation cycles with over 1 billion random patterns were ran before tape-out [14]. The question is whether 5 billion patterns are all necessary to achieve the final quality goal. The test suite used for verifying UltraSPARC I was composed of several different techniques including random patterns and functional patterns. Each of these different techniques are useful during a particular phase of the verification process. When verifying a complex design such as UltraSPARC I with billions of patterns, how do we determine a point where continuing with one test technique will likely result in diminishing return for code coverage, therefore, require a change in test technique? Furthermore, are 5 billion patterns enough for UltraSPARC I? Why not 10 billion patterns? How do we determine when to stop verification (simulation)?

This paper is our attempt to answer some of these questions by presenting a statistical stopping rule and its application to HDL model verification. Section 2 gives the background about the proposed stopping rule and its relationship to stopping rules used in software testing. Section 3 present our stopping rule using the sequential sampling approach. Section 4 summarizes the experimental results. Concluding remarks are given in Section 5.

2. BACKGROUND

Determining when to stop simulation with a given set of patterns (generated according to some strategy) can be done in a variety of ways. Most of the techniques come from software testing. Howden [3] uses a simple binomial distribution to determine probability of finding another coverage element and an associated confidence interval. It is assumed that software test runs are independent of each other. Other approaches similar to [3] that are based on the binomial distribution include [9, 2].

Various statistical techniques have been used to determine the number of software test inputs needed to achieve a particular software test objective (in our case coverage related) [10]. The authors apply their statistical testing formula to a variety of white box software testing techniques. Stopping rules have also been used to attain a given reliability criterion [5, 12]. Poore et al. use statistical testing based on a usage model [6, 11]. This method models usage process and the software testing process as Markov Chains, assuming that the current state (usage as well as failure behavior) completely determines the next state. The model is more accurate than others, because it explicitly models states in the software. On the other hand it assumes memoryless behavior which may not always be the case, thus limiting the model's usability.

Through appropriate reinterpretation of failure events in the models as '*new coverage*' events, one could apply software reliability growth models [19]. However, one has to be careful to consider that while software failures may happen one at a time, coverage elements usually do not; e.g. one input pattern applied to a VHDL model may cause many branches to be covered. Thus, coverage shows a certain clumping effect [7]. This limits the applicability of some of the reliability growth models. An example of a stopping rule that considers clumping is the Compound Poisson Software Reliability Model by Sahinoglu et al., both the time-failure [7] and the clustered (interval) failure models [8]. Dalal et al. [1] also allow general distributions. However, these models are not as successful when the expectations of increase in coverage are low, as may be the case when the coverage is already rather high and further increases in coverage are difficult to achieve. This led us to consider two versions of a sequential sampling rule, because it does not make distributional assumptions nor assumptions about having new coverage when first using a new strategy for inputs (as for example when switching from random inputs with 7 clock cycles to random inputs with 4 clock cycles).

3. A SEQUENTIAL SAMPLING RATE

The sequential sampling approach to determining whether continued simulation of a HDL hardware design is likely to increase coverage or not is based on a related technique in software reliability analysis [4]. This technique, also called (software) reliability demonstration, determines whether the software failure intensity is met with high confidence or not. Key evaluation factors are the *discrimination ratio* and the *supplier and consumer risk*. The discrimination ratio, abbreviated γ, represents the maximum number of input patterns accepted that generate no new coverage. The supplier risk, α, represents false positives. Whereas the consumer risk, β, represents false negatives. When assessing software reliability, the sequential sampling approach considers three decisions: accept, reject, and continue software testing. This results in three regions on the sampling chart: accept, continue software testing, and reject the software.

Figure 1 Original stopping rule plotted against software defect data

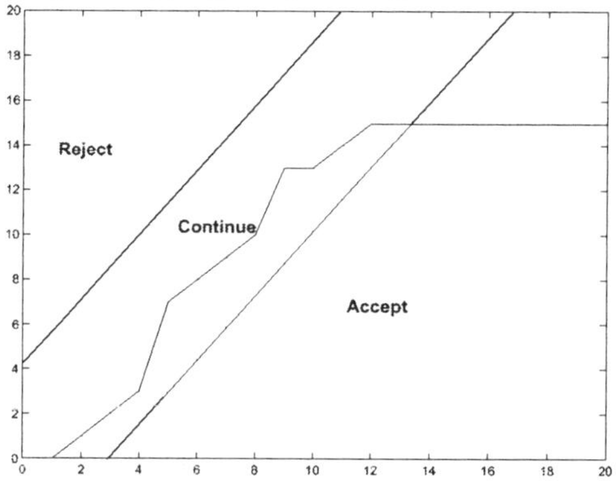

When using the same approach for determining whether continued simulation of a HDL model is likely to yield higher coverage, there are only two regions on the chart: continue, and stop. This requires a modified method. The discrimination ratio is now the ratio of the smallest acceptable branch coverage intensity versus the branch coverage intensity objective. This can also be expressed as the number of inputs one is willing to simulate that are not increasing coverage versus the number of uncovered branches that should still be covered with the given simulation strategy. The consumer and supplier risk have the same meanings. The equation below specifies the boundary between the stop and continue regions on our chart. The x-axis is the number of input patterns applied. The y-axis is the cumulative number of coverage items.

$$f(x) = \frac{B - x * ln(\gamma)}{1 - \gamma} \tag{1.1}$$

where x is the number of input patterns applied, $f(x)$ is the cumulative branch coverage after x inputs have been applied, γ is the discrimination ratio, and B is given by

$$B = ln\left(\frac{1 - \beta}{\alpha}\right) \tag{1.2}$$

α is the supplier risk, β is the consumer risk.

The slope of the boundary line $f(x)$ is determined in large part by the discrimination ratio and to a lesser degree by α and β. The discrimination ratio thus ought to be determined by the expectations of coverage yield. This may be fixed, no matter what strategy is used, or it may be variable. If it is variable, it should be determined both as a function of coverage increase expectations and branches left to cover. We present our sequential sampling technique both ways. When a variable discrimination ratio is used, we assume that a series of strategies are used to increase coverage and that the discrimination ratio is determined for each strategy as follows:

$$\gamma_n = \gamma_{n-1} * ln(\Delta_{n-1}) \quad n > 1 \tag{1.3}$$

where γ_{n-1} is the previously used discrimination ratio, γ_n is the new discrimination ratio to be applied to next data series, Δ_{n-1} is the number of new branches found in the previous technique.

The rationale for this continued growth in the discrimination ratio is based on the following observations: (1) It will get harder to increase coverage as cumulative coverage increases. Thus we must reduce our coverage expectations for successive strategies (the discrimination ratio must increase). (2) The rate of growth in the discrimination ratio is related to how many branches were found in the previous strategy. We chose a logarithmic relationship. Further, if the number of new branches found in strategy n is less than or equal to e, we multiply by a factor of 1. This forces a larger discrimination ratio. This makes the variable discrimination ratio more conservative later in the verification process when yield is likely to be lower.

Sequential sampling with variable discrimination ratio is derived from sequential sampling with a fixed discrimination ratio. Depending on the coverage data, it can behave similar to the fixed model. If, for instance, we have an initially slow rate of finding new branches, the discrimination ratio will not increase by very much, if at all. The more initial coverage generated, the more conservative the next discrimination ratio will be. The less coverage generated, the more closely this statistical sampling method will resemble the fixed sampling method.

4. COMPARATIVE STUDY

We experimented with a VHDL model that implements an image processing algorithm. The model contains 3785 lines of code and 591 branches. The system is organized with several systolic arrays as the processing elements and a global controller. The use of systolic array increases sequential depth, thus, making validation of the model more difficult.

When verifying VHDL models, hardware designers commonly start with a limited number of functional simulations that represent common or typical usages of the design's capabilities. This is then followed by large numbers of random inputs with varying clock cycles. The number of clock cycles accompanying each pattern is determined by sequential depth. Our goal of simulating the model is to have most of the branches covered with a minimum amount of patterns.

We subjected the model to the following five verification strategies: (1) A functional simulation is applied first. It consists of 283 patterns. (2) 5000 random patterns are applied, each was held for 7 clock cycles. (3) 1000 random patterns are applied, each was held for 4 clock cycles. (4) 5000 random patterns are applied, each was held for 2 clock cycles. (5) 5000 random patterns are applied, each was held for 1 clock cycle.

Table 1 Test Coverage Results without Stopping

Test Case	# of patterns	*DCR Branch* *coverage%*	*covered*	*DR Branch* *coverage%*	*covered*
Functional	283	88.66	524	88.66	524
Random H=7cc	5000	93.57	553	96.45	570
Random H=4cc	1000	93.91	555	96.79	572
Random H=2cc	5000	95.26	563	96.79	572
Random H=1cc	5000	96.11	568	97.12	574

We generated random input patterns in two different ways: the first experiment only randomized data signals (marked DR in the results), while the second randomized both data and control signals (marked DCR in the results).

Table 1 shows the number of test patterns a pplied and the cumulative branch coverage at the end of applying each strategy for both the DR and DCR experiment.

Table 2 DCR/DR data using sequential sampling stopping rule $\alpha = .5, \beta = .01, \gamma = 2\ 50$

Testing Strategy	# of patterns	Cumulative coverage	New branches	Patterns saved	Branches missed
Functional	283	524	524	—	—
Random cc7	78/245	536/562	12/38	4922/4755	17/8
Random cc4	250/73	536/563	0/1	750/927	19/9
Random cc2	125/125	538/565	2/2	4875/4875	25/9
Random cc1	125/125	538/565	0/0	4875/4750	30
Total	861/976	538/565		15422/15307	30/9

The first experiment uses a fixed discrimination ratio $\gamma = 250$. This reflects the following yield expectations: we are willing to apply 5,000 patterns to increase coverage by 20 branches. Determination of the discrimination ratio depends on the amount of current coverage (thus, how much more difficult it will be to increase coverage further) and the ability of a strategy (and the patterns it produces) to cover new branches.

Supplier risk is assumed to be medium ($\alpha = .5$; we are not particularly concerned about simulating a little longer than our coverage expectations provide for), while consumer risk must be extremely small ($\beta = .01$). This represents a very conservative approach to stopping.

The results in Table 2 show that for both the D C R and the DR case, the sequential stopping rule saves large amounts of simulation steps while missing few branches. The results of the DCR dataset show that simulating with only 5.3% of the input patterns only causes a loss of 5% in branch coverage. Similarly, the results of the DR dataset show simulating with only 6% of the original input patterns only loses 1.5% of the coverage. Thus, the sequential stopping rule reduces the amount of simulation required by about 95% without large sacrifices in coverage achieved. More importantly, it is able to indicate at which point the chosen strategies for simulation have become inefficient. In this case, simulating with random patterns quickly loses its effectiveness in covering new branches and should be stopped. To illustrate the stopping rule graphically, the Figures in 2 show visually the application of the stopping rule for the random-cc7 and random-cc2 strategies, respectively. The y-axis records the new branches (cf. column 5 in Table 2). The x-axis r epresents patterns applied multiplied by how many clock cycles they were applied. In both cases the stopping rule determined that saturation occured. The probability of new branch coverage is lower than what we established by setting values for γ, α and β.

In the previous section we argued that a variable discrimination ratio might be helpful, indeed, that the rate at which new branches are found should determine the choice of the discrimination ratio. In the VHDL model, SYS7, the rate of finding new branches was very high in the beginning (during the functional simulation). The functional simulation alone covered 88.66 percent of all branches. The relatively high coverage and the low yield expectations for the random inputs lead us to expect that the probability of finding new branches will decrease rapidly with each new test phase.

To adjust for this decreased probability, we multiply γ (the current discrimination ratio) by a logarithmic factor of the number of branches found in the previous phase. Increasing γ forces a shallower slope. It means that we are willing to accept more inputs that generate no new coverage.

The results show that simulating with 24% of the original patterns loses 3.6% of the original coverage for the DCR dataset. For the DR dataset, simulating with 39% of the original patterns loses 1.7% of the original coverage. In this case, there is not a lot of benefit to a more conservative stopping rule, once again emphasizing the limited utility of simulating with random data as a means of increasing coverage.

Table 3 DCR/DR using variable sequential sampling stopping rule $\alpha = .5$, $\beta = .01$, $\gamma = var$.

γ	Testing strategy	# of patterns	Cumulative coverage	Patterns saved	Branches missed
100/100	Functional	283	524	–	–
626/626	Random cc7	195/541	538/563	4805/4459	15/7
1652/2293	Random cc4	305/932	539/563	695/68	16/9
1652/2293	Random cc2	826/2293	543/563	4174/2707	20/11
2290/2293	Random cc1	2290/2293	547/564	2710/2707	21/10
	Total	3899/6342	547/564	12384/9941	21/10

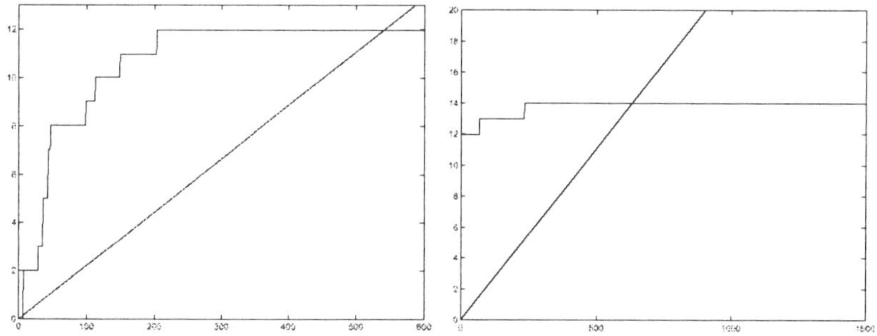

(a) (b)

Figure 2 The stopping rule for random-cc7 and random-cc2 strategies

5. CONCLUSION

This paper presented a stopping rule using the sequential sampling method. Unlike some of the previously proposed stopping rules used mainly for software testing, the advantage of the proposed stopping rule is that it allows the assumption that bug, or code coverage in our case, distribution is not well understood. Furthermore, it takes into account the possibility of very slowly increasing code coverage during the course of validation. Our results show that, using the proposed stopping rule, only 5.3% of the original patterns can cover 95% of the branches and only 6% of original patterns can cover 98.5% of the branches for DCR and DR methods, respectively. We would like to stress here that we are not suggesting that the entire validation process should be stopped altogether after the proposed stopping rule has been met. Rather, the proposed stopping rule is useful in guiding the validation process to determine the point when a different validation phase using a different strategy is warranted. Different validation strategies include applying patterns generated using different methods such as functional, random, and antirandom [15] pattern generation, so that the code coverage can be maximized using the minimum amount of effort. Our future work includes applying the proposed stopping rule to other HDL models to better understand its effectiveness.

6. ACKNOWLEDGMENTS

This research was partially supported by a National Science Foundation Award MIP-9628770.

7. REFERENCES

[1] S. Dalal, C. Mallows, "When should one stop testing software", J. Am. Stat. Assn., v 83, 1988.

[2] D. Hamlet, J. Voas, "Faults on its sleeve: amplifying software reliability testing", Int'l Sym. Software Testing and Analysis, 1993, 89-98.

[3] W. Howden, "Systems testing and statistical test data coverage", COMPSAC 97, 1997, 500-505.

[4] J. Musa, *et. al.*, Software reliability: measurement, prediction, application, McGraw-Hill, 1987.

[5] B. Littlewood, D. Wright, "Some conservative stopping rules for the operational testing of safety-critical software", IEEE-TSE, v 23, n 11, 1997.

[6] J. Poore, *et. al.*, "Statistical testing based on a software usage model", Software Practice and Experience, 1995.

[7] P. Randolph, M. Sahinoglu, "A stopping rule for a compound poisson random variable", Applied Stochastic Models & Data Ana., v 11, 1995.

[8] M. Sahinoglu, U. Can, "Alternative parameter estimation methods for the compound poisson software reliability model with clustered failure data", Software Testing Ver. and Rel., v 7, n 1, 1997.

[9] A. van Schouwen, *et. al.*, "Evaluation of safety-critical software", Comm. of the ACM, v 33, n 6, 1990.

[10] P. Thevenod-Fosse, H. Weselynck, "An investigation of statistical software testing", J. Software Testing Ver., and Rel., v 1, n 2, 1991.

[11] J. Whittaker, M. Thomason, "A markov chain model for statistical software testing", IEEE Trans. Software Eng., v 20, n 10, 1994

[12] M. Yang, "Comparisons of stopping rules and reliability estimation in software testing", SERC-TR-58-F, Purdue Univ., 1992.

[13] P. Bose, "Architectural timing verification and test for super scalar processors", 24th Int. Symp. on Fault-Tolerant Computing, 1994, 256-265.

[14] J. Gateley, "Verifying a million gate processor", Integrated System Design 1997, 19-24.

[15] A. von Mayrhauser, *et. al.*, "Fast antirandom (FAR) test generation", 3rd IEEE High-Assurance Systems Eng. Sym., 1998, 13-14.

Embedded Systems Design And Verification : Reuse Oriented Prototyping Methodologies

S. Raimbault, G. Sassatelli, G. Cambon, M. Robert, S. Pillement, L. Torres

LIRMM,, UMR 5506 – Université Montpellier II / CNRS
161, Rue Ada, 34392 MONTPELLIER CEDEX - FRANCE

Key words: Design Reuse, Co-design, Co-Verification, Prototyping

Abstract: The SIA roadmap plans for 50 millions transistors asics/SOC in 2008 [1]. The design of these chips cannot be achieved in the required *Time-to-Market* constraints without new methodologies. The key solution for saving design time is *Design Reuse*. However, while design reuse solves many design problems, it causes increased verification problems. The complexity of these new designs leads to simulation times that become prohibitive with regard to market pressure. The verification is thus achieved through high-speed emulation and *prototyping* technologies. The scope of this paper is to present these new methodologies. SPW [2] (from Cadence) can handle a wide variety of models in a co-simulation for virtual prototyping. Designs are verified by real prototyping on an Aptix *reconfigurable platform,* using DSP and FPGA components.

1. INTRODUCTION

In a competitive marketplace where many similar products compete for consumer attention, manufacturers must offer noticeably better features and performances. As the integration scale gets larger and larger, these extra features are provided by more and more logic in the design. At the same time, the developement of each unit of the design with the traditionnal methodologies will not meet the time to market constraints. As the required functionalities are often the same from one design to another, design time often is saved by reusing functionals blocks from an older system that has already been validated, or from IP vendors.

The design flow is moving to a higher level of abstraction; concepts such as co-simulation, co-synthesis and co-design are used and applied much more often and, in fact, have become an industrial reality. Around this vocabulary, re-usable cores seem to be the right way to go toward the system on chip and to implement the interface between the system specification and the final physical level. One kind of block which is well suited for design reuse is the processor. Most complex embedded systems are built around one or more processors, with dedicated hardware for intensive tasks, analog parts and interfaces. Therefore, a system is built with hardware and software parts, that have to be designed together (Co-design).

The verification of such systems can be achieved in two ways: formal methods or simulation. Formal verification aims at getting the mathematical proof of the system's predictability. Although this approach can give a high level of confidence in the design, it is very difficult to obtain this result for complex designs, especially if an asynchronous communication scheme is used between the different parts of the system. The simulation approach consists of applying a sequence of stimuli (a testbench) and comparing the results with the expected response. The quality of the validation relies on the testbench's ability to detect the errors. Simulation is powerful for high level design, but the increasing complexity of systems results in prohibitive running time as implementation details become fine. Let's consider a low complexity system which performs edge detection on a 256x256p_ picture with the Sobel algorithm on a 3x3 matrix. Table 1 presents the simulation time for the whole picture. The order of magnitude between these results clearly shows that prototyping is an excellent issue to simulation time.

Description Level	ISS in SPW	VHDL Simulation*	Real world Prototype
Simulation Time	10 mn	20 h	0,5 s

Table 1 : Verification time for a DSP–based edge detection system
*: Vulcan VHDL simulator on a SPARC20 workstation

As systems become heterogeneous (HW/SW, IP/Specific), links between CAD tools and real prototyping seem to be the right way to mixed simulation of heterogeneous modules and/or models (Virtual/Real); designing those systems becomes incremental. We first present an overview of prototyping. The following deals with co-simulating heterogenous systems with heterogenous models in a SPW/Aptix framework.

2. FAST PROTOTYPING: STATE-OF-THE-ART

A prototyping system is actually built with FPGA components. There are 3 main topologies for prototyping systems: specific boards, emulation machines and reconfigurable platforms.

2.1 Specific Boards

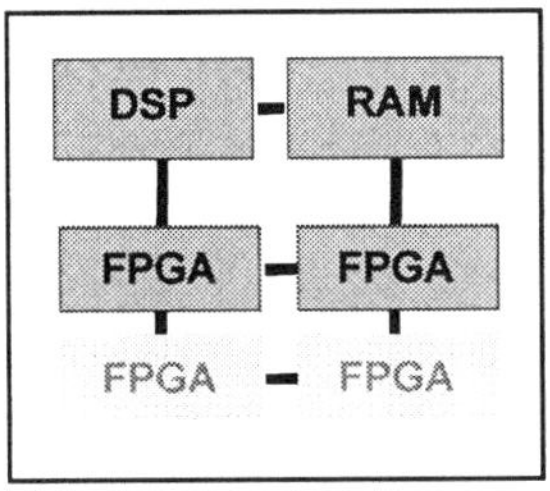

Figure 1. Specific board topolgy

The specific boards [3][4][5] are built for prototyping a given system, or at least systems having a given topology. These boards are actually made of one processor, connected to standard memories and FPGAs for custom logic. It is a low cost solution which allows prototyping of medium complexity HW/SW systems (for example: JPEG algorithm) [6]. The main drawback with these boards is their poor level of reusability.

2.2 Emulation Machines

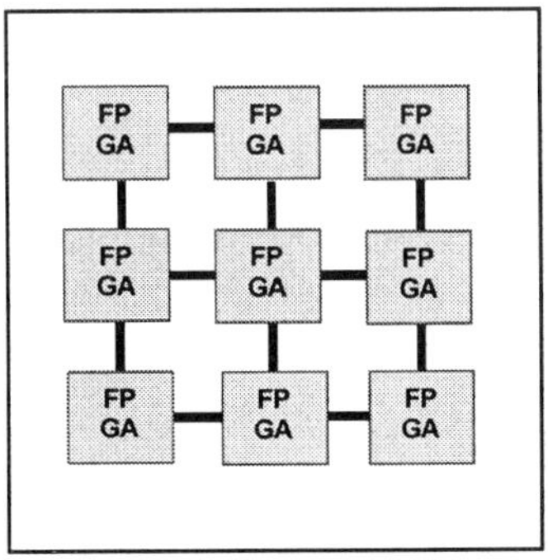

Figure 2. Emulation Board Topology

These machines are built as a network of FPGA, and have the capability to implement large designs [7]. Some of them are using proprietary FPGAs and can thus provide some estimations about the performances of the integrated system [8]. The HDL of the whole system is synthesized for the corresponding FPGA technology. One drawback with this approach is the availabilty of an HDL model for the cores. In the case of an hard-core processor, synthesizing its HDL code may not reflect some implementation details very well. Let's mention that some new emulation machines allow the use of off-the-shelf components, and so the evaluation of hard-cores [7][9]. In [9], the machine is built as an array of specific processors instead of FPGAs.

An other drawback is the price of these machines. They are very expensive, and confront the users with heavy maintenance problems.

2.3 Reconfigurable Platforms

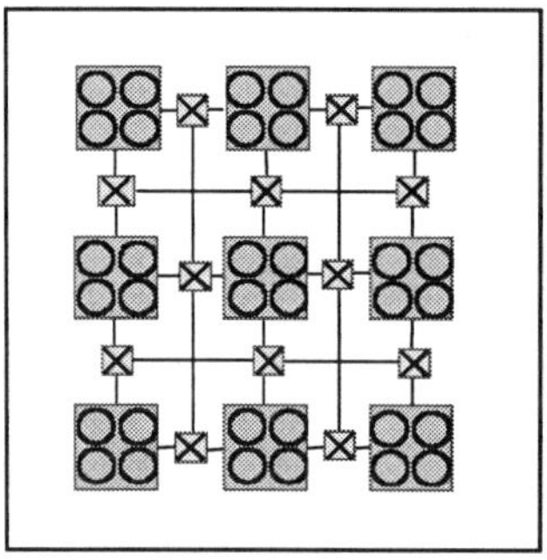

Figure 3. Reconfigurable Platform Topology

This is the most recent way in fast prototyping. These boards are built around Field Programmable Interconnexions Components (FPIC) [10]. The platform consists of a large number of interconnexion holes, which can be connected to each others through FPICs. FPGA or other off-the-shelf components lie on daughter card modules to be plugged into the holes. From a top-level view of the system and a description of each module's pinmap, the associated software automatically performs FPIC and FPGA configurations. The low delay (2-5 ns) introduced by the FPIC devices allows high-speed prototypes. A debug mode is also available in which one can probe any signal from the design into a logic analyzer. This platform has been succesfully used for a demonstrator we built with a d950 DSP (ST-Microelectronics) and an Altera FPGA, presented during the DATE99 exhibition (Aptix booth). A reconfigurable platform is very useful for system validation with a CAD environment, as presented in the following.

3. A GLOBAL PROTOTYPING ENVIRONMENT

Due to the wide variety of models used in the design of complex systems, a global simulation environment is needed. We have made the choice of an industrial frame work with SPW[2] and Aptix MP3 [10]. SPW is proposed as an algorithmic design tool dedicated to signal processing applications. It can handle synchronous or asynchronous dataflows, as well as cycle-based simulation. Overall, this tool offers the ability to co-simulate heterogeneous models. Fig. 4

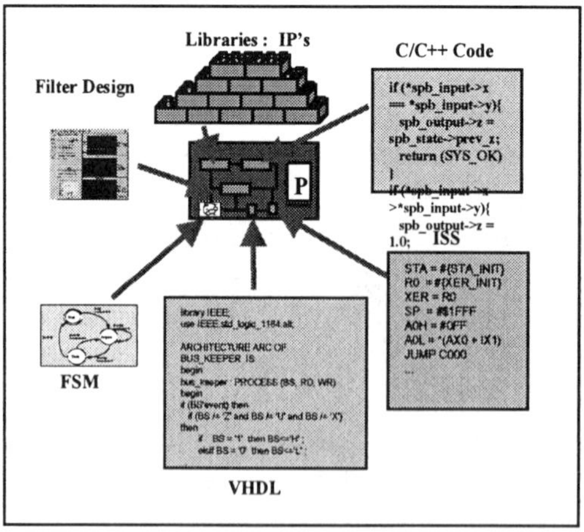

Figure 4. Virtual Prototyping with SPW

presents virtual prototyping with SPW. Within this framework, the system is first designed at a high level of abstraction as a dataflow. The system is built as a set of communicating blocks whose behavior can be specified with an iteration function in C or C++ code. Then an architecture can be developped and refined till the assembly code for software parts and RTL for hardware elements. During these design steps, the whole system is verified by simulations. As previously said, a fine grain of detail will result in long simulation times for complex designs. Therefore, the system is finally verified by running on real world prototype. Fig.5. represents this verification design flow.

3.1 Design space exploration

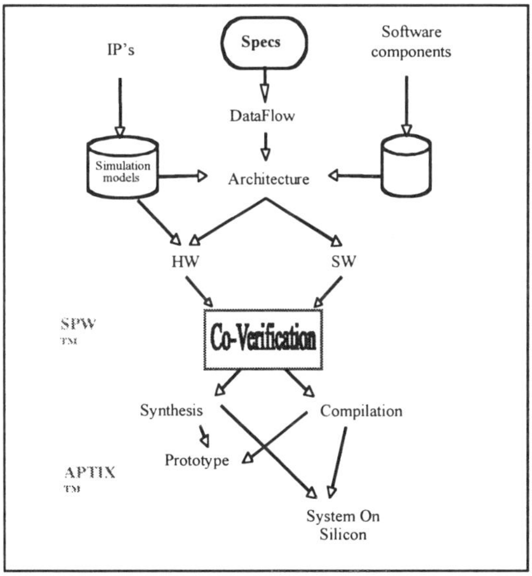

Figure 5. Co-Verification Design Flow

Let's consider the edge detection system described above. The specification was the Sobel algorithm coded in C. We first build a data-flow model of the system, reusing the specification's C code for the computing block. The second step consists of creating the I/O buffering parts, since the original code was using a file system. Here begins the design space exploration. Multiple buffering architectures can be evaluated : using a RAM implies using a communication protocol. Because we need three new points for each pixel, it would mean using either a three times faster clock or three parallel RAMs. On the other side, buffering with FIFOs would solve those problems, but is less reusable (fixed size) and could result in a prohibitive area for larger picture sizes. Note that this is dependant of the target architecture; for a DSP-based system, the processor's memory may be used. It is thus important to complete early evaluations of different solutions.

The system has been first implemented on a DSP and simulated with the processor's ISS. The system was working, but because of the huge data volume of pictures, the image processing rate was not satisfying the specs' constraints.

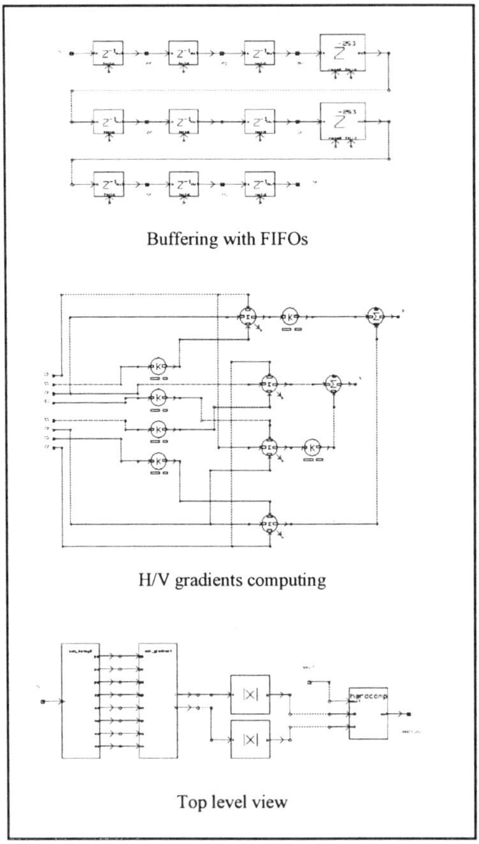

Figure 6. Edge Detection System : Full Hardware Architecture

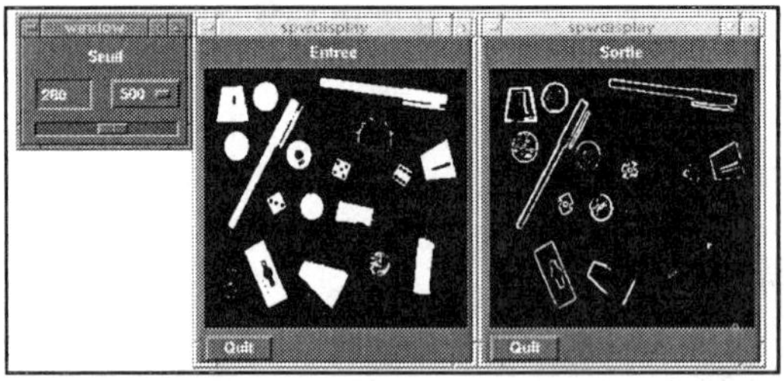

Figure 7. Interactive Parameter Tuning : Subjective Testing

A full hardware solution was then designed with the HDS Library from SPW. The global architecture is given in Fig. 6. The inputs are buffered with 3 FIFOs which can provide the eight required pixels. Then the horizontal and vertical gradients are computed, and their absolute values are added. Comparing this result with a threshold will give the output value for the current pixel. The system was first designed with floating point blocks. The next step was the conversion to the fixed-point system. Overflow and truncation effects are enlightened in order to find the smallest buses sizes (leading to smaller area and power consumption) guaranteeing a correct behavior. At least, the HDL code for the whole system is automatically generated.

The virtual prototype allows verification of course, but also parameter tuning for subjective testing. The threshold constant is determined by using a scroll bar from the Interactive Simulation Library. During simulation, we can tune this parameter according to the subjective quality of the output picture (Fig. 7).

4. TRENDS FOR PROTOTYPING METHODOLOGIES

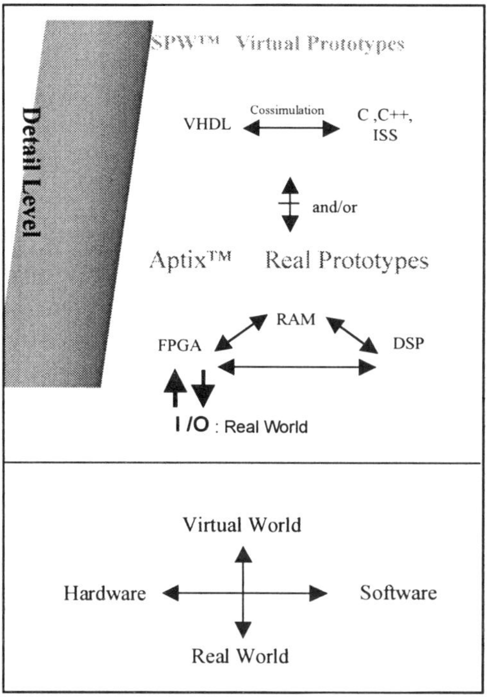

Figure 8. Co-verification Space

Because verification handles heterogeneous models, the verification space can be seen within two dimensions (Fig. 8). Therefore, we believe that an efficient verification environ ment must be able to perform validation with models located anywhere in this verification space. It represents an issue to prohibitive simulation time and also offers new prospects.

First, it allows providing the software team with more and more rapid and accurate proto types. There is another benefit with this approach: the availability of the models. If you plan to use a hard-core, the vendor may not distribute an HDL netlist, but will be able to provide an off-the-shelf component or at least a bounded-out core. In such an open verification environment, it becomes possible to develop the system, even at high level, with high-speed simulation and fine details.

Moreover, it bridges the implementation gap. When a circuit is designed and verified at high level with a virtual prototype, and later globally refined for implementation, there is a lot of work (and thus many possible errors) between them. This problem is more and more frequent with the increasing complexity of systems. When the system does not work after implementation, a lot of work is required to localize and identify the faults sites. An environment which can handle virtual models as well as real models allows the incremental migration from virtual prototype to real world prototype. After the migration of each part of the system, a verification is performed. If the system does not have the required behavior, the last implemented part of the design is analyzed to find the errors. In fact, this approach provides an easier diagnosis of faulty systems during implementation.

5. REFERENCES

[1] http://www.semichips.com

[2] http://www.cadence.com/alta/products/new datasheets/spw.html

[3] J.M Arnold, D.A. Buell, E.G. Davis, *"Splash2"*, Symposium on Parallel Algorithms and Architectures, ACM, June 1992, pp.316-322.

[4] P. Bertin, D. Roncin, J. Vuillemein, *"Introduction to Programmable Active Memories"*, Systolic Array Processor, Prentice Hall, pp. 301-309, 1989.

[5] R.W Hartenstein, J. Becker, R. Kress, H. Reinig, *"High-performance Computing using a Reconfigurable Accelerator"*, CPE Journal, Special issue of concurrency: Practice and experience, John Wiley & Sons Ltd. , 1996.

[6] S. Pillement, L. Torres, M. Robert, G. Cambon, "*Rapid Prototyping – A case study : the JPEG compression algo*rithm", Intl. Conf. On Rapid System Prototyping, Clearwater, Florid, June 15th –17th 1999.

[7] N. Zafar, "*Using emulation to cut Asic and system verification time*", Computer's design, Pennwell Publishing Company, April 1994.

[8] http://www.mentorg.nl/celaro/index.html

[9] http://www.quickturn.com/products/cobalt.htm

[10] http://www.aptix.com

A Virtual CMOS Library Approach for Fast Layout Synthesis

F. Moraes, M. Robert, D. Auvergne

PUCRS - Av. Ipiranga, 6681 - Faculdade de Informática - Prédio 30
90619-900 - Porto Alegre - Brazil
e-mail: moraes@inf.pucrs.br

LIRMM, UMR 5506 CNRS/Uni. Montpellier 2 - 161 rue Ada
34392 - Montpellier - Cedex 5 - France
e-mail: robert@lirmm.fr, auvergne@lirmm.fr

Key words: layout synthesis, physical design, virtual libraries

Abstract: We present a layout synthesis methodology based on the use of virtual CMOS libraries, i.e. using no pre-characterized cells. The proposed methodology is organized around an automatic layout generator, allowing fast on-the-fly implementation of macro-cells. The generator eliminates the need for post-layout compaction procedures and in addition produces parasitic capacitances estimations. Results show that it is possible to quickly generate dense layouts, allowing fast prototyping of logic functions. The proposed method can change the way layout synthesis is seen today, since accurate parasitic evaluation is an important prerequisite for optimized submicronic designs.

1. INTRODUCTION

Shrinking product lifecycles, increased design complexity and time to market are some of the fundamental issues that must be addressed by the IC designers. The recent market trend towards Systems on Silicon (SOS) manufactured with deep submicronic processes has increased the design complexity dramatically, which has lead to the frequent need to reuse IP cores. At the physical level the use of virtual libraries [1] instead of cell libraries, module generators or full custom layout is an alternative for a fast prototyping of these designs on different processes.

In this paper we propose a design methodology where layout synthesis, performance evaluation and optimization are associated in one single design flow as represented in *Figure 1.* The concept of a virtual library is based on using cells available through a layout generator, instead of using a set of pre-characterized cells. The set of available cells is a user defined constraint (number of transistors in series, maximum fanout and output isolation). The technology mapping process starts with a Boolean description and its output file is a netlist of Static CMOS complex gates at the transistor level, which is transferred to a layout generator. The first layout synthesis creates a load file, with routing and diffusion capacitance. Using this load file, with the original Spice file, accurate time analysis [2] and sizing can be performed. The second layout synthesis generates the final layout.

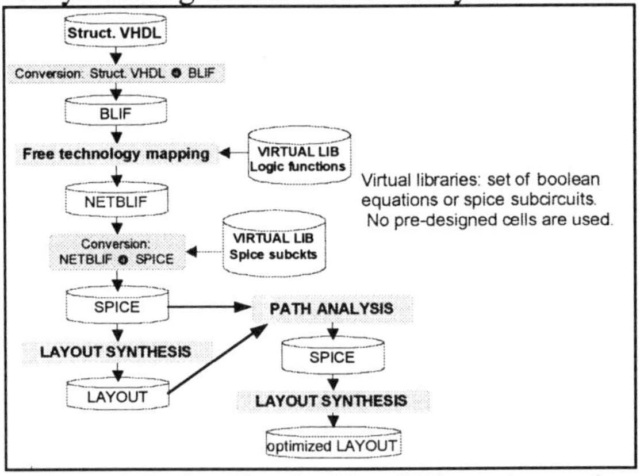

Figure 1. Design Flow for Virtual Library Approach

Using layout synthesis tools [3][4], it is possible to overcome the main limitations of the standard-cell approach: fixed number of functions and fixed transistor sizes. However, a virtual library approach may present the following limitations:

- Cells are not pre-characterized. The cell performance (area, delay, and power) can only be obtained after layout synthesis, since each cell will be generated according to its environment. Layout generation, parasitic capacitance extraction and electrical simulation are CPU time consuming. As will be shown, it is possible to break this cycle, obtaining the parasitic elements immediately after layout synthesis, reducing in this way the amount of CPU necessary to validate a circuit.
- Cell topology is fixed. Usually, only dual gates (same number of N and P transistors) and transmission gates are implemented. When a cell generator implements dynamic logic, like NORA or TSPC, the transistor density is poor [5].

– Automatic layout synthesis design flow requires more steps than standard-cell flow, like layout compaction, detailed extraction and electrical simulation. Layout generation must be entirely transparent and fast to the designer, while efficiently coupled to the logical synthesis step, to technology mapping and to performance optimization tools.

For these reasons, layout synthesis tools are not widely used in the semiconductor industry. However, virtual library methodology has recently grown in importance, since layout synthesis tools may speed up the generation of IP blocks for process migration. Additionally, this can be done without necessarily creating links to specific fabrication process or cell-library vendors.

The approach presented here is original. Our main goal is to quickly synthesize the entire layout of a block, starting from a gate level description (in fact, transistor level), without layout compaction, still maintaining reasonable transistor densities. If layout compaction is employed, more than 3 hours are necessary to translate a symbolic description to the final masks for 5000 transistor blocks, in an Ultra-Sparc 10. Using our approach, we synthesize the layout and calculate the parasitic elements in less than 2 minutes.

As CPU time for layout generation is no more a bottleneck, iterations may be used for optimization. The logic synthesis tool can perform initial iterations to get accurate information on routing length and cell area. This gives facilities for buffer and repeater insertion, since the load of each node is calculated during the layout synthesis.

The technology coding is very simple, requiring only basic rules such as distances, separations and overlapping. For parasitic estimation the data required is the area and the peripheral capacitance of each layer. In addition to the fast synthesis, easy technology migration allows to synthesize circuits in new processes, without waiting for new cell libraries.

This paper is organized as follows. Section 2 presents the layout style at the cell and circuit level, as well as the technology coding. Next, in Section 3, the evaluation of the parasitic elements is introduced. In Section 4 we present our preliminary results, and finally some conclusions and future work.

2. LAYOUT STYLE

The main difference between cell based technology mapping and library free technology mapping concerns the libraries (pre-characterized or virtual) they must cope with. Library free technology mapping implements the functions *directly at the transistor level*, while guaranteeing that the final netlist of complex gates respect some topological constraints (e.g. number of transistors in series). The great number of available complex gates will improve the design space and lead to a minimization of the overall number of transistors, minimizing the design at the transistor level.

The output of the library free technology mapping is a netlist at the transistor level. Two alternatives can be considered at the physical synthesis phase: library or macro-cell generation. During *library generation* ([6] and [7]) a dedicated cell library is created either from symbolic templates or from a regular layout style. These cells have common characteristics (constant height and fixed pin positions), to be treated homogeneously in the same way as with a library in the cell-based approach. After library generation, placement and routing tools for standard-cells are used to obtain the final layout. In *macro-cell generation* ([4]) the complete block will be generated. The initial description is decomposed into leaf cells that will be assembled together using dedicated place and route tools, without constituting a separate library. Two instances of the same logic function can have different layouts, according to its environment. Our approach employs such a method.

Due to the use of regular layout styles for cell or macro-cell synthesis, normally followed by compaction [7], the final transistor density obtained from library free synthesis is smaller than that obtained from the standard-cell approach, where the cells are handcrafted. However, the final area for the library free approach can be *smaller*, due to the use of complex gates, which reduces the overall number of transistors.

2.1 Layout Synthesis

The target technology is CMOS, with 3 metal layers for routing and no restrictions on stacked contacts. Transistors are implemented using the linear-matrix style, where two horizontal diffusion lines are used to implement transistors. Supply lines are placed between transistors, in metal2. The contacts to the substrate (body-ties) are inserted over the supply lines. A metal1 stub is used to connect vcc/gnd nodes to supply lines, requiring a stacked via over body-ties.

The algorithm to implement the layout of a row starts from the left side of each row, processing column by column (*Figure 2*). Each node (drain/gate/source) has its status defined during the data base construction:

- *External row node*: inputs and outputs nodes to be connected to the routing regions. The pin assignment algorithm determines these nodes.
- *Internal row node*: supply, OTC, not connected (e.g., opposite side of a gate) and internal connection (e.g., abutted drains in a nand gate).

Using the design rules and the node status, it is possible to construct the real layout, without intermediate descriptions. The procedure used to implement the rows examines the node status. Only two cases occur: either the node must be connected to the routing region or it must not. If the node must be routed, it will be *aligned* to a grid, with a step defined by the contact rules. Otherwise, the node coordinate will be fixed respecting the minimal distance between transistors. Automatic jog insertion for gates (polysilicon layer) is necessary to reduce the circuit width and the diffusion area.

Over-the-cell nodes will be implemented in metal2, over transistors. The over-the-cell routing is divided into two steps: the first one moves all internal nets of complex gates to the transistor region, and the second one fills each track with nets from the routing region. Experiments over a large set of complex gates [8], up to 4 transistors in series, gives a reduced number of tracks for internal cell routing. In this way, the width of the transistor area is not penalized by this approach.

Different layer directions are used inside row and routing regions. At the row level, diffusion and metal2 are used horizontally; while polysilicon, metal1 and metal3 are used vertically. For the routing regions, metal1 and metal3 are used horizontally; and polysilicon and metal3 are used vertically.

The connection between rows and routing regions is achieved in a dedicated track named "interface line". As can be observed in *Figure 2*, this region will connect the vertical layers. If we have polysilicon connected to polysilicon no contact is needed; otherwise, for metal1-metal3 connection a stacked contact-via1 must be implemented.

The row generation step creates the real layout for each row. Before routing, it is necessary to insert feedthroughs *over* rows. As all rows are transparent to metal3, vertical wires can be placed anywhere over the cells, with the following constraints:

- For any routing region, the number of different signals per vertical column is limited to two.
- Different feedthroughs in two adjacent rows are forbidden.
- Feedthroughs are aligned to the routing grid.

If these restrictions are not observed, vertical cycles can render the routing impossible. Automatic jog insertion over metal3 wires is performed to manage the difference of constraints between column sides.

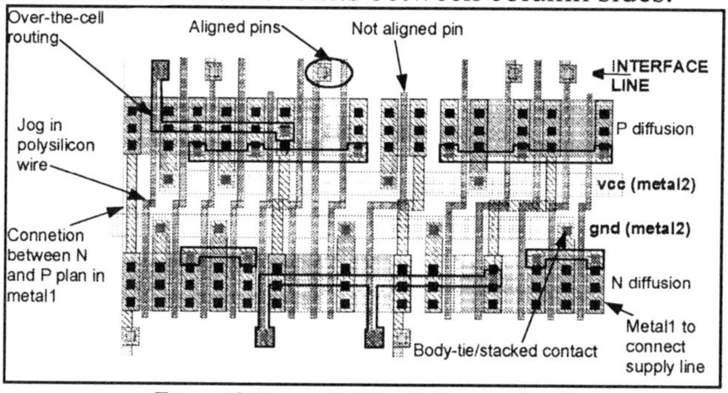

Figure 2. Layout style at the row level

The last step is the routing. A modified left-edge algorithm is employed. The choice of routing layers was made aiming the reduction on the number of contacts in the interface region. If the coupled capacitance between metal1 and metal2 (horizontal layers) is too high it is possible to change the directions to poly-metal1 and metal3-metal2 or to define the maximum superposition length between two wires as a design rule.

2.2 Technology Coding

The input of our layout generator is constituted of only two files: the Spice netlist and the design rules. The outputs are the final layout and a load report for each net. Three sets of rules are necessary:

- *Geometrical rules* (28 rules): for each layer the minimum width and minimum separation for contacts and vias the length, distance and margin to metal; transistor construction rules and power supply width. *Figure 3* illustrates the design rules used for the transistor construction.
- *CIF rules* (15 rules): define the CIF name for each layer of the layout.
- *Electrical rules* (15 rules): for each layer the area and peripheral capacitance, oxide capacitance and "load factor limit" (ratio of load to input capacitance).

The routing grid is defined by equation (1).

$$grid = 2*max(MCTO, MVIA1, MVIA2) + max(LCTO, LVIA1, LVIA2) + max(DM1, DM2, DM3) \quad (1)$$

Where:

- *MCTO, MVIA1, MVIA2*: margin of metal to contact;
- *LCTO, LVIA1, LVIA2*: width of contact;
- *DM1, DM2, DM3*: distance between same metal layer.

Pins assigned to routing are the only elements aligned to this grid. Drains, gates and sources not connected to the routing region (even OTC nodes) are *not* aligned to it. Some experiments were done with gridless routing. This approach was rejected due to the huge number of DRC errors introduced and to the resulting small area gain.

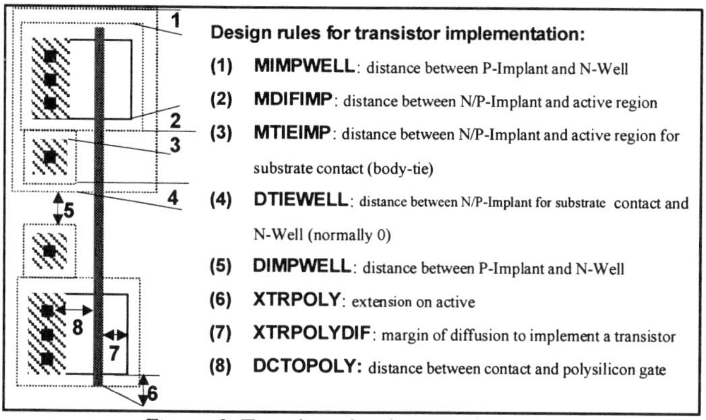

Figure 3. Transistor implementation rules

3. PARASITIC ELEMENTS ESTIMATION

This Section presents the approach used to compute the parasitic elements for each net. We considered only the capacitance to the substrate. No coupled capacitance is computed and a lumped model is used. The goal is to provide a fast evaluation tool, not a precise electrical extractor.

Two procedures can be adopted to compute parasitics:

- compute the wire length during routing,
- compute the wire length after routing, reading the layout file, as an extractor program.

The second approach is CPU time intensive, since it is necessary to find all connected polygons. Despite this problem, we adopted this solution, since this "*wire extractor*" can be used to verify the generated layout, giving information on eventual open connections and short-circuits (import tool during the development of routing algorithms). As we detail in Section 4, for a circuit with 14376 transistors, (257051 polygons), only 228 seconds (Ultra-Sparc 1) were necessary to compute the parasitic elements.

The parasitic capacitance for each net, C_{load}, is given by formula 2. *Figure 4* illustrates the three components defining the load capacitance.

$$C_{load} = C_{active} + C_{route} + C_{diffusion} \tag{2}$$

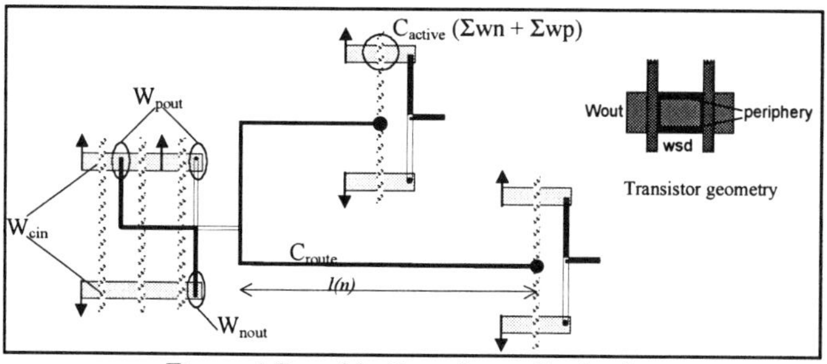

Figure 4. Parasitic Capacitance Evaluation

The active capacitance, C_{active}, is a function of all gate widths connected to the net:

$$C_{active} = (\quad W_n + \quad W_p) . l_{min} . Cox \tag{3}$$

The routing capacitance, C_{route}, is a function of the length of each layer implementing the net. It is important to subtract the polysilicon length over transistors (width) to compute polysilicon capacitance.

$$C_{route} = \qquad (C_{area(n)} . w(n) + 2 . C_{perimeter(n)}) . l(n) \tag{4}$$

$$n = poly, m1, m2, m3$$

The diffusion capacitance, $C_{diffusion}$, corresponds to the number of drain/source regions driving the net. We are assuming worst case behavior: the last transistor in a series branch is switching. The following formula is used to compute $C_{diffusion}$:

$$C_{diffusion} = \frac{\begin{array}{l}(W_{nout} . Cjn + 2 . Cjnsw) . wsd . drainN\# \\ + (W_{pout} . Cjp + 2 . Cjpsw) . wsd . drainP\# + W_{nout} . Cjnsw + W_{pout} . Cjpsw\end{array}}{FDIF} \tag{5}$$

Where:

- *wsd* corresponds to the average drain/source length, and is calculated by the generator tool;
- *Cjn/Cjp* is the area capacitance and Cjnsw/Cjpsw the perimeter capacitance for the diffusion layer;
- *drainN#* and *drainP#* the number of drain/source regions driving the net. This number corresponds to the number of transistors connected to the longest series branch in each plan of the gate;
- the last two factors corresponds to the perimeter length at the end of a diffusion chain.

The input capacitance for a giving gate is given by formula 6.

$$C_{in} = (W_{n\_cin} + W_{p\_cin}) . l_{min} . Cox . drive \tag{6}$$

where:

- W_{n_cin} and W_{p_cin} corresponds respectively to the N and P average width of the gate,
- *drive* is the number of transistor pairs switching together. *Drive* is 1 for nand, nor, inverter and complex gates; and 'k' when we have an inverter with 'k' parallel transistors.

The load factor, $F_{load} = C_{load} / C_{in}$, corresponds to the ratio between the load and the input capacitance. When this factor is greater than the "load factor limit", defined into the design rule file, a warning message is sent to the user. This information can be used to define which gates must be **sized** or where **buffers** can be inserted.

The "wire extractor" supplies three output reports:

- *Flat spice netlist* with parasitic capacitances: this file can be used either by the electrical simulator or by the sizing tool.
- *Parasitic capacitances*:
  - Relative capacitances: C_{ative}/C_{in}, C_{route}/C_{in}, C_{diff}/C_{in} and C_{load}/C_{in}. If C_{load}/C_{in} is greater than the "load factor limit" a warning is printed.
  - Absolute values for capacitances (C_{ative}, C_{route}, C_{diff}, C_{load} and C_{in}).

- Total length per layer and the number of contacts in each net, data used to compute the routing capacitance and (*future work*) resistance.
- Topological data for each gate: number of inputs, fanin, fanout, number of transistors in series, average transistor width and the number of transistors connected to the longest serial branch in each plan (used to compute diffusion capacitance);
- Number of estimated buffers (C_{load}/C_{in} > load factor limit).

– *Wire length histogram.* Running the generator over a set of benchmarks (28 to 15000 transistors, 0.25 μm technology), we got an average value of 89% of connections bellow 200 μm. For this technology, 200 μm corresponds to a C_{route}/C_{in} equal to 2.5 (average transistor width equal to 2 μm). *Figure 5* shows a histogram for a circuit with 14376 transistors and 4764 nets (ISCAS c7552). In this example, 80.5% of the connections are bellow 100 μm, 9.9% between 100 and 200 μm, 7.4 % between 200 and 500 μm and 2.2% (102 connections) have the total length greater than 500 μm. The cells driving these nets must be sized, or buffer insertion must applied.

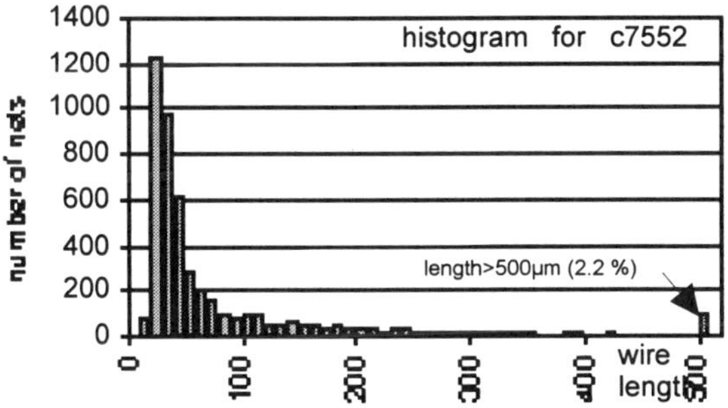

Figure 5. Histogram of wire lengths

The obtained results show that only a small fraction of nets have length larger than 200 μm. A tool for quickly finding these nets is essential in submicronic technologies, in order to guide the sizing and optimization steps. The wire length data can also be used in a second placement iteration to guide the placement of cells inside the critical path of the circuit. The placement algorithm, quadrature with pin propagation, is responsible for this homogeneous length distribution.

4. RESULTS

A method to estimate the silicon area necessary to implement layouts, starting from the number of transistors and the layout design rules, is presented in [9]. The data given in this paper represents an upper bound for the transistor densities. *Table 1* shows the maximum transistor density that can be achieved without routing. Routing area was considered as a constant percentage of layout area, 50% and 25% for two and three metal layers respectively. Then, for a 3-metal layer 0.25 μm process, the transistor density that should be obtained with routing by a layout generator is 94000 transistors/mm^2 (125328*0.75).

Table 1. Transistor Density Roadmap (tr/mm^2) [9]

Process (μm)	Transistor Density
l=0.8	8296
l=0.6	14748
l=0.5	31332
l=0.35	62644
l=0.25	125328
l=0.18	250656

Table 2 gives the area, transistor density, CPU time for layout generation and for "wire extraction" (parasitic capacitance evaluation) obtained with our layout genenerator.

Table 2. Transistor Density and CPU time for layout generation and parasitic capacitance evaluation

Technology: 0.25 μm, 3 metal-layers and stacked contacts, regular sized transistors, w=2 μm, l=0.25 μm. CPU time in milliseconds for an Ultra-Sparc 1

Circuit	Transistors	Nets	Rows	area (mm^2)	Tr. Density (mm^2)	CPU (*ms*) Generation	CPU (*ms*) Par. Eval.
Adder	28	13	1	0.00028	100676	50	170
Addergate	40	15	2	0.00042	94482	180	310
alu	260	94	3	0.00353	73677	440	3970
Alugate	432	117	4	0.00535	80808	500	6050
rip	448	163	4	0.00500	89552	720	8360
cla	528	215	4	0.00700	75439	660	9010
Hdb3	570	191	6	0.00745	76539	420	9810
5xp1	798	308	7	0.01087	73386	790	15310
sao2	930	361	7	0.01321	70388	1030	17770
Mult6	972	308	6	0.01311	74147	1230	15800
9sym	1092	420	8	0.01547	70575	1170	22790
c499	1556	511	7	0.02279	68277	1730	28820
c1355	2244	647	9	0.02975	75437	2390	41460
c1908	3146	990	14	0.04764	66036	3850	63680
Mult2	4512	1239	13	0.07425	60768	5610	66220
c2670	4976	1762	15	0.08408	59181	6360	98230
c75523x3	6164	2101	15	0.13295	46363	14730	101100
c3540	7154	2359	15	0.13449	53193	9100	118810
Mult12	8584	2455	16	0.14757	58169	15110	133850
c6288	10112	2706	18	0.14931	67726	11030	162970
c5315	10656	3429	15	0.24822	42930	24360	167510
c7552	14376	4764	17	0.33096	43437	109700	228220

The average transistor density obtained with this first layout generator version is around 60000 tr/mm^2, to be compared to the ideal density: 94000 tr/mm^2 (0.25 μm technology). The main reason for this difference is the routing approach, which was based on the traditional channel approach (channels and feedthroughs). The number of rows was chose to minimize the area, and fixed as a parameter for the generator. Circuits with transistor density below 58000 tr/mm^2 (c3540, c7552_3x3, c5315 and c7552) have a great number of inverters (at least 55% of the cells). This fact reduces the row transparency, since there is less space to pass metal3 over the cells, reducing in this way the transistor density. One solution to reduce the used silicon area is to use a maze-based algorithm, routing as much as possible over transistors.

5. CONCLUSION

Our primary goal was to show that it was possible to quickly generate dense layouts without compaction. In submicronic technologies, the routing contribution for delay is equivalent to cell delay. We have shown, the capacitance for a wire length with 200 μm is equivalent to 2.5 cells. Then, if only cell delay is used for mapping, sizing and timing analysis, inaccurate results will be obtained. For a submicronic design flow, the layout generation step must give accurate information on parasitic in few minutes, instead hours. This fast layout generation can be used:

- in mapping tools to select which gates will be used (simple gates or and-or-inverters) or where buffers must be inserted;
- in sizing tools to correctly size transistors, using the parasitic capacitance evaluation, or insert buffers not placed in the mapping step;
- in timing analysis tools [2] to get an accurate post-layout performance evaluation.

From this first layout synthesis prototype, two directions can be investigated: (*i*) developing more accurate models for capacitance and resistance evaluation; (*ii*) implementing efficient routing algorithms for this layout style.

REFERENCES

[1] A.Reis; R.Reis; D.Auvergne; M.Robert. "*Library free technology mapping*". VLSI'97 - IX IFIP International Conference on Very Large Scale Integration, August 26-29 1997, Brazil, pp. 303-314.

[2] S.Cremoux; N.Azemard; D.Auvergne. "*Path resizing based on incremental technique*", ISCAS98, Monterey, USA, 1998.

[3] M.Guruswamy; R.L.Maziasz; D.Dulitz; S.Raman; V.Chiluvuri; A.Fernandez; L.G.Jones. "*CELLERITY: A fully automatic layout synthesis system for standard cell libraries*". DAC'97.

[4] CADENCE. "*Virtuoso layout synthesizer - LAS - user guide*". CADENCE™ Version 4.2, October 1991.

[5] S.Rekhi; J.D.Trotter; D.H.Linder. "*Automatic layout synthesis of leaf cells*". DAC'95.

[6] A.J.Velasco; X.Marín; R.Peset; J.Carrabina. "*Performance driven layout synthesis: optimal pairing and chaining*". Fith ACM/SIGDA Physical Design Workshop, April 1996, pp. 176-182.

[7] J.L.Burns; J.A.Feldman. "*C5M - A Control-Logic Layout Synthesis System for High-Performance Microprocessors*". IEEE Transactions on CAD, Vol. 17, no. 1, January 1998, pp. 14-23.

[8] R.L.Maziaz, J.P. Hayes. "*Layout minimization of CMOS Cells*". Kluwer Academic Publishers, 1992, 165p.

[9] F.Moraes; L.Torres; M.Robert; D.Auvergne. "*Estimation of Layout Densities for CMOS Digital Circuits*". PATMOS '98 - International Workshop on Power and Timing Modeling, Optimization and Simulation, October 07-09 1998, Denmark, pp. 61-70.

*The author Fernando Moraes would like to gratefully acknowledge the continued support of the "Conselho Nacional de Desenvolvimento Científico e Tecnológico" (***CNPq***), Brazil, to this work. Research grant 522939/96-1.*

RT-Level Route-and-Place Design Methodology for Interconnect Optimization in DSM Regime

Ananth Durbha
and Srinivas Katkoori
4202, East Fowler Avenue, ENB 118,
Department of Computer Science and Engineering,
University of South Florida, Tampa, FL 33620.
{durbha,katkoori}@csee.usf.edu

Abstract

As the interconnect dominates in DSM regime, we propose a novel *route-and-place* design methodology at RT-level which strongly ties to physical design tasks. Unlike conventional place-and-route, our methodology prioritizes interconnect over devices. At the RT-level, we identify all global nets and carefully design their interconnect topology to achieve desired delay characteristics, followed by the macro-cell placement. Net clustering is carried out to account for net interdependencies as well as to derive the bounding box constraint on a net. Within a macro-cell, as the nets are local in nature, we apply place-and-route to achieve its layout. Experimental results on various datapath circuits resulted in total wirelength reduction varying from 3.7% to 24.6% with an average value of 12.4%.

Keywords: Route & Place Methodology, Interconnect Minimization, Deep Sub Micron Regime.

1. INTRODUCTION

Due to rapid technology scaling, technology generations in the DSM regime give us the capability to realize system-on-a-chip. The accompanied price is the dominance of the interconnect phenomena that were hitherto ignored at the micron and sub-micron technology generations. Therefore, existing CAD algorithms must be revisited and/or new CAD algorithms must be developed to address this paradigm shift. In the 1997 SIA NTRS [1], for example, it is projected that for 100nm technology, the interconnect delay for Al metalization is approximately fifteen times the gate delay. This underlines the severity of the problem. The conventional place-and-route approach is not appropriate for design optimization in DSM regime as the cell placement limits the routing freedom of a net. Thus, in order to make the design methodology interconnect centered, we propose a route and place scheme so that we give priority to net routing over cell placement.

If there are no cells placed, where is one going to route the nets!! We address this problem by *macro-cell clustering* i.e. the cells connected by a bunch of nets are grouped together to give rise to a bounding box constraint for the nets. The clustering is based on the interdependency of nets. Those nets which have many cells in common are put in the same cluster so that the overall wirelength of each net is reasonable. Once clusters are formed, within each cluster we would synthesize the nets that satisfy the delay constraint as well as the bounding box area constraint. It is possible that clusters may share macro-cells. Thus, cluster-level placement is carried out to place the common cells between two clusters at their common boundary. Within each macro-cell, place-and-route methodology is adopted to achieve its layout, as the nets within the cell are local in nature. In summary, this approach gives highest priority to the optimization of the global net characteristics, which is key to success in DSM regime.

The rest of the paper is organized as follows: Section 2. discusses the related work. Section 3. presents the route and place design methodology in detail. Section 4. shows the experimental results. Finally, Section 5. draws conclusions and outlines future work.

2. LITERATURE SURVEY

Although sophisticated models for interconnect delay were developed [2] , their use in synthesis is limited due to the lack of knowledge of geometry of interconnect early enough. So synthesis tools must adopt an approach that first focussed on determining optimal plan for global wiring. Such approaches are known as wire planning.

Given a set of pre-designed modules, which may be standard cells or custom modules, and a netlist describing the connections to be made between them, the objective of module placement stage is to assign each module a unique location on the chip, so that no two modules overlap, and the design constraints are met. Wire planning [3] is one possible source for generating these constraints. There will also be design constraints imposed by the particular design style adopted.

Constructive and Iterative approach. Constructive approach try to imitate the approach taken by a human designer. But the intuitive rules of human designer have proven to be hard to capture into a computer program, resulting in poor quality. Iterative approach involves starting with a seed(initial placement) and improving it iteratively until closure is achieved. For large designs, number of iterations itself can be a bottleneck. Avoiding local minima is also a problem.

Stochastic approach. These algorithms try to avoid the local minima problem noted above. They allow random moves that may increase the cost function temporarily allowing for the solution to proceed in search of a "global minimum" . For example simulated annealing is one such approach [4] . But this approach entails a large amount of computation which tends to be a handicap when dealing with very large circuits.

Min-Cut placement. The most common objective used during above placement process was to minimize the total interconnect length. Such algorithms often lead to non-uniform routings. This can lead to chip area not being minimized even if the wires are short.

Hence min-cut placement techniques [5, 6] were developed which seek in addition, to minimize number of nets crossing a set of horizontal and vertical cut lines crossing the chip. Shorter wirelengths are achieved by finding strongly connected groups of modules.

Timing driven placement. With interconnect delays playing an increasingly significant role in the timing closure problem, it is no longer sufficient to just guarantee routability of a particular placement but needs to satisfy length(delay) constraints. This concern led to work on timing driven placement [7] and routing [8] algorithms. These algorithms can be broadly classified as net-based [9] and path-based [10] approach. Net-based approach usually involves computing timing slack for each net and converting them into wirelength upper-bound constraints or net weights to be used in the cost function during placement. Path-based approach involves modeling path-based constraints as a mathematical programming problem.

The above approaches address the problem in a piece-wise fashion by considering point-to-point length constraints. In DSM regime, however, the entire topology of the net affects the delay. Our approach is global in nature when deriving a routing solution for a given net. The influence of the topology on the delay can be reduced by appropriate buffer insertion through isolation of the downstream capacitance.

3. ROUTE-AND-PLACE DESIGN METHODOLOGY

The design flow of the proposed Route-and-Place design methodology is as shown in Figure 1 The input netlist consists of instances of RT-level components such as n-bit adders, n-bit registers, m-to-1 multiplexors, etc. We assume that the bounding box area of the leaf-cells used to build the macro-cells is known *apriori*. The netlist may be either a hand-written one or the output of a front-end synthesis tool. From the netlist, we eliminate the following global nets: power, ground, clock, and nets such as Global Reset. Specialized net topologies need to be generated for such nets. Grouping of nets that drive the same macro-cells is also carried out.

From the pre-processed netlist, net clusters are formed based on multi-net dependency. For each cluster so formed, the nets in that cluster are prioritized according to their cell-count. Following this priority, an optimum net topology is synthesized for each net. After all nets are routed, macro-cells are then placed around the nets. Within each macro-cell, the traditional place and route is carried out as the nets within a macro-cell are local in nature. Between two or more clusters macro-cells may be shared. To handle this scenario, inter-cluster placement is performed, following which an optimized layout is generated. In the following paragraphs we present a detailed discussion on each phase.

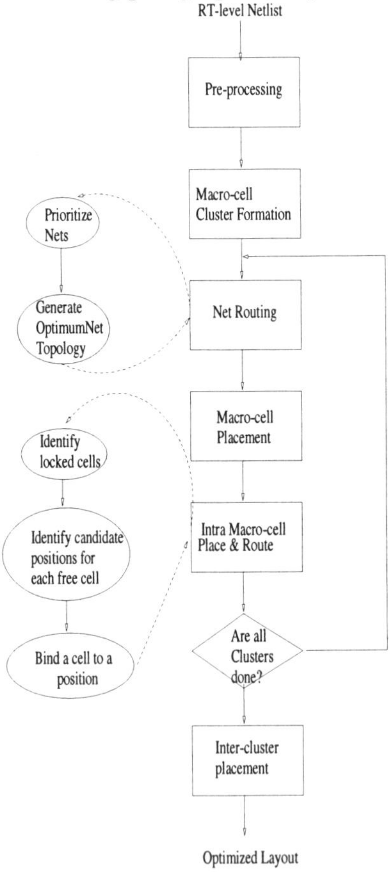

Figure 1 **Route and Place Design Flow**

I. Net Clustering: Two nets are said to be *interdependent* provided they have at-least one common cell. The higher the number of common cells, higher the degree of interdependency. Strongly interdependent nets must be routed in the same *region* in order to find "good" interconnect topologies for all the nets in question. Net clustering is carried out based on the net interdependencies. We employ clique partitioning algorithm to form the clusters.

From the input netlist, a net interdependency graph, G, is constructed where each node represents a net and an edge exists between two nodes if and only if the nets represented by the nodes are interdependent. Each edge is weighted with a weight equal to the number of common cells between the nets. A modified version of clique partitioning algorithm proposed by Tseng and Siewiorek [11] is applied on G_{dual}, the dual of G. The main modification to the original algorithm is to handle a weighted input graph.

The advantage of applying clique partitioning on G_{dual} is that it not only yields cliques of nets but also information as to which nets within the clique needs to closer. In other words, it yields clusters that are hierarchical in nature.

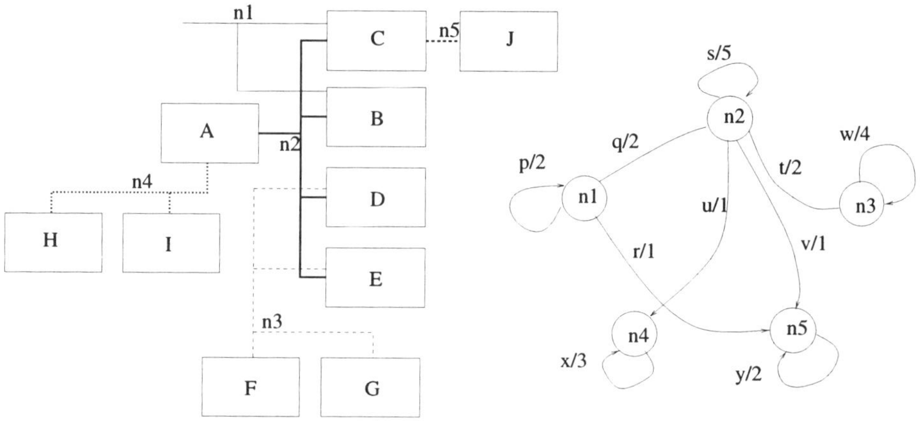

Figure (a) Figure (b)

Figure 2 **(a) A Simple netlist with five nets connecting ten RTL modules; (b) Net interdependency graph**

For a simple RTL netlist shown in Figure 2(a) G and G_{dual} are shown in Figure 2(b) and Figure 3. Clique partitioning on G_{dual} yields minimum number of maximum weighted cliques. The following clique set is obtained from G_{dual}: C={ {b, d, e, f, g}, {a, c}, {i}, {j}, {h}}.

On expanding the clique set into net clusters, we obtain the following: C'={{n1, n2, n3, n4, n5}, {n1, n5}, {n3}, {n4}, {n5}}. As clique partitioning is performed on G_{dual}, a net may occur in more than one cluster. In a given cluster, we can group nets into sub-clusters that identify the nets that are desired to be routed very closely in the same region. A sub-cluster in a cluster arises under the following conditions:
(1) An identical cluster exists by itself.
(2) An identical sub-cluster occurs in one or more other cliques.
The reduction of C' gives rise to C''={{{n1, n5}, n2, n3, n4}}, where {n1, n5} is a sub-cluster that occurs by itself. Thus, we obtain one hierarchical cluster with two levels.

II. Routing: In [12] it is shown that both the total wirelength (i.e., the total interconnect capacitance) as well as the interconnect topology will dictate the interconnect delay. Our methodology gives *near* complete freedom

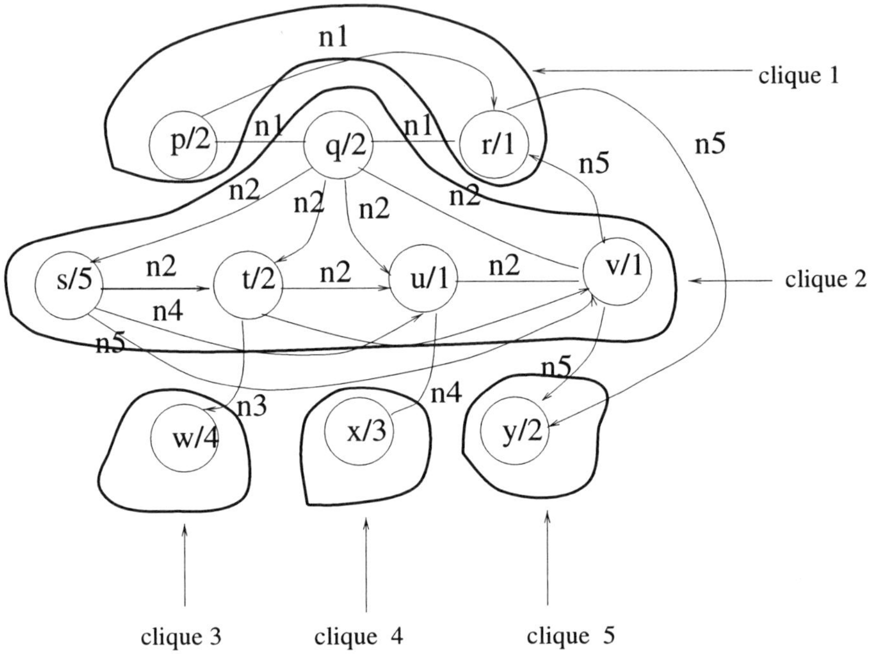

Figure 3 **Dual of the graph shown in Figure 1.2(b)**

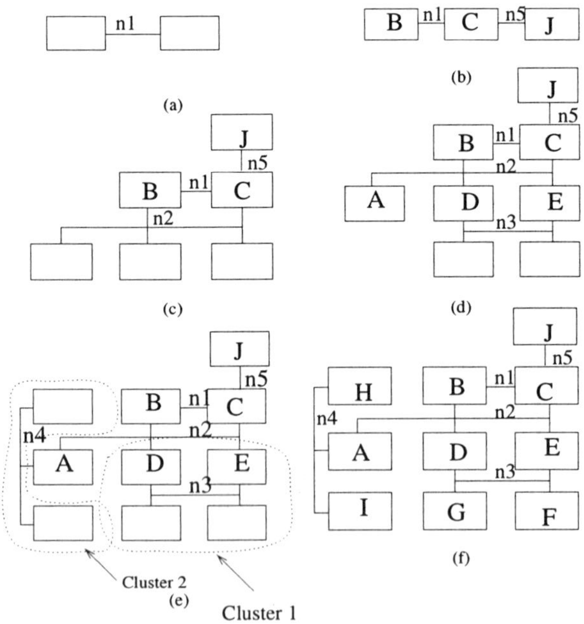

Figure 4 **(a)-(e) Various stages of routing the cluster (f) Final layout**

to optimize both factors for every global net. In this work, we focus on total wirelength minimization. Thus, given a cluster, for every net belonging to the cluster, we aim to obtain an optimum interconnect topology in terms of its total wirelength. We assume that the delay is linearized by repeater insertion and therefore the wirelength is a reasonable delay metric. The delay of a long line that is "quadratic" in nature can be canonically converted into "near-linear" delay by buffer insertion, as a buffer isolates the "downstream" capacitance [13]. Optimal buffer insertion [14] algorithms exists in the literature that can be employed for this purpose.

Work is in progress to compare the actual delays by employing accurate timing analysis tools such as moment-based timing tools [15] and HSPICE.

The generation of a routing solution begins from the innermost sub-cluster(s) and works outwards. For eg., in case of C'', first we generate a routing solution for $\{$n1, n5$\}$ and then continue with nets n2, n3, and n4. For those nets which are at the same level within a cluster, we given preference to larger nets i.e. nets with higher cell count. As the routing solution evolves, macro-cells get locked into certain positions and the end all common macro-cells are locked. Thus a partial placement solution also evolves as a side-effect of this phase. Re-positioning of macro-cells is carried out with the objective of making the layout as square as possible.

Figure 4 shows how the routing solution for the cluster $C\ ''$ evolves. The sub-cluster $\{$n1, n5$\}$ within C'' is routed first. Figure 4(a) shows a simple topology for net n1 with two possible locations for the macro-cells belonging to n1. Figure 4(b) depicts the addition of net n5 to net n1. This fixes the positions of the macro-cells B, C, and J. Between the two possible candidates n2 and n3, we randomly pick n2. With macro-cells B and C already placed, we construct a net topology for n2 as shown in Figure 4(c). There are three possible cell positions that can be occupied by the remaining cells on the net, namely A, D, and E. Routing of net n3 necessitates the placement of macro-cells D and E that are shared with net n2. Thus, macro-cells D and E are placed as shown in Figure 4(d). Finally, net n3 is routed as shown in Figure 4(e).

III. Macro-cell Placement: In the routing solution obtained in the previous phase, the placement of unbound macro-cells is carried out. Generate pin location constraints for each macro-cell. Within a macro-cell (eg. an n-bit adder) since all the nets are local in nature, we can employ the conventional place-and-route approach to obtain an optimized layout. The reason for cluster-level placement is that some clusters may have common macro-cells. In such a case, the common cells need to be placed at the common boundary.

4. EXPERIMENTAL RESULTS

We present results for the following five benchmarks: (1) *Compress*; (2) *Find*; (3) *FIFO*; (4) *TI's Shuffle Exchange Network*; (5) *Traffic Light Controller (TLC)*. These are datapaths are automatically generated by a high level synthesis tool. Pertinent design data of the benchmark examples are shown in Table 1. We present the experimentation procedure below.

(1) *Netlist preprocessing:* From an input RTL netlist, power, ground, and clock nets are eliminated. Nets with identical cell lists are collapsed into one compound net, i.e., they are treated as a single bus. If two nets have no common cells then they are independent and the relative cell placement does not influence each other's total wirelength. In real designs, the degree of interdependency between two arbitrary nets lie between the above two extreme cases.

In a design consisting of N number of nets, the node size of the net interdependency graph (G) is equal to N and the worst-case complexity of the number of edges is O(N^2). Therefore, the worst-case complexity of the number of nodes in G_{dual} is O(N^2). In real life designs, the node-size of G_{dual} is much smaller than the worst-case complexity. Column 3 of Table 1 reports the node-size of G_{dual} for each design.

(2) *Net Clustering:* G_{dual} is provided as an input to Tseng's clique partitioning heuristic [11]. The original algorithm proposed in [11] is modified to handle weighted input graphs. Optimal number of maximum weighted cliques are heuristically obtained. The nodes in the clique identify the cells that need to be in the same neighborhood. From the cliques and from the information about the cells associated with each node, we generate clusters of nets. A net may belong to more than one cluster, as the clique partitioning is performed on G_{dual} rather than G. Recall that by doing so, not only do we find cliques of nets but also information as to which nets within the clique needs to routed closer. The latter information helps in prioritizing the nets when being routed as discussed in Section 3.. For each design, in Table 1 we tabulate the number of cliques obtained (column 5), the size of the largest clique (column 6), and the execution time for clique partitioning (column 7).

(3) *Routing and Macro-cell Placement:* The routing solution for each cluster is generated by hand. Work is in progress to integrate an interconnect synthesis tool that automatically generates the optimized network. We used Flint layout synthesis tool [17] in an interactive mode. We force Flint to generate a desired interconnect topology by generating appropriate channels in the placement. Flint has features to manipulate the macro-cells such as flipping, rotating etc., that are used extensively during the placement of macro-cells.

For each macro-cell, we generate standard cell layouts by employing "Stdcell" layout generator module in Lager silicon compiler. Thus, within each macro-cell cell placement is following by net routing. This approach is justifiable as the nets within the macro-cell are short. From the MAGIC layouts generated by Flint, we measure the wirelength of each net. The nets are routed in two wiring layers (M1 and M2). We sum up the lengths of all M1 and M2 segments of a net to obtain its total wirelength.

4.1 DISCUSSION OF RESULTS

We demonstrate the efficacy of our approach in terms of the following design characteristics: (1) the sum of overall wirelengths of all nets in the design. (2) the bounding box area of the final layout. The reported values of the above measures are in the units of λ, the basic unit of scalable CMOS technology. All designs are targeted for static CMOS technology.

Below, we compare the optimized designs generated by route-and-place (RnP) approach against those generated using Macro-cell (PnR) based layout style as well as Standard-cell (Stdcell) based layout style.

Total Wirelength. Table 2 summarizes the comparison of the total wirelength. The actual values of total wirelength are reported in columns 2-4. Columns 5 and 6 show the percentage change. In comparison with PnR, our approach yielded in significant wirelength reduction between 3.7%–24.6%. This clearly demonstrates the effectiveness of the proposed approach. In comparison with Stdcell, our approach is not effective (but for Find) because of the following reasons: (1) Routing is limited to channels and hence in general the wirelength is higher. This can be alleviated by incorporating over-the-cell routing. (2) Macro-cell shape and port location information is not exploited. Work is underway to improve the approach by addressing these factors.

In Figure 5 we compare the results according the netlength distribution. In each figure, we plotted the net id on the x-axis and its wirelength on y-axis (in units of λ). The nets are sorted in ascending order according to their wirelength. As majority of nets in every design have reduced substantially, it is evident that the proposed methodology is highly effective. Equally important is the fact that there was *no* tradeoff between nets.

Bounding Box Area. The penalty paid for the wirelength reduction is an increase in the bounding box area. Table 3 summarizes the comparison of the bounding box area. The measured values are reported in columns 2-4. Columns 5 and 6 show the percentage change. In comparison with PnR, our approach resulted in an average increase of 25.8%. In comparison with Stdcell, our approach resulted in an average increase of 109.2% !! But, this can be improved by incorporating area in the cost function while generating routing solution for a cluster. Compaction algorithms and over-the-cell routing algorithms will be integrated into the design flow to minimize the area penalty.

Table 1 **Pertinent design data and clique partitioning related data**

Design	*Macro-cells*	*Nets*	*Nodes in G_{dual}*	*No. of Cliques*	*Size of Largest Clique*	*Exec Time seconds*
Compress	35	106	62	14	8	0.39
Find	58	125	130	18	15	3.47
Fifo	63	91	179	33	20	3.48
Shuffle	104	43	353	37	67	52.57
TLC	32	30	86	19	10	0.42

Table 2 **Total Wirelength ("-" stands for reduction)**

Circuit	*PnR*	*RnP*	*Stdcell*	*vs. PnR*	*vs. Stdcell*
Compress	343064	330536	100122	-3.7%	+230.0%
Fifo	1000242	754083	380351	-24.6%	+98%
Find	105256	99956	399962	-5.0%	-75.0%
Shuffle	1115374	873600	164289	-21.7%	+432.0%
TLC	129122	120004	40120	-7.0%	+199.0%

Table 3 **Bounding box area**

Circuit	*PnR*	*RnP*	*Stdcell*	*vs. PnR*	*vs. Stdcell*
Compress	2782 x 2894	3071 x 3435	2459 x 1648	+31%	+160.3%
Fifo	5076 x 4583	3368 x 7529	5735 x 2832	+9%	+56.0%
Find	4520 x 5109	4099 x 6814	5195 x 2528	+21%	+112.7%
Shuffle	6227 x 4916	4883 x 6973	5339 x 3584	+11%	+77.9%
TLC	1963 x 1983	2450 x 2489	1875 x 1360	+57%	+139.1%

5. CONCLUSIONS

We introduced a novel route-and-place design methodology that is applied for interconnect minimization in DSM regime. Full flexibility to design the global nets is the key to high performance design in DSM regime. Extensions to this work includes integrating an interconnect synthesis tool, incorporating area minimization, and delay measurement using DSM-accurate timing tools.

Acknowledgments

We are thankful to Meenakshi Kaul and Ranga Vemuri at the University of Cincinnati for providing us with their clique partitioning implementation and the synthesis benchmarks.

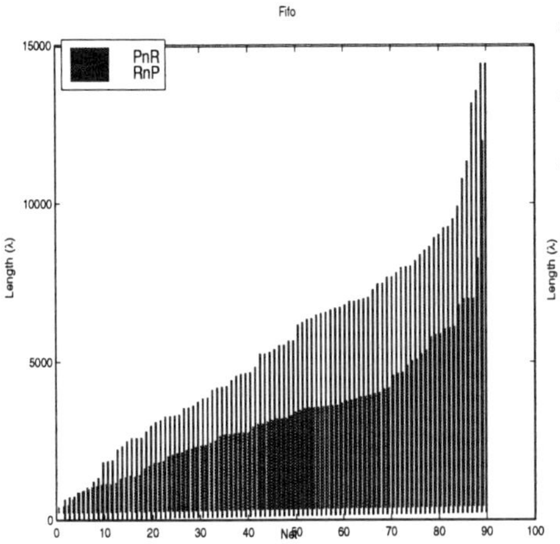

Fifo

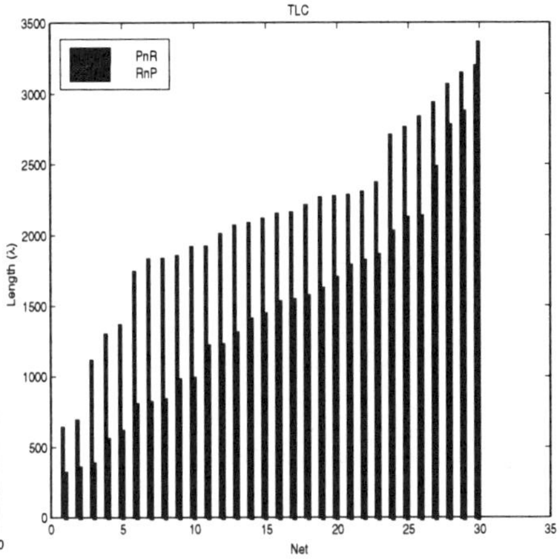

Tlc

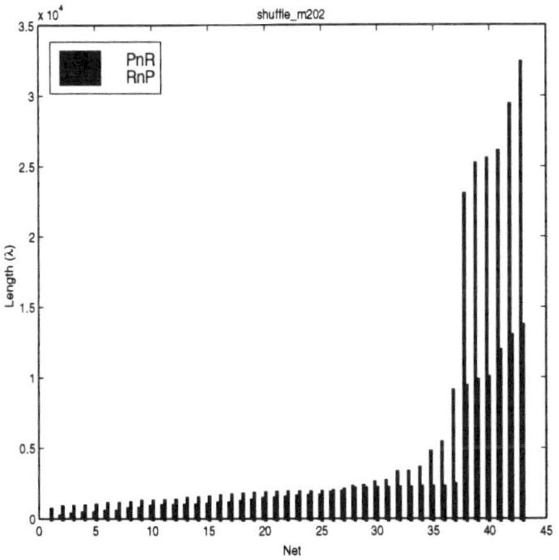

Shuffle

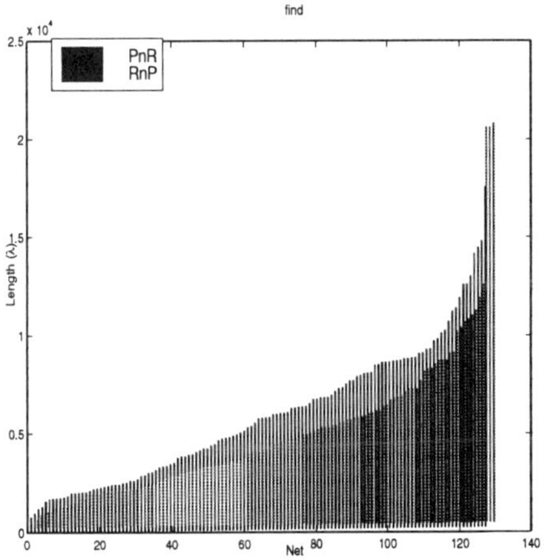

Find

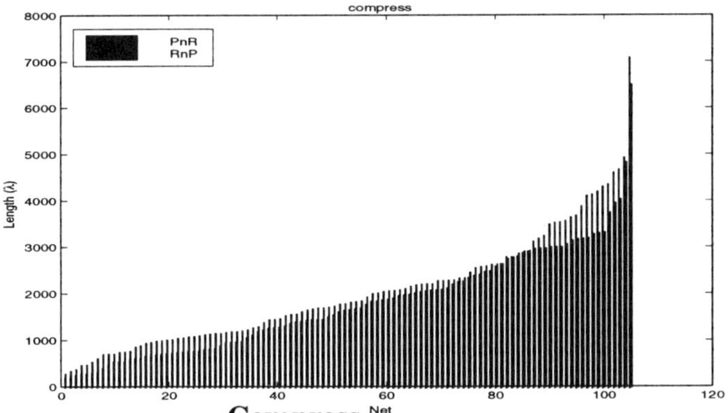

Compress

Figure 5 Comparision of the Wirelength Distribution

6. REFERENCES

[1] –. *"The National Technology Roadmap for Semiconductors"*. SIA (Semiconductor Industry Association), 1997.

[2] Pileggi L. "Delay metrics". In *ISPD*, 1998.

[3] R. H. J. M. Otten and R. K. Brayton. "Planning for Performance". In *DAC*, pages 122–127, June 1998.

[4] Sechen C. et. al. "The TimberWolf placement and routing package". In *Proc. Custom Integrated Circuits Conference*, May 1984.

[5] Breuer M. A. "Min-cut Placement". In *J. Design Automation and Fault-Tolerant Computing*, pages 343–382, 1977.

[6] Kernighan B. Dunlop A. "A procedure for placement of standard-cell VLSI circuits". *IEEE Trans. CAD*, CAD-4:92–98, 1985. Jan.

[7] Chaudhary K et. al. "RITUAL: Performance-Driven Placement of Cell-Based ICs". In *Proc. 3rd Physical Design Workshop*, May 1991.

[8] C-K Cheng J. Huang, X-L Hong and E.S. Kuh. "An Efficient Timing Driven Global Routing Algorithm". In *DAC*, pages 596–600, 1993.

[9] Dunlop A.E. et. al. "Chip layout optimization using critical path weighting". In *DAC*, pages 133 – 136, 1984.

[10] Jackson M.A.B and Kuh E. S. "Performance-driven placement of cell based IC's". In *DAC*, pages 370 – 375, 1989.

[11] C. Tseng and D. P. Siewiorek. "Facet : A Procedure for the Automated Synthesis of Digital Systems". In *20th ACM/IEEE DAC*, pages 490–496, 1983.

[12] J.Cong et. al. Performance-driven interconnect design based on distributed RC delay model. In *DAC*, 1993.

[13] A.B. Kahng et. al. Interconnect tuning strategies for high-performance ics. 1997.

[14] J. Cong et. al. Performance optimization of vlsi interconnect layout. In *Integration, the VLSI Journal*, pages 1–94. 1996.

[15] B.Tutuianu, et. al. An explicit rc-circuit delay approximation based on the first three moments of the impulse response. In *DAC*, pages 611–616, 1996.

[16] Electronic Products. "Novel IC Shuffles Parallel Processing Data". pages 42–50, August 1986.

[17] R. W. Broderson, editor. *Anatomy of a Silicon Compiler*. Kluwer Academic Publishers, 1992.

Designing a Mask Programmable Matrix for Sequential Circuits

Fernanda Lima, Marcelo Johann, José Güntzel, Eduardo D'Avila, Luigi Carro, Ricardo Reis
Universidade Federal do Rio Grande do Sul - Instituto de Informática
Caixa Postal 15064 CEP 91501-970 - Porto Alegre - RS – Brasil
e-mail: <fglima, guntzel, johann, carro, reis>@inf.ufrgs.br
phone: +55 51 316-7036 fax: +55 51 319-1576

Key words: programmable logic, gate arrays, universal logic gates.

Abstract: This paper proposes the use of Universal Logic Gates (ULGs) as basic elements for masked programmable master-slices customizable by the topmost metal layer. This new approach called Maragata combines the efficiency of MPGAs with the flexibility of FPGA architecture. Due to the intensive use of processor-like blocks in current VLSI circuits, ULGs were developed considering the implementation of sequential circuits. A set of ULGs were studied and designed for CMOS technology. Area comparison was accomplished by mapping various combinational and sequential circuits into ULGs master-slices and to a gate array master-slice called Ágata. Results show that significant area gain and connection reduction can be achieved in this new approach.

1. INTRODUCTION

Rapid prototyping is the key to quick turnaround in a product development process. Today's fast paced design cycles require the availability of early silicon and the flexibility of ramping to any volume production. Field Programmable Gate Arrays (FPGAs) are the most popular solution for the time-to-market because they can provide instant manufacturing and low cost prototyping. Since Xilinx Company [XIL98] introduced the FPGA in 1985, many FPGAs have been developed by a number of other Companies like Actel [ACT98], Altera [ALT98].

FPGAs continue to fall short masked gate arrays in performance, density and cost for high volume. Masked Programmable Gate Arrays (MPGAs), on the other hand, have longer turnaround times. New technologies and solutions have emerged to overcome the limitations of FPGAs while maintaining the benefits of traditional gate arrays. One solution is masked gate arrays customizable only by the topmost metal layer [DON93] called Quick Customizable Logic (QCL). Another solution for fast prototyping is the Laser Programmable Gate Array (LPGA). ChipExpress Company [CHI98] offers such programmable gate arrays, which are composed of programmable logic blocks. LPGAs, MPGAs and FPGAs differ significantly in unit price, density, performance and prototyping lead times. Figure 1 shows different logic density and design time tradeoffs.

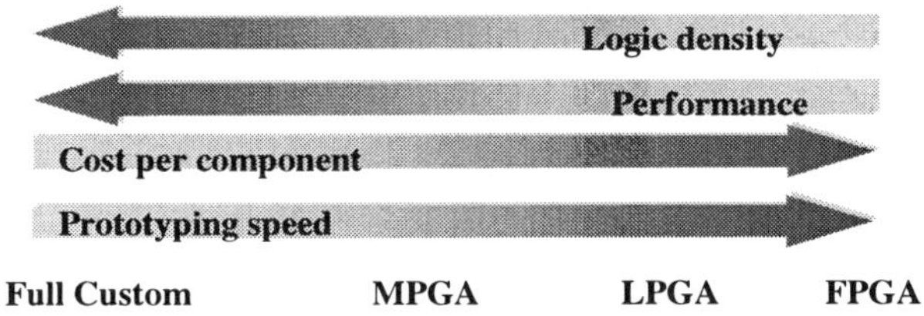

Figure 1. Digital Systems Implementation Options

Present days technology allows the integration of a large number of transistors and the possibility of integrating complete systems on a single chip. In the last 20 years, many technological breakthroughs have led to deep modifications either in our production system, or even in the way people interact with the overall world. Processor-like circuits, on a direct consequence of the integration technology availability, have made a dramatic change in the way systems are designed. The concept of an Application Specific System including microprocessors inside covers a wide range of applications, from portable systems to dedicated embedded control devices or ubiquitous computing.

Aiming at increasing logic density of digital circuits with embedded processors inside, implemented in a programmable matrix, a new methodology based on mask programmable matrix customizable by the top most metal layer is proposed. This new approach is called Maragata [LIM98]. In this methodology the transistor rows are replaced by programmable logic blocks that can be specifically named as Universal Logic Gates (ULGs). Maragata is composed of coarse grain ULGs like in a hard-wired version of a FPGA architecture that combines the efficiency of MPGAs with the flexibility of FPGA architecture. Its ULGs were developed considering the implementation of sequential and processor-like circuits, because these ULGs can implement latches or flip-flops with low area cost.

2. DESIGNING UNIVERSAL LOGIC GATES

The large flexibility of ULGs justifies its use for building up programmable matrix, particularly when customization is performed by using the topmost metal layer. When a more complex cell is used for building MPGAs, it is possible to optimize silicon area by properly sizing its transistors. Moreover, in such approach the transistor connections as well as small connections are already done. For instance, internal cell transistors that do not have to drive large capacitive loads may be smaller or even of minimum size. Overall timing performance of the cell is assured by sizing output cells as buffers by the time the matrix is designed.

A ULG can be defined as a function $f(x_1,\ldots,x_m)$ that can realize some n-input functions $G(y_1,\ldots,y_n)$ connecting x_j to 1, 0, y_i or y_i , or negating the output of G, where m>n. While designing a ULG for a specific programmable methodology, different ULG issues must be carefully analyzed. One issue is the functionality. The number of different Boolean logic functions that can be implemented by the cell defines the ULG functionality. The logic block functionality dominates the matrix logic density. Different ULGs are likely to have different amounts of functionality, and varying costs in terms of area and delay. This issue is very important because it will affect the logic density and the amount of routing resources. Two main considerations when selecting a given structure is to see how many *n*-input functions it can implement, and how easy is to implement a latch or flip-flop using all ULG resources. These logic facts must be combined with layout and electrical issues to ensure good performance and cost.

The topology of a ULG directly influences its functionality. Multiplexors, inverters and simple logic gates may be combined together to build ULGs. A set of topologies was studied analyzing their functionality to achieve an optimal ULG. The goal is to find a ULG capable of realizing as many functions as possible without compromising silicon area and performance.

We have done some research to select a good ULG, looking for low granularity, high flexibility and the availability of a technology mapper. Figure 2 shows some ULGs developed to Maragata approach.

The proposed ULGs to Maragata can implement either combinational logic or sequential logic. Most of FPGAs have logic blocks that can implement combinational logic. To implement sequential logic it is necessary a flip-flop per logic blocks. When this logic block is used only for combinational logic, the flip-flop area is wasted. The ULG3 can implement a flip-flop master-slave (with set and reset) using its multiplexors. It is necessary two ULG3 to implement one flip-flop and only one CLUS2 to implement the same flip-flop.

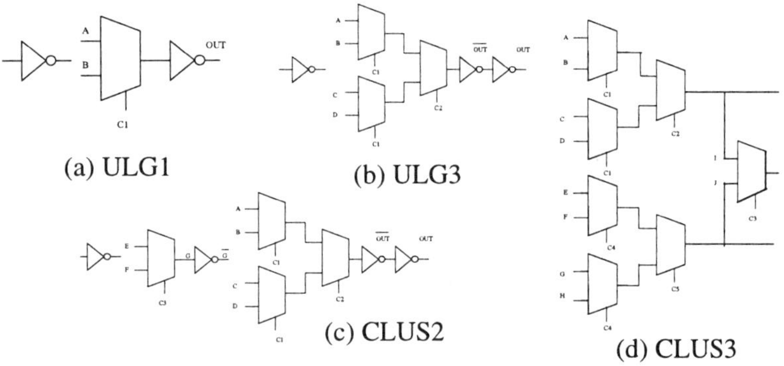

Figure 2. ULGs developed to Maragata

These ULGs developed to Maragata approach were compared to some ULGs developed to commercial FPGAs and LPGAs. Figure 3 shows some ULGs proposed in [LIN97] and in FPGA families like Actel and ChipExpress.

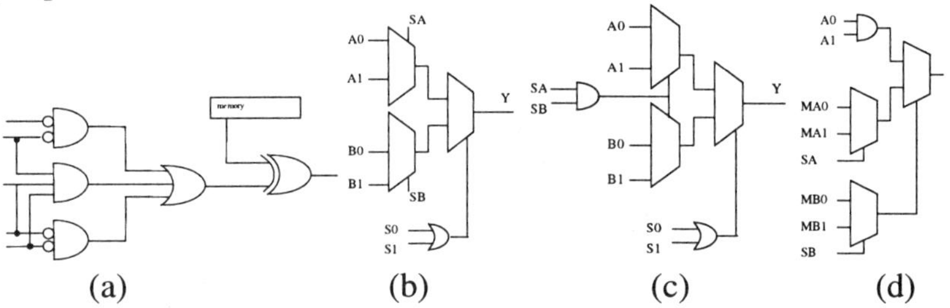

Figure 3. ULGs proposed in [LIN97] (a), Actel (b)(c) and ChipExpress (d) logic blocks

The ULGs presented in figure 2 and in figure 3 were analyzed using the Programa_de_TV tool [LIM99b] and their functionality was obtained. This program is able to realize all the NPN operations (N_I, P and N_O) over any n-input Lookup Tables, where n can be 2, 3 or 4. From the ULG description it is possible to identify all NPN classes that can be realized by this ULG. The graph on figure 4 shows the number of n-input NPN classes implemented by these ULGs. Figure 4 also shows the granularity of ULGs. The relative area indicates this granularity. The number of transistors was calculated based on multiplexors composed of transmission gates. It is necessary to use 6 transistors to implement one 2:1 multiplexor. For example, ULG3 in the x-axis implements all 2-input and 3-input logic functions, few 4-input logic functions and occupies a small fraction of the largest studied ULG area (CLUS3). This one, although with larger area, implements all 4-input logic functions. ULG3 is the best tradeoff among number of logic functions, area and possibility of implementing a full flip-flop.

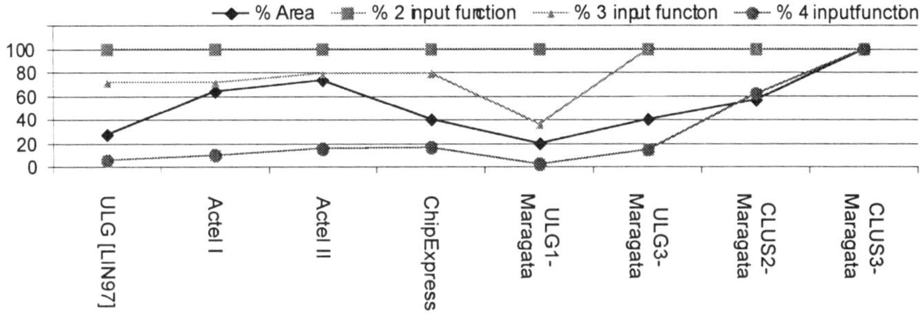

Figure 4. Percentage of number of NPN classes implemented by some ULGs presented in fig. 2 and fig 3, and the percentage of area, respectively

The multiplexers of Maragata ULGs were implemented by using transmission-gates rather than by CMOS static gates, to minimize not only transistor count, but power dissipation as well. In order to achieve minimum layout area, minimum width transistors were used whenever it is possible. In each ULG output transistors were sized to work as buffers. Internal fixed and customizable cell connections may contribute to reduce channel routing complexity.

Table 1 shows the number of transistors and area for all developed ULGs. They have been developed in 0.8 μm double metal layer CMOS technology. All the customizable connections are done over the ULG without using the routing channel. The first metal layer was used for internal connections, while the second one was reserved for customization. Table 1 also presents the area comparison for a master-slave flip-flop implemented into different ULGs. The cell CLUS3 can either implement 1 bit register or a D flip-flop.

Table 1. ULGs Characteristics

ULG	# transistors	Area (μm^2)	# ULGs to implement a flip-flop	Area (μm^2) of a flip-flop
ULG1	10	1057	4	4228
ULG3	22	1922	2	3844
CLUS2	30	3000	1	3000
CLUS3	50	5000	1	5000

Figure 5 presents a circuit layout implemented in the Maragata matrix. The customization is done in metal 2. This matrix is composed of 26 rows, 80 pads and has 1040 ULG3s. The matrix area is about 11.03 mm^2. Its logic density is 2263 tr/mm^2. It is important to notice that the routing channel takes a significant area. By reducing connections one can expect a large reduction in the total matrix area.

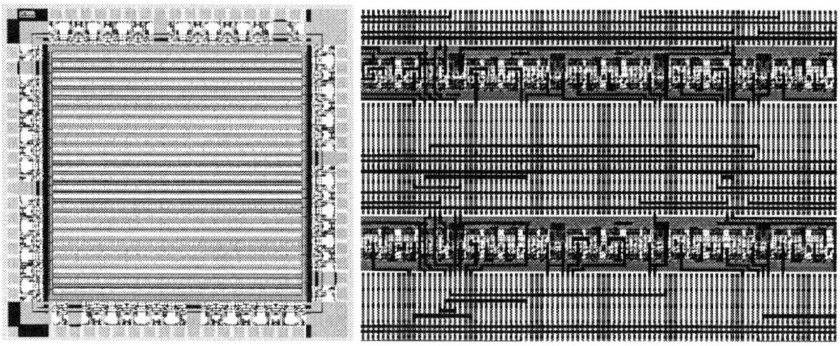

Figure 5. Matrix layout (the routing channel, the ULG rows and the customization in metal 2)

3. AREA AND CONNECTION COMPARISON

In order to evaluate the silicon area gain at Maragata approach, combinational and sequential circuits were mapped into conventional MPGA library elements and to ULGs from Maragata. SIS [SEN92] was used for logic simplification and technology mapping. In case of ULGs, the n-LUT mapper (SIS) was used (xl_part_coll -m -g 2 -n n; xl_coll_ck -n n; xl_partition -m -n n; simplify; xl_imp -n n; xl_partition -t -n n; xl_cover -e 30 -u 200 -n n; xl_coll_ck -n n).

The conventional MPGA approach chosen for benchmarking is a single metal layer customizable, based on transistor rows, refered to as Ágata [CAR96]. It has been designed using the same 0.8μm double metal layer CMOS technology. Its library of personalization patterns has 13 logic functions: inverter, 2-, 3- and 4-input nands and nors, 2-input xor and nxor, 2 to 1 and 4 to 1 multiplexers. There are also two latches and one D master-slave flip-flop with set and reset. In the case of Ágata, it was used the command map -m 0.

Table 2 shows some area and connections results for some combinational and sequential circuits implemented into ULG3. Some of these circuits are from the MCNC'91 benchmark. Previous research [LIM99a] showed that the most appropriated ULG in terms of area and number of connection efficiency is the ULG3.

The area result presents only the amount of active area (cell area without taking account the routing channel) used in each design. Connectivity plays an important role in QCL like designs, since the number of possible connections is fixed. For the Ágata approach there is a complete router working [JOH97], and this same router is used in the Maragata approach. Since the ULG can embed more logic in its core, much less internal connections are needed in the Maragata case when compared to Ágata, as it is showed in table 2.

Table 2. Area comparison between Ágata and Maragata ULGs for combinational and sequential circuits mapped using SIS

Circuits	MAP Ágata		ULG3 (3-LUT)		Comparison	
	CX	Area	Area	CX	%A	%CX
Booth	433	350232.4	282534	257	-19	-41
Booth12	690	545349.2	417074	380	-24	-45
Booth16	936	731685.2	574678	526	-21	-44
Boothp8	829	537555.6	461280	475	-14	-43
Bisect8	919	610075.2	551614	549	-10	-40
Desloca	379	317387.2	251782	224	-21	-41
Descola12	514	764208.8	524706	535	-31	4
8051_control	125	72194.4	78802	92	9	-27
8051_valida	267	150242.4	188356	200	25	-25
8051_stateg	143	108940.0	105710	99	-3	-31
8051_multipl.	885	594790.8	517018	511	-13	-42
Raiz_bs	664	678666.8	526628	383	-22	-42
Raiz_ds	796	790859.4	601586	453	-24	-43
Raiz4q	5102	4733799.6	3865142	3090	-18	-39

Results shows that area reduction may be achieved when the Maragata approach is used to implement a microprocessor (8051) and arithmetic functions like multipliers (booth, boothp, desloca, and bisect) and square extractor (raiz_bs, raiz_ds and raiz4q) using the ULG3 cell. The use of ULGs resulted in area gains around 20% for almost all examples. Only two circuits implemented in Ágata presented a smaller area instead of implementing in Maragata. For almost all circuits implemented using ULG3, the number of internal connections has been reduced around 40%. None of these circuits presented a simultaneous increase in area and connection.

4. CONCLUSIONS

The applicability of ULGs to Quick Customizable Logic (QCL) master-slices has been presented. Comparisons in terms of area gain were developed by mapping some small and medium combinational and sequential circuits. For these circuits, the use of ULGs resulted in area gains around 20% for combinational and sequential circuits. The number of required connections for different examples was also calculated. Results shows that the Maragata approach leads to effective reduction in the number of connections. These area and connection gains can represent a logic density improvement because more connections can be done in the same routing channel.

Current works include the investigation of more appropriate mapping algorithms for ULGs master-slices, to further improve results in both combinational and sequential circuit cases.

5. REFERENCES

[ACT98] ACTEL CORPORATION. Introduction to Actel FPGA Architecture. http://www.actel.com. (Nov.1998).

[ALT98] ALTERA CORPORATION. **Data Sheet.** http://www.altera.com (Nov. 1998).

[BET98] BETZ, Vaughn; ROSE, Jonathan. How Much Logic Should Go in an FPGA Logic Block? **IEEE Design & Test of Computer**. New York, v. 15, n. 1, 1998. http://www.eecg.utoronto.ca/~jayar/ (Aug. 1998).

[CAR96] CARRO, L.; et al. An environment to Design Digital Circuits Based on the Brazilian Gate-Array. In: WORKSHOP IBERCHIP, 2., 1996. **Memorias...**, São Paulo, [s.n.], p.198, Feb. 1996.

[CHI98] CHIP EXPRESS CORPORATION. **Data Sheet.** http://www.chipexpress.com (Nov. 1998).

[DON93] DONG, S. K., et al. Two channel Routing Algorithms for Quick Customized Logic. In: EDAC, 1993. **Proceedings...** Los Alamitos : IEEE Computer Society, 1993. 2 V. p. 122-126, 1993.

[HOP99] HOPKIN, Vince. **Programmable Device or Gate Array?** http://www.isdmag.com (Jan. 1999).

[JOH97] JOHANN, M.; CARRO, L.; REIS, R. Functional Design of GAROTA: Gate Array Router of ÁGATA System. In: Brazilian Symposium on Integrated Circuit Design, 10., 1997, Gramado, RS. p. 21-30. **Proceedings...** Porto Alegre: CPGCC da UFRGS. 1997.

[LIM98] LIMA, Fernanda G.; et al. Improving Logic Density of QCL Masterslices by Using Universal Logic Gates. In: Brazilian Symposium on Integrated Circuit Design, 11., 1998, Búzios, RJ. **Proceedings...** Los Alamitos : IEEE Computer Society, 1998. 250 p.

[LIM99a] LIMA, Fernanda G.; et al. Designing Masked Programmable ULGs for MPGAs. In: WORSHOP IBERCHIP, 4., 1999. **Memorias...** [LIMA : HOZLO S.R.L.], 1999. http://www.inf.ufrgs.br/ ~fglima/maragata.html

[LIM99b] LIMA, Fernanda G.; et al. Programa_de_TV Tool. In: SBCCI, 12., **Proceedings...**, 1999. To be published by IEEE Computer Society.

[LIN97] LIN, Chih-Chang; MAREK-SADOWSKA, Malgorzata; GATLIN, Duane. On Designing Universal Logic Gate and their Application to FPGA Design. **IEEE Transactions on Computer-Aided Design of Integrated Circuits and Systems.** New York, v. 16, n. 5, May 1997.

[SEN92] SENTOVICH, E.; et al. **SIS: A System for Sequential Circuit Synthesis**. Memorandum No. UCB/ERL M92/41. May. 1992.

[XIL98] XILINX CORPORATION. **Devices Data Sheet.** http://www.xilinx.com. (Nov. 1998).

Placement Benchmarks for 3-D VLSI

Stefan Thomas Obenaus
School of Computer Science
McGill University

Ted H. Szymanski
Communications Research Laboratory
McMaster University

Abstract We provide a general definition of the 3-D wirelength placement problem. This definition facilitates comparison of 3-D placement algorithms. Wirelength results using *partitioning placement* are included for the ACM/SIGDA and ISPD98 standard benchmark circuit suites. Further, a wirelength comparison between 2- and 3-D placements is made, and it is shown that larger circuits require 50%-60% less wirelength when utilising the third dimension.

Keywords: 3-D, Benchmarks, Placement, VLSI, Wirelength

1. INTRODUCTION

As research into three dimensional circuit architectures becomes more widespread (Chiricescu and Vai, 1998; Depreitere et al., 1994; Leeser et al., 1998; Leighton and Rosenberg, 1986; Ohmura, 1998; Tong and Wu, 1995), the study of techniques for placing circuits into 3-D structures is emerging. Unlike 2-D placement papers, previous 3-D placement results have been published in isolation (Leeser et al., 1998; Ohmra, 1998) and cannot be compared to each other for lack of benchmark results. In two dimensions wirelengths benchmarks have been a common means of comparing placement algorithms. However, in 3-D no tabulated results for comparison exists.

Our goal is to fill the void and provide wirelength results for comparison of 3-D placement techniques. For this purpose, we have extended the established method of *partitioning placement* (Shahookar and Mazumder, 1991) to three dimensions. Other common placement methods are *simulated annealing* and *force-directed placement* (Shahookar and Mazumder, 1991). We use the currently fastest and best partitioner, hMetis (Karypis et al., 1997). The details of this placement method are described in section 2.. While using a different partitioner, the partitioning placement technique has already been used for 3-D placement for the Rothko architecture (Leeser et al., 1998).

Partitioning placement is fast and has traditionally yielded good results (Shahookar and Mazumder, 1991). For large circuits, which, as we show in section 3.2, profit most from 3-D technology, an asymptotically fast placement algorithm is essential. Asymptotically slower techniques, such as simulated annealing, are not likely to produce high quality results in an acceptable period of time for these circuits as the cooling schedule would have to be drastically shortened.

Given the high quality of the partitioner, and the fundamental soundness of the partitioning placement technique, we have confidence that the placement results presented here provide a good benchmark against which future 3-D placement techniques can be measured.

1.1 PLACEMENT MODEL

In order to provide a convenient basis for benchmark comparisons, the placement model should reflect reality, yet be general and easy to implement. In two dimensions, the *Checkerboard model* (Shahookar and Mazumder, 1991), a.k.a. the *Gate Array Model*, serves this purpose. In this model, circuit elements are of identical size and placed in checkerboard-fashion onto a grid. We extend this model to three dimensions in the natural way. While the purpose of this model is to facilitate comparison of placement algorithms, actual placement methods will have much more specific requirements for the grid size and shape, size of circuit elements, pad placements, etc.

The most common measure of quality of a placement algorithm is the total wirelength required for the circuit (Shahookar and Mazumder, 1991). The most efficient method of connecting points, eg. circuit nodes, in space along horizontal and vertical lines using a minimum of wirelength is to construct a *rectilinear Steiner tree* (Hanan, 1966).

Definition 1 *Rectilinear Steiner Tree*

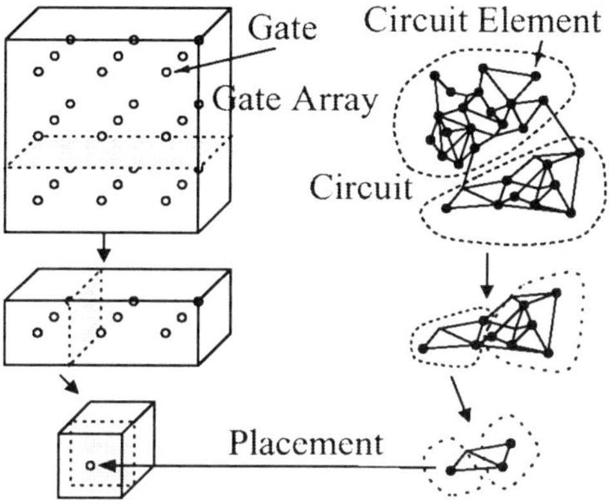

Figure 1.1 Partition Placement Process: Simultaneous Splitting of the Grid and Partitioning of the Circuit

A rectilinear Steiner tree $S(e, f)$ is the shortest tree that connects all nodes $v \in e$ at positions $f(v)$ using only orthogonal segments parallel to the coordinate axes. Its length is $|S(e, f)|$.

However, constructing a rectilinear Steiner tree and measuring its length is difficult. A common approach is to estimate the size of the Steiner tree by adding the width, height, and depth spanned by the nodes in the Steiner tree. This measure is accurate for two and three node nets. We call this approximation the semi-perimeter bounding box approximation $|S^*(e, f)| \approx |S(e, f)|$.

For placement purposes, we reduce all circuit elements to unit-size nodes, and thus we abstract the circuit into a hypergraph $G(V, E)$. $G(V, E)$ is simply the combination of the set of circuit elements, or nodes, V, with the set of nets, or hyperedges, E, where for all $e \in E$, $e \subset V$.

It remains to define the 3-D placement model, as a straight forward extension of the 2-D Checkerboard model, for comparison purposes:

Definition 2 *3-D Placement Model*

Given a circuit $G(V, E)$ find a reversible function

$$f : |V| \rightarrow \{1, \ldots, n_1\} \times \{1, \ldots, n_2\} \times \{1, \ldots, n_3\}.$$

The measure of quality of the placement function is the total wirelength estimate $\sum_{e \in E} |S^*(e, f)|$.

2. PLACEMENT METHOD

Partitioning placement is one of the fundamental placement methods (Shahookar and Mazumder, 1991). When placing a circuit into a two

Variables and predicates:

$V = \{v_1, \dots, v_{\|V\|}\}$	set of circuit nodes
$x[v]$	coordinates of gate for circuit node v
(a_1, a_2, a_3)	coordinates of lower left front corner
(b_1, b_2, b_3)	lengths of sides of gate array box
(n_1, n_2, n_3)	initial size of gate array box

Initial Call:

    place(V,$(0,0,0)$,(n_1, n_2, n_3))

```
place(V,(a1,a2,a3),(b1,b2,b3))
    if |V|=1 then
        x[v1] := (a1,a2,a3)
    else
        find largest side of box
        k := i such that b_i = max(b1,b2,b3)
        split box b into two boxes b1 and b2
        (b1_1,b1_2,b1_3) := (b1,b2,b3)
        b1_k := ⌊b1/2⌋
        (b2_1,b2_2,b2_3) := (b1,b2,b3)
        b2_k := ⌈b1/2⌉
        determine coordinates of lower left front corner of b1 and b2
        (a1_1,a1_2,a1_3) := (a1,a2,a3)
        (a2_1,a2_2,a2_3) := (a1,a2,a3)
        a2_k := a_k + b1_k
        partition V into sub-circuits V1 and V2
        of sizes no more than b1_1·b1_2·b1_3 and b2_1·b2_2·b2_3, respectively
        (V1,V2) := partition(V,b1_1·b1_2·b1_3,b2_1·b2_2·b2_3)
        invoke placement routine on sub-circuits
        place(V1,(a1_1,a1_2,a1_3),(b1_1,b1_2,b1_3))
        place(V2,(a2_1,a2_2,a2_3),(b2_1,b2_2,b2_3))
```

Figure 1.2 Generic Partitioning Placement Algorithm

dimensional gate array, the gates in the gate array are recursively split into smaller sub-arrays. At the same time the circuit is partitioned and the sub-circuits are assigned to the sub-arrays until each circuit element can be assigned to its unique gate. This process extends to three dimensions in the obvious way, as is illustrated in figure 1.1. Figure 1.2 provides the partition placement algorithm.

Leeser et al. (Leeser et al., 1998) used a partitioning placement method for placement in the Rothko architecture. Their partition placement method was based on a 2-D variation of partitioning placement, called *quadrisection* (Shahookar and Mazumder, 1991). In quadrisection, the chip area is recursively split into four quadrants and circuits are recursively partitioned four ways. They extended this method into three dimensions by splitting the chip's volume into eight octants while concurrently partitioning the circuit eight ways. However, no placement results were published.

Circuit	Nodes	Nets	Pins	$\frac{Pins}{Node}$	$\frac{Pins}{Net}$	Circuit	Nodes	Nets	Pins	$\frac{Pins}{Node}$	$\frac{Pins}{Net}$
19ks	2844	3282	10547	3.71	3.21	s15850P	10470	10383	24712	2.36	2.38
avq.large	25178	25384	82751	3.29	3.26	s35932	18148	17828	48145	2.65	2.70
avq.small	21918	22124	76231	3.48	3.45	s38417	23949	23843	57613	2.41	2.42
baluP	801	735	2697	3.37	3.67	s38584	20995	20717	55203	2.63	2.66
biomedP	6514	5742	21040	3.23	3.66	s9234P	5866	5844	14065	2.40	2.41
golem3	103048	144949	338419	3.28	2.33	structP	1952	1920	5471	2.80	2.85
industry2	12637	13419	48158	3.81	3.59	t2	1663	1720	6134	3.69	3.57
industry3	15406	21923	65791	4.27	3.00	t3	1607	1618	5807	3.61	3.59
p1	833	902	2908	3.49	3.22	t4	1515	1658	5975	3.94	3.60
p2	3014	3029	11219	3.72	3.70	t5	2595	2750	10076	3.88	3.66
s13207P	8772	8651	20606	2.35	2.38	t6	1752	1641	6638	3.79	4.05

Table 1.1 The 1993 ACM/SIGDA Benchmark Circuits

As our partitioner, we chose the hMetis hypergraph partitioner developed by Karypis et al. (Karypis et al., 1997). This partitioner is currently the best and fastest partitioner published. Although we use recursive two-way partitioning, we could easily implement 3-D quadrisection with a few modifications to the hMetis library interface. Restrictions in the current hMetis library interface made it necessary to compute a recursive balanced $(k+l)$-way partitioning to achieve a $k{:}l$ split as is sometimes necessary in step 12 of the algorithm in figure 1.2 when an odd number of rows, columns, or layers needs to be split. While this increases run-time and memory requirement, it does not affect the quality of the cut (Karypis, 1999). According to Karypis, the hMetis interface could easily be adapted to allow explicit $k{:}l$ cuts.

In order to compensate for the excessive memory requirement for large $(k+l)$-way partitionings, we restricted the largest dimensions of the gate arrays for the largest circuits to be of even length. Consequently, the largest cuts are balanced two-way cuts which require substantially less memory resources. Further, to obtain accurate estimates of the runtime, assuming the hMetis interface was adapted to allow explicit $k{:}l$ splits, we ran the algorithm while forcing balanced 2-way splits at all levels of recursion.

3. RESULTS

3.1 3-D RESULTS

We performed the benchmark runs on the circuits in the ACM/SIGDA circuit suite (Brglez, 1993), and the newer ISPD98 benchmark circuit suite (Alpert, 1998), cf. tables 1.1 and 1.2. While the ACM/SIGDA

Circuit	Nodes	Nets	Pins	$\frac{Pins}{Node}$	$\frac{Pins}{Net}$	Circuit	Nodes	Nets	Pins	$\frac{Pins}{Node}$	$\frac{Pins}{Net}$
ibm01	12752	14111	50566	3.97	3.58	ibm10	69429	75196	297567	4.29	3.96
ibm02	19601	19584	81199	4.14	4.15	ibm11	70558	81454	280786	3.98	3.45
ibm03	23136	27401	93573	4.04	3.41	ibm12	71076	77240	317760	4.47	4.11
ibm04	27507	31970	105859	3.85	3.31	ibm13	84199	99666	357075	4.24	3.58
ibm05	29347	28446	126308	4.30	4.44	ibm14	147605	152772	546816	3.70	3.58
ibm06	32498	34826	128182	3.94	3.68	ibm15	161570	186608	715823	4.13	3.84
ibm07	45926	48117	175639	3.82	3.65	ibm16	183484	190048	778823	4.24	4.10
ibm08	51309	50513	204890	3.99	4.06	ibm17	185495	189581	860036	4.64	4.54
ibm09	53395	60902	222088	4.16	3.65	ibm18	210613	201920	819697	3.89	4.06

Table 1.2 The 1998 ISPD Benchmark Circuits

Circuit	Grid			Average Wirelength	Standard Deviation	Minimum Wirelength	Runtime (s)
	n_1	n_2	n_3	$\sum\|S^*\|$		$\sum\|S^*\|$	
19ks	15	14	14	14493.3	1.61%	14123	37
avq.large	30	30	28	104104.1	1.22%	101693	303
avq.small	28	28	28	94688.0	0.88%	93823	277
baluP	10	9	9	3263.5	1.77%	3164	13
biomedP	19	19	19	25239.2	0.80%	24839	106
golem3	48	48	45	687104.9	0.68%	679554	1519
industry2	24	23	23	78997.7	0.92%	78052	202
industry3	26	25	24	152962.3	1.05%	149592	301
p1	10	10	9	4156.1	2.17%	4071	14
p2	15	15	14	18562.5	1.64%	18044	43
s13207P	21	21	20	26501.3	2.15%	25617	103
s15850P	22	22	22	30950.7	0.65%	30729	122
s35932	27	26	26	57926.1	1.74%	56120	218
s38417	29	29	29	73282.6	1.21%	72355	253
s38584	28	28	27	72643.9	1.10%	71560	258
s9234P	19	18	18	17670.1	0.89%	17463	86
structP	13	13	12	7064.1	1.41%	6879	23
t2	12	12	12	8501.9	1.44%	8337	22
t3	12	12	12	7828.2	1.42%	7658	22
t4	12	12	11	7375.9	1.63%	7242	22
t5	14	14	14	12568.2	0.57%	12481	36
t6	13	12	12	7968.1	1.51%	7785	23

Table 1.3 3-D Placement Results for the ACM/SIGDA Circuit Suite

Circuit	Grid			Average Wirelength	Standard Deviation	Minimum Wirelength	Runtime (s)
	n_1	n_2	n_3	$\sum \lvert S^* \rvert$		$\sum \lvert S^* \rvert$	
ibm01	24	24	23	92601.5	0.91%	90591	239
ibm02	28	28	26	202121.7	0.65%	199264	391
ibm03	30	30	27	234600.9	0.94%	231824	433
ibm04	32	30	29	301324.1	1.07%	297370	497
ibm05	32	32	30	367004.5	1.27%	360659	554
ibm06	32	32	32	333985.0	3.21%	323919	655
ibm07	36	36	36	473844.3	1.82%	458047	1084
ibm08	38	38	36	531860.7	1.38%	521203	1191
ibm09	38	38	37	617201.7	2.56%	587107	1263
ibm10	42	41	41	832125.1	3.79%	796763	1761
ibm11	42	41	41	872118.2	2.72%	842491	1661
ibm12	42	42	41	991783.9	1.63%	959567	1864
ibm13	44	44	44	1000941.8	2.05%	971330	1864
ibm14	54	54	52	1657408.1	1.12%	1630502	3064
ibm15	56	56	54	1994685.8	1.38%	1957427	3769
ibm16	58	58	57	2222138.0	1.05%	2190510	4030
ibm17	58	58	57	2745042.7	1.16%	2680986	4463
ibm18	60	60	59	2639356.6	4.29%	2504083	4285

Table 1.4 3-D Placement Results for the ISPD98 Circuit Suite

suite is more established, it lacks the larger circuits that can be found in the ISPD98 suite.

Since the hMetis partitioner is randomized, we perform 10 runs for each circuit on a Pentium II/300MHz system. The results are summarised in tables 1.3 and 1.4. The runtimes indicate the runtime for one run.

3.2 FROM 2 TO 3 DIMENSIONS

Besides having a reference platform for 3-D placement wirelengths, it is of interest to know what kind of improvement can be expected from 3-D VLSI. In table 1.5 we have compiled wirelength results for four circuits covering the dynamic range from 800 to 210,000 nodes. For each circuit we computed a placement in $d = 2$, 2 1/3, 2 2/3, and 3 dimensions. By this we mean that the base of the gate array has an edge length of $\sqrt[d]{N}$. Thus a 2-D gate array, is of size $\sqrt{N} \times \sqrt{N} \times 1$, and a 3-D gate array of size $\sqrt[3]{N} \times \sqrt[3]{N} \times \sqrt[3]{N}$. In general, for dimension d and an N-node circuit, we selected a grid size close to $\sqrt[d]{N} \times \sqrt[d]{N} \times N/(\sqrt[d]{N})^2$.

Circuit	Nodes	Grid			Dimen-sions	All Nets Counted		2- and 3-node Nets Only	
	$\|V\|$	n_1	n_2	n_3	d	Length $\sum\|S^*\|$	Change	Length $\sum\|S^*\|$	Change
p1	833	29	29	1	2	6751.5		3800.0	
		17	17	3	2 1/3	4584.2	-32.1%	2727.5	-28.2%
		12	12	6	2 2/3	4175.2	-38.2%	2505.5	-34.1%
		10	10	9	3	4143.5	-38.6%	2498.8	-34.2%
s9234P	5866	78	77	1	2	28353.2		17515.8	
		40	38	4	2 1/3	19192.9	-32.3%	13019.4	-25.7%
		26	26	9	2 2/3	17891.6	-36.9%	12598.2	-28.1%
		19	18	18	3	17640.6	-37.8%	12404.3	-29.2%
ibm06	32498	184	180	1	2	792670.2		224425.6	
		82	82	5	2 1/3	410851.0	-48.2%	125969.6	-43.9%
		50	50	13	2 2/3	372507.4	-53.0%	133579.7	-40.5%
		32	32	32	3	331429.9	-58.2%	116558.5	-48.1%
ibm18	210613	460	460	1	2	7705843.6		2561197.9	
		188	188	6	2 1/3	3597492.0	-53.3%	1258260.2	-50.9%
		98	98	22	2 2/3	2932640.5	-61.9%	1086833.2	-57.6%
		60	60	59	3	2641617.1	-65.7%	969398.1	-62.2%

Table 1.5 Wirelength Improvement from Two to Three Dimensions

Besides the usual total wirelength estimate, table 1.5 also shows the total wirelengths of nets with 2 and 3 nodes. For such nets, the semi-perimeter bounding box approximation of the wirelength is exact.

We notice that a substantial wirelength advantage can be achieved for large circuits even when only a few layers are employed. While small circuits exhibit only a modest wirelength improvement in 3-D, larger circuits clearly benefit from the third dimension. Small circuits `p1` and `s9234P` experience only a 30%-40% improvement in a 3-D placement, whereas `ibm18` saves approximately 50% wirelength in 2 1/3 dimensions and over 60% in 3 dimensions.

4. CONCLUSION

We presented a general definition of the 3-D placement problem. This definition can be used to compare 3-D placement algorithms to each other. On the basis of this definition, we implemented a partitioning placer and generated a first set of comprehensive wirelength results for 3-D placements of circuits in two benchmark circuits suites.

Further, we compared wirelength results of selected circuits for placements varying from two to three dimensions. These results provide empirical evidence that large circuits benefit substantially from utilising the third dimension even if only a small number of layers in the third dimension is used.

Acknowledgments

We thank Prof. Karypis for making the hMetis 1.5.3 library available and providing support for its use.

References

Alpert, C. J. (1998). The ISPD98 circuit benchmark suite. In *Proceedings of the International Symposium on Physical Design*, pages 85–90.

Brglez, F. (1993). ACM/SIGDA design automation benchmarks: Catalyst or anathema? *IEEE Design & Test of Computers*, 10(3):87–91.

Chiricescu, S. M. S. A. and Vai, M. M. (1998). A three-dimensional FPGA with an integrated memory for in-application reconfiguration data. In *1998 IEEE International Symposium on Circuits and Systems*, volume 2, pages 232–235.

Depreitere, J., Neefs, H., Van Marck, H., Van Campenhout, J., Baets, R., Dhoedt, B., Thienpont, H., and Veretennicoff, I. (1994). An optoelectronic 3-d field programmable gate array. In *Field-programmable logic : architectures, synthesis, and applications : 4th International Workshop on Field-Programmable Logic and Applications*, volume 849 of *Lecture notes in computer science*, pages 352–360.

Hanan, M. (1966). On Steiner's problem with rectilinear distance. *SIAM Journal on Applied Mathematics*, 14(2):255–265.

Karypis, G. (1999). private communication.

Karypis, G., Aggarwal, R., Kumar, V., and Shekhar, S. (1997). Multilevel hypergraph partitioning: Application in VLSI domain. In *Proceedings of the 34th ACM/IEEE Design Automation Conference*, pages 526–529.

Leeser, M., Meleis, W. M., Vai, M. M., Chiricescu, S., Xu, W., and Zavracky, P. M. (1998). Rothko: A three-dimensional FPGA. *IEEE Designs and Test of Computers*, 15(1):16–23.

Leighton, F. T. and Rosenberg, A. L. (1986). Three-dimensional circuit layouts. *SIAM Journal on Computing*, 15(3):793–813.

Ohmura, M. (1998). An initial placement algorithm for 3-d VLSI. In *1998 IEEE International Symposium on Circuits and Systems*, volume 6, pages 195–198.

Shahookar, K. and Mazumder, P. (1991). VLSI cell placement techniques. *ACM Computing Surveys*, 23(2):143–220.

Tong, C. C. and Wu, C. (1995). Routing in a three-dimensional chip. *IEEE Transactions on Computers*, 44(1):106–117.

Substrate Noise: Analysis, Models, and Optimization

Edoardo Charbon
Cadence Design Systems, San Jose, CA 95134
charbon@cadence.com

Joel Phillips
Cadence Berkeley Laboratories, Berkeley, CA 94704
jrp@cadence.com

Abstract In this paper we address the main problems posed by substrate noise from two complementary points of view. We look at the effects of substrate noise on performance and reliability in digital, analog and mixed-signal circuits. The mechanisms underlying noise generation, injection, and transport are also analyzed. Solutions to the substrate noise problem using design and layout techniques, as well as accurate analysis and optimization are discussed.

Keywords: Substrate, Optimization, Scaling, Parasitic Extraction, Green's function, Fast Cosine Transform

1. INTRODUCTION

In the past decade, substrate noise has had a constant and significant impact in the design of analog and mixed-signal integrated circuits. Only recently, with the advances in chip miniaturization and innovative circuit design, substrate noise has begun to plague fully digital circuits as well. To combat the effects of substrate noise, heavily over-designed structures are generally adopted, thus seriously limiting the advantages of innovative technologies. For this reason, the modeling of substrate noise is receiving serious attention. Macro-models mimicking substrate noise sources and better simulation of transport mechanisms have been developed, thus allowing designers to detect potential problems before fabrication. Specific guidelines have also been drafted for more aggressive, substrate-aware design practices.

Accurately characterizing substrate noise is problematic for various reasons. The noise results from superposition of a large number of local and remote sources, each attenuated and delayed in a unique way. Modeling signal attenuation and delay individually may be extremely time consuming and would require accurate *ad hoc* characterization of all the sources, which is in itself a hard problem. Substrate noise can be decomposed, at a macroscopic level, in intrinsic and switching noise. Intrinsic noise is a background spurious signal originated in active and passive devices through various physical phenomena, namely thermal, shot and flicker noise. Switching noise originates in digital blocks where frequent state transitions, occurring in gates across the chip, cause current pulses to be absorbed from and transmitted to power and ground buses through direct feedthrough and the charge/discharge of loads. Such pulsing currents are partially injected into the substrate through impact ionization and capacitive coupling.

Switching noise coupled through the substrate is very destructive as it can be broadcasted over great distances and picked up by sensitive circuits, by way of capacitive coupling and body effect. The resulting threshold voltage modulation dynamically changes gate delays locally, thus impacting performance in ways that are difficult to predict. Switching noise has an especially detrimental effect on dynamic logic, memories and embedded analog circuits, such as phase lock loops.

Accurate estimation of substrate noise requires several techniques to model switching noise injection, transport, and reception mechanisms, both at the microscopic and macroscopic level. Moreover, means are necessary to embed substrate noise in optimization loops and trend analysis tools. This is critical to help designers detect substrate related problems as early as possible, tailoring design practices, in order to avoid lengthy redesign cycles.

The paper is organized as follows. In Section 2. the physical mechanisms underlying substrate noise generation, transport and absorption are outlined. Section 3. reviews several substrate noise analysis techniques, while Section 4. presents substrate-aware optimization methods applied to physical design problems.

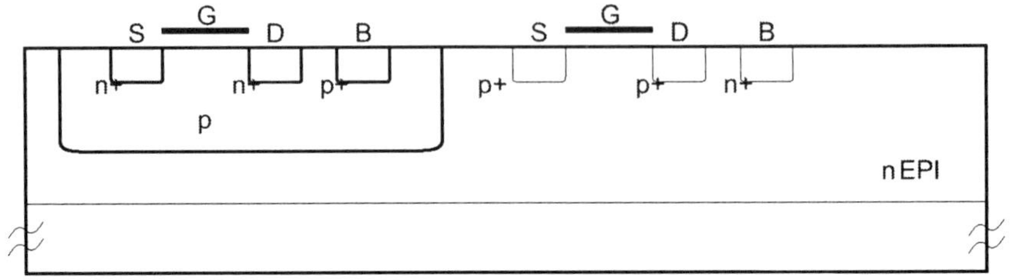

Figure 1 Substrate cross-section of a CMOS inverter

2. NOISE COUPLING MECHANISMS

2.1 Noise Injection and Reception

Substrate noise is caused mainly by the switching activity of fast digital circuits and it is injected into the substrate via *impact ionization* and *capacitive coupling* mechanisms. Figure 1 shows a typical cross-section of substrate epitaxial layer on which a CMOS inverter is integrated. Impact ionization is caused by electron-hole pairs generated in the pinch-off region, when the electric field exceeds a given threshold. The excess holes are collected in the region of substrate under the device and from there they are transported throughout the chip. Impact ionization currents are evaluated as

$$I_{impact} = \int_{E_s}^{E_m} I_d A e^{-B/E(x)} dx, \tag{1.1}$$

where E_s, E_m, $E(x)$ and I_d are source electric field, maximum electric field, local electric field and drain current, respectively. Constants A and B are material related coefficients. Formulae relating these parameters to measurable quantities and the derivation of (1.1) can be found in [1]. Since $E_m \gg E_s$, integral (1.1) can be approximated to

$$I_{impact} \simeq \frac{A}{B} l E_m I_d e^{-B/E_m} = C_1(V_{ds} - V_{dsat}) I_d e^{-\frac{C_2}{(V_{ds} - V_{dsat})}}, \tag{1.2}$$

where l, V_{ds} and V_{dsat} are effective channel length, drain-source voltage and saturation voltage, respectively. C_1 and C_2 are material related coefficients [1]. Equation (1.2) is used by most MOSFET models to represent impact ionization currents [2].

Switching noise can also be coupled into substrate capacitively through reverse biased junctions and metal-to-diffusion capacitances. All spurious currents injected into the substrate travel through the bulk reaching varying depths and resurfacing to be collected by low-resistivity pick-ups.

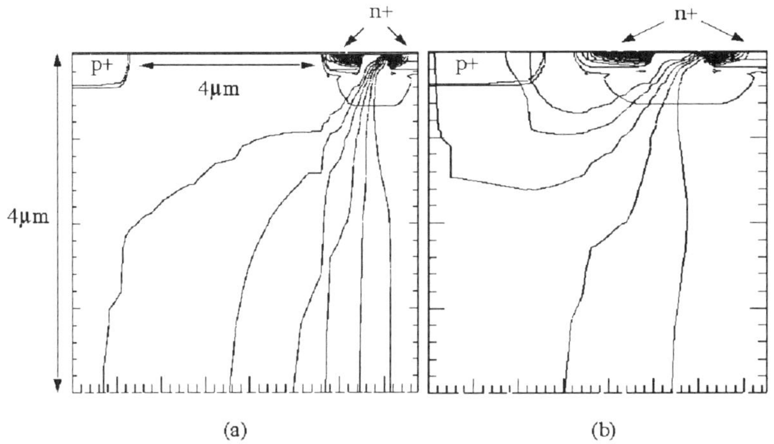

Figure 2 Current flow of substrate noise through low-resistivity substrate with different injector/receptor configurations, assuming grounded backplate

The paths followed by such currents are determined by the relative position of the injector, the pick-up and the other contacts in the circuit, the substrate doping profile, and the backplate potential. Figure 2 shows substrate current flow lines for the case of (a) distant and (b) close injector/receptor systems in a typical high-resistivity substrate. In the whole spectrum of silicon substrates available today, one can recognize two main types: one referred to as *high-resistivity* and the other as *low-resistivity* substrate. In general, the first substrate type is composed of a uniformly doped layer with a resistivity coefficient of $20-50\ \Omega cm$. The second type consists of a thick, high-resistivity epitaxial layer (d $\simeq 10\mu m$, $\rho \simeq 10-15\ \Omega cm$) and a low-resistivity bulk($\rho \simeq 1m\ \Omega cm$). Low-resistivity substrates are generally preferred for their good latch-up suppression properties [1]. High-resistivity ones on the contrary, are better suited to block substrate noise by using guard rings and physical circuit separation. At low and medium frequencies, typically less than 5 GHz, all substrates show a resistive behavior.

Let us consider as an example the CMOS inverter from Figure 1, which we assume has been integrated into a high-resistivity substrate. The plot in Figure 3 shows the input waveform and the resulting injected signals for both High-to-Low and Low-to-High transitions and various slew rates at the input (note the different time scale in the various waveforms). The plot was obtained from SPICE simulations using

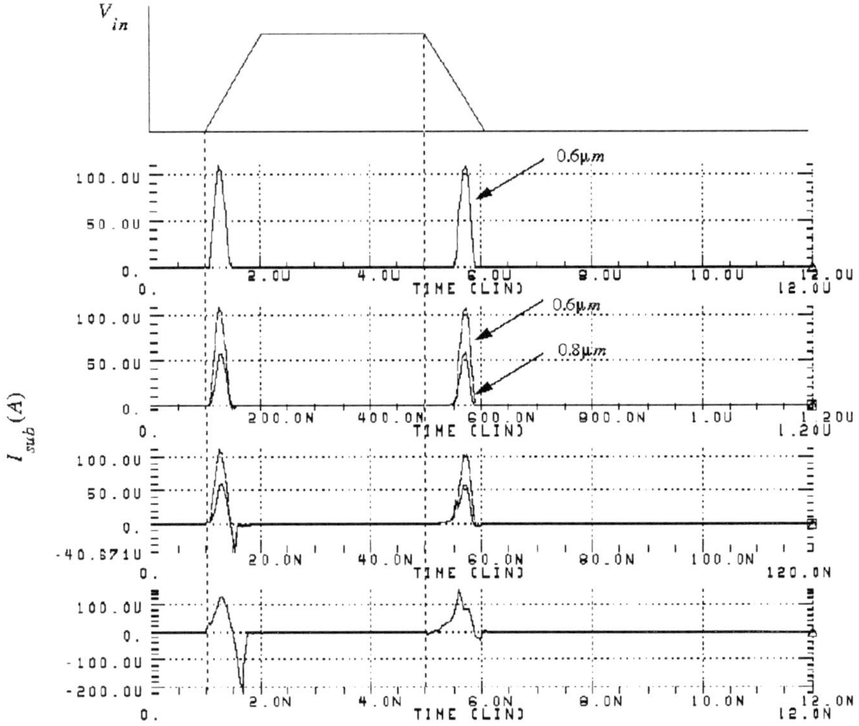

Figure 3 Noise spikes injected into substrate via impact ionization and capacitive coupling (0.8μm and 0.6μm CMOS technologies). The waveforms shown were obtained with input waveforms of varying slew rates: note the different time scales of 2μsec, 200 nsec, 20 nsec, and 2 nsec, while the input waveform is scaled accordingly

custom-fitted device models. Assuming that switching is synchronized with a clock signal, it can be shown that the power has energy components located in a wide spectrum, not necessarily centered at the clock frequency. A significant portion of this energy is usually concentrated around special frequency bands, e.g. at the inverse of the average gate delay. At DC or near DC frequencies, one also observes large spurious currents. This is due to the fact that impact ionization, for its very nature, only generates positive currents. Higher frequency components are due to glitches and fast switching phenomena occurring in large circuits [3].

Impact ionization has the following features: (a) the instant in which the maximum of the waveform is reached depends on the rising and

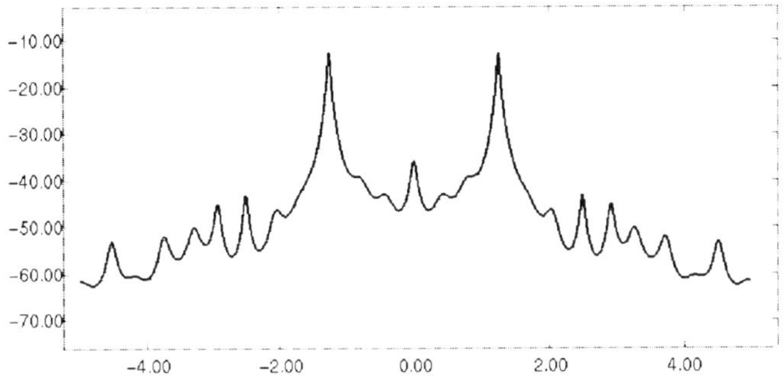

Figure 4 Normalized substrate noise spectrum of C6288 *in dB vs. GHz*

falling time of the input signal; (b) the duration of the pulse depends on the rising and falling time of the input signal; Impact ionization pulses have approximately the same shape in all shown frequencies while only their duration is different (again note the different time scales in Figure 3). Capacitive coupling (drain-substrate and source-substrate reverse-biased junctions) dominates for rising/falling times of less than $10ns$.

In a typical logic circuit of several thousands gates, the effect of switching activity and glitches result in a cumulative injected noise, whose power spectrum represents a significant portion of the total energy absorbed by the circuit. As an illustration consider the MCN91 benchmark C6288. The circuit's spectrum, computed using SubWave[3], is shown in Figure 4. One can recognize the DC, inverse delay and high frequency components. Notably, at the clock frequency, the spectrum is relatively flat. If a circuit with such a large spectrum of injected noise is integrated near sensitive components, then the noise present at the substrate's surface throughout the chip will be unevenly distributed.

2.2 Delay Effects

So far we have outlined the effects of substrate noise coupling on mixed-signal circuit performance. Digital circuits are not immune from substrate noise. The noise is injected by logic gates during switching and glitch transients through impact ionization and capacitive coupling, and it is picked up by active devices via capacitive coupling and body effect. As a result the delay of the datapath may increase, thus possibly exceeding the predefined clock period. Such behavior is known as *delay effect*.

Figure 5 Analogy between interconnect cross-talk and substrate capacitive coupling

Gate delay $t_{50\%}$ is a function of several factors, including fanout, supply voltage, transistor geometry, input waveform, and charge excess caused by charge sharing effects. Ignoring the loading due to interconnect wiring, the gate delay is usually approximated by [4]

$$t_{50\%} = R_{EFF} \times C_G,$$

where $C_G = C_{OX}WL$ is the gate capacitance. R_{EFF}, the effective resistance, is the average transistor resistance during the output voltage swing and is proportional to R_{tr}, with

$$R_{tr} = \frac{L/W}{\mu C_{OX}(V_{DD} - V_T)},$$

where L and W are the dimensions of the transistor, V_{DD} the supply voltage, C_{OX} the gate oxide capacitance, and V_T the threshold voltage. V_T is in turn proportional to the square root of the voltage V_{sb} applied between its substrate contact and the source. Hence,

$$t_{50\%} \propto \frac{1}{\sqrt{V_{sb}}}.$$

The second main contribution to parasitic gate delay is due to capacitive coupling. The analysis of this effect is essentially identical to that of cross-talk between interconnect lines. In this case the aggressor is the substrate underneath the victim interconnect line or device. Figure 5 shows the coupling model equivalence, with coupling capacitances $C_c = C_s$. Resistor R_1 represents the impedance which holds the victim node at ground potential. Using standard analytic charge coupling models [4] one can estimate the charge noise present in the interconnect line due to substrate noise. Figure 6 shows the model proposed for a typical cross-coupling system. A close form analytic model for the response voltage to charge injection $v_1(t)$ was derived in [4]. The voltage at the

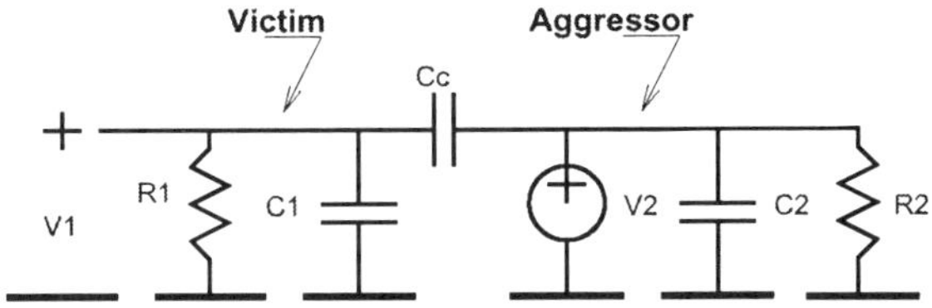

Figure 6 Standard cross-coupling model

peak and the instant at which it occurs are respectively

$$t_p = \begin{cases} \tau_1 \\ \frac{\tau_2\tau_1}{\tau_2-\tau_1} ln(\frac{\tau_1}{\tau_2}) \end{cases} \quad v_1(t_p) = \begin{cases} \frac{C_c}{C_1+C_c} e^{-1} & \text{if } \tau_1 = \tau_2 \\ \frac{C_c}{C_1+C_c} (\frac{\tau_1}{\tau_2})^{\frac{\tau_1}{\tau_2-\tau_1}} & \text{otherwise,} \end{cases}$$

where $\tau_1 = (R_1||R_2)(C_1 + C_c)$ and the waveform of the aggressor node is assumed to be a decaying exponential step with time constant τ_2. From charge noise one can derive the extra delay present on a gate [4].

Empirical delay models based on cross-talk have been proposed in the literature. One such model, relating the length of the parallel running wire l and the average spacing s to the extra delay $\triangle\tau$, was proposed in [5]. The model computes $\triangle\tau$ as

$$\triangle\tau = \frac{\alpha\ l^m}{s^n},$$

where α is a fitting constant, while m and n are empirically observed to be near 2 and 1, respectively.

3. SUBSTRATE ANALYSIS

The goal of substrate analysis is to obtain a compact representation of the interactions of circuit elements that couple through the substrate. An equivalent circuit, or some other model that represents the (possibly frequency-dependent) impedance or admittance matrix describing substrate coupling must be obtained in order to include the substrate effects in circuit simulation or optimization. One approach is to perform experiments on a small number of contacts [6] and fit an empirical model to the results. The other is to address the differential equations that describe substrate transport in a numerical or semi-analytical manner [7, 8, 9, 10, 11].

The basic relation describing substrate transport is the continuity equation

$$\nabla \bullet (\sigma\nabla\Phi(x,y,z,t)) + \frac{\partial}{\partial t}(\nabla \bullet (\epsilon\nabla\Phi(x,y,z,t))) = 0, \tag{1.3}$$

where ϵ and σ are respectively the local dielectric permittivity and conductivity of the substrate. ϵ and σ are potentially spatially varying due to the substrate layer structure, device and well implants, and the presence of other integrated components. If the dielectric relaxation time, $\tau_e = \epsilon/\sigma$, is much smaller than any time-scale of interest, then the second term in Equation (1.3) may be neglected and the substrate treated as purely resistive. A complex conductivity $\sigma' = \sigma + j\omega\epsilon$ may be used to model dynamic effects without change to the basic analysis procedure, though model extraction is complicated since either an equivalent circuit must be fit to data obtained by solving the differential equation at several points in the frequency domain[12, 13] or a model reduction procedure[14, 15, 16] performed.

In uniform material, Equation (1.3) reduces to the Laplace equation

$$\nabla^2\Phi = 0. \tag{1.4}$$

Boundary conditions come from contacts, usually considered equipotential; edges, where zero-normal-current (Neumann) conditions hold; or material interfaces, where the current $\mathbf{J} = \sigma\nabla\Phi$ must be continuous, leading to the boundary condition $\sigma_+ \partial\Phi/\partial n|_+ = \sigma_- \partial\Phi/\partial n|_-$ where $\partial\Phi/\partial n$ refers to the normal derivative, and σ_+, σ_- refer to conductivity on opposite sides of the interface. To extract a column of the impedance matrix corresponding to a specific contact, the potential of that contact is set to a single volt. The currents flowing into each of the other contacts, computed from integrating the normal derivative of the potential over each contact's surface, give the relevant mutual admittances. Solution of the equations governing the substrate requires sophisticated techniques, but, fortunately, the solution of the Laplace equation is one of the most well-studied problems in the applied mathematics literature.

Methods based on differential formulations of the Laplace equation can easily analyze substrates with spatially-varying resistivities. Finite difference[17] or finite element[18] techniques are usually used to discretize Equation (1.3). These methods convert the Laplace equation into a set of sparse algebraic equations, for example by replacing the derivatives of Φ by differences such as

$$\frac{\partial^2\Phi}{\partial^2 x} = \frac{\Phi_{i+1,j,k} - \Phi_{i-1,j,k} - 2\Phi_{i,j,k}}{\Delta_i^2}, \tag{1.5}$$

where Δ_i is the x-directed grid spacing at the grid point indexed by $\{i, j, k\}$. These equations may be solved by standard sparse linear system [19] solution techniques based on on Gaussian elimination but due to the large degree of matrix fill that occurs during the factorization of a matrix that derives from a three-dimensional mesh, iterative solution algorithms often provide better performance in considerably less memory. Modern iterative schemes are usually Krylov-subspace [20] algorithms such as the conjugate-gradient or GMRES methods. For the discretization of elliptic differential equations, preconditioning is required to achieve convergence in a reasonable number of iterations. Incomplete factorization[21, 20] preconditioners are popular, but preconditioners based on multigrid[22] or multiresolutional ideas[23] can be considerably more effective.

When the material conductivities and permittivities are relatively homogeneous (e.g., the resistivity varies along one dimension and/or is piecewise-constant with not too many regions of different conductivity), then integral equation techniques are competitive. By translating the three-dimensional partial differential equation into an integral equation over the two dimensional surfaces that bound the problem domain (usually the substrate contacts), integral equation methods reduce the number of unknown variables that must be analyzed and can provide superior performance if efficient techniques are used to solve the linear equations.

The simplest integral equation encountered in the substrate context is the first-kind equation [24, 25]

$$\Phi(\mathbf{r}) = \int_S \mathbf{G}(\mathbf{r}, \mathbf{r}') j(\mathbf{r}') d^2\mathbf{r}' \tag{1.6}$$

that relates injected contact currents $j(\mathbf{r}')$ to known contact potentials $\Phi(\mathbf{r})$. Physically, $\mathbf{G}(\mathbf{r}, \mathbf{r}')$ represents the potential at $\mathbf{r}$ due to a point charge placed at a point $\mathbf{r}'$ and is called a Green's function. Once the Green's function is known, Equation (1.6) allows one to determine the injected currents, from which the potential and currents at any point in the substrate can be computed.

In the absence of any boundaries, that is in the free-space case, the function $\mathbf{G}(\mathbf{r}, \mathbf{r}')$ reduces to $1/(4\pi\sigma|\mathbf{r} - \mathbf{r}'|)$. In principle the free-space Green's function may be used for substrate extraction calculations, however, the boundary conditions at domain boundaries must be explicitly enforced, which implies discretizing the boundaries[26]. In the substrate analysis problem, it is more convenient to derive a Green's function tailored to the layered media boundary conditions. These Green's functions, which incorporate any effects due to vertically-varying conductivity and possibly finite extent of the substrate, simplify the numerical procedure since the integral equation only needs be written over the multiply-connected surface defined by substrate contacts that are usually in the top layer of the material. However, the Green's function can be expensive to compute. Possible methods for evaluating the Green's function include image-based techniques[27, 28, 10], separation-of-variables (SOV)[29], and spectral domain analysis[30].

In the engineering community, the numerical solution of electromagnetic integral equations is usually done via method-of-moment[31] or boundary-element techniques. The simplest such scheme is to discretize the domain of the integral (in this case, the substrate contacts) into a number of polygonal sections called panels. Given Dirichlet boundary conditions on the panels, the unknowns are the injected currents, and on each panel the injected current is assumed to be constant. The potential $\bar{\Phi}_i$ of a contact panel is defined as the result of summing over the contribution from current injected at every other panel in the domain and averaging the potential over the panel,

$$\bar{\Phi}_i = \sum_j \frac{I_j}{A_j A_i} \int_{A_i} \int_{A_j} G(\mathbf{x}, \mathbf{x}') da_j da_i = \sum_j Z_{ij} I_j \tag{1.7}$$

where the sum runs over all panels j, A_i and A_j are the areas of contacts i and j respectively, and I_j is the current injected from panel j, and the integral is over the panel surfaces. This procedure produces a matrix equation

$$\mathbf{ZI} = \bar{\mathbf{\Phi}} \tag{1.8}$$

where the matrix $\mathbf{Z}$ is *dense*, that is, every entry is non-zero because a normal current injected from any panel induces a potential at *every* other panel in the substrate.

In realistic problems, the matrix in Equation (1.8) can be quite large. Constructing and directly inverting the full $\mathbf{Z}$ matrix for the entire substrate contact configuration can be prohibitively expensive, and so more efficient methods have been sought by many authors. Physically based heuristics involving approximations to the inverse of the $\mathbf{Z}$ matrix[29, 32, 9] can accelerate the matrix solution process as well as the following nonlinear simulation. Numerical stability and error control in these procedures can be difficult to quantify.

More rigorous analysis acceleration techniques typically exploit the analytic properties of the Green's function. For problems with bounded domains, the multilayer Green's function can be computed in $\mathcal{O}(n \log n)$ time using Fast Cosine Transform (FCT) techniques. The FCT can be used to build a technology-dependent table that is used to accelerate the matrix construction procedure for one of the direct techniques. When combined with a matrix simplification procedure and low-rank update techniques[33], the overall procedure can be effective, particularly when embedded in the loop of an optimization procedure.

When direct techniques are no longer feasible, iterative matrix solution algorithms such as GMRES[34] must be used. The dominant computational cost in such an algorithm is the computation of a matrix-vector product with the matrix **Z**. The speed of the FCT can be exploited to directly compute the matrix-vector products in such an iterative procedure [35] in nearly optimal time, if the contacts may be uniformly subdivided and are fairly densely spaced. For complicated contact distributions, several algorithms have been developed that can compute a matrix-vector product in close to $O(n)$ time and memory by approximating the action of the matrix **Z**.

Most of the matrix approximation algorithms are based on the fact that the potential induced by an injected current has a spatially complex profile only near the injection source. Far away from the source it can be easily approximated to within a given error. FCT and FFT related techniques can be applied by using local corrections [36, 37] that remove any constraints on the relation between the FCT/FFT grid and the underlying discretization. The authors of [38] have developed an algorithm that interpolates the Green's function in a hierarchically spatially-decomposed manner, and then uses a procedure similar to the Singular Value Decomposition to further compress the interpolation elements.

Recently algorithms have been developed that combine the matrix approximation with an acceleration of the iterative matrix solution procedure itself. The multigrid method of [39] is based on constructing a hierarchical representation of the irregular problem domain. At each level of hierarchy, a coarser representation of the discretized problem is constructed by using a geometric-moment-matching scheme to approximate the rough features of the finer geometry. The coarser grid problems can be solved relatively cheaply, so the solutions to the coarse grid problems are used to accelerate the iterative solution of the linear systems on the finer levels. The convergence of the iterative solver is extremely rapid, requiring only a few iterations to converge to engineering tolerances. A similar multiresolution approach was described in [40], where a wavelet-like basis for the panel unknowns is constructed by matching moments of the multipole field expansions. The wavelet-like basis is used to perform rapid matrix-vector products and also provides a natural, and very effective, preconditioner.

4. OPTIMIZATION AND SCALING

Optimization phases typically performed in physical design include floorplanning and placement. The algorithms used in floorplanning and placement are based on incremental improvement techniques. Due to its "global" effects felt everywhere in the chip, substrate noise cannot be easily translated into a compact analytical model accounting for the entire substrate area. Hence, even if a small incremental modification is performed on the chip, the whole substrate analysis needs be reevaluated. The traditional approach to this problem consists of using a scheme based on finite difference methods. To reduce the time complexity of the problem the density of the mesh that mimics the substrate bulk is drastically simplified, thus resulting in an accuracy reduction [8, 41]. Another potential problem with this approaches is a strict requirement of alignment between grid and layout objects.

An alternative method is one which transforms the substrate problem into a simpler one, for example using simplified analytical models of contact-to-contact resistances[6, 42]. Moreover, the presence in the design of even relatively small analog circuits complicates the substrate noise analysis problem.

Approaches based on integral equation techniques can make better use of the locality of incremental changes. The key of such techniques is a fast computation of *variations* and *trends* of substrate transport given changes in its physical structure. An often exploited technique is based on the fact that small adjustments in the position and orientation of layout elements results in a small change in the matrix $\mathbf{Z}$. Using the Sherman-Morrison update, $\mathbf{Z}^{-1}$ can be computed in quadratic time complexity. Another method is based on the use of sensitivities to determine the effect on substrate conduction by a small change in the contact organization. Sensitivities can be computed *a priori* efficiently and allow one to obtain a relatively accurate substrate noise map after several component moves.

Re-design generally involves scaling in x- and y- directions, while technology migration involves a three-dimensional scaling. Sensitivity analysis is performed to quantify the effects of small changes in doping profiles and doping concentrations to, for example, a grid of contacts and their associated substrate resistances. Similar regular structures can be designed to test the effects of migrating to a different technology.

Improvements on the performance degradation due to substrate- induced switching noise can be achieved by placing noise injecting and noise sensitive modules at a certain distance or by creating special structures, such as low-resistivity guard-rings, around noise injectors [29]. The first provision is generally implemented in a placement tool using the conventional Simulated Annealing (SA) move-set. The second issue is usually solved by extending the search space, allowing the annealing to choose from a number of alternative implementations for a module, including one with a guard-ring implemented around it.

In order for a placer to be effective in preventing violations to performance specifications, the following features are often implemented in the tool. (1) A model for each noise injecting module, characterizing the *waveform* and the *spatial location* where the noise is injected as precisely as possible. (2) A model of substrate transport for efficient substrate current evaluation, possibly independent of the circuit configuration. (3) A model for substrate noise absorption and its effect on performance.

The evaluation of performance degradation due to substrate noise is generally the most time consuming. In [33] for example, such problem is approached in the following way:

1. generate constraints for each node of noise-sensitive modules
2. generate the resistive network associated with substrate
3. quantify violations to constraints

The calculation of all violations in step 3 to the given constraints is carried out by solving the underlying circuit and evaluating the appropriate parameters at each critical node.

At each stage of the annealing only steps 2 and 3 need be repeated, since step 1 is carried out only once for each chip. The efficiency of a simulator based on integral equation techniques, though high, is still insufficient for such computationally intensive algorithm as SA, hence, appropriate heuristics are generally developed. In SA, at high annealing temperatures, considerable reshuffling is allowed on the components of the layout. Hence, the locations of switching noise generators and receptors can be significantly modified. On the other hand, only when changes in component location reflect a significant change in any performance measure, the entire substrate network should be evaluated along with the estimate of performance degradation. This observation is generally used to create combined heuristics for the evaluation of substrate effects after each tentative annealing move.

The placement algorithm has been proven to converge to a global minimum under the Romeo/Hajek conditions [43] when it is modified to account for noise substrate transport evaluation [44]. A tool based on this algorithm was used extensively in the design of a RAMDAC chip, which was eventually fabricated and successfully tested [45].

5. CONCLUSIONS

The main problem associated with substrate noise is a generalized degradation of performance induced mainly by the appearance of spurious currents generated by the circuit's digital switching activity. The paper focuses on the models of noise transport in view of creating optimized circuits or technologies in mixed-signal and digital designs. Solutions to the substrate noise problem are presented in light of the results obtained from industrial examples.

References

[1] C. Hu, *VLSI Electronics: Microstructure Science*, volume 18, Academic Press, New York, 1981.

[2] Y. Cheng, M. Chan, K. Hui, M. Jeng, Z. Liu, J. Huang, K. Chen, J. Chen, R. Tu, P. K. Ko and C. Hu, *BSIM3v3 Manual*, University of California at Berkeley, Berkeley, CA, 1996.

[3] E. Charbon, P. Miliozzi, L. Carloni, A. Ferrari and A. L. Sangiovanni-Vincentelli, "Modeling Digital Substrate Noise Injection in Mixed-Signal ICs", *Trans. on Computer Aided Design*, vol. CAD-18, n. 3, pp. 301–310, March 1999.

[4] D. A. Kirkpatrick, *The Implications of Deep Sub-micron Technology on the Design of High Performance Digital VLSI Systems*, PhD thesis, University of California at Berkeley, December 1997.

[5] P. Parakh and R. B. Brown, "Crosstalk Constrained Global Route Embedding", in *Proc. ACM International Symposium on Physical Design*, pp. 201–206, April 1999.

[6] D. K. Su, M. Loinaz, S. Masui and B. Wooley, "Experimental Results and Modeling Techniques for Substrate Noise in Mixed-Signal Integrated Circuits", *IEEE Journal of Solid State Circuits*, vol. SC-28, n. 4, pp. 420–430, April 1993.

[7] F. J. R. Clement, E. Zysman M. Kayal and M. Declercq, "LAYIN: Toward a Global Solution for Parasitic Coupling Modeling and Visualization", in *Proc. IEEE Custom Integrated Circuit Conference*, pp. 537–540, May 1994.

[8] B. R. Stanisic, N. K. Verghese, D. J. Allstot, R. A. Rutenbar and L. R. Carley, "Addressing Substrate Coupling in Mixed-Mode ICs: Simulation and Power Distribution Synthesis", *IEEE Journal of Solid State Circuits*, vol. SC-29, n. 3, pp. 226–237, March 1994.

[9] N. K. Verghese, T. Schmerbeck and D. J. Allstot, *Simulation Techniques and Solutions for Mixed-Signal Coupling in ICs*, Kluwer Academic Publ., Boston, MA, 1995.

[10] T. Smedes, N. P. van der Meijs and A. J. van Genderen, "Extraction of Circuit Models for Substrate Cross-Talk", in *Proc. IEEE International Conference on Computer Aided Design*, pp. 199–206, November 1995.

[11] R. Gharpurey and R. G. Meyer, "Modeling and Analysis of Substrate Coupling in ICs", *IEEE Journal of Solid State Circuits*, vol. SC-31, n. 3, pp. 344–353, March 1996.

[12] M. Pfost and H. M. Rein, "Modeling and Measurement of Substrate Coupling in Si-Bipolar ICs up to 40GHz", *IEEE Journal of Solid State Circuits*, vol. SC-33, n. 4, pp. 582–591, 1998.

[13] C. P. Coelho, J. R. Phillips and L. M. Silveira, "Robust Rational Function Approximation Algorithm for Model Generation", in *Proc. IEEE/ACM Design Automation Conference*, pp. 207–212, June 1999.

[14] X. Huang, V. R. Raghavan and R. A. Rohrer, "AWEsim: A Program for the Efficient Analysis of Linearized Circuits", in *Proc. IEEE International Conference on Computer Aided Design*, pp. 534–537, November 1990.

[15] P. Feldmann and R. W. Freund, "Efficient Linear Circuit Analysis by Pade Approximation via the Lanczos Process", *IEEE Trans. on Computer Aided Design*, vol. CAD-14, n. 5, pp. 639–649, May 1995.

[16] L. M. Silveira, M. Kamon, I. Elfadel and J. White, "A Coordinate-Transformed Arnoldi Algorithm for Generating Guaranteed Stable Reduced-Order Models of RLC Circuits", in *Proc. IEEE International Conference on Computer Aided Design*, pp. 288–294, November 1996.

[17] W. F. Ames, *Numerical methods for partial differential equations*, Academic Press, San Diego, 1992.

[18] G. Strang and G. J. Fix, *An Analysis of the Finite Element Method*, Prentice-Hall, Englewood Cliffs,NJ, 1973.

[19] I. S. Duff, A. M. Erisman and J. K. Reid, *Direct Methods for Sparse Matrices*, Clarendon Press, Oxford, 1986.

[20] Y. Saad, *Iterative Methods for Sparse Linear Systems*, PWS Publishing, Boston, MA, 1996.

[21] J. A. Meijerink and H. A. van der Vorst, "An Iterative Solution Method for Linear Systems of which the Coefficient Matrix is a Symmetric M-Matrix", *Mathematics of Computation*, vol. 31, pp. 148–162, January 1977.

[22] W. Hackbusch, *Multigrid Methods and Applications*, volume 4 of *Computational Mathematics*, Springer, Berlin, 1985.

[23] J. H. Bramble, J. E. Pasciak and J. Xu, "Parallel multilevel preconditioners", *Mathematics of Computation*, vol. 55, pp. 1–22, 1990.

[24] D. Colton and R. Kress, *Integral Equations Methods in Scattering Theory*, Krieger, Malabar, Florida, 1992.

[25] K. E. Atkinson, "A Survey of Boundary Integral Equation Methods for the Numerical Solution of Laplace's Equation in Three Dimensions", in M. A. Goldberg, editor, *Numerical Solution of Integral Equations*, pp. 1–34, Plenum Press, New York, 1990.

[26] S. M. Rao, T. K. Sarkar and R. F. Harrington, "The Electrostatic Field of Conducting Bodies in Multiple Dielectric Media", *IEEE Trans. on Microwave Theory and Techniques*, vol. MTT-32, n. 11, pp. 1441–1448, November 1984.

[27] W. T. Weeks, "Calculation of Coefficients of Capacitance of Multiconductor Transmission Lines in the Presence of a Dielectric Interface", *IEEE Trans. on Microwave Theory and Techniques*, vol. MTT-18, n. 1, pp. 35–43, January 1970.

[28] T. Smedes, N. P. van der Mejis and A. J. Genderen, "Boundary Element Methods for 3D Capacitance and Substrate Resistance Calculations in Inhomogeneous Media in a VLSI Layout Verification Package", *Advances in Engineering Software*, vol. 20, n. 1, pp. 19–27, 1994.

[29] R. Gharpurey, *Modeling and Analysis of Substrate Coupling in ICs*, PhD thesis, University of California at Berkeley, May 1995.

[30] R. Crampagne, M. Ahmadpanah and J. L. Guiraud, "A Simple Method for Determining the Green's Function for a Large Class of MIC Lines Having Multilayered Dielectric Structures", *IEEE Trans. on Microwave Theory and Techniques*, vol. MTT-26, pp. 82–87, 1978.

[31] R. F. Harrington, *Field Computation by Moment Methods*, MacMillan, 1968.

[32] A. J. van Genderen, N. P. van der Mejis and T. Smedes, "Fast Computation of Substrate Resistances in Large Circuits", in *Proc. European Design and Test Conference*, pp. 560–565, March 1996.

[33] E. Charbon, R. Gharpurey, R. G. Meyer and A. L. Sangiovanni-Vincentelli, "Analysis and Optimization of Substrate Noise in VLSI ICs", *IEEE Trans. on Computer Aided Design*, vol. CAD-18, n. 2, pp. 172–190, February 1999.

[34] Y. Saad and M. Schultz, "GMRES: A Generalized Minimal Residual Algorithm for Solving Nonsymmetric Linear Systems", *SIAM Journal on Scientific and Statistical Computing*, vol. 7, n. 3, pp. 856–869, July 1986.

[35] J. P. Costa, M. Chou and L. M. Silveira, "Efficient Techniques for Accurate Modeling and Simulation of Substrate coupling in mixed-signal ICs", in *Design, Automation and Test in Europe*, pp. 892–898, February 1998.

[36] J. R. Phillips and J. K. White, "A Precorrected-FFT Method for Electrostatic Analysis of Complicated 3D Structures", *IEEE Trans. on Computer Aided Design*, vol. CAD-16, n. 10, pp. 1059–1072, 1997.

[37] J. P. Costa, M. Chou and L. M. Silveira, "Precorrected-DCT Techniques for Modeling and Simulation of Substrate coupling in mixed-signal IC's", in *Proc. IEEE International Symposium on Circuits and Systems*, volume 6, pp. 358–362, May 1998.

[38] S. Kapur and D. E. Long, "IES^3: Efficient Electrostatic and Electromagnetic Simulation", *IEEE Computational Science and Engineering*, vol. 5, n. 4, pp. 60–67, October-December 1998.

[39] M. Chou and J. White, "Multilevel integral equation methods for the extraction of substrate coupling parameters in mixed-signal IC's", in *Proc. IEEE/ACM Design Automation Conference*, pp. 20–25, June 1998.

[40] J. Tausch and J. White, "A Multiscale Method for Fast Capacitance Extraction", in *Proc. IEEE/ACM Design Automation Conference*, pp. 537–542, June 1999.

[41] S. Mitra, R. A. Rutenbar, L. R. Carley and D. J. Allstot, "Substrate-Aware Mixed-Signal Macro-Cell Placement in WRIGHT", in *Proc. IEEE Custom Integrated Circuit Conference*, pp. 529–532, May 1994.

[42] K. Joardar, "A Simple Approach to Modeling Cross-Talk in Integrated Circuits", *IEEE Journal of Solid State Circuits*, vol. SC-29, n. 10, pp. 1212–1219, October 1994.

[43] F. Romeo, *Simulated Annealing: Theory and Applications to Layout Problems*, PhD thesis, University of California at Berkeley, March 1989.

[44] E. Charbon, *Constraint-Driven Analysis and Synthesis of High-Performance Analog IC Layout*, PhD thesis, University of California at Berkeley, December 1995.

[45] I. Vassiliou, H. Chang, A. Demir, E. Charbon, P. Miliozzi and A. L. Sangiovanni-Vincentelli, "A Video Driver System Designed Using a Top-Down, Constraint-Driven Methodology", in *Proc. IEEE International Conference on Computer Aided Design*, pp. 463–468, November 1996.

Architectural Transformations for Hierarchical Algorithmic Descriptions

Marcio Yukio Teruya, Marius Strum and Wang Jiang Chau
Department of Electronic Enginnering, Escola Politecnica da Universidade de Sao Paulo, Brazil

Key words: Transformational Design, Architectural Synthesis, High Level Synthesis, System Synthesis

Abstract: The use of hierarchy on writing algorithmic descriptions of digital systems allows the implementation of more complex designs since it increases designer's productivity by introducing important features such as modularity, encapsulation and reusability. We are particularly interested in the problem of generating an optimal register transfer logic structure from a hierarchical algorithmic description. It is relatively straightforward to use High Level Synthesis (HLS) tools for producing an implementation from hierarchical algorithmic descriptions; each algorithmic partition is implemented separately and then linked in a following step. In general, the results are sub-optimal due to the large gap existing between the specification and implementation. In this article, we detail a simple architectural model for hierarchical algorithmic descriptions and a set of architectural transformations, which are the core of a methodology, Recursive High Level Synthesis (RHLS), aimed to optimise hierarchical implementations. The transformations are used to reshape the architecture of pre-existing hierarchical algorithmic descriptions in order to provide better synthesis results from HLS. We have implemented a suitable data structure and a set of transformations and tested them over a set of hierarchical algorithmic examples.

1. INTRODUCTION

To cope with growing digital system designs complexity, a well established strategy is to represent them hierarchically. It is well known that hierarchy allows introducing some very desirable features into the design process, such as modularity, encapsulation and reusability, contributing towards increased productivity and making large projects less expensive. Although designers may have their task of specifying and describing a circuit made easier, the results at implementation side may not be as positive. For instance, the hierarchy of a design entry, containing a partitioning that aims at productivity, may show to be inefficient if directly reproduced into the implementation, whose partitioning should be driven by implementation level metrics (timing, power, cost, etc.).

In present methodologies based on higher level descriptions, the design entry presents a widening gap to the final implementation description what increases their aforementioned hierarchical incompatibility. This is very much true for algorithmic level design. Figure 1a shows an example of a hierarchical algorithmic description; **X** is the top level algorithm, **Y** and **Z** are the lower level algorithms that implement operations **opY** and **opZ**, respectively, in algorithm **X**. To produce a structural level implementation, we can use High Level Synthesis (**HLS**) [1] as shown in Figure 1b, where the synthesised structures from algorithms **Y** and **Z** are represented as blocks composed of smaller basic blocks (**A**, **B**, **C** and **D**). Figure 1c shows the synthesised structure from algorithm **X**; it preserves the original algorithmic hierarchy but it presents some redundancy; for instance, the basic block **C** appears 3 times, once at each partition. Depending on the circuit timing requirements, there could be some additional block sharing through a different partitioning, as seen in figure 2.

This example shows that, for the sake of implementation quality, the hierarchical decomposition at design entry level must be in tune with the respective one at implementation level, but, usually, when high-level partitioning is carried out, implementation issues are not visible yet. The problem here is how to keep the mentioned productivity advantages of adopting hierarchical methodologies at design entry level and still deliver efficient implementations. Some attempts have been made to deal with this problem but there is no clear solution yet. In [2][3][4], the problem of generating hierarchical structures was tackled, but the authors did not focus on productivity issues (as possible component reusability) arisen from a hierarchical strategy. Even though, their approaches were capable of producing hierarchical structures containing some optimisation, via restricted rules in partitioning of data-flow [2] or control-flow [3][4] graphs derived from plain algorithmic descriptions. In [5], it is shown a structured methodology capable of dealing with hierarchical algorithmic designs using a HLS system, but they did not show any automatic means for producing optimised implementations. More recently, in [6], the authors presented an extended HLS system, which can allocate components covering different levels of hierarchy to the algorithmic level descriptions. They have proposed an extended model for re-configurable functional units (**FUs**) which includes their behavioural information; at the scheduling phase redundancies are detected and optimised hierarchical structures are produced. The drawback of this approach is its complexity - a whole new set of efficient algorithms must be derived, besides the fact that new generated FUs are problem dependent what restricts their reuse.

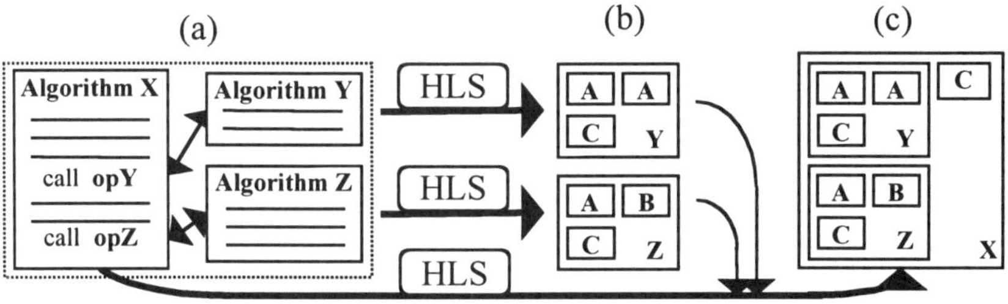

Figure 1. From a hierarchical algorithmic description to a hierarchical structure using HLS.

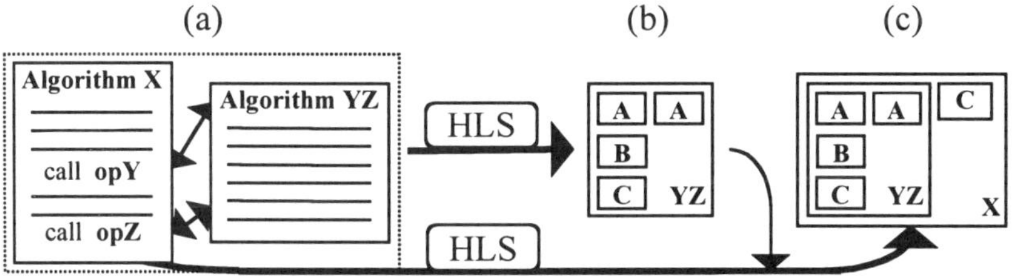

Figure 2. An alternative partitioning for the hierarchical algorithmic description.

In this paper, we follow the **Recursive High Level Synthesis (RHLS)** approach, presented elsewhere[7], to solve the problem of generating an efficient hierarchical register transfer logic (RTL) structure from a hierarchical algorithmic description. It is based on implementation metrics and transformations over hierarchical algorithmic descriptions very much like going from the one in figure 1 to the one in figure 2 - the newly created hierarchical algorithm should then be suitable for generating, via HLS, an optimised structure.

The objective of this article is to present details about the foundations of our methodology for RHLS, which are: an **architectural model for hierarchical algorithmic descriptions** and a set of **Architectural Transformations (ATs)**. Our architectural model establishes a basic framework for designing hierarchical algorithmic descriptions and also establishes a basic vocabulary upon which the transformations will act on. It was designed to be independent of hardware description languages such as the VHDL syntax. The architectural transformations provide new modules which may fulfil better a design implementation level requirements and, since the transformations are defined on modules operation sets, they make a strong contribution for FU reuse environments. It should be noted that the concept of architectural transformations is not new; in [8], the authors presented an implementation based on an architectural model at RTL level. We have adopted a higher abstraction level architecture for two reasons: higher flexibility and simpler implementation. The set of transformations acts on the hierarchical algorithmic descriptions following our model, reshaping them for structural level optimisations, particularly by improving sharing of structural resources, such as FUs, registers and so on, supported by a HLS system, as exemplified previously.

This article is organised as follows. Section 2 gives a general explanation about our architectural model and the transformations. The sections 3 and 4 present details, respectively, of a mathematical formalism for our architecture model and the algorithms we implemented for the set of transformations. In section 5 implementation issues are explained and some results are presented. Finally, section 6 concludes this article.

2. STRATEGY OVERVIEW AND DEFINITIONS

2.1 Recursive High-Level Synthesis

The Recursive High Level Synthesis may be defined as the optimisation process of a hierarchical RTL structure through the application of a sequence of transformations on its functional units (FUs). We say *hierarchical* RTL structure because it contains FUs which are RTL structures themselves. Furthermore, these FUs also have behavioural descriptions from which a RTL structure may be obtained via HLS.

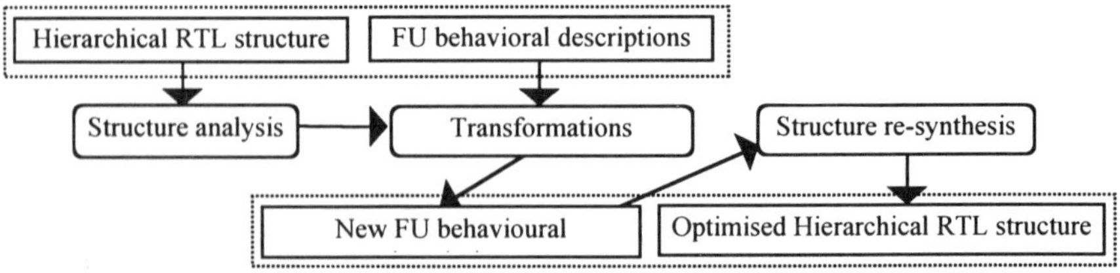

Figure 3. Tasks in RHLS

Figure 3 illustrates the main tasks in RHLS: **Structure analysis** - the hierarchical RTL structure is examined in regard to redundancies or any other inefficiencies through their functional units; **Transformations** - new functional units are created based on existing ones in order to solve problems detected by the structure analysis; and **Structure re-synthesis** - a new hierarchical RTL structure is generated using the newly created functional units. More details about RHLS can be found in [7].

2.2 Architectural Model

Our architectural model establishes the basic framework for writing hierarchical algorithmic descriptions. In other words, it defines the basic components and their basic interrelations for building hierarchical algorithmic descriptions. Therefore, besides providing guidance for writing these descriptions, it states what the transformations can manipulate.

The basic elements of our architectural model are the **behavioural views** (algorithmic descriptions) of **complex functional units**. We understand **functional unit (FU)** as a HLS system library component that is capable of performing one or more operations. **Complex FU** is a FU that holds an algorithmic description for the operation set it is capable of performing. Inside this algorithmic description, there are calls to operations provided by either primitive (non-complex) FUs or other complex FUs. This establishes a relation of hierarchy among complex FUs and, therefore, a relation of hierarchy among algorithmic descriptions. Hence, our architectural model is basically a set of intercommunicating (via operation calls) processes (algorithmic descriptions) hierarchically organised.

```
entity FU_name is
    port ( port declarations );  end FU_name;
architecture behavior of FU_name is begin process
    variable declarations
begin
    case sel is
        when 1 => (instruction set for operation 1)
        ...
        when n => (instruction set for operation n)
    end case;
end process;  end behavior;
```

Figure 4. Behavioural view of a FU represented as a VHDL template.

The architectural model also comprises the architecture of the behavioural views of complex FUs; we have modelled it as a set of **ports**, **variables**, **instruction sets**, **internal operations** and a **list of FUs** (figure 4). The **instruction sets** are the algorithmic descriptions of the operations the FU is capable of performing and there is exactly one instruction set for each operation of the FU. The **internal operations** are the operations used inside the algorithmic description of the FU's behavioural view. The **list of FUs** (not shown in figure 4, but considered part of the behavioural view) refers to the *candidate* FUs to execute the internal operations.

The processes (i.e. the behavioural views of the FUs) are organised in a hierarchical fashion, therefore, the activation of any sibling process or their data exchange must be arbitrated by the parent processes using some kind of communication protocol. Our approach to solve this problem is similar to the one proposed in [5], where intercommunicating protocol is implemented at the parent processes as reproduced in figure 5. The protocol is composed of two operations: the first one (*opX*) starts the operation and sends the related parameters to the sibling process and the second one (*waitX*) runs repeatedly until the result is ready to be used. This protocol is particularly useful in processes whose processing time is data dependent.

```
opX(a,b);                    -- starts the operation with parameters a and b
waitX (c,done);              -- reads the flag (done) for operation completion
while (done /= '1') loop     -- loops until operation completes
     waitX (c,done);         -- and reads a result (c)
end loop;
```

Figure 5. A interprocess protocol in VHDL

2.3 Architectural Transformations (ATs)

We have implemented three ATs, which, essentially, perform modifications on the architecture of complex FU behavioural views, in order to produce new complex FUs with altered operation sets. More elaborate architectural changes can be achieved through combination of these basic transformations: The transformations are:

a) **Merge**: from two pre-existent FUs, **Merge** produces the behavioural view of a FU capable of performing the same operations of the two original FUs. As an optimisation tool, **Merge** is intended to cluster FUs with high probability of resource sharing.
b) **Extract**: from one pre-existent FU, **Extract** produces the behavioural view of a FU with one operation removed from the original operation set. **Extract** is intended to reshape existing FUs removing redundancies so they can better fulfil new requirements.
c) **Promote**: from one pre-existent complex FU (which contains at least one more FU), **Promote** produces the behavioural view of a FU with one extra operation in its operation set (the extra operation is borrowed from a sibling FU). **Promote** is intended to add functionality to existing FUs so they can better fulfil new requirements.

3. MATHEMATICAL MODEL

The architecture of a FU behavioural view is defined as:

$$A = < name, O, P, V, \boldsymbol{I}, O', \boldsymbol{A} >, \text{ where:}$$

- *name* is the name of the FU.
- O is the set of operations executed by the FU; O=<o_1,o_2, ...,o_{nO}>, where o_i is the ith operation; o_i=<*name*> meaning it has a *name*.
- P is the set of ports; P= <p_0,p_1, ...,p_{nP}>, where p_i is the ith port; p_i=<*name*,*width*,*type*> meaning it has a *name*, a *width* (in bits) and a *type* belonging to the set <"*control*","*input*","*output*">; p_0 is of type "*control*", the other ones must be "*input*" or "*output*".
- V is the set of variables; V=<v_1,v_2,...,v_{nV}>, where v_i is the ith variable; v_i=<*name*,*width*> meaning it has a *name* and a *width* (in bits).
- $\boldsymbol{I}$ is the set of instruction sets; $\boldsymbol{I}$ = < I_1, I_2, ..., I_{nI} >, where I_j is the jth instruction set; I_j=<i_1,i_2,...,i_{nIj}>, where i_k is the kth instruction; i_k=<T,B>, where T is the type of instruction, which can be any instruction of the adopted hardware design language's (HDL) instruction set, and B is the body of the instruction which can be a list of arguments (ports, variables, constants) or another instructions.
- O' is the set of internal operations; O'={o_1',o_2' ...,$o_{nO'}'$}, where o_i' is the ith internal operation; o_i=<*name*,*PA*> meaning it has a *name* and a set of parameters PA=<pa_1, pa_2, ..., pa_{nPA}> where pa_i=<*name*,*type*> is the ith parameter; each parameter also has a name and a type belonging to the set <"*input*","*output*">.
- $\boldsymbol{A}$ is the set of architectures (it is the list of FUs mentioned in section 2.2) capable of performing the internal operations in O'; $\boldsymbol{A}$ = {A_1,A_2,...,A_{nA}}, where A_i is the ith architecture.

Observe that the last element of this mathematical model, $\boldsymbol{A}$ (set of architectures), makes the model suitable to represent a hierarchical algorithmic description. The mathematical model has the following assumptions: 1) FUs are capable of executing one or more operations; 2) all ports are bit vectors; 3) all input and output ports can exist in any quantity and there are no width restrictions; 4) bi-directional ports are not allowed; 5) there is only one control port which is designated to activate the FU and to select one desired operation; 6) all variables are bit vectors; 7) variables can exist in any quantity and there are no width restrictions; 8) if the FU is capable of executing more than one operation, then only one operation can be executing at a time; and 9) to each instruction set corresponds one, and only one, operation. Some of these assumptions are restrictions imposed to simplify the implementation of data structures and, as such, they could be relaxed to increase the model's generality. An example of use of the mathematical model is given below for a reciprocal division module.

Listing 1: A behavioural view in algorithmic VHDL.

```
Entity mrep is
    port (input1 : in integer;  sel : in int2bit;  outval : out integer;  outdone : out bit);
end mrep;
architecture behavior of mrep is begin process
    variable A, T, X, Q : integer;
begin
    case sel is
      when 1 => -- reciprocal operation
         T:=constexp20;  X:=0;  A:=input1;  outdone<='0';
         while (X/=20) loop  Q:=shiftleft(T);  X:=X+1;   T:=Q-A;
             if (T>0 and A>0) then  T:=T+1; end if;  T:=qsave(Q,T);
         Q:=selectlsb(T);
      when 2 => -- waiting operation
         outval <= Q;   outdone <= '1';
end case;  end process;  end behavior;
```

Listing 2: The corresponding mathematical model for listing 1.

$Amrep = \langle \text{"mrep"}, Omrep, Pmrep, Vmrep, \mathbf{Imrep}, Omrep', \mathbf{Amrep} \rangle$

$Omrep = \langle \text{"rep"},\text{"wrrep"} \rangle$

$Pmrep = \langle p_0, p_1, p_2, p_3 \rangle, p_0 = \langle \text{"sel"},\text{"control"},2 \rangle, p_1 = \langle \text{"input1"},\text{"input"},32 \rangle,$
$p_2 = \langle \text{"outval"},\text{"output"},32 \rangle, p_3 = \langle \text{"outdone"},\text{"output"},1 \rangle$

$Vmrep = \langle v_1,v_2,v_3,v_4 \rangle, v_1 = \langle \text{"A"},32 \rangle, v_2 = \langle \text{"T"},32 \rangle, v_3 = \langle \text{"X"},32 \rangle, v_4 = \langle \text{"Q"},32 \rangle$

$\mathbf{Imrep} = \langle I_1 \rangle$

$I_1 = \langle I_1,I_2,I_3,I_4,I_5,I_6 \rangle, \; I_2 = \langle I_1, I_2 \rangle$

$I_1.I_1 = \langle ASSIGN,\langle v_2,\langle O_3',\text{null} \rangle\rangle\rangle, I_1.I_2 = \langle ASSIGN,\langle v_3,0 \rangle\rangle,$

$I_1.I_3 = \langle ASSIGN,\langle v_1,p_1 \rangle\rangle, I_1.I_4 = \langle ASSIGN,\langle p_4,0 \rangle\rangle,$

$I_1.I_5 = \langle WHILE,\langle\langle NE,\langle v_3,20 \rangle\rangle, \langle O_4',\langle v_4,v_2 \rangle\rangle, \langle O_1',\langle v_3,v_3,1 \rangle\rangle, \langle O_2',\langle v_2,v_4,v_1 \rangle\rangle,$
$\langle IF, \langle\langle AND,\langle\langle GT,v_2,0 \rangle,\langle GT,v_1,0 \rangle\rangle\rangle,\langle O_1',\langle v_2,v_2,1 \rangle\rangle\rangle\rangle, \langle O_5',\langle v_2,v_4,v_2 \rangle\rangle\rangle\rangle,$

$I_1.I_6 = \langle O_6',\langle v_4,v_2 \rangle\rangle$

$I_2.I_1 = \langle ASSIGN,\langle p_3,v_4 \rangle\rangle, I_2.I_2 = \langle ASSIGN,\langle p_4,1 \rangle\rangle$

$Omrep' = \{ O_1', O_2', O_3', O_4', O_5', O_6' \}$

$O_1' = \langle \text{"+"},\langle pa_1,pa_2 \rangle\rangle, O_2' = \langle \text{"-"},\langle pa_1,pa_2 \rangle\rangle, O_3' = \langle \text{"constexp20"},\text{null} \rangle$

$O_4' = \langle \text{"shiftleft"},\langle pa_1,pa_2 \rangle\rangle, O_5' = \langle \text{"qsave"},\langle pa_1,pa_2,pa_3 \rangle\rangle, O_6' = \langle \text{"selectsb"},\langle pa_1,pa_2 \rangle\rangle,$

$\mathbf{Amrep} = \{ Aalu, Ashift \}$

4. ALGORITHMS

In this section we present the main algorithms we have implemented, all based on the mathematical model just introduced.

The **Merge** AT receives two architectures as inputs: $A1 = \langle name1,O1,P1, V1,\mathbf{I1},O1',\mathbf{A1} \rangle$ and $A2 = \langle name2,O2,P2,V2,\mathbf{I2},O2',\mathbf{A2} \rangle$, and generates a third one: $Am = \langle namem,Om,Pm,Vm,\mathbf{Im},Om',\mathbf{Am} \rangle$, using the following algorithm:

1. *namem* = a different name from those pre-existent
2. $Om' = O1' \cup O2'$
3. $\langle Om,\mathbf{Im} \rangle$ = **Unite**($\langle O1,\mathbf{I1} \rangle,\langle O2,\mathbf{I2} \rangle$)
4. $\langle Pm,\mathbf{Im} \rangle$ = **MergePortSets**($P1,P2,\mathbf{Im}$)
5. $\langle Vm,\mathbf{Im} \rangle$ = **MergeVariableSets**($V1,V2,\mathbf{Im}$)
6. $\mathbf{Am} = \mathbf{A1} \cup \mathbf{A2}$
7. Return $Am = \langle namem,Om,Pm,Vm,\mathbf{Im},Om',\mathbf{Am} \rangle$

First (line1), a name different from any other is given to the architecture, then, the element *Om'* (line 2), the set of internal operations, is generated by simple union of those in the initial architectures; the same occurs for the element ***Am*** (line 6), set of architectures. The elements *Om*, set of operations, and ***Im***, set of instruction sets, are produced by the special function **Unite** (line 3) which works similarly to the connective set union; regarding ***Im***, there is an one-to-one correspondence between the sets *O* and *I*: ***Im*** is generated taking *Om* as guide, i.e., if the first element of *Om* is the first element of *O1*, then the first element of ***Im*** must be the first element of ***I1***, and so on. Finally, the elements *Pm* and *Vm* are generated by algorithms **MergePortSets** and **MergeVariableSets** which perform four similar tasks. **MergePortSets** concatenates the original port sets, minimises them, creates a new control port with a suitable width and updates the set of instruction sets with the new ports. To minimise the port set, an algorithm aimed to share ports among instruction sets is used- given a pair of ports, one can substitute the other if they are of the same type (input or output), have the same width and belong to different instruction sets.

The **Extract** AT has the form **Extract**(*A*,*opname*). It receives one architecture as input, *A*=<*name*,*O*,*P*,*V*,***I***,*O'*,***A***>, and the name of the operation being extracted, *opname*, and it generates a second architecture: *Ae*=<*namee*,*Oe*,*Pe*,*Ve*,***Ie***,*Oe'*,***Ae***>. The undesired operation is extracted from the original operation set, *O*, to produce the new operation set *Oe*; *Ie* is generated in a similar way. The other elements of the new architecture (*Pe*, *Ve*, *Oe* and *Ae*) are derived from the original one by eliminating all the elements that became unused after removing the instruction set corresponding to the extracted operation.

The **Promote** AT has the form **Promote**(*A*,*opname*). It receives one architecture as input, *A*=<*name*,*O*,*P*,*V*,***I***,*O'*,***A***>, and the name of the operation being promoted, *opname*, and it generates a second architecture: *Ap*=<*namep*,*Op*,*Pp*,*Vp*,***Ip***,*Op'*,***Ap***>. The algorithm generates a new operation set, *Op*, appending the new operation to the pre-existing set of operations. The new port set, *Pp*, is generated by concatenating the existing port set, *P*, to a new one, *Pnew*, containing the exact number of ports required by the promoted operation. After this, a new instruction set is created for the promoted operation; it is composed of a simple call to the new operation having the new ports, *Pnew*, as parameters and the created port set, *Pp*, is minimised. A new control port is generated according to the number of operations and it is appended to *Pp*. Finally, the elements *Vp*, *Op'* and ***Ap*** are directly copied from the original architecture. It should be observed that a promoted operation requires a costlier access time compared to a identical non-promoted operation because promotion implies arbitration through FU's top controller (one extra cycle clock in our case).

5. RESULTS

The first set of results (tables 1 and 2) illustrates the AT's *modus operandi*; they were obtained from transformations over three benchmarks (*gcd, mmul* and *mrep*) and from HLS. Table 1 shows the architectural characteristics of the benchmarks and resulting architectures from merging, promoting and extracting. The table shows the operations the FUs are capable of performing (*O*), its ports (*P*), its variables (*V*) and its architecture set (*A*).

Table 1. Architectural characteristics for the benchmarks and the ones produced by the ATs

FU	*O*	*P*	*V*	*A*
mgcd	gcd, wgcd	2 i[32], 1 o[32], 1 o[1], sel[2]	2[32]	alu
mmul	mul, wmul	2 i[32], 1 o[32],1 o[1], sel[2]	5[32]	alubase,shift
mrep	rep, wrep	1 i[32], 1 o[32],1 o[1], sel[2]	4[32]	alu,shift
Merge(*mrep,mmul*)	mul, wmul, rep, wrep	2 i[32], 1 o[32],1 o[1], sel[3]	5[32]	alu,shift,alubase
Promote(*mgcd*,{-})	gcd,wgcd, -	2 i[32], 1 o[32],1 o[1], sel[2]	2[32]	alu
Extract(Promote (*mgcd*,{-}),{-})	gcd, wgcd	2 i[32], 1 o[32],1 o[1], sel[2]	2[32]	alu,alubase

Notes: Column *P*: i designates a input port, o, an output port and sel, a selection port; the number between brackets indicates the port's width in bits. Column *V*: the number between brackets indicates a variable with that many bits. Alu, alubase and shift are primitive FUs.

Table 2 shows characteristics of RTL structures obtained from the architectures in table 1, via a HLS system (we have used AMICAL[9]). Regarding **Merge**, it can be observed a reduction on FU counting: the structure for **Merge**(*mrep,mmul*) uses 1 *alubase* and 2 *shifts*; if *mmul* and *mrep* were taken separately, the total of FUs would be 1 *alubase*, 3 *shifts* and 1 *alu*. The same can be said about the muxes. Regarding number of registers, AMICAL allocates one register for each variable and constant, therefore it closely follows the variable set. The downside is on the controller - it is larger than the sum of the ones in the original FUs, although the number of states, transitions and ios (total sum of controller's inputs and outputs) were smaller as shown in the column 'controller' in table 2. This happened because controller area are non-linearly dependent on the number of states, transitions and ios. Adding data-path and controller, the FU produced by **Merge** is smaller than the sum of the originating ones; the column 'gain' shows the difference. Regarding **Promote**, it forces the creation of paths from ports to FUs; this effect can be observed in table 2, which shows an increased number of muxes and controller hardware when comparing transformed units to the original ones. However, using **Promote** may still be advantageous; for instance, if a data-path contains a *mmul* and an *alu*, and **Promote** is applied in *mmul* generating a new version which is capable of performing all the operations of an *alu*, then the data-path could not need the *alu* anymore. In table 2, the column 'gain' shows the gain in area if the aforementioned optimisation were carried out.

Table 2. Structural characteristics for the benchmarks and for the ones produced by the ATs.

	structural features	area	gain
mgcd	1 FU (alu), 4 Regs, 5 Muxes, (6T,2S,145ios)	4744+604=5348	-
mmul	3 FUs (alubase, shift(2)), 7 Regs, 8 Muxes, (12T,6S,160ios)	12008+1346=13354	-
mrep	2 FUs (shift, alu), 6 Regs, 6 Muxes, (13T,7S,215ios)	8680+1952=10632	-
Merge(*mrep*, *mmul*)	3 FUs (alubase,shift(2)), 7 Regs, 10 Muxes, (24T,12S,94ios)	12712+4927=17639	6347
Promote(*mgcd*, {-})	1 FU (alu), 4 Regs, 6 Muxes, (7T,2S,148ios)	5093+719=5815	597 (467)
Extract(Promote (*mgcd*,{-}),{-})	1 FU (alu), 4 Regs, 5 Muxes, (6T,2S,145ios)	4744+604=5348	-

Notes: Column structural features: lists the main contents of the structure: number and names of FUs (between parenthesis), number of registers, number of multiplexers and features of the controller (number os transitions(T), number os states(S) and the sum of input and output ports (ios) of the controller). Column area: estimations of area in NTBR's - number of transistors - for the data-path, the controller and the total. Column gain(loss): shows potential gains or losses in using the transformed FUs, in NTBRs.

Additional experiments were carried out to observe the changes on hierarchical descriptions when transformations are applied on them. In table 3, the comparison between two hierarchical benchmarks, *pcl* and *pid*, and their transformed versions is shown; only the synthesised structural data are presented. Both transformed versions were obtained from the original ones by adding one extra transformed FU (*mmulmrep*+-=**Promote**(**Merge**(*mmul*, *mrep*),{+-}) into theirs set of architectures. Results in table 3 show that, in all cases, area reductions were obtained; the total area of FUs were always smaller. However, the controller size increased due to the increased complexity in accessing promoted operations, which uses more cycles as commented in section 4; AMICAL created controllers with more transitions due to this fact, it also implies that the new architectures are slower.

Table 3. Structural characteristics for the hierarchical benchmarks

	structural features	**area**	**reduction**
pcl	26Regs, 8FUs {ram,bset,mmul, mrep,alu, bmask(2),shift}, 16Muxes, (110T,48S,196ios)	52242+15254= 67496	
pcl*	26Regs, 6FUs {ram, bset, mmulmrep+-, bmask(2), shiftr}, 15Muxes, (120T,48S,192ios)	44925+16309= 61234	9,3%
pid	14Regs, 4FUs {rom, mrep, mmul, alu}, 8Muxes, (52T,22S,45ios)	40378+1758= 42136	
pid*	14Regs, 3FUs {rom, alu, mmulmrep+-}, 7Muxes, (56T,22S,42ios)	34125+1778= 35903	14,8%

Note: The benchmarks marked with a star are the transformed ones. See also notes in table 2.

6. CONCLUSION

We have presented a set of Architectural Transformations and a suitable modelling for hierarchical algorithmic descriptions. The main application for our ATs is for obtaining optimised hierarchical RTL structures from hierarchical algorithmic descriptions, via HLS. By using such optimising tools, designers are freed to write hierarchical algorithmic descriptions without worrying over some implementation issues like consequences of writing style on structural level quality. We have also presented some experimental results which confirm such a claim.

REFERENCES

[1] D.D.Gajski, N.D.Dutt, A.C.H.Wu e S.Y.L.Lin, "High-Level Synthesis - Introduction to Chip and System Design", Kluwer Academic Publishers, 1992.

[2] W. Geurts, F. Catthoor, H. De Man: Time Constrained Allocation and Assignment Techniques for High Throughput Signal Processing. Proceedings of the 29th Design Automation Conference, pp.124-127,1992.

[3] D. Sreenivasa Rao, F. J. Kurdahi: Controller and Datapath Trade-offs in Hierarchical RT-Level Synthesis. Proceedings of the 7th International Symposium on High-Level Synthesis, 1994.

[4] R. Genevriere, R. Camposano: Partitioning and Restructuring Designs on the Behavioral Level. Technical Report SFB358-B2-11/94, University of Paderborn, Technical University of Dresden, 1994.

[5] P. Kission, H. Ding, A.A.Jerraya: Structured Design Methodology for High-Level Design. Proceedings of the 31th Design Automation Conference, pp.466-471, 1994.

[6] O. Bringmann, W. Rosenstiel: Cross-Level Hierarchical High-Level Synthesis. Proceedings of the 1998 Design Automation and Test in Europe.

[7] J.C.Wang, M.Y.Teruya, J.V.Vale Neto, M. Strum, A.A.Jerraya: A Recursive High Level System. Proceedings of the 5th International Conference on VLSI and CAD - Seoul, South Korea, pp. 412-414, 1997.

[8] B.G.Hald, J.Madsen, A.A.Jerraya: A New Approach to Optimization and Reuse of Hierarchical Architectures. Accepted for publication by IEEE Transactions on VLSI.

[9] A.A.Jerraya, I.Park, K.O'Brien, "AMICAL: An Interactive High-Level Synthesis Environment", Proceedings of European Design Automation Conference, Paris, France, 1993.

An Enhanced Static-List Scheduling Algorithm for Temporal Partitioning onto RPUs

João M. P. Cardoso and Horácio C. Neto jmpc@acm.org, hcn@inesc.pt
University of Algarve/INESC, IST/INESC

Key words: Temporal Partitioning, Reconfigurable Computing, Dynamic Reconfiguration, Field-Programmable Gate-Arrays (FPGAs)

Abstract: This paper presents a novel algorithm for temporal partitioning of graphs representing a behavioral description. The algorithm is based on an extension of the traditional static-list scheduling that tailors it to resolve both scheduling and temporal partitioning. The nodes to be mapped into a partition are selected based on a statically computed cost model. The cost for each node integrates communication effects, the critical path length, and the possibility of the critical path to hide the delay of parallel nodes. In order to alleviate the runtime there is no dynamic update of the costs. A comparison of the algorithm to other schedulers and with close-to-optimum results obtained with a simulated annealing approach is shown. The presented algorithm has been implemented and the results show that it is robust, effective, and efficient, and when compared to other methods finds very good results in small amounts of CPU time.

1. INTRODUCTION

The availability of RPUs (reconfigurable processing units), such as the new FPGAs, with lower reconfigurable times and partial-reconfiguration capability, has made possible the concept of "virtual hardware" [1]: the hardware resources are supposed unlimited and implementations that oversize the RPU area are resolved by temporal partitioning. Then, the partitioned solution is executed by time-sharing the device such that the initial functionality is maintained. This concept promises to be an efficient solution to save silicon area. One of the applications is the switch among functionalities that have mutual exclusiveness on the temporal domain, such as the context-switching between coding/decoding schemes in communication, video or audio systems. However, temporal partitioning algorithms able to exploit efficiently the new concept are needed. They must consider trade-offs among parallelism, communication costs, latency and reconfiguration times. The nodes of a given graph have to be scheduled in time slots to be executed in each temporal partition. Temporal partitioning must preserve the dependencies between nodes (that are already temporal dependencies) such that a node **B** dependent on node **A** cannot be mapped to a partition executed before the partition where node **A** was mapped.

Although, the FPGAs themselves, such as the Xilinx™ XC6200 family [2], do not have mechanisms to implement efficiently temporal partitions and the time of reconfiguration of the overall FPGA is still quite high, the importance of the "virtual hardware" concept has already been demonstrated with computationally complex applications [3]. Industrial efforts are under way to further improve the capability of the devices to handle multiple-configurations by storing several on-chip configurations and permitting the context-switching in few nanoseconds [4][5][6]. The trend to have on-chip configurations instead of more logic cells is explained by the fact that the area of SRAMs to store configurations for each cell is much lower than the cell itself.

Efficient mechanisms of communication between temporal partitions have also been actively researched such as the micro-registers in [5]. The majority of the efforts considers FPGA registers that maintain the same state between contexts whenever wanted.

As referred, research efforts are under way on both new RPU architectures [7] and on the automation of the temporal partitioning process. Our efforts address the temporal partitioning of behaviors during the synthesis steps. This paper presents a new temporal partitioning algorithm that effectively takes into account, among other aspects, the inter-communication costs, while maintaining a small computational complexity. Besides, it is sufficiently flexible to permit the consideration of various target architectures. Results are compared to a number of alternative constructive algorithms and, in order to show how far they are from close-to-optimum solutions, comparisons to a simulated annealing (SA) [8] approach are shown.

From now on we refer to temporal partitions and temporal partitioning simply as partitions or partitioning respectively, since this paper neither considers spatial partitions nor spatial partitioning.

The paper is divided in the following sections. Section 2 summarizes the related work on temporal partitioning. Section 3 describes the computational models considered by the approach proposed in this paper. Section 4 explains the algorithm and the heuristics used. Results are shown in section 5, where the new algorithm is compared to simple heuristics and to results obtained by the SA implementation. Finally, conclusions are enumerated and future work is envisaged.

2. RELATED WORK

The development of temporal partitioning algorithms was firstly considered in [9][1]. The similarities of both scheduling on high-level synthesis [10] and temporal partitioning allow the use of common scheduling schemes for partitioning. However, an important factor that must be considered is the inter-communication (communication among partitions) cost, because it can impose an unacceptable overhead on the overall latency.

In [11] a static-list scheduling (SLS) approach is used for partitioning trying to minimize the number of nets among partitions. It works on the netlists (4-input-LUTs) of combinational circuits and uses the path-to-end's length of each node as a priority function and the size of the fan-out of each node as a tiebreaker. The approach is suitable to RPU architectures with inter-buffering (on-chip buffers that maintain the state among partitions). These RPUs have small inter-communication costs and the overall optimization problem resumes to the minimization of the critical path length. [12] presents an enhanced force direct scheduling algorithm which considers communication costs and which is able to process sequential circuits.

In [13], a variation of the SLS followed by an optimizer is used to perform partitioning of netlists. The algorithm uses three scheduling rules to select among the ready nodes and is tailored to the Time-Multiplexed FPGA [5].

[14][15] present a network-flow based method for multi-way partitioning of netlists. In [14] the algorithm is also targeted to the Time-Multiplexed FPGA and the results out-perform the SLS approach [13] in terms of communication costs (number of nets between partitions). The algorithm uses the max-flow min-cut computation iteratively to find k-partitions. [15] shows improvements over the enhanced force direct scheduling of [12] with respect to communication costs. Results comparing the latency of the solutions are neither presented nor examined.

The above approaches are all based on the netlist of the final circuit previously mapped to the library of the target FPGA. They can be efficient approaches to rapid prototyping but suffer from the impossibility to exploit partitions at the behavioral level. This has more importance when considering the integration of partitioning into the *reconfigware* compilation from behavioral descriptions. Moreover, these approaches suffer from the heavy number of nets and nodes that must be manipulated. At the behavior level the operations encapsulate the tricky connections and only the groups of nets transporting operands are visible by the algorithms.

Some authors, such as in [9] and [16], have considered the partitioning at behavioral levels having in mind the integration of synthesis. In [9], a heuristic based on the SLS enhanced to consider dynamic area constraints is presented. The approach does not consider the inter-communication costs. In [16] the partitioning problem is modeled in a specified 0-1 non-linear programming (NLP) model [17]. Due to the computational complexity of the approach, heuristic methods must be developed to permit feasible executions on large input examples. In [18], a partitioning algorithm based on the levelized nodes obtained by the "as soon as possible" (ASAP) scheduling algorithm is used. The algorithm fills the available area of the RPU in the increasing order of the ASAP levels. The selection of nodes in the same level is arbitrary and the algorithm switches to another partition when it encounters the first node that does not fit in the current partition. In [19], partitioning algorithms based on the extension of the ASAP or "as late as possible" (ALAP) leveling algorithms with the selection of a node, in the same ASAP or ALAP level, by a local priority function based on the nodes' mobilities have been considered. [19] also shows an algorithm that searches recursively in the list of ready nodes so that if a node cannot be mapped to the current partition, other nodes can be considered.

However, all the above approaches do not consider both the inter-communication costs and the latency, and the majority of them are tailored to a specific target architecture. Therefore, new efforts to integrate the inter-communication costs and the latency of the solutions in a temporal partitioning algorithm working at the behavioral level are presented in this paper.

3. RECONFIGURABLE COMPUTING MODEL

Our model assumes that the CPU has access to the *reconfigware* and is responsible for the reconfiguration of the RPU(s) that integrate the *reconfigware* part. We also consider without loss of generality that the CPU has access to the memory attached to the RPU. The partitions are mapped to the *reconfigware* and the CPU store/load primary input/outputs directly to/from the memory, or the FPGA when this is supported. The data transfered between partitions can be stored in the memory by the *reconfigware*, in special registers that are maintained during reconfigurations, or collected by the CPU.

At least three schemes of interface mechanisms between partitions can be considered. The use of registers and a task running on the microprocessor to load and store operands between partitions is one possibility. This scheme is well suited to boards of RPUs with a processor or a micro-controller that can also control the reconfiguration of partitions, or to boards without any buffer scheme to maintain results among partitions. With the advent of the integration of RPUs in processor cores this can be an efficient scheme because of the lower communication overhead on these systems. The second scheme is the use of a set of registers in the RPU when the device can be partially reconfigured. The registers will be configured once in a region of the RPU and shared between partitions (this can include the advent of new RPUs tailored to time-sharing). The third scheme uses a macro-cell to access memory locations controlled by a hardware unit. This is well suited to types of RPUs where there is no processor (only local memory) or the communication between the host and the RPUs is time-consuming.

From the above considerations it is clear that feasible partitioning schemes must consider different inter-communication costs due to different interface mechanisms.

4. PROBLEM FORMULATION

In order to be independent from a particular software language (e.g., C/C++ or Java) the input program is represented as a hierarchy of PDGs (program dependence graphs) in which the bottom level is formed by DAGs (directed acyclic graphs) where each node represents an operation. A node in the PDG can be a group of statements or a single statement. A loop is represented by a special node in the PDG which encapsulates the PDG of the loop body. Herein, we only consider nodes with deterministic delays (known at compile time), and recursive constructions are not allowed. Thus, a behavioral description is represented by a graph, $G = (V, E)$, which is an ordered, directed and acyclic graph with $|V|$ nodes, $\{v_1, v_2, \ldots, v_{|V|}\}$ and $|E|$ edges, where each node v_i represents a single behavior. Each edge $e_{i,j} \in E$ represents a dependency between nodes v_i and v_j. A dependency can be only a precedence-dependency or a transport-dependency due to the transport of data between the two nodes.

The communication cost associated with an edge $e_{i,j}$ representing a transport-dependency is calculated by the number of bytes to transfer ($d_{i,j}$) divided by the maximum bandwidth of each atomic transfer (with the result rounded to the next big integer). In the majority of the communication mechanisms for each connection between different partitions the data must be stored by the partition that defines it and loaded by the partition that uses it. Non-transport dependencies (precedence-dependencies) have a zero $d_{i,j}$.

We assume that an estimation of the execution time and the *reconfigware* size (CLBs, cells, FUs, etc.) of each node is available at compile time.

The objective of the partitioning algorithms is to create partitions in the temporal domain such that the cost is minimum and that each partition fits in the *reconfigware* area physically available. Each partition Π_i is a non-empty subset of V. A graph G partitioned in k subsets is correct if:

- The set of partitions $\wp = \Pi_1 \cup \Pi_2 \cup \ldots \cup \Pi_k \supset V$.
- $\forall\, \Pi_i \in \wp$, $Area(\Pi_i) \leq MaxArea$; Each temporal partition fits in the RPU resources.
- $\forall\, e_{i,j} \in E$, $\wp(v_i) \rightarrow \wp(v_j) \vee \wp(v_i) \equiv \wp(v_j)$; $\rightarrow$ indicates the order of execution. All dependencies are met (necessary condition to obtain the same functionality).

A correct set of partitions guarantees the same behavior of the original graph. However, we are also interested on the minimization of the overall latency. The cost that reflects the latency of a graph in a time-multiplexed RPU can be estimated by the equation (1). τ_{cycles} is the number of clock cycles for each load/store. $In(\Pi_i)$ and $Out(\Pi_i)$ correspond to the number of inter-communications without considering the primary input/outputs. The second term represents the overall critical path delay without inter-communication costs.

$$\Omega(G, RWModel_A) = \sum_{i=0}^{NumPart-1} \left[\tau_{cycles} \times \left(In(\Pi_i) + Out(\Pi_i)\right)\right] + \sum_{i=0}^{NumPart-1} \left[Max_{Delay}(\Pi_i)\right] \quad (1)$$

5. THE ENHANCED-STATIC LIST SCHEDULING ALGORITHM

The partitioning algorithm proposed (ELS), as can be seen in *Figure 1*, is an extension of the SLS algorithm. It starts by computing the ASAP and ALAP values of the nodes in the graph. Then a cost computed with equation (2) is assigned to each node of the input graph. Each term of the equation has a multiplication factor to give more weight to the communication cost (α), to the critical path (β), or to a tradeoff between them (η). The first term (3) gives emphasis to the communication costs. A large difference between the input and output edges of a node assigns a greater priority to that node. Also, a large number of nodes from a given node to the sink can produce more communication costs. The middle term (4) tries to give emphasis to the existent parallelism by assigning more weight to the nodes with lower ASAPs (giving the opportunity to place ready nodes in parallel to the nodes of the critical path already scheduled). This factor increases with the decreasing of the communications' weight. The third term (5) is the starting fine-grain ALAP of each node and permits to sort the nodes by ascending order of their

ALAPs. The scale factor is set to the ratio of the maximum number of levels of the ASAP leveling and the delay of the critical path. Experimentally we have found that the weight η expressed by β/(α+1) generally conduces to good results. However, another independent weight instead of the previous expression can be used to unconstrain the exploitation.

$$W[v_i] = \alpha \cdot \psi_{comm}(v_i) + \eta \cdot \psi_{comm/delay}(v_i) + \beta \cdot \psi_{delay}(v_i) \quad (2)$$

$$\psi_{comm} = scale \cdot [In(v_i) - Out(v_i) - MaxLevels + LevelALAP(v_i)] \quad (3)$$

$$\psi_{comm/delay} = -scale \cdot ASAP_{start}(v_i) \quad (4)$$

$$\psi_{delay} = -scale \cdot ALAP_{start}(v_i) \quad (5)$$

```
ESL(MaxArea, G(V, E), α, β, η) {
  Compute ASAP(G); // compute the fine-grain ASAP for each node
  Compute ALAP(G); // compute the fine-grain ALAP for each node
  int CurrentArea=0, SchedNum=0, NumNode=0, NumSchedNodes=0;
  for each vi ∈ V { Costvi=ComputeCost(vi, α, β, η);} // compute the cost for each node
  SortedList = Create a sorted List of ready nodes to be scheduled; // based on the Costs
  BitSet SchedNodes = new BitSet(|V|); // all nodes marked as unscheduled
  while (NumSchedNodes < |V|) { // while there are unscheduled nodes
    Node A = SortedList.ElementAt(NumNode); // peek a node
    Boolean Fit = ((CurrentArea + Area(A)) <= MaxArea);
    if (Fit) { // the node A fits in the current temporal partition
      SortedList.removeElement(A); // remove A from the SortedList
      // schedule A in the partition, update the current area and mark A as scheduled
      ScheduleElement(A, CurrentArea, SchedNum, SchedNodes);
      NumSchedNodes++; // increment the number of scheduled nodes
      if (SortedList.Update(G, A, SchedNodes)) NumNode = 0;
      else if (NumNode ≥ SortedList.size()) NumNode—;
    } else {
      if (NumNode < SortedList.size()-2) { // try another node in the SortedList
        NumNode++;
      } else {
        SchedNum++; NumNode = 0; CurrentArea = 0; // another temporal partition
      }
    }
  }
}
```

Figure 1. The enhanced static-list scheduling algorithm.

The next step of the algorithm is to compute the nodes ready to be mapped to the current partition and to sort them by descending order of the costs. Then, the algorithm starts the loop by considering each node of the sorted list of ready nodes. The algorithm checks if a node can be mapped. If the node was mapped the algorithm tries to update the list of ready nodes by considering the sink nodes. Then the algorithm starts the loop considering another node. If a node cannot be mapped to the current partition the algorithm searches for a feasible node on the list before the creation of a new partition.

The ASAP and ALAP scheduling algorithms compute with graph traversal and have a runtime complexity of $O(|V|+|E|)$. The ELS algorithm has a worst-case runtime complexity of $O(|V|^2+|E|)$.

To have an idea of how much improvement can be obtained an SA approach has been implemented. It starts from a feasible partitioning solution and tries to improve it by moving nodes (considering the probabilistic selection of randomly valid moves) among adjacent partitions. Moves that can violate the maximum area available on the destination partition, but do not violate any temporal precedence, are considered valid. The approach can exploit results considering more partitions than those of the initial solution by adding empty partitions in the beginning of the execution.

6. RESULTS AND COMMENTS

All the algorithms presented have been implemented with the Java™ language. To permit a statistical comparison a random graph generator has also been implemented.

All the results attributed to SA are close-to-optimum (the best result of several executions with different parameters was collected). Herein Ω is computed with the equation (1) in clock cycles units and Δ_{comm} and Γ_{exec} correspond to the 1st and 2nd terms respectively. All the results neither consider the store/load of primary input/outputs nor the possibility to interleave execution and inter-communications. $\mathbf{S_1}$ refers to the algorithm presented in [18] and $\mathbf{S_2}$ refers to the leveling of the nodes by the ALAP scheme. $\mathbf{S_3}$ and $\mathbf{S_4}$ refer to the algorithm presented in [19] oriented by the ASAP or ALAP levels respectively. $\mathbf{S_5}$ refers to a version of $\mathbf{S_2}$ with the nodes sorted by the ascending ALAP levels and the ascending ASAP levels as a tiebreak. $\mathbf{S_6}$ is a version of $\mathbf{S_1}$ with a list created with the nodes of an ASAP level sorted by the ascending order of their ALAP step time. $\mathbf{S_7}$ refers to an SLS approach with the nodes in the list sorted by the ascending order of their ALAP step time.

In *Table 1* and *Table 2* **E** is the relative improvement cost of the SA over the ELS solution. The constructive approaches obtained each solution in less than 1ms.

The 1st example to be considered is the loop body of the HAL example [20] (all operands with 16-bit width). The example has a total area of 4,384 cells and a critical path delay of 58 cycles. The results presented in *Table 1* show small improvements of the SA over the ELS.

The 2nd example is the AR filter [21]. It has 16 multipliers and 12 adders contributing to a total area of 16,960 cells and a critical path delay of 90 cycles (all operands with 16-bit width). The results are shown in *Table 2*.

Table 1. Results for the HAL example (τ_{cycles}=2; **A**: MaxArea = 2,457; **B**: MaxArea = 4,096).

CASE/#Part	Measure	S_1	S_2	S_3	S_4	S_5	S_6	S_7	ELS, α=2	SA time	E (%)
A/5	Ω	70	86	86	80	80	70	80	66, β=20	62 7.8 s	6.1
B/2		66	86	66	84	84	66	84	66, β=1	60 5.7 s	9.1

Table 2. Results for the AR filter (**A's**: MaxArea =4,096; **B's**: MaxArea =16,384).

CASE/#Part	τ_{cycles}	Measure	S_1	S_2	ELS	(α,β)	SA time	E (%)
A1/5	2	Ω	222	202	202	(2,20)	194 26.4 s	3.9
B1/2	2	Ω	134	142	134	(2, 1)	98 90 s	26.8
A2/5	1	Ω	186	166	166	(1, 1)	162 25.4 s	2.4
B2/2	1	Ω	122	126	122	(1, 1)	94 85 s	22.95
A3/5	0	Γ_{exec}	150	130	118	(0, 1)	118 26 s	0
B3/2	0	Γ_{exec}	110	110	110	(0, 1)	90 68 s	18.18
A4/5	1	Δ_{comm}	36	36	26 (13 edges)	(1, 0)	22 19.2 s	15.38
B4/2	1	Δ_{comm}	12	16	6 (3 edges)	(1, 0)	2 66.7 s	66.7

ELS has produced results always better or equal than those obtained by the other heuristics. An accurate analysis has shown that SA has the capability to map nodes with many connections in the same partition reducing the inter-communications. The ELS approach is unable to balance the last two partitions. This problem has more impact when the number of needed partitions is small and is one of the disadvantages of constructive approaches. The results were not improved by making the SA exploit more partitions than those obtained by the constructive approaches. Also, results from experiments with random graphs confirm that the number of partitions used by the heuristics (e.g., ELS) is close-to-optimum and few cases need more partitions (one) to improve the results. More partitions have higher probability to increase the critical path delay and only in few cases can reduce the overall inter-communication cost.

Table 3 shows results obtained with random graphs. All the graphs have 50 nodes and the output edges of each node vary between 0 and (4 or 10). For each algorithm the median of relative improvements over the ASAP leveling method for different inter-communication's weights are shown. The pairs of rows (2, 3) and (4, 5) present results considering 2 and 1 clock cycles for each inter-communication cost respectively. The results shown in the 6th and 7th rows were obtained considering null inter-communication costs. The 8th and 9th rows show the results obtained when only inter-communication costs are considered. The last column indicates the median of the relative improvements of SA over ELS. The last row shows the median of each column. In these tests the SA has run over an initial solution obtained by ELS.

Table 3. Results for N=100 graphs randomly generated.

	Out-edges	τ_{cycles}	S_2	S_3	S_4	S_5	S_6	S_7	ELS	(α,β)	SA	E_1 (%)
E/N (%)	[0 10]	2	-0.9	9.1	1.1	-2.2	-1.4	9.2	**12.2**	(2,1)	29.7	19.9
	[0 4]	2	-1.0	12.4	1.7	-3.1	-0.8	14.5	**18.5**	(2,1)	38.1	24
	[0 10]	1	-0.3	5.8	0.2	-0.7	0.7	**8.7**	6.8	(1,1)	22.9	17.3
	[0 4]	1	-0.3	7.4	0.7	-1.1	3.3	**13.7**	10.6	(1,1)	28	19.5
	[0 10]	0	2.8	-8.6	-3.5	5.4	8.6	6.0	**9.5**	(0,1)	13	3.9
	[0 4]	0	2.2	-11.1	-2.9	5.8	**16.2**	10.8	10.9	(0,1)	16	5.7
	[0 10]	1	-2.2	15.3	2.7	-5.1	-5.3	10.3	**29.8**	(1,0)	45	21.7
	[0 4]	1	-2.5	22.9	-3.7	-7.1	-9.8	16.0	**48**	(1,0)	57.8	18.8
Tot.	-	-	0.28	6.65	-0.46	-1.0	1.4	11.2	18.3		20.6	16.4

The results show improvements of ELS over the other considered heuristics. On the 4th and 5th rows ELS results would be similar to the S_7 results if only the third term of equation (2) was used. S_3 was the best algorithm of the ASAP/ALAP leveling approaches with respect to communication. The results show better solutions of SLS over the ASAP/ALAP approaches with respect to latency, as was expected due to the opportunity to execute freedom nodes in parallel with the critical path. S_6 has produced the best results of the constructive approaches when only latency has been considered. The fact is due to the consideration of all the nodes of an ASAP level before the addition of nodes from the next level (an SLS tries, for each node scheduled, to update the list of the ready nodes). When the cost of each communication was not high, S_7 has produced better results than the other constructive algorithms (rows 4 and 5).

The close-to-optimum results had about 16% of average relative improvement over the ELS (from 4 to 24%). Better results of ELS can be achieved by exploiting the values of the α, β and η weights on equation (2).

7. CONCLUSIONS & FUTURE WORK

In this paper temporal partitioning techniques have been presented and compared. A novel heuristic extension to the static-list scheduling algorithm was presented. Being an algorithm of the family of the scheduling algorithms it can embody resource constraints without much more complexity. The low complexity of the algorithm makes it applicable on large graphs. The results show improvements over related algorithms and show that simplified algorithms can be used to resolve the temporal partitioning problem at the behavior level. The algorithm can be also applied efficiently to rapid prototyping of DFGs with much better results than the ASAP approach presented in [18] without increasing significantly the execution time.

Although just presented as a comparison term a temporal partitioning approach based on the simulated annealing has been also implemented. Based on the execution time the annealing approach does not seem, at least alone, to be a reasonable choice to *reconfigware* compilers where one of the most important objectives is fast compilation. However, efficient cooling schemes should be studied to improve the efficiency of the annealing.

The close-to-optimum experimental results show that most of times the optimal number of partitions is the minimum as was also stated in [7].

The support of the loop distribution transformation and the possibility to deal with resource sharing will be considered by future temporal partitioning schemes.

ACKNOWLEDGEMENTS

This work has been partially supported by the PRAXIS XXI Program under the scope of Project PRAXIS/2/2.1/TIT/1643/95 and by the Ph.D. program under the PRODEP 5.2 action. The authors would also like to thank Mário P. Véstias for the fruitful comments about the annealing implementation.

REFERENCES

1. X.-P.Long, H. Amano, "WASMII: a Data Driven Computer on a Virtual Hardware," in *Proc. 1st IEEE Workshop on FPGAs for Custom Computing Machines*, Napa Valley, CA, USA, April 5-7, 1993, pp. 33-42.
2. Xilinx Inc., *XC6000 Field Programmable Gate Arrays*, version 1.10, April 24, 1997.
3. R. D. Hudson, D. I. Lehn, and P. M. Athanas, "A Run-Time Reconfigurable Engine for Image Interpolation", in *Proc. 6th IEEE Symposium on Field-Programmable Custom Computing Machines*, Napa Valley, CA, USA, April 15-17, 1998.
4. S. M. Scalera, J. R. Vázquez, "The Design and Implementation of a Context Switching FPGA," in *Proc. 6th IEEE Symposium on Field-Programmable Custom Computing Machines*, Napa Valley, CA, USA, April 15-17, pp. 78-85, 1998.
5. S. Trimberger, D. Carberry, A. Johnson, J. Wong, "A Time-Multiplexed FPGA," in *Proc. 5th IEEE Symposium on Field-Programmable Custom Computing Machines*, Napa Valley, CA, USA, April 16-18, pp. 22-28, 1997.

6. T. Fujii, et al, "A Dynamically Reconfigurable Logic Engine with a Multi-Context/Multi-Mode Unified-Cell Architecture," in *Proc. IEEE Int'l Solid State Circuits Conference*, SA, CA, Feb. 15-17, 1999. See <http://www.nec.co.jp/english/today/newsrel/9902/1502.html>.
7. A. DeHon, *Reconfigurable Architectures for General Purpose Computing*, PhD Thesis, AI Technical Report 1586, MIT, 545 Technology Sq., Cambridge, MA02139, Sept. 1996, <http://www.ai.mit.edu/people/andre/phd.html>.
8. S. Kirkpatrick, C. Gelatt, Jr., and M. Vecchi, "Optimization by Simulated Annealing," Science, Vol. 220, No. 4598, May 1983, pp. 671-680.
9. M. Vasilko, D. Ait-Boudaoud, "Architectural Synthesis Techniques for Dynamically Reconfigurable Logic," in *Proc. 6th Int'l Workshop on Field-Programmable Logic and Applications*, Darmstadt, Germany, Sept. 23-25, 1996, LNCS, vol. 1142, Springer-Verlag, 1996, pp. 290-296.
10. G. De Micheli, *Synthesis and Optimization of Digital Circuits*, McGraw Hill, 1994.
11. D. Chang, M. Marek-Sadowska, "Buffer Minimization and Time-multiplexed I/O on Dynamically Reconfigurable FPGAs," in *Proc. 5th ACM Int'l Symposium on FPGAs*, Monterey, CA, USA, 1997, pp. 142-148.
12. D. Chang, M. Marek-Sadowska, "Partitioning Sequential Circuits on Dynamically Reconfigurable FPGAs," in *Proc. 6th ACM Int'l Symposium on FPGAs*, Monterey, CA, USA, February 22-24, 1998.
13. S. Trimberger, "Scheduling Designs into a Time-Multiplexed FPGA," in *Proc. 6th ACM Int'l Symposium on FPGAs*, Monterey, CA, USA, February 22-24, 1998.
14. H. Liu and D. F. Wong, "Network flow based circuit partitioning for time-multiplexed FPGAs," in *Proc. ACM/IEEE Int'l Conference on CAD*, San Jose, CA, USA, Nov., 1998.
15. H. Liu and D. F. Wong, "Circuit partitioning for dynamically reconfigurable FPGAs," in *Proc.* 7th *ACM Int'l Symposium on FPGAs*, Monterey, CA, USA, Feb 21-23, 1999.
16. I. Ouaiss, et al., "An Integrated Partioning and Synthesis System for Dynamically Reconfigurable Multi-FPGA Architectures," in *Proc. 5th Reconfigurable Architectures Workshop*, Orlando, Florida, USA, March 30, 1998.
17. M. Kaul, R. Vemuri, "Optimal Temporal Partitioning and Synthesis for Reconfigurable Architectures," in *Proc. Design, Automation & Test in Europe,* Paris, France, Feb. 23-26, 1998, pp. 389-396.
18. K. M. GajjalaPurna, and D. Bhatia, "Partitioning in Time: A Paradigm for Reconfigurable Computing", in *Proc. IEEE Int'l Conference on Computer Design*, Austin, Texas, USA, October 5-7, 1998, pp. 340-345.
19. J. M. P. Cardoso, and H. C. Neto, "Macro-Based Hardware Compilation of Java™ Bytecodes into a Dynamic Reconfigurable Computing System, " in *Proc. 7th IEEE Symposium on Field-Programmable Custom Computing Machines*, Napa Valley, CA, USA, April 21-23, 1999, Kenneth Pocek and Jeffrey Arnold, eds., IEEE Computer Society Press.
20. P. G. Paulin, J. P. Knight, and E. F. Girczyc, "HAL: A Multi-Paradigm Approach to Automatic Data Path Synthesis, " in *Proc. 23rd Design Automation Conference*, June 29-July 2, 1986, 263-270.
21. R. Jain, A. Parker, N. Park, "Module Selection for Pipelined Synthesis, " in *Proc. 25th Design Automation Conference*, 1988, 542-547.

Object-Oriented Modeling and Co-Simulation of Embedded Systems

F.R.Wagner, M.Oyamada, L.Carro[1] and M.Kreutz
Universidade Federal do Rio Grande do Sul
Instituto de Informática, [1]Departamento de Engenharia Elétrica
Caixa Postal 15064, 91501-970 Porto Alegre, Brazil
e-mail {flavio,marcio,kreutz}@inf.ufrgs.br, [1]carro@iee.ufrgs.br

Key words: embedded systems, object-oriented modeling, co-simulation

Abstract: This paper presents the modeling and co-simulation capabilities of S^3E^2S, a design environment for electronic systems that can be built as a combination of analog and digital parts and software. S^3E^2S is based on a distributed, object-oriented system model, where abstract objects are initially used to express complex behavior and may be later refined into digital or analog hardware and software. Co-simulation of any heterogeneous model developed during a stepwise refinement process is supported. These capabilities are illustrated by the modeling of a crane and its embedded control.

1. INTRODUCTION

Embedded electronic systems contain a combination of software and hardware, both analog and digital. For simple systems, a single off-the-shelf processor might be sufficient, since there are various available architectures (microcontrollers, DSP, RISC processors) with different cost / performance ratios. However, more complex systems, which represent the current trend in the market, usually have more critical requirements, such as a combination of behaviors (scalar and DSP processing) and tight low-level characteristics (area, speed, power dissipation). Typical examples are voice-modems, cellular phones, and embedded controllers for automotive applications.

The design of complex embedded systems should ideally proceed from an initial, abstract specification, going through a sequence of successive refinements, until a final detailed solution is achieved. Intermediate, heterogeneous descriptions generated during this stepwise refinement must be validated. Co-simulation is the most usual validation tool.

Most existing environments are oriented towards a particular application domain and have a fixed target architecture. They usually follow one of two possible approaches for system specification and co-simulation. In the first approach, system modeling is performed using a single language from a given domain, such as C, VHDL, or a higher-level language, and partitioning is performed in a later stage. In the second approach, the initial specification already considers a given partitioning, and each part is modeled using a language which is more appropriate for the corresponding domain.

The S^3E^2S (Specification, Simulation, and Synthesis of Embedded Electronic Systems) environment combines the advantages of both approaches. Complex systems may be modeled as combinations of objects specified at different domains – abstract object-oriented specification, digital hardware, analog hardware, and software – and at multiple abstraction levels. Co-simulation is performed by coupling different simulation engines, so that the validation of any heterogeneous model developed during a process of stepwise refinement is supported. Furthermore, the environment allows an easy exploration of the design space at a multi-processor level, selecting a combination of processors which best matches the design requirements. This paper covers the modeling and co-simulation features of S^3E^2S. The synthesis capabilities are discussed in detail elsewhere [1].

This paper is organized as follows. An overview of the design environment can be found in Section 2. The modeling and co-simulation capabilities of S^3E^2S are introduced in Section 3. Section 4 presents a case study that fully illustrates the application of the environment. Section 5 compares S^3E^2S to other current approaches for modeling and simulation of embedded systems. Section 6 draws conclusions and discusses future work.

2. AN OVERVIEW OF THE S^3E^2S DESIGN ENVIRONMENT

Figure 1 presents the design flow in S^3E^2S. The initial specification is a set of objects, described as C++ code. A set of support libraries allows not only the sequential modeling of digital hardware and software behavior, but also the modeling of linear or discrete time systems (to model analog domain behavior). If the user wishes, a set of high level VHDL constructs is also available, like a FIR filter, for example.

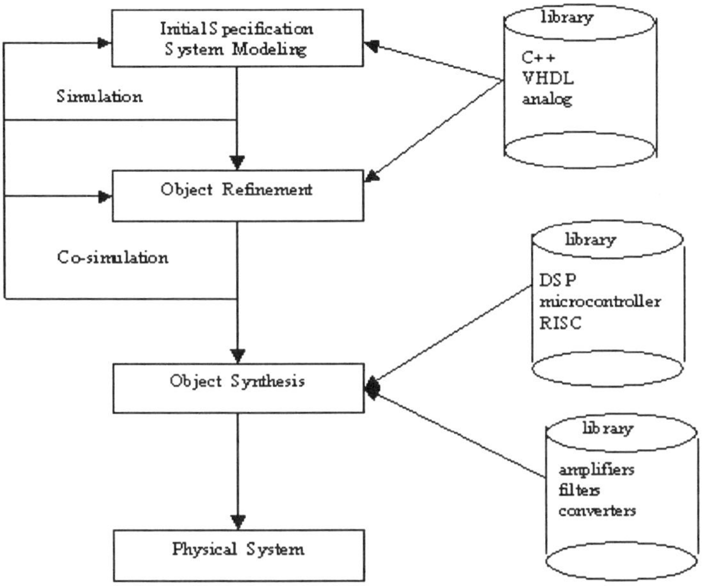

Figure 1. The S^3E^2S design flow

Once the initial specification is defined, it can be validated using the simulation engine behind S^3E^2S. Each object that used a pre-defined set of methods can then be translated into a VHDL component or a set of analog components. Simulation can be done mixing hardware descriptions in VHDL, software descriptions in C++ and analog descriptions developed as a set of analog components (such as operational amplifiers, converters and filters). Each intermediate description obtained during this stepwise refinement process can be co-simulated, whereby each object is simulated by a dedicated simulator for the corresponding abstraction domain.

There are 3 alternatives for the synthesis of the digital parts. The first one is based on the library of VHDL methods used during refinement. If taken from the S^3E^2S library, all methods are synthesizable. Another possibility is the use of the initial C++ specification alone. In this case, an internal compiler analyzes the code of each objet and maps it to an off-the-shelf processor [1]. A library of processors with different characteristics (microcontrollers, digital signal processors, RISC machines) is available. Mapping is done based on the characteristics of the object that best match processor characteristics. S^3E^2S performs the parsing of the C++ object description, obtaining a control-data flow structure. Each function of each object is checked, and the number of memory accesses, arithmetic operations and control instructions is verified. Based on the statistics of resource usage, on the organization of the object code and on the available time to execute a task, a processor best matching these criteria is chosen.

A third synthesis possibility is to let the system map some objects to off-the-shelf processors, leaving some other objects as synthesizable VHDL code. The resulting system is a processor with one or more dedicated ASICs. A typical situation could be a microcontroller and a dedicated filter.

Presently, in order to synthesize analog circuits, the designer must use only blocks that are available from the library, which can mapped to physical components.

3. MODELING AND CO-SIMULATION

S^3E^2S is built on top of SIMOO [2], an integrated environment for object-oriented modeling and simulation of discrete systems. SIMOO is composed of a class library and a model editor. The editor supports the description of the static and dynamic aspects of the model. The static structure is described graphically, while the dynamic structure is described either directly in C++ using the library resources or by means of a state diagram annotated with C++ code. The editor implements extensions to diagrams usually proposed by object-oriented design methodologies, in order to handle simulation-related aspects. From the model description, the editor automatically generates the necessary executable code.

A model is composed of *interface* and *autonomous elements*. Interface elements support tracking of the simulation execution, visualization of simulation results, interactive input of data, and dynamic modification of parameters during the experiments. Autonomous elements, on the other side, are used to model concrete entities. An autonomous element is an active object, i.e., an object with its own execution thread and a message queue. It may interact with other autonomous and interface elements only through messages. The model does not support shared variables, so that it may be also used in distributed environments.

Different objects of the same model may follow different paradigms [3]. A paradigm is defined as a combination of the following modeling approaches: event orientation or process orientation for the description of the object behavior, messages or ports for the communication between objects, and active or passive message handling. These approaches may be extended or specialized by inheritance.

In order to support a progressive replacement of SIMOO objects by VHDL entities or by analog components, and to model interactions with the real analog world, a co-simulation strategy combining SIMOO abstract models, VHDL descriptions, and analog models is needed.

The SIMOO simulation environment is coupled to the VSS simulator from Synopsys. Figure 2(a) shows two SIMOO objects (A and B) that communicate with each other. Object B is to be refined into a VHDL entity. The co-simulation environment then automatically generates necessary interfaces in the SIMOO and VHDL domains. In the SIMOO domain, this includes an interface element. In the VHDL domain, on the other side, the interface specification of the entity corresponding to object B and an interface file written in C are generated, as shown in Figure 2(b). These interfaces in both domains are responsible for data exchange and for synchronizing the simulators.

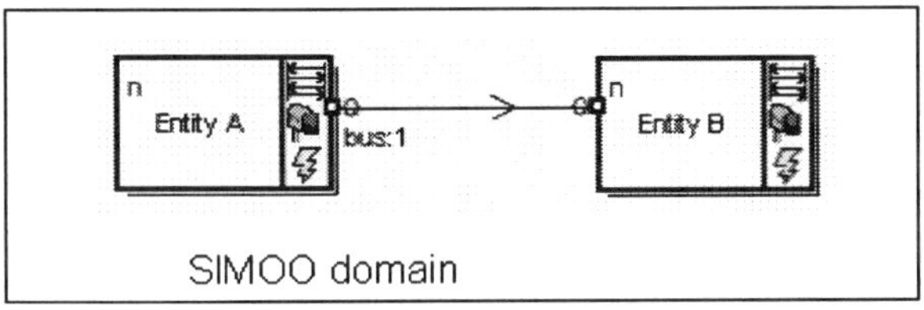

Figure 2(a). An initial specification in the SIMOO domain

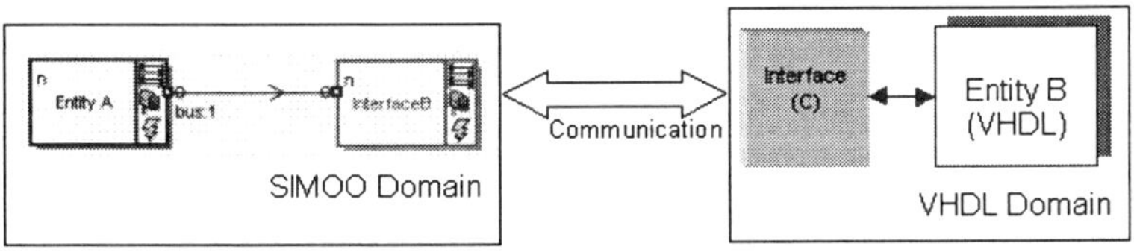

Figure 2(b). The same model, transformed after generation of interfaces for co-simulation

The communication between the SIMOO and VSS simulators is performed via sockets. This allows a distributed solution, where each simulator may run on a different node of a network. In the current version, a conservative approach [4] is adopted for synchronization between the SIMOO and VSS simulators. The base time unit is defined by the VSS simulator, and both simulators advance together their simulation times at each time step. A more efficient implementation of this co-simulation mechanism is currently under investigation. In this more optimistic approach [4], each simulator may run with its own clock and the synchronization is not performed at every clock cycle.

An analog part may be modeled by a set of differential equations that define the object behavior at every possible time value. The co-simulation strategy of the SIMOO environment follows a signal-flow approach, where objects are modeled as mathematical functions from the inputs to the outputs, as opposed to a structural approach, where objects are described as interconnections of analog components. This signal-flow approach is more convenient for systems that don't have a physical implementation yet, or for systems whose design has a stronger emphasis at higher abstraction levels, like PID controllers, converters, filters, transfer functions, etc.

For the integration of an analog part into a SIMOO model, it is thus necessary that the SIMOO object encapsulating this part implements the numerical method needed to solve the differential equations, such as Euler or Runge-Kutta. The SIMOO object must contain attributes such as the time step for the numerical resolution of the equations, the equations themselves, and methods for handling state variables.

4. CASE STUDY

In order to illustrate design possibilities using S^3E^2S, we have modeled a crane and its embedded control. This system has been proposed in [5] as an attempt of benchmarking in the area of system-level modeling and synthesis.

The physical plant is composed of a crane with a load, moving along a track, as depicted in Figure 3. The modeling of the physical system is done by a set of differential equations, which describe the behavior of the crane with a load and external forces being applied. The control of the system involves a set of sequential procedures and the control algorithm itself, which will assure a smooth behavior while the car is moving.

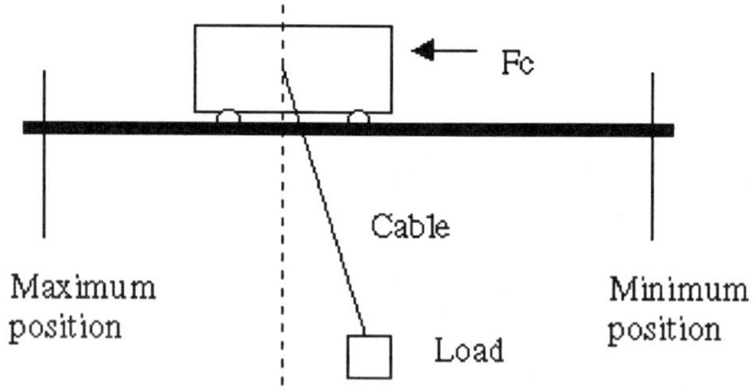

Figure 3. Crane moving along its track with load [5]

A first version of the modeling of the complete system can be seen in Figure 4. Object *Plant_rk* is the physical plant itself. It has been described as a set of differential equations that are solved in continuous time. Object *Actuators* is also modeled in the linear (analog) domain. Object *M_Control* performs the discrete step control algorithm and sensor checking. It receives sensor inputs at each 2 ms, covering the position of the car with some precision, the limits of the displacement for the car, the angle of the cable, and the desired position of the load. All these inputs are used to complete other tasks. For example, the initialization phase of the control algorithm itself and the sensor verification are performed by this major block.

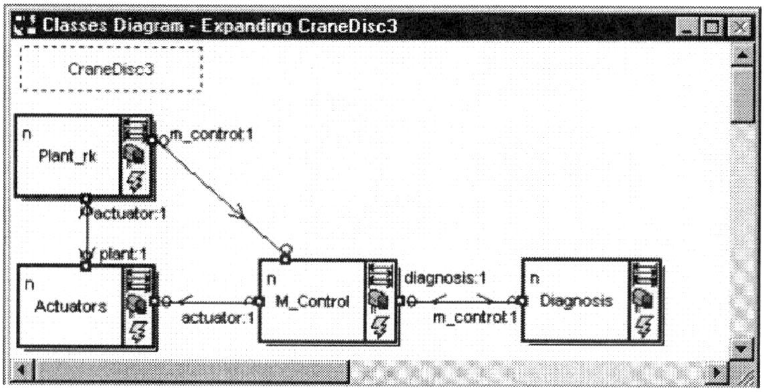

Figure 4. The initial crane and embedded control model

The control algorithm itself is implemented as a discrete computation of the state-variable method. In the control algorithm the goal is to move the crane with a linear displacement, without bumps and oscillations. A set of matrix multiplications must be performed at a fixed time step of 10 ms. If $q_n = [q1_n, q2_n, q3_n, q4_n, q5_n]^T$ is the discrete state vector of the crane, then

$$q_{n+1} = A*q_n + B*[\text{Motor\_Voltage Car\_Position}]^T \quad (1)$$

is the next discrete state of the control algorithm. Coefficient matrixes A and B have dimensions 3x5 and 2x5, respectively.

The object *M_Control* is also responsible for performing sensor checking. When the system enters this mode, the car is driven until extreme positions are found, where sensor inputs should detect its presence. This mode thus allows checking if any sensor function is missing.

Object *Diagnosis* is responsible for continuously checking plausibility of sensor values in parallel to the control algorithm. The position of the car and the value of the angle of the load with regard to the vertical axis are informed by the object *M_Control* and checked for plausible values. In case any discontinuity is verified, the emergency mode is entered immediately.

The control algorithm outputs the value of the force to be applied to the crane, and this is passed to the object *Actuators*. This object is responsible for driving the dc motor that controls the speed, the breaks, and the emergency break, that stops the crane until a power-on-reset is performed.

It is interesting to notice that the proposed benchmark problem allows various modeling solutions, as in almost any real life situation. For example, modeling of the crane behavior should be developed in the analog domain, by solving a set of differential equations that model its behavior. Besides, the control itself could also be developed in the analog world, as some control systems still are. On the other hand, even if one modeled the control circuit as a differential equation in the analog domain, some digital hardware would still be present, due to the finite state machines required to perform sensor checking, system diagnosis, and emergency control. S^3E^2S allows any combination of these different modeling domains and a stepwise refinement of the solution, as it will be described in the next paragraphs.

In Figure 4, which showed our initial modeling of the proposed system, the behavior of all objects is described as C++ methods. Table 1 illustrates the methods implemented for the object *Plant_rk*. Figure 5 shows the main code for this same object.

Table 1. Methods for object *Plant_rk*

Method	Description
Eval	Called in each integration step; invokes functions to solve diff. equations
Rk4_Diff	Implements the Runge-Kutta integration method, to evaluate car distance
Diff	Describe the car position, over the time
Send_sensor	Send the plant state to objects *Control_A* and *AD*

```
/*state[ 0] = xdot, state[ 1] = x,
  state[ 2] = alphadot, state[ 3] = alpha  */

Parameters par;
MSGEA m;                 // port definition
double xdot, x, dxdot_dt, dx_dt;
double alphadot, alpha, dalphadot_dt, dalpha_dt;        // derivatives
xdot = state[ 0] ;       // initial values
x = state[ 1] ;
alphadot = state[ 2] ;
alpha = state[ 3] ;
t = Clock() / 1000;
par.Set(time-t);         // step
m.Set(Id(), actuator, "FORCE", NORMAL_PR, par);
SendReceiveNow(&m);
Fc = m.GetData().GetParAsReal(0);       // get data
dxdot_dt = (fc/mc) + (g*(ml/mc)*alpha) - ((dc/mc)*xdot);
dx_dt= xdot;
dalphadot_dt = ((((dc/mc)-(dl/ml))*(xdot/r))-g/5*(1+(ml/mc))*(alpha))
-((dl/ml)*alphadot) - (fc/(mc*r)) + (fd/(ml*r));    // differ.equations
dalpha_dt = alphadot;
state[ 0] = dxdot_dt;    // next step
state[ 1] = dx_dt;
state[ 2] = dalphadot_dt;
state[ 3] = dalpha_dt;
```

Figure 5. Differential equation describing the crane in object *Plant_rk*

In Figure 6 we have split *M_Control* into two different objects. Since the control algorithm has many arithmetic operations, we described it in a separate object *Control_A*. The finite state machine that performs sensor checking is modeled in object *Job_Control*. All objects of the model are still described using C++ methods.

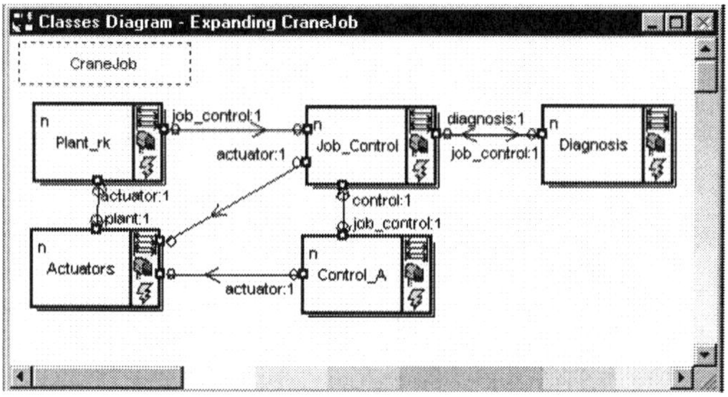

Figure 6. Refining the initial model: control algorithm as a separate object

In order to allow co-simulation and to start a first version of the synthesis of some system parts, *Job_Control* and *Diagnosis* are moved to VHDL. This requires the insertion of analog to digital and digital to analog converters, as shown in Figure 7. Notice that all FSMs, formerly implemented by separate objects *Control_A* and *Diagnosis*, have been grouped in the object *Control_Diag*, although this would not be mandatory. The collapsing of the different FSMs was done because possibly a better synthesis result could be obtained, since all FSMs have many common inputs.

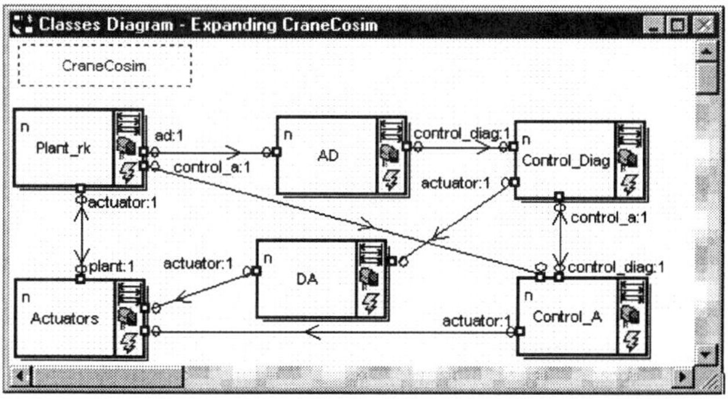

Figure 7. Synthesizing the system: *Control_Diag* as a VHDL object

The synthesis results for the FSMs are a total of 102 logic cells and 41 flip-flops, occupying an area of 8% of an Altera 10k10 FPGA.

5. COMPARISON WITH RELATED WORK

The specification and simulation of application-specific embedded systems is an area of active research. In the case of complex systems, which cannot be implemented by a single processor or controller and its associated software, it is difficult to specify the designer's intention. Many specification languages, or combinations of languages, are being used in industry [6].

The description of complex systems through a single, abstract language has been proposed [7,8]. Some approaches that follow this strategy adopt an object-oriented specification to describe both hardware and software [9-12]. In these cases, design partitioning is left to later design stages. The system is modeled as a set of objects, and each one of them may be later implemented as software or hardware, either digital or analog (as in [13]).

In these object-oriented approaches, system behavior is described by some procedural language, such as C++. The specification of embedded systems using procedural languages is a natural consequence of microprocessor-based system design. An important advantage of this approach is that the specification can be simulated and validated. However, this kind of specification is more appropriate for a software-based synthesis, where the target architecture is fixed and based on a given processor, and the specification is compiled into the processor's code. On the other side, these descriptions have an inherently sequential nature and are not appropriate for describing hardware semantics. Some extensions that allow the description of hardware features like parallelism and communication mechanisms have been developed [14]. Java has been also proposed, not only as a language for describing the abstract system behavior, but also as a basis for a target architecture based on multiple threads [15].

Although the S^3E^2S environment is also based on object-oriented specifications, where abstract behavior is given in C++, the drawbacks of this approach are partially avoided, because heterogeneous specifications mixing C++, VHDL and analog descriptions can be created.

This alternative approach of supporting modeling of heterogeneous systems is also followed by Ptolemy [16], an environment for simulation and prototyping of heterogeneous systems which also uses object-oriented technology. Ptolemy implements the combination of different simulation mechanisms, called domains (such as Synchronous Data Flow, Dynamic Data Flow, discrete event, and analog). Another environment allowing the specification and simulation of heterogeneous systems is described in [17], where a backbone in the operating system implements communication among dedicated simulators that are needed for heterogeneous objects specified in different languages.

S^3E^2S combines the advantages of the multi-language and heterogeneous simulation approach with the abstract, object-oriented specification. Objects can be modeled regardless of their future implementation as digital or analog hardware or software. Then, any object may be specified in any of these domains or refined into any of them, and every possible intermediate model generated during this stepwise refinement process may be co-simulated.

6. CONCLUSIONS AND FUTURE WORK

The automatic design of embedded electronic systems is an open area of research. A design environment must consider important aspects such as system specification, partitioning, validation, and synthesis. Regarding the specification, current design environments either propose a multi-language approach, where each component is modeled in a language which is suitable for a given implementation domain, or a single language approach, where all components must be modeled using a language from a single domain.

This paper presented S^3E^2S, an environment combining the advantages of both approaches in a flexible way. An abstract, object-oriented, initial specification may be created, and each object may be later refined into a different domain, using a suitable language (C, VHDL, differential equations). Any heterogeneous model created during the process of stepwise refinement may be co-simulated for validation. The paper illustrated these modeling and co-simulation capabilities by means of a concrete example.

We are currently implementing the automation of the synthesis capabilities of S^3E^2S. From a manual characterization of available processors and an automatic characterization of the application objects (specified as SIMOO abstract objects), the environment is capable of selecting the best multi-processor platform for implementing the application.

We are also developing a more efficient implementation of the co-simulation strategy. An optimistic synchronization mechanism will allow each object to have its own local time. Each object encapsulating an analog behavior will have a time step most appropriate for the numerical integration of the differential equations, while synchronization among objects will occur only as objects have to communicate events on interface signals.

ACKNOWLEDGMENTS

Special thanks are deserved to B.Copstein, J.F.H.Jornada and C.E.Pereira for the development of the SIMOO environment. Thanks are also due to F.B.Lima, who helped implementing the crane model.

REFERENCES

[1] L.Carro, M.Kreutz, F.R.Wagner and M.Oyamada. "System Synthesis and Processor Selection in the S^3E^2S Environment". In: 12th Symposium on Integrated Circuits and Systems Design. Natal, Brazil, 1999. Proceedings, SBC, 1999.

[2] B.Copstein, F.R.Wagner and C.E.Pereira. "SIMOO – An Environment for the Object-Oriented Discrete Simulation". In: 9th European Simulation Symposium. Passau, Germany, 1997. Proceedings, SCS, 1997. pp 21-25

[3] B.Copstein, C.E.Pereira and F.R.Wagner. "The Object-Oriented Approach and the Event Discrete Simulation Paradigms". In: 10th European Simulation Multiconference. Budapest, Hungary, 1996. Proceedings, SCS, 1996. pp 57-61

[4] R.Fujimoto. "Parallel Discrete Event Simulation". In: Communications of the ACM, Vol. 33, No. 10, Oct. 1990. pp 30-53.

[5] E.Moser and W.Nebel. "Case Study: System Model of Crane and Embedded Control". In: Design, Automation and Test in Europe, 1999. Proceedings, IEEE Computer Society Press, 1999. pp 721-723.

[6] W.Nebel and G.Gorla. "JAVA, VHDL-AMS, ADA or C for System Level Specification?" In: Design, Automation and Test in Europe, 1999. Proceedings, IEEE Computer Society Press, p. 720.

[7] J.K.Adams and D.E.Thomas. "The Design of Mixed Hardware/Software Systems". In: ACM/IEEE Design Automation Conference, 1996. Proceedings, ACM Press, 1996. pp 515-521.

[8] G.Martin. "Design Methodologies". In: Design, Automation and Test in Europe, 1998. Proceedings, IEEE Computer Society Press, 1998. pp 286-289.

[9] J.Böttger et al. "An Object-Oriented Model for Specification, Prototyping, Implementation and Reuse". In: Design, Automation and Test in Europe, 1998. Proceedings, IEEE Computer Society Press, 1998. pp 303-310.

[10] W.Wolf. "An Architectural Co-Synthesis Algorithm for Distributed, Embedded Computing Systems". IEEE Transactions on VLSI Systems, June 1997. pp 218-229.

[11] N.Woo, A.Dunlop and W.Wolf. "Codesign from Cospecification". IEEE Computer, Jan. 1997. pp 42-47.

[12] M.Aiguier et al. "ECOS: A Generic Codesign Environment for the Prototyping of Real Time Applications". In: Hardware/Software Co-Design and Co-Verification. Edited by M.Bergé, O.Levia and J.Rouillard. Kluwer Academic Publishers, 1997. pp 23-58.

[13] J.Z.Pino and K.Kalbasi. "Cosimulating Synchronous DSP Applications with Analog RF Circuits". In: 32nd Asilomar Conference on Signals, Systems and Computers, 1998.

[14] I.Page. "Design of Future Systems". In: Design, Automation and Test in Europe, 1998. Proceedings, IEEE Press, 1998. pp 343-347.

[15] M.Mrva, K.Buchenrieder and R.Kress. "A Scalable Architecture for Multi-Threaded Java Applications". In: Design, Automation and Test in Europe, 1998. Proceedings, IEEE Computer Society Press, 1998. pp 868-874.

[16] A.Kalavade and E.Lee. "Hardware/Software Codesign Using Ptolemy". In: Codesign. Edited by J.Rozenblit and K.Buchenrieder. IEEE Press, 1995. pp 397-413.

[17] A.A.Jerraya and R.Ernst. "Multi-language System Design". In: Design, Automation and Test in Europe, 1999. IEEE Computer Society Press, 1999. pp 696-699.

Architectural Synthesis with Interconnection Cost Control

JEGO Christophe, CASSEAU Emmanuel & MARTIN Eric

LESTER Laboratory, UBS University, France
Tel : (+33) 2.97.87.45.65 Fax : (+33) 2.97.87.45.00
E-mail: {First-name.Surname}@univ-ubs.fr http://lester.univ-ubs.fr:8080/

Abstract:

Architectural synthesis tools map algorithms to architectures under various constraints and quickly provide estimations of area and performance. However, these tools do not take the interconnection cost into account whereas it becomes predominant with the technology decrease and the application complexity increase. A way to control costly interconnections during the architectural process is presented in this paper.

Keywords:

Architectural synthesis, digital ASIC design, sub-micron technologies, interconnection cost

INTRODUCTION

Recent advances in VLSI technology lead to new design methodologies like architectural synthesis, so called behavioral synthesis. Architectural synthesis enables a significant productivity increase by raising the abstraction level of digital designs. This process, which explores the space of possible designs, reaches the "best" architectural solution satisfying a set of constraints such as propagation time, area or power dissipation. However, both technology evolution and application complexity require models to be modified and algorithms that are used during the architectural synthesis process to be adapted. Actually, from the one hand, cost estimation models do not take into account the interconnections that are typically numerous with complex applications. On the other hand, interconnections have become cost effective and critical with deep sub-micron designs. In order to insure performance and reliability of the provided designs, the architectural synthesis flow has to be adapted.

This paper is structured in the following way: section 1 briefly presents the architectural synthesis flow and the GAUT behavioral synthesis tool. Section 2 introduces the interconnection cost problem with VLSI designs and high-level synthesis processes. Previous works about the interconnection cost problem in architectural synthesis tools are presented in section 3.1 and a way to reduce this cost is finally proposed in section 3.2.

1. ARCHITECTURAL SYNTHESIS OVERVIEW

Owing to recent advances in semiconductor technology, application complexity increase and time-to-market constraint, new design methodologies have to be developed. Architectural synthesis promises a significant productivity increase by raising the abstraction level of digital design. Basically, this process maps a behavioral description of an application into a register transfer (RT) level implementation. Since this process quickly provides area and performance estimations, it enables a more efficient exploration of the design space to be done.

1.1 Behavioral synthesis flow

From a behavioral description, an architectural synthesis tool generates an architecture of RT components such as arithmetic operators, registers, interconnection operators (multiplexors, demultiplexors and tristates) and memories, based on a target structural model (register, multiplexor or bus based architecture) [1]. The characteristics of the components (area, propagation time, power dissipation...) are initially given in a library.

Five steps/algorithms are involved during the architectural synthesis process: compilation, allocation, scheduling, binding, and resources optimisation [1,2]. Initially, a compilation step transforms the behavioral description into a control and/or a data flow graph representation. The allocation task determines the arithmetic operators to be used and their number, whereas the scheduling process assigns the flow graph operations to time intervals under a real time constraint. These two tasks are closely linked. If scheduling is performed before allocation, it imposes additional constraints on the operations with respect to allocation. Similarly, if allocation is performed before scheduling, it restricts the scheduling. That is the reason why it is difficult to characterise the quality of a given scheduling without considering the allocation step. Finally, binding maps variables and operations of the scheduled flow graph into the selected components. After the binding of operations to arithmetic operators and variables to storage components, additional algorithms are used for register sharing.

1.2 A behavioral synthesis tool : GAUT

The behavioral synthesis tool we use for this work is called GAUT. This tool has been developed by two French university laboratories: Lester (University of South Brittany, France) and Lasti (University of Rennes, France). GAUT is a pipeline architectural synthesis tool, which is dedicated to signal and image processing applications under real time execution constraints. From one behavioral specification, one mapping technology and one real time constraint, an optimised architecture is synthesised [3,4].

The generic architecture model is composed of four functional units: the processing unit, the control unit, the memory unit and the communication unit. The specification is written in VHDL, at a behavioral level without any architectural directive. After an algorithm compilation, the tool synthesises a data flow graph according to its generic model of architecture (register/multiplexor based architecture) and according to a library that contains the characteristics of components that come from previous logic/physical syntheses. GAUT starts the process with the processing unit synthesis because this unit undergoes the most important constraints for a real time application. Then the memory unit and the communication unit are generated. The control unit is described in order to be synthesised by a finite state machine design tool.

2. INTERCONNECTION COST PROBLEM

With the advancement of the VLSI circuit technology, a rapid scaling of the feature size has been performed. The minimum dimension of a transistor decreased from 2 μm in 1985 to 0.25 μm in 1999. According to the National Technology Roadmap for Semiconducteurs (NTRS) [5], it will further decrease at the rate of 0.7x per generation (consistent with Moore's Law) to reach 0.07 μm by 2010. Table 1 shows the evolution of the design integration features in CMOS technology since 1995 and gives the previsions until 2010.

	1995	1998	2001	2004	2007	2010
technology (μm)	0.35	0.25	0.18	0.13	0.1	0.07
supply voltage (V)	3.3	2.5	1.8	1.5	1.2	1
transistors per chip (M)	10	20	50	110	260	620
metal layers	4-5	5	5-6	6	7	7-8
ASIC area (mm^2)	450	660	750	900	1100	1400
frequency (Mhz)	300	450	600	800	1000	1100

Table 1: Evolution of the design integration features in CMOS technology.

Such scaling implies that the circuit performance will be increasingly determined by the interconnection performance : the wiring delay percentage relative to the cycle time actually becomes more important than the operator propagation time percentage [6,7]. For instance, interconnection contributes 50 percent of total delay in 0.35 μm and is expected to contribute up to 70 percent in 0.25 μm. Whereas this interconnection cost was not of great importance with technologies above 0.7 μm, interconnection design will play the most critical role in achieving of chips with sub-micron technologies.

New applications like multimedia or advance mobile communication systems require complex real time algorithm implementation under constraints such as area and/or power dissipation. Since behavioral synthesis tools map algorithms to architectures and provide fast estimations of area and propagation time, many different architectures can be rapidly explored according to the specified constraints. One of the high-level synthesis characteristics is an "optimal" reusing (sharing) of the operators and the registers. This reusing is performed with interconnection operators (multiplexors, demultiplexors and tristates) and involves interconnection cost (path delay and wiring area). However, the different steps of the synthesis process do not take into account the wiring area and unfortunately the path delay which are difficult to predict. Moreover, these steps are realised on the whole architecture without placement information, which leads to tremendous different wiring lengths when the synthesis of complex applications is concerned.

For instance, the architectural synthesis of the Viterbi algorithm, which is a typically complex application unlike usual synthesis examples like FIR filters etc., has been performed and logic and physical syntheses have been realised afterwards [8]. This work highlighted the problem of interconnection cost (wiring area and path delays) : a great difference may occur between the estimated characteristics of the architectural synthesis and the placed and routed architecture if interconnection cost (wiring) is not efficiently taken into account. In fact, the more complex the architecture is the higher estimation difference.

3. INTERCONNECTION COST CONTROL APPROACH

Since decisions made at the behavioral level may have a pronounced impact on the final design, estimations play a central role in guiding the process to optimal or near-optimal solutions. Typically, there are three major estimations used during the high-level synthesis flow: area, propagation time and power dissipation. These measures can be used at different levels of the process (for instance, they are used to drive the selection of the target architecture, to choose the library type, to select a particular component ...). Since all synthesis decisions depend on these estimations, their accuracy is essential for the generation of high-quality architectures.

Area and timing interconnection costs are known to be difficult to be accurately estimated especially when architecture becomes complex. However, even more than for a logic synthesis, routing performance may be critical for an architectural synthesis. Previous works about the interconnection cost problem in architectural synthesis tools are presented in section 3.1 and a way to control this cost is detailed in section 3.2.

3.1 Previous works

These last years, the interconnection cost problem at the architectural level was the subject of many publications [9-12]. These works are characterised by the techniques used to estimate the interconnection cost (wiring area and/or delay), the methods used to take into account these estimations at a behavioral level and the structural model used.

In [9], an initial solution is generated by partitioning the operations in the design to reduce the interconnection cost. Then a performance driven floorplanning is realised to provide an estimation of the interconnection cost. This estimation is then used for the scheduling of the operations. Afterwards, the solution is optimised by an iterative process that uses design transformations.

Xu [10] detailed a high-level synthesis flow which estimates the layout features with a specific estimation tool before performing the scheduling-binding task. The layout information is then used to guide the scheduling-binding task. The final result is evaluated without actually going through the time consuming phase of placement and routing. When time constraints are met, a structural RTL netlist and it is corresponding physical characteristics are generated.

The estimation of the interconnection length at the architectural level is one of the critical points. Mecha [11] presents a method that takes into account the interconnection wires during the behavioral synthesis without requiring floorplanning to be performed. It is an empirical method which formulates the interconnection length from a study of the routing rules features of Cadence CAD tools. Then, this length enables the interconnection delay to be estimated. In [12], the algorithm that evaluates the interconnection length does not require a complete placement of the components, it uses the topology of the connected components and the interconnections to estimate the distance between each pair of connected components [13]. The interconnection wire delays, computed from the estimated length of the wires, are finally included in an iterative behavioral synthesis.

However, these methods have various disadvantages: either they overestimate the interconnection cost and when the application becomes complex, the architectural solution can not satisfy the constraints (the allocation step selects too many components because the estimations are very pessimistic: worst case approach) or they are based on an iterative process (an initial architectural solution is improved from the estimation of its interconnections) therefore costly in CPU time. Our objective is to quickly provide reliable estimations of the architecture to the designer. We thus propose a different approach that enables the interconnection cost to be controlled all along the synthesis process and that takes care of costly interconnections. The interconnection cost control is carried out by the characterisation of the data flow graph variables. The synthesis also integrates a clustering task in order to insure temporal dependencies.

3.2 A way to control and reduce the interconnection cost

Architectural synthesis tools are based on a generic architecture model. This model is typically composed of four functional units : the processing unit, the control unit, the memory unit and the communication unit. The structural model of the processing unit is based on virtual elementary cells including an arithmetic operator, its connected registers and interconnection operators (figure 1).

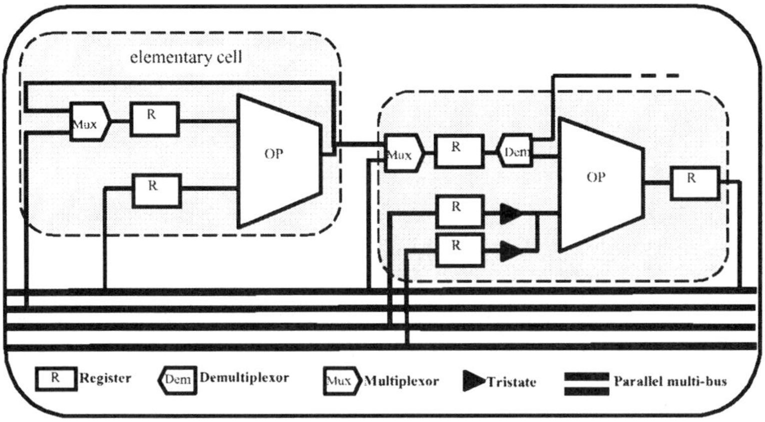

Figure 1: Structural model of the processing unit.

The arithmetic operators perform the data processing whereas the registers are used to temporarily memorise the variables and also to synchronise the data transfers between the processing unit and the memory or communication units. Interconnection operators are multiplexors, demultiplexors and tristates. The multiplexors and demultiplexors are necessary for the register reusing (optimisation step for register sharing) and the tristates make the control of arithmetic operator access possible. Finally, interconnection wires perform the data transfers into/between the elementary cells of the processing unit and a parallel multi-bus is used for the data communications between the four functional units of the architecture. Note that basically each variable of the algorithm is firstly recorded in a register. Different algorithms are next used for the decreasing of the register number during the optimisation step (optimisation by register sharing) : Left Edge [1], Branch&Bound [14], Branch&Bound with heuristics [15].

Consequently, a processing unit associated with this kind of typical structural model is composed of three different types of interconnection wires :

- local to an elementary cell : these interconnections perform the variable transfers into an elementary cell,
- local to the processing unit : these interconnections perform the variable transfers between elementary cells,
- global to the architecture : these interconnections perform the parallel multi-bus access and, by this way, the communication with the other functional units of the architecture.

In fact, the processing unit interconnection cost (wiring area and propagation delay) depends on the type of interconnection wires. On the one hand, this cost is low for interconnection wires that are local to an elementary cell or global to the architecture. On the other hand, for interconnection wires that are local to the processing unit, this cost depends on the complexity of the architecture (related to the algorithm complexity) and the place and route tool performance. Thus, in order to minimise the processing unit interconnection cost, the length and the number of interconnection wires that are local to the processing unit have to be minimised. Our objective is to firstly minimise their number all along the high-level synthesis process (selection-allocation, scheduling, binding and resource optimisation). Then their length will be controlled by a clustering step which has to be inserted between the binding and the register optimisation steps and which provides placement directives.

Since interconnections are associated with data transfers, the idea is to characterise the types of data (temporary processing data, constants or signals) and take advantage of their type. Three categories of data have thus been defined according to the interconnection features :

- category 1 : temporary processing data which are linked to a single arithmetic component,
- category 2 : temporary processing data which are linked to several arithmetic components,
- category 3 : temporary processing data and constants which are stored in the memory unit and input/output signals.

An initial characterisation of the variables is realised from the data flow graph associated with the behavioral description. In fact, in this step, the variables are associated with categories 2 or 3 because the set of arithmetic components is not yet being selected. This characterisation is presented figure 2a for a straightforward example.

The selection algorithm is a basic task of the architectural synthesis process that aims at optimising the cost of dedicated circuits. Its objective is to find the optimal set of components from a given library, for a behavioral description and a set of constraints. Different sets of components can be selected according to the library and the constraints. Then the allocation task determines the minimum number of every selected type of components. An initial automatic selection-allocation step that generates an "optimal" set of components in term of area under a time constraint is first performed. Consequently, the data characterisation may be refined : variables that are linked to one single arithmetic component are thus associated with category 1 whereas variables that are linked to more than one arithmetic component are still associated with category 2. By now, an optimisation phase is performed in order to minimise the number of variables associated with category 2 in favour of category 1 variables (related to component area and propagation time). The designer may take the opportunity of modifying the set of arithmetic components and/or their number to obtain a better set of components in term of data characterisation under a time constraint. Several data characterisations that correspond to different sets of selected components are carried out. The designer can thus choose one of the selection-allocation solutions according to the constraints (propagation time, data characterisation and components area). For example, three sets of components and their characterisation are presented in figure 2 b,c,d. They illustrate the evolution of the variable characterisation. In figure 2b, the architecture is costly in terms of interconnection wires because four variables are associated with category 2. The solutions that are proposed in figures 2c and 2d are more interesting from the interconnection cost point of view (only 2 variables are associated with category 2). However, solution d) is more restricted for temporal data dependencies. Furthermore, these data characterisations may change during the next steps of the synthesis process.

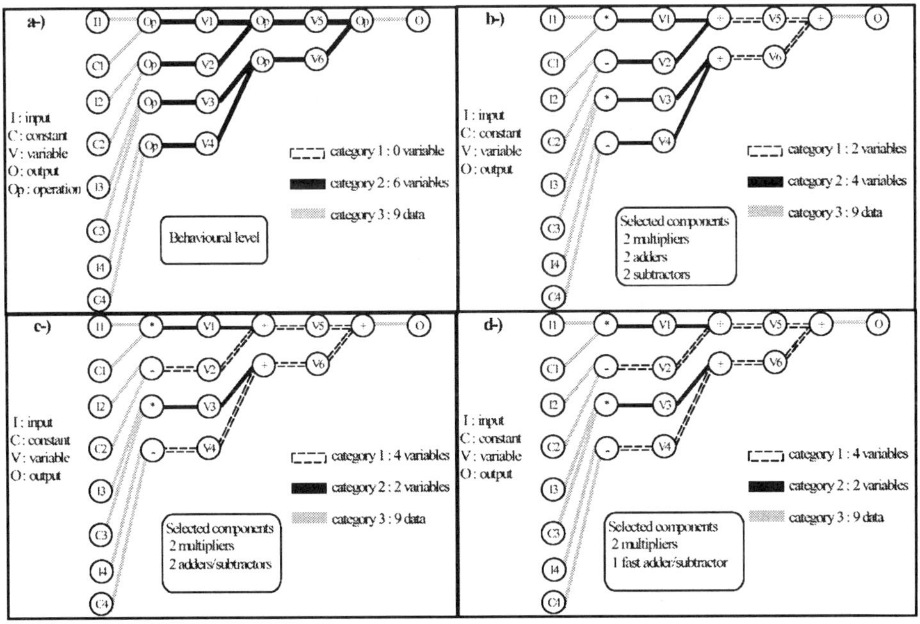

Figure 2: Characterisation of the data flow graph variables for the computation of $O = [(I_1*C_1)+(I_2-C_2)] + [(I_3*C_3)+(I_4-C_4)]$.

Many algorithms can perform the important task of scheduling. However, in order to take into account a given data characterisation, a resource constrained scheduling (List-Based Scheduling) [1] is used in this synthesis process. This algorithm is a generalisation of the ASAP algorithm with the inclusion of constraints. A scheduling priority list is provided according to a priority function. Naturally, the efficiency of this algorithm mainly depends on the priority function used. The priority function used in our approach depends on the mobility of the operations and the data characterisation constraints. For instance, the operations with a small mobility and associated with category 1 variables are scheduled in priority. By this way, the interconnection length is still shorten in the next step.

Finally, a binding algorithm is used to respectively assign the variables and operations of the data flow graph to registers and to the allocated components. Naturally, the variable characterisation is also taken into account in this step. For instance, during the previous steps, category 1 variables were variables linked to a single arithmetic component without regards to the number of this component (the operation performed by the component was only concerned). However, after the binding step, if a variable is linked to two mapped components (even if they carry out exactly the same operation), it becomes a category 2 variable, *i.e.* a potential costly interconnection. In this step, an operation is thus assigned to an available arithmetic component in such a way that the number of variables associated with category 2 is minimised in favour of category 1 variables.

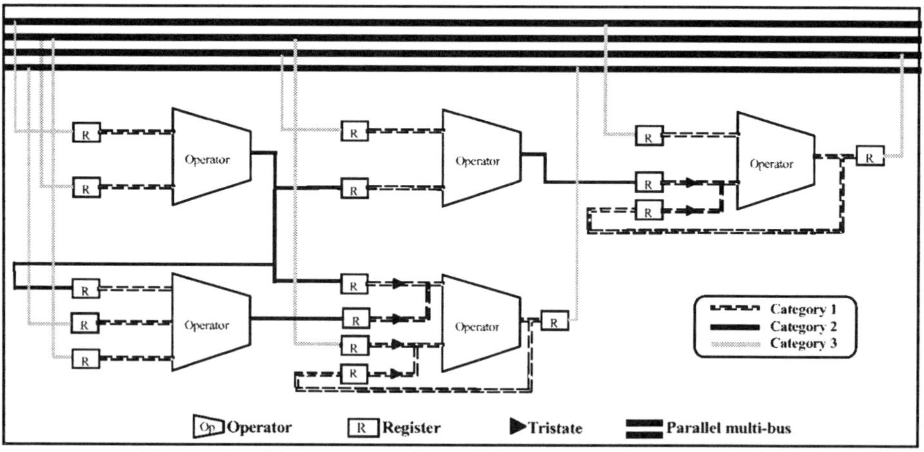

Figure 3: Characterisation of the processing unit wiring.

Obviously, any provided architectural solution will be composed of interconnection wires associated with category 2 variables, like the architecture presented in figure 3. For this reason, a clustering step is necessary to specifically control their cost. Since conventional place and route tools take hierarchical descriptions and placement directives into account, the generation of a hierarchical RTL description enables locally placed components, *i.e.* low cost interconnections. The clustering step of this synthesis flow thus consists in providing hierarchy in the RTL description. It starts with a non-partitioned set of components as provided by the binding process (for instance all the components of the processing unit of figure 3), and places them into clusters according to some component closeness measures. The processing unit is thus partitioned into elementary cells including an arithmetic component, its connected registers and interconnection operators (figure 1). The lengths of the wiring associated with category 1 variables can thus be minimised. Specific cluster placement directives are then provided in order to minimise the inter-cluster interconnection wiring, *i.e.* wiring associated with category 2 variables. Furthermore, when mobility of the operations is not critical, a register can be placed between two clusters to insure against wiring delay and to reduce the cluster placement constraints. Thus, one clock period is specifically dedicated to the data transfer between two clusters. This feature is actually carried out during the scheduling process.

With regards to the register sharing (optimisation step), the optimisation algorithm is applied to each distinct cluster neither on the whole processing unit. Figure 4 presents for instance the architecture obtained after the register optimisation step corresponding to figure 3. This method of register sharing is not as efficient as a register sharing applied to the whole processing unit from the number of registers point of view, however it insures local data transfers and, in this way, the time constraints to be observed.

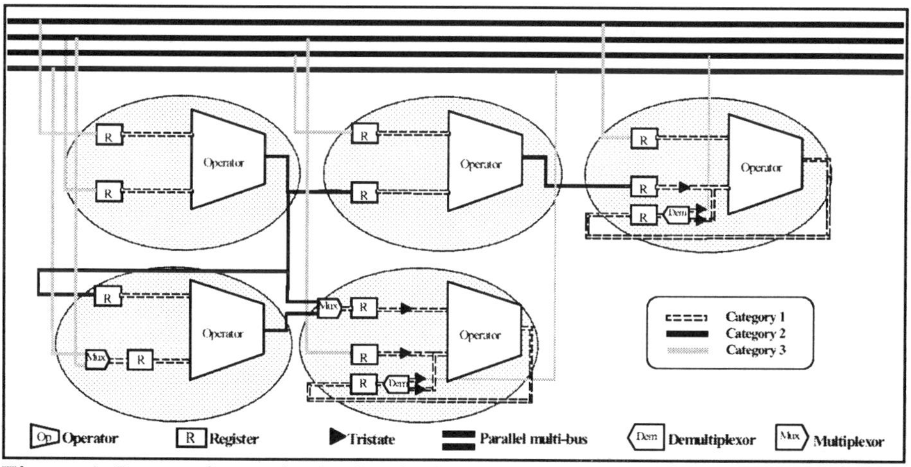

Figure 4: Processing unit obtained after clustering and register optimisation.

Naturally, this synthesis flow may involve a slight increase in the number of resources. However when complex applications are concerned and with sub-micron technologies, the additional resource area is often made up for the interconnection area decrease. Furthermore, the delay control, which is the most important point with VLSI design, can thus be significantly improved. This synthesis flow is currently being tested. The next step of our work is about the control unit, in particular the model of this unit. Actually, it seems that the critical path of the overall architecture is taken back from the processing unit to the control unit. A hierarchical finite state machine, included in the processing unit, may be a solution. This work will be reported later.

4. CONCLUSION

High-level synthesis is said to provide a significant productivity increase by raising the abstraction level of digital designs. Actually, HLS tools attempt to provide RT level design solutions with a quite good trade-off between cost and performance from a behavioral description. However, as interconnection cost becomes predominant, architectural synthesis tools have to take this additional cost into account. An approach that enables this cost to be controlled and that takes care of costly interconnections is proposed in this article.

5. REFERENCES

[1] D.D. Gajski, N. Dutt, A. Wu, S. Lin, "High-level Synthesis", Ed. Kluwer Academic Publisher 1992.

[2] M. C. FcFarland, A. C. Parker, R. Camposano, "The High-Level Synthesis of Digital Systems", In proceedings of IEEE, Vol.78, N°2, pp 301-318, 1990.

[3] E. Martin, O. Sentieys, H. Dubois, J.L. Philippe, "GAUT, an Architecture Synthesis Tool for Dedicated Signal Processors", In proceedings of EURO-DAC 93, pp. 14-19, 1993.

[4] J.L. Philippe, O. Sentieys, J.P. Diguet, E. Martin, " From digital signal processing specification to layout", In Logic and Architecture Synthesis : state-of-the-art and novel approaches, pp. 307-313, Chapman&Hall, 1995.

[5] Semiconductor Industry Association, "The National Technology Roadmap For Semiconductors", 1997.

[6] V. Moshnyaga, K. Tamaru, "Effect of Technology Scaling on Area-Delay Characteristics of RTL Designs: A Case Study", In proceedings of ED&TC'97, pp. 75-79, 1997.

[7] Wayne W.-M. Dai, "Chip Parasitic Extraction and Signal Integrity Verification", In proceedings of DAC 97, p 720-722, 1997.

[8] C. Jégo, E. Casseau, E. Martin, "Architectural Synthesis of a Complex Application : the Viterbi Algorithm", In User Forum proceedings of DATE 99, pp. 69-73, 1999.

[9] V. Moshnyaga, K. Tamaru, "A Floorplan Based Methodology for Data-Path Synthesis of Sub-Micron ASICs", IEICE Trans. on Information and Systems, pp. 1389-1395, Vol. E79-D, N° 10, 1996.

[10] M. Xu, F. J. Kurdahi, "Layout-Driven RTL Binding Techniques for High-Level Synthesis Using Accurate Estimators", ACM Trans. on Design Automation of Electronic Systems, pp. 312-343, Vol. 2, N° 4, 1997.

[11] H. Mecha, M. Fernandez, F. Tirado, J. Septién, D. Mozos, K. Olcoz, "A Method for Area Estimation of Data-Path in High-Level Synthesis", IEEE Trans. on CAD of Integrated Circuits and Systems, pp. 258-265, Vol 15, 1996.

[12] J. Hallberg, Z. Peng, "Estimation and Consideration of Interconnection Delays during High-Level Synthesis", In proceedings of Euromicro'98, pp. 349-356, 1998.

[13] A. Alvandpour and C Svensson, "A Wire Capacitance Estimation Technique for Power Consuming Interconnections at High Levels of Abstraction", In proceedings of Patmos'97, pp. 305-314, 1997.

[14] W. Grass, "A Branch-and-Bound Method for Optimal Transformation of Data Flow Graphs for Observing Hardware Constraints", In proceedings of EDAC 91, pp.73-77, 1991.

[15] J. Septién, D. Mozos, F. Tirado, R. Hermida, M. Fernández, "Heuristics for Branch-and-Bound Global Allocation", In proceedings of EURO-DAC 92, pp. 334-340, 1992.

CAE Environment for Electromechanical Microsystems

R.Lerch, M. Kaltenbacher, H. Landes
Department for Sensor Technology
University of Erlangen-Nuremberg
D-91058 Erlangen

1. INTRODUCTION

Short product lifetime cycles, fast time to market and cost reduction as well as an increasing technical complexity are only some of the challenges developers of electromechanical microsystems are faced with. Since the fabrication of prototypes and experimental based design is a lengthy and costly process, the urgent need for appropriate Computer Aided Engineering (CAE) tools, especially numerical simulation codes, arises. Therefore, we have developed a computer aided engineering software environment for the design of microsystems containing electromechanical sensors and actuators. In this environment we are able to precisely model nearly any type of piezoelectric, electrostatic or electromagnetic transducer including an eventual coupling to a surrounding medium, either a gas, a fluid or a solid. The software even allows the modeling of major nonlinear effects arising in electromechanical transducers, such as dependence of the elastic moduli on deformation, the nonlinear magnetization curves of ferroelectric materials or the geometric nonlinearities due to large displacements.
This software scheme is called CAPA [1] and is based on finite elements (FE), boundary elements (BE) or a coupling of both (FE/BE). Several interfaces for the import or export of data to commercial CAD and numerical modeling packages, like CATIA, IDEAS, ANSYS, have also been developed. Furthermore, interfaces to circuit simulation software are provided (Fig. 1).
For the numerical modeling of electromechanical sensors and actuators one has to consider the mutual interaction of different physical fields:

- Coupling *Electric Field – Mechanical Field*
  This coupling is either based on the electrostrictive or piezoelectric effect or results from the force on an electric charge in an electric field (electrostatic force).

- Coupling *Magnetic Field – Mechanical Field*
  This coupling is twofold. We first have the electromotive force (emf) which describes the generation of an electric field (electric voltage resp. current) when a conductor is moved in a magnetic field and, second, the electromagnetic force.

- Coupling *Mechanical Field – Acoustic Field*
  Very often an electromechanical transducer is surrounded by a fluid or a gaseous medium

in which an acoustic wave is launched (actuator) or is impinging from an outside source towards the receiving transducer.

In the following sections, the basic equations as well as their finite or boundary element implementations are described for these major transducing mechanisms. In each section we introduce one or more practical applications, too.

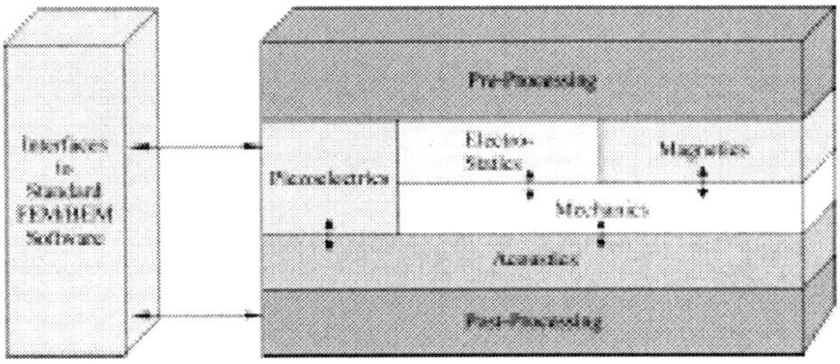

Figure 1 Software System CAPA

2. PIEZOELECTRIC TRANSDUCERS

2.1 BASIC EQUATIONS

The development of piezoelectric finite elements is based on 3 equations, namely the material equations, Newton's law and the potential equation. The material equations, which are the basis of linear piezoelectricity, take into account the piezoelectric effect both in the description of the mechanical stresses as well as the dielectric displacement. According to [2], these equations may be written as

$$\vec{T} = \mathbf{c}^E \vec{S} - \mathbf{e}^T \vec{E} \quad (1.1)$$

$$\vec{D} = \mathbf{e}\vec{S} + \boldsymbol{\varepsilon}^S \vec{E}\,. \quad (1.2)$$

Herein, $\vec{T}$ denotes the mechanical stress, $\mathbf{c}^E$ the mechanical material tensor at constant electric field $\vec{E}$, $\mathbf{e}$ the piezoelectric coupling tensor, $\boldsymbol{\varepsilon}^S$ the dielectric tensor at constant strain $\vec{S}$ and $\vec{D}$ the electrical displacement.

Newton's law, which describes the elastic behaviour of a finite, deformable body can be expressed by the partial differential equation

$$\left(\frac{\partial T_1}{\partial x} + \frac{\partial T_4}{\partial y} + \frac{\partial T_6}{\partial z}\right)\vec{e_1} + \; + \left(\frac{\partial T_2}{\partial y} + \frac{\partial T_4}{\partial x} + \frac{\partial T_5}{\partial z}\right)\vec{e_2}$$
$$+ \left(\frac{\partial T_3}{\partial z} + \frac{\partial T_5}{\partial y} + \frac{\partial T_6}{\partial x}\right)\vec{e_3} = \rho\frac{\partial^2 \vec{u}}{\partial t^2}. \quad (1.3)$$

Herein, ρ denotes the density of the material and $\vec{e_1}, \vec{e_2}$ and $\vec{e_3}$ are unit vectors in the x-, y- and z-direction, respectively.

Using the relation between the electric field $\vec{E}$ and the scalar electric potential Φ according to

$$\vec{E} = -\nabla\Phi\,, \quad (1.4)$$

the constituitive equation for the electric field results in

$$\nabla \cdot \varepsilon \nabla \Phi = q \tag{1.5}$$

where q represents the free electric volume charge.

2.2 FINITE ELEMENT FORMULATION

Combining (1.1)-(1.5) we get a full description of the dynamic behavior of a piezoelectric body.

The application of a finite element discretization scheme to these equations ends up with a linear system of equations, which can be summarized as follows [3, 4]:

$$\begin{pmatrix} \mathbf{M}_{uu} & 0 \\ 0 & 0 \end{pmatrix} \begin{pmatrix} \{\ddot{u}\} \\ \{\ddot{\Phi}\} \end{pmatrix} + \begin{pmatrix} \mathbf{C}_{uu} & 0 \\ 0 & 0 \end{pmatrix} \begin{pmatrix} \{\dot{u}\} \\ \{\dot{\Phi}\} \end{pmatrix} + \begin{pmatrix} \mathbf{K}_{uu} & \mathbf{K}_{u\phi} \\ \mathbf{K}_{u\phi}^T & \mathbf{K}_{\phi\phi} \end{pmatrix} \begin{pmatrix} \{u\} \\ \{\Phi\} \end{pmatrix} = \begin{pmatrix} \{F\} \\ \{Q\} \end{pmatrix} \tag{1.6}$$

Herein $\mathbf{K}_{uu}$, $\mathbf{C}_{uu}$ and $\mathbf{M}_{uu}$ denote the mechanical stiffness, damping and mass matrix, respectively, $\mathbf{K}_{\phi\phi}$ and $\mathbf{K}_{u\phi}$ the dielectric stiffness- and the piezoelectric coupling matrix, $\{F\}$ and $\{Q\}$ the external mechanical forces and electric charges, $\{u\}$ the nodal vector of displacement and $\{\Phi\}$ the nodal vector of scalar electric potential.

Starting with the semidiscrete formulation time stepping procedures based on the Newmark or Hilber-Hughes-Taylor method [5] may be applied. Furthermore, standard algorithms for eigenvalue calculations, like the subspace and Lanczos procedures, as well as harmonic excitations are easily extended.

2.3 ULTRASOUND ARRAYS

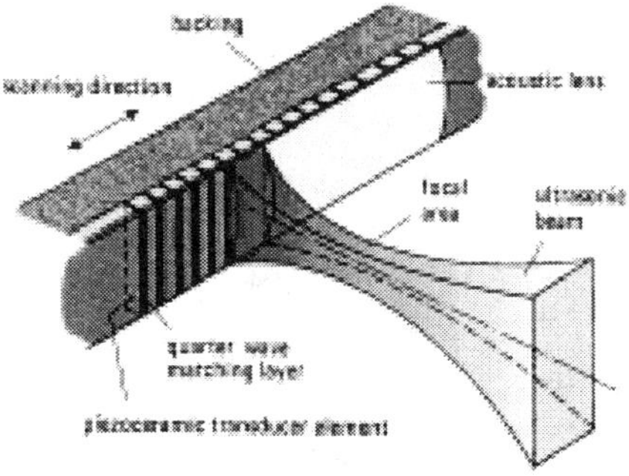

Figure 2 Ultrasound phased array antenna

Ultrasound arrays have found a wide area of applications, ranging from medical imaging to non-destructive testing. The principle of a standard ultrasound phased array antenna, as used in medical imaging applications, is shown in Fig. 2. The backin g is supplied on the backside of the transducer to damp out the vibrations of the antenna and thus generate short ultrasound pulses. The lense on top of the antenna is responsible for the geometric focussing of the

ultrasound beam whereas the matching layer(s) are used to adjust the high acoustic impedance of the piezoelectric transducer to the relatively low impedance of the lense material. Finite element simulations for such antennas have already been reported elsewhere [3, 4, 6].
In recent times, concepts for 2D ultrasound arrays, in which the geometric focussing of the lense is replaced by an addititional subdicing of the array in the length direction, have been developed. A new type of such a 2D ultrasound array has been developed utilizing a new silicon CMOS chip technology [7]. These transducers work simultaneously as ultrasound sensors and as ultrasound transmitters. The transducing mechanism is either piezoelectric or electrostatic. In the piezoelectric case, a thin layer of piezoceramic material is put on top of a micromachined silicon membrane defining a bimorph transducer, while in the electrostatic case, a voltage is applied between this membrane and a base plate, defining a capacitive transmitter (Fig. 3). Detailed results for these devices are reported in [8]

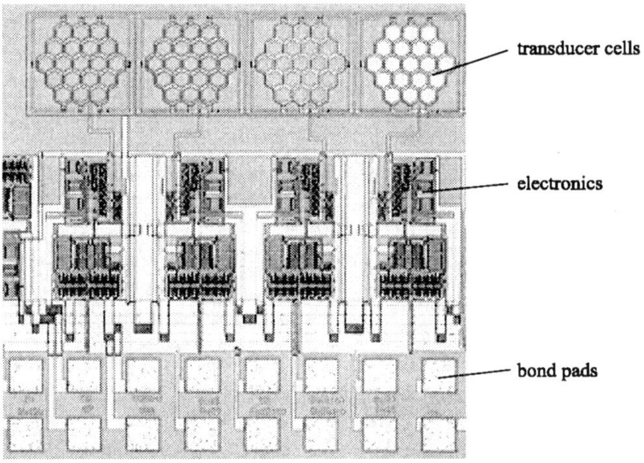

Figure 3 Top view of a CMOS chip with 4 areas, each containing 19 capacitive transducer cells

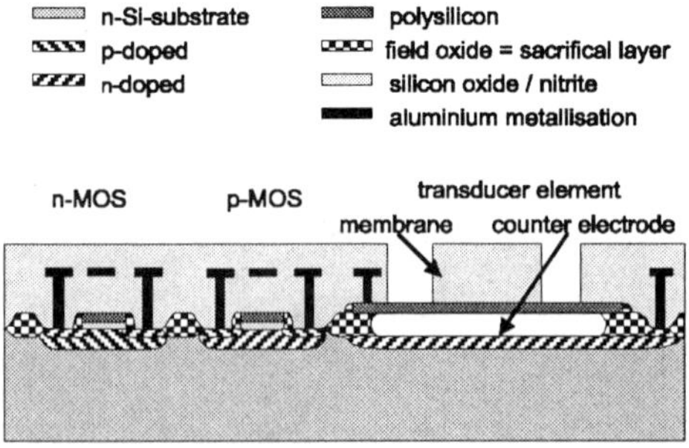

Figure 4 Schematic cross section of an electrostatic ultrasound transducer in CMOS technology

3. ELECTROSTATIC-MECHANICAL-TRANSDUCERS

3.1 BASIC EQUATIONS

In the case of an electrostatic-mechanical-transducer the coupling between the electric and the mechanical field is caused by the electrostatic force between the electrodes. This force is calculated based on the electrostatic force tensor $\mathbf{T}_E$, where $\vec{E} = (E_x,\ E_y,\ E_z)$ denotes the electric field

$$\mathbf{T}_E = \begin{bmatrix} \varepsilon E_x^2 - \frac{1}{2}\varepsilon|E|^2 & \varepsilon E_x E_y & \varepsilon E_x E_z \\ \varepsilon E_y E_x & \varepsilon E_y^2 - \frac{1}{2}\varepsilon|E|^2 & \varepsilon E_y E_z \\ \varepsilon E_z E_x & \varepsilon E_z E_y & \varepsilon E_z^2 - \frac{1}{2}\varepsilon|E|^2 \end{bmatrix}. \tag{1.7}$$

The electrostatic force $\vec{F}_E$ is given by

$$\vec{F}_E = \iint\limits_A \mathbf{T}_E \vec{n}\, dS, \tag{1.8}$$

where $\vec{n}$ is the normal vector on the surface A. The electrostatic force leads to a deformation of the electrodes, which is described by (1.3) and, therefore, introduces a geometric nonlinearity in (1.5).

3.2 FEM-BEM-FORMULATION

The boundary element discretization of (1.5) yields to the following BE-matrix equation

$$\mathbf{H}_\phi\{\Phi\} - \mathbf{G}_\phi\{E_n\} = \{Q\} \tag{1.9}$$

with the two boundary element matrices $\mathbf{H}_\phi$ and $\mathbf{G}_\phi$, the nodal vector of electric charge $\{Q\}$, the nodal vector $\{\Phi\}$ of the scalar electric potential and the nodal vector $\{E_n\}$ of the normal component of the electric field.

Applying the FE-formulation to (1.3) leads to the wellknown matrix equation for the mechanical quantities

$$\mathbf{M}\{\ddot{u}\} + \mathbf{C}\{\dot{u}\} + \mathbf{K}\{u\} - \{F(\Phi, E_n)\} = \{0\} \tag{1.10}$$

as described in section 2.2. Now the nodal force vector $\{F\}$ depends on the values of the scalar electric potential Φ and the normal component of the electric field E_n.

In order to illustrate the procedure of the FE/BE-Method a capacitive acceleration sensor as shown in Fig. 1.5 is used. Finite elements are used to describe the mechanical field in the whole structure, whereas the electric field in the gap between the two electrodes is modeled by boundary elements. This approach has the advantage, that deformations of electrode 2 will not cause a deformation of surrounding finite elements which would be necessary to describe the electric field in the case of a pure FE-modeling. The two boundary element matrices $\mathbf{H}_\phi$ and $\mathbf{G}_\phi$ have to be updated corresponding to the mechanical displacement. The direct coupling of (1.9) and (1.10) leads to a nonlinear system of equations. Using predictor values for the calculation of the electrostatic force, a decoupling into an electric and a mechanical matrix equation can be achieved. To ensure the strong coupling between the electric and mechanical quantities a Predictor/Multicorrector Algorithm as described in [9] has been successfully used.

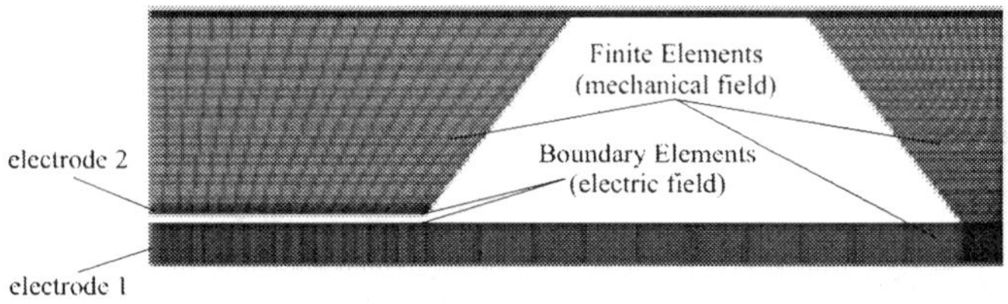

Figure 5 FEM/BEM discretization of a capacitive acceleration sensor

3.3 ACCELERATION SENSOR

The capacitive acceleration sensor (Fig. 1.5) is fabricated by micromachined techniques. Loading the sensor with an acceleration step causes the silicon structure to be deformed (Fig. 1.6) and the change in the capacitance is a direct measure of the acceleration. Without using any controller, the silicon structure oscillates with its eigenfrequency to a new position according to the acceleration step (Fig. 1.7). Applying a PID-controlled voltage to the electrodes, the transient response can be kept to a minimum and the silicon structure moves to the old position (Fig. 1.7) and the controller output voltage (which is applyed to the electrodes) is now a direct measure of the acceleration.

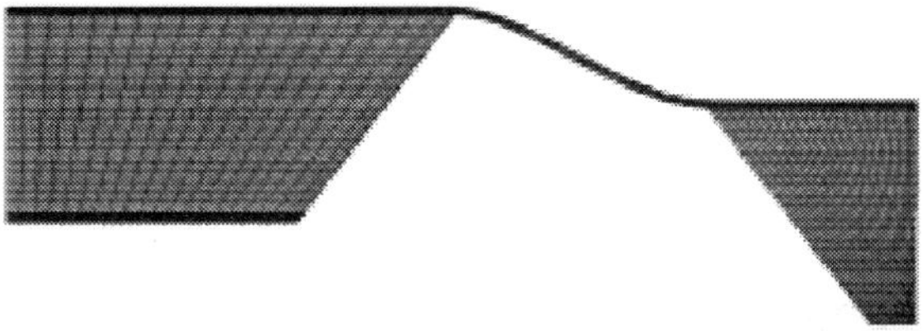

Figure 6 Deformations of the moving part of the acceleration sensor due to an acceleration pulse

3.4 MICROPUMP

A typical microsystem is a micromachined pump, shown in Fig. 1.8. If an electric voltage is applied to the electrodes, the elastic pump diaphragm (electrode 2) is deformed by the electrostatic force and bends towards the counterelectrode (electrode 1). Thereby, fluid will be sucked in through the inlet valve. When the supply voltage is switched off,the relaxation of the diaphragm will push the fluid through the outlet valve. The FEM/BEM discretization of the actuation unit (Fig. 1.8) is performed according to Fig. 1.9. Finite elements are used to describe the mechanical field in the two electrodes, whereas the electric field in the gap between is modeled by boundary elements. This approach has the advantage, that the elastic pump diaphragm (electrode 2) can move towards the stationary counterelectrode (electrode 1) without deforming any finite element which would otherwise (with pure FE-modeling) be necessary to describe the electric field.

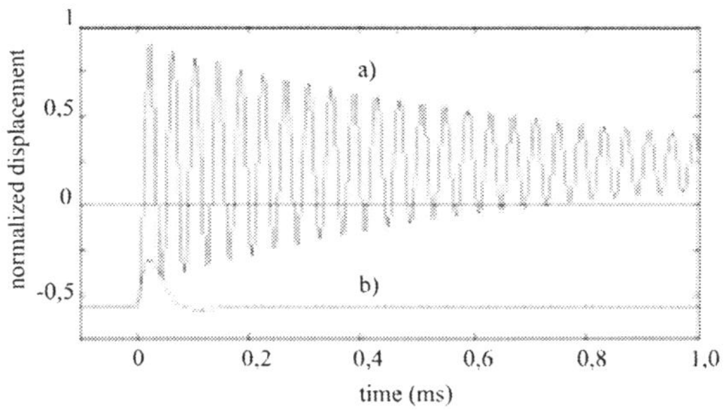

Figure 7 a) Uncontrolled b) controlled dynamic behaviour due to an acceleration step.

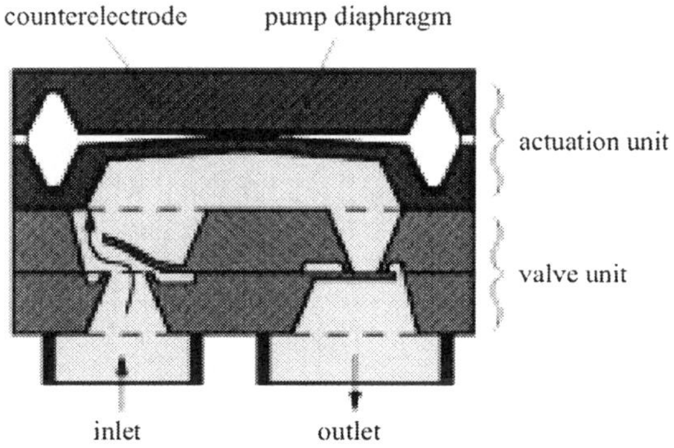

Figure 8 Schematic view of an electrostatically driven micropump

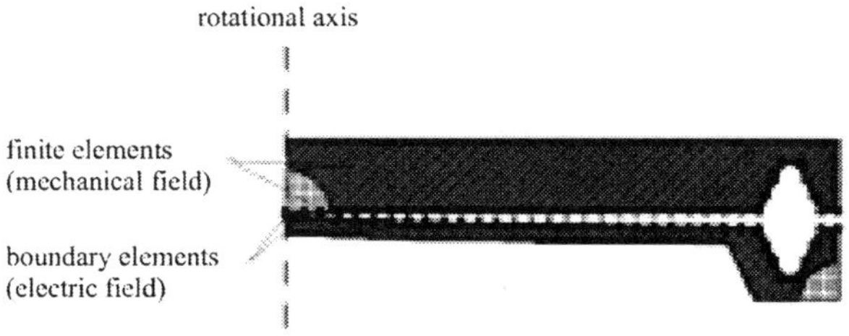

Figure 9 FEM/BEM discretization of the actuation unit

This pump has a diameter of 7 mm, a total height of about 1 mm and a gap thickness of 4 μm between the elastic pump diaphragm and the counterelectrode. Fig. 1.9 shows the mechanical deformations of the actuation unit when a dc voltage is applied. One point of investigation was the nonlinear dynamic response of the micropump. The pump was excited by a sinusoidal voltage with a frequency of 1 kHz and different amplitudes. The dynamic behaviour was analyzed by

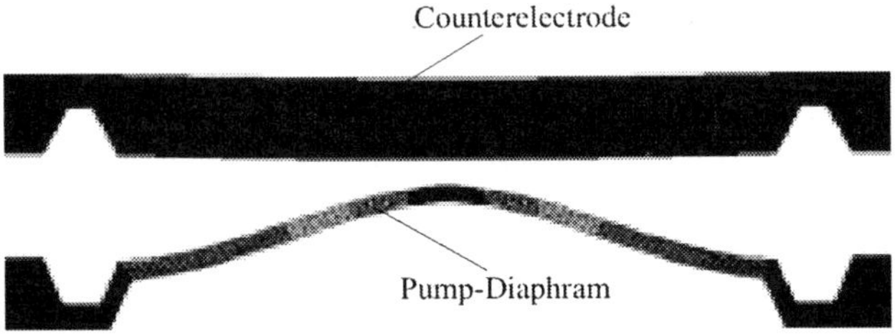

computing the electrostatic-mechanical system with the described calculation scheme. In Fig. 1.10, the center displacement of the elastic pump diaphragm is depicted as a function of the applied voltage. The corresponding frequency spectra can be seen in Fig. 1.11. For a good comparability each curve in Fig. 1.10 and Fig. 1.11 is normalized to its maximum.

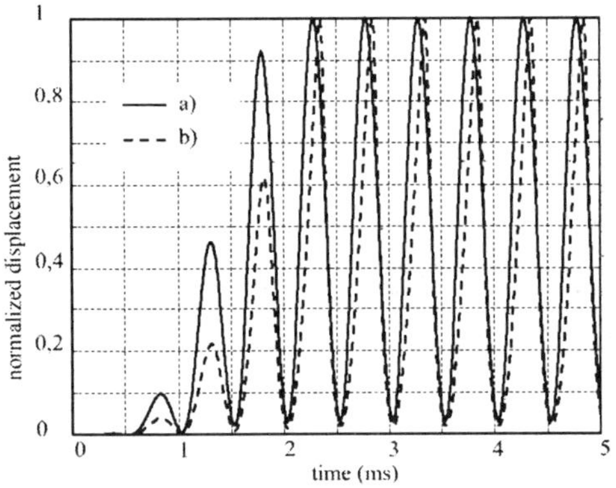

Figure 10 Mechanical displacement in the center of the pump diaphragm. a) Original amplitude of applied voltage. b) Twice the original amplitude of applied voltage

4. MAGNETOMECHANICAL TRANSDUCERS

4.1 BASIC EQUATIONS

For the quasi-stationary case (neglecting displacement currents) the magnetic field is computed by solving Maxwells equations

$$\nabla \times \vec{H} = \vec{J} \tag{1.11}$$

$$\nabla \times \vec{E} = -\frac{\partial \vec{B}}{\partial t} \tag{1.12}$$

$$\vec{B} = \mu \vec{H} \tag{1.13}$$

$$\vec{J} = \gamma \vec{E}\,. \tag{1.14}$$

In (1.11) - (1.14) $\vec{H}$ denotes the magnetic field intensity, $\vec{B}$ the magnetic flux density, $\vec{J}$ the current density, $\vec{E}$ the electric field intensity, μ the permeability and γ the electric conductivity. By introducing the magnetic vector potential $\vec{A}$ as follows

$$\vec{B} = \nabla \times \vec{A} \tag{1.15}$$

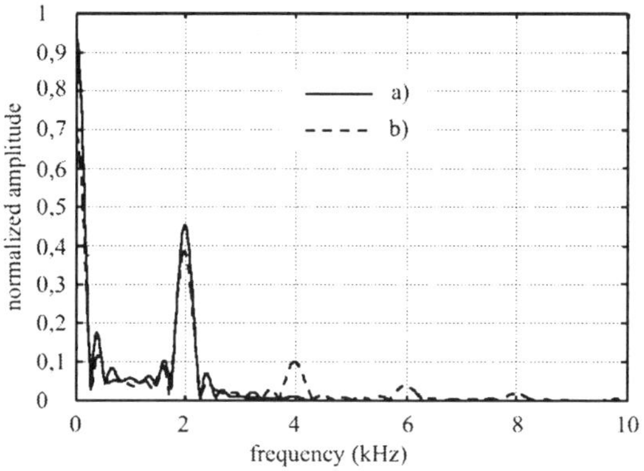

Figure 11 Frequency spectrum of the mechanical displacement in the center of the elastic pump diaphragm. a) Original amplitude of applied voltage. b) Twice the original amplitude of applied voltage

and applying the Coulomb gauging according to [10], the following matrix equation is derived

$$\begin{pmatrix} \mathbf{M}_{AA} & \mathbf{M}_{A\phi} \\ \mathbf{M}_{A\phi}^T & \mathbf{M}_{\phi\phi} \end{pmatrix} \begin{pmatrix} \{\dot{A}\} \\ \{\dot{\Phi}\} \end{pmatrix} + \begin{pmatrix} \mathbf{P}_{AA} & 0 \\ 0 & 0 \end{pmatrix} \begin{pmatrix} \{A\} \\ \{\Phi\} \end{pmatrix} = \begin{pmatrix} \{Q\} \\ \{0\} \end{pmatrix} . \tag{1.16}$$

Herein, $\mathbf{M}_{AA}$ and $\mathbf{M}_{\phi\phi}$ denote the magnetic mass matrix of the magnetic vector potential $\vec{A}$ and scalar electric potential Φ, $\mathbf{M}_{A\phi}$ the coupling matrix between them, $\mathbf{P}_{AA}$ the magnetic stiffness matrix and $\{Q\}$ the nodal vector corresponding to prescribed electric currents. In the case of a voltage loaded coil this equation has to be solved together with the electric circuit equation [11], which determines the current in the coil.

The first coupling term is given by the movement of a coil or conductive body (with velocity $\vec{v}$) in a magnetic field $\vec{B}$. Therewith, an additional current density $\vec{J}_v$ is induced in the coil

$$\vec{J}_v = \gamma \vec{v} \times \vec{B} = \gamma \frac{\partial \vec{u}}{\partial t} \times \vec{B} , \tag{1.17}$$

which has to be added to the electric circuit equation.
Furthermore, the magnetic volume force, resulting from the interaction between the total magnetic field $\vec{B}$ and coil current density $\vec{J}$, expressed by

$$\vec{f}_V = \vec{J} \times \vec{B} , \tag{1.18}$$

has to be added to the nodal force vector of the mechanical equation.

The direct coupling of the magnetic and mechanical matrix equation leads to a nonlinear and unsymmetric system of equations. Using predictor values for the calculation of the magnetic volume force and the mechanical velocity $\vec{v}$ of the coil, a decoupling into a magnetic and mechanical matrix equation can be achieved. To ensure the strong coupling between the magnetic and mechanical quantities a Predictor/Multicorrector Algorithm as described in [12] is used.

4.2 ELECTROMAGNETIC ACOUSTIC TRANSDUCER (EMAT)

A typical setup of an EMAT as used in nondestructive material testing is shown in Fig. 1.12. It can be used either to generate (transmitting mode) or to detect (receiving mode) lamb waves in the material under test [13]

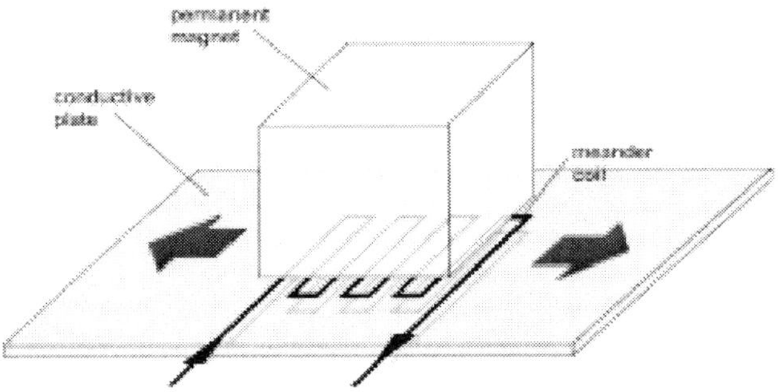

Figure 12 Setup of an electromagnetic acoustic transducer (EMAT)

- Transmitting mode: Loading the coil by a short tone burst signal, the time varying magnetic field induces eddy currents in the conductive material under test. The penetration depth of these eddy currents is given by the classical skin depth δ

$$\delta = \frac{1}{\sqrt{\pi f \gamma \mu}}, \tag{1.19}$$

  where f denotes the frequency of the time varying magnetic field, γ the electrical conductivity and μ the magnetic permeability of the material under test. The interaction of these eddy currents and the overall magnetic field of the permanent magnet results in a lamb wave propagating in the conductive sheet.

- Receiving mode: When the lamb wave passes the region of the receiving EMAT, which is subjected to the static magnetic field of the permanent magnet, locally eddy currents are induced in the conductive metallic sheet. Therewith, the time varying magnetic field of these eddy currents induces a voltage in the meander coil.

The main drawback of conventional FE-discretization is the high solution time due to the large number of unknowns, especially in the 3D case. One reason for the high demands on computer resources is that the generated magnetic FE mesh, describing the magnetic field in the penetration depth δ, which is typically 50 μm deep, has to be extremely fine. Using conventional FE-discretization the mechanical field is also solved at this very fine mesh, which is only necessary for the magnetic field problem.

For this problem a new multigrid approach was utilized which reduces the solution time considerably. This was achieved by using a fine FE grid for the magnetic field and a much coarser one for the mechanical problem (see Fig. 1.13) and on the other hand by applying very fast multigrid methods to solve the linear equation systems at each time step of the transient analysis [14].

For a precise 2D model, the FE-discretization results in a total number of about 300.000 unknowns. A computer run for 200 time steps with direct conventional solution strategies needs

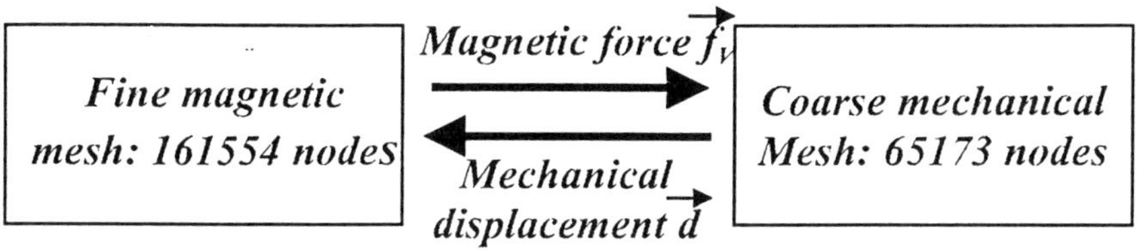

Figure 13 FE grids for the magnetic and mechanical problem

approximately 21 hours on a SGI, Octan 195 MHz. Using multigrid solution techniques the solution time can be reduced to 50 minutes. In the 3D case the advantages of the presented multigrid approach in comparison with standard solvers are even higher.

To show the applicability of the developed calculation scheme for the transient behavior of the EMAT in the transmitting mode as well as in the receiving mode, a set of 2D and 3D simulations has been carried out and compared with experimental data. Fig. 1.14 a) shows the dispersion curves of the group velocity for various lamb modes and Fig. 1.14 b) compares the measured and simulated induced voltage in the meander coil of an EMAT in receiving mode. In both cases a very good agreement between measurement and simulation could be obtained.

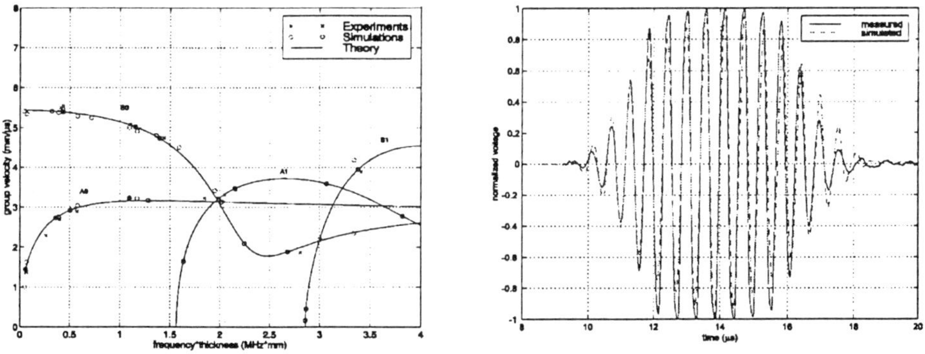

a.) b.)

Figure 14 a) Group velocity curves for A0, A1, SO and S1 lamb wave modes in aluminium plates. b) Normalized induced voltage in the coil of the receiving EMAT

5. REFERENCES

[1] Lerch, R.; Landes, H.; Kaltenbacher, M.: *CAPA User Manual*, Tech. Rept. 3.3, University of Linz, Altenberger Str. 69, A-4040 Linz, 1999.

[2] Mason, W. P. (Ed.): *Physical Acoustics*, Vol. I - Part A, London: Academic Press, 1964

[3] Lerch, R.: *Simulations of Piezoelectric Devices by Two- and Three-Dimensional Finite Elements*, IEEE Trans. on Ultras., Ferroel. and Freq. Control, Vol. 37, No. 3, 1990, pp. 233-247

[4] Lerch, R.: *Finite Element Analysis of Piezoelectric Transducers*, Proc. IEEE Ultrasonics Symposium, 1988, pp. 643-654

[5] Hughes, T. J. R.: *The Finite Element Method*, New Jersey: Prentice-Hall, 1987

[6] Lerch, R.; Landes, H.; Kaarmann, H.: Finite Element *Modeling of the Pulse-Echo Behavior of Ultrasound Transducers*, Proc. IEEE Ultrasonics Symposium, 1994, pp. 1021-1025

[7] Eccardt,P.; Niederer, K; Scheiter, T.; Hierold, C.: *Surface micromachined ultrasound transducer in CMOS technology*, Proc. IEEE Ultrasonics Symposium, 1996, pp. 959-962

[8] Kaltenbacher, M.; Landes, H.; Niederer, K; Lerch, R.: *3D Simulation of Controlled Micromachined Capacitive Ultrasound Transducers*, Proc. IEEE Ultrasonics Symposium, 1999 (to appear)

[9] Kaltenbacher, M.; Landes, H.; Lerch, R.: *A Finite Element / Boundary-Element Method for the Simulation of Coupled Electrostatic-Mechanical Systems*, *J. Phys. III France 7*, 1997, pp. 1975-1982

[10] Biro, O.; Preis, K.: *On the Use of Magnetic Vector Potential in the Finite Element Analysis of Three-Dimensional Eddy Currents*, *IEEE Trans. on Magnetics*, Vol. 25, No. 4, July 1989, pp. 3145-3159

[11] Leonard, P.J.; Rodger, D.: *Voltage forced coils for 3D finite-element electromagnetic models*, *IEEE Transactions on Magnetics*, Vol. 24, No,6, November 1988 pp. 2579-2581

[12] Kaltenbacher, M.; Landes, H.; Lerch, R.: *A Strong Coupling Model for the Simulation of Magnetomechanical Systems using a Predictor/Multicorrector Algorithm*, *Applied Computational Electromagnetics Society Journal*, 1997, Vol. 12, No,2. pp. 102-106

[13] M. Kaltenbacher, R. Lerch, H. Landes, K. Ettinger, B. Tittmann, *Computer Optimization of Electromagnetic Acoustic Transducers*, Proc. IEEE Ultrasonics Symposium, 1999, pp. 1029-1034

[14] M. Jung, U. Langer, *Applications of multilevel methods to practical problems*, Surv. Math. Ind. 1, 1991

Cost Consideration for Application Specific Microsystems' Physical Design Stages

A new approach for microtechnological process design

R. Brück, A. Priebe, K. Hahn
University of Siegen , Institute of Computer Structures
Hölderlinstr. 3, D-57068 Siegen, Germany
Tel. +49 271 740 2472, Fax. +49 271 740 2473
email: hahn@rs.uni-siegen.de

Key words: microsystem physical design, process configuration, process optimisation, design verification, component technology, INTERNET tool access, JAVA

Abstract: The microsystem technology (MST) industry is characterised by small and medium sized enterprises (SMEs) specialised on products for application specific solutions rather than standardised mass volume production. Domains like medicine, automotive sensor technology, etc. are well-known. In this area of business the technology driven design approach known from micro electronics is not appropriate. Instead each design problem aims at its own, specific technology to be used for the solution. The variety of technologies nowadays in use, like Si-surface, Si-bulk, LIGA, laser, precision engineering requires a huge amount of expertise and support for choosing initially the appropriate fabrication technology for specific design requirements. This paper describes methods and tools that support the design of consistent process step configurations that allow to predict in advance the effort assigned to specific process arrangements. Based on formal process description means the microsystems' designer can edit, assess and optimize consistent technology step sequences with regard to costs, time, yield and feasibility providing an interface to later layout verification. The tools presented became recently accessible via the Internet using JavaBeans componentware.

1. MICROSYSTEM PHYSICAL DESIGN VS. MICROELECTRONICS DESIGN

MEMS (so called in the US) or microsystems (so called in Europe) are able to provide the important interfaces between real world parameters and microelectronical information processing. Microstructured components contribute a substantial added value in many innovative products such as medical equipment, home office applications or automobiles.

As a result of some specific properties, the physical design process in microstructure technologies is not sufficiently supported. The design style is problem-oriented - as opposed to the technology oriented design style known from today's classical microelectronic systems design - which turns out to be the appropriate approach for microsystem design. Unlike in microelectronics the dependency between layout design and fabrication technology is very strong as many design properties (e.g. the size of structures in the third dimension) can only be realised choosing specific process parameters like materials, process steps or process resources.

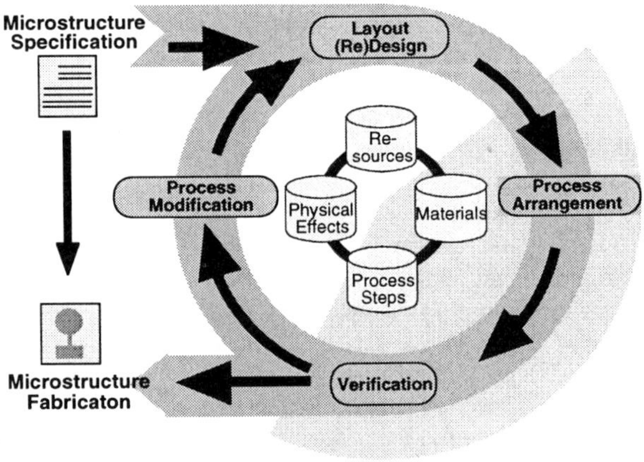

Fig 1. Circular Model of MEMS Design

In contrast to other disciplines of engineering the CAD tools used in this context at the moment were often developed for other purposes such as mechanical engineering or for IC layout design. Although efforts were made to enhance latter tools to deal with microsystem design constraints the application areas for these solutions are restricted. In general the methods and CAD systems available do often not meet the requirements of microsystem designers. [1]

In the area of physical microsystem design that in lithography-based technologies is concerned with the design of mask layout geometry, the typical design cycle is dominated by the fact that the three-dimensional nature of the products calls for a particular sequence of processing steps and parameters to be specified for each design object. This results in the *circle model* for MEMS physical design. It is characterised by the following circular design flow as shown in figure 1.

Generating tasks creating the mask layout geometry for the design object and the specific process step sequence to be performed in order to generate the appropriate orthogonal extension of the design object during production. The process step specification part will be in the focus of this paper.

Checking tasks deriving a consistent set of design rules from the process sequence and applying it to the mask layout in order to find rule violations. If violations are detected, the design will have to be modified.

Modifying tasks used to determine what sort of changes will have to be made in order to turn the design into a correct version. This may include mask layout changes, process specification changes or changes in the higher level design.

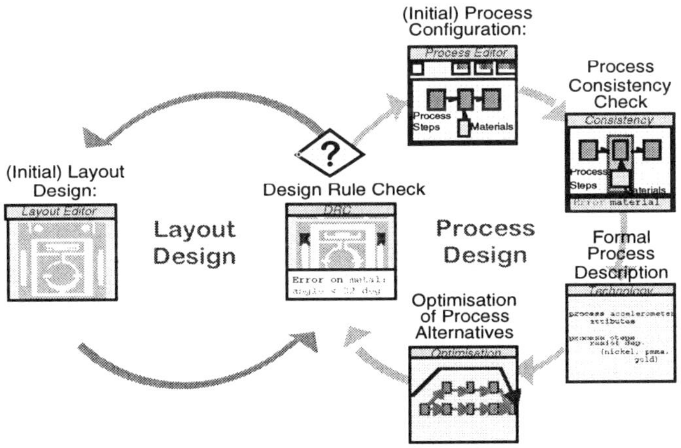

Fig. 2 Concurrent Design of Process and Layout

For more details of the circle model, see [2].

Figure 2 gives an alternative view of this model, showing the design flow as two concurrent, mutual dependent circles of layout and process design. The process editing tool to be presented shortly is used to support the process design cycle.

Each of the steps shown in this figure can be supported by specific design tools. Each of these tools is an independent software module, used to assist the user in a particular subtask of the whole design cycle. The degree of automation provided may differ substantially for each of the tools. The term *assistants* is used to denote this sort of independent software modules. The INTERLIDO-system presented in this paper thus can be regarded as a collection of weakly interacting assistants.

2. INTERLIDO MICROSYSTEM DESIGN SYSTEM

Based on these considerations the INTERLIDO MEMS design system has been developed at the Universities of Dortmund, Jena and Siegen. INTERLIDO supports both the design of appropriate process layouts for microstructure design as well as the construction of consistent and economically optimized process sequences. The link between process design tools and layout is based on a specific process description language LIDO-PDL that allows the specification of all process characteristics that are relevant to layout design including constructs to define economic properties like cost and time of processes.

The system as a whole is composed of a set of assistants. It consists of a common user interface and a common technology and geometry database, both combined in the *INTERLIDO-Manager*. Figure 3 shows a system view of the INTERLIDO system. The user interface offers two major applications: LIDO-*Pedit* - the graphical process configuration editor and LIDO-*Check*. - the microstructure design rule checker.

PEdit allows the configuration of process step sequences without requiring to compose textual LIDO-PDL descriptions. The process sequences can be checked with regard to their internal consistency and the system allows also the optimisation of fabrication processes towards least cost or time consumption. [3]

The consistent and optimised process sequence is used to automatically create a technology file in LIDO-PDL that might e.g. be used to derive geometrical design rules for the design verification module (Check) that checks the mask layout for rule violations. [4].

The system formerly developed in C++ for workstation application, is completely re-implemented as componentware to be accessible via the Internet. The benefits of component based software with Internet access capabilities especially for MEMS design tools will be shown in the end section.

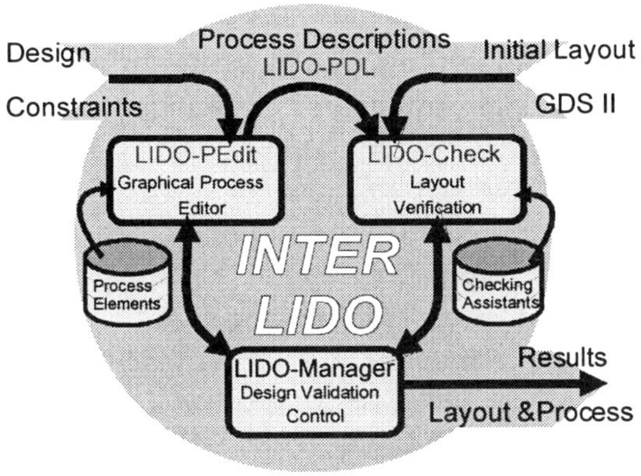

Fig. 3. INTERLIDO System View

3. PROCESS DESCRIPTIONS

The generation of formal description means for lithography-based micromachining processes was a precondition to realize the INTERLIDO system. The realization of the process description language LIDO-PDL is well documented in [5].

LIDO-PDL provides means to describe formally layout rules (with regard to later design verification steps) as well as process parameters. The basic idea for LIDO-PDL is the encapsulation of data (like process rules res. design rules) within objects and the configuration of these objects to complete process sequences. This approach reflects the microstructure physical design methodology with design specific process configurations. Different types of objects can be used. An excerpt of a LIDO-PDL process description is shown further down.

The complete process sequence is defined within a `process` object. A `process` consists of `process steps` carried out in a defined order.

`Material` objects, `resource` objects as well as `layer` objects can be assigned to process step objects. All types of objects can incorporate design rules and process parameters.

As a summary *LIDO-PDL* provides language constructs for the description of

a) Process steps, materials, lithographic masks and useful assemblies of them as well as design rules to be associated with these object types
b) Process step networks taking into account process alternatives and process step concurrency. These process networks form the basis of process

optimisations. Based on economical constraints associated with each process step an optimum utilisation of process resources can be determined making use of parallel process strands and the possibility to select among several alternatives.

c) Economic constraints for each object type. This includes process step execution time, process step resource utilisation, basic and marginal cost for process steps. The information can either be supplied as absolute values or as functions of process parameters like material selections, processing pressures and temperatures, process preparation or setup times and the like. This data is used to perform process optimisations with respect to economics considerations. [6]

```
process step etching (etchant of [KOH, EDP],
        temp of [25°, 70°],
        depth of [100 nm, 3000 nm])
  attributes
    remark: "Anisotropic etching process";
    COST: etch_cost
  attributes end
  calculations
    etch_time = if etchant = KOH
        then 2 * depth * 1 min
        else 0.7  depth * 1 min;
    heating_cost = if temp < 50
          then 10 EUR
          else 25 EUR;
    etch_cost = heating_cost + etch_time * 75 EUR / 1 min
  calculations end
  layout rules
    if etchant = KOH then edge_inclination =
(0°,56.7°,180°,237°);
    length of etching channel <= 300 um;
    spacing > 10 um
  layout rules end
process step end
```

Once all parameters for the particular process are fixed the optimisation procedures of the *LIDO-PDL* compilation system will deliver the most cost-effective path through this network.

4. PEDIT PROCESS EDITING TOOL

INTERLIDO-PEdit provides means to configure a process sequence according to the requirements and to the layout design of the intended microstructure and subsequently to check the current process configuration for consistency. The process arrangement is performed graphically in an editor window. The user selects design process elements like process steps, materials etc. from libraries based on the specific design task he/she is performing. These elements appear as icons that can be placed and graphically be connected to process sequences on the editor window. Each icon is related to process or technology information such as design rules or process configuration rules. Fig. 4 shows the assignment of process description data (that is transparent top the user) to the graphical icon representing process steps. [7]

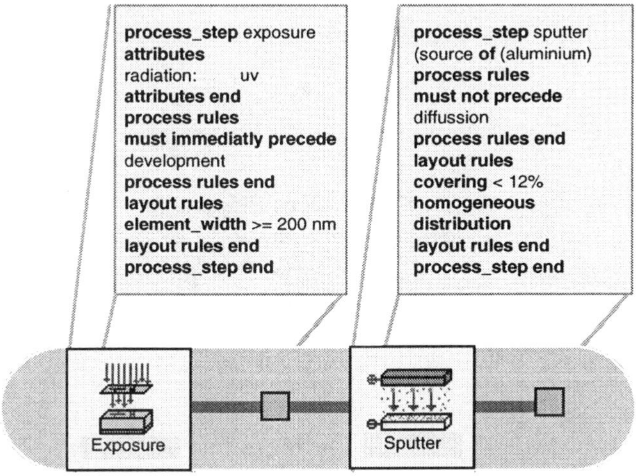

Fig. 4. Assignment of formal description entities to icons

5. COST CONSIDERATION AND PROCESS OPTIMISATION

As process networks with alternative process layouts (like shown in figure 5) can be represented in LIDO-PDL especially designed optimisation methods are implemented to find process sequences with minimum time or cost consumption. [8]

The graphically-based *single-source-shortest-path* algorithm analyses the process network and calculates the process configuration with lowest cost function values. The optimization module INTERLIDO-Opt fulfills the demands of designers wishing to learn about the economic implications of a process layout in early stages of the design flow. The optimisation requires the definition of variables as well as fixed costs within the appropriate

LIDO-PDL objects. PDL offers constructs to declare all resource or process step related costs, time etc. Within the calculation parts of LIDO-PDL descriptions cost statements and function can be provided.

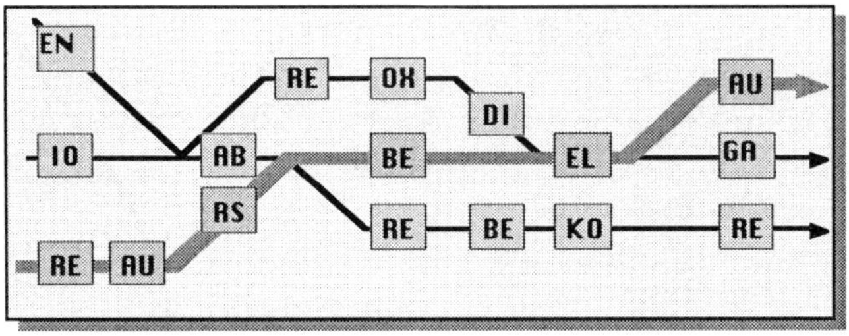

Fig. 5. Process network with optimum path

Based on this information INTERLIDO can pre-calculate the overall expenses assigned to specific process flows in case of alternative process forks. The optimum res. most cost-effective process step sequences can easily be determined before fabricating samples of the microstructures. Time, yield and cost statements can be used as objective functions for the optimisation task.

6. NETWORK ACCESS BY COMPONENT TECHNOLOGY

As has been pointed out in the previous chapters in micro engineering the design process has to be tailored towards every specific design object. As a consequence of this, specific design assistants have to be selected for each design task and especially adapted towards the restrictions to be met in each particular case. To achieve this, component-based software design is currently the method of the choice. It is targeted towards realizing so-called *software components*, executable pieces of software with a clearly defined interface, defined interoperability- and autonomy criteria, as well as the proof of reusability

Componentware is generally understood to be a collection of interacting components. Design assistants in INTERLIDO are realised as componentware based on the component technology *JavaBeans* by SUN Microsystems. JavaBeans is a platform independent component architecture for the Java Application Environment. An important precondition for a useful set of components is that they are well-adapted to each others.

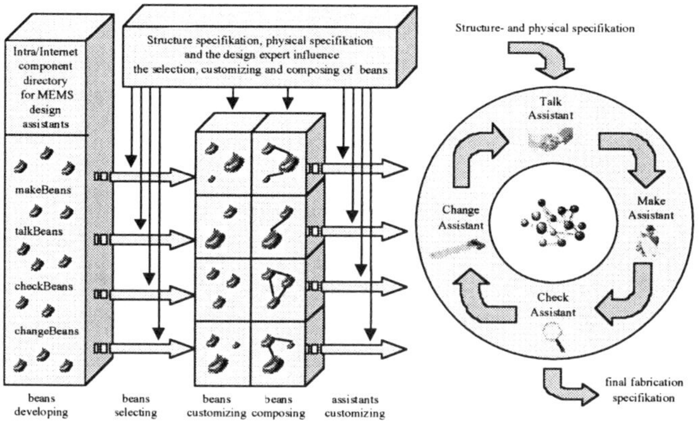

Fig. 6 Component based design assistants

Figure 6 shows an integrated view of the circle model for MEMS design. The Intranet/Internet component directory for physical MEMS design assistants contains various beans for the basic tasks to be performed during design.

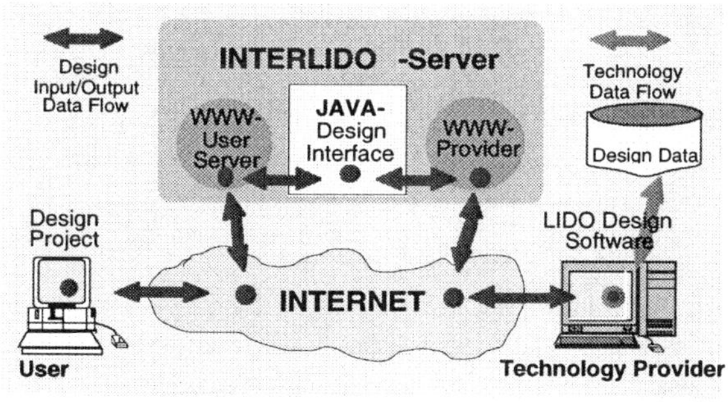

Fig 7: INTERLIDO-network software access

The INTERLIDO is a first prototype of a component-based MEMS physical design tool. At this point of time the dynamic creation of problem-oriented beans has not yet been realised. INTERLIDO is currently used to demonstrate the robustness of a component approach and to show that it can really be useful and not prohibitively limited from a performance point of view, to provide tools that operate merely based on JAVA across the Internet. In addition not all of the LIDO components have been turned into components by now. LIDO-PEdit is the first part that is available based on this new technology. An overview of the INTERLIDO system has already been presented in Figure 3.

Here we only give some remarks on the process of using the INTERLIDO-system via the Internet. Figure 7 shows the principles of Internet-based software access, as they are realised in INTERLIDO. The INTERLIDO-Server is a broker system for user access to tools and technologies across the net. Technology providers will provide their design related knowledge in form of icons and LIDO-PDL descriptions. The LIDO-PEdit functional components will be available on the INTERLIDO-Server and will access the technology related information via the net. The user interface is realized as a JAVA-applet to be run client side. [9]

The system presented so far is currently integrated into a training and working environment to be accessed via the Internet. The TRANSTEC project (No. MM1026) funded by the European Commission aims to implement a training course for different kinds of users teaching them how to take advantage and to solve problems using microsystem technologies.

7. REFERENCES

[1] Karam, J.M. (1998) Microsystems CAD, Proc. MME98 pp. 226-234, Ulvik, Norway

[2] Hahn, K., Brück, R. (1997) An Approach to Layout and Process Verification for Microsystem Physical Design. Journal Microsystem Technologies, pp.82-90

[3] Brück, R., Hahn, K. (1998) MEMS Process/Design and Optimization using Economical Constraints, Proc. MICRO SYSTEM Technologies 98, pp. 479-484

[4] Brück, R., Hahn, K., Reusch, B., (1998) Layout Verification Tools for MEMS Physical Design, Proc MSM 98, April 1998.

[5] Hahn, K., (1999) Methoden und Werkzeuge zur fertigungsnahen Entwurfsverifikation in der Mikrotechnik, Düsseldorf, VDI-Verlag

[6] Lettmann, F. (1997) Prozeßformalisierung zur Unterstützung von Wirtschaftlichkeitsaspekten beim Entwurf von Mikrosystemen. Diploma Thesis, Universität Dortmund

[7]Schneider, C.,Priebe, A., Brück, R., Schumer, C., (1999) Computer aided design tools for economical MEMS fabrication processes, Proc. DTM 99.

[8] Hahn, K., Brück, R., Priebe, A., Schneider, C. (1999) Cost Estimation for LIGA Fabrication Flows using Process Design Methods, Proc. HARMST99 , Kisarazu, June 1999

[9] Brück, R., Schumer, C.(1999) Internet MEMS design tools based on component technology, Proc DTM 99, Paris,.

Moving MEMS into Mainstream Applications : The MEMSCAP Solution

K. Liateni, D. Moulinier, B. Affour, A. Delpoux, M. A. Maher, J.M. Karam
MEMSCAP, 50, allée des Dauphins
38330 St Ismier, France Ph : +33 4 76 52 55 80, Fax : +33 4 76 52 55 81
e-mail : Jean-Michel.Karam@memscap.com, http://www.memscap.com

Key words: MEMS, FEM/BEM, EDA, design, simulation, Intellectual Property, component libraries.

Abstract: This paper presents a fully integrated solution for the development of Micro Electro Mechanical Systems (MEMS) which covers component libraries, design tools and design methodologies which are used in conjunction with conventional design automation tools (EDA). This solution enables system houses in wireless and optical communications and consumer electronics markets to reduce their internal development costs and significantly accelerate their product development cycles.

1. INTRODUCTION

MEMS are the essential link between digital computation and the physical world. They enable the gathering of optical, electromagnetic, mechanical, acoustic, chemical, or thermal information and its conversion into digital data and also serve as the means for controlling the physical environment. Also, in many applications, MEMS serves as an enabling technology for implementing the required system functions in a compact, economical package.

The latest advances in MEMS technology have enabled the design of a new generation of electronic microsystems that are smaller, cheaper, more reliable, and consume less power. These integrated systems integrate numerous analog/mixed signal microelectronics blocks and MEMS functions on a single chip or on two or more chips assembled within an integrated package. A major difficulty in designing these systems resides in the lack of information sharing between designers from different disciplines. With no common interface, such a separatist approach can result in catastrophe when prototype testing reveals a design flaw requiring additional iterations through the design and fabrication cycle. For example, none of the information derived from 3D field solvers on MEMS structures can be automatically transferred to an IC design tool. A secondary hurdle is enabling engineering teams to make full use of existing IP in MEMS. The ability to smoothly integrate cores into a system-on-silicon architecture provides system and IC designers with the latest functionality and process technology and dramatically reduces time-to-market. Until now, designers had to create MEMS by pushing polygons and understanding the fine details about the target fabrication process. Obviously, this approach demanded exceptional engineering skills and expanded design schedules and budgets. The MEMSCAP design environment and IP portfolio open up the playing field by capturing MEMS engineering knowledge to allow automation of MEMS-based designs among the extreme majority of engineers who do not push polygons. They also enable information sharing between system designers, IC designers, process engineers, and MEMS experts from various disciplines thereby reducing development time and cost.

2. A FULLY INTEGRATED SOLUTION

The interdisciplinary nature of MEMS and the significant expertise required to develop the technology has been a significant bottleneck in the timely design of new products incorporating MEMS technology. Until now, designers had to create MEMS by pushing polygons and understanding the fine details about the target fabrication process. Obviously, this approach demands exceptional engineering skills and expanded design schedules and budgets. In many cases it is difficult to justify the R&D investment of both time and money that is necessary to reap the benefits of MEMS technology. The approach used in the MEMS Engineering Kit is based on the following principles and features:

- Define two different flows: the component engineer and the system engineer,
- Ensure the link between the two flows,
- Provide to the system engineer an integrated solution enabling a seamless design flow from front-end to back-end,

- Enable the exchange of data between the different description level: the structural level (FEM/BEM), the system/behavioral level (HDL-A, VHDL-AMS, Verilog-AMS), the physical level (layout),
- Vehicle IP and provide design re-use capabilities.

The key elements of this design environment are a behavioral model to layout generator, a physical layout to 3D solid model translator, and a solid-model to behavioral model translator. These three tools in combination with existing design automation tools enable system designers, IC designers, process engineers, MEMS specialists, and packaging engineers to share critical design and process information in the language most relevant to each contributor. Before, none of the information derived from a given design tool could be automatically converted and transferred to another tool.

2.1 The MEMS Engineering Kit

The MEMS Engineering Kit is a new design paradigm that combines aspects of electronic design automation with mechanical, thermal, and fluidic computer-aided design. This Kit supports a wide range of advanced process technologies at leading MEMS foundries (e.g. MCNC/Cronos). The core software supports best of breed point tools and commonly used design environments (e.g. Mentor Graphics).

The environment contains elements for the device designer, enabling him to design modules, to simulate them, and finally to put the knowledge in the form of characterized standard cells in library. Commercially available optimization and yield management tools, such as OPSIM and ASPIRE have been extended to MEMS technology to enhance the work of the MEMS device engineers.

Figure 1 details the MEMS Engineering Kit design flow.

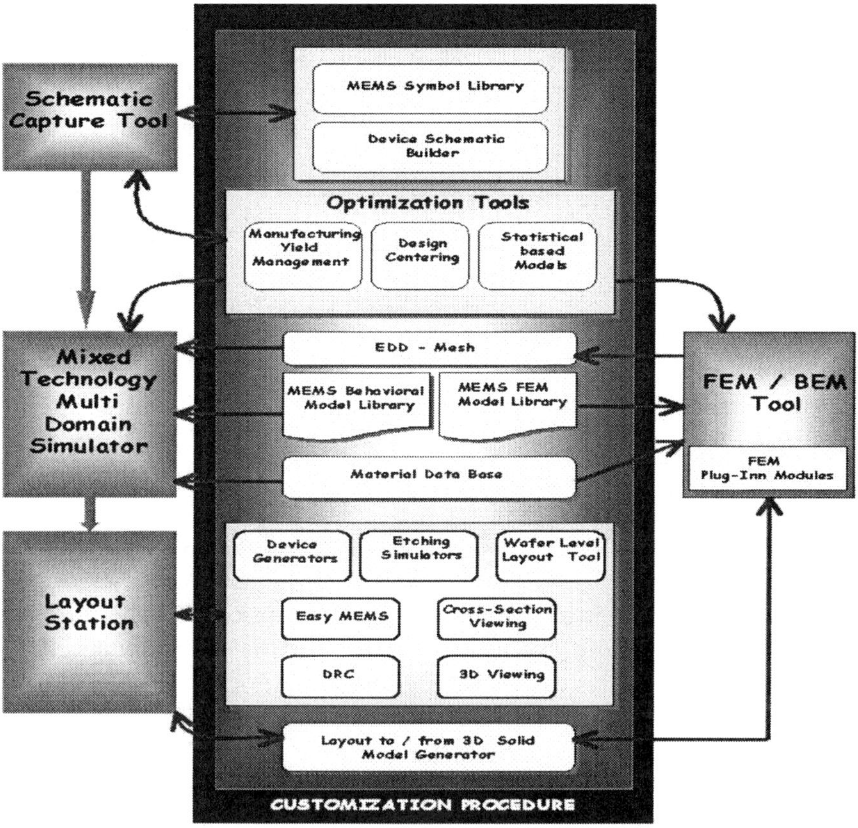

Fig.1. The MEMS Engineering Kit environment

The MEMSCAP MEMS Engineering Kit provides the following features:

- MEMS oriented schematic capture,
- a set of parameterized cells described at different levels (symbolic, behavioral, layout),
- optimization tools (Opsim) very useful for model development as well as specification parameter recycling,
- HDL-A code generator of non-linear behavioral models from lower level description,
- full verification of the design functionality (Continuum/Eldo/HDL-A),
- layout generators for mechanical structures, such as elementary structures (i.e. bridges, cantilevers, membranes, etc.) and application oriented structures coupled to the behavioral models, allowing the schematic driven layout feature,

- the layout generation of the whole MEMS (electronic and non-electronic parts) and a design rule verification,
- an extended design rule checker which can handle monolithic as well as hybrid designs,
- a multi-segment, multi-direction cross-section viewer,
- an anisotropic etching simulator for silicon and gallium arsenide as well as a sacrificial layer etching simulator,
- layout to 3D-solid model generator.

The concept of the environment is based on providing a fully integrated solution, easy to be used and reducing the development cycles of MEMS designs.

The MEMS Engineering Kit includes the Generic Kit and the Foundry Modules:

2.1.1 VULCAIN™ : The Generic Kit

The Generic MEMS Engineering Kit is a customizable design kit (includes a customization procedure to any process technology) and integrates existing third-party design tools (e.g. IC layout, circuit simulator, FEM field solver, etc.) into one common design environment.

Figure 2 shows a schematic capture for a torsional combdrive using building blocks. Each of these bricks encapsulate a behavioral model, written in HDL-A or in VHDL-AMS, making possible the simulation of this system.

A full system functionality can be verified, through a multi-level, mixed-mode, multi-domain behavioral simulation (figure 3). The system can be a MEMS component or a full system including the read-out electronics, whether it is integrated monolithically or in a hybrid assembly.

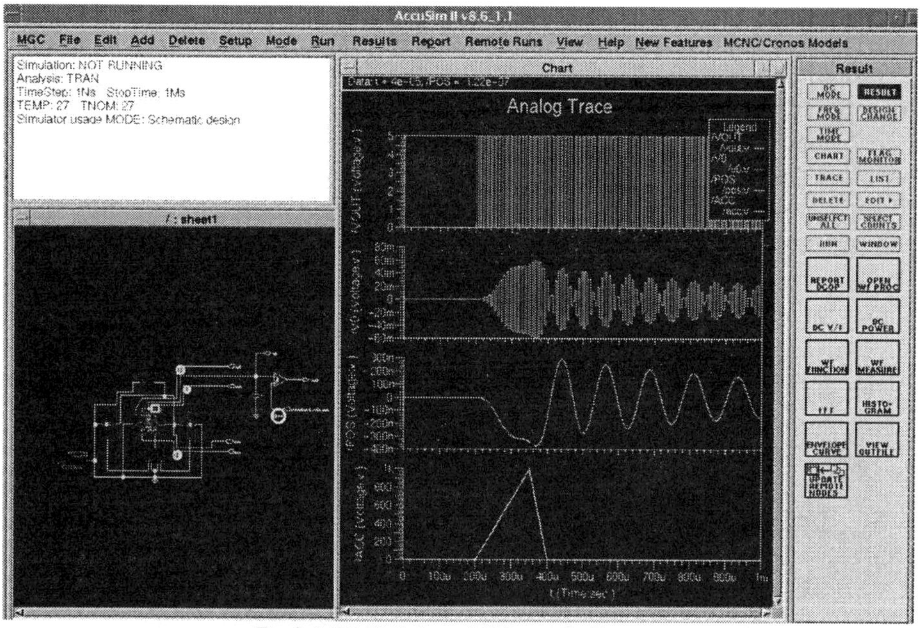

Fig.2. MEMS Oriented Schematic Capture

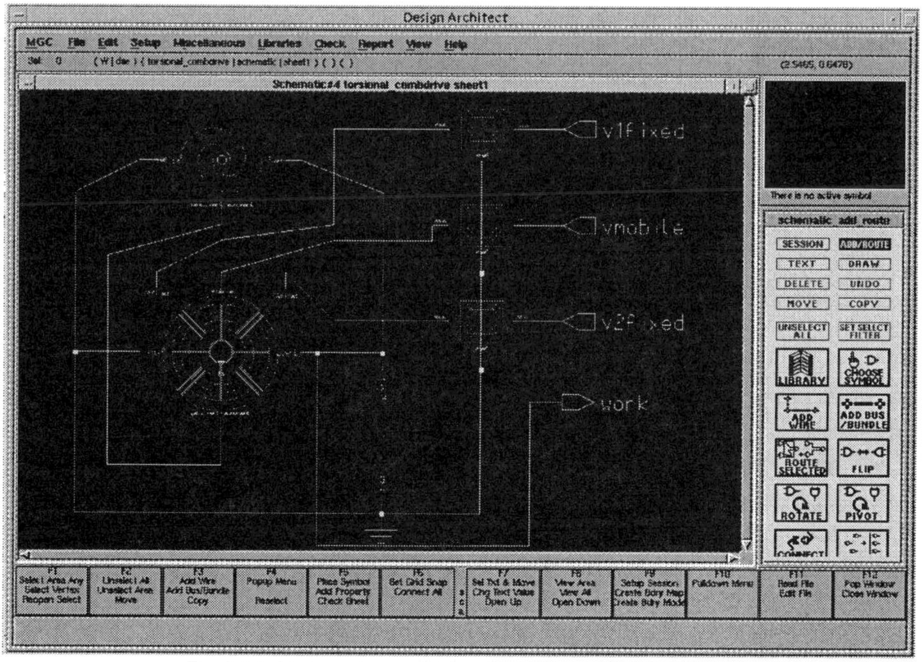

Fig.3. Multi-domain, multi-level, mixed-mode simulation

When simulation is done, the user can perform a schematic driven layout generation (SDL). In this operation, the device generators will consider the parameters fixed at the system level and map them automatically to the layout level keeping properties and connectivities.

Back-end operations, such as design rule checking (which can differentiate the MEMS rules from the electronics rules and verify them simultaneously), multi-segment, multi-angle, multi-direction cross-section viewer, and sacrificial or anisotropic etching simulation, can be realized (figure 4).

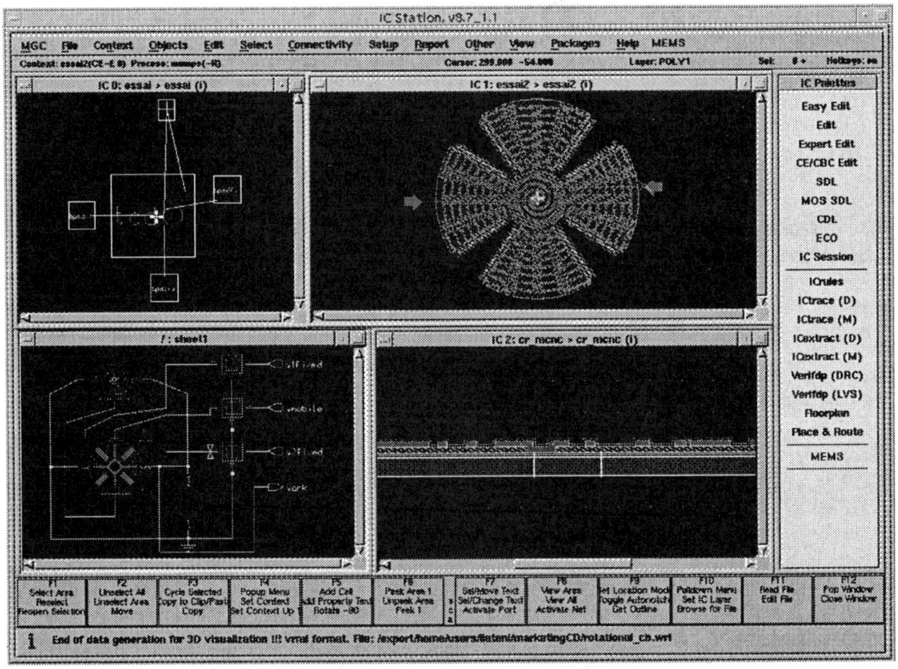

Fig.4. Back-end tools, such as cross-section viewer

This system flow is coupled to the component flow by ensuring the link with the field solvers, such as FEM or BEM tools. For this purpose, and in addition to the FEM/BEM to HDL-A translator described in the following sections, the kit includes a layout to 3D solid model generation which enable the generation of a selected layout area a 3D view (in VRML, or Geomview format) or a FEM input file (for ANSYS and very soon Coyote Systems). The user can select the layers he would like to consider in this generation or simply consider all the structuring layers. Figure 5 shows the 3D view of the torsional combdrive.

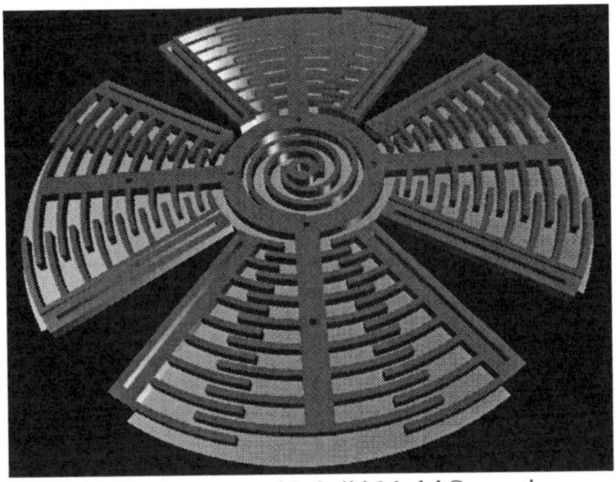

Fig.5. Layout to 3D Solid *Model* Generation

2.1.2 The Foundry Modules

A set of Foundry Specific Modules has been added in order to provide a "faster to fab" solution. In addition to the contents of a customized kit, a foundry specific module includes:

- DRC for the supported technology,
- Component Libraries

The Foundry Specific Modules currently includes, KanagaTM, the MCNC/Cronos Foundry Module. This module ensure a seamless design flow and includes a library of more than fully characterized 75 device generators manufacturable on the MCNC/Cronos production lines.

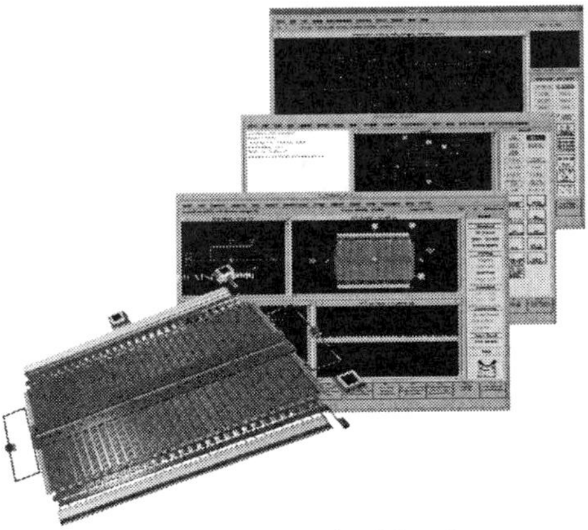

Fig.6. KanagaTM, the MCNC/Cronos Foundry Module, ensures a seamless design flow

2.2 Model Generation Tools: *Edd*™

Simulation of heterogeneous systems on the system level implies the simulation of the whole system and that of particular devices with a special emphasis on points of interests like functional behavior, timing, power consumption and so on. Simulation is very closely connected to modeling because only the system- or device-behavior can be simulated that has been taken into account while developing the models. Functional simulation of microelectromechanical systems (MEMS) and microcomponents can be done efficiently by using a single system description language suitable for a single mixed-mode simulator. The advent of analog HDLs such as HDL-A by Mentor Graphics and the future standard VHDL-AMS offer the possibility to create behavioral models of electrical and non-electrical devices without having the limitations of Spice. The expectations towards VHDL-AMS and commercial AHDLs are high: It is expected that these HDLs simplify the modeling as well as the simulation procedure of analog, analog/digital and other heterogeneous components.

AHDLs will find wide application in the domain of MEMS since modeling of nonlinearities is simplified and supported by almost all AHDLs. In practice, the creation of behavioral models for MEMS components causes several problems:

- The engineer responsible for creating the behavioral models needs excellent theoretical knowledge about the component as well as know-how about the simulation tools. It is difficult to find engineers who are experts in at least two engineering domains including the appropriate simulation tools.
- If the behavioral model is created using first order equations from theoretical work, the accuracy of this model may not be sufficient compared to the realistic 2/3D structural Finite-Element (FEM) model. Therefore often no verification of the behavioral model with regard to the structural 2/3D FEM model is done.
- There is no quantitative number for the accuracy of the behavioral model for other waveforms than the one used during the creation.
- The behavioral models have to obey certain standards concerning system-interfaces, generics etc. It is difficult to manage from the FEM-designers view.

A solution of this dilemma could be a tool that:

- Decouples the process of creating the behavioral model from the simulation process of the 2/3D FEM devices.
- Lowers the required knowledge of the AHDL for creating the behavioral model.

This tool (EddTM) enables the generation of non-linear dynamic behavioral and functional HDL-A models from models on a hierarchical lower level of abstraction (such as Finite-Element or transistor-level description) or measured data. The tool supports the efficient creation of nonlinear behavioral models in HDL-A without requiring deep knowledge of this AHDL. True HDL-A code is generated in a single architecture/entity-unit resulting in a faster simulation time than connected HDL-A models. It has been created as an interface between 2/3D-structural FEM-models and behavioral models in HDL-A (in that case the tool is integrated within the kit, and is called EddTM-Mesh), and it can also be applied for the generation of arbitrary behavioral models from lower level abstractions (e.g. such as analog or analog/digital Spice-descriptions. In that case, it is a stand-alone tool called EddTM-Net).

2.2.1 Edd™-Mesh : FEM/BEM to HDL-A Translator

This CAD-tool provides the following features:

- Simulation of the FEM-model.
- Fixed interface for checking in FEM-models in a CAD-library.
- Support of reuse of already created behavioral models in the library.
- Parameter optimization of the behavioral model.
- Differential adaptation of the behavioral model effects by comparing the waveforms of the FEM and the behavioral model and concluding the insertion or deletion of certain (nonlinear) effects.
- Support of model verification by using up to 10 different input stimuli.
- Check-in of the behavioral model in order to restart the model creation process.
- Creation of a single HDL-A Architecture/Entity pair that has a superior run time behavior than connected HDL-A models.

The basic idea for this CAD-tool is that the new behavioral model should be created in a differential way, i.e. starting with a relatively simple model the more complex one is created by adding missing and/or deleting obsolete effects until the resulting model satisfies a desired quality (the identity of I/O-waveforms for all possible input stimuli). Based on this idea, the tool uses a global optimization scheme, which can be subdivided in parameter and structural optimization. The lacks in the structure of the behavioral model are identified by operators, which are the arguments in a fuzzy-rule based environment for the representation of the knowledge concerning the used component. The basic algorithm is the following:

I. Select a start model from the database.
II. Select the input bitpattern which yields the worst costvalue (the worst correspondence between the real and the behavioral models).
III. Optimize the parammeters by applying simulation procedures (back to II if an internal counter has not exceeded a certain predefined value).
IV. If the targeted accuracy is not met, identify the lacks by operators and insert or delete a certain effect in the behavioral model. Back to II until an internal counter for the number of allowed structural optimizations has not exceeded a certain predefined value.

Figure 7 shows the structure of this CAD-tool in a detailed way.

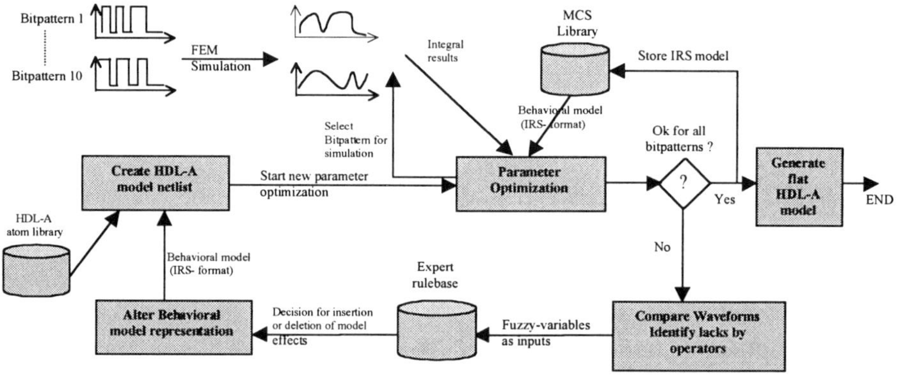

Fig.7. EddTM-Mesh Structure

It should be stressed that with the differential approach used here not all parameter values have to be optimized at each time, but special care may be spent on parameter values which are new (i.e. have been inserted in the last structural optimization step). It can be summarized that using this method more than 12 parameters can be optimized for a behavioral model.

Figure 8 shows the correspondence between the FEM model and the HDL-A generated model of an accelerometer.

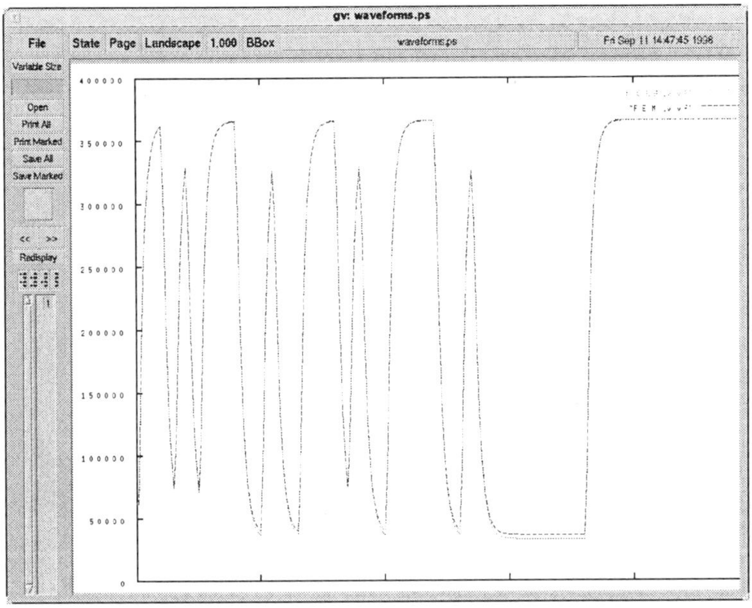

Fig.8. Correspondence between the FEM model and the HDL-A generated model of an accelerometer

3. CONCLUSION

This paper presents a design methodology and an integrated solution for the development of MEMS. It describes the whole design flow from front-end to back-end and details the related tools and functions integrated within a single environment. Finally, model generation tools and methodologies are also discussed and developed.

4. ACKNOWLEDGMENTS

A part of this work is the result of a fruitful cooperation between MEMSCAP, Mentor Graphics and Cronos/MCNC, which is supported by DARPA.

A special acknowledgment is addressed to Dr. Klaus Hofmann for the excellent work he did and transferred to MEMSCAP.

Trends in RF Simulation Algorithms

Joel R. Phillips
Cadence Berkeley Laboratories, Berkeley, CA 94704

Dan Feng
Cadence Design Systems, San Jose, CA 95134

Abstract Continued growth in the market for wireless communication devices has resulted in renewed interest in algorithms for computer-aided design of RF systems. In this article we review some recent progress in the field of circuit-level RF simulation algorithms. Comparisons of the relative advantages of popular simulation approaches are presented, with particular attention to the relation between numerical simulation strategies and the nature of RF circuit problems.

1. INTRODUCTION

Roughly speaking, the primary task of the analog portion of a wireless system is to shift data signals in frequency. A frequency shift of an analog data stream can be accomplished, ideally, by multiplication with an auxiliary periodically time-varying signal. Thus a central characeristic of RF circuits is that they are usually driven with one or more periodic signals. These periodic signals establish a set of frequency ranges about which the data signals cluster in a narrow band. This makes the RF circuit simulation problem difficult, because the response of a circuit over both fast (because of the high frequency carrier signals) and slow (because of the small frequency spacing of the other signals) timescales must be determined. The RF simulation problem is tractable because, though signals must be resolved with very fine frequency spacing, only a small portion of the overall frequency band is involved in the circuit operation. Specialized algorithms for RF simulation have been around for some time. This paper is concerned with recent developments and experience with new numerical techniques. The reader is directed to the book (Kundert et al., 1990) and the reviews (Kundert, 1999; Gilmore and Steer, 1991) for further background and bibliographic material.

2. RF SIMULATION AND THE PERIODIC STEADY-STATE

The simplest non-trivial example of a circuit with a sparse signal spectrum is a circuit that develops a periodic steady-state response to a periodic drive signal. The methods used in solving periodic steady-state problems form the basis of the more complicated analysis methods such as linear time-varying noise analysis and distortion analysis via quasi-periodic steady-state methods.

Formulating steady-state problems. The system of n differential-algebraic equations that describes the RF circuit can be written in the general form

$$\frac{d}{dt}q(v(t)) + i(v(t)) = u(t) \tag{1.1}$$

where $t \in \Re$ is the time variable, $v(t) \in \Re^n$ is the circuit state (such as a collection of node voltages), $q, i : \Re^n \to \Re^n$ represent dynamic and static components in the circuit, and $u(t) \in \Re^n$ represents the effect of external sources. The periodic steady-state problem involves, given an excitation $u(t)$ such that $u(t+T) = u(t)$ for some fundamental period T, finding a solution to Eq. 1.1 such that $v(t+T) = v(t)$ is periodic.

The simplest way to find a steady-state solution is to simply integrate Eq. (1.1) forward in time from a known initial condition until the solution $v(t)$ becomes periodic. Such an approach discards any potential efficiency that could be gained from the sparsity of the signal spectrum, and is impractical since it may take a very long time for the circuit to reach steady-state. Instead what is done is to discretize and solve Equation (1.1) over one period, subject to the periodic boundary condition $v(t+T) = v(t)$.

Discretization. Two methods are presently popular. Finite difference discretizations approximate the state variables $v(t)$ with piecewise-polynomials in t. For example, by using a first-order accurate backward difference approximation to the time-derivative, the backward-Euler integration rule

$$\frac{q(v(t+h)) - q(v(t))}{h} + i(v(t+h)) = u(t+h) \tag{1.2}$$

may be used to solve Equation (1.1). If an equation similar to Equation (1.2) is written at m timepoints in an interval $[t_0, t_0 + T]$ and the periodic boundary condition applied as $v(t_0) = v(t+T) = v(t_m)$, then a nonlinear system of equations in the mn variables $v(t_k), k = 1 \ldots m$ is obtained. Each timepoint will be coupled to $p+1$ others if an order-p polynomial is used to discretize the derivative in Equation (1.1). When $v(t), q(v(t))$, or $i(v(t))$ contain rapid transitions, as happens in highly nonlinear circuits, the local timestep and order of the method can be locally varied to accurately resolve the unusual behavior without an efficiency penalty.

Spectral methods, such as used in harmonic balance, represent the second popular class of discretizations. Spectral discretizations use a high-order, usually global, representation of the functions $v(t), i(v(t)), q(v(t))$. A typical choice is to expand, say, the unknown state variables $v(t)$ in a series of complex exponentials (or, equivalently, sines and cosines), as

$$v(t) = \sum_{k=-K}^{K} c_k e^{ik\omega_0 t} \qquad (1.3)$$

where $\omega_0 = 2\pi/T$ is the fundamental frequency. Note that the periodic boundary condition is automatically satisfied.

If the representation of the function in terms of the frequency-domain coefficients c_k is known, then the a time-derivative can be computed by multiplying the kth Fourier coefficient by $ik\omega_0$. If enough harmonics are chosen to adequately represent $v(t)$, then spectral differentiation is essentially exact. For smooth signals, the approximation error decreases rapidly as the number of terms K in the expansion increases and then spectral methods can achieve high accuracy at low cost. On the other hand, if the voltages or currents, or their low-order derivatives, have sharp or irregular features, whether as a result of circuit operation or due to device models, the spectral representation may converge quite slowly, resulting in an inefficient algorithm.

In modern codes, the nonlinear functions are evaluated in the time-domain, but the time-differentiation operation takes place in the frequency domain, and therefore an efficient means of translating a solution quickly between the domains is needed. The translation operation usually uses the fast Fourier transform (FFT), and this usually imposes constraints on the timestep spacing that create difficulties when dealing with signals containing rapid transitions. Recently, efforts have been made to reformulate the spectral methods to use non-uniform timestep spacing(Nastov and White, 1999).

Harmonic balance methods can naturally include linear distributed elements in simulations. After Fourier transformation the convolution $y(t) = \int_{-\infty}^{t} h(t-\tau)x(\tau)d\tau$ that describes the distributed element becomes the simple multiplication, $Y(\omega) = H(\omega)X(\omega)$ where $Y(\omega), H(\omega), X(\omega)$ are the frequency-domain versions of y, h, and x respectively. Finite-difference based methods must either evaluate the convolution directly in the time domain, an expensive operation, or translate the solution to the frequency domain. It is difficult to achieve efficient time/frequency conversions for signals with non-equispaced timepoints, and so finite-difference based methods are not well suited to analyzing circuits with many distributed elements.

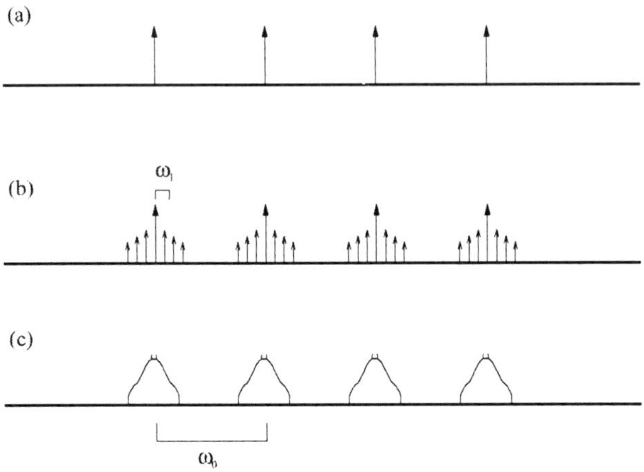

Figure 1 Periodic (a), quasi-periodic (b), and broadband (c) RF data signals.

3. MULTI-FREQUENCY ANALYSES

Most practical problems have signals at multiple frequency scales. Figure (1) shows three common RF data signal profiles. If the frequencies of all the signals are harmonically related, as in Figure 1 (a), then the problem can be solved using periodic steady-state methods, with the fundamental periodic the least common multiple of the periods of the various excitations. However, often signals are present at many unrelated frequencies. If the circuit response can be described by frequencies that are harmonics of several distinct fundamental frequencies, as in Figure 1 (b), the circuit has a *quasi-periodic steady state* that can be solved for directly. For non-quasi-periodic signals that have a spectrum that is dense in a narrow band around a set of (quasi-)periodically spaced frequencies, but unstructured, as in Figure 1 (c), "envelope" techniques can be used. There are various ways of formulating the "multiple-frequency" analyses.

PDE methods. A signal is quasi-periodic if it possesses a Fourier-series expansion in multiple frequencies, such as

$$v(t) = \sum_k \sum_l c_{k,l} e^{ik\omega_0 t} e^{il\omega_1 t} \tag{1.4}$$

for a two-frequency signal with fundamentals w_0, w_1. An obvious way to solve the quasi-periodic problem is by a spectral method : truncating the series (1.4) and solving for the coefficients c_{kl}. When the signals around the carrier frequency have non-periodic character, a representation such as

$$v(t) = \sum_{k=-K_0}^{k=K_0} c_k(t) e^{ik\omega_0 t}. \tag{1.5}$$

may be assumed . A broader class of algorithms can be obtained by making the observation (Brachtendorf et al., 1996; Roychowdhury, 1997) that Equations (1.4) and (1.5) have some of the flavor of a *two-dimensional* Fourier series expansion.

Suppose we define the two-variable quantity

$$V(t_1, t_2) = \sum_{k=-K_0}^{k=K_0} \sum_{l=-K_1}^{K_1} c_{k,l} e^{ik\omega_0 t_1} e^{il\omega_1 t_2}. \tag{1.6}$$

Note that if $V(t_1, t_2)$ is known, then $v(t)$ can be recovered from the "diagonal" as $v(t) = V(t, t)$. If we likewise define multi-dimensional inputs $U(t_1, t_2)$, and replace the time-derivative in Eq.(1.1) by the two-dimensional gradient operator, then, neglecting for the moment distributed terms, the DAE (1.1) becomes a quasi-linear hyperbolic partial differential-algebraic equation

$$\left[\frac{d}{dt_1} + \frac{d}{dt_2}\right] q(V(t_1, t_2)) + i(V(t_1, t_2)) = U(t_1, t_2) \tag{1.7}$$

It can be shown that if $V(t_1, t_2)$ is a solution of the "multivariate partial differential equation" (MPDE), (1.7), then $V(t, t)$, which lies on the "characteristics" of the PDE, is a solution to the DAE (1.1). The original circuit simulation problem can thus be recast as the problem of solving a particular partial differential equation. It should be clear that if periodic boundary conditions are imposed on both variables of the MPDE, and a spectral discretization used in both dimensions, then a representation equivalent to the multi-tone harmonic balance formulation is obtained.

The emergence of the MPDE viewpoint is potentially interesting for three reasons. First, it provides a more formal framework for analyzing methods for solving the "multiple timescale RF simulation problem." Second, it suggests potentially new algorithms. Virtually any scheme that can be used to solve a hyperbolic PDE can be used in the RF simulation context. For example, discretizing the PDE with finite-difference methods along both dimensions produces "time-time"(Roychowdhury, 1997) approaches that are not obviously obtained in other contexts. Third, practical experience with algorithms for solving RF problems could potentially be applied to develop algorithms for solving hyperbolic partial differential equations with multiple-timescale solutions.

Because the PDE formulations embed the one-dimensional quantity $v(t)$ into a higher dimensional space, there is no guarantee of efficiency, and in fact unless the signals in the circuit satisfy certain conditions, the methods can be less efficient than traditional transient simulation. Generally if the signal bandwidth is large relative to the carrier frequency, there is no advantage to the PDE methods. For envelope-like solutions, Equation (1.5), the equivalent statement is that the $c_k(t)$ must vary slowly on a timescale of the order of $1/\omega_0$ for the methods to be both efficient and accurate.

Sampling Methods. The idea behind sampling methods is to remove the difficulties caused by the high-frequency carrier by sampling all the signals in the circuit at multiples of the carrier period $T_0 = 1/\omega_0$. To see how this might work, consider sampling the signal $v(t) = \sum_{k=-K_0}^{k=K_0} c_k(t)e^{ik\omega_0 t}$ at a discrete set of points $t'_n = t_0 + nT_o$ where $t_0 \in [0, T_o)$ and n runs over the integers. The result is a signal $\bar{v}(t'_n) = \sum_{k=-K_0}^{k=K_0} c_k(t'_n)e^{ik\omega_0 t_0}$ defined at the discrete time instants t'_n. Note that the fast time variation due to the high carrier frequency has disappeared. If the signals are concentrated in frequency around the carrier harmonics, then the $c_k(t'_n)$ are slowly varying.

In the quasi-periodic case, where each $c_k(t'_n)$ can be represented by a few terms in a Fourier series, $c_k(t'_n) = \sum_l c_{kl}e^{l\omega_1 t'_n}$ the c_{kl} at consecutive timepoints, t_n and $t_{n+1} = t_n + T_0$ and must be related by the quasi-periodic boundary condition $c_{kl}(t'_{n+1}) = c_{kl}(t'_n)e^{i\omega_1 T_0}$. This boundary condition must relate the signals at the timepoints t'_n and t'_{n+1} that are the beginning and ending points of a single carrier cycle. For envelope-like solutions, boundary conditions are obtained by requiring that the $c_k(t)$ lie on a low-order polynomial. Because of the use of special boundary conditions of this sort, only a few cycles actually need be simulated, and the cost is relatively independent of the relation between the various frequency scales.

When applied to multi-frequency problems, the sampling algorithms can also be thought of as ways to solve the special PDE (1.7). However, they differ from the normal PDE-based implementations in that the solution is actually performed directly on the "characteristics" of the hyperbolic PDE and generally the different cycles on the characteristic are coupled together at a single, common point in the "clock" phase. The sampling methods thus always compute solutions to the DAE (1.1). Error can only be induced as a result of the boundary conditions. Note that, unlike the PDE methods, the sampling schemes are intrinsically one-dimensional. In cases where the PDE methods might become inefficent, the sampling schemes reduce to conventional time-domain simulation algorithms.

4. NONLINEAR SOLUTION TECHNIQUES

Once the circuit equations have been formulated and discretized, a high-dimensional nonlinear system of equations must be solved. The basic Newton method to solve the nonlinear system $f(x) = 0$ is the iterative application of the rule $x^{k+1} = x^k - J(x^k)^{-1} f(x^k)$ for $k = 1, 2, \ldots$ until converged, where $J(x^k)$ is the Jacobian matrix of of $f(x^k)$, $J_{ij}(x^k) = \partial f_i / \partial x^j$. For a discretization of an n-equation DAE with m timepoints, the Jacobian will be a rank-mn matrix. As an example, consider the simple backward-Euler discretization of the periodic-steady-state problem, which has the Jacobian

$$J = \begin{bmatrix} \frac{C_1}{h_1} + G_1 & & & -\frac{C_M}{h_1} \\ -\frac{C_1}{h_2} & \frac{C_2}{h_2} + G_2 & & \\ & \ddots & \ddots & \\ & & -\frac{C_{M-1}}{h_M} & \frac{C_M}{h_M} + G_M \end{bmatrix} \tag{1.8}$$

where

$$C_k = \left. \frac{\partial q(v)}{\partial v} \right|_{v(t_k)}, \qquad G_k = \left. \frac{\partial i(v)}{\partial v} \right|_{v(t_k)}. \tag{1.9}$$

For certain types of finite difference schemes, alternative strategies to the direct Newton updates are available that can prove advantageous. Consider for the moment the periodic-steady-state problem, and suppose Equation 1.1 is discretized using a "causal" low-order difference rule. That is, the rule is such that if an initial condition $v(t_0)$ is known, the rule can be used to propagate the solution to Equation (1.1) forward in time, one timestep at a time. Now imagine that one point on the periodic solution happens to be known. All of the other points on the periodic solution can be easily obtained simply by integrating the DAE (1.1) forward in time.

To make the argument concrete, define the transition function $\phi(v_0, t_k, t_f) = v(t_f) : v(t)$ satisfies equation (1.1) for $t \in [t_k, t_f]$ and $v(t_k) = v_0$. In terms of the transition function, the periodic boundary condition is expressed as $v_0 = \phi(v_0, t_0, t_0 + T)$. The *shooting method* consists of solving the periodic-steady-state problem by solving this alternative equation. To do so we must solve a linear system involving the shooting method Jacobian which is $I - \Phi$ where the matrix Φ is the sensitivity of the transition function to the initial condition,

$$\Phi = \frac{\partial \phi(v_0, t_0, t_0 + T)}{\partial v_0}. \tag{1.10}$$

Consider again Equation (1.8). If we let L denote the lower block-triangular portion of this matrix, and U the upper-block-triangular portion, then $L^{-1}J = I + L^{-1}U$. It turns out that the last size $n \times n$ block in the matrix $L^{-1}U$ is the matrix Φ. Φ can be computed from knowledge of J, and conversely systems involving J can be easily solved if the matrix Φ can be inverted.

The difference between using shooting methods to solve the finite-difference equations and the "normal" Newton iteration is in the way the finite-difference equations are updated after every Newton iteration. In the direct approach, an update to the entire waveform over the period T is made simultaneously. In the shooting method, at first only one point, the first point in the interval, is updated. Then a sequence of local nonlinear problems are solved to be sure that the DAE (1.1) is satisfied at every point in the interval $[t_0, t_0 + T)$. When the end of the interval is reached, the value of the transition function for the new Newton iteration has been computed. If the solution is not yet periodic, another Newton update to the initial condition $v_0 = v(t_0)$ can now take place. Empirically it is observed that the shooting method has better global convergence properties than direct updates of the finite-difference equations. By acting on the function ϕ instead of the individual difference equations, highly nonlinear behavior ocurring in the interior of the periodic interval may be hidden from the Newton iteration.

The shooting method can be naturally extended to solve the equations arising from the sampling schemes, since those schemes depend only on finding the values of the states v at a single point in each of K cycles. The boundary condition that relates the beginning and endpoints of the cycle are expressed using a $K \times K$ matrix D_K, resulting in the the nonlinear system $(D_K \otimes I_n)v_K - \phi_K(v_K) = 0$, where $\otimes$ is the Kronecker product, I_n is the n by n identity matrix, and ϕ_K represents is a multi-cycle transition function that represents the transition functions of the K sample points. The dimension $Kn \times Kn$ Jacobian of the Newton iteration will be $J_K = D_K \otimes I_n - \Phi_K(v_K)$ where Φ_K is a block-diagonal matrix whose blocks are the Jacobians of the individual sensitivity functions. Shooting methods may also be used to solve the equations resulting from finite-difference discretizations of MPDE equation formulations.

For difficult RF simulation problems, continuation methods(Allgower and Georg, 1990) are the procedure of choice to achieve global convergence of the Newton iteration. Continuation methods achieve convergence of the Newton iteration by solving a sequence of nonlinear problems, each of which is easy to solve because the solution of the previous problem provides a starting value for the Newton iteration that is close to the actual solutuion. For example, when performing intermodulation distortion analysis with one of the sampling methods described previously, the Newton iteration may have difficulty converging at high power levels, and so the power is gradually increased from a low level where the solution is known, until the desired operating point is reached (Feng et al., 1999).

5. SMALL SIGNAL ANALYSIS

In some cases, for example, in noise analysis, there are inputs to the RF circuit whose amplitudes are small and who generate small amplitude responses. Performing a linear-time varying analysis of the circuit can be computationally advantageous when there are many such inputs at frequencies unrelated to the larger carrier or RF data signals. The first step in the linear time-varying analysis is to linearize the RF circuit around a time-varying operating point that is set by the large signals in the circuit, for example the local oscillator signal in a mixer. The linearization procedure can be accomplished by separating the excitation and responses in Equation (1.1) into large and small-signal parts and then taking the first term in the Taylor series expansion of the response about the large signal. The resulting time-varying linear system is $\frac{d}{dt}\left[C(v_s(t))v_s(t)\right]+G(v(t))v_s(t) = u_s(t)$ where $u_s(t)$ describes the small-signal inputs and $v_s(t)$ is the small-signal response, and $v(t)$ is the large-signal bias point.

Conversion Matrices. The defining characteristic of linear time-varying operators is that they shift frequency. A small-signal input at a frequency ω will induce a circuit response in a periodically time-varying circuit at frequencies that are offset by multiples of the fundamental frequency. Thus signals at frequencies $w + k\omega_0$, with k an integer, are mutually coupled by the time-variation. We can represent these relations by defining an infinite-dimensional "vector" of the inputs and responses at the various harmonic offsets, $\tilde{v}_s(w) =$

$$\left[v_s^T(\omega - K\omega_0)\cdots v_s^T(\omega)\cdots v_s^T(\omega + K\omega_0)\right]^T, \tilde{u}_s(w) - \left[u_s^T(\omega - K\omega_0)\cdots u_s^T(\omega)\cdots\right]^T$$

that are related by the conversion matrices (Maas, 1988) by $\tilde{v_s}(\omega) = \tilde{H}(\omega)\tilde{u_s}(\omega)$ where

$$\tilde{H} = \begin{bmatrix} \ddots & & & & \\ & H_{-1,-1} & H_{-1,0} & H_{-1,1} & \\ \cdots & H_{0,-1} & H_{0,0} & H_{0,1} & \cdots \\ & H_{1,-1} & H_{1,0} & H_{1,1} & \\ & & & & \ddots \end{bmatrix} \tag{1.11}$$

and each sub-matrix $H_{k,l}$ represents a transfer function matrix that converts signals from freqency $\omega + l\omega_0$ to frequency $\omega + k\omega_0$.

Given an operating point and a linearization, i.e. a Jacobian matrix such as (1.8), computing the conversion matrices is conceptually straightforward. For example, it has been shown in (M. Okumura et al., 1993; Telichevesky et al., 1996) that the conversion matrices associated with finite-difference discretizations are given by (apart from a diagonal scaling) the inverse of the matrix $L + \alpha(\omega)U$ where L and U respectively are again the lower and upper block-triangular pieces of the Jacobian (1.8) and α is a frequency-dependent scalar constant.

Noise Analysis. One of the most important applications of the time-varying small-signal analysis is to compute noise in RF circuits. As the noise signals are stochastic process, we are usually concerned with computing quantities such as power spectral densities or noise correlation matrices. If L is a linear operator and x an element in its domain, then it it a general property of linear systems(Papoulis, 1991) that $E\{Lx\} = LE\{x\}$ and likewise for the adjoint operator L^*. Often we have a system with input x, output y, and we wish to calculated a correlation quantity S_y, $S_y = E\{yy^*\}$. This is easy to do, as $E\{yy^*\} = E\{Lx(Lx)^*\} = E\{Lxx^*L^*\} = LE\{xx^*\}L^*$, and so $S_y = LS_xL^*$. For example, if we let $\tilde{x}$ and $\tilde{y}$ represent the input and output respectively of an RF circuit with conversion matrix $\tilde{H}(\omega)$, then the matrices of (cyclostationary) power spectral densities, $S_x(\omega)$ and $S_y(\omega)$ are related by (Roychowdhury et al., 1998) $S_y(\omega) = \tilde{H}(\omega)S_x(\omega)\tilde{H}(\omega)^H$ where superscript H denotes Hermitian transpose. A similar relation can be derived to relate time-domain representations of noise.

Oscillators and phase noise. The accurate estimation of noise in oscillators is of particular interest to RF designers and requires particular care. In (Kaertner, 1990) a microscopic theory that describes calculation of the stationary part of the phase noise spectrum was presented. Recently, a simple phenomenological time-varying theory has appeared (Hajimiri and Lee, 1998), and a more detailed statistical treatment of the microscopic phase noise analysis problem has been performed (Demir et al., 1998). The time-varying linear noise analysis in Section 5. can be used for phase noise calculations, however, care must be taken when interpreting the results, as the linearized analysis is best interpreted as a specific computational procedure used to perform calculations in the context of the more rigorous theories, rather than an independent theory of phase noise.

6. KRYLOV-SUBSPACE SOLVERS

The major computational obstable in RF simulation is to quickly solve the equations that come from linearizations of the discretized circuit DAEs. One innovation that has enabled the simulation of very large circuits is the use of modern linear algebra techniques, in particular, Krylov-subspace based linear system solution algorithms. These algorithms require only the computation of a matrix-vector product in order to solve a linear system. In both the finite-difference and spectral formulations (Telichevesky et al., 1995; Telichevesky et al., 1996; Melville et al., 1995), the matrix-vector products needed by the Krylov-subspace solvers can be computed in time and memory that is nearly linear in the size of the circuit. However, the performance of Krylov-subspace algorithms can be very problem-dependent, and so their effective use requires close integration of linear algebra techniques with a knowledge of the specific simulation problem and choice of the numerical formulation. This is done by means of a technique called preconditioning.

Convergence of the Krylov-subspace algorithms for solving the matrix equation $Ax = b$ is roughly determined by the location of the eigenvalues of A. In RF analysis, the matrix A is a matrix whose blocks are generated from the small circuit Jacobian matrices C and G. Practical circuits usually have widely-varying physical time constants, and so the matrices assembled from C and G tend to be intrinsically ill-conditioned. Furthermore, forming the large matrices that occur in direct solution of the equations of spectral and finite-difference discretizations tends to exacerbate any underlying ill-conditioning.

Fortunately, the matrices resulting from most RF equation formula have a structure that can be exploited in constructing a preconditioner. A typical approach (Melville et al., 1995; Long et al., 1997) is to approximate the time-varying C and G matrices by constant or piecewise-constant approximations. The resulting preconditioner is block-sparse (block-diagonal in the simplest case) with sparse blocks, and so can be easily inverted. Particularly effective preconditioners can be constructed for the sampling formulations (Feng et al., 1999). In this case, it is the frequency-averaged transition function Jacobian that is used to construct a block-diagonal preconditioner. These preconditioners are not as sensitive to nonlinear behavior as their harmonic-balance counterparts because the dominant nonlinear behavior, the response of the circuit to the carrier signal, is hidden by the transition function.

Optimizations to the Krylov-subspace algorithms can often be made for small-signal problems. The key idea is that the Krylov space of a matrix $\alpha I + A$ is invariant with respect to the complex number α. This means that if a linear system is solved with one α, a second linear system can re-use the matrix-vector products that were computed during the first solve thereby accelerating the iterative solution process (Telichevesky et al., 1996). In principle, all the formulation and discretizations lead to small-signal analyses where block or solvers can be applied, however, they are particularly useful for the shooting-based solution of small-signal equations based on finite-difference discretization of a periodic-steady-state operating point. Essentially the shooting method converts the continuous time problem to a discrete time problem, with system matrix $zI - A$, z being the complex frequency.

7. SUMMARY

Advanced numerical techniques have greatly advanced the state-of-the-art in RF circuit simulation over the past few years. Progress in detailed simulation will continue to derive from deeper understanding of the relation between the generic numerical methods described in Sections 4. and 6. with the RF simulation formulations presented in Sections 3. and 5.

8. ACKNOWLEDGMENTS

The authors would like to acknowledge Ken Kundert, Jess Chen, Baolin Yang, Alper Demir, Jaijeet Roychowdhury and Jacob White for useful discussions that have contributed to the preparation of this article.

9. REFERENCES

Allgower, A. and Georg, K. (1990). *Numerical Continuation Methods*. Springer-Verlag, New York.

Brachtendorf, H. G., Welsch, G., Laur, R., and Bunse-Gerstner, A. (1996). *Electrical Engineering*, 79:103–112.

Demir, A., Mehrotra, S., and Roychowdhury, J. (1998). In *Proc. 35th Design Automation Conference*, pages 26–31, San Francisco, CA.

Feng, D., Phillips, J., Nabors, K., Kundert, K., and White, J. (1999). In *Proc. 36th Design Automation Conference*, pages 635–640, New Orleans, LA.

Gilmore, R. and Steer, M. (1991). *Int. J. Microwave and Millimeter Wave Computer Aided Engineering*, 1(1):159–180.

Hajimiri, A. and Lee, T. (1998). *IEEE J. Sol. State Circuits*, 33:179–193.

Kaertner, F. X. (1990). *Int. J. Circuit Theory and Applications*, 18:485–519.

Kundert, K. (1999). *IEEE J. Sol. State Circuits, to appear September 1999.*

Kundert, K. S., White, J. K., and Sangiovanni-Vincentelli, A. (1990). *Steady-State Methods for Simulating Analog And Microwave Circuits*. Kluwer Academic Publishers, Boston.

Long, D., Melville, R., Ashby, K., and Horton, B. (1997). In *Proc. Custom Integrated Circuits Conference.*

M. Okumura, H. T., Itakura, T., and Sugawara, T. (1993). *IEEE Trans. Circuits and Systems - I*, 40:581–590.

Maas, S. (1988). *Nonlinear microwave circuits.* Artech House, Norwood, MA.

Melville, R., Feldmann, P., and Roychowdhury, J. (1995). In *Proc. Custom Integrated Circuits Conference.*

Nastov, O. and White, J. (1999). In *Proc. 36th Design Automation Conference*, pages 641–646, New Orleans, LA.

Papoulis, A. (1991). *Probability, random variables, and stochastic processes.* McGraw Hill, New York.

Roychowdhury, J. (1997). In *Proc. 34th Design Automation Conference*, pages 269–274, Anaheim, CA.

Roychowdhury, J., Long, D., and Feldmann, P. (1998). *IEEE J. Sol. St. Circuits*, 33:324–336.

Telichevesky, R., White, J., and Kundert, K. (1995). In *Proc. 32rd Design Automation Conference.*

Telichevesky, R., White, J., and Kundert, K. (1996). In *Proc. 33rd Design Automation Conference.*

Device Modeling and Measurement for RF Systems

Franz Sischka
Hewlett-Packard Company,Boeblingen Germany

Abstract: The trend to higher integration and higher speed challenges modelling engineers to develop accurate device models up to tens of Gigahertz. An absolute prerequisite for achieving this goal are accurate RF measurements, and correct de-embedding techniques. Without this, RF modelling can become quite time consuming, with a lot of guesswork and ad-hoc judgements. If, however, the underlying measurements are correct, and the models understood well, RF modelling can provide very accurate design kits.

Since the introduction of the first integrated circuit, consisting of a resistor, a capacitor and a resistor by Kilby in 1958, the number of devices per IC increased steadily. As a consequence, the size of integrated transistors increased as well. Since the early 1970ties, the gate diameter of MOS transistors has decreased by roughly a factor of 100, see fig.1. This implies that the gate area has been reduced by a factor of 100_ = 10000. Since the operating frequency of such transistors is depending mainly on its capacitances, it becomes obvious that the cut-off frequency of modern MOS transistors can be far in the Gigahertz range, see fig. 2.

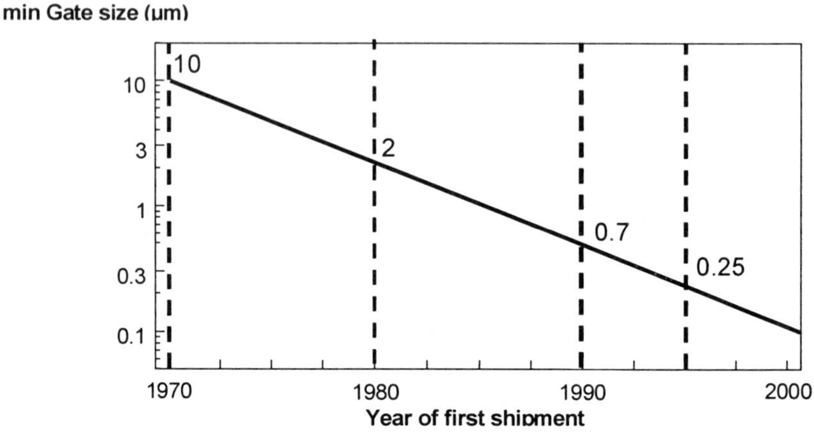

Figure 1: Minimum MOS Circuit Feature Size

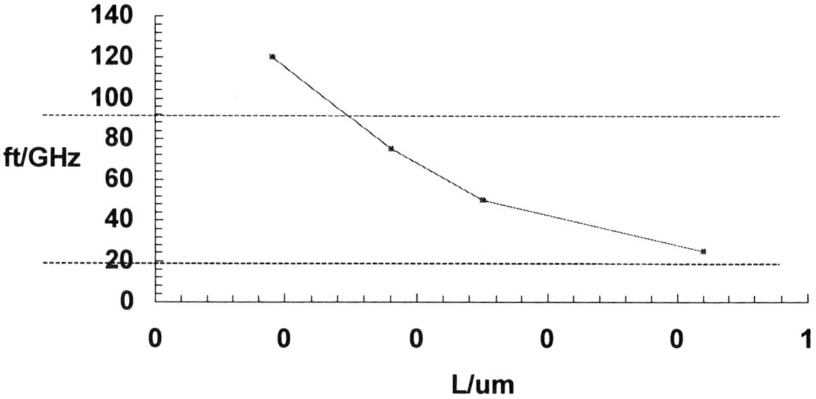

Figure 2: Silicon Transit Frequency Evolution

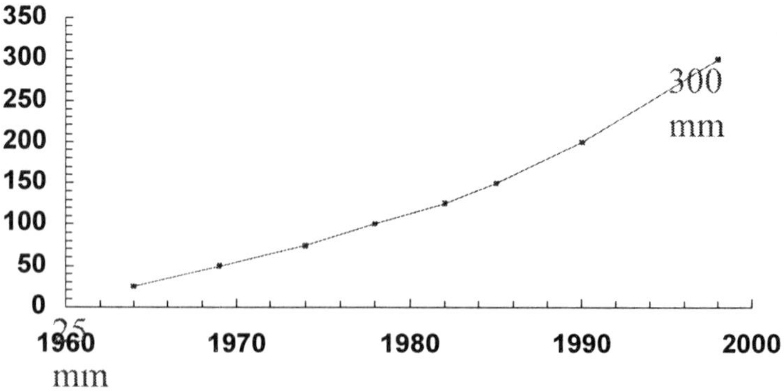

Figure 3: Silicon Substrate Evolution

The interesting effect of this technology evolution is that the silicon wafer diameters increased during this time also drastically (fig.3), what allows today the development of sophisticated high-frequency VLSI chips for and affordable price.

This means: with increasing operating speed, cheaper manufacturing cost, fast turn-around and time-to-market, today's designs become more dependent on good device models. And, as can be seen in figure 4, device models are the base of the complete system design. For the digital design parts, relatively simple yet accurate enough time-dependent models can be developed, which describe basically the delay time of an electrical component, and its input and output characteristics. If, however, the design includes also analogue parts, like RF amplifiers, mixers etc., the weakest part of the design chain becomes the model of the analogue circuit, consisting of transistors and passive components. And, thus, the whole design becomes dependent on good models for these components.

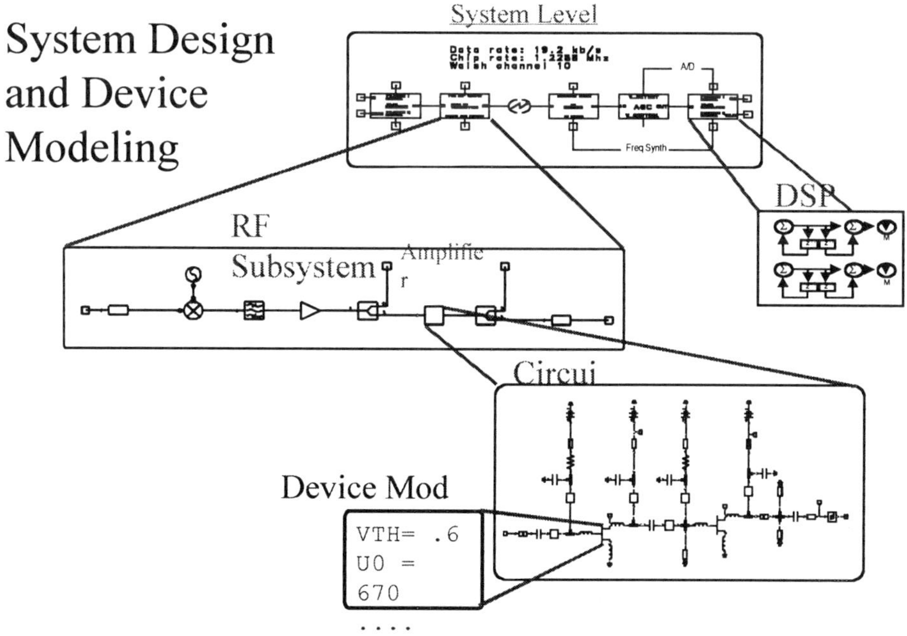

Figure 4: System Design and Device Modelling

In the simplest case, for an ideal ohmic resistor, the device model consists of a single formula, the well-known Ohmic law. And the model parameter is the Ohmic value R, see fig. 5. In order to extract its parameter value, a current is applied to the resistor pins and the voltage drop measured. From a plot of v over i, the calculated slope is equal to the value of R.

$$v = R * i$$

model parameter R

.. is the slope of the resistance curve v(i)

Figure 5: The Device Model of an Ideal Ohmic Resistor

MODEL TYPES:	EXAMPLE:
discrete models:	
physical models empirical models tabular models	individual transistors, diodes, ideal passive components
lumped, composed models:	
sub-circuit models	transistors including parasitics, e.g. npn with paras. pnp
	spiral inductors, varactor diodes, mim capacitors, resistors on silicon
	packages
S-parameter 2-port blocks	if supported by the simulator

Figure 6: Model Types for Electronic Components

For more complex components, the models become more complex too. Figure 6 gives a rough overview. For individual transistors, diodes and ideal passive components, discrete models are available. They can be distinguished between physical models, empirical models and tabular models. Physical models try to include as much as possible equations from device physics. An example is the diode current i=IS*exp(v/vt), characterised by the saturation current IS and the temperature voltage vt. While IS is a model parameter, vt is calculated by the simulator out of the desired device simulation temperature. This physical based type of models can be very accurate for low frequencies, but may become very complex for high frequencies. For this reason, their implementation into simulators is generally a simplification. And this is exactly what causes the problems for RF modelling. The simplification is required in order to allow circuit simulators evaluate the performance of a circuit of thousands of individual circuit components. By this way, the empirical models, based on mathematical equation fitting, come into play. They can also be combined with physical equations. The drawback is, however, that the physical background of a model is lost, and the interpretation of the model parameters becomes doubtful.

On the other hand, tabular models which consist of measured RF data at many bias points, which are interpolated during the simulation, can be a way out of this dilemma. However, they are usually simulator dependent and their migration to another simulator can be pretty difficult.

A commonly used method is to combine discrete models with sub-circuits, when it comes to RF modelling. This way, it is possible to model bipolar transistors including parasitics (e.g. a npn transistor with a parasitic pnp), and MOS transistors using for example the BSIM3 model up to 30GHz and more. In the later case, an additional gate resistor and a small R-L circuitry between drain, bulk and source makes physical sense and helps in achieving this modelling result. Back to the firstly mentioned passive components, i.e. resistors on silicon, spiral inductors, varactor diodes mim capacitors etc., their RF characteristics can be modelled very accurately when using sub-circuits consisting of individual, lumped RLC components. As a side effect, such sub-circuits are simulator independent, and thus can be ported easily to another design environment. This kind of modelling can also be applied to package modelling, a domain which will become more and more important.

Some major RF simulators also allow the import of measured S-parameters for given bias points. This way, cumbersome large sub-circuits which are often required for geometrically big components like packages can be simplified drastically. However, a drawback is the already mentioned reduced portability to other simulators and also the fact that problems with the measurement can easily be hidden behind the S-parameter blocks. Doing modelling the 'conventional' way will in this case lead to either unphysical model parameters or bad fitting performance of the model, and will therefore be detected. Also, a frequency-dependent model parameter in the case of sub-circuit modelling is mostly a hint for a need to improve the equivalent sub-circuit. An experienced modelling engineer can read out of this dependency the required sub-circuit enhancement.

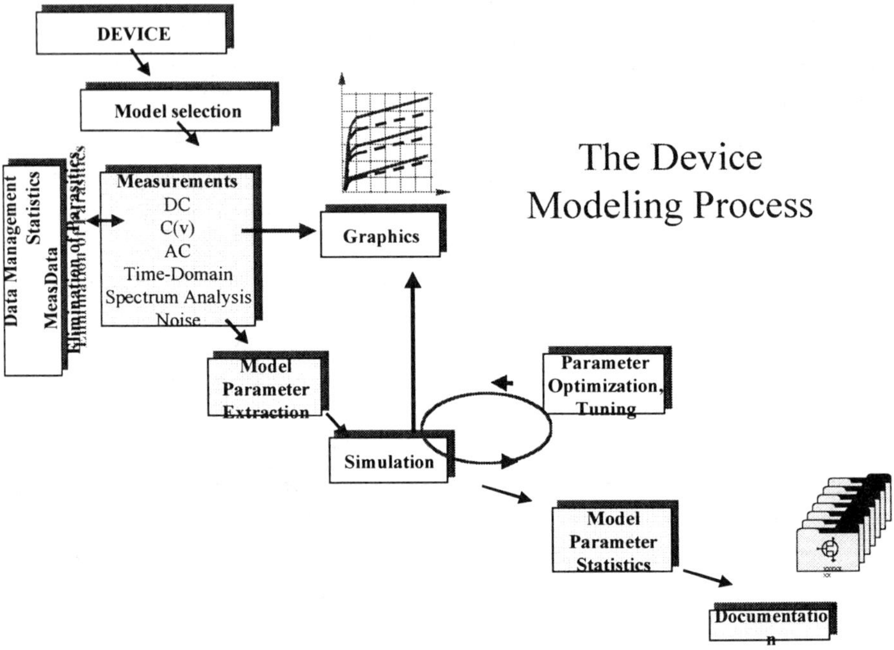

Figure 7: The Device Modelling Process

Figure 7 depicts the global device modelling process. For a given device, a model is selected first. The model equations, which are solved for the model parameters during model parameter extraction, give a clear indication about what kind of measurements and what type of stimulus sweeps are required for characterisation. During the parameter extraction process, the selected model will fit more and more precisely the measured device. This is done by starting with so-called default model parameters, not to be confused with 'typical' parameters. These default parameter values are usually set so that they behave neutral. As an example, a default parameter for a capacitor is zero. In the case of a bipolar transistor, the first model parameters extracted are the DC parasitics RE and RC, followed by the Early voltage and the current amplification beta. The other parameters are still set to default. The so far determined model is a simple Ebers-Moll model, with bias-independent beta, and no frequency dependencies. Step by step, more and more parameters are determined and the model becomes more and more precise with respect to DC and finally RF. Figure 8 depicts this scheme. For a general transistor modelling, the DC performance with respect to its input and output characteristics as well as its transfer function is done first, followed by the so-called CV modelling, i.e. the characterisation of the depletion capacitances. Finally, the S-parameters are measured and modelled by the transit time parameters.

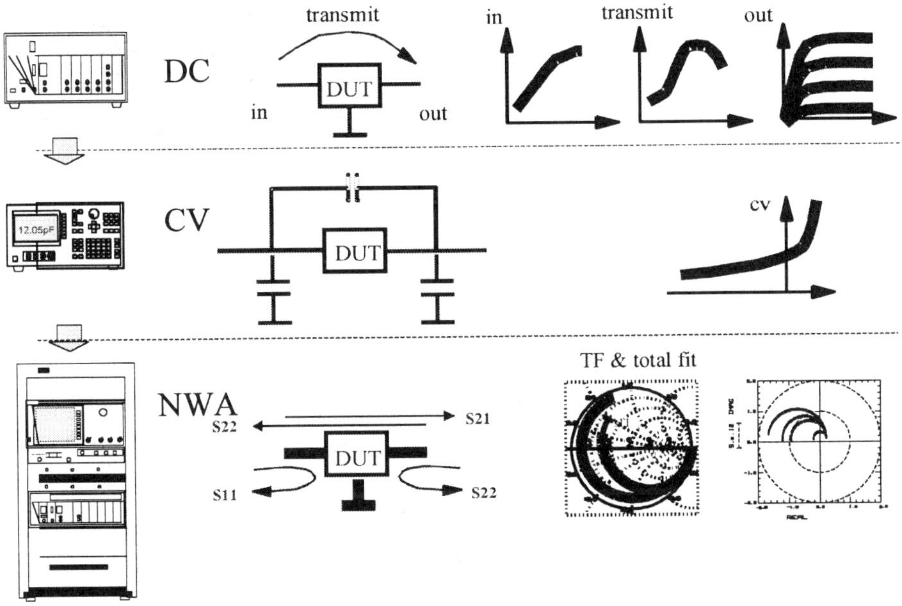

Figure 8: Basic Device Modelling Measurements

Before leading over to the pure RF modelling aspects, figure 9 shows the dilemma to identify the right device for modelling. This means that before an individual component is 'just' selected to serve as a model, it must be assured that this component represents a typical device. Therefore, before doing modelling, some statistical analysis has to be applied to the measured data in all measurement ranges. This is called measurement data management.

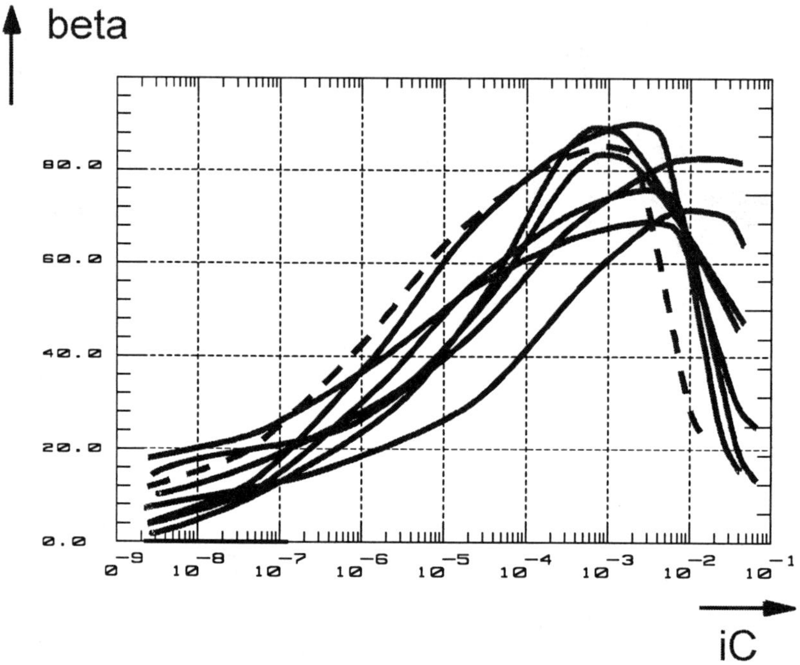

Figure 9: Measured beta traces of bipolar transistors on a wafer

It is also recommendable to perform a measurement for as many DC sweeps as possible and a full frequency S-parameter sweep with as many as possible DC bias conditions. Although this 'over-measurement' may not be required for modelling itself, it is very valuable when comparing the set of simulated data after the model has been established with these many measurement data and to analyse the percentage error over all the bias conditions. Only if this is performed, the developed model can guarantee the desired model accuracy.

Provided the model obtained parameter set as well as the possibly required sub-circuit makes physical sense, the model can be assumed to be accurate also for those operating conditions which may not have been measured for modelling.

RF modelling aspects:

Like sketched in figure 10, S-parameters may easiest be compared to a beam of light, which hits a pair of spectacles. A part of the incident beam is reflected, and the other part is transmitted. With S-parameters, the device under test is tested similarily: first its input characteristics and transfer characteristics from port 1 to port 2 (output) is characterised, and then the same is done backwards. Such a way, it's RF characteristics is completely evaluated.

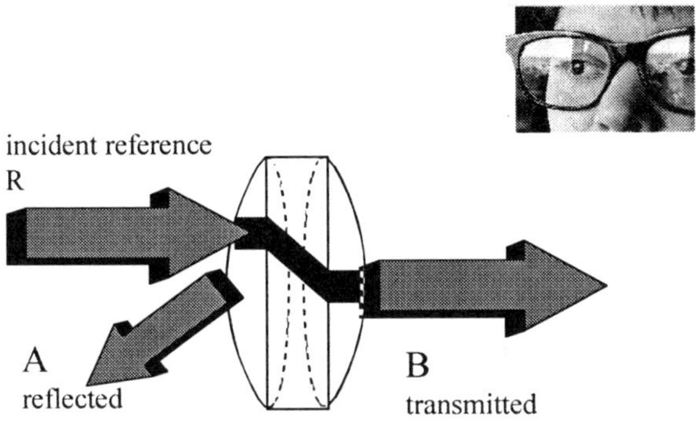

Figure 10: S-Parameter Analogy

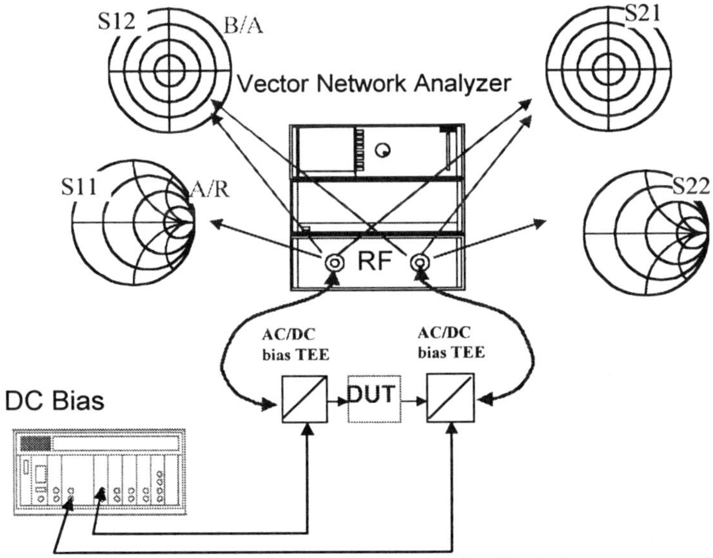

Figure 11: RF Characterisation Techniques

Figure 11 depicts the required measurement setup. The DC and AC signal is combined by the AC/DC bias TEEs. It must be assured that there is no overdrive of the transistor under test, because the network analyzer is a linear system. Therefore, the forward transfer curve must be measured with a much smaller AC signal than the reverse case. From the input reflections of the DUT, S11 is measured as the ratio of incident power to reflected power, i.e. A/R of figure 10. The transfer characteristics is obtained by the network analyzer by calculating the ratio B/A. A second measurement in reverse mode concludes the DUT characterisation. It should be mentioned that for this kind of measurements, special test structures are required. An example is given in fig. 12. Because of the RF frequencies, the test pins also need to include a proper ground connection in order to connect the ground shield of the RF cables properly and with an extremely short way to the DUT. If this is not done properly, self-oscillation of the DUT may occur and the modelling will fail.

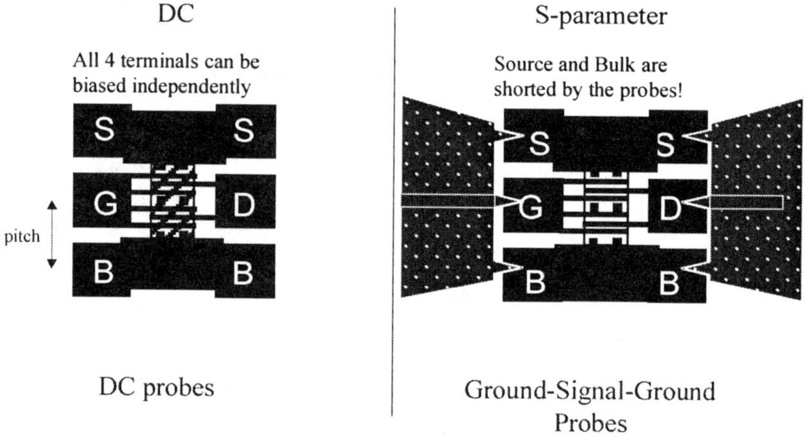

Figure 12: RF modelling measurements require specific test device layout

It is also important to note that the starting points of the S-parameter traces for low frequencies are completely determined by the DC characteristics. For the modelling, this means that the starting points of the simulated traces are in the same way determined by the DC model parameters. The RF transistor parameters only influence the further trace of the S-parameter curves. Therefore, in order to obtain an accurate RF large signal model, a proper DC modelling is a must.

Another prerequisite for good RF models is an accurate de-embedding of the RF parasitics. Firstly, they stem from the network analyzer (cross-talk between the ports, frequency dependence of the port amplifiers etc.), and the delay of the measurement cables. These effects, however, can be eliminated by proper network analyzer calibration. The remaining parasitics, like the on-chip test pad capacitances, as well as parasitic inductances for frequencies above about 10GHz, need to be eliminated by de-embedding. Otherwise, the measured device characteristics is not that of the very inner device alone, but also includes the overlying parasitics. Even if the model fits well, its parameters are physically incorrect, because they reflect both, the very inner device plus the outer parasitics. Since these parasitics will not exist in the later RF design (e.g. the test pads), they need to be stripped off. This procedure is called de-embedding. In most cases, this is done by twoport matrix manipulations. As an example, capacitances, representing basically the test pads, and which are measurable with an OPEN dummy, can be subtracted by subtracting their Y matrix from the total measurement. Consequently, a SHORT dummy measurement, de-embedded first (!) from its outer test pads, represents the delay from the pin contacts to the inner device. Its parasitics can be eliminated by subtracting the SHORT dummy Z matrix from the total measurement. As stated before, the SHORT has to be de-embedded from the outer OPEN parasitics first.

Figure 13 shows the effect of overlying parasitics for a cut-off frequency measurement, and figure 14 sketches the de-embedding process and the required twoport matrix manipulations.

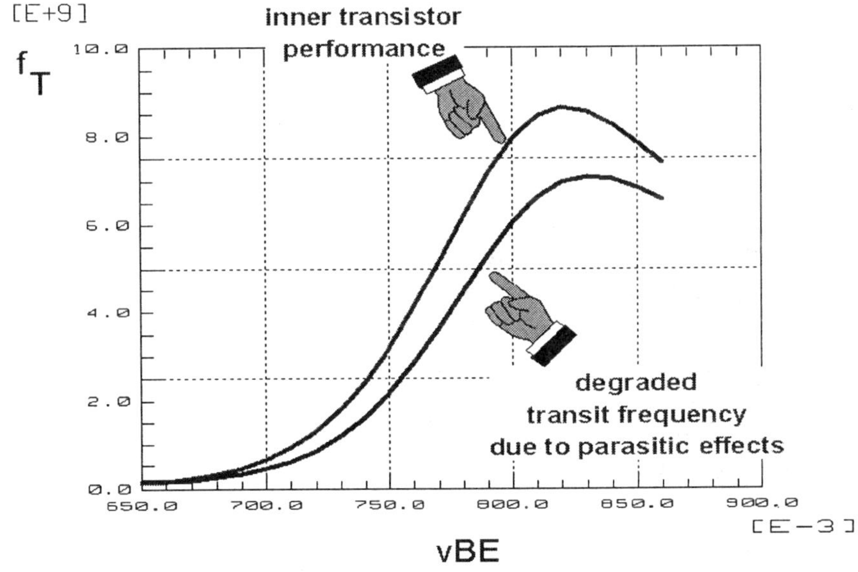

Figure 13: De-embedding Means Off-Stripping of Overlaying, Measurement Related Parasitic Effects.

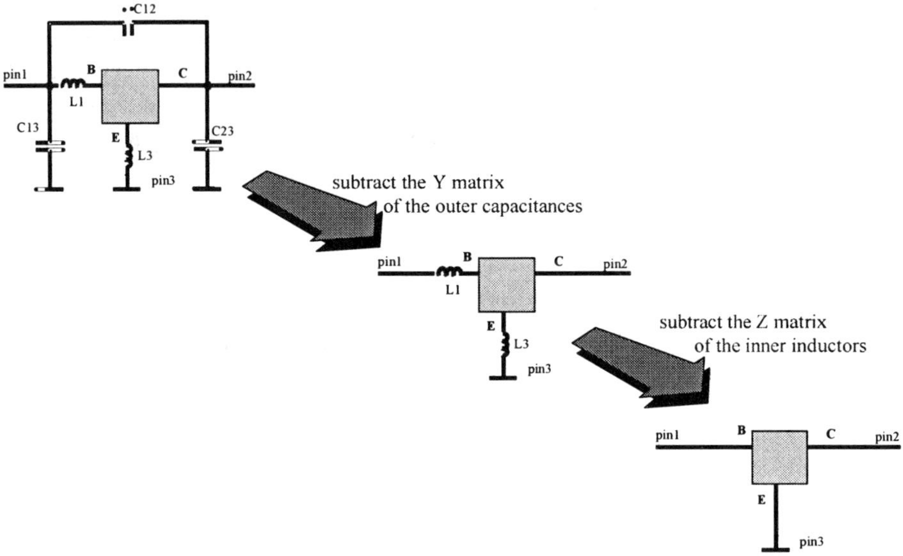

Figure 14: The De-embedding Procedure

As mentioned before, after the de-embedding, the modelling of the RF performance of transistors is basically the correct determination of the transit time model parameters. If the model exhibits some RF model limitations, additional small sub-circuits can be used to match the frequencies in the higher RF range.

If only a transistor model for a fixed DC operating point is required, a small signal model might be sufficient. Figure 15 gives an example. The components of the model branches are defined by evaluating the de-embedded Y matrix of the inner transistor and by displaying the inverse in the complex impedance plain. Such a way, the required branch structure and the parameter values of its components become clearly visible and can be modelled accordingly.

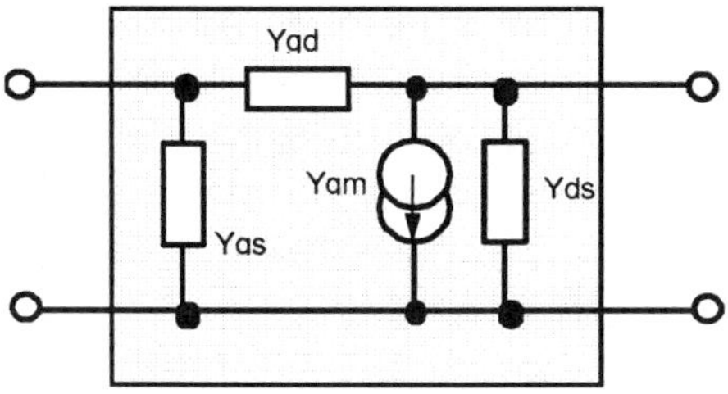

$$\begin{pmatrix} Y11 & Y12 \\ Y21 & Y22 \end{pmatrix} = \begin{pmatrix} Ygs+Ygd & -Ygd \\ Ygm-Ygd & Yds+Ygd \end{pmatrix}$$

Solved for the complex conductances:

$$Ygd = -Y12$$
$$Ygs = Y11 + Y12$$
$$Yds = Y22 + Y12$$
$$Ygm = Y21 - Y12 = gm * \exp(-j\omega TAU)$$

Figure 15: Operating Point RF Modelling for Transistors

When it comes to passive components modelling, be it on-chip or packaged, the standard models of inductors, capacitors and resistors are not sufficient for RF. Therefore, sub-circuit modelling or the development of custom RF models for those components is required. Good RF simulators allow such model development. However, migration of the model to other types of simulators will become difficult. For passive components, the schematic and the parameter values are extracted again from the complex impedance plain.

$$\begin{pmatrix} Y11 & Y12 \\ Y21 & Y22 \end{pmatrix} = \begin{pmatrix} Ygs+Ygd & -Ygd \\ Ygm-Ygd & Yds+Ygd \end{pmatrix}$$

$$\underline{Z}gs = \frac{1}{Y11 + Y12}$$

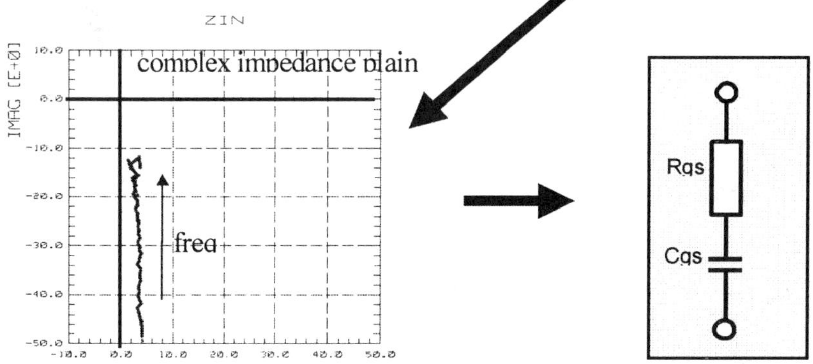

Figure 16: Developing the Schematic and Extracting the Component Parameter Values out of the Complex Impedance Plain

Last not least, the noise performance needs to be modelled. For basically all transistor types, the 1/f noise model is included in the device models and the corresponding parameters can be extracted. Like with the RF modelling, an accurate measurement is a prerequisite for successful noise modelling. Some modern modelling tools also allow the modelling of the non-linear RF performance like 3rd order intersect and 1dB compression point. Such type of modelling ensures the best model accuracy possible.

Conclusions:

Provided the RF measurements and the measurement data handling has been performed accurately, and the DC modelling is correct, experienced modelling engineers can generate accurate device models up to the high RF frequency range. A pre-requisite is good experience in RF measurement techniques and a flexible RF modelling tool, which supports the engineer with the development of a suitable RF model and the extraction of the corresponding model parameters.

Reconfigurable Computing: Viable Applications and Trends

Alexandro M. S. Adário, Sergio Bampi
{adario, bampi}@inf.ufrgs.br
Instituto for Informatics – Federal University at Porto Alegre, Brazil
Av. Bento Gonçalves, 9500 – Porto Alegre(RS) Cx. Postal 15064 – CEP: 91501-970

Key words: Reconfigurable architecture, image processing, FPGA

Abstract: The field of reconfigurable computing has experimented with different architectural and execution models in the last decade. As a radically new architecture approach, reconfiguration has been hindered by a lack of a common development framework and even an unified taxonomy to define the dynamics of the reconfiguration of the hardware. This paper reviews the taxonomy for the field and the previous experiments, which have clearly demonstrated the efficiency of the reconfiguration when, compared with general-purpose processors. The performance and power saving gains possible with dynamic reconfiguration are very high when compared to statically configured architectures. An experiment done with image processors with and without datapath reconfiguration is described, revealing the high area saving achievable in the case of neighborhood processors. The lack of clearly competitive and highly persive applications for the reconfigurable processors has hindered the necessary rethinking of the basic FPGA matrix architectures, as well as the development of a software framework for supporting the reconfiguration paradigm over a wider range of algorithms and applications.

1. INTRODUCTION

In the early 60s, Gerald Estrin proposed the preliminary ideas of the field, which came to be known today as reconfigurable computing [Est63]. Estrin designed a "variable" system, consisting of 3 elements: a general-purpose computer, a variable hardware and a supervisory control unit. The variable hardware in this system allows the implementation of an arbitrary digital function, in a similar way a FPGA (Field-Programmable Gate Array) [Bro92][Fie94]can perform today. The concepts put forward by Estrin were far ahead of the implementation technology available at that time.

Until recently, reconfigurable architectures implemented in user-configurable devices were not feasible, but with increasing levels of FPGA integration (above 200K usable equivalent gates), as well as RISC cores and RAM merging into the reconfigurable arrays, the feasibility picture for such systems has changed dramatically.

This work presents a general overview of the advances in reconfigurable computing. An outline of the main differences between programmable and configurable circuits is initially provided. Dehon sets a useful definition [Deh96], always emphasizing that there are not clear boundaries between the sets of programmable and reconfigurable devices, but it is possible to distinguish the cases at the extremes:

a) **Programmable** – this term refers to architectures which heavily and rapidly reuse a single functional unit for many different functions. The canonical example is a conventional processor, which may perform a different instruction on its ALU every cycle.
b) **Configurable** – in the configurable devices, the active circuitry can perform any of a number of different functions, but the function can be or is changed in successive execution cycles. Once the device is configured to implement a function, it does not change during an operational period. The FPGA devices are the canonical example.

The following section of this paper discusses the motivation for the reconfiguration paradigm, followed by section 3, which gives an account of the possible application areas to explore. Section 4 reviews previous relevant developments and experiments in reconfigurable computing, while section 5 and 6 set the taxonomy applicable to the hardware execution model and the programmability classes, respectively. Finally in section 7 an experiment comparing the application of hardware reconfiguration principles to a previously designed architecture for image processing is presented, showing the potential of this hardware design paradigm.

2. WHY RECONFIGURABLE ARCHITECTURES?

Advanced RISC microprocessors can solve complex computing tasks through a programming paradigm, based on fixed hardware resources. For most computing tasks it is cheaper and faster to develop a program in general-purpose processors (GPPs) specifically to solve them. While GPPs are designed with this aim, focusing on performance and general functionality, total costs of designing and fabricating RISC GPPs are increasing fast. These costs involve three parts:

a) **Hardware costs**: GPPs are larger and more complex than necessary for the execution of a specific task.
b) **Design costs**: functional units that may be rarely used in a given application may be present in GPPs, and may consume substantial part of the design effort. Considerable effort is given to increment the potential instruction level parallelism through superscalar control, so that the design effort is increasingly complicated by the demand for performance in the execution of potential application codes which are not known in advance
c) **Energy costs**: too much power is spent with functional units or blocks not used during a large fraction of the processing time. Solutions like clock slow-down, multiple internal supply voltages, and power-down are implemented in power managers of GPPs.

For specific applications or demanding requirements in terms of power, speed or costs, one may rely on either dedicated processors or reused core processors, which may be well suited to the application or optimized for a given set of performance requirements. In the former case, only the necessary functional units highly optimized for a specific range of problems may be present, which will result in unsurpassed power and area efficiencies for the application-specific algorithm. Until recently, application-specific processors (ASPs) implemented in configurable devices were not feasible, but with increasing levels of FPGA integration, as well as RISC cores and RAM merging into the reconfigurable arrays, the feasibility picture for user-configured ASPs looks excellent, open the opportunity to break away from the general purpose processor paradigm.

Coupling a single one or a set of configurable devices to a GPP makes possible the exploitation of reconfigurable architecture. This structure can also aggregate some local or global-shared memory. The dynamic reconfiguration of the hardware has become a competitive alternative in terms of performance against a GPP software implementation, and it offers significant time-to-market advantage over the conventional ASP approach, in which a full-custom processor or ASIC has to be developed from specification to layout and subsequent fabrication.

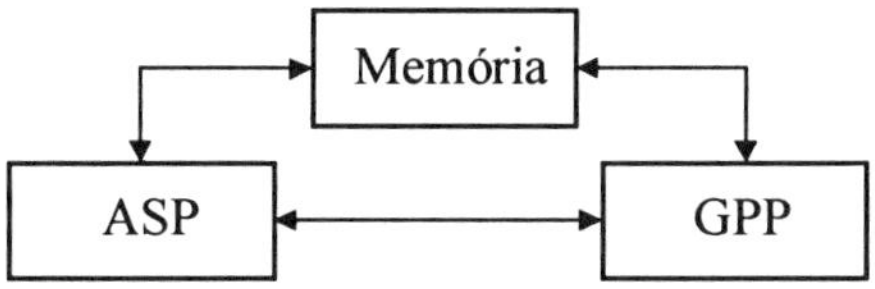

Figure 1. Generic Model of a Reconfigurable Architecture

A reconfigurable architecture allows the designer to create new functions and to perform in customized hardware the operations that would take too many cycles to complete in a GPP. The GPP in *Figure 1* may not include most of the complex functional units often designed-in, as well as considerable power and time can be saved. These units may be implemented on the configurable device, if and when demanded by the application.

3. APPLICATION AREAS AND MARKET CONSIDERATIONS

An important obstacle has hindered the commercial maturity of the reconfigurable hardware platform to date: no killer application demanding a large-volume production was found for this architectural model. A candidate application for a reconfigurable implementation must fulfill some basic characteristics: regularity, high concurrency due to data parallelism, small data operand size, and ability to execute in a pipeline organization with a simplified control.

Considering these characteristics, one may point out that some application which can have some advantage when using reconfigurability are: data encryption, decryption and compression; sequence and string matching; sorting; physical systems analysis and simulation; video and image processing; and specialized arithmetic.

A commercial system based on reconfigurable computing, which may include some of the applications mentioned above, can offer new features to the users and change the way the organizations solve their computing needs:

a) The system can aggregate the most recent trends imposed by consumers, because the design time is reduced and, in general, changes in design specification have less impact than on the traditional design approaches.
b) The savings in design/implementation time considerably reduce the time to market of entire systems.
c) The client support and assistance is easier, once the upgrades and bug fixing can be done remotely at a software level, just sending a new configuration.
d) The development process is more controllable and subject to parameterization, since the design cycle relies on software specification and software tools that follow the previously defined architectural model chosen.
e) Considering the end-user market, the number of interface boards in a conventional system can be reduced, once some applications are mutually exclusive or do not execute at the same time.

4. EXPERIMENTS ON RECONFIGURABLE COMPUTING

Several reconfigurable architectures were designed in the last decade, showing the feasibility of this architectural approach: PAM [Ber89], SPLASH [Gok90], PRISM [Ath93], DISC [Wir95], Garp [Hau97] and NAPA1000 [Nat99].

PAM (Programmable Active Memories) is a project developed by DEC PRL and consists of an array of Xilinx FPGAs (25 XC3020 in the first version, Perle-0, and 24 XC3090 in the Perle-1[Vui96]). With dynamic reconfiguration, it has demonstrated the fastest implementation of RSA cryptography to that date [Sha93]. The set of application implemented using PAM also includes long integer arithmetic, computational biology, resolution of Laplace´s equations, neural networks, video compression and image acquisition, analysis and classification.

SPLASH is a reconfigurable systolic array developed by Supercomputing Research Center in 1988. The basic computing engine of SPLASH is the Xilinx XC3090 FPGA, which composes each one of the 32 stages in the linear array. The second version of SPLASH, Splash 2, is a more general-purpose reconfigurable processor array based on XC4010 FPGA modules. Each module is a processor element in the 16-module basic array boards. Splash2 can include until 16 array boards. The main application tested on the SPLASH architecture was related to computational biology.

PRISM is the acronym for Processor Reconfiguration through Instruction Set Metamorphosis and is a reconfigurable architecture for which specific tools have been developed such that, for each application, new processor instructions are synthesized. The tools for the PRISM environment use some concepts inherited from hardware/software codesign methods. The hardware/software partition process starts with a high-level specification using C, and is user-guided. Two prototypes, PRISM-I and PRISM-II, have been built using Xilinx XC3090 and XC4010 FPGAs, respectively.

The Dynamic Instruction Set Computer, DISC, is a processor that loads complex application-specific instructions as required by a program. DISC uses a National Semiconductor CLAy FPGA and is divided in two parts: a global controller and a custom-instruction space. The instruction loading process is similar to the virtual memory mapping scheme: when a instruction is necessary, the global controller configures it and removes the unnecessary instructions if the custom-instruction space is full. Initially, a library of image processing instructions was created for DISC.

Garp is a reconfigurable architecture that sets a trend to incorporate RISC cores and FPGA arrays in the same die. It incorporates a MIPS-II instruction-set compatible core with a reconfigurable array that may implement co-processor functions as a slave computational unit located on the same die of the processor. Garp simulation results have shown a 24X speed-up over a software implementation in a UltraSparc 1/170 of the DES encryption algorithm. In an image dithering algorithm for a 640x480 pixels frame the speed up obtained by Garp was 9.4 times.

NAPA1000 is an architecture that integrates a 32-bit RISC processor (the NSC CompactRISC), a configurable logic structure, called ALP (Adaptive Logic Processor), memory and a ToggleBus network interface. The ALP is a 2D array of logic cells, dynamically and partially reconfigurable. The network interface allows the NAPA1000 to be connected to 31 other NAPA1000. This architecture tried to combine three relevant architectural paradigms: scalar processing, reconfigurability and parallel computing. The prototyping and commercial introduction of NAPA was eventually discarded due to the killer-application effect.

5. EXECUTION MODELS

Page [Pag96] points out five design strategies by which programs may be embedded in reconfigurable architectures focusing on the relation between algorithm and implemented hardware: pure hardware, application-specific, sequential reuse, multiple simultaneous use and on-demand usage are the possible models for executing in hardware the computational task at hand.

In a **pure hardware model**, the algorithm is converted into a single hardware description, which is loaded into the FPGA. There is no relevant contribution of this model to reconfigurable architectures, since the configuration is fixed at design time and it never changes along the execution of the application. This model can be implemented using conventional HDLs and the currently available synthesis tools. The commercial maturity of FPGAs has thriven under this model, today widely practiced by hardware engineers.

PRISM is an example of an **application-specific microprocessor (ASMP)** execution model. In this system, the algorithm is compiled into two parts (*Figure 2*.a): an abstract machine code and an abstract processor. In the next step, the two are optimized to produce a description of an ASP and the machine code level algorithm implementation.

Very often an algorithm is too large to be implemented on the available devices or the design is area constrained by engineering or economic reasons. To overcome this constraint, the design is splitted into several parts, which are moved in and out of the devices, increasing the hardware density and producing a set of reconfiguration steps (*Figure 2*.b). This model of hardware execution of an algorithm execution is called **sequential reuse**, as only one of each Hw1, Hw2,..., Hwn configuration is present in the FPGA at any given time of the execution.

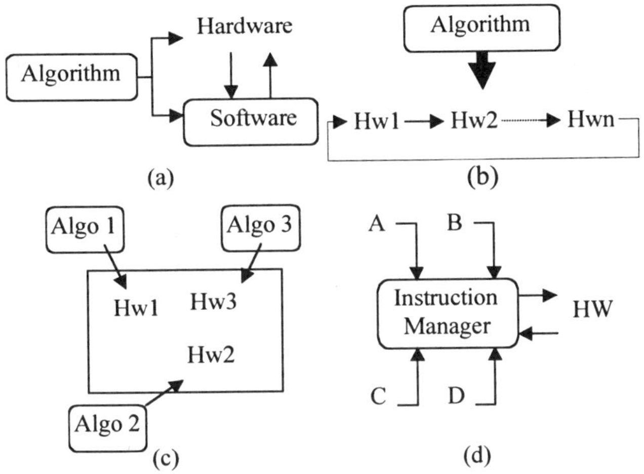

Figure 2. Example of Execution Models

If there is a large availability of configurable devices, many algorithms can be resident and execute simultaneously, interacting with various degrees of coupling (tightened or loose) with the host processor. The **multiple simultaneous use model** (*Figure 2*.c) is less common, requires more area than the sequential reuse, but is certainly very practical and amenable to the growing capacity of the FPGAs and the vastly different levels of demand for hardware of different algorithms. One may argue that most often than not it is the lack of memory capacity that prevents the full exploration of the FPGA resources for a given algorithm.

The last model, **on demand usage** (*Figure 2*.d), is very interesting for reconfigurable computing, and can be applied to a wide range of applications. This model is suitable for real-time systems and systems that have a large number of functions or operations, which are not used concurrently. The proposed dynamic instruction set computer (DISC) follows this strategy of configuring on-demand.

6. PROGRAMMABILITY CLASSES

The execution models presented by Page can be analyzed from the point of view of the reconfigurable architecture design. This classification divides the design models in programmability classes, considering the number of configurations and the time in which reconfiguration takes place:

a) **Static design (SD):** The circuit has a single configuration, which is never changed neither before nor after system reset. The configurable device is fully programmed to perform only one functionality that remains unchanged during system lifetime. This class does not exploit the reconfiguration flexibility, taking advantage only of the implementation/prototipation facilities.
b) **Statically reconfigurable design (SRD):** The circuit has several configurations (N) and the reconfiguration occurs only at the end of each processing task. This can be classified as run-time reconfiguration, depending on the granularity of the tasks performed between two successive reconfigurations. The configurable devices are better used and the circuit can be partitioned in this way, aiming for resources reusability. This class of architecture is called SRA (statically reconfigurable architecture).
c) **Dynamically reconfigurable design (DRD):** The circuit also has N configurations, but the reconfiguration takes place at runtime (RTR, Run-Time Reconfiguration). This kind of design uses more efficiently the reconfigurable architectures. The timing overhead associated to this RTR procedure, during which no useful computation is done, has to be well characterized within the domain of the possible set of run-time configurations. The overall performance will be determined by the overhead-to-computing ratio. The implementation may use partially configurable devices or a set of conventional configurable devices (when one process, the others are reconfigured). The resultant architecture is called DRA (dynamically reconfigurable architecture).

SRD and DRD run-time reconfiguration advantages depend largely on the specific algorithm and its partition in sizable grain tasks. The reconfiguration overhead is heavily dependent on the FPGA microarchitecture, and it will be significantly decreased the convenient – yet costly – integration of FPGA + RISC core + SRAM within the same die. This set of on-chip functions provides for a fertile field of recent [Hau97] and upcoming innovations. The SRD hardware will certainly show better performance when compared to GPP software implementation, given the large time overhead incurred for reconfiguration in current commercial FPGAs. The DRD hardware will benefit the most from innovations in the fast reconfiguration arena, while requiring significant more effort in developing compiler optimization. The set of software support tools (run-time OS, profilers, compilers, loaders) necessary for this DRD represents considerable design effort. An effort already spent for the commercial GPP support, but not yet for the upbringing of the dynamically reconfigurable platforms.

7. A RECONFIGURABLE ARCHITECTURE FOR IMAGE PROCESSING

DRIP (Dynamically Reconfigurable Image Processor) [Ada99] is a reconfigurable architecture based on the pipelined NP9 [Ada97], which is a static-design image processor based on the neighborhood processor architecture proposed by Leite [Lei94]. A neighborhood processor is a special computing device that simulates an array processor, a special class of parallel architecture consisting of simpler processors elements (PE)

The design goal for DRIP is to produce a digital image processing system using a dynamic reconfiguration approach, initially in a SRD scheme. Based on a previous NP9 design, done with conventional FPGAs in a static design scheme, we were expecting to obtain a considerable improvement in performance, with an operating frequency of 32 MHz for real-time processing under a pipeline organization.

7.1 The processor elements

The PEs (*Figure 3*.a) of NP9 are functionally simple and can execute just two basic operations: addition and maximum. Each PE has two inputs (pixels X1 and X2), two weights (W1 and W2) associated to those inputs and one output S. In the current model of the PE, weights can assume only three values: -1, 0 and 1. The DRIP/NP9 PEs are interconnected according to a data flow graph (*Figure 3*.b) which is based on a sorting algorithm..

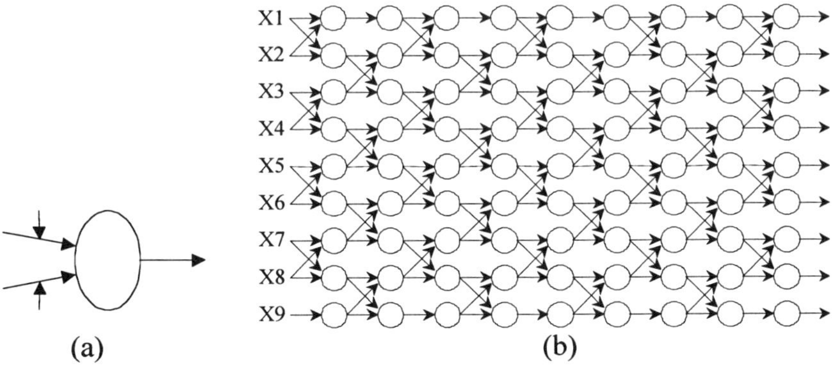

Figure 3. Model of the NP9/DRIP Processor Element(a) and Data Flow Graph (b)

DRIP architecture works with customized PEs. In the set of 18 possible PE functions, one may remove the redundant functions: those are equivalents or symmetrical on its inputs, thus reducing the effective number of functions to 8. Each highly optimized function is a component of a basic function library.

7.2 Design Flow

Typical design flow of the configuration of an algorithm onto DRIP is shown in *Figure 4*. First, an image processing algorithm is specified and simulated to verify its functionality. Later, the algorithm is optimized and translated to an intermediate representation, which matches the DRIP architecture.

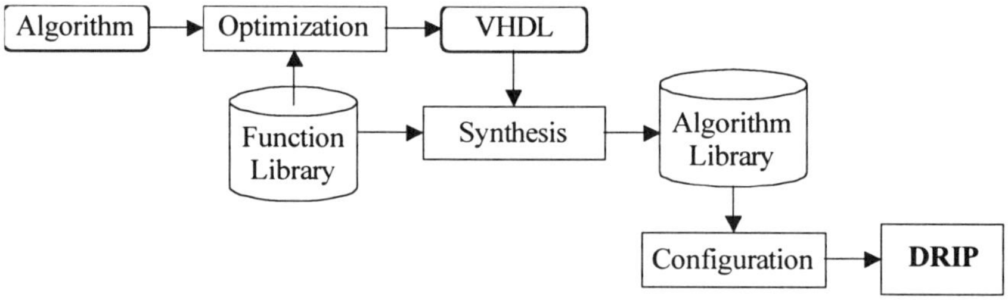

Figure 4. Design Flow of DRIP System

The synthesis process uses the previously designed function library, fully optimized for a better performance. The configuration bitstream that customizes the FPGA for the algorithm implementation is stored in a configuration library. The reuse of the modules of this library is essential for efficient implementation of several image processing functions. Once the configuration bitstream data is stored, it can be used repeatedly and over several modules of the entire architecture (*Figure 5*).

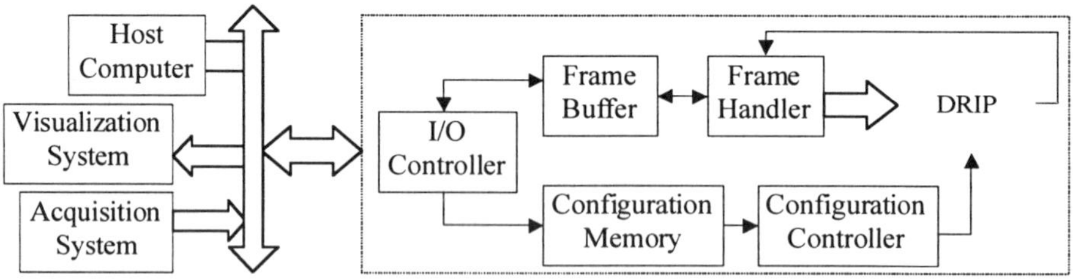

Figure 5. A Digital Image Processing System using DRIP

In the complete digital image processing system, DRIP is part of a more complex system, consisting of two memories (configuration memory and frame buffer, which stores a local copy of the image), the frame handler (which generates the neighborhood), and the configuration and I/O controllers. The most critical issue for the overall performance is bus bandwidth from the frame buffer to DRIP. The best solution is to design a single chip containing DRIP and the buffer handler, also including substantial sub-frame buffer memory.

7.3 DRIP Results with Dynamic Reconfiguration

The resource utilization in the NP9 design was inefficient, and the processor used 6,526 logic elements (LEs) in two FPGA devices (1 Altera Flex10K70 and 1 Flex10K100). DRIP achieved much better results for the same implementation technology, using only 1,113 LEs of a Flex10k30. This is 83% less area (for the worst-case configuration of DRIP) than the NP9 processor.

The target performance of the system is computed for real-time processing, considering 256 gray-level digital images of 1,024 x 1,024 pixels at a rate of 30 frames/s. The preliminary estimated performance of DRIP (51.28 MHz) is 60% greater than the design target performance (32 MHz) and almost 200% faster than the fixed hardware (non-reconfigurable) NP9 implementation (17.3 MHz).

8. CONCLUSIONS

There are plenty of experimental demonstration that the run-time hardware reconfiguration offers a significant performance advantage over the GPP software implementation. The main barriers for the commercial maturity of this technology were identified; on the practical side it is clear that the lack of a killer-application is an obstacle to the introduction of the significant developments that are needed in this field. In particular the pressing needs are: 1) the development of powerful software tools that support the mapping of high level language specifications into a runtime environment that automatically partitions the reconfigurable modules of the hardware. Co-synthesizers, compilers, linkers, and run-time support are complex if geared towards more general application codes; 2) on the hardware side, the development of FPGA families with fast reconfiguration capability, as the DPGA case, as well as considerable more memory mega-blocks to support a larger set of applications which demand both reconfiguration memory as well as large data sets. The image processing DRIP is a case in which the addition of considerable more memory benefits the application. The GPPs themselves are reserving increasing amounts of area to caching data and instructions closer to the decode/execute units; the shortcoming of currently commercial FPGAs is their lack of memory megablocks and of support for fast and/or partial reconfiguration. If FPGAs would incorporate those features, performance of DRD would then become even more appealing when compared to programmable GPP solutions.

9. REFERENCES

[Ada97] ADÁRIO, A. M. S.; CÔRTES, M. L.; LEITE, N. J. "A FPGA Implementation of a Neighbourhood Processor for Digital Image Applications" In: 10 Brazilian Symposium on Integrated Circuit Design, Gramado (RS), Brazil, Aug 1997 p. 125-134.

[Ada99] ADÁRIO, A. M. S.; ROEHE, E. L.; BAMPI; S. "Dynamically Reconfigurable Architecture for Image Processor Applications". In: 36 Design Automation Conference, New Orleans, June 1999. p 623-628.

[Ath93] ATHANAS, P.; SILVERMAN, H. F. "Processor Reconfiguration Through Instruction Set Metamorphosis". IEEE Computer, Mar 1993. p 11-18.

[Bro92] BROWN, Stephen. D.; FRANCIS, Robert J.; ROSE, Jonathan; VRANESIC, Zvonko G. Field-Programmable Gate Arrays. Kluwer Academic Publishers, 1992

[Ber89] BERTIN, P.; RONCIN, D.; VUILLEMIN, J.. Introduction to Programmable Active Memories. Paris: Digital Equipment Corporation, Paris Research Lab, June 1989. (PRL Report 3).

[Deh96] DEHON, A.. Reconfigurable Architectures for General-Purpose Computing. Massachusetts Institute of Technology, 1996. PhD Thesis. (AI Technical Report 1586)

[Deh97b] DEHON, A.. Role of Reconfigurable Computing. Berkeley: University of California. (http://www.reconfig.com/specrept/101596/session1/library/amdrc.htm (1997).

[Est63] ESTRIN, G.; BUSSELL, B.; TURN, R. "Parallel Processing in a Restructurable Computer System". IEEE Transactions on Electronic Computers, vol. EC-12, n. 5, Dec 1963. p 747-755.

[Fie94] FIELD-PROGRAMMABLE GATE ARRAY TECHNOLOGY. Boston: Kluwer Academic, 1994. Editor: TRIMBERGER, Stephen M.

[Gok90] GOKHALE, M. et al. "SPLASH: A Reconfigurable Linear Logic Array." In: International Conference on Parallel Processing, Aug, 1990. p 526-532

[Hau97] HAUSER, J. R.; WAWRZYNECK J. "Garp: A MIPS Processor with a Reconfigurable Coprocessor". In: IEEE Symposium on FPGAs for Custom Computing Machines, 1997. p 24-33.

[Lei94] LEITE, N. J.; BARROS, M. A.; "A Highly Reconfigurable Neighborhood Image Processor Based on Functional Programming". In: IEEE International Conference on Image Processing, Nov 1994. p 659-663.

[Nat99] NATIONAL SEMICONDUCTOR. NAPA1000. http://www.national.com/appinfo/milaero/napa1000 (still available at Aug 1999).

[Pag96] PAGE Ian. Reconfigurable Processor Architectures. Microprocessors and Microsystems, May 1996. (Special Issue on Codesign).

[Sha93] SHAND, M.; VUILLEMIN, J. "Fast Implementations of RSA Cryptography". In: Symposium on Computer Arithmetic, 11., 1993, Los Alamitos, California. p. 252_259.

[Vui96] VUILLEMIN J. et al. "Programmable Active Memories: Reconfigurable Systems Come of Age". IEEE Transactions on VLSI Systems, Mar 1996.

[Wir95] WIRTHLIN, M. J.; HUTCHINGS, B. L. "A Dynamic Instruction Set Computer". In: IEEE Symposium on FPGAs for Custom Computing Machines, Apr. 1995. p 92-103

Hardware Synthesis from Term Rewriting Systems

James C. Hoe and Arvind
Laboratory for Computer Science
Massachusetts Institute of Technology
Cambridge, MA
{jhoe,arvind}@lcs.mit.edu

Abstract Term Rewriting System (TRS) is a good formalism for describing concurrent systems that embody asynchronous and nondeterministic behavior in their specifications. Elsewhere, we have used TRS's to describe speculative micro-architectures and complex cache-coherence protocols, and proven the correctness of these systems. In this paper, we describe the compilation of TRS's into a subset of Verilog that can be simulated and synthesized using commercial tools. TRAC, Term Rewriting Architecture Compiler, enables a new hardware development framework that can match the ease of today's software programming environment. TRAC reduces the time and effort in developing and debugging hardware. For several examples, we compare TRAC-generated RTL's with hand-coded RTL's after they are both compiled for Field Programmable Gate Arrays by Xilinx tools. The circuits generated from TRS are competitive with those described using Verilog RTL, especially for larger designs.

Keywords: Term Rewriting Systems, high level description, high level synthesis, TRAC

1. MOTIVATION

Term Rewriting Systems (TRS's)[Baader and Nipkow, 1998] have been used extensively to give operational semantics of programming languages. More recently, we have used TRS's in computer architecture research and teaching. TRS's have made it possible, for example, to describe a processor with out-of-order and speculative execution succinctly in a page of text[Arvind and Shen, 1999]. Such behavioral descriptions in TRS's are also amenable to formal verification because one can show if two TRS's "simulate" each other. This paper describes hardware synthesis from TRS's.

We describe the Term Rewriting Architecture Compiler (TRAC) that compiles high-level behavioral descriptions in TRS's into a subset of Verilog that can be simulated and synthesized using commercial tools. The TRAC compiler enables a new hardware design framework that can match the ease of today's software programming environment. By supporting a high-level abstraction in design entry, TRAC reduces the level of expertise required for hardware design. By eliminating human involvement in the lower-level implementation tasks, the time and effort for developing and debugging hardware are reduced. These same qualities also make TRAC an attractive tool for experts to prototype large designs.

This paper describes the compilation of TRS into RTL via simple examples. Section 2. presents an introduction to TRS's for hardware descriptions. Section 3. explains how TRAC extracts logic and state from a TRS's type declaration and rewrite rules. Section 4. discusses TRAC's strategy for scheduling rules for concurrent execution to increase hardware performance. Section 5. compares TRAC-generated RTL against hand-coded RTL after each is compiled for Field Programmable Gate Arrays (FPGA) using Xilinx Foundation 1.5i synthesis package. Section 6. surveys related work in high-level hardware description and synthesis. Finally, Section 7. concludes with a few brief remarks.

2. TRS FOR HARDWARE DESCRIPTION

A TRS consists of a set of terms and a set of rewriting rules. The general structure of rewriting rules is:

$$\mathsf{pat}_{lhs}\ \textit{if}\ \mathsf{p} \rightarrow \mathsf{exp}_{rhs}$$

A rule can be used to rewrite a term s if the rule's left-hand-side pattern pat_{lhs} matches s or a subterm in s and the predicate p evaluates to *true*. A successful pattern match binds the free variables of pat_{lhs} to subterms of s. When a rule is applied, the resulting term is determined by evaluating the right-hand-side expression exp_{rhs} in the bindings generated during pattern matching.

In a functional interpretation, a rule is a function which may be expressed as:

$$\begin{array}{l} \lambda\ \mathsf{s}.\ \textit{case}\ \mathsf{s}\ \textit{of} \\ \qquad \mathsf{pat}_{lhs} \Rightarrow \textit{if}\ \mathsf{p}\ \textit{then}\ \mathsf{exp}_{rhs}\ \textit{else}\ \mathsf{s} \\ \qquad\quad \_ \Rightarrow \mathsf{s} \end{array}$$

The function uses a *case* construct with *pattern-matching* semantics in which a list of patterns is checked against s sequentially top-to-bottom until the first successful match is found. A successful match of pat_{lhs} to s creates bindings for the free variables of pat_{lhs}, which are used in the evaluation of the "consequent" expression exp_{rhs}. If pat_{lhs} fails to match to s, the wild-card pattern '_' matches s successfully and the function returns a term identical to s.

In a TRS, the effect of a rewrite is atomic, that is, the whole term is "read" in one step and if the rule is applicable then a new term is returned in the same step. If several rules are applicable, then any one of them is chosen nondeterministically and applied. Afterwards, all rules are re-evaluated for applicability on the new term. Starting from a specially-designated *starting* term, successive rewriting progresses until the term cannot be rewritten using any rule.

Example 1 (GCD): Euclid's Algorithm for finding the greatest common divisor (GCD) of two integers may be written as follows in TRS notation:

GCD Mod Rule
Gcd(a, b) *if* $(a \geq b) \wedge (b \neq 0) \rightarrow$ Gcd($a-b$, b)
GCD Flip Rule
Gcd(a, b) *if* $a < b \rightarrow$ Gcd(b, a)

The terms of this TRS have the form Gcd(a,b), where a and b are positive integers. The answer is the first sub-term of Gcd(a,b) when Gcd(a,b) cannot be reduced any further. For example, the term Gcd(2,4) can be reduced by applying the *Flip* and *Mod* rules to produce the answer 2: Gcd(2,4) $\rightarrow$ Gcd(4,2) $\rightarrow$ Gcd(2,2) $\rightarrow$ Gcd(0,2) $\rightarrow$ Gcd(2,0) □

TRS's for hardware description are often nondeterministic (*"not confluent"* in the programming language parlance) and restricted so that the terms cannot grow. The latter restriction guarantees that a system described by our TRS's can be synthesized using a finite amount of hardware. The nondeterministic aspect of TRS's has a strong flavor of modeling distributed algorithms as *state-transition systems*. (See for example [Manna and Pnueli, 1991, Lamport, 1994, Lynch, 1996, Chandy and Misra, 1988]). The focus of this paper, however, is on automatic synthesis rather than on formal verification of an implementation against a specification.

In the rest of this section we will describe the TRS notation accepted by TRAC. It includes built-in integers, booleans, common arithmetic and logical operators, non-recursive algebraic types and a few abstract datatypes such as arrays and FIFO's. Other user-defined abstract datatype, with both sequential and combinational functionalities, can be included in synthesis by providing an interface declaration and its implementation.

2.1 SIMPLE TYPES

The language accepted by TRAC is strongly typed, that is, every term has a type specified by the user. The complete list of allowed type declarations are:

$$
\begin{array}{rcl}
\text{TYPE} & :: & \text{STYPE} \\
 & [] & \text{CPRODUCT} \\
 & [] & \text{ABSTRACT} \\
\text{CPRODUCT} & :: & \text{CN}_k(\text{TYPE}_1, ..., \text{TYPE}_k)\ where\ k > 0 \\
\text{ABSTRACT} & :: & \text{Array[STYPE}_{idx}]\ \text{STYPE} \\
 & [] & \text{Fifo STYPE} \\
 & [] & \text{Array}_{CAM}[\text{STYPE}_{idx}]\ \text{STYPE}_{key}, \text{STYPE} \\
 & [] & \text{Fifo}_{CAM}\ \text{STYPE}_{key}, \text{STYPE}
\end{array}
$$

We begin by describing simple types (STYPE), which include built-in integer, product and algebraic (disjoint) union types. Product types are designated by a *constructor name* followed by one or more elements. An algebraic union is made up of two or more disjuncts. A disjunct is syntactically similar to a product except a disjunct may have zero elements. An algebraic union with only zero-element disjuncts is also known as an enumerable type. Product and algebraic unions can be composed to construct an arbitrary type hierarchy, but *no recursive types are allowed.*

$$
\begin{array}{rcl}
\text{STYPE} & :: & \text{Bit}[N] \\
 & [] & \text{PRODUCT} \\
 & [] & \text{ALGEBRAIC} \\
\text{PRODUCT} & :: & \text{CN}_k(\text{STYPE}_1, ..., \text{STYPE}_k)\ where\ k > 0 \\
\text{ALGEBRAIC} & :: & \text{DISJUNCT} \parallel \text{DISJUNCT} \\
 & [] & \text{DISJUNCT} \parallel \text{ALGEBRAIC} \\
\text{DISJUNCT} & :: & \text{CN}_k(\text{STYPE}_1, ..., \text{STYPE}_k)\ where\ k \geq 0
\end{array}
$$

The TRS in Example 1 should be accompanied by the type declaration:

Type GCD = Gcd(NUM, NUM)
Type NUM = Bit[32]

Example 2 (GCD_2): We give another implementation of GCD to illustrate some modularity and types issues. Suppose we have the following TRS to implement the *mod* function.

Type VAL = Mod(NUM, NUM) || Val(NUM)
Type NUM = Bit[32]

Mod Iterate Rule
Mod(a, b) *if* $a \geq b \rightarrow$ Mod($a-b$, b)
Mod Done Rule
Mod(a, b) *if* $a < b \rightarrow$ Val(a)

Using this definition of *mod*, GCD can be written as follows:

Type GCD_2 = Gcd_2(VAL, VAL)
GCD_2 Flip&Mod Rule
Gcd_2(Val(a), Val(b)) *if* $b \neq 0 \rightarrow$ Gcd_2(Val(b), Mod(a, b)) □

2.2 ABSTRACT TYPES

Abstract datatypes are defined by their interfaces only and are included to facilitate hardware description and synthesis. An interface can be classified as either *combinational* or *state-transforming*. We discuss array, FIFO and content addressable memory abstract datatypes next.

Array is used to model register files and memories, and has only two operations defined in its interface. Syntactically, if a is an Array then $a[idx]$ represents a combinational "read" operation which gives the value stored in the idx'th location, and $a[idx{:=}v]$, a state-transforming "write" operation gives a new Array identical to a except location idx has been updated to value v. We only support Array of STYPE with an enumerable index type.

Fifo buffers provide the primary means of communication between different modules and pipeline stages. The two main state-transforming operations on Fifo's are *enqueuing* and *dequeuing*. Enqueuing element e to q appears as enq(q,e) while dequeuing the first element from q appears as deq(q). An additional state-transforming interface clr(q) clears the contents of the Fifo. The combinational operation first(q) gives the value of the first element in q. In the description phase, Fifo is abstracted to have a bounded but unspecified size. A rule that makes use of Fifo interfaces has an implied predicate condition that tests whether the Fifo is not empty or not full, as appropriate. We also support access to other Fifo entries with appropriate projection functions. Fifo entries are also restricted to be of STYPE.

Array_{CAM} is similar to Array except its data fields are subdivided into a key field and a normal-data field. The same is true for Fifo_{CAM} and Fifo. The content-associative lookup interface cam(a,key) returns *true* if an entry with a matching key field is found. The content-associative lookup interface camidx(a,key) returns the index of an entry with a matching key field whereas camdata(a,key) returns the data field. The value of camidx(a,key) and camdata(a,key) are undefined when cam(a,key) is *false*.

As can be seen from the definition of TYPE, abstract datatypes are not allowed in algebraic disjuncts. Thus, only a complex product type can have elements of abstract types.

2.3 RULE SYNTAX

Syntactically, a rule is composed of a left-hand-side pattern and a right-hand-side expression. The predicate and *where* bindings are optional. The *where* bindings on the left-hand-side can require pattern matching. Any failure in matching PAT_i to EXP_i in the *where* bindings also deems the rule inapplicable. The expression on the right-hand-side, exp_{rhs}, can also have *where* bindings, but RHS *where* bindings can be made only to simple variables and do not involve pattern matching. In the following '_' represents the "don't care" symbol.

RULE	::	LHS → RHS
LHS	::	PAT_{lhs} [*if* EXP_p] [*where* PAT_1=EXP_1, ..., PAT_n=EXP_n]
PAT	::	_ ▯ *variable* ▯ *constant* ▯ CN_0( ) ▯ CN_k(PAT_1, ..., PAT_k)
RHS	::	EXP_{rhs} [*where* *variable*$_1$=EXP_1, ..., *variable*$_n$=EXP_n]
EXP	::	_ ▯ *variable* ▯ *constant* ▯ CN_0( ) ▯ CN_k(EXP_1, ..., EXP_k)
	▯	Prim-Op (EXP_1, ..., EXP_k)
Prim-Op	::	*Arithmetic* ▯ *Logical* ▯ *Array-Access* ▯ *FIFO-Access*

The type of PAT_{lhs} must be either CPRODUCT or ALGEBRAIC. *In addition, each rule must have PAT_{lhs} and EXP_{rhs} of the same type.* This restriction, together with non-recursive type declaration, guarantees that the size of every term is finite and the size does not change by applying the rewriting rules. In Example 2, VAL is an ALGEBRAIC type with two disjuncts, Val and Mod. It is because of this type declaration that the *Mod Done Rule* does not violate the type discipline - both sides of the rule have the type, VAL.

Example 3 (Single-Cycle RISC Processor): The state of an unpipelined, simple RISC processor is described by its program counter (PC), register file (RF) and memory (MEM). This information is captured in the following type declaration:

Type PROC = Proc_s(PC, RF, MEM)
Type PC = Bit[N]
Type VAL = Bit[N]
Type RF = Array VAL[RNAME]
Type RNAME = Reg0( ) || Reg1( ) || Reg2( ) || . . . . Reg$m$( )
Type MEM = Array INST[PC]
Type INST = Loadc(RNAME,VAL)
|| Loadpc(RNAME)
|| Add(RNAME,RNAME,RNAME)
|| Sub(RNAME,RNAME,RNAME)
|| Bz(RNAME,RNAME)
|| Load(RNAME,RNAME)
|| Store(RNAME,RNAME)

The processor we synthesized in Section 5. has four 32-bit general purpose registers, i.e. N=32, m=4. The behavior of the 7 instructions — move PC to register, load immediate, register-to-register addition and subtraction, branch if zero, memory load and store — can be specified as a TRS by giving a rewrite rule for each instruction. The following rule conveys the execution of the *Add* instruction.

Proc_s(*pc*, *rf*, *mem*)
where Add(*rd*,*r1*,*r2*)=*mem*[*pc*]
→ Proc_s(*pc*+1, *rf*[*rd*:=(*rf*[*r1*]+*rf*[*r2*])], *mem*) □

Example 4 (Pipelined RISC Processor): The processor in Example 3 can be pipelined by introducing FIFO's as pipeline-stage buffers and by systematically splitting each rule into local rules for various pipeline stages. For example, in a two-stage pipeline design, the processing of an instruction can be broken down into separate fetch and execute steps. We model buffers between pipeline stages as a Fifo of an unspecified but finite size. In a behavioral description, it is convenient if the operation of each stage can be described without reference to other stages. FIFO buffers provide this isolation; most pipelined design rules dequeue an input from one FIFO and enqueue the result into another FIFO. In the synthesis phase these FIFO buffers are replaced by a fixed-depth FIFO or simply registers, and flow control logic ensures that a rule does not fire if the destination FIFO is full.

Here, we introduce the pipeline buffer BS in the declaration of the PROC_p term.

Type $\mathsf{PROC}_p = \mathsf{Proc}_p(\mathsf{PC, RF, BS, MEM})$
Type BS = Fifo ITEMP
Type ITEMP = Loadc(RNAME,VAL)
|| Loadpc(RNAME)
|| Add(RNAME,VAL,VAL)
|| Sub(RNAME,VAL,VAL)
|| Bz(VAL,VAL)
|| Load(RNAME,ADDR)
|| Store(ADDR,VAL)

The Add and Bz instruction rules are splitted into *Fetch* and *Execute* stage rules:

Fetch Rule
$\mathsf{Proc}_p(pc, rf, bs, mem)$
$\rightarrow$ $\mathsf{Proc}_p(pc{+}1, rf, \mathrm{enq}(bs, mem[pc]), mem)$

Add Rule
$\mathsf{Proc}_p(pc, rf, bs, mem)$
where $\mathsf{Add}(rd, r1, r2) = \mathrm{first}(bs)$
$\rightarrow$ $\mathsf{Proc}_p(pc, rf[rd := (rf[r1] + rf[r2])], \mathrm{deq}(bs), mem)$

Branch-Taken Rule
$\mathsf{Proc}_p(pc, rf, bs, mem)$
if $rf[rc]=0$ *where* $\mathsf{Bz}(rc, ra) = \mathrm{first}(bs)$
$\rightarrow$ $\mathsf{Proc}_p(rf[ra], rf, \mathrm{clr}(bs), mem)$

Branch-Not-Taken Rule
$\mathsf{Proc}_p(pc, rf, bs, mem)$
if $rf[rc] \neq 0$ *where* $\mathsf{Bz}(rc, ra) = \mathrm{first}(bs)$
$\rightarrow$ $\mathsf{Proc}_p(pc, rf, \mathrm{deq}(bs), mem)$

Notice the *Fetch* rule is always ready to fire. At the same time one of the execute stage rules may be ready to fire as well. This is the first example we have seen where more than one rule can be enabled on a given state. Even though according to TRS semantics, only one rule should be fired in each step, we will see that our compiler tries to fire as many rules in parallel as possible while maintaining correct TRS execution semantics. Without parallel firing of rules we won't get the pipelining effect we want.

Since there is a race to update the *pc* between the *Fetch* and the *Branch Taken* rules, the above rules can exhibit nondeterministic behavior. Specification of microprocessors and cache-coherence protocols often entails nondeterminism, even though a given realization is usually completely deterministic. Our compiler can handle such nondeterministic TRS's. □

In addition to the TRS-to-RTL compilation to be described in Sections 3. and 4., we are developing source-to-source TRS transformations that can achieve the kind of pipelining described in Example 4. The dependence between the rules has to be analyzed carefully to ensure the correctness of all such transformations. Presently, human intervention is required to guide the transformation process at the high level. It is also possible to automatically derive the rules for a superscalar version of the pipelined processor in Example 4 [Arvind and Shen, 1999].

2.4 INPUT AND OUTPUT

Traditionally a TRS describes a closed system, but we are experimenting with new notations and semantics to support description of a system with input and output (I/O) ports. In an approach that only requires minimal deviation from a standard TRS, the designer assigns I/O specific semantics to terms using source code annotations. For example, a wrapper to start and terminate a GCD computation can be given as:

Type TOP = Top(MODE, NUMI, NUMI, NUMO, GCD)
Type MODE = *__iport__* Load( ) || Run( )
Type NUMI = *__iport__* NUM
Type NUMO = *__oport__* NUM

GCD Start
Top(Load( ), x, y, _, _)
→ Top(_, _, _, 0, Gcd(Val(x), Val(y)))

GCD Done
Top(Run( ), _, _, _, Gcd(Val(ans), Val(0)))
→ Top(_, _, _, ans, _)

Ignoring the I/O annotations (*__iport__* and *__oport__*), the type declaration and rules can be interpreted exactly as before. In fact, the combinational logic generated by TRAC is the same irrespective of I/O annotations. The first rule states as long as the first subterm of TOP is Load( ), the GCD term can be rewritten using the second and third subterms of TOP. The second rule states if the first subterm of TOP is Run and the GCD computation is done (when the second subterm of GCD is 0), then copy the first subterm of GCD to the fourth subterm of TOP.

The only effect of annotating the fourth subterm of TOP as an *__oport__* is that TRAC will attach wires to the output of the registers in that subterm and make their content externally visible through an output port. Conversely, the effect of annotating a term as an *__iport__* is that the wires normally connected to the output of the registers in that term are redirected to an input port instead. A rule cannot rewrite a term labeled as an *__iport__* since the value of the term does not correspond to any internal register. From the TRS perspective, an $\_\_iport\_\_$ term may change unexpectedly, but atomically, without any rule application.

By driving the appropriate values on the input ports corresponding to the first three subterms of TOP, a new GCD computation is started. Asserting signals corresponding to Run( ) at the input port enables GCD to execute to completion, and at which point, the answer appears on the output port as a consequence of the *GCD Done* rule.

3. BASIC SYNTHESIS STRATEGY

Although TRS's provide great flexibility in specifying hierarchically organized state and state transitions, a TRS, where recursive types are not allowed and rules are required to have the same type on both sides of $\rightarrow$, can only describe a finite state machine (FSM). TRAC maps a TRS to a synchronous FSM by

- Mapping TRS terms to storage elements (e.g., registers, register files and other abstract datatypes)
- Mapping TRS rules to combinational logic that generates next state values and enable signals for storage elements.

In this section we first describe a functional interpretation of each rule and then derive an "action on state" view of the same rule. The latter view is the starting point for hardware synthesis.

3.1 FUNCTIONAL INTERPRETATION OF A RULE: π AND δ FUNCTIONS

In a functional interpretation, a rule of the form

$$
\begin{array}{ll}
 & \mathsf{pat}_{lhs} \\
 & \quad \textit{if } \mathsf{exp}_p \textit{ where } \mathsf{pat}_1 = \mathsf{exp}_{lhs,1}, \ldots, \mathsf{pat}_n = \mathsf{exp}_{lhs,n} \\
\rightarrow & \mathsf{exp}_{rhs} \\
 & \quad \textit{where } \mathsf{var}_1 = \mathsf{exp}_{rhs,1}, \ldots, \mathsf{var}_m = \mathsf{exp}_{rhs,m}
\end{array}
$$

is a function of *typeof*(pat_{lhs})$\rightarrow$*typeof*(pat_{lhs}), and returns a term identical to the input term if the rule is not applicable. If the rule is applicable, the return value is a new term based on the evaluation of exp_{rhs} using the bindings created during pattern matching.

```
rule = λ s. case s of
            pat_lhs ⇒
              case exp_lhs,1 of
                pat_1 ⇒
                  ...
                    case exp_lhs,n of
                      pat_n ⇒
                          if exp_p then
                            let
                              var_1 = exp_rhs,1, ..., var_m = exp_rhs,m
                            in
                              exp_rhs
                          else
                            s
                      _ ⇒ s
                  ...
                _ ⇒ s
            _ ⇒ s
```

This function can be broken down into its two components: π and δ. The π function determines a rule's applicability to a term and has the type, *typeof*(pat_{lhs})$\rightarrow$*Boolean*. The δ function, on the other hand, determines the new term in case π evaluates to *true*.

```
π = λ s. case s of
        pat_lhs ⇒
          case exp_lhs,1 of
            pat_1 ⇒
              ...
                case exp_lhs,n of
                  pat_n ⇒ exp_p
                  _ ⇒ false
              ...
            _ ⇒ false
        _ ⇒ false
```

```
δ = λ s. let
          pat_lhs = s
          pat_1 = exp_lhs,1, ..., pat_n = exp_lhs,n
          var_1 = exp_rhs,1, ..., var_m = exp_rhs,m
        in
          exp_rhs
```

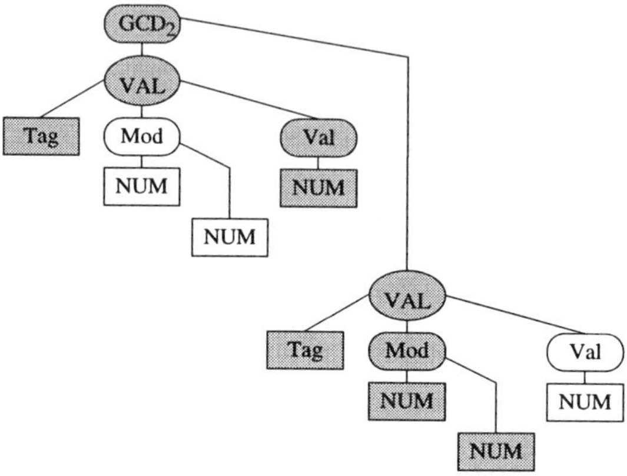

Figure 1 A graph representation of the GCD_2 type structure from Example 2. NUM is treated as a type alias for Bit[32].

Using π and δ, an equivalent functional representation of a rule is

$$rule = \lambda\ \mathsf{s}.\ if\ \pi(\mathsf{s})\ then\ \delta(\mathsf{s})\ else\ \mathsf{s}$$

3.2 A RULE AS A STATE TRANSFORMER

In the architectural context, terms represent state, and rules define how the state can be transformed. If we restrict ourselves to synchronous circuits then each rule "reads" the state at the beginning of the clock cycle and if it can fire, it modifies the state at the end of the same clock cycle. In this "actions on state" view of a rule, one needs to update only those parts of the state that actually change. If two rules are enabled simultaneously and affect disjoint parts of the state then it is possible to execute both rules in the same clock cycle. After discussing the hardware to execute one rule in this section, we will return to the issue of concurrent firings in the next section.

Mapping Terms to Storage Elements: A term can be represented as a tree based on its type. For example, the tree representation of GCD_2 is shown in Figure 1. Algebraic types have an extra branch, Tag, where a register of width $\lceil log_2 d \rceil$ records which of the d disjuncts the term belongs to. An ALGEBRAIC node has a branch for each of the disjuncts, but, at any time, only the branch whose tag matches the content of the tag register holds meaningful data. As an example we have shaded the active portions of the tree corresponding to Gcd(Val(2), Mod(4, 2)) in Figure 1.

We can assign an unique name to each storage element based on its path (also known as projection) from the root. For example, the name for the second (from the left) NUM register in Figure 1 would be "$Proj_1$.Mod.$Proj_2$". The storage implied by a term can be represented as a set of <*proj*, *REG*[*N*]> pairs. For example the storage elements of GCD_2 are represented by the set

$$\{<Proj_1.\mathsf{Tag}, REG[1]>, <Proj_1.\mathsf{Val}.Proj_1, REG[32]>, <Proj_1.\mathsf{Mod}.Proj_1, REG[32]>, <Proj_1.\mathsf{Mod}.Proj_2, REG[32]>, <Proj_2.\mathsf{Tag}, REG[1]>, <Proj_2.\mathsf{Val}.Proj_1, REG[32]>, <Proj_2.\mathsf{Mod}.Proj_1, REG[32]>, <Proj_2.\mathsf{Mod}.Proj_2, REG[32]>\}$$

As an optimization, registers on different disjuncts of an ALGEBRAIC node can share the same physical register. In Figure 1, the registers aligned horizontally are mappable to the same register. This idea can be expressed as allowing multiple pathnames to be associated with a single register state element. In a type structure that includes Array and other abstract datatypes, nodes corresponding to the abstract datatypes appear at the leaves of the tree.

The value embedded in the storage elements of a term can be represented in a similar manner using a set of <*proj*, *value*> pairs. For example the values of storage elements of Gcd(Val(2), Mod(4, 2)) are represented by the set

$$\{<Proj_1.\mathsf{Tag}, \mathsf{Val}>, <Proj_1.\mathsf{Val}.Proj_1, 2>, <Proj_2.\mathsf{Tag}, \mathsf{Mod}>, <Proj_2.\mathsf{Mod}.Proj_1, 4>, <Proj_2.\mathsf{Mod}.Proj_2, 2>\}$$

The procedure *extract-state*(*s*,*proj*) to extract the values of storage elements from term *s* is defined below. Initially it is called with an empty projection ϵ. Since we propose to use this function only at compile time, we assume the representation of a term includes its type structure.

extract-state(*s*,*proj*)=
  case s of
    Bit[N] $\Rightarrow$ {<*proj*, *s*>}
    CPRODUCT: $\mathsf{CN}_k(s_1, ..., s_k) \Rightarrow$
      extract-state(s_1,*proj*.$Proj_1$) $\cup$ ... $\cup$ *extract-state*(s_k,*proj*.$Proj_k$)
    ALGEBRAIC: $\mathsf{CN}_k(s_1, ..., s_k) \Rightarrow$
      extract-state(s_1,*proj*.CN_k.$Proj_1$) $\cup$ ...
        $\cup$ *extract-state*(s_k,*proj*.CN_k.$Proj_k$) $\cup$ {<*proj*.Tag, CN_k>}
    Array: $\Rightarrow$ {<*proj*.Array, s>}
    Fifo: $\Rightarrow$ {<*proj*.Fifo, s>}

Rules as Actions on Storage Elements: Because a rule's pat_{lhs} and exp_{rhs} are required to have the same type, the term resulting from a rewrite must have the same storage structure as the initial term. In other words, beginning with a TRS's starting term and its storage elements, successive rewrite operations never add or delete any storage elements. To implement a TRS, TRAC generates a state structure that is extracted from the starting term, and the rules are implemented as combinational logic that updates the content of the storage elements.

The pattern matching on the left-hand-side of a rule (the π function) essentially tests the values of some of the storage elements. π can also include combinational functions from the interface of abstract datatypes. π for the *Flip&Mod* rule of Example 2 will look like the following:

$$\pi = (Proj_1.\mathsf{Tag}(\mathsf{\textit{state}}) = \mathsf{Val}) \wedge (Proj_2.\mathsf{Tag}(\mathsf{\textit{state}}) = \mathsf{Val}) \\ \wedge (Proj_2.\mathsf{Val}.Proj_1(\mathsf{\textit{state}}) \neq 0)$$

The right-hand-side of a rule (or δ) can be viewed as specifying actions on the storage elements of the input term. The actions can be represented in a set of <*proj*, *action*> pairs. Possible actions include setting a register to a value (*set*(v)) or invoking an abstract datatype's state-transforming interface. The δ of the *Flip&Mod* rule in Example 2 can be viewed as the following set of actions:

$$\{<Proj_1.\mathsf{Tag}, set(\mathsf{Val})]>, <Proj_1.\mathsf{Val}.Proj_1, set(\mathsf{\textit{b}})>, \\ <Proj_2.\mathsf{Tag}, set(\mathsf{Mod})>, <Proj_2.\mathsf{Mod}.Proj_1, set(\mathsf{\textit{a}})>, \\ <Proj_2.\mathsf{Mod}.Proj_2, set(\mathsf{\textit{b}})>\}$$

Recall, a and b refer to some subterms in the initial term s as established by the pattern matching semantics. In cirucit implementation, a and b refer to the initial values of the corresponding storage elements.

Notice, the left-hand-side of the rule requires the first tag register ($Proj_1$.Tag) to be Val when this rule is applicable. Thus we can detele the action <$Proj_1$.Tag, *set*(Val)> without affecting the outcome. Thus, if a compiler can detect that a storage element is assigned the same value as its original content, it can delete that particular action. In general, the necessary actions when a rule fires are

$$\textit{extract-state}(\delta(s),\epsilon) - \textit{extract-state}(s,\epsilon)$$

where '$-$' represents the set difference. In practice, instead of dynamically testing for equality between the next and current state values of a register to eliminate actions, TRAC statically eliminates actions in which a register is updated by a value coming from itself and when a register is updated by the same value that it must have for π to be satisfied.

In another example, consider the pipelined processor of Example 4, whose storage elements are

$$\{<Proj_1, \mathsf{\textit{pc}}>, <Proj_2.\mathsf{Array}, \mathsf{\textit{rf}}>, \\ <Proj_3.\mathsf{Fifo}, \mathsf{\textit{bs}}>, <Proj_4.\mathsf{Array}, \mathsf{\textit{mem}}>\}.$$

The Add rule specifies the following actions on this state:

$$\{<Proj_2.\text{Array}, \text{array-update}(rd, rf[r1]+rf[r2])> \; <Proj_3.\text{Fifo}, \text{deq}(\,)>\}$$

In general, a rule can be applied to a subterm of a whole term. In these cases, *extract-state*(s,proj) is called by a projection, relative to the whole term, that corresponds to the subterm. Furthermore, a rule can be applied to many parts of a term. In these cases, a rule's logic is instantiated multiple times, once for each state sub-structure where the rule is to be applied. In an alternative interpretation, a subterm-applicable rule needs to be lifted to the same type as the TRS's starting term prior to analysis. The effect of applying the lifted rule to the whole term is the same as applying the original rule to the subterm within the whole term. A subterm rule may be applicable to multiple positions in the whole term. A separate lifted version must be created for each possible application. For example, the *Mod Done* rule from GCD_2 in Example 2 could be applicable to both the first and second subterms of a GCD_2 term. The two lifted versions of the *Mod Done* rule are:

$$\begin{array}{ll} & \text{Gcd}_2(\text{Mod}(a, b), t) \; \textit{if} \; a<b \\ \rightarrow & \text{Gcd}_2(\text{Val}(a), t) \end{array}$$

and

$$\begin{array}{ll} & \text{Gcd}_2(t, \text{Mod}(a, b)) \; \textit{if} \; a<b \\ \rightarrow & \text{Gcd}_2(t, \text{Val}(a)) \end{array}$$

3.3 CIRCUIT SYNTHESIS

The π and δ functions for the two GCD rules, *GCD Mod* and *GCD Flip*, in Example 1 are given below. A valid starting term for this TRS has the form Gcd(x, y) where x and y are postive integers. This starting term implies the set of storage elements: $\{<Proj_1, REG[32]>, <Proj_2, REG[32]>\}$. For conciseness, we refer to these registers as a and b in the following definitions:

$$\begin{array}{lcl} \pi_{Mod} & = & a \geq b \wedge b \neq 0 \\ \pi_{Flip} & = & a<b \\ \delta_{Mod,a} & = & set(a-b) \\ \delta_{Flip,a} & = & set(b) \\ \delta_{Flip,b} & = & set(a) \end{array}$$

For hardware synthesis we break down δ into actions on individual storage elements as specified above. Therefore, for each storage element e affected by a rule R, $\delta_{R,e}$ gives its next state value. π_R is the latch-enable signal of all the affected registers. Two state transition circuits corresponding to the two GCD rules, considered indenpendently, is first shown in Figure 2.

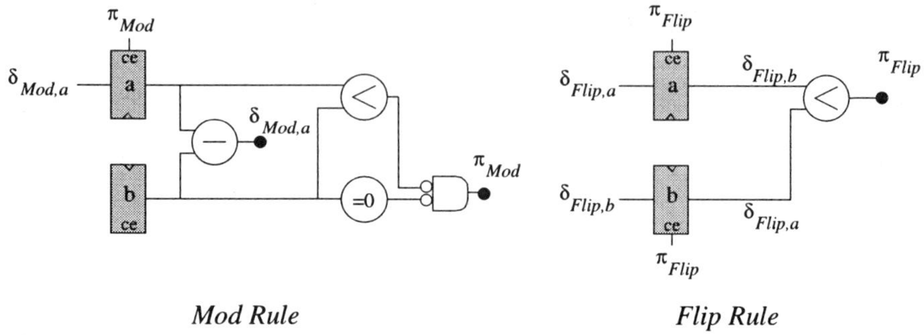

Figure 2 FSM for a TRS with only one rule.

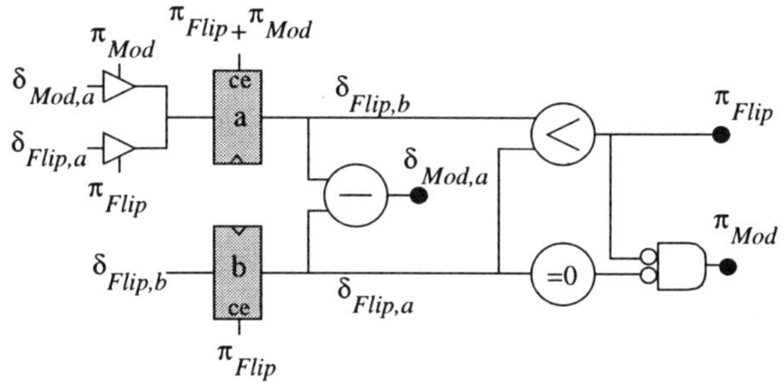

Figure 3 Circuit for computing Gcd(*a*, *b*) from Example 1.

The final circuit is arrived by combining the two circuits. In these cases, both rules affect the storage element *a* but only one of them can actually fire in a given state. When merging the actions from rules with mutally-exclusive firing conditions (π), the final latch enable is simply the logical-OR of their firing conditions (e.g., $\pi_{Mod} + \pi_{Flip}$ in this example), and the next state values are chosen from all of the δ's using a multiplexer where a rule's π enables its own δ. A sample update circuit that merges δ's from two mutually-excluisve rules is illustrated as circuit A in Figure 4. Figure 3 shows the FSM generated by combining the π and δ from both GCD rules.

However, in general, several π's could be asserted, i.e., several rules could be applicable. In the simplest solution, a new set of disjoint triggers $\phi_1, ..., \phi_n$ can be generated using a round-robin priority encoder fed by $\pi_1, ..., \pi_n$. ϕ's, which are mutually exclusive, globally replace π's at all multiplexers and at all latch enable OR-gates. A sample update circuit that merges δ's from two possibly conflicting rules is illustrated as circuit B in Figure 4. This arbitration is simple and correct, but the circuit is inefficient and allows only one rewrite per cycle.

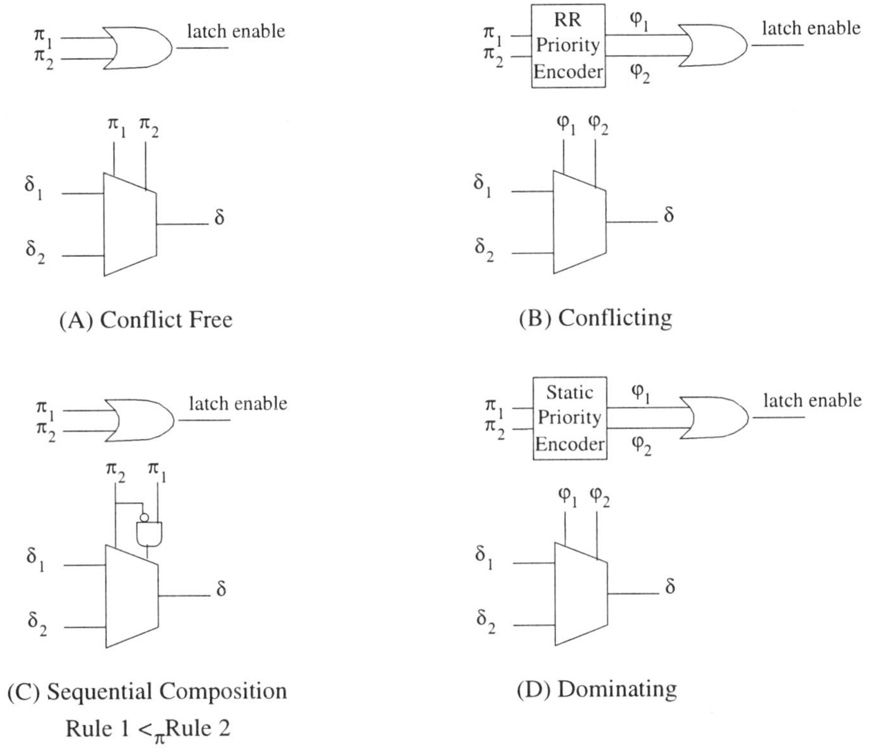

Figure 4 Circuits for combining two rules' δ actions on the same state element.

TRAC does not synthesize any state structures for abstract datatypes. When an abstract datatype is used in a TRS, TRAC instantiates the corresponding Verilog module in the RTL and makes appropriate connections to the interfaces. The user or the library is expected to provide a Verilog module in RTL for each abstract datatype. A state transforming interface has an implied signal driving by π (or ϕ) to enable the state changes when the corresponding rule is fired.

4. EXPLOITING PARALLELISM

According to TRS semantics, if multiple rules can simultaneously become applicable on a given term *s*, one of the rules is chosen nondeterministically and applied atomically to rewrite *s* to *s*'. Next, a new round of rewriting is started from scratch on *s*'. When a TRS exhibits such nondeterminism, multiple behaviors are allowed. Using a scheduler based on a round-robin priority encoder as discussed in Section 3.3, TRAC implements one of the allowed behaviors in a deterministic circuit that fires one rule per clock cycle.

If the simultaneously applicable rules involve mutually disjoint parts of the term, then these rules can be executed in any sequence successively to reach the same final term. In this scenario, although the semantics of a TRS specifies a sequential and atomic term rewriting, a hardware implementation can exploit the underlying parallelism and execute the rules concurrently in the same clock cycle. In general it is not safe to allow two arbitrary applicable rules to execute in the same clock cycle because executing one of them can alter the value of the π or the δ function of the other. This section formalizes the conditions for simultaneous rule execution and suggests a scheduling that improves hardware performance by firing multiple rules in the same clock cycle when allowed.

4.1 TRANSPARENCY

The minimum condition for allowing two simultaneously applicable rules to fire in the same clock cycles is captured by the **π-transparent** relationship.

Definition 1 (π-transparent)
*Rule R_1 is **π-transparent** to rule R_2, denoted as $R_1 <_\pi R_2$, if $\forall s. \pi_1(s) \wedge \pi_2(s) \Rightarrow \pi_2(\delta_1(s))$*

This condition states that if two rules ever become applicable on the same term and $R_1 <_\pi R_2$, then firing R_1 first does not prevent R_2 from firing on the resulting term. Firing in the reverse order may not necessarily be allowed, unless a stronger condition of mutual-transparency (or **π-conflict-free**) is satisfied.

Definition 2 (π-conflict-free)
*Rules R_1 and R_2 are **π-conflict-free** if $(R_1 <_\pi R_2) \wedge (R_2 <_\pi R_1)$*

Given two rules where $R_1 <_\pi R_2$, there are two basic approaches to allow both rules to fire in the same clock cycle. The first approach cascades the combinational logic from the two rules such that R_1 is applied first to the physical state elements, and R_2 is applied to the *effective* state after attempting to apply R_1. In effect, we are creating a composite rule where

$$\begin{aligned}
\lambda\, s\,.\ & \textit{if}\ \pi_1(s)\ \textit{then} \\
& \qquad \textit{if}\ \pi_2(s)\ \textit{then}\ \delta_2(\delta_1(s))\ \textit{else}\ \delta_1(s) \\
& \quad \textit{else if}\ \pi_2(s)\ \textit{then}\ \delta_2(s)\ \textit{else}\ s
\end{aligned}$$

Arbitrary cascading does not always improve circuit performance since cascading combinational logic may lead to a longer cycle time, especially when serveral rules are composed. In a synchronous design, if the clock period increases, every rule firing is penalized, even when at most one rule can fire.

In a more practical approach, the input to the combinational logic from all rules are driven directly by state elements. Two transparent rules are allowed to execute in the same clock cycle only if the correct resulting state can be constructed from independent evaluation of the same current state.

4.2 PARALLEL COMPOSIBILITY

Two rules that do not affect the same storage elements are **parallel composible**, provided allowing them to execute concurrently on the same state produces a behavior that corresponds to at least one ordering of rule-execution in TRS.

Definition 3 (Parallel-Composible Transparency)
*Rule R_1 is **PC-transparent** to rule R_2, denoted as $R_1 <_{PC} R_2$, if*

$$(R_1 <_\pi R_2) \wedge \forall s.(\pi_1(s) \wedge \pi_2(s)) \Rightarrow \delta_2(\delta_1(s)) = \mathbf{PC}(s, \delta_1(s), \delta_2(s))$$

Definition 4 (Parallel Composition)

$\mathbf{PC}(\mathsf{s}, \mathsf{s}_1, \mathsf{s}_2) =$
  case s *of*
    Bit $v \Rightarrow$
      if $\mathsf{s}_1 = \mathsf{s}_2$ *then* s_1
      else if $\mathsf{s}_1 = v$ *then* s_2
      else if $\mathsf{s}_2 = v$ *then* s_1
      else $\perp$
    $\mathsf{CN}_k(\ldots) \Rightarrow$
      if $\mathsf{s}_1 = \mathsf{s}_2$ *then* s_1
      else if $\mathsf{s}_1 = n$ *then* s_2
      else if $\mathsf{s}_2 = n$ *then* s_1
      else if $(\mathsf{Tag}(\mathsf{s}_1) = \mathsf{CN}_k) \wedge (\mathsf{Tag}(\mathsf{s}_2) = \mathsf{CN}_k)$ *then*
        $\mathsf{CN}_k(\mathbf{PC}(Proj_1(s), Proj_1(\mathsf{s}_1), Proj_1(\mathsf{s}_2)), \ldots,$
                $\mathbf{PC}(Proj_k(s), Proj_k(\mathsf{s}_1), Proj_k(\mathsf{s}_2))$)
      else $\perp$
    $\mathsf{Array}_n\ a \Rightarrow \forall_{1 \leq i \leq n}.\ a[i{:=}\ \mathbf{PC}(a[i], \mathsf{s}_1[i], \mathsf{s}_2[\mathrm{i}])]$
    Fifo $f \Rightarrow$
      if (s_1 *is suffix of* f) $\wedge$ (f *is suffix of* s_2) *then*
        chop_prefix(chop_suffix(s_1, f*)*,s_2*)*
      if (s_2 *is suffix of* f) $\wedge$ (f *is suffix of* s_1) *then*
        chop_prefix(chop_suffix(s_2, f*)*,s_1*)*
      else $\perp$

Essentially what this definition says is that, if both rules R_1 and R_2 want to update a register, then they must produce the same value. In the case of an array, if the two rules update different elements of the array, then parallel composition will work assuming the array has multiple write ports. In the case of a FIFO, if one rule enqueues and the other dequeus then they can be combined to execute in the same cycle.

Note $R_1 <_{PC} R_2$ does not imply that the outcome is confluent. Consider the following two rules that operate on four registers:

R_1: $\mathsf{F}(1, r_B, r_C, r_D) \rightarrow \mathsf{F}(1, r_B, 1, r_D)$
R_2: $\mathsf{F}(r_A, 1, r_C, r_D) \rightarrow \mathsf{F}(0, 1, r_C, 1)$

Now consider the starting term F(1,1,r_C,r_D). The effect of executing R_1 after R_2 is F(0,1,1,1). On the other hand if R_2 is executed first the result would be F(0,1,r_C,1) and R_1 will no longer fire.

For two rules to be confluent we need the following stronger condition.

Definition 5 (Conflict-free)
Rules R_1 and R_2 are ***conflict-free*** *if* $(R_1 <_{PC} R_2) \wedge (R_2 <_{PC} R_1)$

If two rules are parallel composible, the δ's do not collide and no special merging circuit is required to arbitrate their actions on the affected storage elements.

4.3 SEQENTIAL COMPOSIBILITY

Even if two rules do affect some common state, by carefully prioritizing the effect of the two rules such that the effects of R_2 overrides R_1 (in case $R_1 <_\pi R_2$), a legal outcome can still be constructed from simultaneous evaluation of the two rules on the same current state.

Definition 6 (Sequentially-Composible Transparency)
Rule R_1 is ***SC-transparent*** *to rule R_2, denoted as $R_1 <_{SC} R_2$, if*

$$(R_1 <_\pi R_2) \wedge \forall s.(\pi_1(s) \wedge \pi_2(s)) \Rightarrow \delta_2(\delta_1(s)) = \mathbf{SC}(s, \delta_1(s), \delta_2(s))$$

Sequential composition that implements the priotization is defined as

Definition 7 (Sequential Composition)

SC(s, s_1, s_2) =
  case s *of*
    Bit $v \Rightarrow$ *if* $s_2 = v$ *then* s_1 *else* s_2
    $\mathsf{CN}_k(...) \Rightarrow$
      if $s_2 = s$ *then* s_1
      else if $(\mathsf{Tag}(s_1) = \mathsf{CN}_k \wedge (\mathsf{Tag}(s_2) = \mathsf{CN}_k)$ *then*
        $\mathsf{CN}_k(\mathbf{SC}(Proj_1(s), Proj_1(s_1), Proj_1(s_2)), ...,$
              $\mathbf{SC}(Proj_k(s), Proj_k(s_1), Proj_k(s_2)))$
      else s_2
    $\mathsf{Array}_n\ a \Rightarrow \forall_{1 \le i \le n}.\ a[i{:=}\ \mathbf{SC}(a[i], s_1[i], s_2[i])]$
    Fifo $f \Rightarrow$
      if $(s_1$ *is suffix of* $f) \wedge (f$ *is suffix of* $s_2)$ *then*
        chop_prefix(chop_suffix(s_1, f*),*s_2*)*
      if $(s_2$ *is suffix of* $f) \wedge (f$ *is suffix of* $s_1)$ *then*
        chop_prefix(chop_suffix(s_2, f*),*s_1*)*
      else $\perp$

If R_1 and R_2 are sequentially composible ($R_1 <_\pi R_2$), then prioritized ϕ_1 and ϕ_2 can be generated. However, instead of applying them globally, they are only used to replace π_1 and π_2 at state elements that are affected by both rules. If a register is only affected by either R_1 or R_2 then π can be used directly. Circuit (C) in Figure 4 illustrates the update circuit for this case.

4.4 DOMINANCE

Definition 8 (Dominance)
*Rule R_2 **dominates** rule R_1, denoted as $R_1 <_D R_2$, if*

$$(R_1 <_\pi R_2) \wedge \forall s.(\pi_1(s) \wedge \pi_2(s)) \Rightarrow \delta_2(\delta_1(s)) = \delta_2(s)$$

If two rules, R_1 and R_2 are conflicting, but R_2 dominates R_1, we can include this information in the priority encoder when generating ϕ's for global replacement of their π's. If π_1 and π_2 are both asserted on a cycle, instead of using a fair round-robin priority encoder, the encoder would statically give priority to π_2. For a two rule circuit, ϕ_2=π_2 and ϕ_1=$\pi_1 \wedge \neg \pi_2$. Circuit (D) in Figure 4 illustrates the update circuit for this case.

4.5 SCHEDULING FOR SIMULTANEOUS FIRING

To conclude this section, we describe a scheduler that is currently implemented in TRAC that makes use of conflict-free (CF) relationships. In general, an exact test for CF relationship between two arbitrary rule instances is expensive (Finding an s such that $\pi_i(s) \wedge \pi_j(s)$ is like solving SAT). Instead, TRAC performs several conservative tests to find as many CF relationships as possible. First, two rule instances that read and write non-overlapping parts of the systems are CF. If two rule instances do not rewrite the same registers, and if none of the registers affected by the δ of one is used by the π and δ of the other, and vice versa, then the two rules are CF since this condition is stronger than the requirement for CF. Lastly, TRAC symbolically analyzes pairs of π's to conservatively determine when a pair can never be satisfied simultaneously and thus are CF by default.

TRAC makes use of certain axioms when analyzing the conflict relationships between rules that reference abstract datatype interfaces. For example,

$(a[idx{:=}v])[idx]=v$
$((a[idx{:=}v])[idx'{:=}v])[idx]=v$ if $idx \neq idx'$

deq(enq(q,e)) = enq(deq(q),e) if q is not empty
first(q) = first(enq(q,e)) if q is not empty

Based on the analysis above and taking into account the properties of FIFO buffers, it can be shown that the rules of Example 4 are CF except for the *Fetch* and the *Branch-Taken* rule. However, it can be shown that the *Branch-Taken* rule dominates the *Fetch* rule in the sense that the effect of applying the *Branch-*

Taken rule after the *Fetch* rule is the same as not applying the *Fetch* rule at all i.e., $(\delta_{BzN}(s) = \delta_{BzN}(\delta_{Fetch}(s)))$. Thus, instead of arbitrating between these two rules, the compiler gives priority to the *Branch-Taken* rule.

After TRAC has establish $\mathcal{C}F$ relationships between as many rule instances as possible, a graph of rule instances can be constructed by adding an edge between each non-$\mathcal{C}F$ pairs. Scheduling groups is formed by partitioning the graph into connected components. Different groups never interfere and can be scheduled independently. For each group, a round-robin priority encoder can be used to map π to ϕ for arbitration. For a small group, an $n \times n$ look-up table can be computed off-line to encode π to ϕ where more than one ϕ can be asserted if the rules of the asserted π's are $\mathcal{C}F$.

5. PERFORMANCE EVALUATION

TRAC generates RTL Verilog that can be synthesized to a variety of technologies by commercial tools like Synopsys and Xilinx hardware compilers. In this paper, we evaluate the quality of the TRAC-generated RTL's against hand-coded RTL when compiled for Xilinx FPGA's.

Synthesis of the GCD Circuit: Both Example 1 and 2 are compiled to RTL by TRAC. The compile time is less than 2 sec on a 166MHz PowerPC604e. As a reference, our colleague, Daniel L. Rosenband, provided a hand-optimized Verilog RTL for GCD that uses only two 32-bit registers, a single subtracter, and simple boolean logic gates. The three RTL's are compiled for XC4010XL-09 FPGA using Xilinx Foundation 1.5i tools. We report the number of flip-flops and the overall utilization of the FPGA. In addition to the maximum clock frequency, we also report the number of clock cycles needed to compute GCD(53857×10957,91159×10957).

Version	FF (bit)	Util. (%)	Freq. (MHz)	Elapse (cyc)
Example 1	64	20	44.2	54
Example 2	102	38	31.5	104
Hand RTL	64	16	53.1	54

The RTL generated by TRAC from Example 2 is significantly worse than the hand-coded RTL because the input TRS maps to a sub-optimal hardware structure. TRAC does not have the same ingenuity that allowed our colleague to realize the high-level transformations that lead to the smaller and simpler circuit of the hand-optimized RTL. However, the necessary information to achieve the same high-level transformation can be expressed at the TRS level. Given Example 1, TRAC produces an RTL that is structurally similar to the hand-coded version and compiles to within 25% of the hand-written RTL in terms of circuit size and 17% in terms of circuit speed.

Synthesis of the Unpipelined Microprocessor: Hand-optimization can often produce much more efficient implementations than machine compilation on small designs. However, as the problem size increases, the pay-back of hand optimizations diminishes while the effort required increases dramatically. This is evident in the synthesis of the simple microprocessor from Example 3. The TRAC generated RTL and a hand-coded Verilog RTL of the unpipelined processor, when targeting an XC4013XL-08 FPGA, are comparable both in size and speed.

Version	FF (bit)	Util. (%)	Freq. (MHz)
Example 3	161	60 %	40.0
Hand RTL	160	50 %	41.0

6. RELATED WORK

A *behavioral* description refers to specifying a component by its input/output behavior without implementation or structural details. In industry, such descriptions are given typically in a sequential language like the behavioral portion of Verilog. Another approach is to extend or adapt a popular software language. Transmorgafier-C[Galloway, 1995] and HardwareC[HardwareC, 1990] compile hardware from a source language based on C. In these systems, some constructs in C are overloaded to convey hardware related information such as clocking and registered storage. In the Programmable Active Memory (PAM) project, Vuillemin, et al. synthesize from an RTL in C++ syntax[Vuillemin et al., 1996]. Algorithms described in data-parallel C languages have been used to program an array of FPGA's in Splash 2 [Gokhale and Minnich, 1993] and CLAy[Gokhale and Gomersall, 1997]. Sequential C and Fortran programs have been parallelized to target an array of simple configurable hardware structures[Babb et al., 1999]. The TRS-based behavioral descriptions are different from these approaches because on one hand TRS terms convey structural information about the hardware, but on the other hand, TRS rules can embody a set of behaviors, including concurrency and nondeterminism. This is not possible to express in any sequential language. TRS also offers a well-understood formalism which is useful in verification.

More related to TRS are hardware description languages that have been developed in the context of formal specification and verification. TRS is perhaps closest to Lamport's TLA's. Windley uses the specification language from the HOL[HOL, 1997] theorem proving system to describe a pipelined processor[Windley, 1995]. Matthews et al. have developed the Hawk language to create executable specifications of processor micro-architectures[Matthews et al., 1998]. However, none of these systems has been used in synthesis to the best of our knowledge. With a somewhat different motivation, Communicating Sequential Processes have been applied to hardware-software co-design by Gupta et al.[Gupta and de Micheli, 1993] and Thomas et al.[Thomas et al., 1993].

7. CONCLUSION

When applied in conjunction with reconfigurable technologies, TRAC can drastically lower the entry cost of taking on a hardware project by people who are not hardware designers by training. Compilers like TRAC have the potential to close the traditional distinction of hardware and software by creating a continuum of trade-offs between development cost and performance. We anticipate the day when all computers are shipped with a FPGA next to the CPU, and developers are just as ready to program the FPGA for a performance critical application as they would program the processor today.

8. ACKNOWLEDGMENTS

This paper describes research done at the MIT Laboratory for Computer Science. Funding for this work is provided in part by the Defense Advanced Research Projects Agency of the Department of Defense under the Ft. Huachuca contract DABT63-95-C-0150 and by the Intel Corporation. James C. Hoe is supported by an Intel Foundation Graduate Fellowship.

9. REFERENCES

[Arvind and Shen, 1999] Arvind and Shen, X. (1999). Design and verification of processors using term rewriting systems. *IEEE Micro Special Issue on Modeling and Validation of Microprocessors.*

[Baader and Nipkow, 1998] Baader, F. and Nipkow, T. (1998). *Term Rewriting and All That.* Cambridge University Press.

[Babb et al., 1999] Babb, J., Rinard, M., Moritz, C. A., Lee, W., Frank, M., Barua, R., and Amarasinghe, S. (1999). Parallelizing applications into silicon. In *Proceedings of the IEEE Workshop on FPGAs for Custom Computing Machines '99*, Napa Valley, CA.

[Chandy and Misra, 1988] Chandy, K. M. and Misra, J. (1988). *Parallel Program Design: A Foundation.* Addison-Wesley.

[Galloway, 1995] Galloway, D. (1995). The Transmogrifier C hardware description language and compiler for FPGAs. In *Proceedings of IEEE Workshop on FPGAs for Custom Computing Machines*, pages 136–144, Napa Valley, CA.

[Gokhale and Gomersall, 1997] Gokhale, M. and Gomersall, E. (1997). High level compilation for fine grained FPGAs. In *Proceedings of the IEEE Workshop on FPGAs for Custom Computing Machines '97*, Napa Valley, CA.

[Gokhale and Minnich, 1993] Gokhale, M. and Minnich, R. (1993). FPGA computing in a data parallel C. In *Proceedings of IEEE Workshop on FPGAs for Custom Computing Machines*, pages 94–101, Napa Valley, CA.

[Gupta and de Micheli, 1993] Gupta, R. and de Micheli, G. (1993). Hardware-software cosynthesis for digital systems. *IEEE Design and Test of Computers*, pages 29–41.

[HardwareC, 1990] HardwareC (1990). *HardwareC – A Language for Hardware Design.* Stanford University.

[HOL, 1997] HOL (1997). *The HOL System Tutorial, Version 2.* SRI International, University of Cambridge.

[Lamport, 1994] Lamport, L. (1994). The temporal logic of actions. *ACM Transactions on Programming Languages and Systems*, 16(3).

[Lynch, 1996] Lynch, N. (1996). *Distributed Algorithms.* Morgan Kaufmann.

[Manna and Pnueli, 1991] Manna, Z. and Pnueli, A. (1991). *The Temporal Logic of Reactive and Concurrent Systems: Specification.* Springer-Verlag.

[Matthews et al., 1998] Matthews, J., Launchbury, J., and Cook, B. (1998). Microprocessor specification in Hawk. In *Proceedings of the 1998 International Conference on Computer Languages*, Chicago, IL.

[Thomas et al., 1993] Thomas, D. E., Adams, J. K., and Schmit, H. (1993). A model and methodology for hardware-software codesign. *IEEE Design and Test of Computers*, pages 6–15.

[Vuillemin et al., 1996] Vuillemin, J., Bertin, P., Roncin, D., Shand, M., Touati, H., and Boucard, P. (1996). Programmable active memories: Reconfigurable systems come of age. *IEEE Transactions on VLSI*, 4(1):56–69.

[Windley, 1995] Windley, P. J. (1995). Verifying pipelined microprocessors. In *Proceedings of the 1995 IFIP Conference on Hardware Description Languages and their Applications (CHDL).*

A Synthesis Algorithm for Modular Design of Pipelined Circuits

Maria-Cristina Marinescu
and Martin Rinard
Laboratory for Computer Science
Massachusetts Institute of Technology
Cambridge, MA 02139
cristina@lcs.mit.edu,rinard@lcs.mit.edu

Abstract: This paper presents a synthesis algorithm for pipelined circuits. The circuit is specified as a collection of independent, loosely-coupled modules connected by queues. The synthesis algorithm transforms this asynchronous, modular specification into a synchronous, tightly-coupled, and fully pipelined circuit in which queues are implemented as finite buffers. Data is read from the buffers at the begining of each clock cycle, new values are computed, then the new results are written back into the buffers at the end of each clock cycle.

We have implemented a prototype synthesizer that is capable of automatically generating synchronous, fully pipelined implementations of modular specifications. This paper presents experimental results from this synthesizer.

1. INTRODUCTION

One successful way to manage the complexity of building very large-scale systems is to specify them as a collection of independent, loosely-coupled modules connected by streams, queues or pipes. Our present work describes a synthesis algorithm for arbitrarily complex and general pipelined circuits, starting from a modular, compact high-level specification.

The designer specifies the circuit as a set of modules connected by queues. The behavior of each module is specified using a set of rewrite rules. Each rule reads data from one or more input queues, uses the data to compute new values, then writes the new values out to the output queues. Conceptually, each module executes independently and asynchronously with respect to the other modules. Because the queues insulate the modules from each other, the designer can use a modular design approach. He or she can first focus on developing each module in isolation, then use queues to connect the modules into a complete specification.

A primary advantage of this approach is that it enables the designer to reason about the behavior and correctness of each module in isolation without worrying about the concurrent behavior of the entire system. This reduces human effort and makes specifications simple, compact, clear, less prone to mistakes, and more easily verified. It also promotes the reuse of existing modules in new specifications. Finally, modular specifications are more suitable for automatic synthesis and simulation than non-modular ones and have good scalability characteristics. This model has proved to be useful in the Unix operating system and in various parallel programming models [Arvind and Nikhil, 1990; Gregory, 1987; Newton and Browne, 1992]. More recently it has been used to successfully model complicated hardware designs, where it has shown great promise in enabling very concise, clear specifications [Arvind and Shen, 1999; Poyneer et al., 1998].

A straightforward synthesis algorithm would implement this model directly in hardware. The problem with this approach is the queue management overhead. If the queues are implemented as asynchronous connections between independently operating modules, the system as a whole suffers from synchronization overhead as modules dynamically handshake to transfer data.

This paper presents an alternative approach: a synthesis algorithm that produces a tightly coupled, fully synchronous implementation of a set of modules connected with queues. The basic idea behind the synthesis algorithm is to automatically compose the module definitions to derive, at the granularity of individual clock cycles, a global schedule for the operations of the entire system, including the removal and insertion of queue elements. The resulting implementation executes in a completely synchronous, pipelined manner. At the beginning of each clock cycle, the modules read their inputs from the input queues and compute the next result. At the end of the clock cycle, the results are written to the output queues, overwriting the inputs from the beginning of the clock cycle. This synthesis algorithm delivers the best of both worlds: it allows the designer to use a modular, high-level specification and obtain an efficient, fully synchronous circuit.

The remainder of the paper is organized as follows. Section 2 presents an example illustrating our synthesis approach. Section 3 presents the synthesis algorithm and Section 4 presents the experimental results. Section 5 discusses related work; we conclude in Section 6.

2. EXAMPLE

We next present an example that shows how to use our approach to synthesize a simple pipelined processor. We use a processor as our example because we expect it will be familiar to a wide audience. Our approach and synthesis algorithms are, of course, generally applicable to wide range of circuits, not just processors.

Our example processor has an instruction memory, a program counter and a register file. Figure 1.1 presents the simplified pipeline that we use to implement the processor. The instruction fetch stage fetches instructions from the instruction memory into the instruction buffer; the register fetch stage moves the instruction from the instruction buffer to the register buffer, replacing the register names in the instruction with the contents of the corresponding registers. The compute and writeback phase computes the results and writes them back into the register file.

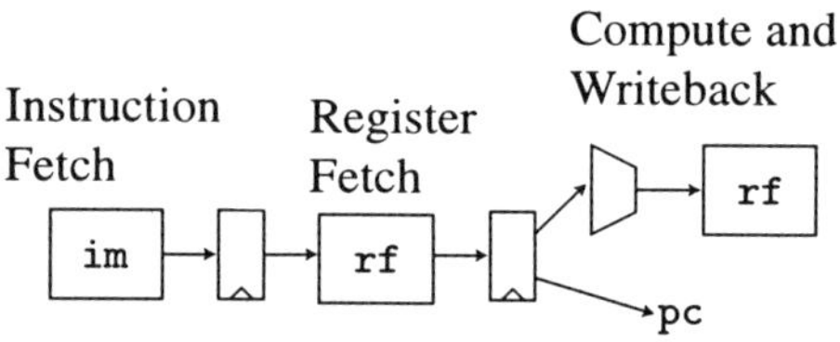

Figure 1.1 Simple Pipeline for Example

2.1 PROCESSOR STATE

Figure 1.2 presents the declaration of the processor state, which consists of the program counter `pc`, the instruction memory `im`, the register file `rf`, and two queues, `iq` and `rq`. Lines `4` and `5` declare the state as a set of state variables; lines `1` through `3` contain the type declarations for these variables. The type declarations include a 3 bit register name type `reg`, an 8 bit integer type `val`, an 8 bit integer type `loc` which represents the locations of instructions in the instruction memory, an instruction type `ins`, and a type `irf` for instructions whose register operands have been fetched from the register file. The instruction type is a tagged union type, similar to those found in ML [Milner et al., 1990] and Haskell [Hudak et al., 1992]. Each instruction can be either an INC instruction, which increments the value in its single register argument, or a JRZ instruction, which tests the value in its register argument and, if the value is zero, jumps to the location in its location argument.

```
type reg = int(3), val = int(8), loc = int(8);
type ins = <INC reg> | <JRZ reg loc>;
type irf = <INC reg val> | <JRZ val loc>;

var pc : loc, im : ins[N], rf : val[8];
var iq = queue(ins), rq = queue(irf);
```

Figure 1.2 State Variables and Type Declarations for Example

2.2 QUEUES

Queues provide buffered, first-in, first-out connections between modules. There are several operations that modules can perform on a queue q:

- `head(q)`: Retrieves the first element in the queue.
- `tail(q)`: The rest of the queue `q` after the first element. Usually used to specify the new value of the queue after removing the first element.
- `insert(q,e)`: The queue `q` after inserting the element `e` at the tail of the queue `q`. Usually used to specify the new value of the queue after inserting a new element.
- `notin(q,e)`: Tests if the element `e` is not in the queue `q`.

Our specification models the pipeline buffers `iq` and `rq` in our example as queues.

2.3 UPDATE RULES

Figure 1.3 presents the code that implements the modules in our example. There are three modules, one for each pipeline stage. Each module is implemented by a set of *update rules*. Each rule has an enabling condition and a set of updates to the state. When the enabling condition evaluates to `true`, the rule is enabled and can execute, in which case its updates are atomically applied to the state. Conceptually, the execution of the system repeatedly chooses an enabled rule and executes it.

This is a standard model of asynchronous execution found, for example, in systems such as Unity [Chandy and Misra, 1988] and term rewriting systems [Baader and Nipkow, 1998].

```
// Instruction Fetch Stage
1: if true then iq = insert(iq,im[pc]); pc = pc+1;

// Register Operand Fetch Stage
2: if <INC r> = head(iq) and notin(rq, <INC r _>) then
      iq = tail(iq); rq = insert(rq, <INC r rf[r]>);
3: if <JRZ r l> = head(iq) and notin(rq, <INC r _>) then
      iq = tail(iq); rq = insert(rq, <INC rf[r] l>);

// Compute and Writeback Stage
4: if <INC r v> = head(rq) then
      rf = rf[r->v+1]; rq = tail(rq);
5: if <JRZ v l> = head(rq) and v = 0 then
      pc = l; iq = nil; rq = nil;
6: if <JRZ v l> = head(rq) and !(v = 0) then
      rq = tail(rq);
```

Figure 1.3 Update Rules for Example

We illustrate the execution of the system by going through the set of rules. The condition for the instruction fetch rule, rule 1, is **true**, which means that it is always enabled. When it executes, it fetches an instruction from the instruction memory and inserts it into the instruction queue iq. It also increments the program counter pc to set up the next fetch.

The two rules in the operand fetch stage, rules 2 and 3, remove instructions from the instruction queue, fetch the register operands, and insert them into the rq. Rule 2 processes INC instructions, and rule 3 processes JRZ instructions. Both rules use a form of pattern matching similar to that found in ML and Haskell. Consider rule 2. The enabling condition is <INC r> = head(iq) and notin(rq, <INC r _>). The first clause of this condition, **<INC r> = head(iq)**, is true if an INC instruction is the first instruction in the instruction queue iq. Furthermore, if there is such an instruction, the clause matches and *binds* the variable r to the register name argument of the INC instruction. The variable r can then be used later in the rule to refer to this operand.

The second clause, `notin(rq, <INC r _>)` uses the binding to check for a read before write hazard. If there is a pending instruction waiting to execute that will write the register `r`, the machine must delay the operand fetch so that it fetches the value after the write. If there is a pending instruction that will write the register `r`, the instruction is in the `rq` queue. The clause `notin(rq, <INC r _>)` checks to make sure that there is no such instruction in `rq`, and the rule as a whole is enabled and can execute only if there is no hazard.

If the rule is enabled, it fetches the register operand and inserts the instruction, along with this operand, into the next queue in the pipeline, the `rq` queue. It also removes the instruction from the instruction queue. The other rules perform similar activities, removing elements from queues, processing the data in the elements to generate results, then inserting the results into the next queue or writing the result back into the register file. In particular, the update `rf = rf[r->v+1]` from the first rule in the compute and writeback stage, rule 4, sets element `r` of the register file `rf` to be `v+1`.

2.4 SYNTHESIS

In the abstract model of computation described above, the modules execute in a completely decoupled way. The rules execute whenever they are enabled, with the queues carrying results between modules. In effect, the queues decouple the modules, enabling the designer to focus on each module in turn. This design methodology scales to very large systems, including systems with hierarchically defined modules. The only problem is that an efficient hardware implementation must be tightly coupled and synchronous. Ideally, the stages of the processor would execute in a strict pipeline, with the queues implemented as hardware buffers and each stage reading the value from the previous stage in the same clock cycle as the new value is written into the register. The next section presents a synthesis algorithm that accomplishes this goal.

3. ALGORITHM

Given a system specification, the synthesis algorithm combines the operations in the rules first into a global schedule, then into a synchronous circuit that implements the specification. The basic approach is, at each clock cycle, to give each rule an opportunity to execute. If a rule is enabled at that cycle, it will execute. The challenge with this approach is to ensure that the final result at the end of the cycle correctly reflects the atomic execution of all of the rules that executed in that cycle. We meet this challenge by symbolically executing the rules in sequence, with each rule operating on the output of the previous rule. The final result is an expression for each state variable. This expression is the new value of the state variable in the next clock cycle, and reflects the combined updates of all the rules that executed in the previous clock cycle.

To avoid the problem of an excessively long clock cycle, the algorithm, when possible, relaxes the enabling condition at each rule so that it is evaluated in the initial state, at the beginning of the clock cycle, rather than in the state produced by the previously executed rule. In particular, this technique ensures that data from state variables moves through at most one module in each clock cycle, which in turn ensures that the critical path of the circuit does not cross module boundaries. The clock cycle time of the system is therefore determined by the modules, not how they are connected together. The algorithm consists of the following phases:

- **Rule Numbering:** The algorithm numbers rules for symbolic execution, determining the intermediate state in which each rule will be evaluated. Figure 1.4 illustrates the numbering of all different versions of the state variables for all the rules in our previous example. As this example shows, the numbering is set up so that each rule reads the version of the state variables produced by the previous rule.

- **Relaxation:** When possible, the algorithm relaxes the calculation of the enabling condition for each rule so that it is evaluated in the initial state, not the intermediate state from the previous rule. This transformation has the effect of limiting the critical path that determines the length of the clock cycle.

- **Queue Finitization:** In the initial specification, the queues have unbounded length. Based on input from the designer, the algorithm chooses a finite length for each queue. It then modifies the rules to ensure that no queue ever exceeds its finite length. The key issue is to ensure that no rule ever executes if there will be no room for its result in the output queues. This is more difficult than it may sound, because each rule must take into account the number of items in the queue at the beginning of the clock cycle, the number of elements inserted and removed by rules before it in the evaluation order, and the number removed by rules after it in the evaluation order.

- **Symbolic Execution:** The algorithm symbolically executes the rules in sequence to obtain an expression for each state variable.

The expression is the value of the variable in the next clock cycle. Because rules may not be enabled in a given state and may therefore not execute, the expressions contain conditionals.

- **Optimizations:** The algorithm optimizes the representation by performing common sub-expression elimination to eliminate any duplication, and mutual exclusion testing to eliminate executions that can never actually occur (i.e. false paths in the circuit).
- **Verilog Generation:** The algorithm generates one or more hardware registers for each state variable, depending on its type. For each state variable, the value in the next clock cycle is determined by the combinational logic implementing the corresponding determined expression.

We next discuss the more complicated phases of the synthesis algorithm.

```
if true then iq1= insert(iq0,im[pc0]); pc1= pc0+1;
if <INC r> = head(iq1) and notin(rq1, <INC r _>) then
   iq2= tail(iq1); rq2=insert(rq1, <INC r rf1[r]>);
if <JRZ r l> = head(iq2) and notin(rq2, <INC r _>) then
   iq3= tail(iq2); rq3 = insert(rq2, <INC rf2[r] l>);
if <INC r v> = head(rq3) then
   rf4= rf3[r->v+1]; rq4= tail(rq3);
if <JRZ v l> = head(rq4) and v = 0 then
   pc5= l; iq5= nil; rq5 = nil;
if <JRZ v l> = head(rq5) and !(v = 0) then
   rq6= tail(rq6);
```

Figure 1.4 Numbered Rules for Example

3.1 RELAXATION

The rule numbering in Figure 1.4 suffers from an excessively long clock cycle. Consider, for example, the system starting out with nothing in any of the queues. The last version of the state variables reflects the entire fetch and execution of the next instruction. Obviously, we would like the fetch and execution to be pipelined over multiple clock cycles. We achieve this goal by relaxing the versions tested in the enabling conditions of each rule — we replace each version of each state variable with the earliest safe version. An earlier version of v_j, name v_k, is safe if the following property holds:

> If the rule's enabling condition, C is true with v_j replaced by v_k, then it is also true with v_j, i.e. $C[v_k/v_j]$ implies C.

This transformation is valid for two reasons:

- **Safety:** After the transformation, each rule is enabled in a subset of the states in which it was enabled before the transformation, and, if enabled, produces the same result as before the transformation. So each execution of the transformed system is also an execution of the original system.

- **Liveness:** The transformation never completely disables a rule — the transformed enabling condition tests the original state, and the rule executes if it is enabled in this state.

Figure 1.5 presents the transformed system in our example. A key property that enables this transformation is that if a rule that tests the element at the head of a queue is enabled, it remains enabled if additional elements are inserted at the tail of the queue. This property makes it possible to relax the rules in the example so that they test the initial version of each queue instead of the version produced by earlier rules.

In many cases, the algorithm can order the rules to perform queue operations in the following order: first checks of the form `notin(q,e)` that test that an element is not in a queue, then insertions into the tail of the queue, then tests that the head of the queue satisfies a given property, then removals from the head of the queue. Being able to put the rules in this order is sufficient (but not necessary) to ensure that the algorithm will be able to relax the enabling conditions so that they all test the initial version of each queue.

```
if true then iq1= insert(iq0,im[pc0]); pc1= pc0+1;
if <INC r> = head(iq0) and notin(rq0, <INC r _>) then
    iq2= tail(iq1); rq2=insert(rq1, <INC r rf1[r]>);
if <JRZ r l> = head(iq0) and notin(rq0, <INC r _>) then
    iq3= tail(iq2); rq3 = insert(rq2, <INC rf2[r] l>);
if <INC r v> = head(rq0) then
    rf4= rf3[r->v+1]; rq4= tail(rq3);
if <JRZ v l> = head(rq0) and v = 0 then
    pc5= l; iq5= nil; rq5 = nil;
if <JRZ v l> = head(rq0) and !(v = 0) then
    rq6= tail(rq6);
```

Figure 1.5 Relaxed Rules for Example

The relaxation algorithm proceeds as follows. It processes the rules of the system in the order in which they are numbered. At each rule, it repeatedly attempts to replace the current version of the state variables in the enabling condition with the previous version. This attempt succeeds if the enabling condition with the previous version of the state variables implies the enabling condition with the current version or if the enabling conditions are mutually exclusive. The implication test and mutual exclusion tests are performed using a combination of resolution [Ballantyne, 1982] and a set of simplification and reduction rules, and operate on the enabling conditions once they have been transformed into conjunctive normal form.

3.2 QUEUE FINITIZATION

When the queues are implemented in hardware, there is a specific number of entries allocated for the queue, and the synthesis algorithm must generate a circuit that does not exceed that length. The algorithm therefore analyzes the rules to determine the circumstances under which a queue may grow beyond its hardware limit. It then modifies the enabling conditions to ensure that the queues never exceed the limit.

Conceptually, the generated circuit maintains several counters for each queue: a counter L_q that contains the number of elements in q at the beginning of the clock cycle, a counter I_q that maintains, for each rule, the net number of elements that preceding rules insert into q (this number is the number of elements inserted minus the number removed), and a counter R_q that maintains, for each rule, the number of elements that succeeding rules remove from q. Both of these counters are dynamically generated using combinational logic, and count only insertions and removals from rules that are enabled in the current clock cycle. There is also the hardware limit $\mathtt{N}_q$ of the maximum number of queue entries.

The basic idea is to augment the enabling condition for each rule that inserts an element into q so that it does not execute unless a subsequent rule clears the queue or $L_q + I_q - R_q < \mathtt{N}_q$. Because the values of the counts depend directly on the enabling conditions, it may be more efficient to simply test combinations of enabling conditions rather than computing the counts explicitly. Figure 1.6 presents our example after the application of the queue finitization algorithm. In this figure, $\mathtt{length}(q) = L_q + I_q$.

Note that because the values of I_q and R_q affect the enabling conditions, it is possible for there to be a cycle of dependences between the different values of these counters. This occurs, for example, when there is a cycle of rules waiting for each other to remove elements from queues. In the worst case, there may simply be no way to avoid deadlock without changing the hardware to add more space in the queues.

```
if length(iq_0) < N_iq or
  (<INC r> = head(iq_0) and notin(rq_0, <INC r _>)) or
  (<JRZ r l> = head(iq_0) and notin(rq_0, <INC r _>)) or
  (<JRZ v l> = head(rq_0) and v = 0) and
  length(rq_0) < N_rq or length(iq_0) < N_iq or
  <INC s _> = head(rq_0) or
  (<JRZ v l> = head(rq_0) and v = 0) or
  (<JRZ v l> = head(rq_0) and !(v = 0) then
    iq_1 = insert(iq_0,im[pc_0]); pc_1 = pc_0+1;
if <INC r> = head(iq_0) and
  notin(rq_0, <INC r _>) and
  length(rq_0) < N_rq or
  <INC s _> = head(rq_0) or
  (<JRZ v l> = head(rq_0) and v = 0) or
  (<JRZ v l> = head(rq_0) and !(v = 0) then
    iq_2 = tail(iq_1); rq_2=insert(rq_1, <INC r rf_1[r]>);
if <JRZ r l> = head(iq_0) and
  notin(rq_0, <INC r _>) and
  length(rq_0) < N_rq or
  <INC s _> = head(rq_0) or
  (<JRZ v l> = head(rq_0) and v = 0) or
  (<JRZ v l> = head(rq_0) and !(v = 0) then
    iq_3 = tail(iq_2); rq_3 = insert(rq_2, <INC rf_2[r] l>);
if <INC r v> = head(rq_0) then
    rf_4 = rf_3[r->v+1]; rq_4 = tail(rq_3);
if <JRZ v l> = head(rq_0) and v = 0 then
    pc_5 = l; iq_5 = nil; rq_5 = nil;
if <JRZ v l> = head(rq_0) and !(v = 0) then
    rq_6 = tail(rq_6);
```

Figure 1.6 Rules in Example After Queue Finitization

But even if there are cycles of rules waiting for each other to remove elements, it may still be possible for the synthesis algorithm to generate a deadlock-free circuit without increasing the queue length.

The key insight in this case is that finitization will not introduce deadlock if there is a way for existing elements to be removed from all of the queues so that there is room for new elements. Assume the sequence of rules $R_j R_{j+1} ... R_{j+m}$ with enabling conditions $C_j C_{j+1} ... C_{j+m}$ creates a cycle for the current rule R_i with queue q, where $\texttt{length}(q) = \texttt{N}_q + R_q$. If C_i implies $\forall 0 \leq l \leq m, C_{j+l}$, then all of the rules in the cycle can execute. Otherwise, none of them can.

3.3 SYMBOLIC EXECUTION

Symbolic execution determines a new value for each state variable at the end of the clock cycle in terms of the values at the start of the clock cycle. It does this by substituting out the intermediate versions of each state variable. The result is an expression, in the original versions of the state variables, for each use of each state variable in the system. The versions at the last rule are latched back into the state variables at the end of the clock cycle, and provide the initial values for the start of the next clock cycle.

3.4 OPTIMIZATIONS

To improve the quality of the synthesized circuit, the compiler optimizes the expressions, using common sub-expression elimination and mutual exclusion testing. If an expression contains a value that will never actually occur in practice because the conditions required to obtain the value are mutually exclusive, the computation of that value is eliminated from the expression. A typical example is a value obtained if both a `JRZ` and an `INC` instruction is at the head of the instruction queue. Obviously, the instruction must be either a `JRZ` instruction or an `INC` instruction, but not both. So such a value will never be computed in the actual circuit. The mutual exclusion testing is implemented using resolution, simplification, and reduction.

We illustrate the symbolic execution and optimizations principle by presenting the final value of the instruction queue `iq`. If there is a taken branch, the instruction queue is cleared. If there is already an instruction at the head of the instruction queue that can go through the register fetch stage, the final result is obtained by inserting the new instruction into the tail of the queue and removing the instruction from the head of the queue. Otherwise, the circuit checks to see if there is an empty entry in the instruction queue. If so, it fetches another instruction; if not, the instruction queue does not change.

Figure 1.7 presents the result of the expression evaluation; note the introduction of the temporary variables `t1`, `t2`, `t3`, and `t4`. These variables will turn directly into combinational logic in the final implementation of the circuit.

```
let
  t1 = <INC r> = head(iq0) and notin(rq0, <INC r _>)
  t2 = <JRZ r l> = head(iq0) and notin(rq0, <INC r _>)
  t3 = insert(iq0, im[pc0])
  t4 = tail(t3)
iq6 =
  if <JRZ v l> = head(rq0) and v = 0 then nil
  else if t1 then t4
  else if t2 then t4
  else if length(iq0) < Niq) then t3
  else iq0
```

Figure 1.7 Result of Symbolic Execution for `iq`

3.5 VERILOG GENERATION

The final step is to generate synthesizable Verilog for the circuit. The basic approach is that each state variable is implemented as one or more hardware registers, with the expressions generated during the symbolic execution providing the new values for the state variables at the end of each clock cycle.

The Verilog generation is straightforward. The algorithm generates combinational logic that computes the value of each expression, then connects the computed values to the inputs of the hardware registers that implement the corresponding state variables. In the future we may explore implementations that use more complicated synthesis algorithms for operations that are expensive to implement in combinational logic.

The compiler currently uses Verilog arrays to implement memories such as the instruction memory in our example. Current memory implementations are single ported, but we are exploring ways to obtain multi-ported memories, either under the control of the designer or automatically as part of the synthesis algorithm. We are also exploring the use of SRAM or DRAM to implement larger memories. Queues are implemented as registers.

Benchmark	Cycle Time	Area (NAND2 gates)	Map Effort	Constraints
Bubblesort	9.59ns	≈ 5121	medium	Clk = 10
Butterfly	9.61ns	≈ 5170	medium	Clk = 10
Processor	10.48ns	≈ 4830	low	none

Table 1.1 Benchmark Characteristics

4. EXPERIMENTAL RESULTS

We have implemented a prototype synthesis system based on the algorithm presented in this paper. We have used this algorithm to generate synthesizable Verilog implementations for the benchmarks described below. These Verilog implementations were tested using the NCVerilog by Cadence, then synthesized using the Synopsys Design Compiler to an industry standard .25 micron standard cell process.

The first benchmark implements Bubblesort for eight 8-bit numbers. The second benchmark implements a butterfly network similar to the ones used in bitonic sorting networks and in FFTs. The last benchmark is an 8-bit pipelined processor specification. The synthesized, gate level model of the processor was then regression tested using the ASIC vendor's simulation libraries in order to confirm correct synthesized functionality. Table 1.1 presents several benchmark characteristics.

5. RELATED WORK

The synthesis of hardware from various description languages has been and continues to be an active area of research [Micheli, 1994]. In this section we discuss systems for specifying custom microprocessors, synchronous dataflow languages, and recent work using term rewriting systems.

There is a large market for embedded processors customized for a specific application. Researchers have proposed to support the development of such systems by providing languages that allow the designer to quickly describe a customized architecture [Pyo et al., 1992; Park and Walker, 1988]. The research presented in our paper, on the other hand, is designed to support the development of arbitrary circuits, not just microprocessors.

Other researchers have proposed a design methodology based on synchronous dataflow [Ho et al., 1998]. While the resulting specifications contain modules, the connections between modules are synchronous, which forces the designer to understand the global timing of the circuit when designing each module. The research presented in our paper uses asynchronous queues to connect modules. The synthesis algorithm automatically derives the global schedule of operations in the circuit, freeing the designer from the need to understand the global timing.

The research most closely related to ours is the work of James Hoe and Arvind on the synthesis of circuits specified as term rewriting systems [Hoe and Arvind, 1999; Hoe et al., 1997]. The basic goal is the same: to synthesize synchronous implementations of modular, queue-based specifications. There are differences, however, in the synthesis algorithms. In particular, their approach executes multiple rewrite rules in the same cycle only if they are completely independent. In our approach, multiple dependent rules may execute in the same clock cycle, with the final result reflecting the combined effect.

6. CONCLUSION

Understanding how to manage the complexity of building large-scale systems is a difficult, challenging, and important problem. This paper presents an approach based on specifying the system as a set of independent, parallel modules connected by queues. This approach enables the designer to control the complexity of the design process by first developing each module in isolation, then using queues to combine the modules and specify the complete system.

The successful use of this design methodology for circuits requires a synthesis algorithm that can translate the asynchronous, loosely-coupled specification into a synchronous, fully pipelined circuit. This paper presents such an algorithm. Our initial experimental results from an implementation of this algorithm provide encouraging evidence that it can be used to deliver efficient pipelined implementations of modular specifications that use queues.

References

Arvind and Nikhil, R. (1990). Executing a program on the MIT tagged-token dataflow architecture. *IEEE Transactions on Computers*, 39(3).

Arvind and Shen, X. (1999). Design and verification of processors using term rewriting systems. Technical Report CSG Memo 419, Laboratory for Computer Science, Massachusetts Institute of Technology, Cambridge, MA.

Baader, F. and Nipkow, T. (1998). *Term Rewriting and All That.* Cambridge University Press.

Ballantyne, M. (1982). Automatic deduction. Technical Report STAN-CS-82-937, Dept. of Computer Science, Stanford Univ., Stanford, Calif.

Chandy, K. M. and Misra, J. (1988). *Parallel Program Design: A Foundation.* Addison-Wesley, Reading, Mass.

Gregory, S. (1987). *Parallel Logic Programming in PARLOG: The Language and Its Implementation.* Addison-Wesley, Reading, Mass.

Ho, W., Lee, E., and Messerschmitt, D. (1998). High level data flow programming for digital signal processing. *VLSI Signal Processing, III*, pages 385–395.

Hoe, J. and Arvind (1999). Hardware synthesis from term rewriting systems. Technical report, Laboratory for Computer Science, Massachusetts Institute of Technology.

Hoe, J., Rinard, M., and Arvind (1997). An exercise in high-level architectural descriptions using a synthesizable subset of term rewriting systems. Technical Report CSG Memo 403, Laboratory for Computer Science, Massachusetts Institute of Technology, Cambridge, MA.

Hudak, P., Peyton-Jones, S., Wadler, P., Boutel, B., Fairbairn, J., Fasel, J., Guzman, M., Hammond, K., Hughes, J., Johnsson, T., Kieburtz, D., Nikhil, R., Partain, W., and Peterson, J. (1992). Report on the programming language Haskell: a non-strict, purely functional language (version 1.2). *SIGPLAN Notices*, 27(5).

Micheli, G. D. (1994). *Synthesis and Optimization of Digital Circuits.* McGraw-Hill.

Milner, R., Tofte, M., and Harper, R. (1990). *The Definition of Standard ML.* The MIT Press, Cambridge, Mass., Cambridge, MA.

Newton, P. and Browne, J. C. (1992). The CODE 2.0 graphical parallel programming language. In *Proceedings of the 1992 ACM International Conference on Supercomputing*, Washington, DC.

Park, N. and Walker, A. (1988). Sehwa: A software package for synthesis of pipelines from behavioral specifications. *IEEE Transactions on Computer-Aided Design*, 7(3).

Poyneer, L., Hoe, J., and Arvind (1998). A TRS model for a modern processor. Technical Report 408, Computation Structures Group, MIT Laboratory for Computer Science.

Pyo, I., Su, C., Huang, I., Pan, K., Koh, Y., Tsui, C., Chen, H., Cheng, G., Liu, S., Wu, S., and Despain, A. M. (1992). Application-driven design automation for microprocessor design. In *Proceedings of 29th Design Automation Conference.*

A Methodology and Associated CAD Tools for Support of Concurrent Design of MEMS

B. F. Romanowicz[1], M. H. Zaman, S. F. Bart, V. L. Rabinovich, I. Tchertkov, C. Hsu, and J. R. Gilbert

Microcosm Technologies, Inc., 215 First Street, Suite 219, Cambridge, MA 02142 U.S.A.

Key words: MEMS, microsystem, CAD, macro-model, simulation

Abstract: Development of micro-electro-mechanical systems (MEMS) products is currently hampered by the need for design aids, which can assist in integration of all domains of the design. The cross-disciplinary character of microsystems requires a top-down approach to system design which, in turn, requires designers from many areas to work together in order to understand the effects of one sub-system on another. This paper describes current research on a methodology and tool-set which directly support such an integrated design process.

1. INTRODUCTION

Considerable progress in technologies for microsystems fabrication has made over the past two decades, resulting in a large variety of commercially successful devices. These products have benefited from the high performance and low manufacturing costs characteristic of MEMS batch fabrication technologies. Though the manufacturing technologies are derived from microelectronic fabrication techniques, most devices require application specific fabrication steps, which must be developed, and characterized. This often results in costly and time consuming prototyping. It

[1]bfr@memcad.com

is recognized that standardization of processes will help reduce the entry cost of microsystems.

MEMS sensors and actuators can be viewed as the electronic interface to the physical world. Some physical quantity of interest is transformed into an electrical quantity that can be measured, or some electrical signal is converted into some action on the environment. As a result of the small dimensions of these transducers, and the semiconductor materials often used, strong field coupling is present, and needs to be taken into account in the modeling of these devices, to accurately capture their behavior. Historically, most of the modeling and simulation effort has gone into developing tools that capture the intricate physics present in MEMS devices.

All MEMS systems have some common layers to their design. These include device design (design a manufacturable component), package design (design a practical package), and system design (design and improve the system the device fits into). The requirement for design aids is best illustrated by considering the MEMS products that have been commercially successful to date. Examples include inkjet printer nozzles, pressure sensors, and a variety of inertial sensors used primarily in the automotive field. In all of these products, the design criteria for each of the individual domains were met successfully in an economic and manufacturable manner.

Development of MEMS products is currently hampered by the need for design aids, which can assist in integration of all domains of the design [1,2]. The cross-disciplinary character requires a top-down approach to system design which, in turn, requires designers from many areas to work together in order to understand the effects of one sub-system on another. What is required is a design methodology based on concurrent design in all required domains. These design domains include the MEMS/MOEMS device, the analog sensing circuitry, the high-level system electronics, the application specific package, and manufacturing sensitivity analysis.

The ability of MEMS devices to be integrated with signal conditioning circuitry and batch fabrication offers an important advantage over their macroscopic counterparts. To ensure proper functioning of such an integrated system, one must perform system-level simulation. Such system-level modeling is extremely useful in determining operation characteristics and verifying performance before the device is actually manufactured. This can reduce the need for prototype fabrication and test iterations and significantly reduce cost and time-to-market.

Performing full 3-D physical simulation within each time step of a typical system simulator (such as SABER™, MATLAB™, or SPICE) is prohibitively time-consuming and numerically impractical. Hence, in order to simulate the appropriate system level dynamic behavior efficiently, a reduced-ordered model or "macro-model" of the MEMS subsystem must be obtained and employed in the system-level simulator. Thus, macro-model construction is a key part of a design methodology [3].

As micro-systems become more complex and the need for models with large numbers of coupled degrees-of-freedom (DOFs) increases, the use of automated tools for generating macro-models becomes increasingly important. Although macro-modeling techniques have been reported by some researchers ([4], [5], [6], [7], [8], [9]), currently there is no systematic method for generating macro-models for MEMS devices in an automatic way.

Design tools to support this methodology must allow passing of models from one group to another to foster communication between them. These tools must also be usable by engineers who are not high level scientists with years of MEMS simulation experience.

In this paper we summarize some of the past three years work on developing methods and tools to enable the concurrent design of MEMS and MST systems. First we describe the concurrent design methodology itself, with particular attention to who does the designing and how different designers interact. We follow that with discussions of a research projects which now provide tools to enable the key (thick) arrow of *Figure 1*. This is a tool for automatic extraction of dynamic macro-models of MEMS devices.

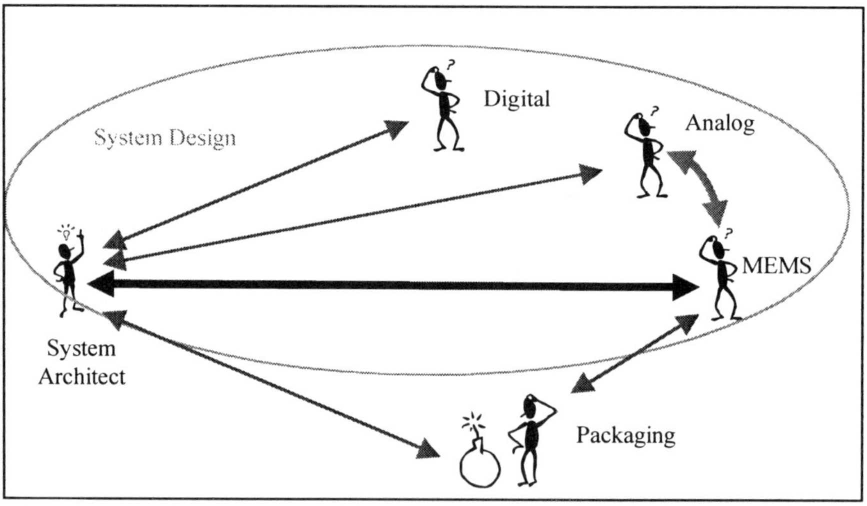

Figure 1. Design groups involved in MEMS design

2. CONCURRENT DESIGN OF MEMS

MEMS design involves several layers of design work, and potentially concurrent engineering among several groups. An "Actor" based view of such concurrent engineering is sketched in *Figure 1.* A System Architect coordinates the design of a product, drawing on the knowledge and experience of design specialists in the digital and analog circuit, MEMS device, and packaging fields. Successful design of systems containing MEMS components requires a top-down approach to system design. This involves supporting the actors (design groups) in Figure 1 with behavioral modeling and simulation at the level of the System Architect, and enabling the System Architect to specify subsystem functions in each area, by specifying behavioral models.

The use of electronic design automation (EDA) tools for the system level simulation of MEMS is attractive, as complete systems may be simulated together with the physical transducers, analog and digital signal processing, compensation and control modules, and external environmental influences. The System Architect can investigate the effects of various design partitions, and trade off the complexity of the various subsystems, and verify if these options are compatible with the required system performance and functions.

2.1 Use of HDLs for Design Intent Communication

Modern analog and mixed signal hardware description languages (HDLs) can advantageously be used to facilitate this communication between the design specialists. In our case, this corresponds to enabling the thick arrows in *Figure 1.* The use of HDLs, as opposed to SPICE models, for system level simulation supports the use of energy-based physical models. This prevents the introduction of spurious energy sources, which can occur,for example, by using SPICE polynomial sources to model nonlinear elements. HDLs also provide a rich set of syntax, constructs, and modeling methods [10]. These are needed to describe the complex coupled physics present in transducers in an accurate and efficient manner. Signal flow modeling, a technique is often used for the analysis of systems containing feedback loops, is supported by HDLs. This is the approach used by tools such as the SimulinkTM (in MatlabTM). The advantage of using a circuit simulator is the ability to seamlessly introduce models of electronics in the feedback loop.

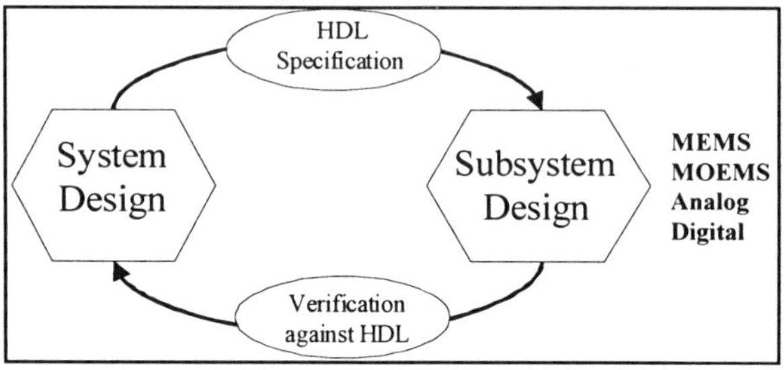

Figure 2. Use of HDLs for specification and validation by system architect

HDLs also support various levels of abstraction, which may be introduced by the System architect as he refines the system from an initial set of components with simplified behavior, to more sophisticated models which contain detailed information about the performance of subsystems, as provided by the various actors.

The System Architect specifies the targeted functionality of each subsystem, as needed to fulfill the required system performance, using VHDL-AMS or a proprietary HDL such as MastTM, VerilogATM, SpectreHDLTM, HDL-ATM, etc. He can perform a system level simulation with these behavioral models to validate the system performance, and validate design partitioning that he has made. For example, it is possible to trade off complexity in the design of a physical transducer against the complexity in the signal processing subsystem. The System Architect then passes each of these behavioral models to the design specialists in each field. These behavioral models are in effect design specifications, which the subsystem designer must now meet. This design process is illustrated in *Figure 2*.

Each of the design specialists may now work independently on his subsystem to insure that it meets the targeted functionality that is required of it. An example of such a design process, and the CAD tools that support it, is presented in *Figure 3*. A MEMS inertial sensor subsystem designer receives a specification from the System Architect. After drafting a set of layouts corresponding to the process he wishes to use in manufacturing the device, he then performs an accurate 3D simulation to capture the physics, which are of importance in his design. He then automatically extracts parameters of importance in this design, to create a reduced order model in a HDL of his choice. This behavioral macro-model can now be simulated and compared to the abstract specification provided by the system architect, to verify if the device design provides the required functionality.

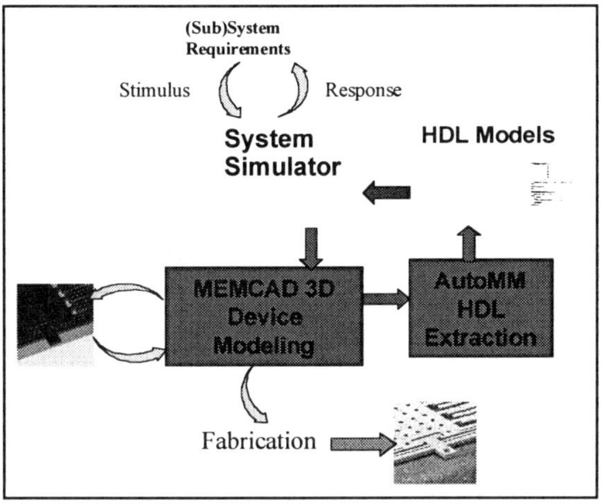

Figure 3. MEMS device design inside the system design loop

The above process is interactive. If any of the actors are unable to meet a targeted specification, they provide the System Architect with a behavioral model of their best design. The System Architect may then change the specification of another subsystem, in order to compensate for the actual performance of a subsystem. This design process is iterated upon until an optimal design is obtained. These decision criteria may include technical specifications such as performance, accuracy, speed, size, power consumption, functionality, but can also include economic and marketing considerations such as cost, reliability, time to market, dimensions. Costly and time consuming prototyping steps are thus avoided.

This top-down approach to MEMS design does place considerable constraints on the CAD tools, as well as on the foundries and plants where the devices are to be manufactured. Without accurate knowledge of parameters such as material properties, residual stresses, manufacturing repeatability, operating conditions, packaging stress, etc. a CAD system can not operate in a predictable manner. In an analogous manner to electronic design automation (EDA), MEMS CAD requires the existence of foundry, process, and even run specific data to predict product performance with an acceptable accuracy. This implies the existence of systematic process characterization through test structures, and the calibration of the CAD tools using constitutive properties extracted from these test structures.

2.2 System Design Partitioning Issues

As indicated in *Figure 1*, MEMS devices are composed of multiple subsystems, which are designed separately and must be integrated. This requires not only communication between the sub-system designers as previously discussed, but also an integrated view of how to partition the sub-systems for optimal cost and performance. Integrated or single chip MEMS are usually manufactured using a set of pre and post CMOS micromachining steps using fab compatible materials. An example of an integrated system is the Analog Devices automobile airbag accelerometer [11]. Integrated devices are attractive because of their generally low unit cost in mass production. The integration of electronics close to sensors also helps noise performance.

Hybrid MST systems offer a greater choice of technologies and materials for the physical transducer, thus offering possible advantages in performance. They are however potentially more costly to package and assemble, and signal conditioning may be problematic as parasitic impedance's are introduced between the sensor and front-end electronics. Ford Microelectronics uses a two-chip hybrid system for their airbag accelerometer [12].

Efficient communication between the actors in Figure 1 is required for both a single chip approach, and for a hybrid multi-chip system solution. The interaction between the designers will be simpler in the second case, as they will not (necessarily) be sharing the same silicon die. The exchange of energy and information between subsystems on different chips will take place through the various interconnects (electrical, fluidic, optical, etc.) connecting the separate substrates. Here, the specifications provided by the System Architect will reflect the structural partitioning provided by the multi-chip approach. In both cases however, the CAD support tools supporting the design flow will be similar.

In both integrated or hybrid MEMS, electronics are used to improve the sensor performance by linearization schemes, active feedback, thermal compensation, chopping to reduce 1/f noise etc. Other system level electronics are often required such as sensor calibration, self-test, programmability and other application specific functions. Additionally design trade-offs are required such as the choice between integrating an on-chip pre-amp versus a complete A/D converter. The partitioning of these sub-systems can have a significant effect on the system performance. System level modeling is critical to optimizing this partitioning.

In inertial measurement systems it is possible to trade complexity between the mechanical and electronic subsystems. A cheap and nonlinear sensing element may be improved through linearization and feedback, but this implies additional IC real estate. For example, a closed-loop system may be used to remove all mechanical and geometric non-linearity's in an accelerometer by sensing the position of the seismic mass and applying a correction force equivalent to the acceleration, thereby immobilizing the mass at it's point of rest. Care must be taken here to ensure that mechanical resonance frequencies are not excited by capacitive sensing means, which requires a mechanical-electronic co-design. This type of force-feedback design was used in initial accelerometer designs [12]. When improved system modeling demonstrated the linearity of the mechanical part, the feedback was discarded in favor of a smaller, cheaper circuit [13].

Another example occurs in resonant sensors such as MEMS gyroscopes. Here it is possible to trade off mechanical complexity in the form of the resonator Q versus the phase accuracy of the sensing electronics. The trade-off point can be very dependent on the specifics of the design (including the fabrication process constraints) and requires modeling to optimize.

All of these design tradeoffs can be taken into account through efficient top-down behavioral model generation and bottom-up validation procedures. Examples of products whose design is supported in MEMCAD today include:

- Inertial Sensors
- Pressure Sensors
- Mirrors, Gratings, Optical Switches
- Electrical/RF Switches
- Thermal Actuators/Sensors
- Packaging Analysis
- Ink Jets
- uTAS and Lab-On-Chip Applications
- Flow Sensors
- Data Storage

3. EXTRACTION OF MACRO-MODELS

In this section, we describe a systematic method for modeling the class of electro-mechanical micro-systems that can be represented as multi-component, lumped, mass-spring-dashpot structures. Examples include accelerometers, gyros, and other structures that have rigid masses and compliant springs. In this lumped modeling assumption, the lumped spring effect originates from mechanical reaction forces and moments of the suspensions (or tethers) holding the proof-mass. Damping forces result from multiple energy loss mechanisms, but are dominated by gas viscosity. In addition, there are electrostatic forces and torques exerted on the dielectrically separated conductors in the system when voltages are applied. The accuracy of the developed method is verified by comparison of two plate-tether MEMS structures to results obtained from the developed models with those from full 3-D physics simulations. Good accuracy is demonstrated in both spatial-domain and frequency-domain dynamic behavior of the models.

3.1 Overview

A semi-automatic and complete modeling procedure that automates the generation of component-level macro-models of MEMS devices has been developed [14]. The user assembles the system-level model by connecting individual component-level macro-models together. For simplicity, the developed method assumes that while the tethers provide mechanical compliance, they are electrostatically inert and massless. It also assumed that the proof mass is electrostatically driven and moves as a rigid body. Devices that do not move as a rigid body, such as membrane devices cannot be accurately modeled with this technique.

The procedure begins by dividing the whole device into sub-components such as mechanical springs, electrostatic elements, dashpots, and proof-masses. These subcomponents are separately meshed and simulated over the desired ranges of operation. These full 3-D physics simulations are done in MEMCAD [15] using hybrid finite element and accelerated boundary element physics. The results of these simulations are fitted to multi-variable polynomials as functions of the desired degrees of freedom (DOFs). The macro-models for each subcomponent are then automatically generated in the behavioral modeling language of a system level simulator (SABER, SPICE, etc.). Finally, the component-level macro-models are assembled into a system-level design to model the behavior of the whole system.

3.2 Implementation

The modeling technique has been implemented in a tool named AutoMM (Auto Macro Modeler). The basic steps involve exploring the device operation space, modeling the data through multi-degree polynomial curve-fitting, and using the polynomial coefficients and other simulation data in dynamic equations. AutoMM consists of several sub-modules that are used to simulate the electrostatic, mechanical and inertial behavior of MEMS components in their operation space as a function of the DOFs.

AutoMM is built around the basic functionalities of the MEMCAD software tool suite [8]. It directly uses the MEMCAD device creation and visualization methods and applies wrappers around the solver modules. AutoMM is constructed as a collection of functional sub-modules. This allows the flexible addition of components with different physical behaviors. It also allows the calculation of the behavioral data to be done in parallel, which reduces the over-all time of macro-model generation.

3.3 Sequence of Operations

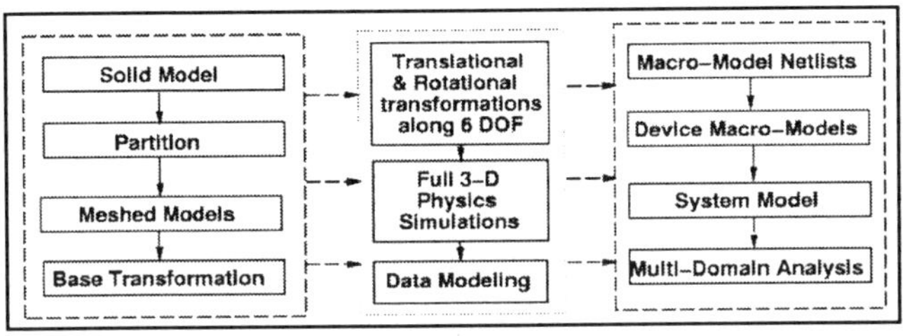

Figure 4. Sequence of operations

Figure 4 shows the sequence of operations that are carried out by the AutoMM module to generate macro-models. The procedure starts by creating the device solid model using the 'MemBuilder' module of MEMCAD [15] from the device process information and the device layout. Then a finite-element mesh is created on the solid model. This meshed solid model is input to the AutoMM module which first carries out a global base transformation on the meshed structure according to the specifications provided by the designer. Examples of such transformations include changing separations of structures, angular orientations, lengths, thicknesses, density, Poisson ratio, stress, and other geometrical and material properties of different subsets of device components. Note that this step can account for the effects of manufacturing variations in the final device macro-model.

The transformed models are then passed to the sub-modules that perform electrostatic, mechanical, and inertial simulations using multi-DOF boundary conditions. The simulation data are then fit to multi-degree polynomial equations (up to fourth order), which are functions of the degrees of freedom over which the device has been simulated. These polynomial fit coeffcients are finally used in system equations to create the device macro-model. Although most of these steps are automated, user interactions and interventions have been allowed in a few cases to include the capability of monitoring the simulation process and specification of user-defined macro-model parameters.

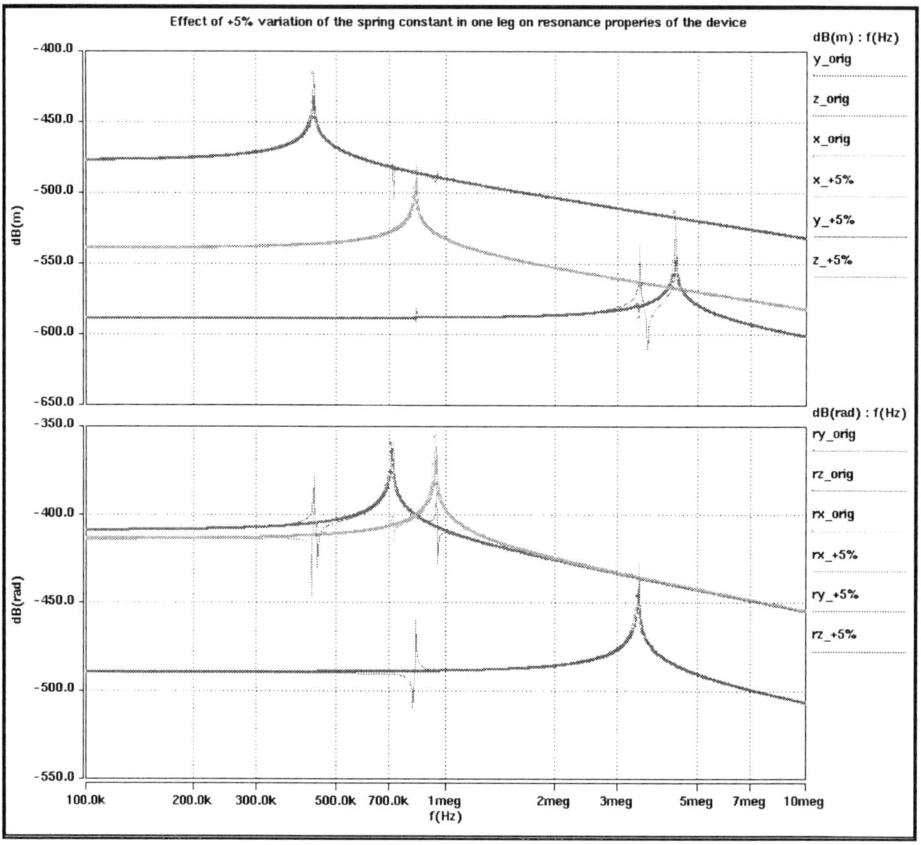

Figure 5. Frequency response in six degrees of freedom of a plate-tether MEMS structure. The thick lines show the nominal response for the 3 translational and 3 rotational DOFs. The thin lines show mode coupling with a 5% variation of thickness of one tether

Figure 5 illustrates the type of results that can be obtained using macro-models generated by AutoMM. The frequency response of a plate-tether structure is compared to that of the same structure with a variation in the thickness of a tether, which breaks the symmetry, and introduces coupling between the translational and rotational degrees of freedom. This is accomplished by modifying the spring constant of a tether, without having to extract another macro-model from a full 3-D simulation.

4. SUMMARY

Development of micro-electro-mechanical systems (MEMS) products is currently hampered by the need for design aids, which can assist in integration of all domains of the design. The cross-disciplinary character of microsystems requires a top-down approach to system design which, in turn, requires designers from many areas to work together in order to understand the effects of one sub-system on another.

In this paper we have summarize some of the past three years work on developing methods and tools to enable the concurrent design of MEMS and MST systems. First we described the concurrent design methodology itself, with particular attention to who does the designing and how different designers interact. Secondly, a modeling procedure that automates the generation of macro-models of MEMS devices and shows good agreement to full 3-D physics simulation has been presented. This tool corresponds to a software realization of the thick arrow between the System Architect, and the MEMS designer in *Figure 1*. The implemented modeling technique is currently limited to the class of devices where the actuating and restoring forces are limited to electrostatic, mechanical (tensile and torsional), damping, and inertial types. Future developments will consider electrostatically active mechanical tethers with nonzero mass and macro-models for other physical forces, such as fluidic pressure, thermal stress, piezo-electric potential, etc.

5. ACKNOWLEDGEMENTS

The authors acknowledge the support of the DARPA CiMEMS Program (Contract No. F30602-96-2-0290). The support of Jonathan Hammond (Motorola SP) and Dennis Greywall (Lucent Technologies) through many discussions is greatly appreciated.

6. REFERENCES

---

1 S. D. Senturia, CAD Challenges for Microsensors, Microactuators and Microsystems, Proc. IEEE, Vol. 86, pp. 1611-1626, 1998.

2 G. Wachutka, "The Art of Modeling Coupled-Field Effects in Microdevices and Microsystems", Technical proceedings of the 1999 International Conference on Modeling and Simulation of Microsystems, MSM 99, San Juan, Puerto Rico, U.S.A., April19-21, 1999, pp.14-19.

3 B. F. Romanowicz, "Methodology for the Modeling and Simulation of Microsystems", Ed. S. D. Senturia, MICROSYSTEMS, Kluwer Academic Publishers, ISBN 0-7923-8306-0, 1998.

4 R. Neul, U. Becker, G. Lorenz, P. Schwarz, J. Hasse, and S. Wünsche, "A Modeling Approach to Include Mechanical Microsystem Componets into the System Simulation", in DATE'98 - Design, Automation and Test in Europe, vol. 1, pp.510-517, Feb, 1998.

5 G. K. Fedder and R. T. Howe, "Multimode Digital Control of a Suspended Polysilicon Microstructure", Journal of Microelectromechanical Systems, pp. 283-297, May, 1996.

6 N. R. Swart, S. F. Bart, M. H. Zaman, M. Mariappan, J. R. Gilbert, and D. Murphy, "AutoMM: Automatic Generation of Dynamic Macromodels for MEMS Devices," Proc. 11th IEEE Intl. Workshop on Micro Electromechanical Systems (MEMS'98), pp. 178-183, Heidelberg, Germany, January 25-29, 1998.

7 S. F. Bart, "CAD for Integrated Surface Micromachined Sensors: Present and Future", in the Proceedings of the Fifth ACM/SIGDA Physical Design Workshop, Reston, VA, April 15-17, 1996, pp. 72-75.

8 G. Lorenz and R. Neul, "Network-Type Modeling of Micromachined Sensor Systems", Proceedings of 1998 International Conference on Modeling and Simulation of Microsystems, Semiconductors, Sensors and Actuators (MSM'98), pp. 233-238, Santa Clara, California, USA, April 6-8, 1998.

9 S. F. Bart, N. R. Swart, M. Mariappan, M. H. Zaman, and J. R. Gilbert, "An Environment for MEMS Design and Verification," Proceedings of 1998 International Conference on Modeling and Simulation of Microsystems, Semiconductors, Sensors and Actuators (MSM'98), pp. 386-391, Santa Clara, California, USA, April 6-8, 1998.

10 B. F. Romanowicz, et al., "VHDL 1076.1 Modeling Examples for Microsystem Simulation," published in "Current Issues in Electronic Modeling", Issue No. 10, Berg et al. Editors, Kluwer Academic Publishers, July 1997, ISBN 0-7923-9875-0.

11 S. Sherman, et al., "A Low Cost Monolithic Accelerometer," Proc. IEEE 1992 Int. Elec. Dev. Meeting, San Francisco, CA, Dec. 13-16, 1992, pp. 19.1.1-19.1.4.

12 L. Spangler and C. J. Kemp, "ISAAC – Integrated Silicon Automotive Accelerometer," Proc. of the 8th International Conf. Of Solid-State Sensors and Actuators, June 25-29, 1995, pp. 585-588.

13 S. F. Bart and H. R. Samuels, "Design Techniques for Minimizing Manufacturing Variations in Surface Micromachined Accelerometers," Microelectromechanical Systems, ASME, DSC-Vol. 59, 1996, pp. 427-433.

14 M. H. Zaman, S. F. Bart, V.L. Rabinovich, C.K. Ghaddar, I. Tchertkov, and J. R. Gilbert, "A Technique for Extraction of Macro-Models in System Level Simulation of Inertial Electro-Mechanical Micro-Systems," Technical Proceedings of 1999 International Conference on Modeling and Simulation of Microsystems, MSM 99, pp. 163-167,San Juan, Puerto Rico, USA, April 19-21, 1999.

15 M. H. Zaman, S. F. Bart, V.L. Rabinovich, C.K. Ghaddar, I. Tchertkov, and J. R. Gilbert, "A Technique for Extraction of Macro-Models in System Level Simulation of Inertial Electro-Mechanical Micro-Systems," Technical Proceedings of 1999 International Conference on Modeling and Simulation of Microsystems, MSM 99, pp. 163-167,San Juan, Puerto Rico, USA, April 19-21, 1999.

SIPPs, Why Do We Need a New Standard for Interconnect Process Parameters ?

Dr Martin G. Walker, Dr Keh-Jeng (KJ) Chang, Dr Christophe J. Bianchi
Frequency Technology, Inc., Technology Station, 469 El Camino Real, Santa Clara, CA 95050 (408) 961-2300 (main line), (408) 961-2323 (FAX), christophe@frequency.com (principal contact)

Key Words: SIPPs, interconnect, characterisation, process, extraction, modelling

Abstract:

The overall performance of deep sub-micron integrated circuits is dominated by the interconnect. Standard Interconnect Process Parameters (SIPPs) [1] provides an abstract model of the interconnect system, as well as a methodology to derive interconnects electrical parameters from a process description.

SIPPs also enables measurement of process variations and their impact on interconnect performance.

This paper describes in details the SIPPs model, as well as highlights some of the benefits of using a standard for interconnect parameters.

SIPPs is now an SI2 (Silicon Integration Initiative) [2] endorsed standardisation effort.

1. INTRODUCTION

The overall performance of deep sub-micron integrated circuits is dominated by the interconnect. One of the most difficult challenges of producing new, complex chips is providing process experts with tools to communicate the capabilities of their technology to designers. These two communities quite often use different terminologies. The adoption of SIPPs will bridge this gap and should result in higher performance devices coming out in shorter periods of time. Momentum is building among industry leaders for a program that can reduce or eliminate tool-specific interconnect characterization and provide highly accurate performance estimates as foundries implement very deep sub-micron processes.

Silicon Integration Initiative (Si2) has taken over leadership of the SIPPs development effort and is committed to going through an assessment of the technology. Si2 expects to complete SIPPs standardization within one year.

1.1 Definition of SIPPs

A set of parameters has been proposed that predicts the electrical performance of on-chip interconnect for multilevel-metal, VLSI processes. The proposed parameters (SIPPs) are a minimal set of parameters needed to accurately predict the resistance and capacitance of the state-of-the-art interconnect systems. SIPPs describe a simple physical model, so they can be used as a common interconnect parameter set for all design calculations, including R/C extraction tools.

The proposed SIPPs are based on a model for the interconnect system. The cross-section of the model, shown in Figure 1, comprises: rectangular polygons representing conductors, planar dielectrics filling the volumes between them, and cylindrical vias (not shown).

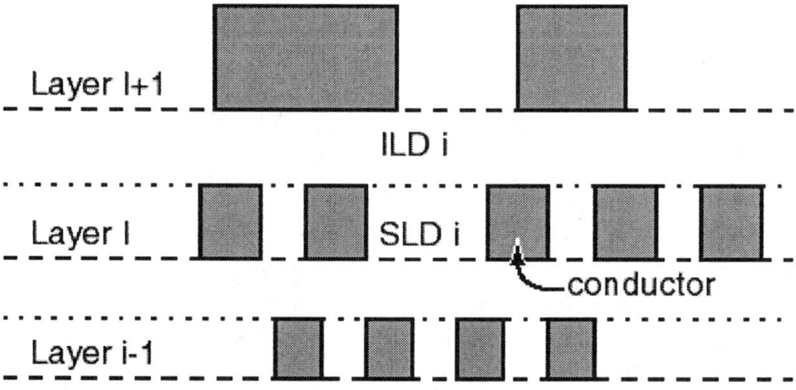

Figure 1. Cross section of a SIPPs model

For each interconnect layer, there are two planar dielectrics: one lies between the conducting lines on the same layer (same-layer dielectric or SLD), and the other is between the conductors of a layer and the layer above (inter-layer dielectric or ILD). The dielectric constants of the two need not to be the same. A via is modeled as a cylinder connecting two conductor polygons at adjacent layers. The height of the via is the thickness for ILD.

SIPPs are electrically predictive parameters, not an exact replica of interconnect geometry. The model closely approximates, but does not exactly replicate, the interconnect geometry of current IC processes. Instead, it intentionally simplifies the actual physical geometry. Nonetheless, the model provides an *electrically equivalent* model for the interconnect system when dimensioned with properly chosen SIPPs. It should accurately predict the electrical performance of the interconnect system. For example, using a properly dimensioned model, a field solver operating on suitable test structures will predict Rs or Cs whose values match those measured from silicon.

As an example of a typical simplification, fabricated aluminum conductors often consist of a sandwich of barrier metal surrounding an aluminum core. The cross section of such a wire may look I-shaped, due to different etch rates between aluminum and the barrier metal, or trapezoidal, due to different etch rates at different depths. The model accounts for all of these effects with equivalent rectangular polygons that predict wire resistance and capacitance as accurately as possible.

Similarly, a complex multi-material dielectric is usually deposited between two metal levels. It is not represented explicitly in the SIPPs model, but is summarized by the planar dielectrics and the conductor polygons.

SIPPs

1.2 Description for each parameter in SIPPs

For each interconnect layer in an interconnect system, the SIPPs based on the above model are shown in Figure 2.

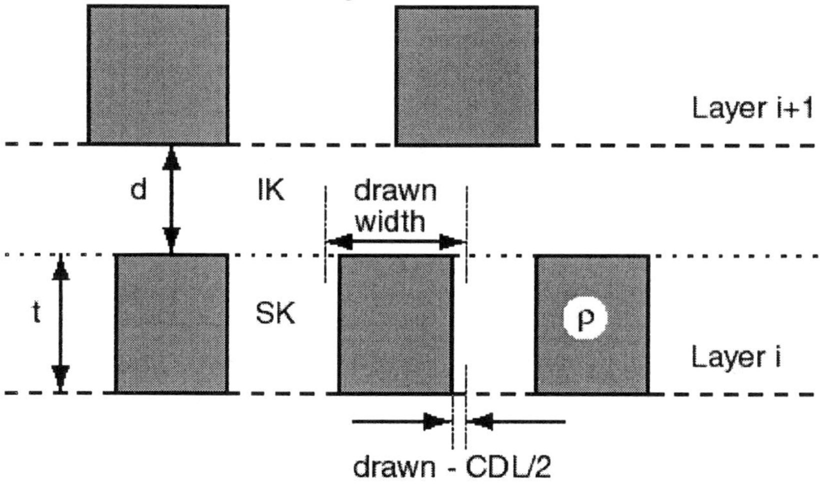

Figure 2. SIPPs associated with a layer.

1. **Sheet resistance (ρ):** is the sheet resistance of the conductor layer. For example, sheet resistance for Metal 2 may be 0.05 Ω /square. Sheet resistance is a piece-wise linear function of drawn width.
2. **Critical dimension loss (CDL):** is the loss in the conductor width compared with drawn width. For example, the actual width of a Metal 2 wire of 0.5μm drawn width and 0.1μm CDL is 0.4μm. CDL is a two-dimensional, piece-wise linear function of drawn width and drawn spacing.
3. **Conductor thickness (t)**: is the thickness of the rectangular conductor polygons in the model. For example, conductor thickness for Metal 2 may be 0.6μm.
4. **Dielectric constant for same-layer dielectric (SK):** is the dielectric constant for same-layer dielectric (SLD), the planar dielectric between conductor polygons of the same layer. For example, SK may be 3.9 for the dielectric between Metal 2 polygons. SK is a piece-wise linear function of drawn spacing.
5. **Dielectric constant for inter-layer dielectric (IK):** is the dielectric constant for inter-layer dielectric (ILD), the planar dielectric between conductor polygons of this interconnect layer and the layer above. For example, IK may be 4.1 for the dielectric between Metal 2 and Metal 3 layers.
6. **Inter-layer dielectric thickness (d):** is the distance between conductor polygons of this interconnect layer and the layer above. For example, d may be 0.8μm between Metal 2 and Metal 3 layers. d is a two-dimensional, piece-wise linear function of drawn width and drawn spacing of the underlying conductor polygons.
7. **Via resistance (R_{via}) and via diameter (d_{via}):** are the resistance and diameter of the via cylinder. For example, R_{via} may be 3 Ω and d_{via} may be 0.5μm for Metal 2 ñ Metal 3 vias.

1.3 Description of how to incorporate variations into SIPPs

Instead of a single number, each parameter in a SIPPs set is defined by a pair of numbers: mean and standard deviation. SIPPs assume that the probability distribution of each parameter is Gaussian. Thus, SIPPs represent process variations, and they allow you to determine performance corners for the interconnect.

Note that SIPPs do not include basic design rules, such as minimal width and spacing for conductors. However, these design rules would ordinarily accompany the SIPPs to provide a complete picture of the interconnect.

2. USING SIPPS

The SIPPs described here are used to accurately predict resistance and capacitance of the interconnect system from layout. Simple situations may require only hand calculations; complex ones generally require field solvers, such as Raphael™, or QuickCap™ or other software.

SIPP	Substrate	Poly	Metal1	Metal2	Passivation
d	0.3 ±0.5%	0.5 ±10%	0.8 ±10%	2.0 ±5%	2.0 ±5%
IK	3.9 ±1%	3.9 ±1%	3.9 ±1%	3.9 ±1%	4.0 ±1%
t		2.0 ±0.2	0.6 ±0.05	0.6 ±2%	
SK		3.9 ±1%	4.1 ±1%	4.1 ±1%	
CDL		0.05 ±0.5%	0.1 ±0.02	0.1 ±2%	

Figure 3. SIPPs for the 2M1P example process

Figure 3 shows the SIPPs for a hypothetical 2M1P process, i.e., a process with two metal and one poly layers. For this process, constants adequately model all SIPPs, without need for piece-wise linear functional descriptions.

2.1 Example: using SIPPs for predicting resistance

Considering calculating the resistance of a Metal 2 wire of 200μm drawn length and 0.5μm drawn width for this example interconnect system. Metal 2 has a sheet resistance of 0.05Ω /o and a CDL of 0.1μm. Therefore, the wire will have resistance of:

200μm/(0.5μm - 0.1μm) * 0.05W /o = 25Ω .

Similarly, if CDL were 0, the final line width on the silicon equals the 0.5μm drawn width, and the resistance drops to 20Ω .

2.2 Example: using SIPPs for predicting capacitance

Figure 4 shows a structure consisting of 3 parallel Metal 2 wires running over a regular array of Metal 1 wires. Consider calculating the capacitance the middle Metal 2 wire per pitch of the underlying Metal 1 wire. Assuming widths and spacings of 0.5μm and 1.0μm for both the Metal 1 and the Metal 2 wires.

Accurate calculation of this capacitance requires a field solver simulation. All off the SIPPs listed in Table 1 except CDL_p affect the capacitance. Appendix A shows the Raphael™[3] input and output for an appropriate field solution for the structure. The desired capacitance is 0.19fF per Metal 1 pitch, or equivalently, 0.13fF/μm.

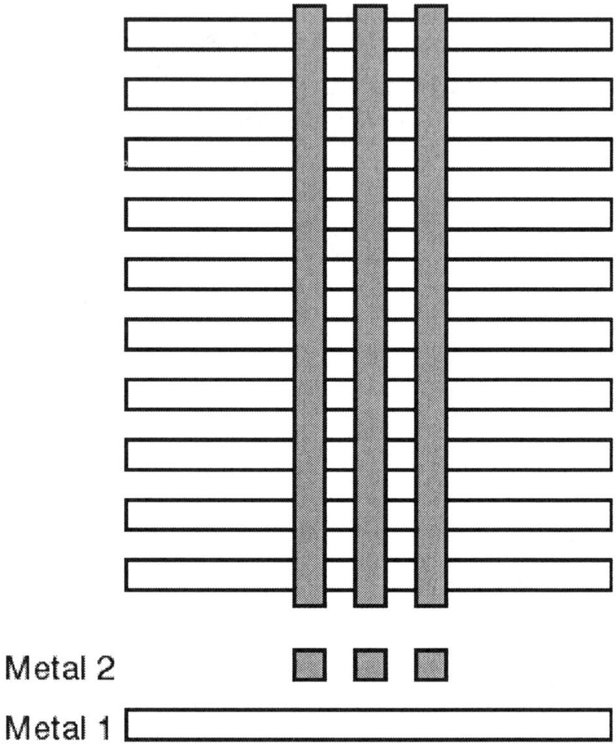

Figure 4. Structure for capacitance calculation, in top view and cross section.

Three Metal 2 wires cross above an array of ten Metal 1 wires. Determine the total capacitance of the center Metal 2 wire.

2.3 Characterisation of SIPPs parameters

SIPPs parameters can be verified on silicon by characterisation of test structures, electrical measurements, SEM measurements and end-to-end electrical performance checking [5].

By verifying SIPPs parameters, process-induced effects can be more accurately mapped into the design flow, allowing better modelling of RC parasitics [6].

APPENDIX A. RAPHAEL EXAMPLE FOR CAPACITANCE

This appendix contains the input and output for one possible Raphael [3] simulation of the cross-over structure of Figure 4. In the code, "0" refers to everything below Metal 1, "1" refers to the Metal 1 layer, "2" refers to the Metal 2 layer, and "3" refers to the passivation above the Metal 2 layer. The total capacitance for M2line2, the center Metal 2 line, is 1.9fF, so its capacitance per Metal 1 pitch is 0.19fF.

```
*****************************************************
*** RAPHAEL RC3 ***
*** Version 4.1.1 ***
*** Copyright (C) 1991 - 1997 ***
*** Technology Modeling Associates, Inc. ***
*** All Rights Reserved ***
*****************************************************
* File: SIPPs
param
ILD0thick=1.0;
vILD0=3.9;
param
M1CDL=0.1;
M1thick=0.6;
vILD1s=4.1;
vILD1=3.9;
ILD1thick=0.8;
param
M2CDL=0.1;
M2thick=0.6;
vILD2s=4.1;
vILD2=3.9;
ILD2thick=0.8;
param
ILD3thick=2.0;
vILD3=4;
*
param
M1w=0.5;
M1s=1.0;
M2w=0.5;
M2s=1.0;
M1width=M1w-M1CDL;
```

```
M1space=M1s+M1CDL;
M2width=M2w-M2CDL;
M2space=M2s+M2CDL;
Sthick=0.9;
*
param
M1pitch=M1width+M1space;
M2pitch=M2width+M2space;
Xtotal=10*M1pitch;
Ytotal=10*M2pitch;
Ztotal=Sthick+ILD0thick+M1thick+ILD0thick+M2thick+ILD2thick+ILD3thick;

block name=ILD0; width=Xtotal; length=Ytotal; diel=vILD0;
v1=0,0,Sthick;
v2=0,0,Sthick+ILD0thick;

block name=ILD1s; width=Xtotal; length=Ytotal; diel=vILD1s;
v1=0,0,Sthick+ILD0thick;
v2=0,0,Sthick+ILD0thick+M1thick;

block name=ILD1; width=Xtotal; length=Ytotal; diel=vILD1;
v1=0,0,Sthick+ILD0thick+M1thick;
v2=0,0,Sthick+ILD0thick+M1thick+ILD1thick;

block name=ILD2s; width=Xtotal; length=Ytotal; diel=vILD2s;
v1=0,0,Sthick+ILD0thick+M1thick+ILD1thick;
v2=0,0,Sthick+ILD0thick+M1thick+ILD1thick+M2thick;

block name=ILD2; width=Xtotal; length=Ytotal; diel=vILD2;
v1=0,0,Sthick+ILD0thick+M1thick+ILD1thick+M2thick;
v2=0,0,Sthick+ILD0thick+M1thick+ILD1thick+M2thick+ILD2thick;

block name=ILD3lanket; width=Xtotal; length=Ytotal; diel=vILD3;
v1=0,0,Sthick+ILD0thick+M1thick+ILD1thick+M2thick+ILD2thick;
v2=0,0,Sthick+ILD0thick+M1thick+ILD1thick+M2thick+ILD2thick+ILD3thick;

block name=M1line10; width=M1width; length=Ytotal; volt=0;
v1=4.5*M1pitch,0,Sthick+ILD0thick;
v2=4.5*M1pitch,0,Sthick+ILD0thick+M1thick;

copy3d from=M1line10; to=M1line9; direction=-1*M1pitch,0,0; volt=0;
```

```
copy3d from=M1line9; to=M1line8; direction=-1*M1pitch,0,0; volt=0;
copy3d from=M1line8; to=M1line7; direction=-1*M1pitch,0,0; volt=0;
copy3d from=M1line7; to=M1line6; direction=-1*M1pitch,0,0; volt=0;
copy3d from=M1line6; to=M1line5; direction=-1*M1pitch,0,0; volt=0;
copy3d from=M1line5; to=M1line4; direction=-1*M1pitch,0,0; volt=0;
copy3d from=M1line4; to=M1line3; direction=-1*M1pitch,0,0; volt=0;
copy3d from=M1line3; to=M1line2; direction=-1*M1pitch,0,0; volt=0;
copy3d from=M1line2; to=M1line1; direction=-1*M1pitch,0,0; volt=0;
block name=M2line3; width=Xtotal; length=M2width; volt=0;
v1=0,1.5*M2pitch,Sthick+ILD0thick+M1thick+ILD1thick;
v2=0,1.5*M2pitch,Sthick+ILD0thick+M1thick+ILD1thick+M2thick;
copy3d from=M2line3; to=M2line2; direction=0,-1*M2pitch,0; volt=1;
copy3d from=M2line2; to=M2line1; direction=0,-1*M2pitch,0; volt=0;
block name=Substrate; width=Xtotal; length=Ytotal; volt=0;
v1=0,0,0;
v2=0,0,Sthick;
window3d v1=-0.5*Xtotal,-0.5*Ytotal,0.0;
v2=0.5*Xtotal,0.5*Ytotal,Ztotal;
options set_grid=500000;
potential
*** POTENTIAL CALCULATION [Coulombs]
Charge on M1line10 = -8.337879e-17
Charge on M1line9 = -8.476684e-17
Charge on M1line8 = -8.471164e-17
Charge on M1line7 = -8.482564e-17
Charge on M1line6 = -8.438169e-17
Charge on M1line5 = -8.432124e-17
Charge on M1line4 = -8.441266e-17
```

Charge on M1line3 = -8.464931e-17
Charge on M1line2 = -8.582391e-17
Charge on M1line1 = -8.443481e-17
Charge on M2line3 = -5.231111e-16
Charge on M2line2 = 1.905325e-15
Charge on M2line1 = -5.206713e-16
Charge on Substrate = -1.609720e-17

3. REFERENCES:

[1] Open SIPPs International " Proposal for SIPPs, Standard Interconnect Performance Parameters", California, USA, Sep. 1998

[2] SI2 "Silicon Integration Initiative", *World-wide-web site* http://www.si2.org., Texas, USA, 1998

[3] Raphael™ v4.0, TMA Inc, Sunnyvale, CA, USA, 1997

[4] QuickCap™ v1.1, RLC, Fairfax, VA, USA, 1996

[5] A.Chou, *et al.*, "Characterization and Application of Interconnect Process Parameters", *Dig. ICMTS,* Japan, Mar. 1998.

[6] L.-F. Chang, *et al.*, "Incorporating process induced effects into RC extraction", *Dig. VLSI 99*, India, January 1999.

ILP-Based Board-Level Routing of Multi-Terminal Nets for Prototyping Reconfigurable Interconnect

A. Kirschbaum, J. Becker, M. Glesner
Institute of Microelectronic Systems
Darmstadt University of Technologie
{kirschbaum|becker|glesner} @mes.tu-darmstadt.de

Abstract For the board-level routing of intermodule connections with reprogrammable devices in a prototyping environment we present an Integer Linear Programming (ILP) model. It is solvable in polynomial time for architectures consisting of two routing devices even for multi-terminal nets. A net decomposition strategy is presented to handle all remaining infeasible problems. In contrast to previous work this approach allocates a minimum of additional port resources and thus significantly improves the routability of multi-terminal net dominated designs.

Keywords: Reconfigurable Interconnect, Rapid-Prototyping, Board-Level Routing, ILP-Model

1. INTRODUCTION

Advances in integrated circuit technology resulted in large and fast reconfigurable devices such as field-programmable gate arrays (FPGAs) and field-programmable interconnection components (FPICs) and have enabled designers to create a new class of open-architecture reconfigurable hardware platforms. Whereas FPGAs are mainly used for prototyping digital logic designs, FPICs allow the arbitrary interconnection of almost any kind of digital system modules.

A typical prototyping system consists of a set of communicating processing elements such as processors, custom logic modules in multiple FPGAs, memory devices, and I/O circuits. In order to preserve flexibility, the system topology is often personalized with a set of FPICs. The interconnection of the system modules and consequently the configuration of the FPICs is hereinafter referred to as board-level routing problem.

Chan and Schlag examined the related problem of net routing in an FPGA-based computing system for the special case of prototyping platforms which are more restrictive with respect to the number of communication links than the approach presented in the following Chan and Schlag, 1993.

Mak and Wong Mak and Wong, 1995b investigated the board-level routing problem for FPGA-based logic emulation systems Varghese et al., 1993. For the case where all nets are two-terminal nets, they presented an $O(n^2)$-time optimal algorithm (n = number of intermodule nets) for configuring the FPICs. This is in contrast to previous approaches which are based on greedy heuristics Slimane-Kadi et al., 1994. The algorithm, which basically finds an Euler circuit in a multigraph, guarantees 100% routing completion if the number of intermodule nets does not exceed the I/O-capacity of the FPGAs. Furthermore, they proved that the BLRP with multi-terminal nets in the presence of k routing devices ($k > 2$) is *NP*-complete. In Mak and Wong, 1995a the routing of m-terminal nets is relaxed by decomposing them into $m - 1$ two-terminal nets. A network flow-based algorithm is used to determine a decomposition whenever a feasible one exists. The routing of the resulting two-terminal nets is then completed with the optimal algorithm of Mak and Wong, 1995b. Obviously, this approach suffers from the extensive allocation of additional port resources at the FPGAs and FPICs. Unfortunately, this property becomes crucial for the feasibility of the BLRP in presence of many multi-terminal nets.

In this paper we will investigate the BLRP for system architectures with coarse grained modules which occur for example in embedded systems with bus-dominated interconnection topologies Yen and Wolf, 1995 Ortega and Borriello, 1997 Gasteier and M., 1996. The coarse granularity as opposed to that found in FPGA-based logic emulation systems allows us to reduce the number of independent switching devices in our prototyping environment to two. It can be shown that the BLRP can be solved in polynomial time for this instance of the problem even in the presence of multi-terminal nets. We will present an Integer Linear Programming (ILP) model Lawler, 1976 for determining a feasible router configuration whenever one exists without splitting any net. For cases where the problem turns out to be infeasible due to resource conflicts we propose an iterative net decomposition strategy. This two phases approach consists of a heuristic to find a suitable net for splitting and a computationally efficient ILP formulation for determining the module at which the net is to be split as well as a valid router configuration. Again, this approach allocates only as much resources as necessary for solving the BLRP which will significantly increase routability for a particular prototyping environment compared to the approach from Mak and Wong, 1995a.

In Section 2. we define a generic system model as basis for the BLRP and introduce our prototyping environment. The formal definition of the BLRP is given in Section 3.. The ILP-based algorithm for routing multi-terminal nets without splitting is presented in Section 4.. A detailed discussion of the net decomposition strategy which is applied if the primary BLRP turns out to be infeasible follows in Section 5. . Some experimental results and remarks about the complexity of the presented algorithms are given in Section 6.. Section 7. concludes the paper.

2. A GENERIC PROTOTYPING ENVIRONMENT

The board-level routing problem discussed in this paper is not specific for a particular prototyping environment but also occurs in many switch-based system architectures. Therefore we introduce a generic architecture of a reconfigurable system as basis for the board-level routing problem.

Such an execution unit $EU(M, P, K)$ consists of a set of functional modules M, a set of ports P, and a set of routers K (Figure 1 . 1). The architecture is completely described by the parameters M, P, K from which the following can be derived: Each module $m \in M$ contains $B = P/M$ ports which are used for intermodule communication. They are partitioned into $|K|$ equally sized disjoint sets $P_m^1 \dots P_m^{|K|}$ which denotes the ith portset at module m. Let $P_m = \bigcup_{i=1}^{|K|} P_m^i$ the set of all ports of module m and $P^i = \bigcup_{m=1}^{|M|} P_m^i$ the set of all ports with a physical connection to router i.

Figure 1 . *1* Generic Architecture of an Execution Unit EU(MR).

Intermodule connections can be established between any two ports $p_x, p_y \in P^i$ via the non-blocking switching matrix of router i. As there is no physical interconnection *between* the routers, it is impossible to connect any ports p_x, p_y with $p_x \in P^i$, $p_y \in P^j$ and $i \neq j$. Additionally, any two ports $p_x, p_y \in P_m$ can also be connected via non-blocking *intramodule* switching resources (e.g. router, FPGA).

A rapid prototyping system which complies with the generic architecture of an execution unit is *REPLICA* [1] Kirschbaum and Glesner, 1997, a Rapid Prototyping System focusing on inter-/processor-module communication emulation. The system is integrated in a design environment for embedded mixed hardware/software systems (*DICE* Gasteier et al., 1996). Due to the coarse granularity of the target system modules (processors, memories, ASICs etc.) which are to be connected and the large number of available I/O ports of the switching devices (320 ports icu, 1996), the prototyping system consists of only two independent routers. Thus *REPLICA* can be modeled as an execution unit $EU(6, 636, 2)$. This prototyping environment facilitates design space exploration and validation of communication links. The reprogrammable system architecture allows prototyping of different topologies, communication types and protocols and is supported by a powerful toolset for automatic system configuration. A reconfigurable hardware monitor Kirschbaum et al., 1998 is available for non-intrusive observation and extraction of real-time data about I/O-channel activities.

3. PROBLEM DESCRIPTION

To prototype a system, it has to be mapped to the execution unit. However, the interconnection of the system modules can be established by different switching devices, i.e. appropriate routers have to be selected for the communication links of the system. Thus, an important task is to find a mapping of the target system topology to the reconfigurable interconnection architecture of the EU which avoids router congestions. In the following section we give some basic definitions of the underlying system model, and a formal description of the above-mentioned board-level routing problem.

The topology of the target system, which is to be mapped onto an execution unit EU, can be described with an undirected graph $NG(V, E)$ termed *netgraph.* The nodes V of this graph represent a set of terminals which are connected via hyperedges E termed *nets.* A *net* e consists of a set of terminals $V_e \subset V$. It represents a physical connection of two or more different modules of a set $M_e \subset M$ of an execution unit $EU(M, P, K)$. Each terminal is unambiguously mapped to a module by $\psi : V \to M$ which is determined by the target system architecture. Thus, our task is to assigns each *terminal* of a net to a *port* of a module. We can formulate the BLRP as follows:

Definition Feasible Board-Level Routing Problem: Given an execution unit $EU(M, P, K)$ and a netgraph $NG(V, E)$: Does there exist an assignment ϕ: $V \to P$ of terminals to ports, i.e. a mapping of the netgraph NG onto the execution unit EU, such that the following constraints are not violated:

1. **Unambiguousness:** Each port p of a module m can only be allocated to at most one terminal v:

$$\forall v_x, v_y \in V : \phi(v_x) \neq \phi(v_y). \qquad (1.1)$$

2. **Connectivity:** The connectivity must be preserved:

$$\forall e \in E \, \forall v \in e : \phi(v) \in P_{\psi(v)}. \qquad (1.2)$$

   That means that a terminal which maps onto a module m obviously has to be connected to a port from P_m.
   Also because there is no physical connection between the routers K, all terminals v of a net e must be connected via the same router:

$$\forall e \in E \, \forall v_x, v_y \in e \, \exists i \in \{1, \ldots, |K|\} : \\ \phi(v_x) \in P^i \wedge \phi(v_y) \in P^i. \qquad (1.3)$$

Obviously, there are several equivalent solutions for the same board-level routing problem (if a solution exists at all), because the connectivity constraint forces only the allocation of a portset but not of a specific port within that set. Moreover, there are no costs associated with a particular mapping, because all routing resources within the generic architecture are equivalent. The described problem is a pure feasibility problem and thus there exists no cost function.

4. BOARD-LEVEL ROUTING

Board-level routing in our prototyping environment is relaxed with respect to the general case, because of the architecture of *REPLICA*. Since the system consists of only two routers ($K = 2$), the terminals have to be assigned to only two different sets of ports.

4.1 ILP FORMULATION

Given a netgraph $NG(V, E)$ and an execution unit $EU(M, P, 2)$, we present a computationally efficient ILP formulation for solving the board-level routing problem.

The mapping ψ of the terminals to the two different sets of ports can easily be modeled with a set of binary decision variables $X = \{x_i | v_i \in V\}$. It is $x_i = 0$, if terminal v_i is assigned to a port $p \in P^1$ and $x_i = 1$, if $p \in P^2$. In order to guarantee the semantic correctness of the mapping the following constraints have to be fulfilled:

- For preserving the connectivity of the target system we require

$$\forall e \in E\ \forall v_i, v_j \in e : x_i = x_j. \tag{1.4}$$

That means all terminals v_i of a net e are connected via the same router.

- The number of terminals assigned to port sets P_m^1, P_m^2 must not exceed their cardinality. Thus, the resource constraints are

  - router 1:

$$\forall m \in M : \sum_{i:\psi(v_i)=m} (1 - x_i) \leq |P_m^1| = \frac{|P|}{2} \tag{1.5}$$

  - router 2:

$$\forall m \in M : \sum_{i:\psi(v_i)=m} x_i \leq |P_m^2| = \frac{|P|}{2}. \tag{1.6}$$

Each assignment $\phi : V \to P$ which does not violate these constraints is termed *feasible.*

5. SPLITTING OF MULTI-TERMINAL NETS

It turns out that the ILP problem, as defined in the previous section, may be infeasible due to router congestions. This means that the original netgraph $NG(V, E)$ cannot be mapped to the execution unit due to resource conflicts.

The netgraph shown in Figure 1.2 consisting of one 4-terminal net, one 3-terminal net, and one 2-terminal net, for example, leads to an infeasible ILP problem. This is non-obvious: although the overall number of terminals per module is not exceeded, there are router resource problems due to the fact that all terminals of a net must be connected via the same router.

Splitting of multi-terminal nets is a technique to enable the mapping of netgraphs even if the ILP-problem turned out to be infeasible. The specific property of *intramodule connectivity* of the execution unit is used for that purpose: All ports p of a module m ($p \in P_m$) can be internally connected to each other via a non-blocking switching matrix without allocating resources outside of the module. When splitting a net, a terminal v_x is duplicated with a terminal $v_{x'}$ at the splitting module m_s. The associated ports $p_x = \phi(v_x)$ and $p_{x'} = \phi(v_{x'})$, respectively, are connected via the internal routing resources such that $p_x \in P^i$ and $p_{x'} \in P^j$ with $i \neq j$. Hence, net e may be accessed from router i *and* j ($i, j \in \{1, \ldots, |K|\}$) which reduces the connectivity constraint of Equation (1.3) to

$$\forall e \in E\ \forall v_x, v_y \in e\ \exists i, j \in \{1 \ldots |K|\} : \\ \phi(v_x) \in P^i \cup P^j \wedge \phi(v_y) \in P^i \cup P^j. \tag{1.7}$$

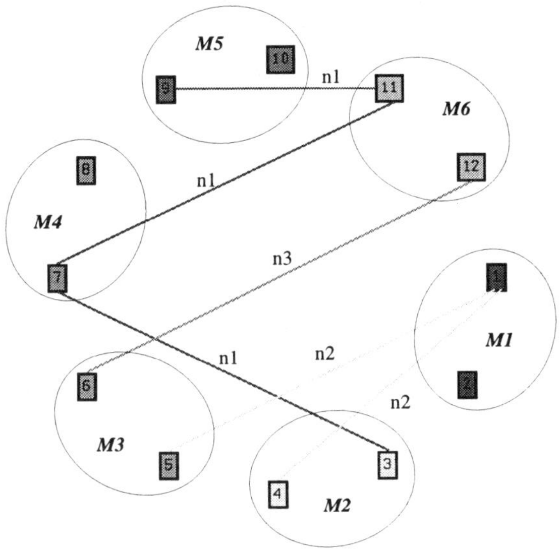

Figure 1.2 Netgraph for which no feasible router configuration exists without splitting.

Our proposed net decomposition strategy consists of two main phases: *net selection* and *net splitting*. First a heuristic is applied for finding a suitable net for splitting. Starting with all multi-terminal nets which can be split, i.e. nets attached to modules with free port resources, the set of candidates is ordered by the degree of mobility MD. MD is defined as the sum of free ports of all modules m to which net e is attached. We assume that choosing the net with the lowest degree of mobility for splitting will result in the best relaxation of the connectivity constraint for the overall system. After having selected the net to be split we derive a computationally efficient ILP model to determine the splitting module as well as a valid router configuration. The optimality of the ILP approach guarantees that such a router configuration will be found whenever one exists. If this problem turns out to be infeasible, we try to split another net without merging the first one again. In this case we select manually the module m_s with the most free port resources as splitting node and reformulate the board-level routing problem. This decomposition strategy is iterated until there are no more free port resources or nets to split, or the board-level routing problem becomes feasible with the relaxed constraints.

Nevertheless, splitting also introduces some cost. Depending on the chosen splitting module m_s, some additional delay may be incurred on the net as well as additional port resources are allocated in the splitting module itself. Therefore, in the presence of many multi-terminal nets as for example in bus-dominated communication architectures of embedded systems, the number of multi-terminal nets which are split has to be minimized. Otherwise the target system's clock frequency will be reduced unnecessarily or even worse, the limited resources of the execution unit can easily be exceeded. The iterative algorithm proposed above terminates as soon as a feasible solution has been found, rather than splitting all nets as proposed in Mak and Wong, 1995a.

5.1 ILP FORMULATION

For finding the splitting node m_s as well as a feasible allocation $\phi : V \to P$ we again use an ILP-formulation. Let $NG(V, E)$ be a netgraph, $EU(M, P, 2)$ an execution unit, e_s a net which is to be split, and M_s a set of modules at which e_s can be split.

The mapping of terminals to ports is again modeled with a set of binary decision variables $X = \{x_i | v_i \in V\}$ with $x_i = 0$, if terminal v_i is assigned to a port $p \in P^1$ and $x_i = 1$, if $p \in P^2$. Additionally, we define a new set of binary decision variables $S = \{s_1, \ldots, s_{|M_s|}\}$ in order to determine the splitting module m_s. It is $s_m = 1$, if net e_s is split at module m, otherwise let $s_m = 0$. The problem is feasible, if it complies with the following constraints:

- For preserving the connectivity of the target system we require

$$\forall e \in E \setminus e_s \; \forall v_i, v_j \in e : x_i = x_j. \tag{1.8}$$

  That means all terminals v_i of a net are connected via the same router. This is not true for net e_s which is available at both routers due to the splitting. Thus, as *REPLICA* is an execution unit of type $EU(M, K, 2)$ the connectivity constraint given in Equation (1.7) can be eliminated for net e_s completely.

- Net e_s should be split at exactly one node:

$$\sum_{m \in M_s} s_m = 1. \tag{1.9}$$

- For the sake of a more compact notation, we define two new decision variables for the total number of terminals at module m which are assigned to router 1 and router 2 respectively:

$$W_m^1 = \sum_{i:\psi(v_i)=m} (1 - x_i)$$
$$W_m^2 = \sum_{i:\psi(v_i)=m} x_i.$$

For all modules which could not serve as splitting module for net e_s we only apply the port constraints already introduced in Section 4.1:

$$\forall m \in M :$$
$$W_m^1 \leq |P_m^1| = \frac{|P|}{2} \quad (1.10)$$
$$W_m^2 \leq |P_m^2| = \frac{|P|}{2}. \quad (1.11)$$

Let v_s be the node at which net e_s is split. We have to allocate two ports (one in each port set) for net e_s at the splitting module v_s. Thus, the resulting port constraints are

$$\forall m \in M_s \; \exists v_s \in e_s : \psi(v_s) = m :$$
$$W_m^1 - (1 - x_s) + s_m \leq |P_m^1| \quad (1.12)$$
$$W_m^2 - x_s + s_m \leq |P_m^2|. \quad (1.13)$$

It should be noted that for all $m \in M_s$ with $s_m = 0$ the port constraints of equations (1.10) and (1.11) are more restrictive than those of equations (1.12) and (1.13) respectively and therefore are effective.

In contrast to the feasibility problem in Section 4.1 particular solutions of this ILP formulation are now rated according to a cost function. Let us define g_m as the weight of module m. We assign g_m with the number of used ports at module m and sum it for all modules net e_s is attached to:

$$\text{min:} \sum_{m \in M_s} g_m \cdot s_m. \quad (1.14)$$

When minimizing this sum, the module with the most free ports will be chosen as splitting node. Considering the iterative nature of our decomposition strategy this will keep the cardinality of M_s for subsequent splitting operations as high as possible. This in turn will relax constraints for v_s and improve routability. It should be noted that g_m is a constant and can be directly derived from the netgraph.

6. EXPERIMENTAL RESULTS

For evaluating the runtime behavior of the ILP formulation of Section 4.1 a benchmark of 21200 different board-level routing problems has been generated. A random generator was used to create different classes of netgraphs $NG(V, E)$

with the cardinality $|V|$ as distinguishing feature. Each class consists of 400 different netgraphs. In order to maintain a maximally balanced utilization of the hardware resources the execution unit was also scaled with the cardinality $|V|$. Thus, for all problems an execution unit of type $EU(6, |V|, 2)$ is assumed.

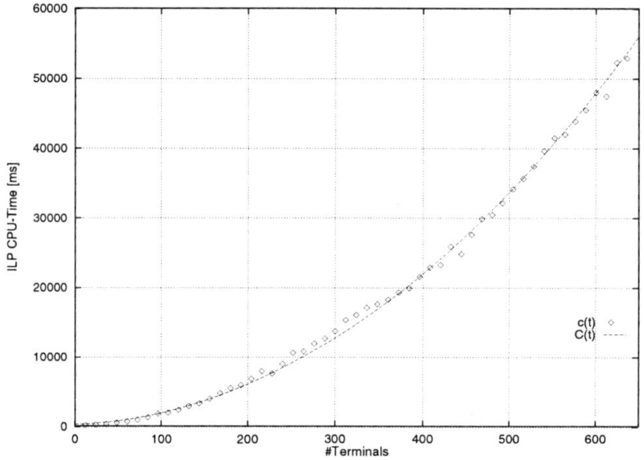

Figure 1.3 Runtime Analysis.

In Figure 1.3 the dependence of the execution time on the cardinality of the investigated netgraph (i.e. the number of terminals) is displayed. The measurements were taken on a *Sun UltraSparc (167 MHz, 562 MB RAM)* with *lp_solve V2.2* Berkelaar, 1997. as ILP-solver. It can be shown that the computational complexity of the particular instance of the BLRP presented in this paper is polynomially bounded. In fact, Figure 1.3 shows that for the benchmark circuits the runtime was bounded by a second degree polynomial.

Figure 1.4 shows the number of infeasible ILP problems in the set of benchmark circuits. It turns out that for execution units of type $EU(6, P, 2)$ almost all problems are feasible for large values of P. The investigation of another set of benchmarks with 6041 different netgraphs and an execution unit of type $EU(6, 636, 2)$ concluded with similar results: only 8 problems (0.13%) were infeasible. The decomposition strategy described in Section 5. was applied to the remaining problems, all of which were solved after the first iteration of the algorithm.

7. CONCLUSIONS

In this paper we have addressed the board-level routing problem of multi-terminal nets in reconfigurable interconnection architectures with coarse granularity. We have presented an ILP-based algorithm which finds a feasible router configuration for a prototyping system consisting of two independent switching

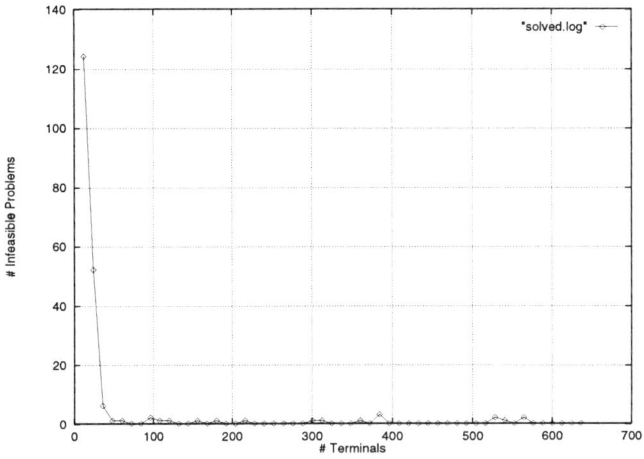

Figure 1.4 Problems which are infeasible without splitting.

devices whenever one exists in polynomial time. Moreover, we have presented an iterative net decomposition strategy for netgraphs which initially cannot be mapped to the execution unit. Instead of decomposing all multi-terminal nets as suggested by previous research, our approach allocates minimal additional port resources to increase routability in the presence of many multi-terminal nets significantly. Given a net to be decomposed, we have also proposed an ILP-based algorithm for finding a splitting node as well as a feasible routing scheme, whenever one exists.

8. ACKNOWLEDGMENTS

The work presented here is part of the research program *Rapid Prototyping of Embedded Systems with Hard Time Constraints.* It is supported by the German Research Foundation *Deutsche Forschungsgemeinschaft (DFG)* under grant Gl 144/14-1.

Notes

1. Realtime Execution Platform for Intermodule Communication Architectures

9. REFERENCES

(1996). *IQX Family Data Sheet.* I-Cube Inc.

Berkelaar, M. (1997). lp_solve. Available via anonymous ftp from `ftp.es.ele.tue.nl`. TU Eindhoven.

Chan, P. and Schlag, M. (1993). Architectural Tradeoffs in Field-Programmable-Device-Based Computing Systems. In *IEEE Workshop on FPGAs for Custom Computing Machines*, pages 138–141.

Gasteier, M. et al. (1996). An Interactive Approach to Hardware/Software Co-Design. In *International Workshop on Logic and Architecture Synthesis*, pages 211–218, Grenoble, France.

Gasteier, M. and M., G. (1996). Bus-Based Communication Synthesis on System-Level. In *9th Int. Symposium on System Synthesis*, pages 65–70, La Jolla, CA.

Kirschbaum, A., Becker, J., and Glesner, M. (1998). A Reconfigurable Hardware-Monitor for Communication Analysis in Distributed Real-Time Systems. In *5th Reconfigurable Architectures Workshop RAW98*, Orlando, USA.

Kirschbaum, A. and Glesner, M. (1997). Rapid Prototyping of Communication Architectures. In *IEEE Workshop on Rapid System Prototyping*, pages 136–141, Chapel Hill, USA.

Lawler, E. (1976). *Combinational Optimization: Networks and Matroids*. Holt, Rinehart and Winston, New York.

Mak, W. and Wong, D. (1995a). Board-Level Multi-Terminal Net Routing for FPGA-based Logic Emulation. In *IEEE/ACM Int. Conference on Computer Aided Design*, pages 339–344.

Mak, W. and Wong, D. (1995b). On Optimal Board-Level Routing for FPGA-based Logic Emulation. In *ACM/IEEE Proc. 32nd Design Automation Conference*, pages 552–556.

Ortega, R. and Borriello, G. (1997). Communication Synthesis for Embedded Systems with Global Considerations. In *Fifth Int. Workshop on Hardware/Software Codesign*, pages 69–73.

Slimane-Kadi, M., Brasen, D., and Saucier, G. (1994). A Fast-FPGA Prototyping System that uses inexpensive High-performance FPIC. In *ACM/SIGDA Int. Workshop on Field-Programmable Gate Arrays*.

Varghese, J. et al. (1993). An Efficient Logic Emulation System. *IEEE Transactions on VLSI*, 1(2):171–174.

Yen, T.-Y. and Wolf, W. (1995). Communication Synthesis for Distributed Embedded Systems. In *International Conference on Computer Aided Design*, pages 288–294. IEEE Computer Society Press.

INDEX OF CONTRIBUTORS

INDEX OF KEYWORDS